Grundlehren der mathematischen Wissenschaften 297

A Series of Comprehensive Studies in Mathematics

Editors

M. Artin S. S. Chern J. Coates J. M. Fröhlich
H. Hironaka F. Hirzebruch L. Hörmander
C. C. Moore J. K. Moser M. Nagata W. Schmidt
D. S. Scott Ya. G. Sinai J. Tits M. Waldschmidt
S. Watanabe

Managing Editors

M. Berger B. Eckmann S. R. S. Varadhan

Leonid Pastur   Alexander Figotin

Spectra of Random and Almost-Periodic Operators

Springer-Verlag

Berlin Heidelberg New York
London Paris Tokyo
Hong Kong Barcelona
Budapest

Leonid Pastur
Mathematical Division
Institute for Low Temperature Physics and Engineering
Academy of Sciences of the Ukrainian SSR
47 Lenin Avenue
310164 Kharkov, USSR

Alexander Figotin
Department of Mathematics
University of Nevada
Reno, NV 89557-0045, USA

Mathematics Subject Classification (1991): 47-02, 60H25, 35J10

ISBN 3-540-50622-5 Springer-Verlag Berlin Heidelberg New York
ISBN 0-387-50622-5 Springer-Verlag New York Berlin Heidelberg

Library of Congress Cataloging-in-Publication Data
Pastur, L. A. (Leonid Andreevich)
Spectra of random and almost-periodic operators/Leonid Pastur, Alexander Figotin.
p. cm. – (Grundlehren der mathematischen Wissenschaften; 297)
Includes bibliographical references and index.
ISBN 0-387-50622-5
1. Random operators. 2. Almost periodic operators. 3. Spectral theory (Mathematics)
I. Figotin, Alexander, 1954–, II. Title. III. Series.
QA274.28.P37 1992 519.2–dc20 91-38704

Printed in Germany

41/3140-543210 – Printed on acid-free paper

Contents

Introduction

In the last fifteen years the spectral properties of the Schrödinger equation and of other differential and finite-difference operators with random and almost-periodic coefficients have attracted considerable and ever increasing interest. This is so not only because of the subject's position at the intersection of operator spectral theory, probability theory and mathematical physics, but also because of its importance to theoretical physics, and particularly to the theory of disordered condensed systems.

It was the requirements of this theory that motivated the initial study of differential operators with random coefficients in the fifties and sixties, by the physicists Anderson, I. Lifshitz and Mott; and today the same theory still exerts a strong influence on the discipline into which this study has evolved, and which will occupy us here. The theory of disordered condensed systems tries to describe, in the so-called one-particle approximation, the properties of condensed media whose atomic structure exhibits no long-range order. Examples of such media are crystals with chaotically distributed impurities, amorphous substances, biopolymers, and so on. It is natural to describe the location of atoms and other characteristics of such media probabilistically, in such a way that the characteristics of a region do not depend on the region's position, and the characteristics of regions far apart are correlated only very weakly. An appropriate model for such a medium is a homogeneous and ergodic, that is, metrically transitive, random field.

We are thus faced with the problem of studying the spectral properties of the Schrödinger operator with a metrically transitive potential, just as the study of ideal crystals (ordered systems) led to the spectral analysis of this operator with a periodic potential. Similar problems also arise in some other branches of quantum and wave physics.

The last fifteen years have brought to light a variety of remarkable new properties of differential and finite-difference operators with metrically transitive coefficients, and witnessed the development of many probabilistic methods unknown to traditional spectral theory and their application to a range of profound problems associated with these operators. Thus metrically transitive operators stand out as important mathematical objects not only in themselves, but also because their study leads to new standpoints and results in many areas of spectral theory. In particular, we now understand much better the structure of the spectrum of Schrödinger operators with bounded potentials (that is, potentials that do not increase or decrease at infinity in a regular way), one of least well studied classes of operators in traditional spectral theory.

The present monograph attempts to summary in a systematic way the results in this field. It develops the spectral theory of metrically transitive operators under a unified point of view, including both differential and finite-difference operators, and (among others) random and almost-periodic coefficients.

The detailed contents of the book can be gleaned from the Table of Contents and from the introductions to the chapters (except for Chapter II, which is introduced by Section 3). We will therefore restrict ourselves to some general remarks.

The notion of a metrically transitive field is very broad, and includes both "random" fields whose values at sufficiently distant points are correlated very weakly, if at all (for example, fields on $\mathbf{Z}^d$ independent at distinct points, or Markov processes with exponential mixing), and "regularly behaved" fields, such as the hull of an almost-periodic function on $\mathbf{R}$ (the closure of the set of its shifts in the topology of uniform convergence on the whole axis).

Accordingly, the book can be divided into three parts. The first includes Sections 1–5, 8, 11, 12 and parts of Sections 6 and 13; it contains results derived under the weakest possible assumptions on the metrically transitive coefficients, often simply boundedness conditions expressed in terms of finiteness of the moments. These sections focus on the study of new objects and problems arising in the spectral theory of metrically transitive operators. The results obtained are quite general and lead to a number of important conclusions on the structure of the spectrum, especially in the one-dimensional case.

However, to obtain a more detailed picture of the spectrum, including its characterization as absolutely continuous, point, etc., and of the generalized eigenfunctions, including their behavior at infinity and other analytic properties, we must restrict our attention to particular subclasses of metrically transitive fields, which enjoy a sufficiently rich set of properties. At present, the best understood subclasses are those of random and of almost-periodic operators. These two classes, corresponding in some sense to the two possible extremes in the degree of randomness of the coefficients, are the focus of the two other parts of the book. From the point of view of solid-state physics, random operators describe disordered structures (impure crystals, amorphous media, and so on), while almost-periodic operators correspond to incommensurate structures, in particular the recently discovered quasicrystals.

The second part, then, covering Sections 6–7, 9–10 and 14–15, discusses random operators whose coefficients admit a fairly concrete description. Under these conditions, one can carry out the constructive spectral analysis of the operators pretty far. In particular, we study the integrated density of states in detail, and establish the remarkable property of Anderson localization, that is, the existence of a dense point spectrum with exponentially decaying eigenfunctions. In one dimension, the absolutely continuous and singular continuous components are absent, so in fact in this case the typical random operator has a pure point, dense spectrum with multiplicity one (just as operators with periodic coefficients typically have absolutely continuous spectra with multiplicity two, whose eigenfunctions have the well-known Floquet–Bloch form). In the multidimensional case, the same behavior is observed near the the so-called fluctuation spectral boundaries (Sections 6, 9, 10 and 15).

The third part, encompassing Sections 16–18, deals with almost-periodic operators. Here the structure of the spectrum is, even in the one-dimensional case, very diversified, and the spectral classification of almost-periodic operators in terms of the coefficients is far from complete. However, for some subclasses—especially quasiperiodic operators and limit-periodic operators very well approximated by periodic ones—one can carry out the spectral analysis in a fair amount of detail. Unlike the case of periodic operators, where the absolutely continuity of the spectrum is determined essentially by the algebraic fact of their commuting with the operator of shift by a period, the character of the spectrum of almost-periodic operators depends strongly on the analytic properties of the coefficients: their amplitude, smoothness, the arithmetic properties of the almost-periods, and so on.

The spectral theory of metrically transitive operators represents the result of a nontrivial synthesis of ideas and constructions from spectral theory, probability theory, mathematical and theoretical physics. Therefore it employs a wide range of often specialized concepts and methods, belonging to branches of mathematics that have been traditionally only very tenuously related. For the same reason, it attracts the interest of scientists from many branches of mathematics and related sciences. Finally, this theory has, in its development, accumulated a good deal of profound facts and interrelations between them, some of which have not yet been rigorously proved in the generality in which they are believed to be true.

All this makes it difficult to present the theory in a book of reasonable size, especially if one uses the traditional style of mathematical writing, where everything is proved in detail, in the interests of completeness and self-containedness. We therefore depart sometimes from this style, in the three following ways:

A few results—including some that are profound and central to the subject—are given without proof, or with just the sketch of a proof. Such statements are called propositions, in contrast with theorems and lemmas, which are proved in full.

Other results, especially those obtained very recently, are formulated in detail, but their proofs, which are as a rule cumbersome and technically complicated, are reduced to a discussion of the main ideas involved. Results of this type are either presented in the form of remarks, or in special subsection that are more survey-like than the rest of the book: for example, 6.F, 15.B, part of 15.C, 16.B and 16.D.

Finally, we thought it appropriate to provide a heuristic discussion of certain problems before they are considered rigorously (see 9.A and 13.A).

We are aware that this type of presentation may not satisfy everyone, but hope that our sincere goal of covering the subject as completely as possible will serve as at least a partial justification.

We now turn briefly to some interesting and important open questions in, or related to, the spectral theory of metrically transitive operators.

1. The best studied objects in the spectral theory of metrically transitive operators, as in traditional spectral theory, are one-dimensional differential and finite-difference operators of second order. They are given considerable attention in this book. But even for this class of operators there exist a variety of questions requiring serious further study, including "inverse problems" where one tries to obtain information about the operator's coefficients from the spectrum.

For the Schrödinger operator, the case of a periodic potential (the Hill operator) has been exhaustively studied by Marchenko and Ostrovskii [1975, 1980]. But other than that, the only case for which more or less complete results are available is that of a limit-periodic potential that can be approximated very well by periodic functions (Section 17).

For some other classes of metrically transitive operators, the inverse problem is not even entirely well posed yet. However, some results for the class discussed in [Kotani 1985b, 1987b] hold the promise of deep analytic and probabilistic questions.

There is also a substantial disparity between our knowledge of the singular spectra of one-dimensional finite-difference operators with almost-periodic coefficients and their differential counterparts. While for the former we know the class of coefficients for which there is no absolutely continuous spectrum, and interesting cases have been studied that lead to a pure point or a purely singular continuous spectrum (see Sections 16 and 18), there are practically no such results for the continuous case (see, however, [Molchanov and Chulaevskii 1984]).

2. In higher dimensions there is a very wide range of interesting open problems. In recent years a first remarkable advance has been made (see 15.C): it was proved that, for a discrete Schrödinger operator with a potential of the form $gQ(x)$, where $Q(x)$ is a metrically transitive field on $\mathbf{Z}^d$, independent at distinct points, the spectrum is pure point near a fluctuation boundary, if the amplitude g is large. The size of the neighborhood where the spectrum is pure point depends on g, and in some cases it is the whole spectrum for g large enough.

However, for a region of the spectrum far enough from any fluctuation boundaries, the spectrum must be absolutely continuous in dimension $d \geqslant 3$, as is commonly believed in the physics literature. In particular, such a situation should obtain near a so-called stable boundary ($\lambda = \infty$ for the Schrödinger operator and $\lambda = 0$ for the divergent elliptic operator (8.1), where λ is the spectral parameter). A mathematical proof of this seemingly universal property of multidimensional metrically transitive operators is one of the most important open problems.

After that, one must tackle the subtle question of the structure of the spectrum near the point separating the absolutely continuous and pure point components, known in the physics literature as the mobility edge. There is at present no proof even of the existence of this point, that is, there is no rigorous statement giving conditions that exclude the coexistence (intersection) of these two spectral components. In Section 18.D we consider an operator—the Schrödinger operator with a nonlocal quasiperiodic potential—for which there is coexistence, so any such statement cannot be absolute. But it is believed that for "random enough" metrically transitive operators—even abstract ones—the mobility edge exists, and there cannot be a singular continuous spectrum. These questions have been mathematically studied very little; we mention the work of Kunz and Souillard [1983], who treated the case of a Schrödinger operator on the Bethe lattice (Cayley tree) with a random, smoothly distributed potential independent at different points.

As numerical calculations and theoretical physics arguments show, generalized eigenfunctions in the vicinity of the mobility edge must have a very complicated behavior, essentially self-similar for a large number of different scales. Such behavior may also characterize the generalized eigenfunctions of one-dimensional almost-periodic operators, whose spectrum can change in character as a result of a change in the amplitude of the potential, rather than in the spectral parameter [Sokoloff 1985]. Another example is that of coefficients having a recurrent structure [Bellissard et al. 1982], as is the case when the associated metrically transitive transformation is characterized by what is called "dynamic chaos": see, for example, [Zaslavskii 1984].

3. An interesting, "intermediate" behavior is expected in two dimensions. Here the spectrum of random operators, as in the one-dimensional case, is expected to be pure point, but possibly with polynomial rather than exponential decay of eigenfunctions at large distances, in some regions of the spectrum (for large values of the spectral parameter, in the case of the Schrödinger operator).

A new class of problems arises for two-dimensional Schrödinger operators with a random potential and a constant or random magnetic field. In this case there should occur subtle effects associated with the eigenvalues of infinite multiplicity (Landau levels), including, for example, an absolutely continuous spectrum in their vicinity. These problems are closely related to the important physical questions of weak localization [Altshuler et al. 1982] and the quantum Hall effect [Thouless 1981, Joynt and Prange 1984], among others.

4. A different approach from the one presented in this book can be pursued in the study of the spectral properties of random matrices. This approach arose in connection with problems of nuclear physics [Landau and Smorodinskii 1957] and was developed by Wigner and Dyson, among others [Wigner 1967; Dyson 1963; Mehta 1967; Pastur 1973; Brody et al. 1981]; it has proved fruitful also in the study of what is called "quantum chaos" [Casati 1985], multidimensional statistical analysis [Girko 1988], and other areas.

The starting point of this approach is a symmetric matrix h_L of order L, all entries of which are of the same order of magnitude (for example, independent, identically distributed random variables). The limit of such an operator for $L \to \infty$ does not exist, but there exist, under certain normalizations, limits for various spectral quantities, and in particular for the eigenvalue distribution function. An important feature of this approach is to obtain, for such asymptotic quantities, closed functional equations that can often be solved explicitly [Marchenko and Pastur 1966; Pastur 1972a]. The relationship between this approach and the one offered in this book was explained by Wegner [1979], but the mathematical status of the problems arising therefrom is still far from clear (but see Appendix B). It is also of great interest to investigate the statistical properties of the distances between nearest eigenvalues (spacings) as $L \to \infty$.

5. Many applications dictate the need to consider not only the traditional problems of spectral theory, but also other ones, associated with the study of solutions of equations with random coefficients. An example is the problem of wave and particle propagation through disordered media. A theoretical physics analysis [Lifshitz et al. 1982a,b] reveals a considerable amount of variation in the behavior of the solutions (including behaviors involving resonance) and in the transmission coefficient. But so far only a few cases have been considered rigorously (see, for example, [Marchenko and Pastur 1985]), and further progress in this area seems unlikely without the development of new mathematical methods.

Many quantities that are functionals of solutions of equations with random coefficients—including those constructed from the eigenvalues and eigenfunctions, as in equations (5.33) and (5.34)—arise naturally in the thermodynamics and kinetics of disordered condensed systems. Examples are the density of states, the coefficient of interband light absorption, the conductivity, and various correlation functions [Bonch-Bruevich et al. 1981; Lifshitz et al. 1982b; Efros and Shklovski 1984; Altshuler et al. 1982; Efetov 1983; Lee and Ramakrishnan 1985]. These quantities are as a rule very complicated, and therefore the basic problem here is to carry out their asymptotic or approximate analysis in various ranges of the parameters and for various classes of metrically transitive coefficients.

Theoretical physics has developed many methods for this analysis, as witnessed by the papers cited in the previous paragraph; but rigorous asymptotic results and even estimates are scarce. We mention [Kunz and Souillard 1980; Martinelli and Scoppola 1987; Figotin and Pastur 1984b]. Similarly, not much has been done in terms of applying the techniques of the quasi-classical approach—which plays an important role in the quantum theory of solids, in radiophysics and in optics—to equations with random coefficients, and the role the percolation theory might play in these problems is also mathematically not well understood. For a discussion of these questions in the context of theoretical physics, see, for example, [Efros and Shklovski 1984].

6. A central problem of theoretical physics is the question of the evolution of a system for very large times, that is, in mathematical terms, the behavior of solutions of Cauchy problems for nonstationary Schrödinger equations or for wave equations. When the characteristics of the disordered medium are independent of time, this question can be formulated, and often solved, in spectral terms. But for a variety of situations in astrophysics, magnetic hydrodynamics, radiophysics, and so on, the coefficients vary with time.

It is known that the solutions in this case typically display a complex hierarchic behavior, called intermittency and characterized by sharp, high peaks, whose separation and height tends to cluster around certain values of widely different orders of magnitude [Molchanov et al. 1985, Zel'dovich et al. 1987]. Any attempt to describe this behavior in detail, including a consistent derivation and analysis of equations for moments of solutions and other, more complicated functionals, involves many difficult new problems.

7. Similar problems arise in the study of nonlinear equations with random coefficients. In theoretical physics, one passes from linear to nonlinear equations when one takes into consideration interactions between particles and waves in a random environment. In this case a variety of new effects are to be expected: even a relatively simple problem like the one-dimensional scattering problem for a random barrier displays new phenomena, such as polynomial, rather than exponential, decay of solutions [Devillard and Souillard 1986]. It is reasonable to expect various regimes of behaviors for the solutions, depending on the relative magnitudes of the random and nonlinear terms, on the formulation of the problem (which can vary much more than in the linear case, where all formulations can be reduced to one by a simple argument), and on many other factors.

Even from this fragmentary discussion it is obvious that the subject of this book falls within a wide class of problems of modern mathematics and related sciences. A mathematician wishing to gain insight into the subject should keep in mind that it may require considerable and unusual efforts to read the related physics literature, and, what is more, he or she must be inclined to listen sympathetically to nonrigorous, but often deep, statements. The reward will be a multitude of interesting and fresh mathematical problems, as well as new and effective approaches to their solution, often requiring serious mathematical understanding.

Chapter I

Metrically Transitive Operators

This introductory chapter starts with a presentation of some general facts of probability theory and spectral operator theory that will be used throughout the book. This material can be found in many textbooks; we have used [Gihman and Skorohod I–III] and [Reed and Simon I–IV], and generally follow their notation. Our presentation is, on the whole, conventional, but we occasionally use different examples, among other reasons because metrically transitive fields and processes that occur often in spectral theory do not necessarily in traditional treatments of probability. Such is the case, for instance, with Poisson fields (1.29) and the field modeling a random alloy (1.31). Our treatment also allows us to consider almost periodic, and even periodic, functions as realizations of metrically transitive fields (Example 1.15(g)).

After that we define metrically transitive operators and establish their simplest and most general spectral properties, such as the nonrandomness of a spectrum and of its components as closed sets on the axis, its dependence on the support of the probability measure, the nonrandomness of multiplicities, and the density of the set of eigenvalues.

1 Basic Definitions and Examples

1.A Random Variables, Functions and Fields

By a *probability space* we mean, as usual, a triple $(\Omega, \mathcal{F}, \mathsf{P})$, where Ω is a set, $\mathcal{F}$ is a σ-algebra of subsets of Ω, and P is a *probability measure*, that is, a positive measure on Ω such that $\mathsf{P}(\Omega) = 1$. The elements of Ω are called *realizations* and those of $\mathcal{F}$ are called *events*. A real-valued $\mathcal{F}$-measurable function on Ω is called a *random variable* if it takes finite values on a subset of Ω of P-measure 1. The expression "on a subset of Ω of P-measure 1" is often abbreviated "with probability 1."

The *mathematical expectation* of an object defined on a probability space is denoted by E, that is, $\mathsf{E}\{\dots\} = \int_\Omega \dots \, d\mathsf{P}$. By $\mathcal{L}^p$, for $1 \leqslant p \leqslant \infty$, we denote the Banach space of random variables ξ with finite p-norm $\|\xi\|_p = \mathsf{E}\{|\xi|^p\}^{1/p}$.

A real-valued random variable ξ can be seen as a map from the measure space $(\Omega, \mathcal{F})$ into the measure space $(\mathbf{R}, \mathcal{B})$, where $\mathcal{B}$ is the σ-algebra of Borel sets. In this way it induces, together with P, a probability measure P_ξ on $(\mathbf{R}, \mathcal{B})$, called the *distribution* of ξ:

$$P_\xi(X) = \mathsf{P}\{\xi \in X\}$$

for $X \in \mathcal{B}$. In the same way, a finite family of real-valued random variables $\xi_1, \ldots, \xi_n$ induces a probability measure $\mathsf{P}_{\xi_1,\ldots,\xi_n}$ on the measure space $(\mathbf{R}^n, \mathcal{B}^n)$, where $\mathcal{B}^n$ is the σ-algebra of Borel sets:

$$P_{\xi_1,\ldots,\xi_n}(X_n) = \mathsf{P}\{(\xi_1, \ldots, \xi_n) \in X_n\}$$

for $X_n \in \mathcal{B}^n$, called the *joint distribution* of the random variables $\xi_1, \ldots, \xi_n$. When dealing with probabilistic problems involving only the random variables $\xi_1, \ldots, \xi_n$, we can restrict our attention to $(\mathbf{R}^n, \mathcal{B}^n, P_{\xi_1,\ldots,\xi_n})$, instead of $(\Omega, \mathcal{F}, \mathsf{P})$.

A real-valued *random function* on a set G is a function $f(g;\omega)$, for $g \in G$ and $\omega \in \Omega$, such that $f(g;\omega)$ is a random variable for any fixed g. For each finite family $g_1, \ldots, g_n$ of elements of G, we can consider the joint distributions of the family of random variables $f(g_1), \ldots, f(g_n)$, denoted by $\mathsf{P}_{g_1,\ldots,g_n}$:

$$P_{g_1,\ldots,g_n}(X_n) = \mathsf{P}\{f(g_1), \ldots, f(g_n) \in X_n\}$$

for $X_n \in \mathcal{B}^n$. These finite-dimensional distributions clearly satisfy the following conditions:

(a) for any $Y_1, \ldots, Y_n \in \mathcal{B}$, any $g_1, \ldots, g_n \in G$ and any permutation $i_1, \ldots, i_n$ of the numbers $1, \ldots, n$,

$$P_{g_{i_1},\ldots,g_{i_n}}(Y_{i_1} \times \cdots \times Y_{i_n}) = P_{g_1,\ldots,g_n}(Y_1 \times \cdots \times Y_n);$$

(b) for any $Y_1, \ldots, Y_n \in \mathcal{B}$ and any $g_1, \ldots, g_n, g_{n+1} \in G$,

$$P_{g_1,\ldots,g_n}(Y_1 \times \cdots \times Y_n) = P_{g_1,\ldots,g_n,g_{n+1}}(Y_1 \times \cdots \times Y_n \times \mathbf{R}).$$

Conditions (a) and (b) are called the *consistency conditions* of a system of finite-dimensional distributions.

We have just seen that the system of finite-dimensional distributions of a random function is consistent; the converse turns out to be true as well:

(1.1) Proposition [Gihman and Skorohod I]. *A system of finite-dimensional distributions is the system of distributions of a random function if and only if it is consistent.* □

We will not prove this important statement, but merely show how, starting from a consistent system of finite-dimensional distributions, the corresponding probability space and random function can be constructed. Given a system of finite-dimensional distributions, indexed by the elements of a set G and satisfying conditions (a) and (b) above, we choose for Ω the set $\mathbf{R}^G$ of all real-valued functions on G. A *cylinder* in $\mathbf{R}^G$ is a set of the form

$$\{f \in \mathbf{R}^G : (f(g_1), \dots, f(g_n)) \in X^n\},$$

for $X^n \in \mathcal{B}^n$ and $g_1, \dots, g_n \in G$; cylinders clearly form an algebra of subsets of $\mathbf{R}^G$. Define a function P_G on this algebra by setting $P_G(\hat{X}) = P_{g_1,\dots,g_n}(\hat{X})$ for $\hat{X}$ of the form $\{f \in \mathbf{R}^G : (f(g_1), \dots, f(g_n)) \in X^n\}$. The consistency conditions imply that P_G is a finite additive function on the algebra of cylinders in $\mathbf{R}^G$, and obviously $P_G(\mathbf{R}^G) = 1$. Extending P_G to the minimal σ-algebra $\mathcal{B}^G$ containing all cylinders, we obtain a σ-additive function on $\mathcal{B}^G$, that is, a probability measure [Gihman and Skorohod I]. On the probability space $(\mathbf{R}^G, \mathcal{B}^G, P_G)$ we define a random function

$$f(g; \omega) = \omega(g), \tag{1.1}$$

for $g \in G$ and $\omega \in \mathbf{R}^G$, which is easily seen to have the prescribed system of finite-dimensional distributions.

Now consider an arbitrary real-valued random function f on a probability space $(\Omega, \mathcal{F}, \mathsf{P})$, and construct from its finite-dimensional distributions, as above, a probability space $(\mathbf{R}^G, \mathcal{B}^G, P_{f,G})$. The probability measure $P_{f,G}$ on the measure space $(\mathbf{R}^G, \mathcal{B}^G)$ plays the same role as the joint distribution of $P_{\xi_1,\dots,\xi_n}$ on $(\mathbf{R}^n, \mathcal{B}^n)$ in the case of a finite set $\xi_1, \dots, \xi_n$ of random variables (which, by the way, may be regarded as a random function on the set $G = \{1, \dots, n\}$). Thus, if we are only interested in probabilistic questions about f, we can disregard the original probability space $(\Omega, \mathcal{F}, \mathsf{P})$ and think instead in terms of $(\mathbf{R}^G, \mathcal{B}^G, P_{f,G})$, with f given by the canonical map (1.1). In any case, a real-valued random function is often given by its finite-dimensional distributions, that is, by the measure $P_{f,G}$ and the canonical map (1.1). We discuss these well-known facts because below we will often tacitly identify random functions f on G with their corresponding spaces $(\mathbf{R}^G, \mathcal{B}^G, P_{f,G})$.

In this book the set G is generally either $\mathbf{R}^d$ or $\mathbf{Z}^d$, for $d \geqslant 1$. Random functions defined on commutative groups are called *random fields*, and one-dimensional random fields are called *random processes*. The relationship between the probability and group structures, which is of central importance for the rest of the book, will be established in 1.C by the introduction of the concept of metric transitivity, roughly a sort of invariance of the probability space under the action of the group.

(1.2) Examples

(a) Sequences of independent, identically distributed random variables. Let F be a probability measure on $(\mathbf{R}, \mathcal{B})$, and consider the system of finite-dimensional distributions determined by

$$P_{g_1,\ldots,g_n}(X_1 \times \cdots \times X_n) = F(X_1) \cdots F(X_n)$$

for any $X_1, \ldots, X_n \in \mathcal{B}$ and $g_1, \ldots, g_n \in \mathbf{Z}$. This system is obviously consistent, and therefore gives rise to a probability space $(\mathbf{R}^{\mathbf{Z}}, \mathcal{B}^{\mathbf{Z}}, P_{f,\mathbf{Z}})$, where the space of realizations is the set $\mathbf{R}^{\mathbf{Z}}$ of doubly infinite sequences of real numbers and $f(g;\omega) = \omega(g)$ for $\omega \in \mathbf{R}^{\mathbf{Z}}$ and $g \in \mathbf{Z}$.

This construction can also be carried out with $G = \mathbf{Z}^d$, for any $d \geqslant 1$, giving a random field on the lattice $\mathbf{Z}^d$, with independent values at each point.

(b) Gaussian processes. Let $b : \mathbf{R} \to \mathbf{R}_+$ be a positive function, and consider the finite-dimensional distributions indexed by $G = \mathbf{R}$ and defined by

$$P_{g_1,\ldots,g_n}(X^n) = \left((2\pi)^n \det B_{g_1,\ldots,g_n}\right)^{-1/2} \times \int_{X^n} \exp\left(-\tfrac{1}{2}(B^{-1}_{g_1,\ldots,g_n})_{ij} x_i x_j\right) dx_1 \cdots dx_n,$$

where $g_1, \ldots, g_n \in \mathbf{R}$, $X^n \in \mathcal{B}^n$ and the $n \times n$ positive definite matrix $B_{g_1,\ldots,g_n}$ has entries $b_{ij} = b(g_i - g_j)$. This system is consistent and so gives rise to a probability space $(\mathbf{R}^{\mathbf{R}}, \mathcal{B}^{\mathbf{R}}, P_{q,\mathbf{R}})$, where the space of realizations is the set $\mathbf{R}^{\mathbf{R}}$ of real-valued functions on $\mathbf{R}$ and $q(x;\omega) = \omega(x)$ for $\omega \in \mathbf{R}^{\mathbf{R}}$ and $x \in \mathbf{R}$. The random field $q(x;\omega)$, also written just $q(x)$, has zero mathematical expectation, $\mathsf{E}\{q(x)\} = 0$, and its correlation function is b, that is, $\mathsf{E}\{q(x)q(y)\} = b(x-y)$. These two conditions uniquely characterize the field.

A similar construction in the case of $\mathbf{R}^d$, for $d > 1$, results in a Gaussian random field $q(x)$ on $\mathbf{R}^d$, also with zero mathematical expectation and correlation function $b(x)$.

When dealing with a random function $f(g)$, for $g \in G$, one often considers sets of realizations of the form

$$\{\omega : f(g,\omega) \in F \text{ for all } g \in G_1\}, \tag{1.2}$$

where F is closed in $\mathbf{R}$ and G_1 is a subset of G. When G_1 is uncountable, this set can fail to be an event, that is, it can lie outside the σ-algebra $\mathcal{F}$. This difficulty is usually overcome by "correcting" the random function f on sets of P-measure zero. More specifically, let G be a separable metric space and I a countable and dense subset of G. Consider the set

(1.3) $$\{\omega : f(g,\omega) \in F \text{ for all } g \in G_1 \cap I\}.$$

The real-valued random function f will be called *separable* if for any closed set F in $\mathbf{R}$ and any $G_1 \subset G$ the symmetric difference between sets (1.2) and (1.3) is contained in a set in $\mathcal{F}$ of P-measure zero.

Two real-valued random functions f_1 and f_2 are called *stochastically equivalent* if $\mathsf{P}\{f_1(g) \neq f_2(g)\} = 0$.

(1.3) Proposition [Doob 1952]. *For any real-valued random function $f(g)$, $g \in G$, where G is a separable metric space, there exists a separable random function $\tilde{f}(G)$, for $g \in G$, with values in $\mathbf{R} \cup \{\infty\}$, that is stochastically equivalent to f.* □

Doob's result shows that without loss of generality we can restrict our attention to separable random functions, and we will do so from now on.

All that we've said above for real-valued random functions holds also for complex-valued ones, and, in general, for random functions with values in a complete, separable metric space [Gihman and Skorohod I].

1.B Random Vectors and Operators

In this book we concentrate mainly on self-adjoint matrix and differential operators. But a wide range of problems in spectral theory, which arose initially in the study of these special, if very important, operators, can be formulated in the more general framework of abstract operators. This section describes this framework and some of the results and examples that can be couched in it. Other definitions of random vectors and operators, and the interrelations among them, can be found in [Gihman and Skorohod I] and [Skorohod 1984].

All our Hilbert spaces are assumed separable, that is, endowed with a countable dense subset. We will mostly denote them by $\mathcal{H}$, and the inner product in them by $(\cdot\,,\,\cdot)$.

Definition. *A random vector h in a Hilbert space $\mathcal{H}$ is a map $h : \Omega \to \mathcal{H}$ such that (u,h) is a random variable for any $u \in \mathcal{H}$.*

Definition. *Let $\mathcal{D}$ be a dense linear subspace of $\mathcal{H}$. A random* operator *with domain $\mathcal{D}$ is a map A from the set of realizations Ω into the set of linear operators on $\mathcal{H}$, such that $\mathcal{D}$ lies in the domain of A with probability 1 and that Au is a random vector in $\mathcal{H}$ for all $u \in \mathcal{D}$. If, in addition, $(Au,v) = (u,Av)$ for all $u,v \in \mathcal{D}$, we say that A is* symmetric.

(1.4) Examples of random symmetric operators

(a) Jacobi matrices. For $\mathcal{H} = \ell_2(\mathbf{Z})$, let $\mathcal{D}$ be the set of sequences with compact support in $\mathcal{H}$. If $s(x)$ and $q(x)$, for $x \in \mathbf{Z}$, are two sequences of random variables, the operator J given by

$$(J\psi)(x) = -s(x)\psi(x-1) - s(x+1)\psi(x+1) + q(x)\psi(x), \tag{1.4}$$

for $\psi \in \mathcal{D}$ and $x \in \mathbf{R}$, is a random symmetric operator on $\ell_2(\mathbf{Z})$ with domain $\mathcal{D}$.

(b) Schrödinger and Sturm–Liouville operators. Here $\mathcal{H} = L^2(\mathbf{R})$, $\mathcal{D} = C_0^\infty(\mathbf{R})$ is the subspace of smooth functions with compact support, and $q(x)$ is a real-valued random process whose realizations, with probability 1, are bounded on each finite interval. The operator H given by

$$(H\psi)(x) = -\psi''(x) + q(x)\psi(x), \tag{1.5}$$

for $\psi \in \mathcal{D}$ and $x \in \mathbf{R}$, is a random operator, called the one-dimensional Schrödinger or Sturm–Liouville operator with random potential $q(x)$ [Levitan and Sargsyan 1970; Hartman 1964].

The name Sturm–Liouville is often used for operators of a more general form, defined by the equation

$$-\big(s(x)\psi'(x)\big)' + q(x)\psi(x) = \lambda r(x)\psi(x), \tag{1.6}$$

where $r(x)$, as well as $q(x)$, is bounded in each finite interval, $s(x)$ is continuously differentiable and $s(x) \geqslant 1$, $r(x) \geqslant 1$. Such an operator may be either considered on the space of square-integrable functions with weights $r(x)$, or reduced to the Schrödinger operator by a suitable change of variables.

We have confined ourselves here to examples of one-dimensional finite-difference and differential operators of the second order, which are the most frequent and interesting in many respects. We could have included one-dimensional finite-difference and differential operators of arbitrary finite order, but we will consider such operators in the more general multidimensional case.

(c) Multidimensional matrix operators. For $\mathcal{H} = \ell_2(\mathbf{Z}^d)$, let $\mathcal{D}$ be the set of sequences with finite support in $\mathcal{H}$, and $\{a(x,y)\}$, for $x, y \in \mathbf{Z}^d$, a family of complex-valued random variables with the following properties:

(1) $a(x,y) = a^*(y,x)$ for $x, y \in \mathbf{Z}^d$;

(2) $\sum_{x \in \mathbf{Z}^d} |a(x,y)|^2 < \infty$ for all $y \in \mathbf{Z}^d$ with probability 1.

Then there is a symmetric operator A in $\mathcal{H}$ given by

$$(A\psi)(x) = \sum_{y\in\mathbf{Z}^d} a(x,y)\psi(y) \tag{1.7}$$

for $\psi \in \mathcal{D}$ and $x \in \mathbf{Z}^d$.

A special case is the operator A of the form

$$A = A_0 + Q, \tag{1.8}$$

where A_0 is a multidimensional finite-difference operator, that is, $a_0(x,y) = a_0(x-y)$ with $\sum_{x\in\mathbf{Z}^d}|a_0(x)| < \infty$, and Q is the operator of multiplication by a random field $q(x)$, for $x \in \mathbf{Z}^d$. This operator is common in physics; if $a_0(x) = 0$ for $|x| > 1$ and $q(x)$ is independent it is called the *Anderson model.*

(d) Integral operators. Let G be a locally compact, metrizable abelian group, with Haar measure dg. Set $\mathcal{H} = L^2(G, dg)$, let D be the set of complex-valued continuous functions on G with compact support, and $a(x,y)$ a random complex-valued field on $G \times G$ such that:

(1) $a(x,y) = a^*(y,x)$ for $x, y \in G$;

(2) $\int_G |a(x,y)|^2\, dx$ is a random variable dg-almost everywhere;

(3) $b(y) = \int_G |a(x,y)|^2\, dx$, for $y \in G$, is a locally integrable field.

Then there is a symmetric operator A on $\mathcal{H}$ given by

$$(A\psi)(x) = \int_G a(x,y)\psi(y)\, dy$$

for $x \in G$ and $\psi \in \mathcal{D}$. The operator defined in (c) is clearly a special case of this, with $G = \mathbf{Z}^d$.

(e) Multidimensional Schrödinger operators. The Anderson model considered in (c) is easily seen to be the finite-difference analogue of the following well-known and intensively studied operator. Set $\mathcal{H} = L^2(\mathbf{R}^d)$, let $\mathcal{D} = C_0^\infty(\mathbf{R}^d)$ be the set of smooth functions with compact support on $\mathbf{R}^d$, and $g(x)$, for $x \in \mathbf{R}^d$, a real-valued random field whose realizations are locally square-integrable with probability 1. A symmetric random operator H, called the *multidimensional Schrödinger operator*, is defined by the equality

$$(H\psi)(x) = -\Delta\psi(x) + q(x)\psi(x)$$

for $\psi \in \mathcal{D}$ and $x \in \mathbf{R}^d$, where Δ is the Laplace operator in $L^2(\mathbf{R}^d)$. For $d = 1$ we obtain the operator (1.5) from part (b).

A more general class of operators, including multidimensional Schrödinger operators as a particular case, is the following:

(f) Divergence-type elliptic differential operators. Let $\mathcal{H} = L^2(\mathbf{R}^d)$ and $\mathcal{D} = C_0^\infty(\mathbf{R}^d)$ be as in the preceding example. Set

$$\partial_k = \frac{\partial}{\partial x_k} \quad \text{and} \quad \partial^\alpha = \prod_{k=1}^d \partial_k^{\alpha_k},$$

where $\alpha = (\alpha_1, \dots, \alpha_d) \in \mathbf{Z}_+^d$ is a multiindex with $|\alpha| = \sum_{k=1}^d \alpha_k$. Assume that for each multiindex α with $|\alpha| \leqslant p$, where p is a natural number, we are given random fields $a_\alpha(x)$, for $x \in \mathbf{R}^d$, such that $a_\alpha \in C^{|\alpha|}(\mathbf{R}^d)$ with probability 1 and $a_\alpha(x)$ is real for $|\alpha| = p$. The random symmetric operator A is

$$A = \sum_{|\alpha| \leqslant p} i^{|\alpha|} a_\alpha(x) \partial^\alpha, \tag{1.9}$$

and we assume that A is formally self-adjoint and elliptic, that is, p is even and

$$a_p(x, \xi) = \sum_{|\alpha| = p} a_\alpha(x) \xi^\alpha \geqslant \varepsilon(x) |\xi|^p$$

for $\varepsilon(x) > 0$ and $\xi \in \mathbf{R}^d$.

(1.5) Theorem. *Let h and h_1 be random vectors in $\mathcal{H}$ and ξ a random variable. Then*
(i) *ξh and $h + h_1$ are random vectors;*
(ii) *$\|h\|$ and (h, h_1) are random variables.*

Proof. Part (i) follows immediately from the definition of a random vector. Part (ii) follows from the equality $(h, h_1) = \sum_i (h, e_i)(e_i, h_i)$ and the similar equality for $\|h\|^2$, where $\{e_i\}$ is the orthonormal basis of $\mathcal{H}$ that guarantees the $\mathcal{F}$-measurability of the functions of ω and their boundedness for P-almost every ω. □

Definition. (a) *A random projection π in a Hilbert space $\mathcal{H}$ is a random operator in $\mathcal{H}$ whose realizations are orthogonal projections in $\mathcal{H}$.*
(b) *A subspace X of a Hilbert space $\mathcal{H}$ is called random if the orthogonal projection on X, which we denote by $\pi(X)$, is random.*

(1.6) Theorem. *If $\{h_i\}_1^\infty$ is a set of random vectors in $\mathcal{H}$, the subspace X spanned by $\{h_i\}_1^\infty$ is random in $\mathcal{H}$.*

Proof. We apply the Gram–Schmidt orthogonalization process to the sequence $\{h_i\}_1^\infty$, defining a new sequence $\{u_l\}_1^\infty$ by the recurrence relation

$$u_1 = \lim_{n\to\infty} \frac{h_1}{n^{-1} + \|h_1\|}, \quad u_l = \lim_{n\to\infty} \frac{h_l - \sum_{k=1}^{l-1}(h_l, u_k)u_k}{n^{-1} + \left\|h_l - \sum_{k=1}^{l-1}(h_l, u_k)u_k\right\|}.$$

Using Theorem 1.5, we verify that the vectors $\{u_l\}_1^\infty$ are random, pairwise orthogonal, and, for each realization of ω, either have norm 1 or are equal to 0; it is this second possibility that necessitates taking the limit with respect to n in the definition of u_l. Also, $\{u_l\}_1^\infty$ and $\{h_i\}_1^\infty$ clearly span the same subspace X. It follows that the projection $\pi(X)$ onto X is given by

$$\pi(X)v = \sum_{l=1}^{\infty}(v, u_l)u_l$$

for $v \in \mathcal{H}$, showing that $\pi(X)$, and hence X, are random. □

1.C Metrically Transitive Random Fields

We start by introducing metric transitivity in the case of a real-valued random field on the set of integers $\mathbf{Z}$, that is, on a sequence of real-valued random variables. There are two conditions. The first is that the field $f(g)$, for $g \in \mathbf{Z}$, be *homogeneous* (or *stationary*, since in many applications g is interpreted as time). Homogeneity is here understood as invariance under translation in $\mathbf{Z}$ of the finite-dimensional distributions associated with the field. In symbols,

$$P_{g_1,\ldots,g_n} = P_{g_1+g,\ldots,g_n+g} \tag{1.10}$$

for any $g, g_1, \ldots, g_n \in \mathbf{Z}$.

We can also formulate the concept of homogeneity without direct reference to finite-dimensional distributions, and will do so in view of subsequent generalizations. Consider the probability space $(\mathbf{R}^{\mathbf{Z}}, \mathcal{B}^{\mathbf{Z}}, P_{f,\mathbf{Z}})$ constructed in 1.A from the finite-dimensional distributions of a random field f. Define a group of shifts T_g, for $g \in \mathbf{Z}$, in the space of realizations $\mathbf{R}^{\mathbf{Z}}$:

$$(T_g\omega)(g_1) = \omega(g + g_1) \tag{1.11}$$

for $g, g_1 \in \mathbf{Z}$ and $\omega \in \mathbf{R}^{\mathbf{Z}}$. Each operator T_g, for $g \in \mathbf{Z}$, is a one-to-map map from $\mathbf{R}^{\mathbf{Z}}$ (or $\mathcal{B}^{\mathbf{Z}}$) into itself, and the homogeneity condition (1.10) will take the form of a measure-preservation condition:

$$P_{f,\mathbf{Z}}(T_gX) = P_{f,\mathbf{Z}}(X) \tag{1.12}$$

for $g \in \mathbf{Z}$ and $X \in \mathcal{B}^{\mathbf{Z}}$.

Notice that the group $\{T_g : g \in \mathbf{Z}\}$, is isomorphic to $\mathbf{Z}$, and $T_g = (T_1)^g$. In view of (1.1) and (1.11), we have

(1.13) $$f(g+g_1,\omega)=\omega(g_1+g)=(T_{g_1}\omega)(g)=f(g,T_{g_1}\omega).$$

Using (1.12) and (1.13) we can now characterize a homogeneous random field without reference to finite-dimensional distributions:

Definition. *A random field $f(g)$, for $g\in\mathbf{Z}$, is called homogeneous if there is a group $\mathcal{T}=\{T_g : g\in\mathbf{Z}\}$ of automorphisms of the probability space $(\Omega,\mathcal{F},\mathsf{P})$ associated with f, such that:*
(a) *$\mathcal{T}$ is isomorphic to $\mathbf{Z}$, that is, $T_{g+g_1}=T_gT_{g_1}$ for $g,g_1\in\mathbf{Z}$, and T_g is the identity if and only if $g=0$;*
(b) *$\mathsf{P}(T_gX)=\mathsf{P}(X)$ for all $X\in\mathcal{F}$;*
(c) *$f(g+g_1,\omega)=f(g,T_{g_1}\omega)$ for all $\omega\in\Omega$ and $g,g_1\in\mathbf{Z}$.*

The second condition for metric transitivity of a random field is the vanishing of the correlations between field values in regions that are infinitely far from one another. This notion can be formalized in several ways; we start with the weakest one, *ergodicity*. Keeping in mind that $T_g=T^g$, for $T=T_1$, the ergodicity condition is
(1.14)
$$\lim_{n\to\pm\infty}\frac{1}{|n|}\sum_{k=0}^{n-1}\mathsf{P}(T^kX\cap Y)=\lim_{n\to\infty}\frac{1}{2n+1}\sum_{k=-n}^{n}\mathsf{P}(T^kX\cap Y)=\mathsf{P}(X)\mathsf{P}(Y)$$

for all X and Y in the σ-algebra $\mathcal{F}$. From this condition follows an important property of the automorphism T, which we use in the definition of metrical transitivity. Assuming that $X\in\mathcal{F}$ is an event invariant under T, and making $Y=X$ in (1.14), we get $\mathsf{P}^2(X)=\mathsf{P}(X)$, that is, $\mathsf{P}(X)$ equals either 0 or 1.

Definition. *A measure-preserving automorphism T of a probability space is called metrically transitive if any T-invariant event has measure 0 or 1. In this case, the group generated by the integer powers of T is also called metrically transitive.*

It turns out that the metric transitivity of an automorphism T not only follows from ergodicity, but is in fact equivalent to it:

(1.7) Proposition (Birkhoff–Khintchine ergodic theorem [Gihman and Skorohod I]). *Let T be a metrically transitive automorphism of a probability space and ξ a random variable having a finite mathematical expectation. With probability* 1, *we have*

(1.15) $$\lim_{n\to\pm\infty}\frac{1}{|n|}\sum_{k=0}^{n-1}\xi(T^k\omega)=\lim_{n\to\infty}\frac{1}{2n+1}\sum_{k=-n}^{n}\xi(T^k\omega)=\mathsf{E}\{\xi\}.\qquad\square$$

A stronger condition than ergodicity for an automorphism T of a probability space is the following:

Definition. *We call a measure-preserving automorphism of a probability space* $(\Omega, \mathcal{F}, \mathsf{P})$ *mixing if it satisfies*

$$\lim_{|n|\to\infty} \mathsf{P}(T^n X \cap Y) = \mathsf{P}(X)\mathsf{P}(Y) \tag{1.16}$$

for $X, Y \in \mathcal{F}$.

Condition (1.16) clearly implies (1.14), so a mixing automorphism is automatically metrically transitive. It is also easy to see that (1.16) is equivalent to the following relation between two random variables ξ and η in the space $\mathcal{L}^2$ of P-square-integrable functions:

$$\lim_{|n|\to\infty} \mathsf{E}\{\xi(T^n\omega)\eta(\omega)\} = E\{\xi\}\mathsf{E}\{\eta\}. \tag{1.17}$$

Definition. *Let* $f(g)$, *for* $g \in \mathbf{Z}$, *be a random field. Denote by* $\mathcal{F}_n$ *the* σ-*algebra generated by the random variables* $f(g)$, *for* $g \in \mathbf{Z}$ *such that* $|g| \geqslant n\}$, *and by* $\mathcal{F}_\infty$ *the intersection* $\bigcap_{n\geqslant 1} \mathcal{F}_n$. *We say that* $\mathcal{F}_n$ *satisfies the zero-one law if* $\mathsf{P}(X)$ *is* 0 *or* 1 *for all* $X \in \mathcal{F}_\infty$.

(1.8) Proposition [Gihman and Skorohod I]. *If a homogeneous field* $f(g)$ *satisfies the zero-one law, the corresponding measure-preserving automorphism* $T = T_1$ *(from the definition of homogeneity) is mixing and therefore metrically transitive.* □

Equation (1.17) and Proposition 1.8 are useful in checking the validity of the mixing condition.

An even stronger formulation of the vanishing of correlations with increasing distance is the following. Let $\mathcal{F}_n^\infty$ and $\mathcal{F}_{-\infty}^n$ be the σ-algebras generated by the random variables $f(g)$ for $g \geqslant n$ and $g \leqslant n$, respectively. A measure-preserving automorphism of the probability space T is said to be *strongly mixing* if

$$\lim_{|m-n|\to\infty} \sup_{\substack{X\in\mathcal{F}_{-\infty}^n \\ Y\in\mathcal{F}_m^\infty}} \left|\mathsf{P}(X\cap Y) - \mathsf{P}(X)\mathsf{P}(Y)\right| = 0. \tag{1.18}$$

This condition clearly implies (1.16) and (1.17). Both mixing conditions (1.16) and (1.18) mean that the points of the set $T^n X$ are, for $|n| \to \infty$, "uniformly distributed," or mixed, over the entire space Ω.

There is a somewhat different classification of random fields in terms of the decay of their correlation function; we shall consider it in one dimension only.

Let $f(g)$, for $g \in G$, be a random field on either $G = \mathbf{Z}$ or $G = \mathbf{R}$. Denote by $\mathcal{F}(f)$ the σ-algebra generated by the random variables $f(g)$, for $g \in G$; by $\mathcal{F}^n_{-\infty}(f)$ the σ-algebra generated by $f(g)$, for $g \in G$ with $g \leqslant n$; by $\mathcal{F}_{-\infty}(f)$ the intersection

$$\mathcal{F}^{\infty}(f) = \bigcap_n \mathcal{F}^n_{-\infty}(f) = \lim_{n \to -\infty} \mathcal{F}^n_{-\infty}(f),$$

called a *tail algebra*; and by $\mathcal{F}^{(0)}(f)$ the σ-algebra $\{X \in \mathcal{F}_{-\infty}(f) : \mathsf{P}(X) = 0 \text{ or } 1\}$. We clearly have $\mathcal{F}(f) \supset \mathcal{F}_{-\infty}(f) \supset \mathcal{F}^{(0)}(f)$.

Definition. *A field $f(g)$, for $g \in G$, is called:*
(a) *singular, or deterministic, if $\mathcal{F}_{-\infty}(f) = \mathcal{F}(f)$;*
(b) *nondeterministic if $\mathcal{F}_{-\infty}(f)$ is strictly included in $\mathcal{F}(f)$;*
(c) *regular, or completely nondeterministic, if $\mathcal{F}_{-\infty}(f) = \mathcal{F}^{(0)}(f)$.*

Regularity can be expressed by a condition similar to the ones defining ergodicity (1.14), mixing (1.16) and strong mixing (1.18). The condition is

$$\lim_{n \to -\infty} \sup_{Y \in \mathcal{F}^n_{-\infty}} \left|\mathsf{P}(X \cap Y) - \mathsf{P}(X)\mathsf{P}(Y)\right| = 0$$

for all $X \in \mathcal{F}(f)$. This relation suggests that regularity lies between mixing (1.16) and strong mixing (1.18). Strong mixing is sometimes called *complete regularity.*

The concept of a regular random process arose naturally from the study of the problem of predicting, or extrapolating, the future behavior of a time series from its past behavior. In this framework, a situation is considered regular when the prediction gets worse as it advances more and more into the future, because it is natural to assume that for a "truly random" process information about past events should be gradually "forgotten." Therefore the possibility of accurate prediction of a process can be associated with its level of "degeneracy," from a probabilistic point of view.

An example of a regular process is a sequence of independent, identically distributed random variables ($G = \mathbf{Z}$); a singular process is exemplified by periodic and almost periodic functions ($G = \mathbf{R}$). See Examples 1.15(a) and 1.15(g) below.

We now define metric transitivity in the general case of a locally compact commutative group G that is also a separable metric space. In view of subsequent applications, we will want the realizations of a random field

$f(g,\omega)$ for $g \in G$, when $G = \mathbf{R}$ or $\mathbf{R}^d$, to be measurable functions of G for P-almost every ω. Therefore, our definition of metric transitivity will include the requirement of stochastic continuity:

Definition. *A topological group $\mathcal{T}$ of automorphisms of probability spaces is called stochastically continuous if, for any $X \in \mathcal{F}$ and $T \in \mathcal{T}$, we have*

$$\lim_{T_1 \to T} P(T_1 X \bigtriangleup TX) = 0,$$

where $\bigtriangleup$ denotes the symmetric difference of sets: $X \bigtriangleup Y = (X \cup Y) \backslash (X \cap Y)$.

Definition. *A stochastically continuous group $\mathcal{T}$ of measure-preserving automorphisms of a probability space is called metrically transitive if any $\mathcal{T}$-invariant set from the σ-algebra $\mathcal{F}$ has probability 0 or 1. In other words, any $X \in \mathcal{F}$ such that $TX = X$ for all $T \in \mathcal{T}$ satisfies $\mathsf{P}(X) = 0$ or 1.*

Definition. *A random field $f(g)$, for $g \in G$, is called metrically transitive if there exists a stochastically continuous metrically transitive group $\{T_g : g \in G\}$ of automorphisms of the associated probability space $(\Omega, \mathcal{F}, \mathsf{P})$ that is a homomorphic image of G. In symbols,*

$$f(g + g_1, \omega) = f(g, T_{g_1}\omega)$$

for all $g, g_1 \in G$ and $\omega \in \Omega$.

In the canonical representation of the random field f by means of a probability space $(\mathbf{R}^G, \mathcal{B}^G, P_{f,G})$, the group $\{T_g : g \in G\}$ is given by equation (1.11).

(1.9) Proposition. *The realizations of a metrically transitive random field $f(g)$, for $g \in G$, are measurable functions of g with probability 1.*

Proof. By definition, a metrically transitive random field $f(g)$, for $g \in G$, is stochastically continuous, that is,

$$\lim_{g_1 \to g} \mathsf{P}\{|f(g_1) - f(g)| > \varepsilon\} = 0 \tag{1.19}$$

for any $\varepsilon > 0$. The proposition follows easily [Gihman and Skorohod I]. □

It is easy to construct metrically transitive fields on a metrically transitive commutative group $\mathcal{T}$: just choose an arbitrary random variable ξ, that is, a measurable function on $(\Omega, \mathcal{F}, \mathsf{P})$, and set

$$f(T, \omega) = \xi(T\omega) \qquad \text{for } T \in \mathcal{T}. \tag{1.20}$$

A random variable ξ on the probability space $(\Omega, \mathcal{F}, \mathsf{P})$ is *invariant* under an automorphism T of this space if $\xi(T\omega) = \xi(\omega)$ for all $\omega \in \Omega$. The following simple result is left as an exercise to the reader:

(1.10) Theorem. *A group $\mathcal{T}$ of measure-preserving automorphisms of a probability space is metrically transitive if and only if the set of random variables invariant with respect to all $T \in \mathcal{T}$ reduces to a set of constants.* □

We also leave to the reader the straightforward reformulation of the mixing conditions (1.16) and (1.17) for the case of the groups $\mathbf{Z}^d$ and $\mathbf{R}^d$. Similarly, the ergodicity property (1.14), which is equivalent to metric transitivity, can be extended to the cases $G = \mathbf{Z}^d, \mathbf{R}^d$, with $d \geqslant 1$. We have the following results:

(1.11) Proposition [Dunford and Schwartz 1962]. *If $\{T_x : x \in \mathbf{Z}^d\}$ is a metrically transitive group, and $\xi \in \mathcal{L}^1$ a random variable, the following relations hold with probability* 1:
(1.21)
$$\lim_{|n| \to \infty} |n|^{-d} \sum_{x_1, \ldots, x_d = 0}^{n-1} \xi(T_x \omega) = \lim_{n \to \infty} (2n+1)^{-d} \sum_{x_1, \ldots, x_d = -n}^{n} \xi(T_x \omega) = \mathsf{E}\{\xi\}.$$

If, moreover, $\xi \in \mathcal{L}^p$, for some $1 \leqslant p < \infty$, the same convergence takes place in $\mathcal{L}^p$. □

(1.12) Proposition [Dunford and Schwartz 1962]. *If $\{T_x : x \in \mathbf{R}^d\}$ is a metrically transitive group, and $\xi \in \mathcal{L}^1$ a random variable, the following relations hold with probability* 1:

$$(1.22) \quad \lim_{|t| \to \infty} |t|^{-d} \int_0^t \cdots \int_0^t \xi(T_x \omega)\, dx_1 \cdots dx_d$$
$$= \lim_{t \to \infty} |2t|^{-d} \int_{-t}^t \cdots \int_{-t}^t \xi(T_x \omega)\, dx_1 \cdots dx_d = \mathsf{E}\{\xi\}.$$

If, moreover, $\xi \in \mathcal{L}^p$, for some $1 \leqslant p < \infty$, the same convergence takes place in $\mathcal{L}^p$. □

The ergodic theorems 1.7, 1.11 and 1.12 can obviously be rephrased in terms of metrically transitive fields.

(1.13) Proposition. *Let G be $\mathbf{Z}^d$ or $\mathbf{R}^d$, for $d \geqslant 1$, and let $f(g)$, for $g \in G$, be a metrically transitive field such that $\mathsf{E}\{|f(0)|\} < \infty$. Denoting by Λ a cube of volume V centered at the origin, we have, with probability* 1:

$$\lim_{V\to\infty} V^{-1} \int_{\Lambda} f(g)\,dg = \mathsf{E}\{f(0)\}, \tag{1.23}$$

where for $G = \mathbf{Z}^d$ integration should be understood as summation. If, moreover, $f(0) \in \mathcal{L}^p$ for some $1 \leqslant p < \infty$, the same convergence takes place in $\mathcal{L}^p$. □

(1.14) Examples of metrically transitive groups

(a) For G a compact abelian group, with Haar measure dg normalized to 1, consider the probability space $(G, \mathcal{B}(G), dg)$, where $\mathcal{B}(G)$ the Borel σ-algebra on G. Let $\mathcal{T}$ be the group of automorphisms of this probability space defined by $T_g\omega = \omega + g$ for $\omega, g \in G$. One verifies readily that this group is metrically transitive. This example will be important when we consider almost periodic functions in Example 1.15(g).

(b) Consider the probability space $(\mathbf{R}^{\mathbf{Z}}, \mathcal{B}^{\mathbf{Z}}, P_{f,\mathbf{Z}})$ from Example 1.2(a), corresponding to a sequence of independent, identically distributed random variables. Define the group $\mathcal{T} = \{T_g : g \in G\}$ by (1.11); we show that $\mathcal{T}$ is metrically transitive. It is enough to prove that the automorphism T_1 is mixing. To do this, we could check the validity of the zero-one law and use Proposition 1.8; but in this case there is a more straightforward way. Take arbitrary cylinders X, Y. If $|g|$ is sufficiently large, $T_1^g X$ and Y are independent, that is,

$$\mathsf{P}(T_1^g X \cap Y) = \mathsf{P}(T_1^g X)\mathsf{P}(Y) = \mathsf{P}(X)\mathsf{P}(Y),$$

which shows (1.16) in this case. Since every $X \in \mathcal{B}^{\mathbf{Z}}$ can be approximated by cylinders, that is, for every $\varepsilon > 0$ one can find a cylinder X_ε such that $P(X \triangle X_\varepsilon) < \varepsilon$ [Gihman and Skorohod I], it follows that (1.16) holds for all $X, Y \in \mathcal{B}^{\mathbf{Z}}$.

(1.15) Examples of metrically transitive random fields

(a) Independent, identically distributed random variables. This field was constructed in Example 1.2(a), for $G = \mathbf{Z}$ or $\mathbf{Z}^d$. Its metric transitivity follows from the metric transitivity of the corresponding group of automorphisms, which was proved in Example 1.12(b).

(b) Random Markov processes. We start by recalling the definition of a Markov process in our context [Gihman and Skorohod II]. A real-valued random process $q(x)$, for $x \in \mathbf{R}$, is called a *homogeneous Markov process* if there is, on the set Ω of functions $q : \mathbf{R} \to \mathbf{R}$, a family of measures $\mathsf{P}_{x,q}$ with the following properties:

(i) if $\mathcal{F}_x^y$ denotes the σ-algebra generated by the random variables $\{q(z) : x \leqslant z \leqslant y\}$, then $\mathsf{P}_{x,q}$ is a probability measure on the measure space $(\Omega, \mathcal{F}_x^\infty)$, and $\mathsf{P}_{x,q}\{q(x) = q\} = 1$;

(ii) for any $y \geqslant x \geqslant 0$ and $Q \in \mathcal{B}$ the function

$$P(x, q; y, Q) = \mathsf{P}_{x,q}\{q(y) \in Q\}$$

is $\mathcal{B}(\mathbf{R})$-measurable with respect to q;

(iii) if $P\{\cdot \mid \mathcal{F}_1\}$ denotes the conditional probability with respect to a fixed σ-subalgebra $\mathcal{F}_1$, we have, for any $z \geqslant y \geqslant x$ and $Q \in \mathcal{B}$:

$$\mathsf{P}_{x,q}\{q(z) \in Q \mid \mathcal{F}_x^y\} = \mathsf{P}_{y,q(y)}\{q(z) \in Q\};$$

(iv) for any $z \geqslant 0$, $y \geqslant x$ and $Q \in \mathcal{B}$ we have

$$\mathsf{P}_{x,q}\{q(y) \in Q\} = \mathsf{P}_{x+z,q}\{q(y+z) \in Q\}.$$

Property (iii) of the process $q(x)$ means that the corresponding transition probability satisfies the Chapman–Kolmogorov equation

$$P(x, q; y, Q) = \int P(x, q; y, dq_1) P(y, q_1; z, Q), \tag{1.24}$$

and property (iv) means homogeneity. The homogeneity implies, in particular, that the transition probability $P(x, q; y, dq_1)$, for $x \leqslant y$, defined in (ii) depends only on the difference $y - x$. In view of this we can define a transition probability $P(x, q; dq_1)$, for $x \geqslant 0$, by

$$P(x, q; y, dq_1) = P(y - x, q; dq_1)$$

for all $x \leqslant y$.

This definition is for Markov processes that take values in a subset of $\mathbf{R}$, but it can be extended without significant changes to processes with values in $\mathbf{R}^d$ [Gihman and Skorohod II]. It is clear from the definition that the central datum in specifying a Markov process is the transition probability $P(x, q; dq_1)$.

An important example of a Markov process is the Wiener process in $\mathbf{R}^d$, which we will often use as a technical tool. It is defined by the following transition probability:

$$P(x, q; dq_1) = (4\pi x)^{-d/2} e^{-(q-q_1)^2/(4x)}\, dq_1 \tag{1.25}$$

for $x > 0$.

Here's a procedure to get a stationary process from any Markov process $q(x)$, for $x \in \mathbf{R}$, with conditional probabilities $\mathsf{P}_{x,q}$. Let $F(dq)$ be a probability measure on $\mathbf{R}$ invariant with respect to the transition probability of $q(x)$, that is, satisfying

$$F(dq_1) = \int_{\mathbf{R}} F(dq)P(x,q;dq_1)$$

for $x \geqslant 0$. Introduce a probability measure $\mathsf{P} = \int_{\mathbf{R}} F(dq)\mathsf{P}_{x,q}$ on the measure space $(\Omega, F_{-\infty}^{\infty})$. This definition really gives P on each σ-algebra F_x^{∞}, but since, by the invariance of F, the result is independent of x, we see that P is well-defined on $F_{-\infty}^{\infty}$. This probability measure P describes a stationary Markov process $q(x)$ with the prescribed transition probability $P(x,q;dq_1)$ and distribution $\mathsf{P}\{q(x) \in dq\} = F(dq)$ for $x \in \mathbf{R}$.

This construction of a stationary Markov process is universal, in the sense that for any stationary Markov process $q(x)$, for $x \in \mathbf{R}$, the distribution $\mathsf{P}\{q(x) \in dq\} = F(dq)$ is invariant with respect to the corresponding transition probability, and the representation $\mathsf{P} = \int_{\mathbf{R}} F(dq)\mathsf{P}_{x,q}$ holds.

The stationary Markov process obtained by this construction will be metrically transitive if the set of distributions $\mathcal{M}_0$ invariant under the transition probability $P(x,q;dq_1)$ has exactly one element. If $\mathcal{M}_0$ has more than one element, it is clearly convex, and a metrically transitive process is obtained if and only if we choose a distribution $\mathsf{P}\{q(x) \in dq\}$ that is an extremal point of $\mathcal{M}_0$.

As an illustration, consider the Markov process $r(x)$, for $x \in \mathbf{R}$, taking values in $\{0,1\}$. Define the transition probability by

$$P(x,r,r') = \begin{cases} (n_{1-r} + n_r e^{-(n_0+n_1)x})/(n_0+n_1), & \text{if } r = r'; \\ n_r(1-e^{-(n_0+n_1)x})/(n_0+n_1), & \text{if } r \neq r', \end{cases}$$

where $r = 0,1$ and $n_0, n_1 > 0$ are fixed positive real numbers. It is easy to see that such a transition probability satisfies the Chapman–Kolmogorov equation

$$P(x+y,r,r') = \sum_{r''=0,1} P(x,r,r')P(y,r'',r');$$

that it determines a homogeneous Markov process; and that it has a unique invariant distribution, concentrated at points $r = 0,1$ and given by

$$F(r) = \mathsf{P}\{r(x) = r\} = \frac{n_{1-r}}{(n_0+n_1)}$$

for $r = 0,1$. The Markov process thus determined is metrically transitive; it is often called a *dichotomous process*, or a *random telegraph signal.*

Clearly, the only invariant distribution for a Wiener process, defined by the transition probability (1.25), is the Lebesgue measure, and therefore a Wiener process cannot be converted into a metrically transitive one. But in the case $d = 1$, we can consider, instead of (1.25), the distribution

$$P(x;q,dq_1) = \big(2\pi D(x)\big)^{-1/2} e^{-(q_1-qe^{-x/2})^2/\left(2D(x)\right)}\, dq_1$$

for $x \geqslant 0$, where $D(x) = 1 - e^{-x}$. This is seen to satisfy the Chapman–Kolmogorov equation (1.24), and to have an invariant distribution

$$F(dq) = (2\pi)^{-1/2} e^{-q^2/2}\, dq.$$

The corresponding Markov process is called the *Ornstein–Uhlenbeck process* (see also Exercise I.12).

We have discussed above Markov processes with a continuous argument. The same considerations apply to process with a discrete argument $x \in \mathbf{Z}$, called *Markov chains.*

(c) Gaussian random fields. Consider a homogeneous Gaussian random field $g(x)$, for $x \in \mathbf{R}$, as in Example 1.2(a); as mentioned there, its mathematical expectation $\mathsf{E}\{g(x)\}$ is zero, and its correlation function $\mathsf{E}\{g(x+y)g(y)\}$ coincides with the function $b(x)$ that occurs in the definition of the field's finite-dimensional distributions. If $b(x)$ is continuous at $x = 0$ and approaches 0 as $|x| \to \infty$, the corresponding group of automorphisms $\{T_x : x \in \mathbf{R}\}$ will be stochastically continuous and mixing, which is more than enough to guarantee that it is metrically transitive. If $b(x)$ has compact support, the field values will be independent at long distances, and the field will satisfy all our criteria for "randomness." A similar situation obtains in the discrete case.

However, the nondeterministic character of a Gaussian field cannot be guaranteed solely on the basis of the rate of decay of its correlation function. This is particularly easy to see in the continuous case: already for $d = 1$, there exist Gaussian processes almost all of whose realizations are analytic functions on some strip of the complex plane, or even entire functions, as long as the correlation function has the same property [Belyaev 1959]. A simple example is given by the correlation function $b(x) = \sigma^2 e^{-\delta^2 x^2}$, for $x \in \mathbf{R}$. Clearly, such a process, while mixing, is at the same time deterministic (singular), since almost all its realizations can be reconstructed from their values on any semiaxis $x \leqslant -n$ or $x \geqslant n$, for $n = 1, 2, \ldots$

The properties of a Gaussian field are conveniently formulated in terms of its spectral measure $\hat{B}(dk)$, given by the Fourier series of its correlation function:

$$b(x) = \int_{\mathbf{R}^d} e^{ikx} \hat{B}(dk).$$

By the well-known Bochner–Khintchine theorem, $\hat{B}(dk)$ is nonnegative. It turns out [Gihman and Skorohod I] that in the one-dimensional case, a Gaussian process is metrically transitive if and only if $\hat{B}$ is free of atoms; it is nondeterministic if and only if

$$\int_{\mathbf{R}} \frac{\log \hat{b}_{ac}(k)}{1+k^2}\, dk > -\infty,$$

where $\hat{b}_{ac}(k)$ is the derivative of the absolutely continuous component of the measure $\hat{B}$; and it is regular if and only if $\hat{B}$ has a density $\hat{b}$ for which the above condition holds.

(d) Poisson random fields. Let $m(dx)$ be a *Poisson random measure* on $\mathbf{R}^d$. This means that, if $(B_1, \dots, B_n)$ are pairwise disjoint Borel sets in $\mathbf{R}^d$, the random variables $m(B_1), \dots, m(B_m)$ are independent, and if B is a bounded Borel set, the random variable $m(B)$ has the distribution

$$\mathsf{P}\{m(B) = n\} = e^{-c|B|}\frac{(c|B|)^n}{n!}$$

for $n = 0, 1, \dots$, where $|B|$ denotes the volume of B and c is a positive constant called the concentration, because obviously $\mathsf{E}\{m(B)\} = c|B|$.

Let $u : \mathbf{R}^d \to \mathbf{R}$, be a nonrandom function with compact support. (A weaker condition on u will be considered in subsequent chapters.) The field $q(x)$, for $x \in \mathbf{R}^d$, given by

$$q(x) = \int_{\mathbf{R}^d} u(x - y)m(dy) \tag{1.28}$$

is called a *Poisson random field.* Poisson random fields, like Gaussian random fields, are metrically transitive and mixing. If $u(x)$ has compact support, the Poisson random field $q(x)$, for $x \in \mathbf{R}^d$, satisfies even the strong mixing condition (1.18). But this property, or the property of being nondeterministic (regular) cannot be ensured solely by the rate of decay of $u(x)$, so that this intuitively "typical" random field can turn out to be deterministic. For instance, in the one-dimensional case, if the function $u(x)$, decreasing sufficiently rapidly as $|x| \to \infty$, is analytically continued onto a strip of the complex plane, almost all realizations of process (1.28) will be analytic functions and therefore will be uniquely determined by their values on any semiaxis $x \leqslant -n$ or $x \geqslant n$, for $n = 1, 2, \dots$ For details, see [Kirsch et al. 1985].

If we denote by x_j the atoms of the Poisson measure, the field (1.28) can also be written as

$$q(x) = \sum_j u(x - x_j). \tag{1.29}$$

Thus the Poisson field is a random field of "summation" type, consisting of a sum of independent contributions due to various regions in space. Taken as a random potential for the Schrödinger operator of Example 1.4(e), it provides a model for the effect of a field of identically, randomly distributed, very heavy atoms (say, an amorphous solid or a liquid) on a light, quantum-mechanical particle (say, an electron). If the medium has atoms of several kinds, it is natural to consider, instead of the Poisson potential, a more general one

$$q(x) = \sum_j u_j(x - x_j), \tag{1.30}$$

where the points x_j are still distributed according to the Poisson law and the functions $u_j(x)$ are selected from a fixed set, independently of the x_j and of one another. For example, in the case of atoms of two types, A and B, we set $u_j(x)$ to $u_A(x)$ with probability p_A, and to $u_B(x)$ with probability $p_B = 1 - p_A$, where p_A is the concentration of atoms of type A. Another example is obtained by taking $u_j(x)$ as $\xi_j u(x)$, where $u(x)$ is a nonrandom function and the ξ_j are random charges, that is, independent, identically distributed random variables. Finally, if we consider not an amorphous medium but a so-called *substitution alloy*, where all the atoms are at the sites of a regular crystal lattice $\mathbf{Z}^d$, but are of several kinds, the corresponding random potential [Lifshitz et al. 1982b] is given by:

$$q(x) = \sum_{a \in \mathbf{Z}^d} u_a(x - a), \tag{1.31}$$

where the $u_a(x)$ are random functions like the ones above. For example, we can take $u_a(x) = \xi_a u(x)$, where the ξ_a, for $a \in \mathbf{Z}^d$, are independent, identically distributed random variables, or, more generally, a metrically transitive field on the lattice $\mathbf{Z}^d$.

Unlike the previous examples of metrically transitive fields on $\mathbf{R}^d$, here the metrically transitive group T is isomorphic to $\mathbf{Z}^d$, rather than to $\mathbf{R}^d$. Therefore, rigorously, the field (1.31) is not subject to the definitions and constructions given above for metrically transitive fields. But this proves to be of little importance from our point of view, in particular for the definition of abstract metrically transitive operators and for the study of their spectral properties.

Moreover, there is a simple method [Stratonovich 1961; Kirsch 1985] to reduce such periodic random fields to homogeneous ones. Indeed, the finite-dimensional distributions $P_{x_1,\ldots,x_n}(B_1,\ldots,B_n)$ of the field (1.31) are only invariant under shifts $x_i \mapsto x_i + \xi$, for $\xi \in \mathbf{Z}^d$, rather than under shifts by any vector $a \in \mathbf{R}^d$. Let $q(x)$ be an arbitrary random field with this property. Consider a random variable θ independent of the field $q(x)$ and homogeneously distributed in the unit cube $C \subset \mathbf{R}^d$ with opposite faces identified, so that for any Borel set $B \subset C$ we have $P\{\theta \in B\} = |B|$. Now define a new field $\tilde{q}(x)$ by $\tilde{q}(x) = q(x + \theta)$. One can easily see that the finite-dimensional distributions $\tilde{P}_{x_1,\ldots,x_n}(B_1 \times \cdots \times B_n)$ of $\tilde{q}(x)$ are obtained from the finite-dimensional distributions of $q(x)$ by the relation

$$\tilde{P}_{x_1,\ldots,x_n}(B_1 \times \cdots \times B_n) = \int_C P_{x_1+t,\ldots,x_n+t}(B_1 \times \cdots \times B_n)\,dt,$$

and therefore are invariant under any shift in $\mathbf{R}^d$. Thus, $\tilde{q}(x)$ is a homogeneous field on $\mathbf{R}^d$. Its probability space $(\tilde{\Omega}, \tilde{\mathcal{F}}, \tilde{\mathsf{P}})$ is a product of the probability space $(\Omega, \mathcal{F}, \mathsf{P})$ of the initial periodic field $q(x)$ with the probability space $(\Omega_0, \mathcal{F}_0, \mathsf{P}_0)$ of a random variable θ, where $\Omega_0 = C$, $\mathcal{F}_0$ is the σ-algebra of Borel sets and P_0 is the Lebesgue measure. The points of $\tilde{\Omega}$ are pairs (ω, τ), for $\omega \in \Omega$ and $\tau \in \Omega_0$, and the shift operator $\tilde{T}_x$ is defined as

$$\tilde{T}_x(\omega, \tau) = \big(T_{[x+\tau]}\omega, \{x + \tau\}\big),$$

where $[x]$ and $\{x\}$ indicate the decomposition of a vector $x \in \mathbf{R}^d$ as a sum of an integer-valued vector $[x] \in \mathbf{Z}^d$ and a vector $\{x\} \in C$. The group $\{\tilde{T}_x : x \in \mathbf{R}^d\}$ is metrically transitive in $\tilde{\Omega}$ if $\{T_\xi, \xi \in \mathbf{Z}^d\}$ is metrically transitive in Ω.

The simplest example of a periodic random field is a periodic function, which plays here a role analogous to that of a constant in the homogeneous case. Applying this procedure to a periodic function results in a periodic function with a random origin, uniformly distributed over the period. This view of a periodic function as a homogeneous random field will be discussed again when we consider almost periodic functions in Example 1.15(g).

(e) Generalized Poisson random fields. A random measure $m(dx)$ on $\mathbf{R}^d$ is called a *generalized Poisson measure* if it satisfies the following conditions:

(i) $m(B)$ is a random variable for any bounded Borel set B, and we have $\operatorname{ess\,inf} m(B) = 0$ and $\mathsf{E}\{m(B)\} < \infty$;

(ii) if $B_1, \ldots, B_n$ are bounded, pairwise disjoint Borel sets, the random variables $m(B_1), \ldots, m(B_n)$ are independent;

(iii) the random measure m is metrically transitive, that is, there exists a metrically transitive group $\{T_x : x \in \mathbf{R}^d\}$ of automorphisms of probability spaces such that $m(B, T_x\omega) = m(B + x, \omega)$ for any bounded Borel set B.

It is easy to see that for a generalized Poisson measure the distribution of the random variable $m(B)$ depends only on $|B|$, so that such measures may be defined by the distribution function $F_V(m)$, for $V > 0$ and $m \in \mathbf{R}$, corresponding to the random variable $m(B)$ for some B with $|B| = V$. The only requirement to be satisfied by the functions $F_V(m)$, besides the obvious condition $F_V(m) = 0$ for $m \leqslant 0$, is that

$$\int_{\mathbf{R}} F_{V_1}(m - m_1) F_{V_2}(dm_1) = F_{V_1+V_2}(m) \tag{1.32}$$

for $V_1, V_2 > 0$. In particular, we get the Poisson measure with parameter c when we put

(1.33) $$F_V(m) = e^{-cV} \sum_{k=0}^{m} \frac{(cV)^k}{k!},$$

for $m \in \mathbf{Z}_+$.

Another example of a generalized Poisson measure is the measure generated by Γ-distributions [Feller 1958]:

(1.34) $$F_v(dm) = \frac{1}{\Gamma(V)} c^V m^{V-1} e^{-mc}\, dm$$

for $m \geqslant 0$, where Γ is the gamma function and c is a positive parameter. The functions in (1.33) and (1.34) are easily seen to satisfy (1.32).

A *generalized Poisson random field* is constructed from a generalized Poisson random measure m and a nonrandom function u on $\mathbf{R}^d$ with compact support, as follows:

(1.35) $$q(x) = \int_{\mathbf{R}^d} u(x-y) m(dy).$$

(f) Smoothed square of the Gaussian random field. This field is obtained from (1.35) with $g^2(y)\, dy$ as the random measure, where $g(y)$, for $y \in \mathbf{R}^d$, is the Gaussian random field from Example 1.15(c).

(g) Periodic and quasiperiodic functions. Let $f(x)$ be a continuous function on $\mathbf{R}$, periodic with period 1. Let the probability space $(\Omega, \mathcal{F}, \mathsf{P})$ be a circle of unit circumference, together with its σ-algebra of Borel sets and its Lebesgue measure; and the group T, the group of rotations of the circle, with respect to which Ω is obviously metrically indecomposable and the measure, invariant. Then the family of functions

(1.36) $$q(x, \omega) = f(x + \omega),$$

for $x \in \mathbf{R}$ and $\omega \in \Omega$, is a metrically transitive process, which can be seen, as it were, as the result from adding to the argument of the 1-periodic function a random variable ω uniformly distributed over the circle.

Clearly, a similar construction is possible in a d-dimensional space, where it associates with each periodic function a metrically transitive field on $\mathbf{R}^d$.

We now consider almost periodic functions, restricting ourselves, for simplicity, to the one-dimensional case.

Definition [Levitan 1953]. *A bounded, continuous function $f(x)$ on $\mathbf{R}$ is called almost periodic if any subset of the set of its possible shifts $f_\xi(x) = f(x+\xi)$ contains a sequence that converges uniformly on the entire axis, that is, is precompact in the topology $C(\mathbf{R})$.*

(These functions are called *uniform* or *Bohr*-almost periodic. For other classes of almost periodic functions, see chapter VIII.)

The closure of the set of shifts $\{f_\xi(x) : \xi \in \mathbf{R}\}$ in the metric $C(\mathbf{R})$ induced by the $\|\cdot\|_\infty$-norm is compact in the space of bounded and uniformly continuous functions on the axis, and is called the *hull* Ω_f of the function $f(x)$. This set has a natural commutative group structure, defined as follows: if $\omega = \lim_{n\to\infty} f_{\xi_n}$ and $\omega' = \lim_{n\to\infty} f_{\xi'_n}$ are two elements of Ω_f, the sequence $f_{\xi_n+\xi'_n}$ is also convergent and we set $\omega+\omega'$ to its limit in Ω_f. Checking that this addition is well-defined, satisfies the properties of a group law and is continuous is straightforward.

As a compact topological group, Ω_f possesses an invariant Haar measure which, if normalized to unity, makes Ω_f into a probability space. The group $\mathcal{T}$ of measure-preserving automorphisms is the usual group of shifts in Ω_f (see, for example, Example 1.14(a)). Since $\mathbf{R}$ is a dense subgroup in Ω_f, it follows that $\mathcal{T}$ is a metrically transitive group.

Thus an almost periodic function, like a periodic function, gives rise to a metrically transitive process. A particularly illustrative result of this construction is the case of the so-called quasiperiodic functions. Here Ω_f is an n-dimensional torus $\mathbf{T}^n$, the Haar measure is the Lebesgue measure, and

$$f(x) = F(\alpha_1 x, \ldots, \alpha_n x),$$

where $F(x_1, \ldots, x_n)$ is periodic of period 1 on every variable (that is, a function on $\mathbf{T}^n$) and the numbers $\alpha_1, \ldots, \alpha_n$ are *rationally independent*, that is, any relation $\alpha_1 r_1 + \cdots + \alpha_n r_n = 0$ with rational coefficients $r_1, \ldots, r_n$ implies that all the r_i are 0. In this case (cf. (1.13) and (1.20)) we have

$$q(x,\omega) = F(\alpha_1 x + \omega_1, \ldots, \alpha_n x + \omega_n). \tag{1.37}$$

A metrically transitive process constructed from a quasiperiodic function can be thought of as a function on $\mathbf{T}^n$, describing in $\mathbf{T}^n$ a movement of uniform velocity whose components are rationally independent, and having started at a random initial point.

In contrast with the previous examples of metrically transitive random fields, which were, under very natural assumptions, nondeterministic and even regular for $G = \mathbf{Z}$ or $\mathbf{R}$, the metrically transitive process we have just constructed is singular. This is because an almost periodic function $f(x)$ is determined [Levitan 1953] by its Fourier coefficients c_λ, calculated as the limits

$$c_\lambda = \lim_{T\to\infty} \frac{1}{T} \int_a^{a+T} f(x) e^{ix\lambda}\, dx = \lim_{T\to\infty} \frac{1}{T} \int_{a-T}^{a} f(x) e^{ix\lambda}\, dx,$$

for any $a \in \mathbf{R}$. Thus, $f(x)$ is determined by its values on any semiaxis $(-\infty, n)$ or (n, ∞), for $n = 0, 1, \ldots$ But then any of the σ-algebras $\mathcal{F}_n(f)$ in the definition of a singular process coincides with the σ-algebra $\mathcal{F}(f)$, that is, $\mathcal{F}_\infty(f) = \mathcal{F}(f)$.

Statistical information about the finite-dimensional distributions associated with a random field $q(x)$ can often be conveniently specified and employed in the form of the so-called *characteristic functional* $\chi_F(\phi)$, defined by

$$\chi_F(\phi) = \mathsf{E}\left\{\exp\left(i\int_{\mathbf{R}^d} \phi(x)q(x)\,dx\right)\right\}, \tag{1.38}$$

for some class of functions $\phi : \mathbf{R}^d \to \mathbf{R}$. The characteristic functional is a natural generalization of the characteristic functions from probability theory.

For example, let $g(x)$ be a homogeneous Gaussian field in $\mathbf{R}^d$, with $\mathsf{E}\{g(x)\} = 0$ and $\mathsf{E}\{g(0)g(x)\} = b(x)$. Because any linear combination of Gaussian random variables is also one, the integral $\int_{\mathbf{R}^d} \phi(x)g(x)\,dx$, defined for any bounded function $\phi(x)$ with compact support, has a Gaussian distribution. Therefore, for a homogeneous Gaussian field,

$$\chi_F(\phi) = \exp\left(-\frac{1}{2}\int_{\mathbf{R}^{2d}} b(x-y)\phi(x)\phi(y)\,dx\,dy\right). \tag{1.39}$$

For the Poisson random field (1.29),

$$\chi_F(\phi) = \exp\left(c\int_{\mathbf{R}^d}\left(\exp\left(i\int_{\mathbf{R}^d} \phi(x-y)u(y)\,dy\right) - 1\right)dx\right). \tag{1.40}$$

Besides the characteristic functional (1.38), which generalizes the Fourier transform of the probability distribution function, the *Laplace characteristic functional*

$$\chi_L(\phi) = \mathsf{E}\left\{\exp\left(-\int_{\mathbf{R}^d} \phi(x)q(x)\,dx\right)\right\} \tag{1.41}$$

can be useful. This expression, unlike (1.38), is often far from being defined, even if $\phi(x)$ is smooth and has compact support. In the particular case of Gaussian and Poisson fields, however, it is defined and we have $\chi_L(\phi) = \chi_F(i\phi)$, as can be seen by direct computation.

Here are simple conditions guaranteeing the existence of $\chi_L(\phi)$:

(1.16) Proposition. *The Laplace characteristic functional of a homogeneous random field whose realizations are, with probability* 1, *locally summable, is defined on all bounded, measurable functions with compact support if*

$$\mathsf{E}\{e^{-tq(0)}\} < \infty \tag{1.42}$$

for all $t \in \mathbf{R}$. *If, in addition,* $\phi(x) \geqslant 0$ *for* $x \in \mathbf{R}$, *we have* $\chi_L(\phi) < \infty$ *if*

$$\mathsf{E}\{e^{-t\max\{0,-q(0)\}}\} < \infty \tag{1.43}$$

for all $t > 0$.

Proof. Write ϕ in the form $\phi_+ + \phi_-$, where $\phi_\pm = \max\{0, \pm x\}$. Applying the Schwartz inequality, we have

$$\chi_L^2(\phi) \leqslant \mathsf{E}\{e^{-2\int \phi_+ q\,dx}\}\mathsf{E}\{e^{2\int \phi_- q\,dx}\}.$$

Now the functions ϕ_+ and ϕ_-, being nonnegative, can be written as $\phi_\pm(x) = \hat{\phi}_\pm \alpha_\pm(x)$, where the $\hat{\phi}_\pm$ are nonnegative numbers and the $\alpha_\pm$ are nonnegative functions whose integral is 1. According to Jensen's inequality, the product of expectations in the previous display is bounded above by

$$\int_{\mathbf{R}^d} \alpha_+(x)\mathsf{E}\{e^{-2\hat{\phi}_+ q(x)}\}\,dx \int_{\mathbf{R}^d} \alpha_-(x)\mathsf{E}\{e^{2\hat{\phi}_- q(x)}\}\,dx$$
$$= \mathsf{E}\{e^{-2\hat{\phi}_+ q(0)}\}\mathsf{E}\{e^{2\hat{\phi}_- q(0)}\} < \infty,$$

where the equality is based on the homogeneity of $q(x)$. The second statement in the proposition follows similarly. □

1.D Metrically Transitive Operators

As typical and most important examples of metrically transitive operators, we can cite the Schrödinger operator of Example 1.4(b) with a metrically transitive field as the potential $q(x)$, and Jacobi matrices (1.4) when $q(x)$ and $s(x)$ are both metrically transitive sequences—for example, sequences of independent, identically distributed random variables.

The general definition of metrically transitive operators is formulated naturally in terms of metrically transitive groups, defined in 1.C, and of random operators, defined in 1.B.

Definition. *A random operator A on a Hilbert space $\mathcal{H}$ is called metrically transitive if there exists a homomorphism from a metrically transitive group $\mathcal{T}$ of automorphisms of the probability space $(\Omega, \mathcal{F}, \mathsf{P})$ into a group $\mathcal{U} = \{U_T : T \in \mathcal{T}\}$ of unitary operators on $\mathcal{H}$, such that*

$$A(T\omega) = U_T A(\omega) U_T^{-1}. \tag{1.44}$$

If, moreover, the operator A is symmetric, its domain $\mathcal{D}$ is invariant under the group $\mathcal{U}$, that is, $U\mathcal{D} \subset \mathcal{D}$ for $U \in \mathcal{U}$.

(1.17) Examples of metrically transitive operators

(a) Jacobi matrices. Here $\big(s(x), q(x)\big)$, for $x \in \mathbf{Z}$, is a metrically transitive random field on $\mathbf{Z}$ with values in $\mathbf{R}^2$ and group $\mathcal{T} = \{T_a : a \in \mathbf{Z}\}$, and the homomorphism $\mathcal{T} \to \mathcal{U} = \{U_T : T \in \mathcal{T}\}$ is defined by

$$(U_a\psi)(x) = \psi(x+a)$$

for $\psi \in \ell_2(\mathbf{Z})$ and $x, a \in \mathbf{Z}$ (with $U_a = U_{T_a}$).

(b) One-dimensional Schrödinger operator. Here the potential $q(x)$, for $x \in \mathbf{R}$, is a metrically transitive real-valued random field, with group $\mathcal{T} = \{T_a : a \in \mathbf{Z}\}$, and the homomorphism is defined by

$$(U_a\psi)(x) = \psi(x+a)$$

for $\phi \in L^2(\mathbf{R})$ and $x, a \in \mathbf{R}$.

(c) Multidimensional matrix operators. Here $a(x,y,\omega)$, for $(x,y) \in \mathbf{Z}^d \times \mathbf{Z}^d$, is a metrically transitive complex-valued random field on $\mathbf{Z}^d \times \mathbf{Z}^d$. The group of automorphisms is isomorphic to the diagonal subgroup of $\mathbf{Z}^d \times \mathbf{Z}^d$, hence to $\mathbf{Z}^d$. It has a natural parametrization $\mathcal{T} = \{T_x : x \in \mathbf{Z}^d\}$. The condition for the field to metrically transitive is

$$a(x,y,T_z\omega) = a(x+z, y+z, \omega). \tag{1.45}$$

The homomorphism $\mathcal{T} \to \mathcal{U}$ is defined by

$$(U_x\psi)(y) = \psi(x+y)$$

for $\psi \in \ell_2(\mathbf{Z}^d)$ and $x, y \in \mathbf{Z}^d$.

(d) Integral operators. The kernel $a(x,y)$, for $(x,y) \in G \times G$, is a metrically transitive, complex-valued random field on G. Otherwise everything is the same as in the previous example, with $\mathbf{Z}^d$ replaced by G.

(e) Multidimensional Schrödinger operators. The potential $q(x)$, for $x \in \mathbf{R}^d$, is a metrically transitive real-valued random field, and

$$(U_a\psi)(x) = \psi(x+a)$$

for $\psi \in L^2(\mathbf{R}^d)$ and $x, a \in \mathbf{R}^d$.

(f) Elliptic differential operators. For every $\alpha \in \mathbf{Z}^d_+$ such that $|\alpha| \leqslant p$ the functions $a_\alpha(x)$, for $x \in \mathbf{R}^d$, are metrically transitive real-valued random fields of the same type as the above field $q(x)$.

We see that the class of metrically transitive operators is quite wide, even if we restrict ourselves to the above examples. Already in the case of differential operators, we obtain all operators with constant coefficients if Ω consists of a single point, all operators with periodic coefficients if Ω is the torus $\mathbf{T}^d$, and, for a higher-dimensional torus, we get operators with quasiperiodic coefficients.

From the definition of metric transitivity for operators (1.44) it follows that all scalar spectral characteristics of an operator, being unitary invariants, are, as random variables, invariant under the corresponding metrically transitive group of automorphisms of the probability space. Therefore, by theorem 1.8, they are nonrandom, that is, independent of ω. These spectral characteristics include the integrated density of states (see Chapter II) or the spectrum as a set, because the latter may be described by scalar quantities.

However, for the analysis of more detailed spectral properties, such as the point and absolutely continuous components (see 2.B), the notion of metric transitivity turns out to be too coarse. For instance, the structure of the spectrum depends on whether the metrically transitive field is deterministic or not. Consider the example of one-dimensional Schrödinger operators: if the potential is taken as a random periodic function (1.36)—an example of a deterministic random field—the spectrum of the resulting operator is absolutely continuous with probability 1 [Reed and Simon IV]. If, on the other hand, the potential is a nondeterministic random process, the spectrum has no absolutely continuous component, again with probability 1 (Section 12). If the potential is a smooth Markov process that has good mixing properties, the spectrum is pure point with probability 1 (Section 15). In the intermediate situation of an almost periodic potential, there are quite a variety of spectral types, including purely singular continuous spectra (Chapter VII).

Remark. An important point in the definition of abstract random operators, and in particular of metrically transitive operators, is to get the right mix of unboundedness and measurability, so that, on the one hand, traditional objects of spectral theory, such as the resolvent, the resolution of the identity, and so on, are measurable if the operator is, and, on the other hand, the most important classes of operators, such as differential and matrix operators with random and almost periodic coefficients, are covered by a general definition whose conditions can be verified as simply and efficiently as possible.

The definition we gave for metrically transitive operators seems to satisfy these requirements. As we will see in Section 2, it implies measurability of the spectral projections and of the resolvent. And checking the conditions in the definition, in particular measurability, for differential and matrix operators with metrically transitive coefficients, turns out to be quite simply in many important cases, for example, for elliptic differential operators with bounded and sufficiently smooth coefficients.

A different possible approach in defining metrically transitive operators is to include the most technically complicated property of measurability of the resolvent or resolution of the identity in the definition. According to this definition, an operator is metrically transitive if its resolvent (or resolution of the identity or some other function of the operator bounded *a priori*, such as the semigroup e^{-tA} for $A \geqslant 0$) is a measurable operator satisfying (1.44). To be constructive, such a definition must be complemented with sufficiently general and readily verifiable criteria for measurability.

These two approaches to definition metrically transitive operators are clearly equivalent in essence, but the first seems more convenient.

These same questions are addressed, under slightly different points of view, in [Skorohod 1984; Carmona 1984; Bellissard 1985].

2 Simple Spectral Properties of Metrically Transitive Operators

2.A Deficiency Indices

Definition. *A function f defined on Ω and with values on an arbitrary set is called* nonrandom *if there is a set $\tilde{\Omega} \subset \Omega$ such that $\mathsf{P}\{\tilde{\Omega}\} = 1$ and $f(\omega_1) = f(\omega_2)$ for all $\omega_1, \omega_2 \in \tilde{\Omega}$.*

The following simple theorem has a variety of applications in the spectral analysis of metrically transitive operators.

(2.1) Theorem. *The rank of a metrically transitive orthogonal projection π operating on a Hilbert space $\mathcal{H}$ is a nonrandom function. If $\mathcal{T}$ and $\mathcal{U} = \{U_T : T \in \mathcal{T}\}$ are the metrically transitive group and the group of unitary operators, respectively, associated with π, and $\mathcal{U}$ has no nontrivial finite-dimensional invariant subspaces in $\mathcal{H}$, the rank of π is either* 0 *or* ∞ *with probability* 1.

Proof. Since the rank of a projection is a unitary invariant, that is, it does not change when the operator is replaced by a unitarily equivalent one, it follows immediately from the metric transitivity of π that the rank of π is an invariant random variable. By theorem 1.10, there exists $c \in [0, \infty]$ such that the rank of π equals c with probability 1. Since the rank of an orthogonal projection coincides with its trace, we have $\operatorname{Tr} \pi = c$ with probability 1.

For the second part, consider the operator $\pi_1 = \mathsf{E}\{\pi\}$, which is easily seen to be self-adjoint and nonnegative, and to have norm no greater than 1. Then, with probability 1, $\operatorname{rank} \pi = \operatorname{Tr} \pi_1$, so what we have to prove is that

$\operatorname{Tr}\pi_1 = 0$ or ∞. Since π is a metrically transitive random projection, we have

$$\pi_1 = \mathsf{E}\{\pi(T\omega)\} = \mathsf{E}\{U_T \pi U_T^{-1}\} = U_T \pi_1 U_T^{-1}$$

for all $T \in \mathcal{T}$. But this means that π_1 commutes with the U_T, so the resolution of the identity $E_{\pi_1}(d\lambda)$ of π_1 (see 2.B) also commutes with the U_T. Thus, for any $\varepsilon \in [0,1]$, the space $E_{\pi_1}([\varepsilon,1])\mathcal{H}$ is invariant under the U_T, and, by the condition of the theorem, its dimension is either 0 or ∞. This easily implies that $\operatorname{Tr}\pi_1 = 0$ or ∞. □

The verification of the theorem's condition is facilitated by the following simple result:

(2.2) Proposition. *If, for any finite family of vectors $h_j \in \mathcal{H}$, for $j = 0, \dots, n$, we have*

$$\inf_{T\in\mathcal{T}} \sup_j \left|(h_j, U_T h_0)\right| = 0,$$

the group of unitary operators $\mathcal{U} = \{U_T : T \in \mathcal{T}\}$ has no nontrivial finite-dimensional invariant subspaces.

Proof. Assume to the contrary that $\mathcal{H}_1$ is an n-dimensional subspace of $\mathcal{H}$ with $n < \infty$ and $U_T\mathcal{H}_1 = \mathcal{H}_1$ for all $T \in \mathcal{T}$. Choose an orthonormal basis $\{h_j\}_1^n$ for $\mathcal{H}_1$ and take an arbitrary vector $h_0 \in \mathcal{H}_1$ of norm 1. For any $T \in \mathcal{T}$, we have

$$\sum_{j=1}^{n} \left|(h_j, U_T h_0)\right|^2 = \|U_T h_0\|^2 = 1.$$

This obviously contradicts the condition of the proposition. □

(2.3) Corollary. *The following groups of unitary operators have no nontrivial finite-dimensional invariant subspaces:*

(i) $\mathcal{H} = L^2(\mathbf{R}^d)$ *and* U_x, *for* $x \in \mathbf{R}^d$, *given by* $(U_x\psi)(y) = \psi(x+y)$ *for* $y \in \mathbf{R}^d$ *and* $\psi \in L^2(\mathbf{R}^d)$;
(ii) $\mathcal{H} = L^2(\mathbf{R}^d)$ *and* U_x, *for* $x \in \mathbf{Z}^d$, *given by* $(U_x\psi)(y) = \psi(x+y)$ *for* $y \in \mathbf{R}^d$ *and* $\psi \in L^2(\mathbf{R}^d)$;
(iii) $\mathcal{H} = \ell_2(\mathbf{Z}^d)$ *and* U_x, *for* $x \in \mathbf{Z}^d$, *given by* $(U_x\psi)(y) = \psi(x+y)$ *for* $y \in \mathbf{Z}^d$ *and* $\psi \in \ell_2(\mathbf{Z}^d)$. □

(2.4) Theorem [Figotin and Pastur 1983]. *Let A be a symmetric metrically transitive operator with domain $\mathcal{D}$, satisfying the following conditions:*

(a) *the group of unitary operators $\mathcal{U} = \{U_T : T \in \mathcal{T}\}$ associated with A has no nontrivial finite-dimensional invariant subspaces in $\mathcal{H}$;*

(b) *there exists a countable set* $\mathcal{D}_1 \subset \mathcal{D}$, *dense in* $\mathcal{H}$, *such that the closure of the restriction* $A\big|_{\mathcal{D}_1}$ *of the random operator* A *to* D_1 *coincides, with probability* 1, *with the closure* $\bar{A}$ *of* A *(in other words, for any* $\varepsilon > 0$ *and* $x \in \mathcal{D}$ *there exists* $y \in \mathcal{D}_1$ *such that* $\|x - y\| + \|A(x - y)\| < \varepsilon$*).*

Then the deficiency indices of the operator A *are nonrandom and equal to either* 0 *or* ∞.

Proof. The deficiency subspace of the symmetric operator A coincides, by definition, with the orthogonal complement of the closure of the linear subspace $(A - zI)\mathcal{D}$, for $z \in \mathbf{C}$ such that $\operatorname{Im} z \neq 0$. Denote the projection on the deficiency subspace by π. From condition (ii) it follows that the closure of $(A - zI)\mathcal{D}$ coincides with the subspace of $\mathcal{H}$ spanned by the countable family of vectors $(A - zI)\mathcal{D}_1$, and therefore, by Theorem 1.6, is a random subspace in $\mathcal{H}$. Hence π is a random projection in $\mathcal{H}$. Since A is a metrically transitive operator, π is clearly one. The deficiency index of A is the dimension of $\pi\mathcal{H}$, that is, the rank of π. But π satisfies the conditions of theorem 2.1, implying our result. □

Theorem 2.4 and Corollary 2.3 have the following consequence:

(2.5) Corollary. *Symmetric, metrically transitive random matrix and differential operators (Examples* 1.15(*c*,*e*,*f*)*) have nonrandom deficiency indices equal to either* 0 *or* ∞.

Proof. We need only check condition (b) in theorem 2.4. In the case of matrix operators, this condition is evidently satisfied by the set $\mathcal{D}_1$ of entries of $\mathcal{D}$ having rational coordinates. For differential operators, notice that the set $C_0^\infty(\mathbf{R}^d)$ is separable in the appropriate topology [Reed and Simon I]. □

In the spectral theory of multidimensional Schrödinger operators, there was a conjecture [Glazman 1965] to the effect that the deficiency indices of the corresponding symmetric operator may only be 0 or ∞. Although we now know that this cannot be true in general (see, for example, [Piepenbrink and Rejto 1974]), it remains an important problem to find conditions on the coefficients under which the conjecture is true. An example of such conditions is given by Corollarly 2.5, which says that for a metrically transitive potential, having, say, a finite first moment, the deficiency index must be 0 or ∞. Because of Exercise I.5, this statement may be seen as a positive answer to this hypothesis for "typical" locally summable potentials.

As we shall see in Section 5, under fairly natural and general conditions the deficiency indices of the Schrödinger operator are zero, that is, the operator is essentially self-adjoint.

(2.6) Corollary. *Metrically transitive Jacobi matrices (Example 1.17(a)), under the additional condition that* $\mathsf{P}\{|s(0)| = 0\} = 0$, *and one-dimensional Schrödinger operators (Example 1.17(b)), under the additional condition that the potential q is locally square-integrable with probability 1, are essentially self-adjoint with probability 1.*

Proof. This follows immediately from Theorem 2.5, since the indices cannot be more than 2. □

We now recall an important notion from spectral theory; for details, see [Reed and Simon I].

Let A be a self-adjoint operator on a Hilbert space $\mathcal{H}$. The *resolution of the identity* of A, denoted by $E_A(\lambda)$, for $\lambda \in \mathbf{R}$, is a nondecreasing, strongly left-continuous family of projection operators such that $E_A(-\infty) = 0$, $E_A(\infty) = I$,

$$E_A(\lambda)E_A(\mu) = E_A(\min\{\lambda, \mu\}),$$

$A = \int_{\mathbf{R}} \lambda E_A(d\lambda)$, and the domain $\mathcal{D}_A$ of A contains exactly those vectors $h \in \mathcal{H}$ satisfying

$$\int_{\mathbf{R}} \lambda^2 (E_A(d\lambda)h, h) < \infty,$$

where $E_A(d\lambda)$ denotes, as usual, the operator-valued measure on $\mathbf{R}$ associated with the nondecreasing function $E_A(\lambda)$.

(2.7) Theorem. *Let A be a symmetric metrically transitive operator with domain* $\mathcal{D}$ *that satisfies condition (b) of Theorem 2.4 and is essentially self-adjoint with probability 1. If* $E_A(\lambda)$, *for* $\lambda \in \mathbf{R}$, *is its resolution of the identity,* $E_A(\lambda)$ *is a metrically transitive projection for every* $\lambda \in \mathbf{R}$.

Proof. The metric transitivity of $E_A(\lambda)$, that is, the validity of equation (1.44), follows immediately from the validity of the same equation for A. Thus we just need to show the randomness, that is, the measurability, of $E_A(\lambda)$. (A similar fact is proved in [Skorohod 1984].) To do this, we need an auxiliary result:

(2.8) Lemma. *If A satisfies the conditions of theorem 2.7,* $(A + iI)^{-1}$ *is a random operator on* $\mathcal{H}$.

Proof. Construct an orthogonal basis $\{e_n\}_1^\infty$ for $\mathcal{H}$ by applying Gram–Schmidt to the set $\mathcal{D}_1$ of condition (b) of theorem 2.4, and let $\mathcal{D}_2$ be the span of $\mathcal{D}_1$. Also, let Q_n be the projection onto the span of $\{e_1, \ldots, e_n\}$, and consider the sequence of random operators $A_n = Q_n A Q_n$ and $B_n = A_n + iI$, for $n \in \mathbf{Z}_+$. The measurability of the operators B_n^{-1} is equivalent to the measurability of the entries of the matrices $\left((A_n + iI)|_{Q_n\mathcal{H}}\right)^{-1}$, which is obvious. Therefore, if $B = A + iI$, all that remains to show is that

$$\text{w-lim}_{n\to\infty} B_n^{-1} = B^{-1}, \tag{2.1}$$

where w-lim denotes weak convergence. But, for any $u \in \mathcal{D}_2$ and any $v \in \mathcal{H}$, we have

$$(B^*u, B_n^{-1}v) = (u, v) + \left((B^* - B_n^*)u, B_n^{-1}v\right),$$

and, for n sufficiently large, $Q_n u = u$. Therefore s-$\lim_{n\to\infty} B_n^*(u) = B^*(u)$, where s-lim denotes strong convergence. Hence, because $\|B_n^{-1}\| \leqslant 1$ and by (2.1), we have

$$\lim_{n\to\infty} (B^*u, B_n^{-1}v) = (u, v),$$

which can be rewritten as

$$\lim_{n\to\infty} (w, B_n^{-1}v) = ({B^*}^{-1}w, v) = (w, B^{-1}v),$$

where $w \in B^*\mathcal{D}_2$ and $v \in \mathcal{H}$. Since A is self-adjoint, $B^*\mathcal{D}_2$ is dense in $\mathcal{H}$. This, together with the inequalities $\|B_n^{-1}\| \leqslant 1$ and $\|B^{-1}\| \leqslant 1$, shows that the limit holds, in fact, for any $w \in \mathcal{H}$. From this follows (2.1) and with it the lemma. □

The lemma implies that the unitary operator $U = (A - iI)(A + iI)^{-1}$ on $\mathcal{H}$ is $\mathcal{F}$-measurable. Consider the resolution of the identity $F(\phi)$, for $\phi \in [0, 2\pi)$, of this unitary operator:

$$U = \int_0^{2\pi} e^{i\phi}\, F(d\phi).$$

For any $\phi \in [0, 2\pi)$, there is a sequence of polynomials p_n such that s-$\lim_{n\to\infty} p_n(U) = F(\phi)$; whence, in view of the $\mathcal{F}$-measurability of U, follows that of $F(\phi)$. But, since $E_A(\lambda) = F(\phi)$, for $\lambda = \cot(\phi/2)$, we see that $E_A(\lambda)$ is also $\mathcal{F}$-measurable, as we wished to show. □

(2.9) Corollary. *For a metrically transitive matrix or differential operator (Examples* 1.17(c,e,f)*) that is essentially self-adjoint with probability* 1, *the resolution of the identity* E_λ *is a metrically transitive operator.*

Proof. This follows immediately from the preceding theorem, because, as was shown in the proof of Corollary 2.5, condition (b) of Theorem 2.4 is fulfilled for operators of these types. □

2.B Nonrandomness of the Spectrum and of its Components

We recall some more spectral notions; see, for example, [Reed and Simon I].

Let A be a self-adjoint operator on a Hibert space $\mathcal{H}$ and $E_A(\lambda)$, for $\lambda \in \mathbf{R}$, its resolution of the identity. The *spectrum* $\sigma(A)$ of A is the closed set of growth points of $E_A(\lambda)$:

$$\sigma(A) = \{\lambda \in \mathbf{R} : E_A(\lambda+\varepsilon) - E_A(\lambda-\varepsilon) \neq 0 \text{ for any } \varepsilon > 0\}.$$

A discontinuity point $\lambda \in \sigma(A)$ of $E_A(\lambda)$ is an eigenvalue of A, and the operator

$$P(\lambda) = \lim_{\lambda' \downarrow \lambda} E_A(\lambda') - E_A(\lambda), \tag{2.2}$$

is the orthogonal projection onto the eigenspace corresponding to the eigenvalue λ. The set of all eigenvalues is the *point spectrum* of A, and is denoted by $\sigma_p(A)$. For a separable Hilbert space—the only kind we're dealing with—the point spectrum is at most a countable set.

Denote by $\mathcal{H}_p$ the closed linear subspace generated by all the subspaces $P(\lambda)\mathcal{H}$, for $\lambda \in \sigma_p(A)$. If $\mathcal{H}_p = \mathcal{H}$, the spectrum of A is *pure point*. In this case, $\sigma(A)$ consists of the set $\sigma_p(A)$ of eigenvalues of A, plus the limit points of this set: $\sigma(A) = \overline{\sigma_p(A)}$. If $\mathcal{H}_p = 0$ the spectrum is *purely continuous*, and $E_A(\lambda)$ is strongly continuous with respect to λ. In general, $\mathcal{H}$ can be represented as an orthogonal sum of two A-invariant subspaces $\mathcal{H}_p$ and $\mathcal{H}_c$, such that the spectrum of the restriction of A to $\mathcal{H}_p$ is pure point and coincides with $\overline{\sigma_p}$, while the spectrum of the restriction of A to $\mathcal{H}_c$ has no eigenvalues. The latter is called the *continuous spectrum* $\sigma_c(A)$ of A. Thus $\sigma(A) = \overline{\sigma_p(A)} \cup \sigma_c(A)$, and the two components may intersect. The subspaces $\mathcal{H}_p$ and $\mathcal{H}_c$ can also be defined as the set of vectors $h \in \mathcal{H}$ for which the measure $\mu_h(d\lambda) = (E_A(d\lambda)h, h)$ is atomic and continuous, respectively.

$\mathcal{H}_c$ can in turn be represented as an orthogonal sum of two invariant subspaces $\mathcal{H}_{ac}$ and $\mathcal{H}_{sc}$, consisting of the vectors $h \in \mathcal{H}$ for which the measure μ_h is absolutely continuous and singular continuous with respect to the Lebesgue measure. The spectra of the restrictions of A to $\mathcal{H}_{ac}$ and $\mathcal{H}_{sc}$ are called the *absolutely continuous* and *singular continuous* components of the spectrum of A, and denoted by $\sigma_{ac}(A)$ and $\sigma_{sc}(A)$. Thus we have $\mathcal{H} = \mathcal{H}_p \oplus \mathcal{H}_{ac} \oplus \mathcal{H}_{sc}$ and $\sigma(A) = \overline{\sigma_p(A)} \cup \sigma_{ac} \cup \sigma_{sc}$.

The points of the spectrum of a self-adjoint operator can be classified in other ways. For example, a point $\lambda \in \mathbf{R}$ is said to belong to the *essential* or *limit spectrum* $\sigma_e(A)$ of A if the projection $E_A(\lambda+\varepsilon) - E_A(\lambda-\varepsilon)$ is infinite-dimensional for any $\varepsilon > 0$, and λ belongs to the *discrete spectrum* $\sigma_d(A)$ if, for some $\varepsilon > 0$, this projection is finite-dimensional. The set σ_e is closed and consists of the points in the continuous spectrum, the limit points of the closed spectrum and the eigenvalues of infinite multiplicity; while σ_d consists of isolated eigenvalues of finite multiplicity:

$$\sigma_d = \{\lambda : (\lambda-\varepsilon, \lambda+\varepsilon)\cap\sigma = \{\lambda\} \text{ for every } \varepsilon < \varepsilon_0 \text{ and } 0 < \operatorname{rank} P(\lambda) < \infty\}.$$

Contrary to the preceding decomposition, this one splits the spectrum into two disjoint sets.

Theorem 2.7 implies two simple results about the spectrum of a metrically transitive operator:

(2.10) Theorem. *Let A be a metrically transitive operator, self-adjoint with probability* 1, *and satisfying the conditions of Theorem* 2.4. *The number of points in the spectrum of A falling inside a prescribed interval $\Delta \subset \mathbf{R}$ is, with probability* 1, *equal to* 0 *or* ∞, *depending on whether the operator* $\mathsf{E}\{E_A(\Delta)\}$ *is zero or not.* □

Here is another formulation of the same result:

(2.11) Theorem. *The spectrum of a self-adjoint metrically transitive operator satisfying the conditions of Theorem* 2.4 *is essential, with probability* 1: $\sigma = \sigma_e$. *Equivalently, such an operator has, with probability* 1, *no isolated eigenvalues of finite multiplicity*: $\sigma_d = \varnothing$.

(2.12) Theorem. *The probability of any given point $\lambda \in \mathbf{R}$ being an eigenvalue of finite multiplicity of a metrically transitive operator is zero.*

Proof. Since the projection $P(\lambda)$ onto the corresponding subspace is a strong limit, for $\varepsilon \downarrow 0$, of the projections $E_A(\lambda+\varepsilon) - E_A(\lambda-\varepsilon)$, it follows from Theorem 2.7 that $P(\lambda)$ is metrically transitive. By Theorem 2.1, this implies that $\operatorname{rank} P(\lambda) = 0$ or ∞ with probability 1, and since we're assuming $\operatorname{rank} P(\lambda) < \infty$, this rank must be zero. □

Theorems 2.10 and 2.12 do not preclude the existence of a point component in the spectrum of a metrically transitive operator. It is just that, on the one hand, this part of the spectrum is the most sensitive to changes in ω, because, generically, every eigenvalue is shifted by perturbation; and, on the other, the point spectrum is at most a countable set. Thus, the probability of a point being in this relatively "thin" and mobile set turns out to be zero.

(2.13) Example. We illustrate Theorem 2.12, as well as several other facts to be discussed in this section, by considering the following simple metrically transitive operator. Let $\xi(x)$, for $x \in \mathbf{R}^d$, be a family of real-valued, independent, identically distributed random variables with common distribution function $F(\xi)$. The operator A on $\ell_2(\mathbf{Z}^d)$ acts on each vector x by

$$(A\psi)(x) = \xi(x)\psi(x). \tag{2.3}$$

Since it consists of multiplication by real quantities $\xi(x)$, this operator is self-adjoint, and the metric transitivity of the sequence $\xi(x)$, shown in Example 1.15(a), implies that A is also metrically transitive. Now the vectors e_y with components $e_y(x) = \delta_{xy}$, where δ_{xy} is Kronecker's symbol, are eigenvectors of A, for $Ae_y = \xi(y)e_y$ for all $y \in \mathbf{Z}^d$. Furthermore, these vectors form a basis for $\ell_2(\mathbf{Z}^d)$, so the spectrum of A is pure point, that is, $\sigma(A) = \overline{\sigma_{\mathrm{P}}(A)}$, and coincides with the closure of the set of values of random variables $\xi(x)$, that is, with the set Φ of the growth points of $F(\xi)$. The operator $E_A(\Delta)$ is in this case

$$E(\Delta; x, y) = \chi_\Delta\big(\xi(x)\big)\delta_{xy},$$

where $\chi_\Delta(\xi)$ is the characteristic function of the interval Δ.

Therefore, the rank of $E_A(\Delta)$ is, with probability 1,

$$\operatorname{rank} E_A(\Delta) = \operatorname{Tr} E_A(\Delta) = \sum_{x\in\mathbf{Z}^d} \chi_\Delta\big(\xi(x)\big) = \begin{cases} 0 & \text{if } \Delta\cap\Phi = \varnothing, \\ \infty & \text{if } \Delta\cap\Phi \neq \varnothing, \end{cases}$$

because for any sequence of cubes $\{\Lambda\}$ exhausting $\mathbf{Z}^d$ we have, by the ergodic theorem,

$$\lim_V \frac{1}{V}\sum_{x\in\Lambda} \chi_\Delta\big(\xi(x)\big) = \mathsf{P}\{\xi(0)\in\Delta\}$$

with probability 1, where V is the volume of Λ. On the other hand,

$$\big(E_A(\Delta)\psi, \psi\big) = \sum_{x\in\mathbf{Z}^d} \chi_\Delta\big(\xi(x)\big)\left|\psi(x)\right|^2,$$

and therefore

$$\mathsf{E}\big\{\big(E_A(\Delta)\psi,\psi\big)\big\} = \mathsf{P}\{\xi(0)\in\Delta\}\left\|\psi\right\|^2.$$

Finally,

$$\operatorname{Tr} P(\lambda) = \sum_{x\in\mathbf{Z}^d} \chi_{\{\lambda\}}\big(\xi(x)\big).$$

Therefore, if λ is a discontinuity point of $F(\xi)$, the right-hand side of this last equality is equal to infinity, and if λ is a continuity point, the right-hand side is zero. Thus, in our example, Theorem 2.12 is equivalent to the following trivial fact: the probability of a random variable assuming a value which is a continuity point of its distribution function is zero.

However, if we consider intervals, rather than individual points, on the spectral axis, the point spectrum, like the other components of the spectrum of a metrically transitive operator, has, with probability 1, a property similar to the property of the entire spectrum formulated in Theorem 2.10.

We denote by π_ε the projection on $\mathcal{H}_\varepsilon$, for $\varepsilon = \mathrm{p}, \mathrm{ac}, \mathrm{sc}$.

(2.14) Lemma. π_{p}, π_{ac} *and* π_{sc} *are metrically transitive random projections.*

Proof. It is clear that (1.44) holds for π_ε, for $\varepsilon = \mathrm{p}, \mathrm{ac}, \mathrm{sc}$, if it holds for A. Thus we just have to prove the randomness, that is, the measurability, of π_ε. Denote by μ_u the random measure on $\mathbf{R}$ associated with the distribution function $\big(E_\mu(\lambda)u, u\big)$, where $E_A(\lambda)$ is the resolution of the identity of A, and $u \in \mathcal{H}$. By Theorem 2.7, μ_u is a random measure. But all the components of a random measure $\nu(d\lambda)$ in the Lebesgue decomposition $\nu = \nu_{\mathrm{p}} + \nu_{\mathrm{ac}} + \nu_{\mathrm{sc}}$ are also random measures; this follows from the relations

$$\nu_{\mathrm{ac}} = \nu - \nu_{\mathrm{p}} - \nu_{\mathrm{sc}},$$

$$\nu_{\mathrm{p}}\big((-\infty, \lambda]\big) = \lim_{\alpha \downarrow 1} \lim_{n \to \infty} \sum_{k \in \mathbf{Z}} \big(\nu\big((\tfrac{k}{n}, \tfrac{k+1}{n}) \cap (-\infty, \lambda]\big)\big)^2,$$

$$\nu_{\mathrm{sc}}\big((-\infty, \lambda]\big) = \lim_{n \to \infty} \sup_{\substack{B \in \mathcal{B}_Q \\ |B| \leqslant n^{-1}}} (\nu - \nu_{\mathrm{p}})\big(B \cap (-\infty, \lambda]\big),$$

where $\mathcal{B}_Q$ is a countable family of Borel sets on $\mathbf{R}$ of the form $\bigcup_{j=1}^k (p_j, q_j)$, for $k < \infty$, and the $p_j < q_j$ are all rational numbers.

Therefore, $(\pi_\varepsilon u, u) = \mu_{u,\varepsilon}(\mathbf{R})$ is a random variable, hence $(\pi_\varepsilon u, v)$, for $u, v \in \mathcal{H}$, is also a random variable. □

(2.15) Theorem [Pastur 1974a]. *Let A be a metrically transitive operator that is essentially self-adjoint with probability* 1 *and satisfies the conditions of Theorem* 2.4. *For any interval $\Delta \subset \mathbf{R}$ the number of eigenvalues of A that lie in Δ is either* 0 *or* ∞, *depending on whether or not the operator* $\mathsf{E}\{\pi_{\mathrm{p}} E_A(\Delta)\}$ *vanishes.*

Proof. Since π_{p} commutes with $E_A(\Delta)$, and because of Theorem 2.7 and Lemma 2.14, $\pi_{\mathrm{p}} E_A(\Delta)$ is a metrically transitive random projection. By Theorem 2.1, the number of eigenvalues of A that lie in Δ is, with probability 1, equal to $\mathsf{E}\{\operatorname{Tr} \pi_{\mathrm{p}} E_A(\Delta)\}$, which proves the theorem. □

There are criteria for the existence of the point spectrum other than the one given by this theorem. These criteria, which can also be used to find out whether the spectrum is pure point, are considered in Exercise I.15 and in Sections 13 and 15.

Theorems 2.10 and 2.16 effectively say that the spectrum of a metrically transitive operator and the closure of its point component are nonrandom sets. The next theorem says the same thing for all the components of the spectrum.

(2.16) Theorem [Pastur 1974a, 1980; Kunz and Souillard, 1980; Kirsch and Martinelli, 1982a]. *Let A be a metrically transitive operator that is essentially self-adjoint with probability* 1 *and satisfies the conditions of Theorem* 2.4. *The spectrum $\sigma(A)$ of A and its components $\overline{\sigma_{\mathrm{p}}(A)}$, $\sigma_{\mathrm{c}}(A)$, $\sigma_{\mathrm{ac}}(A)$ and $\sigma_{\mathrm{sc}}(A)$ are all nonrandom sets.*

Proof. Since $\sigma = \overline{\sigma_{\mathrm{p}}} \cup \sigma_{\mathrm{ac}} \cup \sigma_{\mathrm{sc}}$ and $\sigma_{\mathrm{c}} = \sigma_{\mathrm{ac}} \cup \sigma_{\mathrm{sc}}$, it is enough to prove the statement for $\overline{\sigma_{\mathrm{p}}}$, σ_{ac} and σ_{sc}. Each of these sets is closed and coincides with the set of growth points of the corresponding projection-valued functions $\pi_{\mathrm{p}} E_A(\lambda)$, $\pi_{\mathrm{ac}} E_A(\lambda)$ and $\pi_{\mathrm{sc}} E_A(\lambda)$, which are orthogonal projections since π_{p}, π_{ac} and π_{sc} commute with $E_A(\lambda)$.

Denote by σ_ε the component $\overline{\sigma_{\mathrm{p}}}$, σ_{ac} and σ_{sc} for $\varepsilon = p, ac, sc$, respectively, and by Q_2 the set of intervals of the form $(p, q]$, for $p, q \in \mathbf{Q}$. For each $(p, q] \in Q_2$, the function

$$n_{\sigma_\varepsilon}\big((p,q]\big) = \big\|\pi_\varepsilon(E_A(q) - E_A(p))\big\|$$

is an invariant random variable, because the projections $E_A(\lambda)$ and π_ε are metrically transitive, by Theorem 2.7 and Lemma 2.14. By Proposition 1.4, $n_{\sigma_\varepsilon}\big((p,q]\big)$ is nonrandom; but since Q_2 is countable, there is in fact a set $\tilde{\Omega} \subset \Omega$ of full measure such that $n_{\sigma_\varepsilon}\big((p,q]\big)$ is independent of Ω for all $(p,q] \in Q_2$. The theorem now follows from the observation that, if C_1, C_2 are closed sets on the axis and $n_{C_i}\big((p,q]\big)$, for $(p,q] \in Q_2$ and $i = 0, 1$, is defined by

$$n_{C_i}\big((p,q]\big) = \begin{cases} 1 & \text{if } (p,q] \cap C_i \neq \varnothing, \\ 0 & \text{if } (p,q] \cap C_i = \varnothing, \end{cases}$$

we have $n_{C_1} = n_{C_2}$ if and only if $C_1 = C_2$. □

We emphasize again that the nonrandomness of the spectrum of a metrically transitive operator, and of its point component in particular, results from the fact that these sets are closed. In particular, this nonrandomness is consistent with the fact that the spectrum is pure point in cases when Theorem 2.12 applies, as this theorem bespeaks the high degree of "mobility" of the point component.

In view of this, it is natural to think that the bigger the set of realizations of a metrically transitive operator, the bigger its spectrum, since the support of the corresponding probability measure is also bigger. However, an attempt to formalize this intuitively obvious observation shows at once that the probability-theory definition of the support of a measure as any set $S \in \mathcal{F}$ such that $\mathsf{P}(S) = 1$ turns out to be too coarse for this purpose, inasmuch as by virtue of the metric transitivity condition, the support of a measure P' corresponding to a random operator with a smaller set of realizations will, as a rule, have zero P-measure.

This is clear already in the very simple case of Example 2.13, for which the spectrum obviously depends monotonically on the support. Indeed, if $\xi(x)$ and $\xi'(x)$, for $x \in \mathbf{Z}^d$, are fields of independent random variables, with corresponding probability measures P and P′ and distribution functions supported on Φ and Φ', any set of realizations of ξ' will have zero P-measure if $\Phi' \subsetneq \Phi$.

We therefore introduce a more suitable definition for the support of a probability measure. We consider probability spaces for which the space of realizations Ω is a topological space, the σ-algebra is the Borel σ-algebra $\mathcal{B}_\Omega$ of Ω (that is, the σ-algebra generated by all sets open in the topology of Ω), and the probability measure P on the space $(\Omega, \mathcal{B}_\Omega)$ is regular [Reed and Simon I]. Then the *topological support*, or simply *support*, of P, denoted by supp P, is the set of points $\omega \in \Omega$ such that the P-measure of any neighborhood of ω is positive. It is easy to see that the topological support of P is the smallest closed set in $\mathcal{B}_\Omega$ whose P-measure is 1.

For the next result, whose statement includes continuity conditions for certain functions in Ω, it is best to assume that Ω is a σ-compact (or, as a special case, compact) metrizable topological space, for which there is a metric compatible with the topology making it into a complete space.

(2.17) Theorem [Kotani 1985a]. *Let Ω be a space as specified in the preceding paragraph, and let $A(\omega)$ be a function on Ω taking values in the set of self-adjoint operators on a Hilbert space $\mathcal{H}$. Assume that for each $\psi \in \mathcal{H}$ the measure $(E_{A(\omega)}(d\lambda)\psi, \psi)$, where $E_{A(\omega)}(\lambda)$ is the resolution of the identity of $A(\omega)$, is a weakly continuous function of ω, that is,* $\text{w-lim}_{\omega_n \to \omega}(E_{A(\omega_n)}(d\lambda)\psi, \psi) = (E_{A(\omega)}(d\lambda)\psi, \psi)$.

Assume given on $(\Omega, \mathcal{B}_\Omega)$ two regular metrically transitive probability measures P_1 and P_2 which define, together with $A(\omega)$, two self-adjoint metrically transitive operators $A_1(\omega)$ and $A_2(\omega)$. If $\operatorname{supp}\mathsf{P}_1 \subset \operatorname{supp}\mathsf{P}_2$, *the spectra of A_1 and A_2 satisfy $\sigma(A_1) \subset \sigma(A_2)$.*

Proof. Consider an open interval $\Delta \subset \mathbf{R}$ not intersecting $\sigma(A_2)$. By Theorem 2.16, for P_2-almost every ω we have $(E_{A(\omega)}(\Delta)\psi, \psi) = 0$ for all $\psi \in \mathcal{H}$. By the assumptions of the theorem and by continuity, this equality extends to all $\omega \in \operatorname{supp} P_2$. In particular, it is valid on $\operatorname{supp}\mathsf{P}_1$, so that $\Delta \subset \sigma(A_1) = \varnothing$, proving the theorem. □

Sections 4 and 5 show how this theorem can be applied to the problem of finding the location of the spectrum of matrix and differential operators. From the results there it follows, for example, that if the potential $q(x)$ of a Schrödinger operator is bounded and nonnegative, and its topological support (in the topological space of generalized functions) contains a function that vanishes identically, the spectrum of the operator is the semiaxis $[0, \infty)$.

We will see in Section 12 that for one-dimensional and finite-difference operators of second order, absolutely continuous components of spectra are related by inclusion in the opposite way as in Theorem 2.17.

2.C Nonrandomness of Multiplicities

Let A be an essentially self-adjoint operator on a Hilbert space $\mathcal{H}$. A subspace $\mathcal{H}_0$ is called a *generating subspace* for A if the closure of the linear span of the vectors $E_A(\Delta)h$, where Δ runs over all intervals in $\mathbf{R}$ and h runs over $\mathcal{H}_0$, equals $\mathcal{H}$.

Definition. *The spectral multiplicity of an essentially self-adjoint operator A, denoted by $\kappa(A)$ or simply κ, is the minimum of the dimensions of generating subspaces for A.*

Any essentially self-adjoint operator A is unitarily equivalent to the operator $\tilde{A}$ of multiplication by λ on the Hilbert space

$$\tilde{\mathcal{H}} = \sum_{\alpha} \oplus \kappa_\alpha L^2(\mathbf{R}, \rho_\alpha),$$

where there are at most a countable number of α, the measures ρ_α are pairwise singular, and κ_α indicates how many times $L^2(\mathbf{R}, \rho_\alpha)$ occurs in the sum (see [Reed and Simon I]). The multiplicity κ is equal to $\sup_\alpha \kappa_\alpha$.

If a subspace $\mathcal{H}_1$ reduces an essentially self-adjoint operator A, the spectral multiplicity of the restriction $A\big|_{\mathcal{H}_1}$ does not exceed the spectral multiplicity of A, because the orthogonal projection onto $\mathcal{H}_1$ of any generating subspace for A is a generating subspace for the restriction. This equips us for the definition of the spectral multiplicity at a point λ:

Definition. *The spectral multiplicity of an essentially self-adjoint operator A at a point λ, denoted by $\kappa(A; \lambda)$, is the limit of the multiplicities of the restrictions $A\big|_{E_A(\Delta_n)\mathcal{H}}$, where $\{\Delta_n\}$ is a monotonically decreasing sequence of intervals containing λ in their interior, such that $\bigcap_n \Delta_n = \{\lambda\}$.*

This limit always exists and does not depend on the sequence $\{\Delta_n\}$.

We say that a random subspace $\mathcal{H}_1$ of $\mathcal{H}$ is metrically transitive if the orthogonal projection onto $\mathcal{H}_1$ is.

(2.18) Theorem. *If A is a metrically transitive, essentially self-adjoint operator satisfying the conditions of Theorem 2.4, and $\mathcal{H}_1$ a random metrically transitive A-reducing subspace of $\mathcal{H}$, the multiplicity $\kappa(A|_{\mathcal{H}_1})$ is nonrandom.*

Proof. Assume that $\kappa(A|_{\mathcal{H}_1})$ is a random variable. Since A and the projection onto $\mathcal{H}_1$ are metrically transitive operators, it follows from (1.44) that $\kappa(A|_{\mathcal{H}_1})$ is an invariant random variable and, by Theorem 1.10, nonrandom. Thus we'll be done if we can show that $\kappa(A|_{\mathcal{H}_1})$ is $\mathcal{F}$-measurable.

Let $h_1, \ldots, h_n \in \mathcal{H}$. Denote by $\mathcal{H}(h_1, \ldots, h_n)$ the closure of the span of the vectors $E_A(\Delta)h_1, \ldots, E_A(\Delta)h_n$, where Δ runs over all intervals in $\mathbf{R}$, and by $\pi(h_1, \ldots, h_n)$ the orthogonal projection onto $\mathcal{H}(h_1, \ldots, h_n)$. If Q_2 is set of all intervals in $\mathbf{R}$ with rational endpoints, $\mathcal{H}(h_1, \ldots, h_n)$ coincides with the closure of the linear span of the vectors $E_A(\Delta)h_1, \ldots, E_A(\Delta)h_n$, for $\Delta \in Q_2$. Hence, by Theorems 1.6 and 2.7, $\mathcal{H}(h_1, \ldots, h_n)$ is a random subspace and $\pi(h_1, \ldots, h_n)$ a random projection.

To prove that $\kappa(A|_{\mathcal{H}_1})$ is $\mathcal{F}$-measurable, we have to express it in terms of at most a countable number of objects known to be measurable. We do this in the following way: Let A be an essentially self-adjoint operator on $\mathcal{H}$, and let $\mathcal{H}_1$ be a subspace that reduces A. Introduce the metric of strong convergence of uniformly bounded operators in $\mathcal{H}$: namely, let $\{u_m\}_1^\infty$ be a countable, dense family of vectors in $\mathcal{H}$, and set

$$\rho(B) = \sum_{m=1}^{\infty} 2^{-m} \frac{\|Bu_m\|}{1 + \|Bu_m\|}. \tag{2.4}$$

Denote by $\pi_{\mathcal{H}_1}$ the orthogonal projection onto $\mathcal{H}_1$ and by $\mathcal{D}$ an arbitrary countable subspace dense in $\mathcal{H}$. Define a sequence of numbers ζ_k, for $k \in \mathbf{Z}_+$, related to the multiplicity of the operator $A|_{\mathcal{H}_1}$, by the equalities

$$\zeta_k(A, \mathcal{H}_1) = \inf_{h_1,\ldots,h_n \in \mathcal{D}} \rho(\pi_{\mathcal{H}_1} - \pi(\pi_{\mathcal{H}_1}h_1, \ldots, \pi_{\mathcal{H}_1}h_n)). \tag{2.5}$$

(2.19) Lemma. *With the notations above, $\zeta_k(A, \mathcal{H}_1) > 0$ if $k < \kappa(A|_{\mathcal{H}_1})$.*

Proof. We first show that we can reduce the problem to the case $\mathcal{H}_1 = \mathcal{H}$. Indeed, the sets $\mathcal{D}_1 = \pi_{\mathcal{H}_1}\mathcal{D}$ and $\{u_m\}_1^\infty = \{\pi_{\mathcal{H}_1}u_m\}_1^\infty$ are obviously countable and dense in $\mathcal{H}_1$. Therefore, if I_1 is the identity operator on $\mathcal{H}_1$ and ρ_1 denotes the metric in (2.4), with $\{\tilde{u}_m\}_1^\infty$ playing the role of $\{u_m\}_1^\infty$, we can rewrite $\zeta_k(A, \mathcal{H}_1)$ as follows:

$$\zeta_k(A, \mathcal{H}_1) = \inf_{h_1,\ldots,h_k \in \mathcal{D}_1} \rho_1(I_1 - \pi(h_1, \ldots, h_k)) = \zeta_k(A|_{\mathcal{H}_1}, \mathcal{H}_1).$$

This clearly reduces the case of arbitrary $\mathcal{H}_1$ to that of $\mathcal{H}_1 = \mathcal{H}$. Now put $\zeta_k(A\big|_{\mathcal{H}}, \mathcal{H}) = \zeta(A)$; what we have to show is that for $k < \kappa(A)$ we have

$$\zeta_k(A) = \inf_{h_1,\dots,h_k\in\mathcal{D}} \rho\big(I - \pi(h_1,\dots,h_k)\big) > 0. \tag{2.6}$$

Let us assume to the contrary that for some $k < \kappa$ (that is, for k finite) $\zeta_k(A) = 0$. This means there is a sequence of families $\{(h_1^{(n)},\dots,h_k^{(n)})\}$, for $n \in \mathbf{Z}_+$, such that $\text{s-lim}\, \pi(h_1^{(n)},\dots,h_k^{(n)}) = I$. Thus any vector in $\mathcal{H}$ can be approximated, to any prescribed accuracy, by a vector from $\mathcal{H}(h_1^{(n)},\dots,h_k^{(n)})$, as long as n is sufficiently large.

Consider the above canonical representation $\tilde{A}$ of A. An arbitrary vector $\tilde{f} \in \tilde{\mathcal{H}}$ is defined by a family of functions $\{f_{\alpha,l}(\lambda)\}$, for $1 \leqslant l \leqslant \kappa_\alpha$, where $f_{\alpha,l} \in L^2(\mathbf{R},\rho_\alpha)$. Since $k < \kappa(A) = \max_\alpha \kappa_\alpha$, there exists β such that $k < m\beta$, and we specify $k+1$ vectors $\tilde{f}^{(j)}$, for $1 \leqslant j \leqslant k+1$, by the equalities

$$\tilde{f}_{\alpha,l}^{(j)}(\lambda) = \begin{cases} 0 & \text{if } \alpha \neq \beta, \\ \delta_{j,l} & \text{if } \alpha = \beta \text{ and } 1 \leqslant l \leqslant k+1, \\ 0 & \text{if } \alpha = \beta \text{ and } l \leqslant k+1 < m\beta. \end{cases} \tag{2.7}$$

Now look at the images $\tilde{\mathcal{H}}(\tilde{h}_1^{(n)},\dots,\tilde{h}_k^{(n)})$ of the spaces $\mathcal{H}(h_1^{(n)},\dots,h_k^{(n)})$. There exist vectors $\tilde{g}^{(j,n)} \in \tilde{\mathcal{H}}(\tilde{h}_1^{(n)},\dots,\tilde{h}_k^{(n)})$ such that $\|\tilde{g}^{(j,n)} - \tilde{f}^{(j)}\| \to 0$ as $n \to \infty$; setting $\tilde{\eta}^{(j,n)} = \tilde{f}^{(j)} - \tilde{g}^{(j,n)}$, we have $\|\tilde{\eta}^{(j,n)}\| \to 0$ as $n \to \infty$. There exist also Borel functions $\psi_r^{(j,n)}(\lambda)$ such that

$$\tilde{g}^{(j,n)} = \sum_{r=1}^{k} \psi_r^{(j,n)}(\tilde{A})\tilde{h}_r^{(n)};$$

in other words,

$$\tilde{f}_{\alpha,l}^{(j)}(\lambda) = \tilde{\eta}_{\alpha,l}^{(j,n)}(\lambda) + \sum_{r=1}^{k} \psi_r^{(j,n)}(\lambda)(\tilde{h}_r^{(n)})_{\alpha,l}(\lambda).$$

By (2.7), this implies, for $\alpha = \beta$ and $1 \leqslant l \leqslant k+1$, that

$$\delta_{j,l} - \tilde{\eta}_{\beta,l}^{(j,n)}(\lambda) = \sum_{r=1}^{k} \psi_r^{(j,n)}(\lambda)(\tilde{h}_r^{(n)})_{\beta,l}(\lambda). \tag{2.8}$$

In addition,

$$\lim_{n\to\infty} \int_{\mathbf{R}} \left|\tilde{\eta}_{\beta,l}^{(j,n)}(\lambda)\right|^2 \rho_\beta(d\lambda) = 0.$$

This implies that, for any $\delta > 0$, there exists a set $\Delta_\delta \subset \mathbf{R}$ and a number $n(\delta)$ such that $\rho_\beta(\Delta_\delta) > 0$ and

$$|\tilde{\eta}_{\beta,l}^{(j,n(\delta))}(\lambda)| \leqslant \delta$$

for any $\lambda \in \Delta_\delta$.

We now set $\delta = (2k+2)^{-1}$ and consider the equalities (2.8), with $n = n(\delta)$ and $\lambda \in \Delta_\delta$. They may be seen as an equality between two matrices, having the two sides of (2.8) as entries for various j and l. The left-hand side matrix is clearly nondegenerate, while the right-hand side one is degenerate for λ in the set Δ_δ, whose ρ_β-measure is positive. This contradiction proves (2.6) and concludes the proof of the lemma. □

(2.20) Lemma. *If $\kappa(A|_{\mathcal{H}_1}) < \infty$ and $k \geqslant \kappa(A|_{\mathcal{H}_1})$, we have $\zeta_k(A, \mathcal{H}_1) = 0$.*

Proof. It is clear that $\zeta_k(A, \mathcal{H}_1)$ is a nonincreasing function of k. Therefore, it is enough to show that $\zeta_\kappa(A, \mathcal{H}_1) = 0$, where $\kappa = \kappa(A|_{\mathcal{H}_1})$. As in the proof of the preceding lemma, we can reduce to the case $\mathcal{H}_1 = \mathcal{H}$, so the task is to prove that $\zeta_\kappa(A) = 0$. Now there exist vectors $f_1, \dots, f_\kappa \in \mathcal{H}$ such that $\pi(f_1, \dots, f_\kappa) = I$. Take a sequence of families $\{(h_1^{(n)}, \dots, h_\kappa^{(n)})\}$, for $n \in \mathbf{Z}_+$, such that $h_j^{(n)} \in \mathcal{D}$ for $1 \leqslant j \leqslant \kappa$ and that $\|h_j^{(n)} - f_j\| \to 0$ as $n \to \infty$. We show that $\text{s-lim}_{n\to\infty} \pi(h_1^{(n)}, \dots, h_\kappa^{(n)}) = I$, which clearly implies $\zeta_\kappa(A) = 0$.

Consider the set $\mathcal{H}' \subset \mathcal{H}$ of the vectors $h \in \mathcal{H}$ that can be written as

$$h = \sum_{j=1}^{\kappa} \psi_j(A) f_j, \tag{2.9}$$

where the ψ_j are bounded Borel functions. Since $\mathcal{H}'$ is clearly dense in $\mathcal{H}$, it is enough to show that $\text{s-lim}_{n\to\infty} \pi(h_1^{(n)}, \dots, h_\kappa^{(n)}) = I$ for any $h \in \mathcal{H}'$. We do this by exhibiting a sequence of vectors $\{g^{(n)}\}_1^\infty$ in $\mathcal{H}$ such that $g^{(n)} \in \mathcal{H}(h_1^{(n)}, \dots, h_\kappa^{(n)})$ and $\|g^{(n)} - h\| \to 0$ as $n \to \infty$, as follows. Define the functions ψ_j by writing $h \in \mathcal{H}'$ as in equation (2.9), and let $g^{(n)}$ be also given by (2.9), but with f_j replaced by $h_j^{(n)}$. Clearly $g^{(n)} \in \mathcal{H}(h_1^{(n)}, \dots, h_\kappa^{(n)})$, and, since the ψ_j are bounded, $\|g^{(n)} - h\| \to 0$ as $n \to \infty$. □

We continue with the proof of Theorem 2.18. Let $\chi(t)$, for $t \in \mathbf{R}$, be 1 for $t \neq 0$ and 0 for $t = 0$. By the definition of the multiplicity $\kappa(A|_{\mathcal{H}_1})$ and by Lemmas 2.19 and 2.20, we have

$$\kappa(A|_{\mathcal{H}_1}) = \sup_{k \in \mathbf{Z}_+} \{(k+1)\chi(\zeta_k(A, \mathcal{H}_1))\}. \tag{2.10}$$

Fix the sets $\{u_m\}_1^\infty$ and $\mathcal{D}$ that appear in (2.4) and (2.5). Then $\zeta_k(A, \mathcal{H}_1)$ is a random variable, which means, in view of (2.10), that the multiplicity $\kappa(A|_{\mathcal{H}_1})$ is also one. □

Theorem 2.18 leads to the following statement:

(2.21) Theorem. *If A is a metrically transitive, essentially self-adjoint operator satisfying the conditions of Theorem 2.4, the spectral multiplicities of A, $A|_{\mathcal{H}_p}$, $A|_{\mathcal{H}_{ac}}$, $A|_{\mathcal{H}_{sc}}$ and, $A|_{\mathcal{H}_c}$ are nonrandom. The spectral multiplicities of the same operators at any point $\lambda \in \mathbf{R}$ are also nonrandom.*

Proof. We just have to prove the nonrandomness of $\kappa(A, \lambda)$ and $\kappa(A|_{\mathcal{H}_\varepsilon}, \lambda)$, for $\varepsilon = p, ac, sc, c$. But the definitions of these functions include only the projections $E_A(\Delta)$ corresponding to intervals Δ with rational endpoints. The nonrandomness of these functions is therefore a corollary of Theorems 2.7 and 2.18. □

Problems

1. Prove that, in the probability space $(S^1, \mathcal{B}, d\phi/2\pi)$ consisting of the unit circle with the Lebesgue measure, a rotation by an angle ϕ such that $\phi/2\pi$ is irrational is a one-to-one metrically transitive transformation.

2. Prove that, for almost periodic functions, the ergodic theorem holds not just for almost every point, but for every point.

Hint. Use the approximation theorem [Levitan 1953] that says that every almost periodic function is the uniform limit of polynomials in $C(\mathbf{R})$.

3. Let G be either $\mathbf{R}^d$ or $\mathbf{Z}^d$, for $d \geqslant 1$. Prove that, with probability 1, a realization of a metrically transitive field $f(g)$, for $g \in G$:

(a) does not belong to $L_2(G, dg)$, where dg is the Haar measure on G;
(b) does not have compact support;
(c) is such that $\inf_{g\in G} f(g)$ and $\sup_{g\in G} f(g)$ are nonrandom.

If $G = \mathbf{R}^d$ and the realizations of f are differentiable with probability 1, prove that the mathematical expectation of any of its partial derivatives is zero.

4. Let $f(g)$ and $f_1(g), \ldots, f_n(g)$, for $g \in G$, be metrically transitive fields, with $G = \mathbf{R}^d$ or $\mathbf{Z}^d$.

(a) If $c_1, \ldots, c_n$ are constants, $c_1 f_1 + \cdots + c_n f_n$ is a metrically transitive field.
(b) If $L(x_1, \ldots, x_n)$ is a Baire function on $\mathbf{R}^n$, $L(f_1, \ldots, f_n)$ is a metrically transitive field.
(c) If $h(g) \in L_1(G)$ is a nonrandom function and $\mathsf{E}\{|f(0)|\}$ is finite,

$$F(g) = (h * f)(g) = \int_G h(g - g_1) f(g_1) dg_1$$

is a metrically transitive field and $\mathsf{E}\{|F(0)|\} < \infty$.

5. If $Q(x)$, for $x \in \mathbf{R}^d$, is a metrically transitive field such that $\mathsf{E}\{|Q(0)|^p\} < \infty$ for some positive p, show that:

(a) $Q \in L_p^{\mathrm{loc}}(\mathbf{R}^d)$ with probability 1;

(b) $\lim_{x\to 0} \mathsf{E}\{|Q(x) - Q(0)|^p\} = 0$.

Hint. For (b), remember that the notion of metric transitivity includes stochastic continuity.

6. Let $g(x)$, for $x \in \mathbf{R}$, be a homogeneous Gaussian process with zero mean and correlation function $b(x) = \mathsf{E}\{g(x)g(0)\}$. By the Bochner–Khintchine theorem,

$$b(x) = \int_{\mathbf{R}} e^{ikx} \hat{B}(dk),$$

where $\hat{B}(dk)$ is a nonnegative finite measure on $\mathbf{R}$. Show that, if $g(x)$ is metrically transitive, $\hat{B}$ has no atoms.

Hint. Use the ergodic theorem (1.23) to show that

$$\lim_{L\to\infty} \mathsf{E}\left\{ \left(L^{-1} \int_0^L g^2(x)\, dx - \mathsf{E}\{g^2(0)\} \right)^2 \right\} = 0.$$

Then calculate explicitly the mathematical expectation, and use Wiener's theorem, according to which if $m(x) = \int e^{ikx} \mu(dk)$, where μ is a finite measure on $\mathbf{R}^d$ with atoms μ_n,

$$\lim_{L\to\infty} L^{-1} \int_0^L |m(x)|^2\, dx = \sum_n \mu_n^2.$$

7. Let $g(x)$, for $x \in \mathbf{Z}^d$, be a stationary Gaussian field with zero mean and correlation function $b(x) = \mathsf{E}\{g(x)g(0)\}$ whose Fourier transform $\hat{b}(k)$, for $k \in \mathbf{T}^d$, satisfies $\int_{\mathbf{T}^d} \hat{b}^{-1}(k)\, dk < \infty$. Let $\mathcal{F}(0)$ denote the σ-algebra generated by intersections of cylinders $\{g(x) \in \Delta\}$, for $x \neq 0$ and Δ an interval. Show that the conditional probability

$$F(dg \mid \mathcal{F}(0)) = \mathsf{P}\{g(0) \in dg \mid \mathcal{F}(0)\}$$

has a density bounded by a nonrandom constant.

Hint. Find the conditional mathematical expectation $\mathsf{E}\{g(0) \mid \mathcal{F}(0)\}$ and show that $F(dg \mid \mathcal{F}(0))$ is the Gaussian measure coinciding with the distribution of $g(0) - \mathsf{E}\{g(0) \mid \mathcal{F}(0)\}$, which is independent of ω.

8. Find the characteristic functional (1.38) for

(a) a random field on $\mathbf{R}^d$ that is the square of a Gaussian field;

(b) a random field on $\mathbf{R}^1$ (or random process) generated by the quasiperiodic function $\sum_{k=1}^{n} a_k \cos \alpha_k x$, that is, having the form

$$q(x, \omega) = \sum_{k=1}^{r} a_k \cos(\alpha_k x + \omega_k)$$

(see equation (1.37)), where the a_k and α_k are real, and the α_k are rationally independent.

Answer. (a) $\chi_F(\phi) = D^{-1/2}(\phi)$, where $D(\phi)$ is the Fredholm determinant of the following integral equation considered on the support Φ of ϕ:

$$f(x) - 2i \int_{\Phi} b(x-y)\phi(y)f(y)\, dy = h(x)$$

for $x \in \Phi$.

(b) $\chi_F(\phi) = \prod_{k=1}^{n} J_0\big(\big|a_k \int \phi(x) e^{i\alpha_k x}\, dx\big|\big)$, where J_0 is the Bessel function of order zero.

9. Using the explicit forms (1.39) and (1.40) of the characteristic functionals of the Gaussian and Poisson fields, show that, for a Poisson field of concentration c corrected by its mathematical expectation, that is,

$$\sum_j u(x - x_j) - c \int u(x)\, dx$$

the finite-dimensional distributions converge, as $c \to \infty$, to the finite-dimensional distributions of a Gaussian field with correlation function $b(x) = cu_0^2 \int w(x-y)w(y)\, dy$, where $u(x) = u_0 w(x)$ for $w(x)$ a fixed summable function and u_0 a real number that tends to 0 in such a way that cu_0^2 tends to a constant.

10. If $q(x)$ is a homogeneous Markov process assuming a finite number of values $q_1, \dots, q_r$, the probability that $q(x)$ has the value q_k for an interval of length at least l is of the form $F_k(l) = e^{-n_k l}$, with $n_k > 0$, for $k = 1, \dots, n$.

Hint. Denote by $h_k(x_1, x_2)$ the probability that $q(x)$ keeps the value q_k for an interval of length at least x_2, given that it has had the same value during the preceding interval of length x_1. By the theorem on multiplication of probabilities, $h(0, x_1 + x_2) = h(x_1, x_2)h(0, x_1)$.

11. Let $\{l_j\}_0^\infty$ be a sequence of nonnegative, independent random variables, identically distributed for even and odd j, with distributions

$$\mathsf{P}\{l_{r+2k} \leqslant l\} = \int_0^l f_r(l')\,dl' \qquad \text{for } r = 0, 1,$$

where $f_r(l) > 0$. For fixed $\xi' \geqslant 0$ and $r' = 0, 1$, lay down end-to-end, starting at $-\xi'$ and continuing to the right, random intervals of lengths $l_{r'}, l_{r'+1}, \ldots$, considering only realizations such that $l_{r'} \geqslant \xi'$. Set $x_n = \sum_{k=r'}^n l_k$, for $n \geqslant r'$, and

$$q(x) = \begin{cases} r' & \text{if } -\xi' \leqslant x \leqslant -\xi' + x_{r'}, \\ 1 - r' & \text{if } -\xi' + x'_r \leqslant x \leqslant -\xi' + x_{r'+1}, \\ r' & \text{if } -\xi' + x_{r'+1} \leqslant x \leqslant -\xi' + x_{r'+2}, \end{cases}$$

and so on. Verify that

(a) $q(x)$ is a component of the two-component homogeneous Markov process $Q(x) = (q(x), \xi(x))$, where $\xi(x)$ is the distance from x to the lower endpoint of the interval where x is contained;

(b) the invariant distribution of the process $Q(x)$ is

$$F_r(d\xi) = \frac{1}{n_0 + n_1} \int_\xi^\infty f_r(l)\,dl\,d\xi$$

for $r = 0, 1$, where $n_r = \left(\int_0^\infty l\,f_r(l)\,dl\right)^{-1}$;

(c) the density $p(x; r', \xi'; r, \xi)$ of the transition probability $P(x; r', \xi'; r, \xi) = p(x; r', \xi'; r, \xi)\,d\xi$ satisfies the *Fokker–Planck–Kolmogorov (FPK) equation*

$$\frac{\partial p}{\partial x} = -\frac{\partial p}{\partial \xi} - \nu_r(\xi) p,$$

where $\nu_r(\xi) = f_r(\xi)\left(\int_\xi^\infty f_r(l)\,dl\right)^{-1}$, and that

$$p(x; r', \xi; r, 0) = \int_0^\infty \nu_{1-r}(\eta') p(x; r', \xi'; 1 - r, \eta)\,d\eta$$

$$p\big|_{x=0} = \delta_{r,r'} \delta(\xi - \xi');$$

(d) the density of the invariant distribution is a stationary (i.e., independent of x) solution of the FPK equation;

(e) For $f_r(l) = n_r e^{-n_r l}$, the solution of the FPK equation coincides with (1.27) after integration over ξ.

12. Show that if a stationary Gaussian process $g(x)$ is Markovian, its correlation function $b(x)$ is

$$b(x) = D(2\beta)^{-1}e^{-\beta|x|},$$

where D and β are positive constants. Find, by direct calculation, the transition probability of this process. In the physics literature this is usually called the *Ornstein–Uhlenbeck process*; when $\beta = 0$, it coincides with the Wiener process.

13. Show that, if a self-adjoint operator A acting on a Hilbert space $\mathcal{H}$ is an orthogonal sum of self-adjoint operators A_1 and A_2 acting on $\mathcal{H}_1$ and $\mathcal{H}_2$, with $\mathcal{H} = \mathcal{H}_1 \oplus \mathcal{H}_2$, we have $\sigma(A) = \sigma(A_1) \cup \sigma(A_2)$ and $\sigma_{\mathrm{ac}}(A) = \sigma_{\mathrm{ac}}(A_1) \cup \sigma_{\mathrm{ac}}(A_2)$.

14. Prove that the norm of a metrically transitive operator is nonrandom.

15. Let A be a self-adjoint metrically transitive operator acting on a Hilbert space $\mathcal{H}$. For any vectors $h, g \in \mathcal{H}$, any interval $\Delta \subset \mathbf{R}$ and any $t > 0$, set

$$B_{\Delta,t}(h,g) = \int_\Delta e^{it\lambda}(E(d\lambda)h, g).$$

(a) The limit

$$\lim_{T\to\infty} \frac{1}{2T}\int_{-T}^{T} \left|B_{\Delta,t}(h,g)\right|^2 dt = b_\Delta(h,g)$$

exists and is equal to $\sum_{\lambda\in\sigma_{\mathrm{p}}\cap\Delta}\left|(P(\lambda)h,g)\right|^2$.

(b) Δ contains the point spectrum of A with probability 1 if and only if

$$\sup_{\|h\|=1} \mathsf{E}\{b_\Delta(h,h)\} > 0.$$

(c) The spectrum of A within Δ is pure point if and only if

$$\mathsf{E}\{b_\Delta(h)\} = \mathsf{E}\{(E(\Delta)h,h)\}$$

for any $h \in \mathcal{H}$, where $b_\Delta(h) = \sum_k b_\Delta(h, e_k)$ and $\{e_k\}$ is an orthonormal basis in $\mathcal{H}$.

(d) If $R(z) = (A - zI)^{-1}$, for $\operatorname{Im} z \neq 0$, is the resolvent to A, we have

$$b_\Delta(h,g) = \lim_{\varepsilon\downarrow 0}\frac{\varepsilon}{\pi}\int_\Delta \left|(R(\lambda+i\varepsilon)h,g)\right|^2 d\lambda.$$

Remark. The quantity $b_{\mathbf{R}}(h,g)$ has a simple quantum-mechanical interpretation, as the probability that a particle which was in state h at the origin of time will be, after an infinitely long time, in state g. Thus, saying that an operator A has a point spectrum is saying that at least some part of a wave packet will remain in a bounded region of space after an infinitely long time if its evolution is governed by a Hamiltonian given by A.

Hint. To prove (a), use Wiener's theorem [Reed and Simon III]; to prove (c), recall that $b_\Delta(h) = \big(E_p(\Delta)h, h\big)$.

16. According to (1.20), the metrically transitive field $q(x,\omega)$ on $\mathbf{Z}^d(\mathbf{R}^d)$, which may be regarded as a "diagonal" metrically transitive operator (the multiplication operator), can be represented as $q(x,\omega) = Q(T_x\omega)$, where $Q(\omega)$ is a measurable function on the space of elementary events Ω, and T_x is the corresponding shift operator. Prove that, for the kernels of matrix and integral operators given in Examples 1.17(c,d), the following representation generalizing (1.20) obtains:

$$a(x,y;\omega) = A(x-y, T_x\omega),$$

where $A(x,\omega)$, for every $x \in \mathbf{Z}^d(\mathbf{R}^d)$, is a measurable function on Ω.

For a self-adjoint operator, $A^*(-x,\omega) = A(x,T_x\omega)$. Give examples of such functions.

17. If $q_1(x)$ and $q_2(x)$, for $x \in \mathbf{Z}$, are mutually independent metrically transitive fields, and the random variables $q_2(x)$ are also independent, the metrically transitive field $q_1(x) + q_2(x)$ is nondeterministic.

18. Let $U(x)$, for $x \in \mathbf{R}$, be the group of shift operators in $L_2(\mathbf{R})$, that is, $\big(U(x)\psi\big)(y) = \psi(x+y)$. Call a bounded operator A on $L_2(\mathbf{R})$ almost periodic if the closure Ω of the family of operators $\{U(x)AU^*(x) : x \in \mathbf{R}\}$ is compact in the space of bounded linear operators in $L_2(\mathbf{R})$. The operator A can be regarded as a metrically transitive operator if Ω is endowed with a probability measure P coinciding with the Haar measure and we set $A(\omega) = \omega$. Show that, with probability 1, the spectrum σ of $A(\omega)$ is

$$\sigma = \bigcup_{\omega \in \operatorname{supp} P} \sigma\big(A(\omega)\big).$$

19. Let A be an almost periodic operator (see Problem 18) and $C^*(A)$ the minimal C^*-algebra of linear bounded operators in $L_2(\mathbf{R})$ containing A. Prove that any operator in $C^*(A)$ is almost periodic.

20. We say that a random process $f(x)$ on $\mathbf{R}$ or $\mathbf{Z}$ is *reversible* if all its finite-dimensional distributions $\mathsf{P}\{f(x_1) \in X_1, \dots, f(x_n) \in X_n\}$ are invariant under the transformation $x_j \mapsto -x_j$, for $j = 1, \dots, n$. Show that the following processes are reversible:

(a) a homogeneous Gaussian process;
(b) a Poisson process (1.28) with $u(x)$ an even function;
(c) a random telegraph signal (1.27);
(d) a homogeneous ergodic Markov process (possibly multicomponent, that is, taking values in $\mathbf{R}^d$) whose infinitesimal operator is self-adjoint in

the space of functions on the phase space of the process that are square-integrable with weights given by the invariant distribution; in particular, the one-dimensional diffusion process whose infinitesimal operator is

$$(\mathfrak{A}g)(x) = a\frac{dg}{dx} + b + b\frac{d^2g}{dx^2};$$

(e) the quasiperiodic function (1.37) with $Q(x_1, \ldots, x_n)$ an even function.

Chapter II

Asymptotic Properties of Metrically Transitive Matrix and Differential Operators

3 Review of Basic Results

The preceding chapter was devoted to the study of those properties of abstract symmetric and essentially self-adjoint operators that follow from metric transitivity. For example, we proved that their deficiency indices are nonrandom and equal either 0 or ∞, and their their spectra, spectral components and multiplicities are all nonrandom.

Of course, these results are true, in particular, for matrix and differential operators; but many natural and important questions about the commonest types of matrix and Schrödinger operators cannot be answered by a general theory that relies solely on the notion of metric transitivity. For example, the crucial question of essential self-adjointness for general multidimensional matrix and differential operators remains unanswered, because Theorem 2.4 states only that these operators have deficiency indices equal to 0 or ∞. Therefore, essential self-adjointness follows only for one-dimensional finite-difference and differential operators of finite order, such as Jacobi matrices and Schrödinger operators on the line (Corollary 2.6), or for operators whose deficiency indices are known to be finite.

In this chapter we present some results about the essential self-adjointness of metrically transitive multidimensional matrix and Schrödinger operators. We show that a matrix operator $A = \{a(x,y)\}$, for $x, y \in \mathbf{Z}^d$, is essentially self-adjoint with probability 1 if, for instance, the matrix A is quasidiagonal—that is, if $a(x,y) = 0$ for $|x - y| > R$, where R is a nonrandom positive number—or if the distribution functions $\mathsf{P}\{|a(x,y)| > t\}$ vanish sufficiently fast as $t \to \infty$.

For Schrödinger operators $-\Delta + q(x)$, for $x \in \mathbf{R}^d$, there are two criteria implying essential self-adjointness, which in a certain sense are complementary. One, Theorem 5.8, guarantees essential self-adjointness with probability 1 under the sole condition $\mathsf{E}\{|q(0)|^p\} < \infty$, where p is a particular natural number. To prove this criterion, one must take the limit from the case of smooth, bounded $q(x)$ to the general case. The other criterion, Theorem 5.15, does not impose any restrictions on the moments of the field $q(x)$, but requires that $q(x)$ be continuous and that its radius of correlation be finite (by definition, the σ-algebras generated by $q(x)$ for values of x whose distance is greater than the radius of correlation are independent). Its proof depends on an important criterion of essential self-adjointness for nonrandom operators and on recent results about the structure of local maxima of random fields.

The question of essential self-adjointness and, more generally, questions about the deficiency indices of matrix and differential operators on $\ell_2(\mathbf{Z}^d)$ and $L_2(\mathbf{R}^d)$, may be considered in terms of the asymptotic behavior of sequences of self-adjoint operators defined on bounded regions Λ, as Λ grows to exhaust all of $\mathbf{Z}^d$ or $\mathbf{R}^d$.

For a matrix operator A, this sequence of finite-order matrices is obtained by substituting zeros for the entries of A in positions outside $\Lambda \subset \mathbf{Z}^d$. For a differential operator, the corresponding sequence of operators on $\Lambda \subset \mathbf{R}^d$ is specified by preserving the differential operation and imposing certain self-adjoint boundary conditions on $\partial\Lambda$. According to this interpretation, which goes back to H. Weyl, essential self-adjointness amounts to a kind of independence of the "limiting" operator on $L_2(\mathbf{R}^d)$ from the choice of a sequence of expanding regions and of boundary conditions on their surfaces.

This asymptotic approach can be applied to any matrix or differential operator and is very widespread in spectral analysis [Levitan and Sargsyan 1970; Berezanskii 1968]. It turns out to be equally fruitful in the study of a variety of other properties of metrically transitive operators, especially the distribution of eigenvalues. It commends itself both in terms of the intrinsic logic of the spectral analysis of such operators—their structure and the results about them that it leads to—and in view of the historical roots of the discipline in theoretical physics, where the asymptotic approach is closely associated with the so-called thermodynamic limit and the self-averageability of physical quantities [Lifshitz et al. 1982b].

We now turn to some simple, but typical, examples that illustrate the class of problems associated with the asymptotic approach and properties. We will call such properties ergodic, because they have much in common, in form and essence, with the ergodic theorem. The main assumption made in these examples is metric transitivity, whereas in the results of Sections 4 and 5 more technical conditions will intervene.

Suppose that J is a metrically transitive random Jacobi matrix, that is,

$$(J\psi)(x) = -s(x+1)\psi(x+1) - s(x)\psi(x-1) + q(x)\psi(x).$$

Consider the sequence of matrices J_n, for $n \in \mathbf{Z}_+$, obtained by truncating J: namely, if $J = \{a(x,y)\}$ for $x, y \in \mathbf{Z}$, we set $J_n = \{a(x,y)\}$ for $|x|, |y| \leqslant n$. The order of J_n is evidently $2n+1$. For each J_n, we let $N_n(\lambda)$ be $(2n+1)^{-1}$

times the number of eigenvalues not exceeding λ, and $N_n(d\lambda)$ the measure on $\mathbf{R}$ associated with the normalized distribution function $N_n(\lambda)$.

We will show that there is a nonrandom measure $N(J, d\lambda)$ on $\mathbf{R}$ such that, with probability 1,

$$\lim_{n\to\infty} N_n(d\lambda) = N(J, d\lambda), \tag{3.1}$$

where the limit is in the sense of weak convergence of measures [Gihman and Skorohod I]. Moreover, the limiting measure $N(J, d\lambda)$ is related to the resolution of the identity $E_J(d\lambda)$ of J by

$$N(J, d\lambda) = \mathsf{E}\{(e_0, E_J(d\lambda)e_0)\}, \tag{3.2}$$

where the vector e_0 has components $e_0(x) = \delta_{0,x}$. The nondecreasing, left-continuous distribution function $N(J, \lambda)$ defined by the relation

$$N(J, \lambda) = N(J, (-\infty, \lambda])$$

is the normalized limiting distribution function of eigenvalues of the operator J acting on $\ell_2(\mathbf{Z})$. The derivative of $N(J, \lambda)$ is known in the physics literature as the operator's density of states, and $N(J, \lambda)$ itself as the number of states or the *integrated* (or *cumulative*) density of states [Lifshitz et al. 1982b].

In two special cases, relations (3.1) and (3.2) can be interpreted as well-known mathematical facts:

If the probability space Ω consists of a single point, a matrix operator A is metrically transitive if and only if it is Toeplitz, that is, its entries depend only on the difference between indices: $a(x, y) = a(x - y)$. In this case, (3.1) and (3.2) express results of Szegö on the asymptotic distribution of eigenvalues of Toeplitz matrices of increasing order. See, for example, [Grenander and Szegö 1958], which also presents similar results for integral operators whose kernels depend only on the difference between arguments, considered on a sequence of expanding intervals.

If, on the other hand, Ω is arbitrary but the matrix A is diagonal (Example 2.13) and its diagonal entries $\xi(x)$, for $x \in \mathbf{Z}$, are independent, identically distributed random variables with distribution function $F(\xi)$, relation (3.1) is a version of the *Glivenko–Cantelli theorem* stating that the so-called empirical distribution function $F_n(\xi)$ of a random variable $\xi(x)$, for n independent and identical tests, converges, with probability 1, to $F(\xi)$, where $F(\xi)$ is a distribution function given *a priori*.

Now let $H = -\Delta + q(x)$, for $x \in \mathbf{R}^d$, be a metrically transitive, essentially self-adjoint Schrödinger operator, the metrically transitive field $q(x)$ being assumed continuous with probability 1. Consider an arbitrary cube Λ, centered at the origin, with volume V and boundary ∂V. Consider the Schrödinger operator H_Λ on Λ,

$$(H_\Lambda\psi)(x) = -\Delta\psi + q(x)\psi, \qquad \text{for } x \in \Lambda,$$

with boundary conditions

$$\psi + \sigma\frac{\partial\psi}{\partial\nu}\bigg|_{\partial\Lambda} = 0,$$

where σ is a fixed nonnegative constant and the differential represents the derivative in the direction of the external normal to $\partial\Lambda$. Since we know that the spectrum of H_Λ consists of isolated eigenvalues of finite multiplicity, and the only limiting point of the spectrum is ∞, we can, as in the preceding case, define a function $N_\Lambda(\lambda)$ equal to V^{-1} times the number of eigenvalues of H_Λ not exceeding λ. Let $N_\Lambda(d\lambda)$ be the measure associated with $N_\Lambda(\lambda)$. We will prove that there exists a nonrandom measure $N(H, d\lambda)$ on $\mathbf{R}$ such that

$$\lim_{\nu\to\infty} N_\Lambda(d\lambda) = N(H, d\lambda) \tag{3.3}$$

with probability 1, where the limit is in the sense of weak convergence of measures as the cubes Λ expand to exhaust $\mathbf{R}^d$. The nonrandom measure $N(H, d\lambda)$ is related to the kernel $E_H(d\lambda; x, y)$ of the resolution of the identity $E_H(d\lambda)$ of H by the equation

$$N(H, d\lambda) = \mathsf{E}\{E_H(d\lambda; 0, 0)\}. \tag{3.4}$$

The nondecreasing function $N(H, \lambda)$ corresponding to $N(H, d\lambda)$ and normalized so that $N(H, -\infty) = 0$ will be called, as in the discrete case, the *integrated density of states* of the operator H.

The proof of relations (3.1) to (3.4) goes roughly as follows. Let A stand for the operators J and H, and $N_\Lambda(\lambda)$ for both $N_\Lambda(\lambda)$ and $N_n(\lambda)$. To show that $N_\Lambda(\lambda)$ converges as $V \to \infty$ it is clearly enough to verify the convergence as $V \to \infty$ of the integrals $\int_{\mathbf{R}} f(\lambda)N_\Lambda(d\lambda)$, for all f in a sufficiently large family $\mathcal{E}$ of continuous functions. Rewrite the integral as

$$\int_{\mathbf{R}} f(\lambda)N_\Lambda(d\lambda) = \frac{1}{V}\operatorname{Tr} f(A_\Lambda) = \frac{1}{V}\int_\Lambda f(A_\Lambda)(x, x)\, dx, \tag{3.5}$$

it being understood that, in the matrix case, the latter integral stands for a sum.

Suppose that, when $V \to \infty$, the operator A_Λ on the right-hand side of (3.5) can be replaced by A; the justification of this assumption will largely dictate the choice of $\mathcal{E}$ and will impose certain restrictions on A. Since A is a metrically transitive operator, its resolution of the identity $E_A(d\lambda)$ consists of metrically transitive projections, as was shown in Section 2; therefore $f(A)$ is a metrically transitive operator, and $f(A_\Lambda)(x, x)$ is a metrically transitive field. By applying the ergodic theorems (1.21) and (1.22) to the right-hand side of (3.5), we have, with probability 1:

$$\lim_{V\to\infty} \int_{\mathbf{R}} f(\lambda) N_\Lambda(d\lambda) = \mathsf{E}\{f(A)(0,0)\}. \tag{3.6}$$

The right-hand side of this equation is easily transformed into

$$\int_{\mathbf{R}} f(\lambda)\, \mathsf{E}\{E_A(d\lambda;0,0)\}, \tag{3.7}$$

where $E_A(d\lambda;x,y)$ is the kernel of the resolution of the identity of A; in the case of matrices, $E_A(d\lambda;0,0)$ obviously coincides with $\big(e_0, E_A(d\lambda)e_0\big)$. If $\mathcal{E}$ is a broad enough class of functions, (3.6) and (3.7) imply equations (3.1) to (3.4) with probability 1.

Arguments of this type allow us to prove equations (3.1) to (3.4) relatively effortlessly for an initial class of metrically transitive operators, consisting of matrices with only a finite number of nonzero diagonals, all bounded, or of differential operators with bounded coefficients. We then bootstrap the process by extending the original class under a variety of limiting processes, the details of which will be found in Sections 4 and 5.

In the case of Jacobi matrices, the family $\mathcal{E}$ is chosen as the set of polynomials $\mathcal{E} = \{\lambda^p : p = 0,1,\ldots\}$, as the traces of the powers of J_n are fairly easy to relate to the diagonal entries of the powers of J. The same $\mathcal{E}$ is also used for general metrically transitive matrix operators whose entries tend to zero fast enough away from the main diagonal.

In the case of Schrödinger operators, or, more generally, of elliptic differential operators A for which the A_Λ are unbounded, it is more convenient to use for $\mathcal{E}$ the family $\{e^{-t\lambda} : t > 0\}$. Then our objects of study are the operators e^{-tA_Λ} and e^{-tA}, whose kernels are solutions of well-studied parabolic equations, and, in the case of Schrödinger operators, relate explicitly to the potential $q(x)$ through the Wiener integral. This yields a relatively simple proof for (3.3) and (3.4), and even an efficient representation for the Laplace transforms of the limiting measure $N(H,d\lambda)$, which is then used to study the properties of $N(H,d\lambda)$.

However, the family $\{e^{-t\lambda}\}$ can be used only when the random coefficients in A have a small enough probability of taking on negative values. For example, in the case of Schrödinger operators, we must assume that $\mathsf{E}\{e^{-tq(0)}\} < \infty$ for all $t > 0$: see Theorem 5.18. To obtain (3.3) and (3.4) under weaker restrictions on the coefficients, we can take $\mathcal{E} = \big\{(\lambda - z)^{-p} : \operatorname{Im} z \neq 0\big\}$, that is, we use the powers of the resolvent $(A-z)^{-1}$ instead of the semigroup e^{-tA}. Then the restriction (for Schrödinger operators) becomes that the order-p moment $\mathsf{E}\big\{|q(0)|^p\big\}$ of the potential be finite, where p is the smallest even number larger than $d/2$: see Theorem 5.20.

In the study of differential and finite-difference operators whose coefficients are metrically transitive fields of a general form, all the convergence results obtained for $N_\Lambda(d\lambda)$ hold with probability 1, as was the case in the previous chapter with respect to a number of spectral properties of metrically transitive operators. Clearly, this limitation—the "collective" character of these statements, so to speak—is a fundamental aspect of our theory, and, in a sense, cannot be improved on: as in probability theory and in ergodic theory, probability 1 is the highest "accuracy" that we can hope for. This is due both to the methods of the proofs, which are based, at some point or another, on the ergodic theorem, and to the very essence of our problems: indeed, even for a one-dimensional Schrödinger operator, it is easy to construct examples of potentials for which the limit of $N_\Lambda(d\lambda)$ for $V \to \infty$ does not exist.

Yet there are some classes of metrically transitive operators that stand out because for them we can prove statements that hold for every, not just almost every, representative of the operator ensemble in question. An example is the class of almost periodic differential or finite-difference operators, that is, those whose coefficients are almost periodic in $\mathbf{R}^d$ or $\mathbf{Z}^d$. The special character of these operators lies in that for almost periodic functions the ergodic theorem is valid for every $\omega \in \Omega$ (Problem I.2), which in turn is due to a rather strong topology on Ω, induced by the uniform topology. For such operators, then, the relevant quantities turn out to be, as a rule, continuous functions on Ω (in a fairly good sense), and this allows us to get rid of the "with probability 1" clause.

Similar, but weaker, continuity properties underlie all the limiting processes mentioned above for extending the existence of the integrated density of states $N(d\lambda)$ to successively broader classes of metrically transitive operators.

To illustrate this, we consider the integrated density of states of a Schrödinger operator with an almost periodic potential. As explained in Example 1.15(g), the probability space Ω can be identified with the hull of the almost periodic function $f(x)$, that is, the closure of the set of all its shifts $\{f(x+\xi) : \xi \in \mathbf{R}^d\}$ [Levitan 1953]. Therefore Ω is an abelian topological group, and two points ω_1 and ω_2 are close in this topology if the potentials $q(x,\omega_1)$ and $q(x,\omega_2)$ are close in the uniform metric. But if two potentials q_1 and q_2 are such that $\sup_{x\in\mathbf{R}^d}|q_1(x) - q_2(x)| < \varepsilon$, that is, if the norm of their difference is no larger than ε, it follows from general inequalities for eigenvalues of two operators that the corresponding functions $N_\Lambda^{(1)}(\lambda)$ and $N_\Lambda^{(2)}(\lambda)$ satisfy

$$N_\Lambda^{(1)}(\lambda-\varepsilon) \leqslant N_\Lambda^{(2)}(\lambda) \leqslant N_\Lambda^{(1)}(\lambda+\varepsilon). \tag{3.8}$$

We have already seen that, in any case, $N_A(\lambda)$ converges with probability 1, as $V \to \infty$, to the nonrandom limit $N(\lambda)$ of equation (3.4), at all points where $N(\lambda)$ is continuous. From this and from (3.8) it is clear that, for an almost periodic potential, equation (3.3) holds for all $\omega \in \Omega$ at every continuity point of $N(\lambda)$.

In Chapter VII we will study almost periodic operators more thoroughly. For now let's turn to some simple properties enjoyed by $N(\lambda)$ whenever it exists.

(3.1) Theorem [Pastur 1974a, 1980]. *If A is a metrically transitive operator satisfying relations* (3.1)–(3.2) *or* (3.3)–(3.4), *the spectrum of A coincides, with probability* 1, *with the set of growth points of the integrated density of states $N(\lambda)$.*

Proof. We consider only the case that A is a matrix operator; the case of a Schrödinger operator is treated similarly. By the conditions of the theorem, for any interval $\Delta \subset \mathbf{R}$ we have

$$N(A, \Delta) = \mathsf{E}\{(e_0, E_A(\Delta)e_0)\}. \tag{3.9}$$

We show that $N(A, \Delta) = 0$ if and only if $E_A(\Delta) = 0$ with probability 1, as the theorem claims. The if part is clear. Conversely, if $N(A, \Delta) = 0$, it follows from (3.9) that $(e_0, E_A(\Delta)e_0) = 0$ with probability 1. Now $E_A(\Delta)$ is a metrically transitive projection by Theorem 2.6 and Corollary 2.7, so $(e_x, E_A(\Delta)e_x) = 0$ for all $x \in \mathbf{Z}^d$ with probability 1. But this says that $E_A(\Delta) = 0$. □

Remark. Denote by $\operatorname{supp}\mu$ the support of a measure $\mu(d\lambda)$, or of the associated nondecreasing function $\mu(\lambda) = \mu((-\infty, \lambda])$; in other words, $\operatorname{supp}\mu$ is the set of $\lambda \in \mathbf{R}$ such that $\mu((\lambda - \varepsilon, \lambda + \varepsilon))$ for any $\varepsilon > 0$. Then the conclusion of Theorem 3.1 can be expressed simply as

$$\sigma(A) = \operatorname{supp} N(A, \lambda). \tag{3.10}$$

(3.2) Theorem [Pastur 1974a, 1980]. *If A is either a metrically transitive one-dimensional Schrödinger operator satisfying* (3.3)–(3.4) *or a metrically transitive Jacobi matrix satisfying* (3.1)–(3.2), *the corresponding integrated density of states $N(A, \lambda)$ is continuous.*

In fact, the validity of (3.1)–(3.4) follows from the metric transitivity of A: see theorems 4.8, 5.20 and 6.3.

Proof. Let λ be an arbitrary point on the spectral axis. By theorem 2.1, $E_A(\{\lambda\}) = 0$ with probability 1, since the rank of $E_A(\{\lambda\}) = 0$ can never exceed 2. From this, and from the equality

(3.11) $$N(A,\{\lambda\}) = \mathsf{E}\{E_A(\{\lambda\},0,0)\},$$

which is a special case of formulas 3.2 and 3.4 for $\Delta = \{\lambda\}$, it follows that $N(A,\{\lambda\}) = 0$. Thus, the measures $N(A, d\lambda)$ corresponding to one-dimensional and finite-difference metrically transitive operators of the second order have no point component. □

In some of the cases below, under additional conditions, the integrated density of states is absolutely continuous and even has smooth density: see Chapter III, Theorem 4.21, and Problems II.21, II.31 and III.14. This property turns out to be important in the spectral analysis of metrically transitive operators. The meaning of Theorem 3.2 is that the continuity of $N(A,\lambda)$ for one-dimensional Jacobi matrices and Schrödinger operators follows almost immediately from their metric transitivity.

Theorem 3.2 is also true for metrically transitive ordinary differential operators of arbitrary finite order and for one-dimensional quasidiagonal matrix operators, that is, those with a finite number of nonzero diagonals. Such operators obviously have finite spectral multiplicities, so the above proof applies for them as well.

Theorem 3.2 is related to theorem 2.12, which gives conditions for the absence, with probability 1, of immobile eigenvalues, that is, eigenvalues independent of ω on a set of positive measure. This relationship is clarified by the following simple result:

(3.3) Theorem [Craig and Simon, 1983a]. *If A is a metrically transitive multidimensional matrix operator of finite order (that is, having a finite number of nonzero diagonals), or a metrically transitive elliptic differential operator with smooth coefficients (Examples 1.4(c,f)), the corresponding integrated density of states $N(A,\lambda)$ is continuous at a point λ if and only if the probability of λ being an eigenvalue of A is zero.*

Proof. The if part follows from (3.11), which is valid under the conditions of the theorem by the results of Sections 4 and 5. (In fact, more general equalities such as (3.2) and (3.4), and therefore the statement of Theorem 3.3, are true under much weaker conditions. The conditions given here are used for the sake of simplicity.)

We now prove the converse. In the discrete case, this also follows from (3.11), read backwards, and from the metric transitivity of the projection $E_A(\{\lambda\})$, which implies that all its diagonal entries $E_A(\{\lambda\},x,x)$, for $x \in \mathbf{Z}^d$, equal zero with probability 1. But then this projection is zero, that is, with probability 1, λ is not an eigenvalue of A.

In the continuous case, it follows from (3.11) that all the integrals

$$\int_{C_n} E_A(\{\lambda\}, x, x)\, dx,$$

for C_n ranging over all cubes in $\mathbf{R}^d$ with rational center coordinates and edge lengths, equal zero. After that we use the smoothness of the spectral kernel with respect to the spatial variables: see, for example, [Berezanskii 1968]. □

This result reduces the problem of continuity of the integrated density of states of a metrically transitive operator to the existence of immobile eigenvalues, and therefore, together with Theorem 2.12, it implies Theorem 3.2.

Theorem 3.3 suggests that the problem of continuity of the integrated density of states in the multidimensional case is not at all simple, because, for example, even for a multidimensional Schrödinger operator with a periodic potential, the absence of eigenvalues is a delicate result, following from a nontrivial theorem of Thomas [Thomas 1973; Dyakin and Petrukhnovskii 1982]. This being so, it is remarkable that the discrete multidimensional analogue of Theorem 3.2 can easily be proved using the same ideas above:

(3.4) Theorem [Delyon and Souillard 1984]. *If A is a metrically transitive multidimensional matrix operator of finite order, its integrated density of states is continuous.*

Proof. We use formula (3.11) again. By the ergodic theorem, its right-hand side can be written as

$$\lim_{V\to\infty} \frac{1}{V} \sum_{x\in\Lambda} \left(e_x, E_A(\{\lambda\})e_x\right) = \lim_{V\to\infty} \frac{1}{V} \operatorname{Tr} \pi_\Lambda E_A(\{\lambda\}),$$

where π_Λ is the projection on the space of functions in $\ell_2(\mathbf{Z}^d)$ with support in Λ. But the range of the operator $\pi_\Lambda E_A\big(\{\lambda\}\big)$ consists of the restrictions to Λ of the solutions of the equation $A\psi = \lambda\psi$. Since A is of finite order, any such solution is uniquely defined by its values in a shell of thickness equal to the order of the equation, adjacent to $\partial\Lambda$; therefore the rank of $\pi_\Lambda E_A\big(\{\lambda\}\big)$ is of order $|\partial\Lambda|$ for $V \to \infty$. It follows that

$$N\big(A, \{\lambda\}\big) \leqslant \lim_{V\to\infty} \frac{1}{V} \operatorname{rank} \pi_\Lambda E_A\big(\{\lambda\}\big) = \lim_{V\to\infty} \frac{|\partial\Lambda|}{V} = 0. \qquad \square$$

The type of the measure $N(A, d\lambda)$ for a metrically transitive operator A does not, in general, coincide with the types of the measures $\big(E_A(d\lambda)h, h\big)$. Indeed, by (3.2) and (3.4), $N(A, d\lambda)$ is the integral of $\big(E_A(d\lambda)h, h\big)$ over the probability measure P, so already in the simple example of the multiplication operator (2.3) we get different types: the operator's spectrum is always pure point, while $N(A,\lambda)$ coincides with $F(\lambda)$, the distribution function of the random variables $\xi(x)$, and therefore may be arbitrarily smooth.

In the less trivial case of one-dimensional Schrödinger operator and Jacobi matrices, $N(A,\lambda)$ is, by Theorem 3.2, at least continuous, while the spectrum turns out to be pure point if the coefficients have a sufficiently smooth Markovian structure (see Sections 13 and 15), or are quasiperiodic functions whose frequencies are not very well approximated by rational numbers (Sections 16 and 18). If the frequencies are sufficiently well approximated by rational numbers, the spectrum is singular continuous. In the Markov case, $N(A,\lambda)$ has a differentiable density; and in certain quasiperiodic cases, $N(A,\lambda)$ is analytic, irrespective of the arithmetic properties of the frequencies (Section 18).

Further properties of the integrated density of states are discussed in Subsections 4.C and 5.C, and in subsequent chapters.

We conclude this section with formulas for the integrated density of states of matrix and differential operators with constant coefficients, that is, operators of the form (1.7) with $a(x,y) = a_0(x-y)$ for $x, y \in \mathbf{Z}^d$, or of the form (1.9) with a_m independent of $x \in \mathbf{R}^d$ for all $|m| \leqslant p$. The formulas can be obtained, for example, from (3.2) and (3.4) (see Theorems 4.4 and 5.18) and from Fourier analysis. For matrix operators we have

$$(3.12) \qquad N(A, d\lambda) = (2\pi)^{-d} \operatorname{meas}\Big\{p \in \mathbf{T}^d : \sum_{x \in \mathbf{Z}^d} a_0(x) e^{2\pi i p x} \in d\lambda\Big\},$$

while differential operators satisfy

$$(3.13) \qquad N(A, d\lambda) = (2\pi)^{-d} \operatorname{meas}\Big\{p \in \mathbf{R}^d : \sum_{|m| \leqslant p} a_m \prod_1^d p_k^{2m_k} \in d\lambda\Big\}.$$

4 Matrix Operators on $\ell_2(\mathbf{Z}^d)$

4.A Essential Self-Adjointness

In this section we consider metrically transitive random matrices $\{a(x,y)\}$, for $x, y \in \mathbf{Z}^d$, and the corresponding random operators of example 1.17(c), given by

$$(4.1) \qquad (A\psi)(x) = \sum_{y \in \mathbf{Z}^d} a(x,y)\psi(y).$$

It is convenient to single out certain classes of random operators. Denote by $\mathcal{L}_{1,1}$, $\mathcal{L}_{2,1}$ and $\mathcal{L}_{2,2}$ the Banach spaces of metrically transitive matrices A with norms

$$\|A\|_{1,1} = \mathsf{E}\Big\{\sum_{x\in\mathbf{Z}^d} |a(x,0)|\Big\}, \tag{4.2}$$

$$\|A\|_{2,1}^2 = \mathsf{E}\Big\{\Big(\sum_{x\in\mathbf{Z}^d} |a(x,0)|\Big)^2\Big\}, \tag{4.3}$$

$$\|A\|_{2,2}^2 = \mathsf{E}\Big\{\sum_{x\in\mathbf{Z}^d} |a(x,0)|^2\Big\}. \tag{4.4}$$

That these are Banach spaces can be readily checked. In particular, in all three cases the norm of the bounded operator and the norm of its adjoint operator coincide. The metrically transitive matrices for which the norm $\|A\|_{2,2} = \mathsf{E}\{\|Ae_0\|^2\}^{1/2}$ is finite are analogous to Hilbert–Schmidt operators. This analogy is supported by the fact that the norm $\|\cdot\|_{2,2}$ may, by the ergodic theorem, be represented as

$$\|A\|_{2,2}^2 = \lim_{V\to\infty} \frac{1}{V} \operatorname{Tr} \pi_\Lambda A^* A \pi_\Lambda,$$

where Λ represents cubes of volume V centered at the origin and exhausting $\mathbf{Z}^d$, and π_Λ is the projection onto the space $\ell_2(\mathbf{Z}^d)$ of functions with support in Λ.

Furthermore, since the norm of a metrically transitive operator is non-random (Problem I.14), the result of multiplying a metrically transitive operator in $\mathcal{L}_{2,2}$ by a bounded metrically transitive operator is still in $\mathcal{L}_{2,2}$. More generally:

(4.1) Proposition. *Let A and B be metrically transitive matrix operators with $\|B\| < \infty$, and, for $\Lambda \subset \mathbf{Z}^d$, let π_Λ be the orthogonal projection onto the space of functions of $\ell_2(\mathbf{Z}^d)$ supported in Λ. Then*

(i) $\|A\|_{1,1} \leqslant \|A\|_{2,1}$ *and* $\|A\|_{2,2} \leqslant \|A\|_{2,1}$, *so* $\mathcal{L}_{2,1} \subset \mathcal{L}_{1,1}$ *and* $\mathcal{L}_{2,1} \subset \mathcal{L}_{2,2}$;

(ii) *If $A \in \mathcal{L}_{1,1}$ then $A^* \in \mathcal{L}_{1,1}$, with $\|A\|_{1,1} = \|A^*\|_{1,1}$, and similarly for $A \in \mathcal{L}_{2,2}$;*

(iii) *If $A^* \in \mathcal{L}_{2,1}$ and $\{e_x, x \in \mathbf{Z}^d\}$ is a basis for $\ell_2(\mathbf{Z}^d)$, that is, $e_x(y) = \delta_{x,y}$ for $x, y \in \mathbf{Z}^d$, we have*

$$\mathsf{E}\{\|A\pi_\Lambda B e_x\|^2\} \leqslant \|B\|^2 \|A^*\|_{2,1}^2$$

for all $x \in \mathbf{Z}^d$.

Proof. Parts (i) and (ii) are obvious. Part (iii) follows from the metric transitivity of A and B and the inequalities

$$\begin{aligned}
\mathsf{E}\{\|A\pi_\Lambda Be_x\|^2\} &= \mathsf{E}\Big\{\sum_{y,z\in\Lambda} B^*(x,y)(A^*A)(y,z)B(z,x)\Big\} \\
&\leqslant \sum_{y,z\in\Lambda} \mathsf{E}\{|(A^*A)(y,z)|\,|B^*(x,y)B(z,x)|\} \\
&= \sum_{y,z} \mathsf{E}\{|(A^*A)(y-z,0)|\,|B^*(x-z,y-z)B(0,x-z)|\} \\
&= \sum_{y} \mathsf{E}\Big\{|(A^*A)(y,0)|\sum_z |B^*(z,y)B^*(z,0)|\Big\} \\
&\leqslant \sum_{y} \mathsf{E}\{|(A^*A)(y,0)|\,\|B^*e_y\|\,\|B^*e_0\|\} \\
&\leqslant \|B\|^2 \sum_{y,z} \mathsf{E}\{|a^*(y-z,0)a(0,-z)|\} \\
&= \|B\|^2 \mathsf{E}\Big\{\Big(\sum_y |a^*(y,0)|\Big)^2\Big\} = \|B\|^2\|A^*\|_{2,1}^2. \qquad \square
\end{aligned}$$

(4.2) Theorem [Figotin 1987]. *A metrically transitive Hermitian matrix $A \in \mathcal{L}_{2,1}$ defines, with probability* 1, *a symmetric operator A on the set $\mathcal{D}$ of sequences in $\ell_2(\mathbf{Z}^d)$ with finite support. This operators is self-adjoint on $\mathcal{D}$.*

Proof. Denote by e_x, for $x \in \mathbf{Z}^d$, a basis for the space $\ell_2(\mathbf{Z}^d)$, that is, $e_x(y) = \delta_{x,y}$. From 4.1, and owing to the metric transitivity of A and to the fact that it belongs to $\mathcal{L}_{2,1}$, we obtain, from proposition 4.1(i), that

$$\mathsf{E}\{\|Ae_x\|^2\} = \|A\|_{2,2}^2 \leqslant \|A\|_{2,1}^2 < \infty$$

for all $x \in \mathbf{Z}^d$. It follows with probability 1 that the operator A is defined on the vector space spanned by the e_x, which coincides with $\mathcal{D}$.

The proof that A is essentially self-adjoint will consist in verifying that the closure of $(A - zI)\mathcal{D}$, for $\operatorname{Im} z \neq 0$, equals $\ell_2(\mathbf{Z}^d)$ with probability 1, or, equivalently, that it contains, with probability 1, all the basis vectors e_x. Toward this purpose we approximate $A = \{a(x,y)\}$ by a sequence of bounded self-adjoint metrically transitive operators $A^{(p)} = \{a_p(x,y)\}$, for $p = 0, 1, \ldots$, defined as follows:

$$a_p(x,y) = \begin{cases} 0 & \text{if } |x-y| > p, \\ \chi_{[0,p]}\big(|a(x,y)|\big)a(x,y) & \text{if } |x-y| \leqslant p, \end{cases} \tag{4.5}$$

where $\chi_{[0,p]}$ is the characteristic function of the interval $[0,p]$. Clearly,

(4.6)
$$\lim_{p\to\infty} \|A - A^{(p)}\|_{2,1} = 0.$$

Now consider the resolvents $R^{(p)} = (A^{(p)} - zI)^{-1}$, for $\operatorname{Im} z \neq 0$; and denote by π_l^+ and π_l^- the projections π_Λ for $\Lambda = \{x : |x| \leqslant l\}$ and $\Lambda = \{x : |x| > l\}$, respectively. Clearly, $\pi_l^+ + \pi_l^- = I$. Using Proposition 4.1(c) and the inequality $\|R^{(p)}\| \leqslant |\operatorname{Im} z|^{-1}$, we get

(4.7)
$$\begin{aligned} \mathsf{E}\{\|(A - zI)\pi_l^+ R^{(p)} e_x - e_x\|\}^2 &\leqslant 2\mathsf{E}\{\|(A - A^{(p)})\pi_l^+ R^{(p)} e_x\|^2 + \|(A^{(p)} - zI)\pi_l^- R^{(p)} e_x\|^2\} \\ &\leqslant 2\,|\operatorname{Im} z|^{-2}\,\|A - A^{(p)}\|_{2,1}^2 + 2C^2(p)\mathsf{E}\{\|\pi_l^- R^{(p)} e_x\|^2\}, \end{aligned}$$

where $C(p)$ is a nonrandom bound for the norm of the operators $A^{(p)} - zI$, whose existence follows immediately from the definition of $A^{(p)}$. Since $\|\pi_l^- R^{(p)} e_x\| \leqslant |\operatorname{Im} z|^{-1}$ and $\lim_{l\to\infty} \|\pi_l^- R^{(p)} e_x\| = 0$ for fixed p, we conclude that

$$\lim_{l\to\infty} \mathsf{E}\{\|\pi_l^- R^{(p)} e_x\|^2\} = 0.$$

Therefore we can choose $l = l(p, x)$ such that

(4.8)
$$C^2(p)\mathsf{E}\{\|\pi_l^- R^{(p)} e_x\|^2\} \leqslant p^{-1}.$$

If we now set $f_x^{(p)} = \pi_{l(p,x)}^+ R^{(p)} e_x$, it follows from (4.6), (4.7) and (4.8) that

$$\lim_{p\to\infty} \mathsf{E}\{\|(A - zI) f_x^{(p)} - e_x\|^2\} = 0.$$

Hence, for some sequence of natural numbers p_n that tends to ∞ as n increases, it is true with probability 1 that

(4.9)
$$\lim_{n\to\infty} \|(A - zI) f_x^{(p_n)} - e_x\| = 0.$$

Because $f_x^{(p)} \in \mathcal{D}$, this means that the closure of $(A - zI)\mathcal{D}$ contains all the vectors e_x, for $x \in \mathbf{Z}^d$, with probability 1. □

Theorem 4.2 has a simple consequence:

(4.3) Corollary. *If $A = \{a(x, y)\}$ is a metrically transitive Hermitian matrix satisfying*

(4.10)
$$\sum_{x\in\mathbf{Z}^d} \left(\mathsf{E}\{|a(x,0)|^2\}\right)^{1/2} < \infty,$$

the corresponding operator is defined, with probability 1, on the set $\mathcal{D}$ of sequences in $\ell_2(\mathbf{Z}^d)$ with finite support, and is essentially self-adjoint on $\mathcal{D}$.

Proof. Condition (4.10) and metric transitivity imply that $A \in \mathcal{L}_{2,1}$, because

$$\mathsf{E}\Big\{\Big(\sum_{x\in\mathbf{Z}^d}|a(x,0)|\Big)^2\Big\} = \sum_{x,y\in\mathbf{Z}^d}\mathsf{E}\{|a(x,0)a(y,0)|\}$$
$$= \Big(\sum_{x\in\mathbf{Z}^d}\big(\mathsf{E}\{|a(x,0)|^2\}\big)^{1/2}\Big)^2 < \infty. \qquad \square$$

4.B Existence of the Integrated Density of States and Other Ergodic Properties

Let A be a symmetric metrically transitive operator on $\ell_2(\mathbf{Z}^d)$ (Example 1.17(c)):

$$(A\psi)(x) = \sum_{y\in\mathbf{Z}^d} a(x,y)\psi(y).$$

Consider the sequence of Hermitian matrices $A_\Lambda = \{a(x,y)\}$, for $x, y \in \Lambda$, where Λ is a cube in $\mathbf{Z}^d$ centered at the origin. The order of A_Λ is clearly the volume V of Λ. Introduce a nondecreasing function $N_\Lambda(\lambda)$ equal to V^{-1} times the number of eigenvalues of A_Λ not exceeding λ, and denote the measure arising from $N_\Lambda(\lambda)$ by $N_\Lambda(d\lambda)$.

In this section, as in the previous one, we will understand convergence of measures to mean weak convergence [Gihman and Skorohod I].

Following the plan outlined in the last section to prove relations of the type (3.1) and (3.2), we start by considering a restricted class of metrically transitive matrices, namely those with a finite number of nonzero diagonals with bounded entries. For such matrices it is easy to prove (3.1) and (3.2), by comparing the moments of the associated measures: this is done in Lemma 4.5. We then extend the relations to other matrices by taking various kinds of limits.

(4.4) Theorem [Figotin and Pastur 1983]. *Let A be a symmetric metrically transitive operator on $\ell_2(\mathbf{Z}^d)$. Assume there exists a sequence $\{c(r) : r \in \mathbf{Z}^d\}$, with $0 \leqslant c(r) \leqslant \infty$, such that*

(a) $c(r) = c(-r)$;

(b) $\sum_{|r|\geqslant \rho} c(r) < \infty$ *for some constant* ρ;

(c) $\sum_{|r|\in\mathbf{Z}^d} \mathsf{P}\{|a(r,0)| > c(r)\} < \infty$.

Then there exists a nonrandom positive measure such that $N(A, d\lambda) \leqslant 1$ *and, with probability* 1,

$$\lim_{V\to\infty} N_\Lambda(d\lambda) = N(A, d\lambda). \tag{4.11}$$

If, in addition, A is essentially self-adjoint with probability 1 and $E_A(d\lambda)$ is its resolution of the identity, we also have

$$N(A, d\lambda) = \mathsf{E}\{(e_0, E_A(d\lambda)e_0)\}, \tag{4.12}$$

where $e_0(x) = \delta_{x,0}$.

We start the proof with the special case discussed above.

(4.5) Lemma. *Let $A = \{a(x,y)\}$, for $x, y \in \mathbf{Z}^d$, be a symmetric metrically transitive operator such that $|a(x,y)| \leqslant C$, where C is a nonrandom positive constant, and $a(x,y) = 0$ whenever $|x - y|$ is greater than some fixed bound R. Then A is self-adjoint, and satisfies* (4.11) *and* (4.12).

Proof. We immediately verify that

$$\|A_\Lambda\| \leqslant \|A\| \leqslant C(2R+1)^d = C_1.$$

Thus A is self-adjoint, being a bounded operator. Furthermore the spectra of A_Λ and A, the measures $N_\Lambda(d\lambda)$ and the projection-valued measures $E_{A_\Lambda}(d\lambda)$ are all concentrated in the interval $[-C_1, C_1]$. To prove that the measures $N_\Lambda(d\lambda)$ converge weakly, then, it is enough to prove that their moments converge. We rewrite the moments as

$$V \int_{-C_1}^{C_1} \lambda^p N_\Lambda(d\lambda) = \operatorname{Tr} A_\Lambda^p = \sum_{x_1,\ldots,x_p \in \Lambda} a(x, x_1) \ldots a(x_{p-1}, x), \tag{4.13}$$

for $p \in \mathbf{Z}_+$. The conditions on A imply that, as $V \to \infty$, the right-hand side of (4.13) is

$$\sum_{\substack{x \in \Lambda \\ x_2,\ldots,x_{p-1} \in \mathbf{Z}^d}} a(x, x_1) \ldots a(x_{p-1}, x) + O(V^{1-1/d}) = \sum_{x \in \Lambda} A^p(x,x) + O(V^{1-1/d}).$$

From this and from (4.13) we now obtain

$$\lim_{V \to \infty} \int_{-C_1}^{C_1} \lambda^p N_\Lambda(d\lambda) = \lim_{V \to \infty} \frac{1}{V} \sum_{x \in \Lambda} A^p(x,x), \tag{4.14}$$

as long as the limit on the right exists. Since A is a bounded metrically transitive operator, so is A^p, and therefore the random field $A^p(x,x)$ on $\mathbf{Z}^d$ is metrically transitive and bounded by the nonrandom constant C_1^p. By applying the ergodic theorem 1.13 to this field, we conclude from (4.14) that, with probability 1,

$$\lim_{V\to\infty}\int_{-C_1}^{C_1}\lambda^p N_\Lambda(d\lambda)=\mathsf{E}\{A^p(0,0)\}.$$

By Corollary 2.9, the resolution of the identity $E_A(d\lambda)$ of A is a metrically transitive projection, so the preceding equality may be rewritten as

$$\lim_{V\to\infty}\int_{-C_1}^{C_1}\lambda^p N_\Lambda(d\lambda)=\int_{-C_1}^{C_1}\lambda^p\mathsf{E}\{(e_0,E_A(d\lambda)e_0)\}.$$

Here p is any positive integer, but the same equality is obviously true for $p=0$. This implies the lemma. □

To extend the validity of (4.11) and (4.12) we need two lemmas.

(4.6) Lemma. *Let $A^{(p)}$, for $p\in\mathbf{Z}_+$, be a sequence of symmetric metrically transitive operators in $\ell_2(\mathbf{Z}^d)$, and $N_\Lambda^{(p)}(d\lambda)$ the corresponding prelimit functions. If, for each p, the limit*

$$\lim_{V\to\infty}N_\Lambda^{(p)}(d\lambda)=N(A^{(p)},d\lambda)$$

exists, and if there exists a symmetric metrically transitive operator A in $\ell_2(\mathbf{Z}^d)$ such that $\lim_{p\to\infty}\|A-A^{(p)}\|=0$, the sequence $N_\Lambda(d\lambda)$ corresponding to A converges to $N(A,d\lambda)=\lim_{p\to\infty}N\big(A^{(p)},d\lambda\big)$.

Proof. The conditions on $A^{(p)}$ and the definition of the integrated density of states show that

$$N_\Lambda^{(p)}(\lambda-\varepsilon_p)\leqslant N_\Lambda(\lambda)\leqslant N_\Lambda^{(p)}(\lambda+\varepsilon_p),$$

where $\varepsilon_p=\|A-A^{(p)}\|$. The lemma follows readily. □

(4.7) Lemma. *Suppose a symmetric metrically transitive operator $A=\{a(x,y)\}$, for $x,y\in\mathbf{Z}^d$, is such that, with probability 1, $|a(x,y)|\leqslant c(x-y)$, where $\{c(r):r\in\mathbf{Z}^d\}$ is a sequence of nonrandom nonnegative numbers with $\sum_{r\in\mathbf{Z}^d}c(r)\leqslant\infty$. Then, with probability 1, $\|A\|\leqslant\sum_{r\in\mathbf{Z}^d}c(r)$ and*

$$\lim_{V\to\infty}N_\Lambda(d\lambda)=\mathsf{E}\{(e_0,E_A(d\lambda)e_0)\}.$$

Proof. This is a straightforward corollary of Lemmas 4.5 and 4.6. □

Proof of Theorem 4.4. For each $p\in\mathbf{Z}_+$, we define a sequence $\{c_p(r):r\in\mathbf{Z}^d\}$ by setting $c_p(r)=p$ if $|r|\leqslant p$ and $c_p(r)=c(r)$ otherwise. We also approximate A by operators $A^{(p)}=\{a_p(x,y)\}$, for $p\in\mathbf{Z}_+$, given by

$$a_p(x,y)=\chi_{(-\infty,c_p(x-y)]}\big(|a(x,y)|\big)a(x,y),$$

where χ represents, as usual, the characteristic function. The operators $A^{(p)}$ satisfy the conditions of Lemma 4.6, because $|a_p(x,y)| \leqslant c_p(x-y)$ and $\sum_{r\in\mathbf{Z}^d} c_p(r) < \infty$. Therefore, there exist nonrandom measures $N(A^{(p)}, d\lambda)$ such that, with probability 1,

$$\lim_{V\to\infty} N_\Lambda^{(p)}(d\lambda) = N(A^{(p)}, d\lambda) = \mathsf{E}\{(e_0, E_{A^{(p)}}(d\lambda)e_0)\}. \tag{4.15}$$

We now use the following simple statement: If A_1 and A_2 are Hermitian matrices of order n and $N_1(\lambda)$ and $N_2(\lambda)$ are their integrated densities of states,

$$|N_1(\lambda) - N_2(\lambda)| \leqslant \frac{1}{n}\,\mathrm{rank}(A_1 - A_2).$$

Since the rank of a matrix cannot exceed the number of nonzero entries, we have

$$\begin{aligned} |N_\Lambda(\lambda) - N_\Lambda^{(p)}(\lambda)| &\leqslant \frac{1}{V}\sum_{x,y\in\Lambda} \chi_{[c_p(x-y),\infty)}(|a(x,y)|) \\ &\leqslant \frac{1}{V}\sum_{x\in\Lambda}\sum_{r\in\mathbf{Z}^d} \chi_{[c_p(r),\infty)}(|a(x,x-r)|). \end{aligned} \tag{4.16}$$

The inner sum is a metrically transitive field, and, by condition (c) of Theorem 4.4, we have

$$\mathsf{E}\Big\{\sum_{r\in\mathbf{Z}^d} \chi_{[c_p(r),\infty)}(|a(0,-r)|)\Big\} = \sum_{r\in\mathbf{Z}^d} \mathsf{P}\{|a(r,0)| > c_p(r)\} < \infty. \tag{4.17}$$

Now let $\varepsilon_\Lambda^{(p)}$ be the right-hand side of (4.16) and ε_p the left-hand side of (4.17). By the ergodic theorem 1.13 and condition (c) of Theorem 4.4,

$$\lim_{V\to\infty} \varepsilon_\Lambda^{(p)} = \varepsilon_p, \qquad \lim_{p\to\infty} \varepsilon_p = 0. \tag{4.18}$$

By passing to a subsequence, is necessary, we may also assume that there is a nondecreasing function $N(\lambda)$ such that

$$\lim_{p\to\infty} N(A^{(p)}, \lambda) = N(\lambda) \tag{4.19}$$

for all λ outside a countable set. Let Ω_1 be the set of $\omega \in \Omega$ where (4.15) and the first equality in (4.18) hold, and set $N^{(p)}(\lambda) = N(A^{(p)}, \lambda)$. Fix $\varepsilon > 0$, $\omega \in \Omega_1$, and a point λ where N and all the $N(p)$ are continuous. By the second equality in (4.18) and by (4.19), there exists p_0 such that $\varepsilon_p \leqslant \varepsilon/4$ and $|N^{(p)}(\lambda) - N(\lambda)| \leqslant \varepsilon/3$, whenever $p \geqslant p_0$. By (4.15), (4.16) and the first equality in (4.18), there exists Λ_0 such that, for $\Lambda \supset \Lambda_0$,

$$\left|N_\Lambda(\lambda) - N_\Lambda^{(p_0)}(\lambda)\right| \leqslant \varepsilon_\Lambda^{(p_0)} \leqslant \frac{\varepsilon}{3}, \qquad \left|N_\Lambda^{(p_0)}(\lambda) - N^{(p_0)}(\lambda)\right| \leqslant \frac{\varepsilon}{3}.$$

Therefore, for $\Lambda \supset \Lambda_0$,

$$\begin{aligned}&\left|N_\Lambda(\lambda) - N(\lambda)\right| \\ &\quad \leqslant \left|N_\Lambda(\lambda) - N_\Lambda^{(p_0)}(\lambda)\right| + \left|N_\Lambda^{(p_0)}(\lambda) - N^{(p_0)}(\lambda)\right| + \left|N^{(p_0)}(\lambda) - N(\lambda)\right| \leqslant \varepsilon.\end{aligned}$$

This shows that, for $\omega \in \Omega_1$, we have $\lim_{V\to\infty} N_\Lambda(\lambda) = N(\lambda)$ if λ is a continuity point of N and of all the $N^{(p)}$; in other words, (4.11) holds with probability 1. Since the limit measure $N(d\lambda)$ exists, it is clear that $N(\mathbf{R}) \leqslant 1$ because the prelimit measures $N_\Lambda(d\lambda)$ enjoy the same property.

We now prove (4.12) on the assumption that A is self-adjoint. First notice that, for each $x \in \mathbf{Z}^d$, the series $\sum_{y\in\mathbf{Z}^d} |a(y,x)|^2$ converges with probability 1: for $x = 0$, this follows from conditions (b) and (c) of the theorem, which imply that $\left|a(r,0)\right| \leqslant c(r)$ with probability 1 for r large enough; and for arbitrary x, it follows because A is metrically transitive.

By the definitions of $c_p(x)$ and of the approximating operators $A^{(p)}$, the expression

$$\left\|(A - A^{(p)})e_x\right\|^2 = \sum_{y\in\mathbf{Z}^d} \left|a(y,x)\right|^2 \left(1 - \chi_{(-\infty, c_p(x-y)]}(|a(y,x)|)\right)^2$$

tends to 0 with probability 1 as p goes to ∞. This shows that

(4.20) $$\operatorname*{s-lim}_{p\to\infty} A^{(p)}(\omega)\psi = A(\omega)\psi$$

with probability 1 for ψ a basis vector e_x, and consequently for all ψ in the set $\mathcal{D}$ of sequences in $\ell_2(\mathbf{Z}^d)$ with finite support. Equation (4.20), in turn, implies that the resolutions of the identity $E_{A^{(p)}}(\lambda)$ converge with probability 1 to $E_A(\lambda)$ at all continuity points of the latter [Reed and Simon I]. Therefore

$$\lim_{p\to\infty} \mathsf{E}\left\{(e_0, E_{A^{(p)}}(d\lambda)e_0)\right\} = \mathsf{E}\left\{(e_0, E_A(d\lambda)e_0)\right\}.$$

Equation (4.12) follows immediately from (4.15), (4.19) and (4.20). This concludes the proof of Theorem 4.4. □

The following theorems can be considered, in a way, as generalizations of Theorem 4.4.

(4.8) Theorem [Figotin 1987]. *Let $A = \{a(x,y)\}$ be a metrically transitive Hermitian matrix belonging to $\mathcal{L}_{1,1}$, that is, such that*

(4.21) $$\sum_{x\in\mathbf{Z}^d} \mathsf{E}\left\{|a(x,0)|\right\} < \infty.$$

The matrix A defines, with probability 1, a symmetric metrically transitive operator on the set $\mathcal{D}$ of sequences in $\ell_2(\mathbf{Z}^d)$ with finite support, and there exists a nonrandom positive measure $N(A, d\lambda)$ such that $N(A, \mathbf{R}) = 1$ and, with probability 1,

$$\lim_{V \to \infty} N_\Lambda(d\lambda) = N(A, d\lambda). \tag{4.22}$$

If, moreover, the operator A is essentially self-adjoint with probability 1 and $E_A(d\lambda)$ is its resolution of the identity, this measure satisfies

$$N(A, d\lambda) = \mathsf{E}\{(e_0, E_A(d\lambda)e_0)\}. \tag{4.23}$$

In particular, any symmetric operator $A \in \mathcal{L}_{2,1}$ satisfies (4.22) and (4.23).

Proof. The last assertion follows from the others by Proposition 4.1 and Theorem 4.2. To show that A defines an operator on $\mathcal{D}$ with probability 1, we notice that, by (4.2), $\sum_{y \in \mathbf{Z}^d} |a(y,x)| < \infty$ with probability 1 for all $x \in \mathbf{Z}^d$, and also that

$$\|Ae_x\|^2 = \sum_{y \in \mathbf{Z}^d} |a(y,x)|^2 < \infty. \tag{4.24}$$

The operator thus defined is obviously symmetric, because A is Hermitian.

We prove (4.22) and (4.23) by investigating the resolvents of the sequence of bounded, self-adjoint operators $A^{(p)}$, for $p \in \mathbf{Z}_+$, defined in the proof of Theorem 4.2 (equation (4.5)). We denote by $R_A(z)$, where $\operatorname{Im} z \neq 0$, the resolvent $(A - zI)^{-1}$ of A. We also set

$$r_{A_\Lambda}(z) = \int_{\mathbf{R}} (\lambda - z)^{-1} N_\Lambda(d\lambda),$$
$$r^{(p)}(z) = \int_{\mathbf{R}} (\lambda - z)^{-1} N^{(p)}(d\lambda),$$

where $N^{(p)}(d\lambda)$ is the integrated density of states of $A^{(p)}$. Clearly, we have

$$r_{A_\Lambda}(z) = \frac{1}{V} \operatorname{Tr} R_{A_\Lambda}(z) \tag{4.25}$$

$$r^{(p)}(z) = \mathsf{E}\{(e_0, R^{(p)}(z)e_0)\} \tag{4.26}$$

if $\operatorname{Im} z \neq 0$; the second equation follows from Lemma 4.5.

We will need the following simple relation between the resolvents of any two operators A and B:

$$R_A(z) - R_B(z) = -R_A(z)(A - B)R_B(z), \tag{4.27}$$

and two intermediate results:

(4.9) Proposition. *Let $\mathcal{M}$ be the set of nonnegative measures $m(d\lambda)$ on the real line such that $m(\mathbf{R}) \leqslant 1$.*

(i) *If $m_1, m_2 \in \mathcal{M}$ satisfy $\int_{\mathbf{R}}(\lambda - z)^{-1} m_1(d\lambda) = \int_{\mathbf{R}}(\lambda - z)^{-1} m_2(d\lambda)$ for all z such that $\operatorname{Im} z \neq 0$, the two measures are equal.*

(ii) *If $\{m_n : n \in \mathbf{Z}_+\}$ is a sequence of measures in $\mathcal{M}$ and the limit*

$$\lim_{n\to\infty} \int_{\mathbf{R}} (\lambda - z)^{-1} m_n(d\lambda) = r(z) \tag{4.28}$$

exists for all z such that $\operatorname{Im} z \neq 0$, there exists a positive measure $m \in \mathcal{M}$ such that

$$\lim_{n\to\infty} m_n(d\lambda) = m(d\lambda) \tag{4.29}$$

and

$$r(z) = \int_R (\lambda - z)^{-1} m(d\lambda). \tag{4.30}$$

(iii) *If, conversely, m and $\{m_n\}$ are measures in $\mathcal{M}$ satisfying (4.29), and $r(z)$ is defined by (4.30), equality (4.28) is also satisfied.*

Proof. Straightforward. □

(4.10) Lemma. *The following relations hold between $A^{(p)}$, A_Λ and $A_\Lambda^{(p)}$:*

(i) $\left|r_{A_\Lambda^{(p)}}(z) - r_{A_\Lambda}(z)\right| \leqslant V^{-1} |\operatorname{Im} z|^{-2} \sum_{x\in\Lambda} \sum_{t\in\mathbf{Z}^d} \left|(a - a^{(p)})(t,x)\right|$;

(ii) $\left|r_{A^{(p)}}(z) - r_{A^{(q)}}(z)\right| \leqslant |\operatorname{Im} z|^{-2} \left\|A^{(p)} - A^{(q)}\right\|_{1,1}$;

(iii) *for $\varepsilon > 0$ we have $\varepsilon \left|r^{(p)}(i\varepsilon)\right| \geqslant 1 - \|A\|_{1,1}/\varepsilon$.*

Proof. From (4.25), (4.27) and the relation $\|R_{A_\Lambda}(z)\| \leqslant |\operatorname{Im} z|^{-1}$ we have

$$\begin{aligned}\left|r_{A_\Lambda^{(p)}}(z) - r_{A_\Lambda}(z)\right| &= \frac{1}{V}\left|\operatorname{Tr} R_{A_\Lambda}(z)(A_\Lambda - A_\Lambda^{(p)}) R_{A_\Lambda^{(p)}}(z)\right| \\ &\leqslant \frac{1}{V} |\operatorname{Im} z|^{-2} \sum_{x,y\in\Lambda} \left|(a - a^{(p)})(x,y)\right|.\end{aligned}$$

This obviously implies (i), and (ii) is proved in a similar way. To prove (iii), we use the following simple equality:

$$(e_0, (A^{(p)} - i\varepsilon I) R_{A^{(p)}}(i\varepsilon) e_0) = 1.$$

From this, using (4.26) and the metric transitivity of $A^{(p)}$, we get

$$\begin{aligned}\varepsilon\left|r_{A^{(p)}}(i\varepsilon)\right| &= \mathsf{E}\left|\{(e_0, i\varepsilon R_{A^{(p)}}(i\varepsilon)e_0)\}\right| \\ &\geqslant 1 - \mathsf{E}\left\{\sum_{t\in\mathbf{Z}^d}|a^{(p)}(0,t)|\left|(R_{A^{(p)}}(i\varepsilon)e_0, e_t)\right|\right\} \\ &\geqslant 1 - \frac{1}{\varepsilon}\|A^{(p)}\|_{1,1} \geqslant 1 - \frac{1}{\varepsilon}\|A\|_{1,1}. \qquad \square\end{aligned}$$

We continue with the proof of Theorem 4.8. Because $A^{(p)}$ converges to A in the norm $\|\cdot\|_{1,1}$, Proposition 4.9 and Lemma 4.10(ii) imply the existence of a positive measure $N(d\lambda)$ such that $N(\mathbf{R}) \leqslant 1$ and

$$\lim_{p\to\infty}\int_{\mathbf{R}}(\lambda - z)^{-1}N^{(p)}(d\lambda) = \int_{\mathbf{R}}(\lambda - z)^{-1}N(d\lambda), \tag{4.31}$$

$$\lim_{p\to\infty} N^{(p)}(d\lambda) = N(d\lambda). \tag{4.32}$$

Because $N^{(p)}$ is nonrandom, so is N. Also, we have from (4.31) and Lemma 4.10(iii) the inequalities

$$1 \geqslant N(\mathbf{R}) = \lim_{\varepsilon\to\infty}\left|\int_{\mathbf{R}}\varepsilon(\lambda - i\varepsilon)^{-1}N(d\lambda)\right| \geqslant 1,$$

so that $N(\mathbf{R}) = 1$. On the other hand, Lemma 4.10(i), Lemma 4.5 and the ergodic theorem 1.13 show that

$$\limsup_{V\to\infty}\left|r_{A^{(p)}_\Lambda}(z) = r_{A_\Lambda}(z)\right| \leqslant |\operatorname{Im} z|^{-2}\|A - A^{(p)}\|_{1,1}$$

with probability 1 for all $p \in \mathbf{Z}_+$. Hence, for any z such that $\operatorname{Im} z \neq 0$, we conclude from (4.31) and from the fact that $A^{(p)}$ tends to A in the $\|\cdot\|_{1,1}$ norm, that

$$\lim_{V\to\infty} r_{A_\Lambda}(z) = \int_{\mathbf{R}}(\lambda - z)^{-1}N(d\lambda) \tag{4.33}$$

with probability 1.

The family of functions $\int_{\mathbf{R}}(\lambda - z)^{-1}m(d\lambda)$ for $m \in \mathcal{M}$ is equicontinuous with respect to z on any subset of $\mathbf{C}$ of the form $|\operatorname{Im} z| \geqslant \varepsilon > 0$, so the validity of (4.33) with probability 1 for any z such that $\operatorname{Im} z \neq 0$ implies its validity with probability 1 for *all* such z. This, combined with Proposition 4.9, shows that (4.22) holds with probability 1, and we have already seen that N satisfies $N(\mathbf{R}) = 1$ and (4.32).

There remains to show (4.23) when A is, with probability 1, an essentially self-adjoint operator. By (4.24), the arguments used to prove (4.22) also apply here. This concludes the proof of Theorem 4.8.

(4.11) Theorem [Figotin 1987]. *Let $A = \{a(x,y)\}$ be a metrically transitive Hermitian matrix belonging to $\mathcal{L}_{2,2}$, that is, such that*

$$\sum_{x\in\mathbf{Z}^d} \mathsf{E}\{|a(x,0)|^2\} < \infty. \tag{4.34}$$

The matrix A defines, with probability 1, a symmetric metrically transitive operator on the set $\mathcal{D}$ of sequences in $\ell_2(\mathbf{Z}^d)$ with finite support, and this operator has a metrically transitive self-adjoint extension $\tilde{A}$. Furthermore, there exists a nonrandom positive measure $N(A,d\lambda)$ such that, with probability 1,

$$\lim_{V\to\infty} N_\Lambda(d\lambda) = N(A,d\lambda), \tag{4.35}$$

$$N(A,d\lambda) = \mathsf{E}\{(e_0, E_{\tilde{A}}(d\lambda)e_0)\}. \tag{4.36}$$

Proof. The proof is along the same lines as that of Theorem 4.8. First, (4.24) is still valid, so A defines a symmetric operator on $\mathcal{D}$ with probability 1. The analogue of Lemma 4.10 is the following:

(4.12) Lemma. *The following relations hold between $A^{(p)}$, A_Λ and $A_\Lambda^{(p)}$:*

(i) $\left|r_{A_\Lambda^{(p)}}(z) - r_{A_\Lambda}(z)\right| |\mathrm{Im}\, z|^{-2} \leqslant \left(V^{-1}\sum_{x\in\Lambda}\sum_{t\in\mathbf{Z}^d}\left|(a-a^{(p)})(t,x)\right|^2\right)^{1/2}$;

(ii) $\left|r_{A^{(p)}}(z) - r_{A^{(q)}}(z)\right| \leqslant |\mathrm{Im}\, z|^{-2}\left\|A^{(p)} - A^{(q)}\right\|_{2,2}$.

Proof. From (4.25), (4.27) and the inequality $\|R_{A_\Lambda}(z)\| \leqslant |\mathrm{Im}\, z|^{-1}$, we have

$$\begin{aligned}\left|r_{A_\Lambda^{(p)}} - r_{A_\Lambda}\right| &\leqslant \frac{1}{V}\left|\mathrm{Tr}\, R_{A_\Lambda}(z)(A_\Lambda - A_\Lambda^{(p)})R_{A_\Lambda^{(p)}}(z)\right| \\ &\leqslant \frac{1}{V}\left(V\,|\mathrm{Im}\, z|^{-4}\sum_{x,y\in\Lambda}\left|(a-a^{(p)})(x,y)\right|^2\right)^{1/2}.\end{aligned}$$

This shows part (i); part (ii) is showed in a similar way. □

We can now prove the existence of $N(A,d\lambda)$ and the validity of (4.35) with probability 1 in exactly the same way we did in the proof of Theorem 4.8. In particular, equation (4.32) is true.

There remains to prove the existence of the metrically transitive self-adjoint extension $\tilde{A}$ of A, and that it satisfies (4.36). To do this, we show that a certain subsequence of the sequence $A^{(p)}$ has a weak graph limit (see [Reed and Simon I], for example) that satisfies the conditions for $\tilde{A}$. In fact, it is enough to show that the full sequence $A^{(p)}$ has a densely defined weak graph limit and that, for some subsequence, the resolvents $R_{A^{(p)}}(z)$ converge strongly for every z with $\operatorname{Im} z \neq 0$ (see [Reed and Simon I] or Proposition 5.6 below).

The existence of a densely defined weak graph limit $\tilde{A}$ follows immediately from the equality $\lim_{p\to\infty} A^{(p)} e_x = e_x$, which is true with probability 1 for every $x \in \mathbf{Z}^d$ in view of (4.24). The strong convergence of $R_{A^{(p)}}(z)$ is equivalent to the convergence of $R_{A^{(p)}}(z)e_x$ for all basis vectors e_x, because $\|R_{A^{(p)}}(z)\| \leqslant |\operatorname{Im} z|^{-1}$. Taking (4.27) into account, we can write

$$(4.37)\quad \mathsf{E}\{\|(R_{A^{(p)}}(z) - R_{A^{(q)}}(z))e_x\|\} \\ \leqslant |\operatorname{Im} z|^{-1}(\mathsf{E}\{\|(A^{(p)} - A^{(q)})R_{A^{(p)}}(z)e_x\|^2\})^{1/2}.$$

But if a metrically transitive operator B belongs to $\mathcal{L}_{2,2}$, we clearly have

$$(4.38)\qquad \mathsf{E}\{\|Be_x\|^2\} = \mathsf{E}\{\|B^* e_x\|^2\} = \|B\|_{2,2}^2,$$

so (4.37) implies that

$$(4.39)\qquad \mathsf{E}\{\|(R_{A^{(p)}}(z) - R_{A^{(q)}}(z))e_x\|\} \leqslant |\operatorname{Im} z|^{-2}\|A^{(p)} - A^{(q)}\|_{2,2}.$$

We choose the subsequence $\{A^{(p_n)}\}$ in such a way that

$$(4.40)\qquad \sum_{n\geqslant 1} \|A^{(p_{n+1})} - A^{(p_n)}\|_{2,2} < \infty;$$

this is possible because $\lim_{p\to\infty} \|A^{(p)} - A\|_{2,2} = 0$. It follows from (4.39) and (4.40) that

$$\mathsf{E}\Big\{\sum_{n\geqslant 1} \|(R_{A^{(p_{n+1})}}(z) - R_{A^{(p_n)}}(z))e_x\|\Big\} < \infty,$$

so the expression in braces is finite with probability 1, and the vectors $R_{A^{(p_n)}}(z)e_x$ converge with probability 1. This shows that $\tilde{A}$, as the weak graph limit of $A^{(p_n)}$ as $n \to \infty$, is self-adjoint with probability 1, and the $A^{(p_n)}$ converge to $\tilde{A}$ in the strong resolvent sense. Clearly $\tilde{A}$ is metrically transitive, being a limit of metrically transitive operators.

Finally, $\tilde{A}$ satisfies (4.36) because the $A^{(p)}$ satisfy an analogous formula (by Lemma 4.5) and $E_{A^{(p_n)}}(d\lambda) \to E_{\tilde{A}}(d\lambda)$ (since $A^{(p_n)} \to \tilde{A}$ in the strong resolvent sense).

(4.13) Theorem. *Let A be a metrically transitive Jacobi matrix such that* $\mathsf{P}\{s(0) = 0\} = 0$, *in the notation of Example* 1.17(*a*). *There exists a nonramdom measure $N(A, d\lambda)$ such that, with probability* 1,

$$\lim_{V\to\infty} N_A(d\lambda) = N(A, d\lambda) = \mathsf{E}\{(e_0, E_A(d\lambda)e_0)\}.$$

Proof. Corollary 2.6 implies that A is essentially self-adjoint with probability 1, and the conditions of Theorem 4.4 will be satisfied with $c(r) = \infty$ for $|r| \leqslant 1$ and $c(r) = 0$ otherwise. □

Remarks. 1. The proofs of Theorems 4.1, 4.4, 4.8 and 4.11 show that these results apply to block matrix operators $A = \{a(x,y)\}$ whose entries $a(x,y)$ are square matrices of a fixed finite order, and to metrically transitive integral operators (Example 1.17(d)) with $G = \mathbf{R}^d$.

2. The approximating operators A_Λ above were based on cubes centered at the origin, whose volume goes to infinity. With slight refinements, the same proofs would work for a broader class of sets, tending to infinity in the so-called sense of van Hove. This concept is defined as follows: if $a = (a_1, \ldots, a_d)$ is a vector in $\mathbf{Z}^d$ with strictly positive components, let $\Lambda(a)$ be the parallelepiped $\{(k_1, \ldots, k_d) \in \mathbf{Z}^d : 0 \leqslant k_j \leqslant a_j\}$. For a finite subset $\Lambda \subset \mathbf{Z}^d$, let $\nu_a^+(\Lambda)$ be the number of translates $\Lambda(a) + na$, where $n = (n_1, \ldots, n_d) \in \mathbf{Z}^d$ and $na = \{n_1 a_1, \ldots, n_d a_d\}$, which have a nonempty intersection with Λ. Similarly, let $\nu_a^-(\Lambda)$ be the number of such translates which are contained in Λ. A sequence $\{\Lambda_i\}$ of finite subsets of $\mathbf{Z}^d$ tends to infinity in the sense of van Hove if, for any $a \in \mathbf{Z}^d$, we have

$$\lim_{i\to\infty} \nu_a^-(\Lambda_i) = \infty \quad \text{and} \quad \lim_{i\to\infty} \frac{\nu_a^-(\Lambda_i)}{\nu_a^+(\Lambda_i)} = 1.$$

All the above theorems on the existence of $\lim N_{\Lambda_i}(d\lambda)$ are true for any sequence $\{\Lambda_i\}$ that tends to infinity in this sense, and the limit does not depend on the choice of $\{\Lambda_i\}$. Such sequences are used, for example, in statistical mechanics: see [Ruelle 1969b].

4.C Simple Properties of the Integrated Density of States and of the Spectra of Metrically Transitive Matrix Operators

We continue the investigation of matrix operators started in Section 3. Recall that the integrated density of states $N(\lambda)$ is left-continuous and normalized by the condition $N(-\infty) = 0$. If A and B are self-adjoint operators (or Hermitian matrices), we write $A \geqslant B$ if $A - B \geqslant 0$, that is, $A - B$ is positive semidefinite.

(4.14) Theorem. *If A and B are self-adjoint metrically transitive matrix operators on $\ell_2(\mathbf{Z}^d)$,*

(4.41) $$N(A,\lambda) \leqslant N(B,\lambda) \quad \textit{if } A \geqslant B$$

and

(4.42) $$N(A, \lambda - \|B\|) \leqslant N(A+B, \lambda) \leqslant N(A, \lambda + \|B\|).$$

Proof. The second inequality is an immediate corollary of the first, which, in turn, follows from the definition of the integrated density of states and the fact that if $A \geqslant B$, where A and B are Hermitian matrices of finite order, the eigenvalues of A are greater than or equal than those of B: see [Reed and Simon I]. □

Inequality (4.42) is useful in describing the asymptotic behavior of $N(\lambda)$ for $\lambda \to \pm\infty$, in some simple cases. For example:

(4.15) Theorem. *Let A and B be self-adjoint metrically transitive matrix operators on $\ell_2(\mathbf{Z}^d)$, and assume that B is bounded. If $N(A, d\lambda)$ satisfies*

(4.43)
$$\lim_{\lambda \to \pm\infty} \frac{N(A, \lambda + a)}{N(A, \lambda)} = 1$$

for all $a \in \mathbf{R}$, we have

(4.44)
$$N(A + B, \lambda) = N(A, \lambda)(1 + o(1))$$

as $\lambda \to \pm\infty$. Similarly, if

(4.45)
$$\lim_{\lambda \to \pm\infty} \frac{\ln N(A, \lambda + a)}{\ln N(A, \lambda)} = 1$$

for $a \in \mathbf{R}$, we have

(4.46)
$$\ln N(A + B, \lambda) = \ln N(A, \lambda)(1 + o(1))$$

as $\lambda \to \pm\infty$. □

We will often consider the case of metrically transitive operators on $\ell_2(\mathbf{Z}^d)$ of the form

(4.47a)
$$A = A_0 + Q,$$

where Q is the operator of multiplication by a metrically transitive random field $q(x)$ and $A_0 = \{a_0(x, y)\}$ is a nonrandom translationally invariant (Toeplitz) matrix operator, that is,

(4.47b)
$$a_0(x, y) = a_0(x - y) \qquad \text{with } \sum_{x \in \mathbf{Z}^d} |a_0(x)| < \infty.$$

When A_0 is a multidimensional finite-difference operator of the second order, that is, $a_0(x) = 0$ for $|x| > 1$, this is the discrete analogue of the multidimensional Schrödinger operator. If, in addition, $q(x)$ is independent at different points, A is usually referred to in the physics literature as the Anderson model [Lifshitz et al. 1982b]. See Example 1.4(c,e).

(4.16) Theorem. *Consider a symmetric metrically transitive operator A of the form* (4.47), *with $a_0(x) = a_0(-x)$ and $\mathsf{E}\{|q(0)|^2\} < \infty$. If $N(A,\lambda)$ and $N(A_0,\lambda)$ coincide, $Q = 0$.*

Proof. Notice that $N(A,\lambda)$ and $N(A_0,\lambda)$ exist by Theorem 4.4, which applies to A and A_0 because A_0 is bounded, so A is a self-adjoint operator. Theorem 4.4, equation (4.12) and the equality $N(A,\lambda) = N(A_0,\lambda)$ imply that

$$\mathsf{E}\{(e_0, Ae_0)\} = \int_{\mathbf{R}} \lambda N(A, d\lambda) = \int_{\mathbf{R}} \lambda N(A_0, d\lambda) = (e_0, A_0 e_0),$$
$$\mathsf{E}\{(e_0, A^2 e_0)\} = (e_0, A^2 e_0).$$

Together with the conditions on A, this shows that $\mathsf{E}\{q(0)\} = 0$ and $\mathsf{E}\{q^2(0)\} = 0$, and the theorem follows. □

(4.17) Theorem. *Let A be a symmetric metrically transitive operator of the form* (4.47). *If the integrated density of states $N(Q,\lambda) = \mathsf{P}\{q(0) \leqslant \lambda\}$ satisfies* (4.43), *we have*

$$N(A,\lambda) = \mathsf{P}\{q(0) \leqslant \lambda\}(1 + o(1)) \tag{4.48}$$

as $\lambda \to \pm\infty$.

Proof. This follows directly from Theorem 4.15. □

(4.18) Theorem [Figotin 1987]. *Denote by $\mathcal{L}_{1,1}^{(s)}$ and $\mathcal{L}_{2,2}^{(s)}$ the Banach spaces of metrically transitive, Hermitian random matrix in $\mathcal{L}_{1,1}$ and $\mathcal{L}_{2,2}$, respectively, and by $\mathcal{N}$ the set of probability measures $N(d\lambda)$ on $\mathbf{R}$ endowed with the topology of weak convergence of measures. The mapping $N : A \to N(A, d\lambda)$ from $\mathcal{L}_{1,1}^{(s)}$ or $\mathcal{L}_{2,2}^{(s)}$ into $\mathcal{N}$ is continuous.*

Proof. Consider first $\mathcal{L}_{1,1}^{(s)}$. If $A, B \in \mathcal{L}_{1,1}^{(s)}$, we see from the proof of Lemma 4.10(b) that

$$\left|r_{A^{(p)}}(z) - r_{B^{(p)}}(z)\right| \leqslant |\operatorname{Im} z|^{-2} \|A - B\|_{1,1},$$

where $r_A(z) = \int_{\mathbf{R}} (\lambda - z)^{-1} N(A, d\lambda)$. From this and from equality (4.31), we get

$$\left|r_A(z) - r_B(z)\right| \leqslant |\operatorname{Im} z|^{-2} \|A - B\|_{1,1}. \tag{4.49}$$

The theorem now follows for $\mathcal{L}_{1,1}^{(s)}$ by Proposition 4.9. The case of $\mathcal{L}_{2,2}^{(s)}$ is treated similarly, based on the inequality

$$\left|r_A(z) - r_B(z)\right| \leqslant |\operatorname{Im} z|^{-2} \|A - B\|_{2,2}$$

derived from the proof of Lemma 4.12(b). □

Here is a simple special case of Theorem 4.18:

(4.19) Corollary. *Let A_n, for $n \in \mathbf{Z}_+$, be a sequence of metrically transitive operators of the form (4.47). Let their potentials be $q_n(x)$, for $x \in \mathbf{Z}^d$. If A is an operator of the same form with potential $q(x)$ such that $\mathsf{E}\{|q(0)|^2\} < \infty$ and $\lim_{n\to\infty} \mathsf{E}\{|q_n(0) - q(0)|^2\} = 0$, the measures $N(A_n, d\lambda)$ converge to $N(A, d\lambda)$ as $n \to \infty$.* □

In the spectral analysis of metrically transitive operators, it is often interesting to know if the integrated density of states $N(A, d\lambda)$ is absolutely continuous. We present now a useful theorem, first proved by Wegner [1981], that is relevant to this question. The proof given here differs from the original one, and uses the following simple lemma:

(4.20) Lemma. *Let A_1 and A_2 be self-adjoint operators on an abstract Hilbert space $\mathcal{H}$, whose difference is a one-dimensional operator, that is, $A_2 - A_1 = \tau P$, where $\tau \in \mathbf{R}$ and P is the orthogonal projection $P\psi = (e, \psi)e$ onto some unit vector $e \in \mathcal{H}$. For any z such that $\operatorname{Im} z \neq 0$, the resolvents R_1 and R_2 of A_1 and A_2 are related by*

$$R_2 - R_1 = -R_1 P R_1 \tau \big(1 + \tau(R_1 e, e)\big)^{-1}.$$

Proof. We can rewrite (4.27) in the form $R_2 = R_1\big(I + (A_2 - A_1)R_1\big)^{-1}$. But it is easy to see that for any bounded operator A such that $1 + (Ae, e)$ is nonzero, $(I + PA)^{-1} = I - PA\big(1 + (Ae, e)\big)^{-1}$. This proves the lemma. □

(4.21) Theorem [Wegner 1981]. *Let A be an operator of the form (4.47), where the potential $q(x)$ consists of a family of independent, identically distributed random variables whose distribution function has a bounded density: $F(dq) = f(q)dq$, with $\sup_{q\in\mathbf{R}} f(q) = f_0 < \infty$. The integrated density of states $N(A, d\lambda)$ is absolutely continuous, and the density of states $n(\lambda)$, given by $N(A, d\lambda) = n(\lambda)\, d\lambda$, is bounded by f_0.*

Proof. We use Lemma 4.20, with $A_1 = A\big|_{q(0)=0}$ and A playing the roles of A_1 and A_2. The role of τP is played by $q(0)P_0$, where P_0 is the projection onto the unit vector e_0. Then, by Theorem 4.4,

$$\int_R \frac{N(A, d\lambda)}{\lambda - z} = \mathsf{E}\{(e_0, R_A(z)e_0)\} = \mathsf{E}\Big\{\int_{\mathbf{R}} \frac{f(q)dq}{q - \zeta}\Big\}, \tag{4.51}$$

where $-\zeta^{-1} = (e_0, R_A(z)e_0)\big|_{q(0)=0}$. Since the diagonal element of the resolvent of a self-adjoint operator is a Nevanlinna function of z, that is, it has positive imaginary part on the upper half-plane, ζ has the same property: $\zeta = \zeta_1 + i\zeta_2$ with $\zeta_2 > 0$ if $\operatorname{Im} z > 0$. Then we get

$$\operatorname{Im} \int_{\mathbf{R}} \frac{f(q)\,dq}{q-\zeta} = \zeta_2 \int_{\mathbf{R}} \frac{f(q)\,dq}{(q-\zeta_1)^2+\zeta_2^2} \leqslant f_0 \int_{\mathbf{R}} \frac{\zeta_2\,dq}{(q-\zeta_1)^2+\zeta_2^2} = \pi f_0.$$

The theorem now follows from the Stieltjes–Perron inversion formula A.4 of the Appendix. □

Remarks. 1. Problems II.18 to II.25 present extensions of Theorem 4.21 to more general matrix operators.

2. Problem II.42 and [Kotani and Simon 1986] discuss an analogue of Theorem 4.21 for Schrödinger operators with random potentials of a certain form.

4.D Location of the Spectrum

We again consider metrically transitive operators of the form (4.47). When the correlation between the potential function $q(x)$ at distant points is weak enough, fairly simple methods are sufficient to yield information about the geometric disposition of the spectrum.

For every metrically transitive field on $\mathbf{Z}^d$, we consider two subsets of $\mathbf{R}$:

(4.52a) $\sigma_1(q) = \{\alpha : \mathsf{P}\{|q(x)-\alpha| < \varepsilon\} > 0 \text{ for all } \varepsilon > 0\}$,

(4.52b) $\sigma_2(q) = \{\alpha : \mathsf{P}\{\sup_{x\in\Lambda} |q(x)-\alpha| < \varepsilon\} > 0 \text{ for all } \varepsilon > 0 \text{ and } \Lambda \in L\}$,

where L is the set of cubes in $\mathbf{Z}^d$ centered at the origin.

The first set is simply the set of growth points of the one-dimensional distribution $F(dq) = \mathsf{P}\{q(x) \in dq\}$ of the field $q(x)$. The second set, $\sigma_2(q)$, is always contained in $\sigma_1(q)$, and the inclusion is strict only when there is essential correlation between the values of $q(x)$ at different points. The two sets are equal when $q(x)$ is independently distributed. For the one-dimensional field $q(x) = \cos(2\pi\alpha x + \omega)$, where α is irrational, we have $\sigma_1(q) = [-1, 1]$ and $\sigma_2(q) = \varnothing$.

If we consider in the space of realizations of the potential $q(x)$ the topology of pointwise convergence, $\sigma_2(q)$ can be seen as the set of $s \in \mathbf{R}$ such that the topological support of the probability measure P of the field $q(x)$ contains the constant function s.

In the statement of the following theorem, $X+Y$ denotes the arithmetic sum of X and Y, that is, the set of $\xi + \eta$ for some $\xi \in X$ and $\eta \in Y$.

(4.22) Theorem [Kunz and Souillard 1980]. *Let A be a metrically transitive operator of the form (4.47a,b), and let $\hat{a}_0(k) = \sum_x a_0(x)e^{2\pi ikx}$, for $k \in \mathbf{R}^d$, be the Fourier transform of the matrix of its translationally invariant part A_0. Set $a_+ = \sup_k \hat{a}_0(k)$ and $a_- = \inf_k \hat{a}_0(k)$. Then*

(i) $\sigma(A) \subset [a_-, a_+] + \sigma_1(q)$, *and*
(ii) $\sigma(A) \supset [a_-, a_+] + \sigma_2(q)$.

Proof. We clearly have $\sigma(A_0) = [a_-, a_+]$ and $\sigma(Q) = \sigma_1(q)$. Therefore the first inclusion is a corollary of the well-known fact that for a self-adjoint operator Q to which we add a perturbation A_0, the spectrum of $Q + A_0$ is contained in $\sigma(Q) + \sigma(A_0)$.

To prove the second inclusion we assume first that the potential $q(x)$ is bounded. The inclusion then follows from Theorem 2.17 and the characterization of $\sigma_2(q)$ given above: if $\alpha \in \sigma_2(q)$, we have $q(x) \equiv \alpha \in \operatorname{supp} \mathsf{P}$. If $q(x)$ is unbounded we truncate it to $q_a(x) = \chi_{(-a,a)} q(x)$ and take the limit as $a \to \infty$. □

(4.23) Corollary. *Let A be a metrically transitive operator of the form* (4.47), *the field $q(x)$ being given by independent, identically distributed random variables with distribution function $F(q)$. Then*

$$\sigma(A) = \sigma(A_0) + \operatorname{supp} F,$$

where $\sigma(A_0)$ is the set of values of the function $\hat{a}_0(k) = \sum_{x \in \mathbf{Z}^d} e^{2\pi i k x} a_0(x)$, for $k \in \mathbf{R}^d$. In particular, for the Anderson model with $a_0(x) = -1$ for $|x| = 1$ and $a_0(x) = 0$ otherwise, we have

$$\sigma(A) = [-2d, 2d] + \operatorname{supp} F.$$

Proof. If $q(x)$ is a sequence of independent, identically distributed random variables, it is clear that $\sigma_2(q) = \sigma_1(q) = \operatorname{supp} F$. □

Remark. If Q is the operator of multiplication by $q(x)$, it is always true that $\sigma(Q) = \sigma_1(q)$, so the proof of the corollary applies whenever $\sigma_1(q) = \sigma_2(q)$, to give the relation

$$\sigma(A) = \sigma(A_0) + \sigma(Q).$$

This implies, in particular, that if the support of the distribution F splits into several components sufficiently far apart, there will be also gaps in the spectrum of A (see also subsection 5.C).

5 Schrödinger Operators and Elliptic Differential Operators on $L_2(\mathbf{R}^d)$

5.A Criteria for Essential Self-Adjointness

The methods we will use to obtain results on essential self-adjointness mostly apply equally well to the study of the ergodic properties of elliptic metrically transitive operators, and, in particular, of Schrödinger operators. For this reason many auxiliary results in this subsection are considered in somewhat more generality than would be necessary if we were interested only in essential self-adjointness; this way we can use them later, in the study of ergodic properties.

Consider the Schrödinger operator $H = -\Delta + q(x)$, for $x \in \mathbf{R}^d$, where $q(x)$ is a metrically transitive field. We assume the realizations of $q(x)$ to be Borel functions. As explained in Section 3, and in analogy with the study of matrix operators in Section 4, we will introduce for each cube Λ centered at the origin and each function $a : \mathbf{R}^n \to [0,1]$ an approximating operator $H_{\Lambda,a}$, defined by the same operation $-\Delta + q(x)$ in Λ and subject to the self-adjoint boundary condition

$$(5.1) \qquad (1-a)\psi + a\frac{\partial\psi}{\partial\nu}\bigg|_{\partial\Lambda} = 0,$$

where $\partial\Lambda$ is the boundary of Λ and the differential represents the derivative in the direction of the external normal to $\partial\Lambda$. We will also denote $H_{\Lambda,a}$ by $(-\Delta + q)_{\Lambda,a}$. The cases $a \equiv 0$ and $a \equiv 1$ correspond to the Dirichlet and Neumann boundary problems and will play a special role later on.

We use expanding cubes to define the approximating operators because this simplifies the techniques of the proofs and allows the treatment of periodic boundary conditions. But the same results are valid for more general sequences of regions, such as regions tending to infinity in the Fisher sense: see [Shubin 1979], and compare the end of Subsection 5.B.

Denote by $C_a^\infty(\Lambda)$ the set of functions ψ on Λ that are differentiable infinitely often up to the boundary and such that ψ and $\Delta^n\psi$, for all $n \in \mathbf{Z}_+$, satisfy the boundary conditions (5.1).

Before we can consider H on the whole space, we need some information about $H_{\Lambda,a}$. We denote by p the smallest even number strictly larger than $d/2$.

(5.1) Theorem. *If $q(x)$ be a metrically transitive field with $\mathsf{E}\{|q(0)|^{p+1}\} < \infty$,*

(i) *the operator $(-\Delta+q)_{\Lambda,a}$ is essentially self-adjoint on $C_a^\infty(\Lambda)$ with probability 1;*

(ii) *the resolvent* $R_{\Lambda,a} = (-\Delta + q - z)^{-1}_{\Lambda,a}$, *where* $\operatorname{Im} z \neq 0$, *satisfies*

$$\mathsf{E}\{\|R^p_{\Lambda,a}\|_1\} < \infty,$$

where $\|A\|_1 = \operatorname{Tr}(A^*A)^{1/2}$.

Before proving this theorem we state some auxiliary facts.

(5.2) Proposition. *Let A be a compact linear operator on a Hilbert space, and* $\|A\|_t = (\operatorname{Tr}(A^*A)^{t/2})^{1/t}$. *The norm $\|\cdot\|_\infty$ coincides with the usual (uniform) operator norm, which is smaller than the norm $\|\cdot\|_t$ for $t > 1$. Furthermore,*

(i) $\|AB\|_t \leqslant \|B\|\,\|A\|_t$ *and* $\|A\|_t = \|A^*\|_t$;
(ii) $\|AB\|_1 \leqslant \|A\|_2\,\|B\|_2$;
(iii) $\|A + B\|_t \leqslant \|A\|_t + \|B\|_t$.

Proof. See [Reed and Simon II]. □

For $0 < \varepsilon \leqslant \pi/2$ and $\alpha > 0$, set

$$\Gamma_{\varepsilon,\alpha} = \{z : |\arg z| \geqslant \varepsilon \text{ and } |\operatorname{Im} z| > \alpha\}.$$

(5.3) Lemma. *Let V be the volume of the cube Λ, and assume $V \geqslant 1$. Take an integer $r \geqslant p$, complex numbers $z_j \in \Gamma_{\varepsilon,\alpha}$, for $1 \leqslant j \leqslant r$, each equal to either z or $\bar z$, and let $R_{0,j}(x, y)$, for $x, y \in \Lambda$, be the kernel of the resolvent $R_{0,j} = (-\Delta - z_j)^{-1}_{\Lambda,a}$. There exists a constant c_ε, independent of Λ, a and α, such that, for every $x \in \Lambda$,*

$$\int_{\Lambda^{r-1}} \left|R_{0,1}(x, y_1)R_{0,2}(y_1, y_2)\dots R_{0,r}(y_{r-1}, x)\right| dy_1 \dots dy_{r-1} \leqslant c_\varepsilon\,|z|^{d/2-r}.$$

Proof. Denote by $R_{\lambda,a}(z; x, y)$ the kernel of the resolvent $R_{\Lambda,a} = (-\Delta - z)^{-1}_{\Lambda,a}$. By the scaling properties of the resolvent,

$$R_{\Lambda,a}(z; x, y) = L^{2-d} R_{C,aL^{-1}(1-a-aL^{-1})^{-1}}(zL^2, xL^{-1}, yL^{-1}),$$

where C is the cube of volume 1 centered at the origin and $L = V^{1/d}$. Thus, it is sufficient to proved the lemma for $\Lambda = C$. In this case it follows from a suitable representation of the kernel $R_{C,a}(z; x, y)$: see, for example, [Titchmarsh 1958]. □

Let $|f|_t$ denote the expression $\left(\int_\Lambda |f(x)|^t\,dx\right)^{1/t}$. As usual, χ_Λ is the characteristic function of Λ.

(5.4) Lemma. *Let $g_1, g_2 \in C(\mathbf{R}^d)$ and $g_0 = 0$. For z_j as in Lemma 5.3, let* $R_{k,j} = (-\Delta + g_k - z_j)^{-1}_{\Lambda,a}$.

(i) *There is a positive constant* $c^{(1)}_{\varepsilon,\alpha}$ *independent of* Λ, a *and* g_k *and such that*

$$\left\| \prod_{j=1}^{p-s} R_{2,j} \prod_{j=p-s+1}^{p} R_{0,j} - \prod_{j=1}^{p-s} R_{1,j} \prod_{j=p-s+1}^{p} R_{0,j} \right\|_1 \leqslant c^{(1)}_{\varepsilon,\alpha} |z|^{d/2-p} \big|\chi_\Lambda(g_1 - g_2)\big|_{p+1} \big(\max_{k=1,2} |\chi_\Lambda(|g_k| + 1)|_{p+1} \big)^p.$$

(ii) *There is a positive constant* $c^{(2)}_{\varepsilon,\alpha}$ *independent of* Λ, a *and* g_k *and such that*

$$\left\| \prod_{j=1}^{p-s} R_{1,j} \prod_{j=p-s+1}^{p} R_{0,j} \right\|_1 \leqslant c^{(2)}_{\varepsilon,\alpha} |z|^{d/2-p} \big|\chi_\Lambda(|g_1| + 1)\big|_p^p.$$

(iii) *If* $z = \pm i$, $f(x)$ *is a Borel function on* $\mathbf{R}^d$ *and* f *is the operator of multiplication by* $f(x)$, *there is a positive constant* c_1 *independent of* Λ, a *and* g_1 *such that, for* $V \geqslant 1$,

$$\|f R_1 R_0^{p-1}\|_1 \leqslant c_1 |\chi_\Lambda f|_{p+1} \big|\chi_\Lambda(|g_1| + 1)\big|_{p+1}^p.$$

(iv) *If* $r \geqslant p+1$ *and* $z = -t + i$, *where* $t \geqslant 0$, *there is a positive constant* c_2 *independent of* Λ, a *and* g_1 *such that, for* $V \geqslant 1$,

$$\|R_1^r\| \leqslant c_2 \big|\chi_\Lambda(|g_1| + 1)\big|_r^r (t^2 + 1)^{(d/2-r)/2}.$$

Proof. We start by setting $\delta g = g_2 - g_1$, and noticing that

$$R_2 - R_1 = R_2\, \delta g\, R_1 = R_1\, \delta g\, R_2, \tag{5.2}$$

$$R_k - R_0 = -R_k g_k R_0, \tag{5.3a}$$

$$R_k - R_0 = -R_0 g_k R_k. \tag{5.3b}$$

Applying (5.2) repeatedly we can write

$$\prod_{j=1}^{p-s} R_{2,j} - \prod_{j=1}^{p-s} R_{1,j} = -\sum_{k=1}^{p-s} \Big(\prod_{j=1}^{k} R_{2,j}\, \delta g \prod_{j=k}^{p-s} R_{1,j} \Big). \tag{5.4}$$

Hence, in view of Proposition 5.2(iii), the bound in part (i) can be obtained by estimating in a suitable way the expression

$$\left\| \prod_{j=1}^{k} R_{2,j}\, \delta g \prod_{j=k}^{p-s} R_{1,j} \prod_{j=p-s+1}^{p-1} R_{0,j} \right\|_1,$$

for $1 \leqslant k \leqslant p - s$. Applying (5.3a) to R_2 and (5.3b) to R_1 and using Proposition 5.2(iii), we reduce the problem to estimating

$$\left\| B_2 \prod_{j=1}^{k} (a_j \tilde{R}_{0,j})\, \delta g \prod_{j=k+1}^{p} (\tilde{R}_{0,j} b_j) B_1 \right\|_1,$$

where $\tilde{R}_{0,j}$ equals $R_{0,j'}$ for some j'; a_j and b_j equal either 1 or g_2 and g_1; and B_1 and B_2 are products of factors, each coinciding with $R_{0,j}$ for some j or with $R_{2,j}$ or $R_{1,j}$, respectively, for some other j. Since $\|R\| \leqslant |\operatorname{Im} z|^{-1} \leqslant \alpha^{-1}$, we can apply Proposition 5.2(i) and further reduce the problem to estimating an expression of the from

$$\|a_1 \tilde{R}_{0,1} a_2 \tilde{R}_{0,2} \dots a_k \tilde{R}_{0,k} \dots a_p \tilde{R}_{0,p} a_{p+1}\|_1, \tag{5.5}$$

where a_j equals g_1, g_2, δg or 1, and δg appears among the a_j exactly once. Since p is even, say $p = 2m$, we can split this product as follows:

$$(a_1 \tilde{R}_{0,1} \dots a_m \tilde{R}_{0,m} \sqrt{|a_{m+1}|})(\operatorname{sign} a_{m+1} \sqrt{|a_{m+1}|} \tilde{R}_{0,m+1} \dots a_p \tilde{R}_{0,p} a_{p+1}).$$

Applying parts (i) and (ii) of Proposition 5.2, we obtain

$$\begin{aligned} \|a_1 \tilde{R}_{0,1} \dots \tilde{R}_{0,p} a_{p+1}\| &\leqslant \left(\operatorname{Tr} a_1 \tilde{R}_{0,1} \dots \tilde{R}_{0,m} |a_{m+1}| \tilde{R}^*_{0,m} \dots \tilde{R}^*_{0,1} a_1\right)^{1/2} \\ &\quad \times \left(\operatorname{Tr} a_{p+1} \tilde{R}_{0,p} \dots \tilde{R}_{0,m+1} |a_{m+1}| \tilde{R}^*_{0,m+1} \dots \tilde{R}^*_{0,p} a_{p+1}\right)^{1/2}. \end{aligned} \tag{5.6}$$

We now use Hölder's inequality and Lemma 5.3 to estimate the first radical in (5.6); the second is treated in the same way.

$$\begin{aligned} &\operatorname{Tr} a_1 \tilde{R}_{0,1} \dots \tilde{R}_{0,m} |a_{m+1}| \tilde{R}^*_{0,m} \dots \tilde{R}^*_{0,1} a_1 \\ &\leqslant \left(\int_{\Lambda^p} |a_1(x_1)|^{p+1} |R_{0,1}(x_1, x_2) \dots R_{0,p}(x_p, x_1)|\, dx_1 \dots dx_p \right)^{\frac{1}{(p+1)}} \times \cdots \\ &\leqslant c_{\varepsilon,\alpha} |z|^{d/2-p} \prod_{j=1}^{m} |\chi_\Lambda a_j|^2_{p+1} |\chi_\Lambda a_{m+1}|_{p+1}. \end{aligned}$$

Plugging this into (5.6), we get

$$\|a_1 \tilde{R}_{0,1} \dots \tilde{R}_{0,p} a_{p+1}\| \leqslant c_{\varepsilon,\alpha} |z|^{d/2-p} \prod_{j=1}^{p+1} |\chi_\Lambda a_j|_{p+1}.$$

Taking into account the definition of a_j, we obtain part (i) of the Lemma. Parts (ii) and (iii) are proved similarly, starting from equations (5.3) and Proposition 5.2.

The proof of (iv) is also similar, but the following comments may be made. As in the case of (i), it is necessary to find an upper bound for $\left\|\prod_{j=1}^r (a_j R_0)\right\|_1$, where a_j equals g_1 or 1. If r is even, say $r = 2s$, we can use the estimate

$$\|a_1R_0\dots a_rR_0\|_1 \leqslant (\operatorname{Tr} a_1R_0\dots a_sR_0R_0^*a_s\dots R_0^*a_1)^{1/2} \times (\operatorname{Tr} a_{s+1}R_0\dots a_rR_0R_0^*a_r\dots R_0^*a_{s+1})^{1/2}$$

If, on the other hand, $r = 2s+1$, we write instead

$$\begin{aligned}\|a_1R_0\dots a_rR_0\|_1 &= \left\|(a_1R_0\dots R_0\sqrt{|a_{s+1}|})(\operatorname{sign} a_{s+1}\sqrt{|a_{s+1}|}R_0\dots a_rR_0)\right\| \\ &\leqslant (\operatorname{Tr} a_1R_0\dots a_sR_0|a_{s+1}|R_0^*a_s\dots R_0^*a_1)^{1/2} \\ &\quad\times (\operatorname{Tr} R_0^*a_r\dots R_0^*a_{s+2}R_0^*|a_{s+1}|R_0a_{s+2}R_0\dots a_rR_0)^{1/2}\end{aligned}$$

After that we continue as before, using Lemma 5.3 and the appropriate Hölder inequalities. □

(5.5) Lemma. *Suppose that $g \in C^\infty(\Lambda)$ is zero in a neighborhood of the boundary of Λ. If R is the resolvent $(-\Delta + g - z)^{-1}_{\Lambda,a}$, where $\operatorname{Im} z \neq 0$, we have $RC_a^\infty(\Lambda) = C_a^\infty(\Lambda)$.*

Proof. The proof is rather lengthy, so we limit ourselves to a sketch of it.

(1) The eigenfunctions e_k and eigenvalues λ_k of the operator $(-\Delta)_{\Lambda,a}$ are indexed by $k \in \mathbf{Z}^d$ and have the following properties: λ_k is nonnegative, $\lim_{|k|\to\infty} |k|^{-2}\lambda_k$ is a positive constant, $e_k \in C_a^\infty$, and $\sup_{k\in\mathbf{Z}^d, x\in\Lambda}|e_k(x)| < \infty$. For the proof, see [Titchmarsh 1958].

(2) If $\psi \in L^2(\Lambda)$, let $\psi_k = (e_k, \psi)$ be the coefficients of the expansion of ψ with respect to the basis $\{e_k : k \in \mathbf{Z}^d\}$. Then $\psi \in C_a^\infty(\Lambda)$ if and only if $\lim_{|k|\to\infty} |k|^n|\psi_k| = 0$ for all $n \in \mathbf{Z}_+$. Moreover, if $\psi \in C_a^\infty(\Lambda)$, the series $\psi = \sum_k \psi_k e_k$ can be differentiated term by term. If we denote by $\mathcal{D}(A)$ the domain of A, we have $C_a^\infty(\Lambda) = \bigcap_{m\geqslant 1}\mathcal{D}(\Delta^n_{\Lambda,a})$, and $\mathcal{D}(\Delta^n_{\Lambda,a}) \supset \mathcal{D}(\Delta^{n+1}_{\Lambda,a})$. This latter inclusion follows immediately from Hölder's inequality $\|A^n\psi\| \leqslant \|A^{n+1}\psi\|^{n/(n+1)}$, which is true for any self-adjoint operator A.

(3) If $R_0 = (-\Delta - z)^{-1}_{\Lambda,a}$, we have $R_0C_a^\infty(\Lambda) = C_a^\infty(\Lambda)$, and $R_0\mathcal{D}(\Delta^{n+1}_{\Lambda,a}) = \mathcal{D}(\Delta^n_{\Lambda,a})$.

(4) We have $g\mathcal{D}(\Delta^n_{\Lambda,a}) \subset \mathcal{D}(\Delta^n_{\Lambda,a})$. This is proved by induction on n, using step (2) and the fact that the self-adjoint operator $\Delta^n_{\Lambda,a}$, with domain $\mathcal{D}(\Delta^n_{\Lambda,a})$, is the closure of its own restriction to the linear subspace $C_a^\infty(\Lambda)$.

(5) The inclusion $RC_a^\infty(\Lambda) \supset C_a^\infty(\Lambda)$ follows from the obvious relation $(-\Delta + g - z)_{\Lambda,a}C_a^\infty(\Lambda) \subset C_a^\infty$. The reverse is a consequence of the inclusion $RC_a^\infty(\Lambda) \subset \mathcal{D}(\Delta^n_{\Lambda,a})$, for all $n \in \mathbf{Z}_+$, which follows from (5.3b) and from steps (1)–(4), by induction on n. □

(5.6) Proposition. *Let $\{A_n\}_1^\infty$ be a sequence of self-adjoint operators on a Hilbert space. Assume there are operators R_+ and R_- such that $(A_n \pm i)^{-1} \to$*

$R_\pm$ strongly, and that the weak graph limit A of $\{A_n\}_1^\infty$ is densely defined. Then A is self-adjoint, $(A \pm i)^{-1} = R_\pm$, and A_n converges to A in the strong resolvent sense.

Proof. See [Reed and Simon II]. □

We now take a nonnegative function $\rho(x) \in C_0^\infty(\mathbf{R}^d)$, with support in a ball of radius $\frac{1}{2}$ and such that $\int_{\mathbf{R}^d} \rho(x)\,dx = 1$. We define a sequence of smooth, bounded potential $q_n(x)$ approximating $q(x)$ by taking the convolution of the functions $\rho_n(x) = n^d \rho(nx)$ and $\tilde{q}_n(x) = \operatorname{sign} q(x) \min\{n, |q(x)|\}$:

$$q_n(x) = \rho_n * \tilde{q}_n = n^d \int_{\mathbf{R}^d} \rho(n(x-y))\tilde{q}_n(y)\,dy. \tag{5.7}$$

(5.7) Lemma. *Assume either that the functions g_1 and g_2 belong to $L^{p+1}(\Lambda)$ or that $g_1 \in L^r$, where p, as usual, is the smallest even number strictly greater than $d/2$ and $r \geqslant p+1$. Under these conditions, parts (i), (ii) and (iv) of Lemma 5.4 are true.*

Proof. We first show, using Proposition 5.6, that if $q \in L^{p+1}(\Lambda)$ the operators $(-\Delta + q_n)_{\Lambda,a}$ converge in the strong resolvent sense to the self-adjoint operator $(-\Delta + q)_{\Lambda,a}$. Applying Lemma 5.4(i) to q_n and q_l, for $z_j = z$ and $s = p-1$, we get $\lim_{n,l\to\infty} \|(R_n - R_l)R_0^{p-1}h\|_1 = 0$ for any $h \in L^2(\Lambda)$. Since $\|(R_n - R_l)R_0^{p-1}h\| \leqslant \|(R_n - R_l)R_0^{p-1}\|_1 \|h\|$, we see that the strong limit $\operatorname{s-lim}_{n\to\infty} R_n R_0^{p-1} h$ exists. Since $R_0^{p-1}L^2(\Lambda)$ is dense in $L^2(\Lambda)$ and $\|R_n\| \leqslant |\operatorname{Im} z|^{-1}$, this means that $\operatorname{s-lim}_{n\to\infty} R_n$ also exists, showing that the first condition in Proposition 5.6 is fulfilled.

Now note that $\lim_{n\to\infty} |\chi_\Lambda(q - q_n)|_{p+1} = 0$, so, for every $\phi \in C_a^\infty(\Lambda)$ we have $\operatorname{s-lim}_{n\to\infty}(-\Delta + q_n)\phi = (-\Delta + q)\phi$, and the second condition is also satisfied. Therefore $\operatorname{s-lim}_{n\to\infty} R_n = R$. Moreover,

$$\lim_{n\to\infty} \|B_n - B\|_1 = 0, \quad B_n = \prod_{j=1}^{p-s} R_{n,j} \prod_{j=p-s+1}^{p} R_{0,j}, \quad B = \prod_{j=1}^{p-s} R_j \prod_{j=p-s+1}^{p} R_{0,j}. \tag{5.8}$$

Indeed, since $\lim_{n\to\infty} |\chi_\Lambda(q - q_n)|_{p+1} = 0$, parts (i) and (ii) of Lemma 5.4 imply that the sequence B_n is fundamental in the $\|\cdot\|_1$ metric and therefore converges in this metric to some operator $\tilde{B}$ with a finite $\|\cdot\|_1$-norm. On the other hand, from the strong convergence of R_n to R and the inequalities $\|R\| \leqslant |\operatorname{Im} z|^{-1}$ and $\|R_n\| \leqslant |\operatorname{Im} z|^{-1}$, we see that $\tilde{B} = B$.

Using (5.8) it is straightforward to prove that the inequalities in parts (i) and (ii) of Lemma 5.4 hold under the weakened conditions of Lemma 5.7. Part (iv) of Lemma 5.4 is treated in a similar way. □

Proof of Theorem 5.1. Since the operator $(-\Delta+q)_{\Lambda,a}$ is symmetric on the space $C_a^\infty(\Lambda)$ with probability 1, part (i) will follow if we show that the set $(-\Delta+q-z)_{\Lambda,a}C_a^\infty(\Lambda)$ is dense in $L^2(\infty)$ with probability 1. The inequality $\mathsf{E}\{|q(0)|^{p+1}\}<\infty$ implies that $q\in L^{p+1}(\Lambda)$ with probability 1. Fix a realization of q and choose a sequence of functions $g_n\in C_a^\infty(\Lambda)$ such that each g_n is zero in some neighborhood of Λ, and that $\lim_{n\to\infty}|\chi_\Lambda(g-g_n)|_{p+1}=0$. Take an arbitrary $h\in C_a^\infty(\Lambda)$ and set $R_n=(-\Delta+q-z)^{-1}_{\Lambda,a}$. Then, by Lemma 5.5, $\phi_n=R_nh$ is in $C_a^\infty(\Lambda)$, and $h\in R_0^{p-1}h_1$, for some $h_1\in C_a^\infty(\Lambda)$.

Now consider the sequence

$$(-\Delta+q-z)_{\Lambda,a}\phi_n=h+(q-g_n)R_nR_0^{p-1}h_1.$$

Using Lemma 5.4(iii), we see that

$$\lim_{n\to\infty}\left\|(q-g_n)R_nR_0^{p-1}h_1\right\|\leqslant\lim_{n\to\infty}\left\|(q-g_n)R_nR_0^{p-1}h_1\right\|_1\|h_1\|=0.$$

Clearly this implies that $\lim_{n\to\infty}(-\Delta+q-z)_{\Lambda,a}\phi_n=h$. Since $h\in C_a^\infty(\Lambda)$ and $\phi_n\in C_a^\infty(\Lambda)$ are arbitrary, it follows that $(-\Delta+q-z)_{\Lambda,a}C_a^\infty(\Lambda)$ is dense in $L^2(\infty)$, as we wished to prove.

Part (ii) follows from Lemmas 5.4 and 5.7 and the Hölder inequality for mathematical expectations, because

$$\mathsf{E}\{\|R^p\|_1\}\leqslant c_{\varepsilon,\alpha}|z|^{d/2-p}V\mathsf{E}\{(|q(0)|+1)^p\}.\qquad\square$$

We now formulate the first of the two main criteria for essential self-adjointness of Schrödinger operators used in this book. For other similar criteria, see [Kirsch and Martinelli 1983b].

As usual, we let p stand for the smallest even number strictly larger than $d/2$.

(5.8) Theorem [Figotin 1983]. *Let q be a metrically transitive field.*

(i) *If $\mathsf{E}\{|q(0)|^{p+1}\}<\infty$, there exists a subsequence of $\{-\Delta+q_n : n\in\mathbf{Z}_+\}$ that converges in the strong resolvent sense, with probability 1, to a self-adjoint metrically transitive operator. We denote this limit by $-\Delta+q$.*

(ii) *If $\mathsf{E}\{|q(0)|^{p+2}\}<\infty$, the limit $-\Delta+q$ is, with probability 1, an essentially self-adjoint operator on $C_0^\infty(\mathbf{R}^d)$.*

(5.9) Lemma. *Take $z\in\Gamma_{\varepsilon,\alpha}$ and complex numbers z_j, for $1\leqslant j\leqslant p$, each equal to either z or $\bar{z}$, and let $R_{0,j}(x,y)$, for $x,y\in\mathbf{R}^d$, be the kernel of the resolvent $R_{0,j}=(-\Delta-z_j)^{-1}$. There exists a positive constant c_ε, independent of α, such that, for every $x\in\mathbf{R}^d$,*

$$(5.9)\quad \int_{(\mathbf{R}^d)^{r-1}} \left|R_{0,1}(x,y_1)R_{0,2}(y_1,y_2)\dots R_{0,p}(y_{p-1},x)\right| dy_1\dots dy_{p-1} \leqslant c_\varepsilon\, |z|^{d/2-p}.$$

Proof. By the scaling properties of the resolvent R_0 we have

$$R_0(z;x,y) = L^{2-d}R_0(zL^2, xL^{-1}, yL^{-1})$$

for any $L>0$. Taking $L=|z|^{-1/2}$, we reduce the proof to the case $|z|=1$. But this follows from the easily checked fact that the left-hand side of (5.9) is continuous on the intersection of the unit circle with the set $\{|\arg z| \geqslant \varepsilon\}$. □

(5.10) Lemma. *Let $A(x,y)$, for $x,y\in\mathbf{R}^d$, be the kernel of the metrically transitive operator A in $L^2(\mathbf{R}^d)$, that is, $A(x,y,T_a\omega)=A(x+a,y+a,\omega)$.*

(i) $\mathsf{E}\{\operatorname{Tr}\chi_\Lambda AA^*\chi_\Lambda\} = \mathsf{E}\{\operatorname{Tr}\chi_\Lambda A^*A\chi_\Lambda\}$.

(ii) *If B is a metrically transitive operator on $L^2(\mathbf{R}^d)$ such that $\|B\|\leqslant c$ for a nonrandom constant c, and if $(BA)(x,y)$ is the kernel of BA, we have*

$$\mathsf{E}\{\operatorname{Tr}\chi_\Lambda BAA^*\chi_\Lambda\} \leqslant c^2\mathsf{E}\{\operatorname{Tr}\chi_\Lambda AA^*\chi_\Lambda\}.$$

Proof. Since A is a metrically transitive operator,

$$\begin{aligned}\mathsf{E}\{\operatorname{Tr}\chi_\Lambda AA^*\chi_\Lambda\} &= V\int_{\mathbf{R}^d}\mathsf{E}\{|A(0,y)|^2\}\,dy = V\int_{\mathbf{R}^d}\mathsf{E}\{|A(y,0)|^2\}\,dy\\ &= V\int_{\mathbf{R}^d}\mathsf{E}\{|A^*(0,y)|^2\}\,dy = \mathsf{E}\{\operatorname{Tr}\chi_\Lambda A^*A\chi_\Lambda\}.\end{aligned}$$

Part (ii) follows from (i) and the inequality $A^*B^*BA\leqslant c^2A^*A$. □

(5.11) Lemma. *For $1\leqslant j\leqslant r$, where $r>p$, let $g_j(x)$, for $x\in\mathbf{R}^d$, be metrically transitive fields.*

(i) *If $R_{0,j}$ is as in Lemma 5.3 and t_j, for $1\leqslant j\leqslant p+1$, are real numbers such that $\sum_{j=1}^{p+1}t_j^{-1}=1$, we have*

$$\mathsf{E}\{\|\chi_\Lambda g_1R_{0,1}\dots R_{0,p}g_{p+1}\chi_\Lambda\|_1\}c_\varepsilon\,|z|^{d/2}\,V\prod_{j=1}^{p+1}\mathsf{E}^{1/t_j}\{|g_j(0)|^{t_j}\}.$$

(ii) *Assume the fields $g_j(x)$ are uniformly bounded, and continuous with probability 1. Take $z\in\Gamma_{\varepsilon,\alpha}$ and complex numbers z_j, for $1\leqslant j\leqslant r$, each equal to either z or $\bar z$, and let $R_{0,j}(x,y)$, for $x,y\in\mathbf{R}^d$, be the kernel of the resolvent $R_{0,j}=(-\Delta+g_j-z_j)^{-1}$. If $f(x)$, for $x\in\mathbf{R}^d$, is a metrically transitive field and $t_1,t_2\geqslant 1$ are real numbers with $t_1^{-1}+pt_2^{-1}=1$, there exists a positive constant $c_{\varepsilon,\alpha}$ such that, for every $s\in[0,r]$,*

$$\mathsf{E}\{\|\chi_\Lambda R_1 \dots R_s f R_{s+1} \dots R_r \chi_\Lambda\|_1\}$$
$$\leqslant c_{\varepsilon,\alpha} |z|^{d/2-p} V \mathsf{E}^{1/t_1}\{|f(0)|^{t_1}\} \sup_{1\leqslant j\leqslant r} \mathsf{E}^{p/t_2}\{(|g_j(0)|+1)^{t_2}\}.$$

Proof. Part (i) follows easily from Lemma 5.9, inequality (5.6), where $\chi_\Lambda g_1$ and $g_{p+1}\chi_\Lambda$ play the roles of a_1 and a_{p+1}, and the Hölder inequality for mathematical expectations.

The proof of (ii) is carried out by successively applying (5.3a) to the operators $R_1, \dots, R_s$ and (5.3b) to $R_{s+1}, \dots, R_r$, until the number of R_0's is equal to p. In view of 5.2(iii), it is necessary to estimate expressions of the form

$$\mathsf{E}\{\|\chi_\Lambda B_1 a_1 \tilde{R}_{0,1} \dots \tilde{R}_{0,p} a_{p+1} B_2 \chi \Lambda\|_1\},$$

where B_1 and B_2 are products of resolvents taken from among $R_1, \dots, R_s$ and $R_{s+1}, \dots, R_r$, respectively; $\tilde{R}_{0,j}$ is taken from among $R_{0,1}, \dots, R_{0,r}$; and a_j takes values from among $1, g_1, \dots, g_r, f$, with f appearing exactly once. The same argument used in the proof of (5.6) gives

$$\mathsf{E}\{\|\chi_\Lambda B_1 a_1 \tilde{R}_{0,1} \dots \tilde{R}_{0,p} a_{p+1} B_2 \chi \Lambda\|_1\}$$
$$\leqslant \mathsf{E}^{1/2}\{\operatorname{Tr} \chi_\Lambda B_1 a_1 \tilde{R}_{0,1} \dots \tilde{R}_{0,m} |a_{m+1}| \tilde{R}^*_{0,m} \dots \tilde{R}^*_{0,1} a_1 B_1^* \chi_\Lambda\}$$
$$\times \mathsf{E}^{1/2}\{\operatorname{Tr} \chi_\Lambda B_2^* a_{p+1} \tilde{R}^*_{0,p} \dots \tilde{R}^*_{0,m+1} |a_{m+1}| \tilde{R}_{0,m+1} \dots \tilde{R}_{0,p} a_{p+1} B_2 \chi_\Lambda\}.$$

Applying Lemma 5.10(ii) to the right-hand side, and taking into account that $\|R_j\| \leqslant |\operatorname{Im} z|^{-1}$ and part (i) of the lemma, we obtain the desired estimate. □

(5.12) Lemma. *Take $z \in \Gamma_{\varepsilon,\alpha}$ and complex numbers z_j, for $1 \leqslant j \leqslant p$, each equal to either z or $\bar{z}$, and let $R_{n,j} = (\Delta + q_n - z_j)^{-1}$. If $\mathsf{E}\{|q(0)|\}^{p+1} < \infty$, there exist positive constants $c^{(u)}_{\varepsilon,\alpha}$, for $u = 1, 2, 3$, such that:*

(i) *for any s such that $0 \leqslant s \leqslant p-1$,*

$$\mathsf{E}\{\|\chi_\Lambda (\prod_{j=1}^{p-s} R_{n,j} - \prod_{j=1}^{p-s} R_{l,j}) \prod_{j=p-s+1}^{p} R_{0,j}\chi_\Lambda\|_1\}$$
$$\leqslant c^{(1)}_{\varepsilon,\alpha} |z|^{d/2-p} V \mathsf{E}^{1/(p+1)}\{|q_n(0) - q_l(0)|^{p+1}\}$$

(ii) $\mathsf{E}\{\|\chi_\Lambda \prod_{j=1}^p R_{n,j}\chi_\Lambda\|_1\} \leqslant c^{(2)}_{\varepsilon,\alpha} |z|^{d/2-p} V$;

(iii) $\mathsf{E}\{\|(R_n - R_l) R_0^p \chi_\Lambda\|_2^2\} \leqslant c^{(3)}_{\varepsilon,\alpha} |z|^{d/2-p} V \mathsf{E}^{1/(p+1)}\{|q_n(0) - q_l(0)|^{p+1}\}$.

(iv) *If, in addition, $\mathsf{E}\{|q(0)|\}^{p+2} < \infty$, there exists $c^{(4)}_{\varepsilon,\alpha}$ such that*

$$\mathsf{E}\{\|(q - q_n) R_n R_0^p \chi_\Lambda\|_2^2\} \leqslant c^{(4)}_{\varepsilon,\alpha} |z|^{d/2-p} V \mathsf{E}^{1/(p+2)}\{|q(0) - q_n(0)|^{p+2}\}.$$

Proof. Part (i) follows from (5.4) and Lemma 5.11(ii) by setting $t_1 = t_2 = p+1$ and using the fact that $\mathsf{E}\{|q(0)|\}^{p+1} < \infty$. Part (ii) also follows from Lemma 5.11(ii) with $f \equiv 1$ and $t_1 = t_2 = p+1$. To prove (iii) and (iv), we notice that $\|A\|_2^2 = \|A^*A\|_1$ and use again Lemma 5.11(ii): in the case of (iii), with $f = q_n - q_l$ and $t_1 = t_2 = p+1$, and using the fact that

$$(R_n - R_l)^*(R_n - R_l) = -R_n^* R_n(q_n - q_l)R_l + R_l^* R_n(q_n - q_l)R_l;$$

and in the case of (iv), with $f = (q - q_n)^2$, $t_1 = (p+2)/2$ and $t_2 = p+2$. □

(5.13) Lemma. *If* $\mathsf{E}\{|q(0)|\}^t < \infty$ *for some* $t \geqslant 1$, *we have*

$$\lim_{n\to\infty} \mathsf{E}\{|q(0) - q_n(0)|^t\} = 0.$$

Proof. Let C_1 and C_2 be cubes centered at the origin and having sides 1 and 2, respectively. Since q is a metrically transitive field,

$$\mathsf{E}\{|q(0) - q_n(0)|^t\} = \mathsf{E}\left\{\int_{C_1} |q(x) - q_n(x)|^t\, dx\right\}.$$

For any $f \in L_+^{\mathrm{loc}}(\mathbf{R}^d)$ Young's inequality [Reed and Simon II] implies that $|\chi_{C_1}(\rho_n * f)|_t \leqslant |\chi_{C_2} f|_t$, for all n. Therefore the integral $\int_{C_1} |q(x) - q_n(x)|^t\, dx$ has the P-integrable upper bound $2^t |\chi_{C_2} q|_t^t$; but this tends to zero as n increases, which proves the lemma. □

Proof of Theorem 5.8. We shall use Proposition 5.6. From the condition $\mathsf{E}\{|q(0)|^{p+2}\} < \infty$ it follows that, with probability 1, $q \in L_{p+1}(\Lambda)$ for any cube Λ. Hence, using Young's inequality (see the proof of Lemma 5.13), we find $\lim_{n\to\infty} |\chi_\Lambda(q - q_n)|_{p+1} = 0$. Therefore, with probability 1, we have s-$\lim_{n\to\infty}(-\Delta + q_n)\phi = (-\Delta + q)\phi$ for all $\phi \in C_0^\infty(\mathbf{R}^d)$. This means that the second condition in Proposition 5.6 is satisfied.

We use Lemma 5.12(iii) to choose a sequence of integers $\{n_i\}$ converging to ∞ and such that, with probability 1, $\lim_{i,j\to\infty} \|(R_{n_i} - R_{n_j})R_0^p \chi_\Lambda\|_2^2 = 0$ for any cube Λ. For an arbitrary $\phi \in C_0^\infty(\mathbf{R}^d)$, we choose Λ so that $\chi_\Lambda \phi = \pi$. Then

$$\lim_{i,j\to\infty} \|(R_{n_i} - R_{n_j})R_0^p \chi_\Lambda\| \leqslant \|\phi\| \lim_{i,j\to\infty} \|(R_{n_i} - R_{n_j})R_0^p \chi_\Lambda\|_2 = 0.$$

Thus, the limit s-$\lim_{i\to\infty} R_{n_i} R_0^p \phi$ exists, with probability 1, for every $\phi \in C_0^\infty(\mathbf{R}^d)$. Since $R_0^p C_0^\infty(\mathbf{R}^d)$ is dense in $L_2(\mathbf{R}^d)$, this implies that the first condition in Proposition 5.6 is also satisfied, with probability 1, for the subsequence $-\Delta + q_{n_i}$. Applying the proposition, we conclude that this subsequence converges to a self-adjoint operator which, like the prelimit operators, is metrically transitive. This proves the first part of the theorem.

To prove the second part, the self-adjointness criterion, we use Lemma 5.12(iv). Choose a sequence $\{n_i\}$ converging to ∞ and such that, with probability 1, $\lim_{i\to\infty} \|(q - q_{n_i})R_{n_i}R_0^p\chi_\Lambda\|_2^2 = 0$ for any Λ, and then consider only those realizations of q for which the limit is true. To prove that $-\Delta + q$ is essentially self-adjoint, it is enough to show that $(-\Delta + q - z)C_0^\infty(\mathbf{R}^d)$ is dense in $L^2(\mathbf{R}^d)$ if $\operatorname{Im} z \neq 0$, or, equivalently, that $R_0^p\phi$ is a cluster point of $(-\Delta + q - z)C_0^\infty(\mathbf{R}^d)$ for any $\phi \in C_0^\infty(\mathbf{R}^d)$. Fix such a ϕ and define $\tilde{\psi}_{n_i} = R_{n_i}R_0^p\phi$, so that $(-\Delta + q_{n_i} - z)\tilde{\psi}_{n_i} = R_0^p\phi$. We have $\tilde{\psi}_{n_i} \in C_0^\infty(\mathbf{R}^d)$ [Berezanskii 1968]. Choose nonnegative functions $s_{n_i} \in C_0^\infty(\mathbf{R}^d)$ so that $s_{n_i}(x) \leqslant 1$ and $\|(-\Delta + q_{n_i} - z)s_{n_i}\tilde{\psi}_{n_i} - R_0^p\phi\| \leqslant n_i^{-1}$. Consider the action of $-\Delta + q - z$ on the sequence of functions $\psi_{n_i} = s_{n_i}\tilde{\psi}_{n_i}$:

$$(-\Delta + q - z)\psi_{n_i} = (q - q_{n_i})s_{n_i}\tilde{\psi}_{n_i} + (-\Delta + q_{n_i} - z)s_{n_i}\tilde{\psi}_{n_i}.$$

If we choose Λ so that $\chi_\Lambda\phi = \phi$, we have

$$\begin{aligned}\|(-\Delta + q - z)\psi_{n_i} - R_0^p\phi\| &\leqslant n_i^{-1} + \|(q - q_{n_i})s_{n_i}R_{n_i}R_0^p\phi\| \\ &\leqslant n_i^{-1} + \||q - q_{n_i}|\,R_{n_i}R_0^p\chi_\Lambda\|_2\,\|\phi\|,\end{aligned}$$

and this last expression clearly tends to zero as i increases. □

The following fact will be needed below.

(5.14) Lemma. *If q is as in part (i) of Theorem 5.8, parts (i) and (ii) of Lemma 5.12 remain valid if we replace q_n and $R_{n,j}$ by q and $R_j = (-\Delta + q - z_j)$*

Proof. Analogous to the proof 5.7. □

We now turn to another criterion for essential self-adjointness, which complements in a certain sense the criterion of Theorem 5.8. We say that a random field has a finite correlation radius ρ if the σ-algebras generated by the field at points separated by more than ρ are independent.

(5.15) Theorem [Grenkova 1981]. *If $d \geqslant 2$ and $q(x)$, for $x \in \mathbf{R}^d$, is a continuous, metrically transitive random field with finite correlation radius, the Schrödinger operator $-\Delta + q$ is essentially self-adjoint on $C_0^\infty(\mathbf{R}^d)$ with probability 1.*

Proof. This will follow immediately from Proposition 5.16, an important criterion for essential self-adjointness of Schrödinger operators whose proof will not be given here, and from Lemma 5.17. □

(5.16) Proposition [Eastham et al. 1976]. *Let $d \geqslant 2$ and $q(x)$, for $x \in \mathbf{R}^d$, be a continuous real-valued function. Assume there exist real numbers d_n and*

$Q_n \geqslant 1$ and compact sets Γ_n^+ and Γ_n^-, for $n \in \mathbf{Z}_+$, that satisfy the following properties:

(a) *$\Gamma_n^- \subset \Gamma_n^+$ and $\Gamma_n^+ \subset \Gamma_{n+1}^-$, for $n \in \mathbf{Z}_+$, and $\bigcup_{n=1}^{\infty} \Gamma_n^+ = \mathbf{R}^d$;*
(b) *Γ_n^+ and Γ_n^- have smooth boundary and $\min_{x \in \partial\Gamma_n^-, y \in \partial\Gamma_n^+} |x - y| = d_n$, for $n \in \mathbf{Z}_+$;*
(c) *$q(x) \geqslant -Q_n$ for $x \in \Gamma_n^+ \setminus \Gamma_n^-$.*

If, in addition, $\sum_n d_n(Q_n + d_n^{-2})^{-1/2} = \infty$, the operator $-\Delta + q$ is essentially self-adjoint on $C_0^\infty(\mathbf{R}^d)$. □

(5.17) Lemma. *Let $d \geqslant 2$ and $q(x)$, for $x \in \mathbf{R}^d$, be a metrically transitive field with a finite correlation radius. If b_0 is sufficiently large, there exist for any $b > b_0$, with probability 1, compact sets $\Gamma_n^\pm$ such that:*

(i) *conditions (a) and (b) of Proposition 5.16 are satisfied with $d_n \geqslant 1$, and*
(ii) *$q(x) \geqslant -b$ in the layer $S_n = \Gamma_n^+ \setminus \Gamma_n^-$.*

Proof. The proof is based on the ideas in [Molchanov and Stepanov 1983], which studies the structure of sufficiently high peaks of homogeneous random fields. It goes basically like this: one chooses an increasing sequence of cubes $\Lambda_n^\pm$, for $n \in \mathbf{Z}_+$, so that the layers $\Lambda_n^+ \setminus \Lambda_n^-$ are thick enough. Then one considers the sequence of events Ω_n consisting of the existence of compact sets Γ_n^+ and Γ_n^- satisfying the conditions of Proposition 5.16 for $d_n = 1$ and $Q_n = b$, where b is big enough. In view of the Borel–Cantelli lemma, the problem is reduced to proving the convergence of the series $\sum_n \mathsf{P}\{\Omega_n^c\}$, which requires estimating the probability of the event Ω_n^c. If $\mathbf{R}^d$ is divided into congruent cubes, the field $q(x)$ can be "transformed" in a natural way into a metrically transitive field $\tilde{q}(n)$, for $n \in \mathbf{Z}^d$, with a finite correlation radius. In terms of this field, we can define a related sequence of events $\tilde{\Omega}_n \supset \Omega_n^c$, whose probability is easier to estimate. If $\Lambda_n^+ \setminus \Lambda_n^-$ is thick enough in comparison with the correlation radius of $\tilde{q}(n)$, the series $\sum_n \mathsf{P}\{\tilde{\Omega}_n\}$, and hence also the series $\sum_n \mathsf{P}\{\Omega_n^c\}$, converge. □

5.B Ergodic Properties

We consider again a metrically transitive Schrödinger operator $H = -\Delta + q(x)$, for $x \in \mathbf{R}^d$, and the associated family of operators $H_{\Lambda,a} = (-\Delta + q)_{\Lambda,a}$ defined in 5.A. We denote by $N_{\Lambda,a}(\lambda)$ the normalized distribution function of the eigenvalues of $H_{\Lambda,a}$, that is, V^{-1} times the number of eigenvalues of $H_{\Lambda,a}$ not exceeding λ, and by $N_{\Lambda,a}(d\lambda)$ the corresponding measure on $\mathbf{R}$.

Here again, we will be talking about weak convergence when we write limits of measures.

We start with results for bounded, smooth potentials $q(x)$, then extend them to more general cases by taking limits.

(5.18) Theorem [Pastur 1971a]. *Assume the realizations of the metrically transitive field $q(x)$, for $x \in \mathbf{R}^d$, are bounded, smooth functions on $\mathbf{R}^d$, and let a be as in equation* (5.1). *There exists a nonrandom measure $N(d\lambda)$ independent of a and such that, with probability* 1,

$$\lim_{V\to\infty} N_{\Lambda,a}(d\lambda) = N(d\lambda) \tag{5.10}$$

and

$$N(d\lambda) = \mathsf{E}\{E_H(d\lambda; 0, 0)\}, \tag{5.11}$$

where $E_H(d\lambda; x, y)$ is the kernel of the resolution of the identity $E_H(d\lambda)$ of the operator $H = -\Delta + q$.

(5.19) Proposition. *Suppose the positive measures $N_n(d\lambda)$, for $n \in \mathbf{Z}_+$, have finite Laplace transform $\tilde{N}_n(t) = \int_0^\infty e^{-\lambda t} N_n(d\lambda)$ for all $t > 0$, and that the limit $\lim_{n\to\infty} \tilde{N}_n(t) = \tilde{N}(t)$ exists for all $t > 0$. There exists a nonnegative measure $N(d\lambda)$ such that*

$$\lim_{n\to\infty} N_n(d\lambda) = N(d\lambda)$$

and

$$\tilde{N}(t) = \int_0^\infty e^{-\lambda t} N(d\lambda).$$

Proof. This is a simple generalization of the well-known continuity theorem of probability theorem [Feller 1966] for Laplace transforms of the distribution functions of nonnegative random variables. □

Proof of Theorem 5.18. Notice that, because the realizations of $q(x)$ are smooth and bounded, the spectrum of $H_{\Lambda,a}$ is discrete; therefore the functions $N_{\Lambda,a}(\lambda)$ are well-defined. The right-hand side of (5.11) is also well-defined, in view of the measurability of the resolution of the identity $E_H(d\lambda)$ (Corollary 2.9) and of the continuity in x and y of the kernel $E_H(d\lambda; x, y)$ [Berezanskii 1968]. The metric transitivity of $q(x)$ implies that $\inf_{x\in\mathbf{R}^d} q(x)$ is invariant, and therefore nonrandom, by Theorem 1.10. We may as well assume this infimum to be zero, by subtracting if from $q(x)$, so that the measures $N_{\Lambda,a}$ are concentrated on the positive semiaxis $[0,\infty)$.

We consider first the case where the parameter a in (5.1) is zero, that is, the Dirichlet problem. By Proposition 5.19, equations (5.10) and (5.11) will be proved if we can show that, for any $t \geqslant 0$, we have, with probability 1,

$$\lim_{V\to\infty} \tilde{N}_{\Lambda,0}(t) = \int_0^\infty e^{-\lambda t} \mathsf{E}\{E_H(d\lambda; 0, 0)\} \tag{5.12}$$

Write $\tilde{N}_{\Lambda,0}(t)$ in the form

(5.13) $$\tilde{N}_{\Lambda,0}(t) = \frac{1}{V}\operatorname{Tr}\exp(-t(-\Delta+q)_{\Lambda,0}) = \frac{1}{V}\int_{\Lambda} K_{\Lambda,0}(t;x,x)\,dx,$$

where $K_{\Lambda,0}(t;x,y)$ is the Green's function of the parabolic equation

$$\frac{\partial}{\partial t}K_{\Lambda,0} = \Delta_x K_{\Lambda,0} - qK_{\Lambda,0}, \quad K_{\Lambda,0}\big|_{t=0} = \delta(x-y), \quad K_{\Lambda,0}\big|_{\partial\Lambda} = 0.$$

For this function, which is the kernel of the semigroup $\exp(-tH_{\Lambda,0})$, we have the following *Feynman-Kac* representation [Reed and Simon II]:
(5.14)
$$K_{\Lambda,0}(t;x,y) = k(t;x-y)\mathsf{W}\{e^{-\int_0^t q(x(s)+x)\,ds}\chi_{\Lambda,t}(x(\cdot)+x) \mid x(t)=y-x\},$$

where $\chi_{\Lambda,t}(f)$ is 1 if $f([0,t]) \subset \Lambda$ and 0 if not, $k(t;x) = (4\pi t)^{-d/2}e^{-x^2/(4t)}$, and $\mathsf{W}\{\cdot \mid x(t)=x\}$ denotes integration with respect to the measure concentrated on the trajectories of the Wiener process that satisfies the conditions $x(0)=0$, $x(t)=x$ (see Example 1.15(b)).

Plugging (5.14) into (5.13) we get
(5.15)
$$\tilde{N}_{\Lambda,0}(t) = \frac{1}{(4\pi t)^{d/2}V}\int_{\Lambda} dx\,\mathsf{W}\{E^{-\int_0^t q(x(s)+x)ds}\chi_{\Lambda,t}(x(\cdot)+x) \mid x(t)=0\}.$$

On the other hand, the function

$$R(x) = \mathsf{W}\{e^{-\int_0^t q(x(s)+x)ds} \mid x(t)=0\}$$

is easily seen to be a metrically transitive random field. So if it were not for the $\chi_{\Lambda,t}$ in (5.15), which accounts for the zero boundary conditions, the convergence of (5.15) with probability 1 to the nonrandom limit $(4\pi t)^{-d/2}\mathsf{E}\{R(0)\}$ would follow simply from the ergodic theorem (1.22). Thus, we have to make sure that

$$\lim_{V\to\infty}\frac{1}{V}\int_{\Lambda} dx\,\mathsf{W}\{e^{-\int_0^t q(x(s)+x)ds}(1-\chi_{\Lambda,t}(x(\cdot)+x)) \mid x(t)=0\} = 0.$$

Since $q \geqslant 0$, it is sufficient to verify that, for each fixed trajectory,

$$\lim_{V\to\infty}\frac{1}{V}\int_{\Lambda}\chi_{\Lambda,t}(x(\cdot)+x) = 1,$$

which can be done directly. Consequently, with probability 1, there exists the limit

(5.16) $$\lim_{V\to\infty}\tilde{N}_{\Lambda,0}(t) = \tilde{N}(t)$$

and

$$(5.17)\quad \tilde{N}(t) = (4\pi t)^{-d/2}\mathsf{E}\{\mathsf{W}\{e^{-\int_0^\infty q(x(s))\,ds} \mid x(t)=0\}\} = \mathsf{E}\{K(t;0,0)\},$$

where $K(t;x,y)$ is the fundamental solution of the parabolic equation

$$\frac{\partial K}{\partial t} = \Delta K - q(x)K, \qquad K\big|_{t=0} = \delta(x-y).$$

In writing equality (5.17), we again used the Feynman-Kac formula, this time for $K(t;x,y)$. Since this function is the kernel of the semigroup e^{-tH}, we can use the spectral theorem for differential operators with smooth coefficients [Berezanskii 1968] to rewrite (5.17) in the form

$$\tilde{N}(t) = \int_0^\infty e^{-\lambda t}\mathsf{E}\{E_H(d\lambda;0,0)\}.$$

From this equation, equation (5.16) and Proposition 5.19, (5.10) and (5.11) follow for $a=0$.

Now consider the case of a arbitrary, and $q \geqslant 0$ as before. Denote by $K_{\Lambda,a}(t;x,y)$, for $t>0$ and $x,y\in\Lambda$, the Green's function for the parabolic equation

$$\frac{\partial K_{\Lambda,a}}{\partial t} = \Delta_x K_{\Lambda,a} - qK_{\Lambda,a}, \qquad K_{\Lambda,a}\big|_{t=0} = \delta(x-y),$$
$$(1-a)K_{\Lambda,a} + a\frac{\partial}{\partial \nu_x}K_{\Lambda,a}\big|_{\partial\Lambda} = 0,$$

and by $k_{\Lambda,a}(t;x,y)$ the Green's function for the same equation with $q=0$. These functions are nonnegative, symmetric with respect to x and y, and satisfy the following inequalities, which may be obtained either from Green's formulas or from the maximum principle [Friedman 1964]:

$$(5.18)\quad \begin{aligned} K_{\Lambda,1}(t;x,y) \geqslant K_{\Lambda,a}(t;x,y) \geqslant K_{\Lambda,0}(t;x,y),\\ k_{\Lambda,1}(t;x,y) \geqslant k_{\Lambda,a}(t;x,y) \geqslant k_{\Lambda,0}(t;x,y).\end{aligned}$$

These properties also have a simple probabilistic interpretation. Indeed, $K_{\Lambda,a}(t;x,y)$, for $q\geqslant 0$, is the transition probability of the d-dimensional Markov process representing Brownian motion of a particle inside a volume Λ, under the condition that at its boundary the particle will be absorbed with probability $1-a$ and reflected with probability a, and inside the volume the probability of its absorption in the vicinity of a point x is proportional to $q(x)\,dx$: see [Gihman and Skorohod III]. It is clear, therefore, that any such probability is nonnegative and nondecreasing as $a(x)$ decreases at each point of the boundary. From the same considerations we get $K_{\Lambda,a}(t;x,y) \leqslant k_{\Lambda,a}(t;x,y)$. But then, from (5.18) and the integral equation

$$K_{\Lambda,a}(t) = k_{\Lambda,a}(t) - \int_0^t k_{\Lambda,a}(t-\tau)q(\tau)K_{\Lambda,a}(\tau)\,d\tau$$

(which also implies $K_{\Lambda,0} \leqslant k_{\Lambda,0}$), we immediately get the relation

$$(5.19) \qquad 0 \leqslant K_{\Lambda,1}(t;x,y) - K_{\Lambda,0}(t;x,y) \leqslant k_{\Lambda,1}(t;x,y) - k_{\Lambda,0}(t;x,y).$$

To finish the proof there remains only to check that

$$\lim_{V\to\infty} \frac{1}{V} \int_\Lambda (k_{\Lambda,1}(t;x,x) - k_{\Lambda,0}(t;x,x))\,dx,$$

which can be done by explicit calculation, since $k_{\Lambda,1}$ and $k_{\Lambda,0}$ are known for the cube [Feller 1966]. □

The method we have used, based on an analysis of the properties of the kernel of the semigroup generated by the Schrödinger operator, can be applied with some modifications to the more general case of a random potential unbounded from below, provided that $\mathsf{E}\{e^{-tq(0)}\} < \infty$ for all $t > 0$. The same method may also be used to prove ergodic properties (i.e., the existence of nonrandom limits as $V \to \infty$) of random quantities other than $N_{\Lambda,a}(\lambda)$, constructed from the eigenvalues and eigenfunctions of the operator $(-\Delta + q)_{\Lambda,a}$, and to prove similar statements for elliptic operators of arbitrary order with random metrically transitive coefficients (see Theorem 5.23).

If, however, we assume the finiteness of only power-law, rather than exponential, moments of the negative part of the random potential, we must work with a characteristic set consisting not of functions $e^{-t\lambda}$, but of powers of $(\lambda - z)$, for $\operatorname{Im} z \neq 0$. In other words, we must use not the semigroup, but the resolvent of the operator. We have already used this technique in the first part of this section, in the analysis of essential self-adjointness. With this modification, the following result can be proved. (For related results, see [Kirsch and Martinelli 1982c].)

(5.20) Theorem [Figotin 1983]. *Let p be the smallest even number strictly greater than $d/2$, and suppose that $\mathsf{E}\{|q(0)|^r\} < \infty$, where $r \geqslant p+1$. There exists a nonrandom measure $N(d\lambda)$, independent of the function a in* (5.1), *such that, with probability* 1,

$$(5.20) \qquad \lim_{V\to\infty} N_{\Lambda,a}(d\lambda) = N(d\lambda)$$

and

$$(5.21) \qquad N(d\lambda) = \mathsf{E}\{\operatorname{Tr} \chi_C E_H(d\lambda) \chi_C\},$$

where $E_H(d\lambda)$ is the resolution of the identity of the self-adjoint operator $H = -\Delta + q$, C is the cube of volume 1 and centered at the origin, and χ_C is the operator of multiplication by the characteristic function of C.

Note that, by Theorem 5.1, the operators $H_{\Lambda,a} = (-\Delta + q)_{\Lambda,a}$ are, with probability 1, essentially self-adjoint and have a compact resolvent, so the functions $N_{\Lambda,a}(\lambda)$ are well-defined.

(5.21) Lemma. *Let $u < p$ be a positive number and $m(d\lambda)$ a positive measure on* $\mathbf{R}$.

(i) *If $\int_{\mathbf{R}} |\lambda - i\tau|^{-p} m(d\lambda) \leqslant c\tau^{u-p}$ for $\tau \geqslant 1$, we have $m([-\lambda, \lambda]) \leqslant 2^{p/2} c\lambda^u$ for $\lambda \geqslant 1$ and $\int_{|\lambda| \geqslant \alpha} |\lambda - z|^{-p} m(d\lambda) \leqslant 4^p c\alpha^{u-p}$ for $\alpha \geqslant \max\{2|z|, 1\}$;*

(ii) *If $\int_{\mathbf{R}} \big((\lambda+t)^2 + 1\big)^{-r/2} m(d\lambda) \leqslant ct^{u-r}$ for $\tau \geqslant 1$, we have $m\big((-\infty, \lambda]\big) \leqslant c_1 \lambda^{u-r+1}$ for $\lambda \geqslant 1$, where c_1 depends on c, u and r, but not on m.*

Proof. From the conditions in (i) we have

$$m\big([-\lambda, \lambda]\big)(\lambda^2 + \tau^2)^{p/2} \leqslant c\tau^{u-p},$$

whence the estimate in the case $\tau = \lambda$. For arbitrary $\tau \geqslant 1$, we have

$$\begin{aligned}
\int_{|\lambda| \geqslant \alpha} |\lambda - z|^{-p} m(d\lambda) &= \int_{|\lambda| \geqslant \alpha} \big|(\lambda - i\tau)(\lambda - z)^{-1}\big|^p |\lambda - i\tau|^{-p} m(d\lambda) \\
&\leqslant \sup_{|\lambda| \geqslant \alpha} \big\{(1 + |z - i\tau|/|\lambda - z|)^p\big\} c\tau^{u-p} \\
&\leqslant \left(1 + \frac{|z| + \tau}{\alpha - |z|}\right)^p c\tau^{u-p} \leqslant \left(\frac{1 + \tau + \alpha/2}{\alpha/2}\right)^p c\tau^{u-p}.
\end{aligned}$$

Setting $\tau = \alpha$ in this latter expression, we get the second inequality in (i).

As for part (ii), its conditions imply that

$$ct^{u-r} \geqslant \int_{-\infty}^{0} \big((\mu + t)^2 + 1\big)^{-r/2} m(d\mu).$$

By integrating this inequality with respect to t over $[\lambda, \infty)$, we get

$$\begin{aligned}
c(u - r + 1)^{-1} \lambda^{u-r+1} &\geqslant \int_{-\infty}^{0} \int_{\lambda}^{\infty} \big((\mu + t)^2 + 1\big)^{-r/2} dt \\
&\geqslant \int_{-\infty}^{-\lambda} m(d\mu) \int_{\lambda+\mu}^{\infty} (t^2 + 1)^{-r/2} dt \\
&\geqslant ((-\infty, -\lambda)) \int_{0}^{\infty} (t^2 + 1)^{-r/2} dt.
\end{aligned}$$

□

(5.22) Lemma. (i) *If $m_1(d\lambda)$ and $m_2(d\lambda)$ are positive measures on* $\mathbf{R}$ *satisfying the condition of Lemma* 5.21(*i*), *and*

$$\int |\lambda - z|^{-p} m_1(d\lambda) = \int |\lambda - z|^{-p} m_2(d\lambda)$$

for every z with $\operatorname{Im} z \neq 0$, *the two measures are equal.*

(ii) *If $m(d\lambda)$ and $m_n(d\lambda)$, for $n \in \mathbf{Z}_+$, are positive measures on* $\mathbf{R}$ *satisfying the conditions of Lemma* 5.21(*i*) *with a common constant, and*

$$\lim_{n\to\infty} \int |\lambda - z|^{-p} m_n(d\lambda) = \int |\lambda - z|^{-p} m(d\lambda)$$

for every z with $\operatorname{Im} z \neq 0$, *the sequence $m_n(d\lambda)$ converges to $m(d\lambda)$ as $n \to \infty$.*

Proof. If m_1 and m_2 have continuous densities $m_1'(\lambda)$ and $m_2'(\lambda)$ with respect to Lebesgue measure, an application of Lemma 5.21(i) in the limit $\operatorname{Im} z \to 0$ leads to $m_1'(\operatorname{Re} z)$ and $m_2'(\operatorname{Re} z)$. This in turn gives $m_1'(\lambda) = m_2'(\lambda)$, that is, $m_1 = m_2$, since z is arbitrary. The case of m_1 and m_2 arbitrary can be easily reduced to the above.

Lemma 5.21(i) also implies that the sequence $\{m_n : n \in \mathbf{Z}_+\}$, is precompact. But if a measure $\tilde{m}(d\lambda)$ is a cluster point for this sequence, m and $\tilde{m}$ satisfy the conditions of part (i), so that $m = \tilde{m}$. □

Proof of Theorem 5.20. We will consider, along with the essentially self-adjoint operators $H_{\Lambda,a} = (-\Delta + q)_{\Lambda,a}$ and $H = -\Delta + q$, the essentially self-adjoint operators $H_{\Lambda,a}^{(n)} = (-\Delta + q_n)_{\Lambda,a}$ and $H^{(n)} = -\Delta + q_n$ with potentials $q_n(x)$ given by (5.7), and their respective measures $N_{\Lambda,a}^{(n)}(d\lambda)$ and $N^{(n)}(d\lambda)$. In this case, by Theorem 5.18, the limiting measure $N(H^{(n)}, d\lambda)$ exists, is nonrandom and, in view of equations (5.10) and (5.11),

$$N(H^{(n)}, d\lambda) = \mathsf{E}\{E_{H^{(n)}}(d\lambda; 0, 0)\} = \mathsf{E}\{\operatorname{Tr} \chi_C E_{H^{(n)}}(d\lambda) \chi_C\}. \tag{5.22}$$

We define the measure $N(d\lambda)$ by relation (5.21) and set

$$r^{(n)}(z) = \int |\lambda - z|^{-p} N(H^{(n)}, d\lambda), \quad r(z) = \int |\lambda - z|^{-p} N(H, d\lambda). \tag{5.23}$$

We show that, if $p = 2m$ and $\operatorname{Im} z \neq 0$, we have $\lim_{n\to\infty} r^{(n)}(z) = r(z)$. Indeed, according to (5.21), (5.22) and (5.23),

$$r^{(n)}(z) = \{\|\chi_C R_n^m (R_n^*)^m \chi_C\|_1\}, \quad r(z) = \{\|\chi_C R^m (R^*)^m \chi_C\|_1\}, \tag{5.24}$$

where $R_n = (H^{(n)} - zI)^{-1}$ and $R = (H - zI)^{-1}$. This gives the desired limit by (5.24), Lemma 5.12, and parts (i) and (ii) of Lemma 5.14, with $s = 0$, $z_j = z$ for $1 \leqslant j \leqslant m$ and $z_j = \bar{z}$ for $m + 1 \leqslant j \leqslant p$.

To prove (5.20), we introduce the functions $r_{\Lambda,a}^{(n)}(z)$ and $r_{\Lambda,a}(z)$, for $\operatorname{Im} z \neq 0$. Their definition is the same as that of $r^{(n)}(z)$ and $r(z)$ in equation (5.23), but with $N(H^{(n)}, d\lambda)$ and $N(H, d\lambda)$ replaced by $N_{\Lambda,a}^{(n)}(d\lambda)$ and $N_{\Lambda,a}(d\lambda)$. The following relations, similar to (5.24), are obvious:

(5.25)
$$r_{\Lambda,a}^{(n)}(z) = 1/V\big\|(R_{\Lambda,a}^{(n)})^m((R_{\Lambda,a}^{(n)})^*)^m\big\|_1, \qquad r_{\Lambda,a}(z) = 1/V\big\|R_{\Lambda,a}^m(R_{\Lambda,a}^*)^m\big\|_1.$$

From Lemmas 5.4 and 5.7(i), for $s = 0$, we have

$$\begin{aligned}(5.26)\quad & \big|r_{\Lambda,a}^{(n)}(z) - r_{\Lambda,a}(z)\big| \\ & \leqslant c\left(\frac{1}{V}\int_\lambda |q(x) - q_n(x)|^{p+1}dx\right)^{\frac{1}{p+1}}\left(\frac{1}{V}\int_\lambda (|q(x)|+1)^{p+1}dx\right)^{\frac{p}{p+1}},\end{aligned}$$

where c depends only on z, for $\operatorname{Im} z \neq 0$. In view of the ergodic theorem 1.13, this inequality yields, with probability 1,

$$\begin{aligned}&\limsup_{V,V_1\to\infty}\big|r_{\Lambda,a}^{(n)}(z) - r_{\Lambda,a}(z)\big| \\ &\qquad \leqslant 2c\mathsf{E}^{1/(p+1)}\{|q(0) - q_n(0)|^{p+1}\}\mathsf{E}^{p/(p+1)}\{(|q(0)|+1)^{p+1}\},\end{aligned}$$

whence, by Lemma 5.13, follows the existence of the limit $\lim_{V\to\infty} r_{\Lambda,a}(z)$. From this and from (5.26), after taking the limits $V \to \infty$ and $n \to \infty$, we have, with probability 1,

$$(5.27)\qquad \lim_{V\to\infty} r_\Lambda(z) = \lim_{n\to\infty} r^{(n)}(z) = r(z),$$

if $\operatorname{Im} z \neq 0$. Here we used the fact that $\lim_{n\to\infty} r_{\Lambda,a}^{(n)}(z) = r^{(n)}(z)$, which follows from the convergence of $N_{\Lambda,a}^{(n)}(d\lambda)$ to $N_{\Lambda,a}(d\lambda)$, Theorem 5.18, and Lemmas 5.4(ii) and 5.21(i).

Now Lemmas 5.4 and 5.7(ii), applied to the measures $N_{\Lambda,a}(d\lambda)$, with $V \geqslant 1$, imply that the conditions of Lemma 5.21(i) are fulfilled with probability 1 with a single common constant c. This, together with (5.27), allows us to apply Lemma 5.22(ii), which leads directly to equation (5.20). □

Remarks. 1. As in the discrete case, a result similar to Theorem 5.18 is valid for more general sequences of expanding regions than the cubes we've been considering. But here we define the convergence of a sequence of bounded regions to infinity in a somewhat different way [Ruelle 1969b]. Given a bounded set $\Lambda \subset \mathbf{R}^d$, let $V(\Lambda)$ be its Lebesgue measure, $d(\Lambda)$ its diameter, Λ_h the set of points within a distance h from the boundary of Λ, and V_h the Lebesgue measure of Λ_h. We say that a sequence $\{\Lambda_i\}$ tends to infinity in the sense of Fisher, and write $\Lambda_i \to \infty$ if $\lim V(\Lambda_i) = \infty$ and for any $\varepsilon > 0$ there exists $\delta > 0$ independent of i and such that $V_{\delta d(\Lambda_i)}/V(\Lambda_i) < \varepsilon$.

This definition is less general than the definition of convergence in the sense of van Hove (see Remark 2 after Theorem 4.11), which would be equivalent to $\lim V(\Lambda_i) = \infty$ and $\lim V_h(\Lambda_i)/V(\Lambda_i) = 0$ for any fixed $h > 0$. However, it is still invariant under linear inhomogeneous transformations of $\mathbf{R}^d$. Since $V_{\delta d(\Lambda)}$ is larger than the volume of a ball of radius $\delta d(\Lambda)$, we see that for a sequence that converges to infinity in the sense of Fisher there exists a constant $c > 0$ such that $V(\Lambda) > c(d(\Lambda))^d$.

2. Results similar to Theorem 5.18 can be obtained also for elliptic difference equations of arbitrary order with metrically transitive, bounded, smooth random coefficients. More exactly, consider the operator of Example 1.4(f):

$$A = \sum_{|\alpha| \leqslant p} i^{|\alpha|} a_\alpha(x) \partial^\alpha, \tag{5.28}$$

with $a_\alpha(x) \in C^\infty(\mathbf{R}^d)$. Assume A is formally self-adjoint and uniformly elliptic, that is, its order p is even and its principal symbol

$$a_p(x, \xi) = \sum_{|\alpha| = p} a_\alpha(x) \xi^\alpha, \tag{5.29}$$

(the highest-degree homogeneous component, considered as a function of an auxiliary variable $\xi \in \mathbf{R}^d$) satisfies $a_p(x, \xi) \geqslant c|\xi|^p$, where $c > 0$ doesn't depend on x and ξ. If the coefficients $a_\alpha(x)$ are smooth and bounded, A is essentially self-adjoint on the set $C^\infty(\mathbf{R}^d)$ of smooth functions with compact support [Berezanskii 1968]. Suppose then that there is a sequence $\{\Lambda\}$ of bounded domains tending to infinity in the sense of Fisher, and that we impose on the boundary of each domain self-adjoint boundary conditions of the from $B_j u\big|_{\partial\Lambda} = 0$, for $j = 1, \ldots, p/2$ (or periodic boundary conditions, if the domains Λ are rectangular parallelepipeds). Suppose also that the boundary operators B_j have order not exceeding $p-1$ and smooth, bounded coefficients.

Under these assumptions, the operator A_Λ defined by (5.28) and the boundary condition $B_j u\big|_{\partial\Lambda} = 0$ is self-adjoint, its domain is the respective Sobolev space, and its spectrum is discrete, bounded from below and has a single limit point $+\infty$. Therefore, as in the case of the Schrödinger operator with conditions (5.1), the distribution function $N_\Lambda(\lambda)$ and the corresponding measures $N_\Lambda(d\lambda)$ can be defined.

(5.23) Theorem [Gusev 1977; Shubin 1979]. *Suppose the operators A and A_Λ are as in the previous paragraphs, and Λ is a domain that tends to infinity in the sense of Fisher. Then, with probability* 1, *there exists a nonrandom limit*

$$\lim_{V \to \infty} N_\Lambda(d\lambda) = N(A, d\lambda) = \mathsf{E}\{E_A(d\lambda; x, x)\},$$

where $E_A(d\lambda; x, x)$ *is the kernel of the resolution of the identity* $E_A(d\lambda)$ *of the operator* A, *from equation* (5.28).

Proof. The proof goes along the same lines as that of Theorem 5.18. One considers the Laplace transform $\tilde{N}_\Lambda(t)$ of $N_\Lambda(d\lambda)$ and expresses it in terms of the trace of the integral operator $K_\Lambda(t; x, y) = (e^{-tA_\Lambda})(x, y)$ which is the Green's function for the respective parabolic boundary problem. But unlike the case of a Schrödinger operator, where the Feynman-Kac formula and the simple estimates (5.18) and (5.19) sufficed to compare the kernels of e^{-tA_Λ} and e^{-tA}, here we must use the theorem of existence and uniqueness of a fundamental solution of the parabolic equation $\partial K/\partial t = -AK$ [Friedman 1964], from which it follows that this solution $K(t; x, y; \omega)$ for every fixed $t > 0$ is a metrically transitive integral operator in the sense of Example 1.17(d), and that

$$K(t; x, y) \leqslant c_1 t^{-d/p} e^{-c_2(|x-y|/t)^{1/(p-1)}} \tag{5.30}$$

[Friedman 1964] and

$$\left|K_\Lambda(t; x, y) - K(t; x, y)\right| \leqslant c_3 t^{-d/p} e^{-c_4((|x-y|+d(y,\partial\Lambda))^p/t)^{1/(p-1)}}, \tag{5.31}$$

[Eidel'man and Ivasishen 1970], where the c_i do not depend on Λ and $d(y, \partial\Lambda)$ is the distance from a point $y \in \Lambda$ to the boundary.

In the case of periodic boundary conditions, where

$$K_\Lambda(t; x, y) = \sum_{z \in \mathbf{Z}_\Lambda} K(t; x+z, y),$$

where $\mathbf{Z}_\Lambda$ is the lattice generation by periodic translations of Λ, (5.31) follows from (5.30).

Given these facts from the theory of parabolic equations, the proof of the relation $\lim_{V\to\infty} \tilde{N}_\Lambda(t) = \mathsf{E}\{K(t; x, x)\}$, which is clearly equivalent to the statement of the theorem, is quite straightforward. Indeed, by splitting the integral

$$\int_\Lambda \left|K_\Lambda(t; x, x) - K(t; x, x)\right| dx$$

into two, with domains $\Lambda_{\delta d(\Lambda)}$ and $\Lambda \setminus \Lambda_{\delta d(\Lambda)}$, and using (5.30) and (5.31), we find that the integral is bounded from above by

$$O\left(e^{-c(\delta d(\Lambda))^{p/(p-1)}}\right) + O\left(V(\Lambda_{\delta d(\Lambda)})\right),$$

which, after division by $V(\Lambda)$, tends to 0 as Λ tends to infinity, as we wished to show. □

The scheme of the proof of Theorems 5.18 and 5.20 may also be used to prove the existence of nonrandom limits of more complicated random quantities constructed from the eigenvalues λ_l and eigenfunctions ψ_l of the operators A_Λ. Noting that

$$N_\Lambda(\lambda) = \frac{1}{V} \sum_{\lambda_l<\lambda} (\psi_l, \psi_l)$$

we see that it is natural to consider quantities like

$$\frac{1}{V} \sum_{\lambda_l<\lambda} (\partial^\alpha \psi_l, \psi_l) \tag{5.32}$$

or

$$\frac{1}{V} \sum_{\lambda_l<\lambda} (f, \psi_l)(\psi_l, g), \tag{5.33}$$

where $f(x)$ and $g(x)$ are metrically transitive almost periodic or random functions. Such quantities appear in theoretical physics [Lifshitz et al. 1982b], where the role of f and g is played, for example, by e^{ikx} for $k \in \mathbf{R}^d$. Finally, we may consider multiple sums over the spectrum of A_Λ, such as

$$\frac{1}{V} \sum_{\substack{\lambda_k<\lambda \\ \lambda_l<\mu}} (\partial^{\alpha_1} \psi_l, \psi_k)(\partial^{\alpha_2} \psi_l, \psi_k). \tag{5.34}$$

Quantities of this kind express important characteristics of solid state systems, such as the conductivity of a disordered system. More general quantities of the same nature can also be considered. In all these cases, the proof of the existence with probability 1 of nonrandom limits for $\Lambda \to \infty$ is made by passing to the Laplace transform (which is two-dimensional in the case of quantities like (5.21)), expressing it in terms of integral of the Green's function $K_\Lambda(t; x, y)$ of the corresponding parabolic problems, and proving the allowability of replacing $K_\Lambda(t; x, y)$ by $K(t; x, y)$ by means of estimates such as (5.18), (5.19), (5.30) and (5.31). The exact statements and detailed proofs can be found in [Pastur 1971a; Gusev 1977; Shubin 1979].

5.C Some Properties of the Integrated Density of States

We now consider some properties of the function $N(\lambda)$ for metrically transitive differential operators, which are the most general and, in a sense, also the simplest. A more detailed study of this function, including finding asymptotic formulas in various regions of the spectrum and for various classes of random operators, will be presented in Chapters III and IV; see also [Lifshitz et al. 1982b].

Some properties of $N(\lambda)$ for differential operators are similar to the corresponding properties for the matrix operators described in Subsection 4.C. For example, it is easy to see that the arguments used in the proofs of Theorems 4.14 and 4.15 also hold in the continuous case, so we have the following counterparts:

(5.24) Theorem. *Let A and B be self-adjoint metrically transitive operators on $L_2(\mathbf{R}^d)$ such that the integrated densities of states $N(A,\lambda)$, $N(B,\lambda)$ and $N(A+B,\lambda)$ exist. Then:*

(i) $N(A,\lambda) \leqslant N(B,\lambda)$ *if* $A \geqslant B$;

(ii) *if* $\beta_+ = \sup_{\|\psi\|=1}(B\psi,\psi)$ *and* $\beta_- = \inf_{\|\psi\|=1}(B\psi,\psi)$ *are the upper and lower bounds of* B, *we have*

$$N(A,\lambda-\beta_+) \leqslant N(A+B,\lambda) \leqslant N(A,\lambda-\beta_-);$$

(iii) *if* $\lim_{\lambda\to\infty} N(A,\lambda+a)/N(A,\lambda) = 1$ *and* B *is a bounded metrically transitive operator,*

$$N(A+B,\lambda) = N(A,\lambda)(1+o(1))$$

as $\lambda \to \infty$. □

Part (iii) of this theorem is more significant than the corresponding statement in Theorem 4.15, because, unlike matrix operators, elliptic differential operators are always unbounded from above, so the study of the asymptotic behavior of their integrated density of states as $\lambda \to \infty$ is less trivial. In the comparatively simple case of a Schrödinger operator with a bounded, metrically transitive potential, this asymptotic behavior is prescribed by Theorem 5.24, including an error estimate. Indeed, since for the Laplace operator the integrated density of states $N_0(\lambda)$, according to (3.10), is

$$N_0(\lambda) = \begin{cases} c_d\lambda^{d/2} & \text{for } \lambda \geqslant 0, \\ 0 & \text{for } \lambda < 0, \end{cases} \tag{5.35}$$

where $c_d^{-1} = (4\pi)^{d/2}\Gamma(d/2+1)$, the condition in (iii) for $\lambda \to \infty$ is evidently fulfilled for this function. Therefore, if the potential $q(x)$ is bounded with probability 1, that is, if $|q(x)| \leqslant C < \infty$, we apply parts (ii) and (iii) of Theorem 5.24, with $A = -\Delta$ and $B = q$, to find the following asymptotic formula for the integrated density of states of $-\Delta + q$, as $\lambda \to \infty$:

$$N(\lambda) = c_d\lambda^{d/2}(1+O(1/\lambda)) \tag{5.36}$$

Later we will discuss in more detail the asymptotics of the integrated density of states of Schrödinger operators and more general elliptic metrically transitive operators as $\lambda \to \infty$. For now we note that for the Schrödinger operator $H = -\Delta + q$, and, more generally, for self-adjoint, metrically transitive elliptic differential operators of second order, many other inequalities in addition to those in Theorem 5.24 can be obtained. They relate $N(\lambda)$ with the prelimit functions $N_{\Lambda,a}(\lambda)$, for $a = 0, 1$ corresponding to the Dirichlet and Neumann problems in the domain Λ, and are based on well-known Courant inequalities for the eigenvalues of these problems [Courant and Hilbert 1968]. These inequalities say that if $H_{\Lambda,a}$ is the operator defined on the domain Λ by $-\Delta+q$ and the boundary conditions (5.1), and $\lambda_n^{(a)}(\Lambda)$ are its eigenvalues, numbered in nondecreasing order, we have

$$\lambda_n^{(1)}(\Lambda) \leqslant \lambda_n^{(a)}(\Lambda) \leqslant \lambda_n^{(0)}(\Lambda). \tag{5.37}$$

Furthermore, if Λ is divided into two disjoint subregions Λ_1 and Λ_2 with union Λ, and the same boundary condition is made to apply at the separating surface as at the outer surfaces, the eigenvalues of the resulting operator in $\Lambda_1 \cup \Lambda_2$ do not decrease for the Dirichlet problem and do not increase for the Neumann problem. This observation is embodied in the so-called *Weyl–Courant inequalities* (see [Titchmarsh 1958]):

$$N_{\Lambda,0}(\lambda) \leqslant N_{\Lambda,a}(\lambda) \leqslant N_{\Lambda,1}(\lambda) \tag{5.38}$$

$$(V_1 + V_2)N_{\Lambda_1\cup\Lambda_2,0}(\lambda) \geqslant V_1 N_{\Lambda_1,0}(\lambda) + V_2 N_{\Lambda_2,0}(\lambda) \tag{5.39a}$$

$$(V_1 + V_2)N_{\Lambda_1\cup\Lambda_2,1}(\lambda) \leqslant V_1 N_{\Lambda_1,1}(\lambda) + V_2 N_{\Lambda_2,1}(\lambda) \tag{5.39b}$$

The last two inequalities imply that $VN_{\Lambda,0}(\lambda)$ is a nondecreasing, and $VN_{\Lambda,1}(\lambda)$ a nonincreasing, function of Λ: if $\Lambda' \subset \Lambda$,

$$VN_{\Lambda,0}(\lambda) \geqslant V'N_{\Lambda',0}(\lambda) \quad \text{and} \quad VN_{\Lambda,1}(\lambda) \leqslant V'N_{\Lambda',1}(\lambda). \tag{5.40}$$

(5.25) Theorem [Pastur 1971b]. *Assume that, for a metrically transitive Schrödinger operator, the integrated density of states $N(\lambda)$ exists and is independent of the form of the boundary conditions of the prelimit problems. Then, for any cube $\Lambda \subset \mathbf{R}^d$,*

$$\mathsf{E}\{N_{\Lambda,0}(\lambda)\} \leqslant N(\lambda) \leqslant \mathsf{E}\{N_{\Lambda,1}(\lambda)\} \tag{5.41}$$

Proof. Divide $\mathbf{R}^d$ into cubes congruent to Λ and consider a sequence of regions tending to infinity, each consisting of a union of such cubes. Then apply (5.39). □

Here is an analogue or Theorem 4.16:

(5.26) Theorem. *Let $H = -\Delta + q(x)$, for $x \in \mathbf{R}^d$, be a metrically transitive Schrödinger operator such that $\mathsf{E}\{|q(0)|^{2+\delta}\} < \infty$ for some $\delta > 0$ and*

$\mathsf{E}\{e^{-tq(0)}\}$ *for all* $t > 0$. *If* $N(H,\lambda) = N_0(\lambda)$, *where* $N_0(\lambda) = c_d\lambda^{d/2}$ *is the integrated density of states* (5.35) *for the operator* $-\Delta$, *we have* $q(x) \equiv 0$ *with probability* 1.

Proof. If $N(H,\lambda) = N_0(\lambda)$ for all λ, the Laplace transforms $\tilde{N}(H,t)$ and $\tilde{N}_0(t)$ are also equal. By (5.17) this is equivalent to

$$\mathsf{E}\{\mathsf{W}\{e^{-\int_0^t q(x(s))\,ds} \mid x(t)=0\}\} = 1. \tag{5.42}$$

Denote the operation $\mathsf{E}\{\mathsf{W}\{\cdot \mid x(t)=0\}\}$ by $\mathsf{M}\{\cdot\}$. Using the identity

$$1 - e^{-a} = a - a^2 \int_0^1\int_0^1 e^{-\tau_1\tau_2 a}\tau_1\,d\tau_1\,d\tau_2,$$

we obtain

$$\mathsf{E}\{q(0)\} = \frac{1}{t}\mathsf{M}\Big\{\Big(\int_0^t q(x(s))\,ds\Big)^2 \int_0^1\int_0^1 e^{-\tau_1\tau_2\int_0^t q(x(s))\,ds}\tau_1\,d\tau_1\,d\tau_2\Big\}. \tag{5.43}$$

This expression clearly admits the following estimate, for any $\beta \geqslant 1$:

$$\big|\mathsf{E}\{q(0)\}\big| \leqslant \mathsf{M}\Big\{\int_0^t \big|q(x(s))\big|^2\,ds\Big(\int_0^1\int_0^1 e^{-\tau_1\tau_2\beta\int_0^t q(x(s))\,ds}\,d\tau_1\,d\tau_2\Big)^{\frac{1}{\beta}}\Big\}.$$

Now Hölder's inequality further permits us to write, if $\alpha^{-1}+\beta^{-1}=1$,

$$\begin{aligned}\big|\mathsf{E}\{q(0)\}\big| &\leqslant \mathsf{W}\Big\{\int_0^t \mathsf{E}^{1/\alpha}\{|q(x(s))|^{2\alpha}\}\,ds \\ &\qquad\times \mathsf{E}^{1/\beta}\Big\{\int_0^1\int_0^1 e^{-\tau_1\tau_2\beta\int_0^t q(x(s))\,ds}\,d\tau_1\,d\tau_2\,\Big|\,x(t)=0\Big\}\Big\} \\ &\leqslant t\mathsf{E}^{1/\alpha}\{|q(0)|^{2\alpha}\}\mathsf{M}^{1/\beta}\Big\{\int_0^1\int_0^1 e^{-\tau_1\tau_2\beta\int_0^t q(x(s))\,ds}\,d\tau_1\,d\tau_2\Big\}\end{aligned}$$

By Jensen's inequality,

$$e^{-a\int_0^t q(x(s))\,ds} \leqslant \frac{1}{t}\int_0^t e^{-aq(x(s))},$$

so the previous estimate, together with the metric transitivity of $q(x)$, implies

$$\big|\mathsf{E}\{q(0)\}\big| \leqslant t\mathsf{E}^{1/\alpha}\{|q(0)|^{2\alpha}\}\Big(\int_0^1\int_0^1 d\tau_1\,d\tau_2\mathsf{E}\{e^{-\beta\tau_1\tau_2 tq(0)}\}\Big)^{\frac{1}{\beta}}.$$

Since for any $0 \leqslant t \leqslant 1$ and $a \in \mathbf{R}$ we have $e^{-at} \leqslant 1 + e^{-a}$, the estimate becomes, for $0 \leqslant t \leqslant 1$:

$$\left|E\{q(0)\}\right| \leqslant t\mathsf{E}^{1/\alpha}\{|q(0)|^{2\alpha}\}\left(1+\mathsf{E}\{e^{-\beta q(0)}\}\right)^{1/\beta}.$$

Now picking $\alpha = 1 + \delta/2$ and using the conditions of Theorem 5.26, we find that $\mathsf{E}\{q(0)\} = 0$ as $t \to \infty$. From this and from (5.43), it follows that

$$\mathsf{M}\left\{\left(\int_0^t q(x(s))\,ds\right)^2\right\} = 0,$$

which clearly means that $q(x) \equiv 0$ with probability 1. □

(5.27) Theorem. *Denote by $\mathcal{L}$ the family of metrically transitive random Schrödinger operators $H = -\Delta + q(x)$, for $x \in \mathbf{R}^d$, that satisfy the condition $\mathsf{E}\{|q(0)|^{p+1}\} < \infty$, where p is the smallest even number larger than $d/2$. Denote by $\mathcal{N}$ the set of measures $N(d\lambda)$ on $\mathbf{R}$ together with the weak topology. If $\mathcal{L}$ is given a metric ρ by*

$$\rho(H_1, H_2) = \mathsf{E}^{1/(p+1)}\{|q_{H_1}(0) - q_{H_2}(0)|^{p+1}\}$$

for $H_1, H_2 \in \mathcal{L}$, the mapping $H \mapsto N(H, d\lambda)$ from $\mathcal{L}$ into $\mathcal{N}$ is continuous.

Proof. Notice first that $N(H,\lambda)$ exists for all $H \in \mathcal{L}$ by Theorem 5.20. Now set

$$r_H(z) = \int |\lambda - z|^{-p} N(H, d\lambda)$$

if $\operatorname{Im} z \neq 0$; by (5.21), we have

$$r_H(z) = \mathsf{E}\{\operatorname{Tr} \chi_C R_H^{p/2} (R_H^*)^{p/2} \chi_C\},$$

where $R_H = (H - zI)^{-1}$. Hence, by Lemmas 5.12 and 5.14(i), we have

$$\left|r_{H_1}(z) - r_{H_2}(z)\right| \leqslant c(z)\rho(H_1, H_2) \tag{5.44}$$

if $\operatorname{Im} z \neq 0$, where $c(z)$ is a positive constant depending on z. The desired continuity property easily follows from (5.44) and Lemma 5.22. □

Theorem 5.27 implies the following statement:

(5.28) Corollary. *Suppose $\mathsf{E}\{|q(0)|^{p+1}\} < \infty$, where p is the smallest even number larger than $d/2$, and*

$$\lim_{n\to\infty} \mathsf{E}\{|q_n(0) - q(0)|^{p+1}\} = 0.$$

Then $\lim_{n\to\infty} N(H_n, d\lambda) = N(H, d\lambda)$. □

(5.29) Theorem [Figotin 1983]. *Let the metrically transitive potential $q(x)$ be such that $\mathsf{E}\{|q(0)|^r\} < \infty$, where $r \geqslant p+1$, p still being the smallest even number larger than $d/2$. Then*

$$N(\lambda) \leqslant c|\lambda|^{d/2+1-r} \quad \textit{for } \lambda \leqslant -1 \tag{5.45a}$$

for some positive constant c, and

$$N(\lambda) = N_0(\lambda)\big(1+o(1)\big) \quad \textit{for } \lambda \to \infty, \tag{5.45b}$$

where $N_0(\lambda) = c_d\lambda^{d/2}$, c_d being the volume of the d-ball of unit radius.

A similar result is also found in [Kirsch and Martinelli 1982c].

Proof. From (5.27), and using arguments similar to those in the proof of Theorem 5.20, we have

$$\lim_{n\to\infty} N_n(d\lambda) = N(d\lambda) \tag{5.46}$$

where $N_n(d\lambda)$ is the integrated density of states of the operator $-\Delta + q_n$ with potential specified in (5.7).

To prove (5.45a), note that the measures $N_\Lambda(d\lambda)$, by Lemma 5.4 and parts (ii) and (iv) of Lemma 5.7, satisfy the conditions of Lemma 5.21(i,ii), with the common constants c and c_1 for all Λ such that $V \geqslant 1$. From this and from (5.20) it follows that $N(d\lambda)$ satisfies the conditions of Lemma 5.21(ii), which leads to inequality (5.45a). Further, it follows from Lemmas 5.12 and 5.14(ii) and from the definition of $N(d\lambda)$ and $N_n(d\lambda)$ that the latter satisfy the conditions of Lemma 5.21(i) for $u = d/2$, with common constant c. Therefore, for some constant c,

$$N_n\big([0,\lambda]\big) \leqslant c(\lambda^{d/2}+1) \quad \text{and} \quad N\big([0,\lambda]\big) \leqslant c(\lambda^{d/2}+1) \tag{5.47}$$

for $\lambda \geqslant 0$, as we wished to prove.

The proof of (5.45b) is based on some auxiliary statements.

(5.30) Lemma. *If $q(x)$, for $x \in \mathbf{R}^d$, is a metrically transitive field satisfying the conditions of Theorem 5.20, and if $q(x) \geqslant c$ with probability 1 for some constant c, equation (5.45b) is true.*

Proof. By the conditions of the lemma, the supports of $N(d\lambda)$ and $N_n(d\lambda)$ are contained in the interval $[c,\infty)$. Hence, and from (5.17),

$$\int_{-\infty}^{\infty} e^{-\lambda t} N_n(d\lambda) = (4\pi t)^{-d/2}\mathsf{M}\{e^{-\int_0^t q_n(x(s))\,ds}\}.$$

In view of (5.46) and (5.47) and the condition $q_n \geqslant c$, we can pass to the limit $n \to \infty$, whence

(5.48)
$$\int_{-\infty}^{\infty} e^{-\lambda t} N(d\lambda) = (4\pi t)^{-d/2} \mathsf{M}\{e^{-\int_0^t q(x(s))\,ds}\}.$$

Therefore,

$$\int_{-\infty}^{\infty} e^{-\lambda t} N(d\lambda) = (4\pi t)^{-d/2}(1 + o(1))$$

as $t \downarrow 0$. Equality (5.45b) now follows from an application of the classical Tauberian theorems [Feller 1966]. □

We continue with the proof of (5.45b) in Theorem 5.29. Set $q_+(x) = \max\{q(x), 0\}$ and let $N_+(d\lambda)$ denote the nonrandom measure corresponding to the operator $-\Delta + q_+$. Theorems 5.24 and inequalities (4.41) give

(5.49)
$$N(\lambda) \geqslant N_+(\lambda).$$

Applying Lemma 5.30, we find that

(5.50)
$$\lim_{\lambda \to \infty} N(\lambda)\lambda^{-d/2} \geqslant c_d.$$

To obtain an upper bound for $N(\lambda)$, we will use equalities (5.24), (5.25) and Lemmas 5.12(i) and 5.14(i), with $s = 0$, $z_1 = \cdots = z_{p/2} = i\tau$, $z_{p/2+1} = \cdots = z_p = -i\tau$, $\tau \geqslant 1$. This gives

(5.51)
$$\left| \int_{-\infty}^{\infty} (\lambda^2 + \tau^2)^{-p/2} (N(d\lambda) - N_n(d\lambda)) \right| \leqslant \varepsilon_n \tau^{d/2-p},$$

where ε_n is a constant times $\mathsf{E}^{1/(p+1)}\{|q(0) - q_n(0)|^{p+1}\}$. Applying (5.45a) to $-\Delta + q$ and $-\Delta + q_n$, and then (5.46), we get

$$\left| \int_0^{\infty} p\lambda(\lambda^2 + \tau^2)^{-p/2-1} (N(\lambda) - N_n(\lambda))\, d\lambda \right| \leqslant \varepsilon_n \tau^{d/2-p}.$$

Replacing λ by $t\lambda$, we arrive at

(5.52)
$$\int_0^{\infty} \tau^{-d/2} N(\tau\lambda) p\lambda(\lambda^2 + 1)^{-p/2-1}\, d\lambda$$
$$\leqslant \varepsilon_n + \int_0^{\infty} \tau^{-d/2} N_n(\tau\lambda) p\lambda(\lambda^2 + 1)^{-p/2-1}\, d\lambda.$$

Taking an arbitrary $\delta \in (0, 1)$, we find a lower bound for the left-hand side of (5.52), using (5.49) and the monotonicity of $N(\lambda)$:

$$\int_0^\infty \tau^{-d/2} N(\tau\lambda) p\lambda(\lambda^2+1)^{-p/2-1} d\lambda$$
$$= \int_{J_1} \tau^{-d/2} N(\tau\lambda) p\lambda(\lambda^2+1)^{-p/2-1} d\lambda + \int_{J_2} \tau^{-d/2} N(\tau\lambda) p\lambda(\lambda^2+1)^{-p/2-1} d\lambda$$
$$\geqslant \tau^{-d/2} N(\tau) \int_{J_1} p\lambda(\lambda^2+1)^{-p/2-1} d\lambda + \int_{J_2} \tau^{-d/2} N_+(\tau\lambda) p\lambda(\lambda^2+1)^{-p/2-1} d\lambda,$$

where $J_1 = [1, 1+\delta]$ and $J_2 = [0, \infty) \setminus J_1$. Combining this with (5.47), (5.49), (5.52), and applying Lemma 5.30 to $N_n(\lambda)$ and $N_+(\lambda)$, we get:

$$\limsup_{\tau\to\infty} \tau^{-d/2} N(\tau) \int_{J_1} p\lambda(\lambda^2+1)^{-p/2-1}\, d\lambda \leqslant c_d \int_{J_1} p\lambda^{d/2+1}(\lambda^2+1)^{-p/2-1} d\lambda + \varepsilon_n.$$

In view of (5.51) and Lemma 5.13, $\lim_{n\to\infty} \varepsilon_n = 0$. Therefore, by making $n \to \infty$ in the preceding inequality, then dividing both sides by δ and making δ approach 0, we get

$$\limsup_{\tau\to\infty} \tau^{-d/2} N(\tau) \leqslant c_d.$$

This, together with (5.50), gives (5.45b). □

Formula (5.45b), being an analogue of the well-known asymptotic Weyl formula $N_\Lambda = c_d \lambda^{d/2}(1+o(1))$, as $\lambda \to \infty$, for the prelimit integrated densities of state, and suggesting that the limits $\lambda \to \infty$ and $\Lambda \to \infty$ for $N_\Lambda(\lambda)$ are interchangeable, shows also that $N(\lambda)$ has the same behavior, as $\lambda \to \infty$, as in the case of the Laplace operator, when its value is $c_d \lambda^{d/2}$. In this sense, the behavior of $N(\lambda)$ as $\lambda \to \infty$ is universal, that is, it doesn't depend (to leading order) on the form of the random potential. As we will see in subsequent chapters, in other regions of the spectrum, especially near its left edge, the situation is different: the asymptotic behavior of $N(\lambda)$ there depends essentially on the random potential, and in some cases is entirely dictated by it (Chapters III and IV).

The behavior of $N(\lambda)$ for $\lambda \to \infty$ is also universal from a different point of view. It turns out that in the very general situation of an elliptic differential operator of arbitrary order p with metrically transitive coefficients, this asymptotic behavior is determined, as in the case of a Schrödinger operator, by the term with highest-order derivatives, that is, by the operator's principal symbol. More exactly, we have the following result:

(5.31) Theorem [Gusev 1977]. *If an elliptic operator satisfy the conditions of Theorem* 5.23, *its integrated density of states satisfies*

$$N(\lambda) = \frac{1}{d(2\pi)^d}\eta(a)\lambda^{d/p}(1 + O(\lambda^{-1/p}))$$

as $\lambda \to \infty$, *where*

$$\eta(a) = \int_{|\xi|=1} \mathsf{E}\{a_p^{-d/p}(0,\xi)\}\, dS_\xi,$$

$a(x,\xi)$ *is the principal symbol of* A, *and* dS_ξ *is the surface element of the unit sphere in* $\mathbf{R}^d$.

The proof of this result uses techniques from pseudo-differential operator theory and is bears little relation with the special nature of the metrically transitive operators. Therefore we will not present it. □

5.D Location of the Spectrum of a Metrically Transitive Schrödinger Operator

The geometry of the spectrum of metrically transitive finite-difference and differential operators is very closely related to the behavior of their integrated density of states because, by Theorem 3.1, the former coincides with probability 1 with the set of growth points of the latter. If we have, as above, an asymptotic formula for the integrated density of states of a metrically transitive operator that estimates the order of the remainder as $\lambda \to \infty$, we also have asymptotic bounds on the lengths of gaps in the spectrum.

(5.32) Theorem [Shubin 1979]. *Let* A *be a metrically transitive operator with spectrum unbounded from above. If the integrated density of states of* A *has asymptotic behavior*

(5.53) $$N(A,\lambda) = \text{const}\,\lambda^\alpha(1 + O(\lambda^{-\beta}))$$

as $\lambda \to \infty$, *where* α *and* β *are nonnegative constants, and the interval* (λ_1, λ_2), *with* $\lambda_1 \geqslant 1$, *does not intersect the spectrum of* A, *we have*

(5.54) $$|\lambda_2 - \lambda_1| \leqslant \text{const}\,\lambda_1^{1-\beta}.$$

Proof. By Theorem 3.1, $N(A,\lambda_1) = N(A,\lambda_2)$, so (5.53) can only be satisfied if (5.54) is. □

When applied to Schrödinger operators with bounded potentials, which, as we know, satisfy (5.36), this theorem yields a result obvious even from elementary perturbation theory, namely, that the lengths of gaps in the spectrum are bounded. In fact, the theorem becomes tautological, because (4.42), which we used to obtain formula (5.36), follows from the same elementary facts of perturbation theory. However, in the case of a general elliptic metrically transitive operator of order p, where the exponent β in (5.54) equals $1/p$ (see Theorem 5.31, for example), the bound for gap lengths provided by (5.54) is nontrivial. And if we can get an asymptotic formula better than (5.36), the same reasoning used in Theorem 5.32 gives correspondingly better estimates for gap lengths. Thus, in the one-dimensional case, if a random potential has finite variance, $\mathsf{E}\{q^2(x)\} < \infty$, we have (see Theorem 6.5)

$$N(\lambda) = \frac{\sqrt{\lambda}}{\pi}\left(1 - \frac{1}{2\lambda}\mathsf{E}\{q(0)\} + o(\lambda^{-1})\right) \tag{5.55}$$

as $\lambda \to \infty$, and therefore the gap lengths tend to zero. If, in addition, the realizations of the potential are smooth enough or random enough, the asymptotic formula can be improved (Section 6), and the decrease in gap lengths can be estimated more accurately.

The problem of the decrease in gap lengths for one-dimensional differential operators has been studied for quite a while (see, for example, [Glazman 1965]), and the lengths are known to go to zero even when the potential is only bounded, no smoothness properties being required [Gordon 1979]. Under certain smoothness conditions, fairly accurate estimates of the rate of decrease are available [Marchenko 1986].

It is interesting that for operators with metrically transitive random potentials, similar results are true, under different conditions. Thus, in the asymptotic formula (5.55), the condition for the gap lengths to tend to zero was simply that the potential have finite variance, regardless of whether or not the potential's realizations have singularities or, if locally continuous, are bounded.

For a Schrödinger operator H with a metrically transitive potential, the analogue of Theorem 4.22 relating the spectrum of H with the spectrum of the translationally invariant part $-\Delta$ and the random diagonal part $q(x)$ is also valid, and can be shown using the same arguments—namely, Theorem 2.17:

(5.33) Theorem. *The spectrum $\sigma(H)$ of a self-adjoint Schrödinger operator $H = -\Delta + q$ in $L_2(\mathbf{R}^d)$ with metrically transitive potential $q(x)$ satisfies*

$$\sigma(H) \subset [0,\infty) + \sigma_1(q),$$
$$\sigma(H) \supset [0,\infty) + \sigma_2(q),$$

where $\sigma_1(q)$ and $\sigma_2(q)$ are defined in the same way as in the discrete case (formulas (4.52)*).* □

This has the following consequences:

(5.34) Theorem. *Let $H = -\Delta + q$ be a Schrödinger operator.*

(i) *If q is a Gaussian random field whose correlation function goes to 0 as $|x| \to \infty$, or a Poisson field (Example 1.15(d)) with $u(x)$ a bounded, summable function that tends to zero as $|x| \to \infty$ and takes negative values on a set of positive Lebesgue measure, we have $\sigma(H) = \mathbf{R}$.*

(ii) *If q is the square (or smoothed square, see Example 1.15(f)) of a Gaussian field or Poisson field with a nonnegative $u(x)$ that tends to zero as $|x| \to \infty$, we have $\sigma(H) = \mathbf{R}_+$.*

Proof. Part (i) for a Gaussian potential is immediate, because we always have $\sigma_1(q) = \sigma_2(q)$. For a Poisson potential with a nonpositive $u(x)$, we check that $\sigma_2(q)$ contains negative numbers of arbitrarily large absolute value: this will prove the claim, since $\{a\} + [0, \infty) = [a, \infty)$. Given a cube Λ, we consider realizations of the Poisson points x_j of the following form: a number of points lie so densely within and around Λ that the sum of the corresponding terms is equal, within a prescribed accuracy, to the given negative number; and the rest of the points is so far away from Λ that their contribution to $q(x)$ is negligible (remember that $\lim_{|x|\to\infty} = 0$). The set of such realizations clearly has positive (although exponentially decreasing) probability, since it is defined by inequalities.

To prove part (ii), notice that $\sigma(q) = \mathbf{R}_+$ in both cases, so it is sufficient to check that $\sigma_2(q)$ contains arbitrarily small positive numbers. This is all but obvious for both potentials. □

The arguments used in this proof suggest that the geometry of the spectrum might be different if in the additive potential $q(x) = \sum_j u_j(x - x_j)$ of (1.30) the points x_j are distributed not according to Poisson's law (independently of each other, regardless of distance), but in such a way that the distance between them can be neither too large nor too small. Indeed, in this case, $\inf \sigma_2(q)$ does not necessarily equal $-\infty$ for nonpositive $u_j(x)$ and 0 for nonnegative $u_j(x)$. A typical example is the random potential (1.31) that models a disordered *substitutional alloy*

$$q(x) = \sum_{y \in \mathbf{Z}^d} \xi_y u(x - y), \tag{5.56}$$

where $u(x)$ is bounded with compact support, and the ξ_y, for $y \in \mathbf{Z}^d$, are independent, identically distributed random variables whose distribution function $F(q)$ has compact support $[a, b]$, for $-\infty < a < b < \infty$. These conditions are imposed for the sake of simplicity; a more general potential is discussed in [Kirsch and Martinelli 1982b].

Suppose that in (5.56) all the ξ_y are equal, or, more generally, that there are d linearly independent vectors $a_1, \ldots, a_d \in \mathbf{Z}^d$ such that $\xi_y + a_j = \xi y$ for

$j = 1, \dots, d$, so that $q(x)$ is a periodic function, with fundamental domain $\Gamma = \{s_1 a_1 + \cdots + s_d a_d : 0 \leqslant s_i \leqslant 1\}$. Under the fairly weak condition that $q(x) \in L_p(\Gamma)$, where $p = 2$ for $d \leqslant 3$, $p > 2$ for $d = 4$ and $p = d/2$ for $d \geqslant 5$, it can be shown [Reed and Simon IV] that the spectrum of the corresponding Schrödinger operator is the union of the spectra of the operators H_k prescribed by

$$H_k = -\Delta + 2i(k, \nabla) + k^2 + q(x) \tag{5.57}$$

in Γ, and by periodic conditions on $\partial\Gamma$. Because of periodicity, we can take k to range over a fundamental domain B of a lattice reciprocal to the lattice of periods, called the *first Brillouin zone.* The spectrum of H_k is discrete, consisting of isolated eigenvalues $\lambda_n(k)$ of finite multiplicity, which depend continuously on $k \in B$. If Δ_n denotes the image of $\lambda_n(k)$, a closed interval, we can write

$$\sigma(H) = \bigcup_{k \in B} \sigma(H_k) = \bigcup_n \Delta_n.$$

Thus, the spectrum of a Schrödinger operator with a periodic potential has a *band structure*: it consists of closed intervals (*allowed bands*) separated by gaps (*forbidden bands*).

In the one-dimensional case, neighboring intervals Δ_n are either disjoint or intersect in one point. Therefore, generically, $\sigma(H)$ consists of an infinite number of allowed bands, and the gaps between them lie near the points $(\pi n/a)^2$; their lengths tend to zero as $n \to \infty$, the more rapidly the smoother the potential (see [Marchenko 1986] and Problem III.7).

Not all gaps need be present; when all but a finite number are absent, or closed, we talk of a *finite-zone potential.* In the one-dimensional case, finite-zone potentials form a dense set in the set of all periodic potentials of a fixed period, under the metric induced by a certain Sobolev norm on the period [Marchenko 1986]. In the multidimensional case, under fairly general conditions, the number of gaps is finite [Skriganov 1982; Veliev 1983].

Interestingly enough, for the potential (5.56), which is a natural generalization of the periodic potential, corresponding to equal amplitudes ξ_y, the spectrum of the Schrödinger operator is determined by the spectra of Schrödinger operators with periodic potentials associated with (5.57).

Definition. *A nonrandom, periodic function $w(x)$ is called an admissible potential corresponding to the random potential* (5.56) *if it has the form*

$$w(x) = \sum_{y \in \mathbf{Z}^d} c_y u(x - y), \tag{5.58}$$

where $c_y \in \operatorname{supp} F$. We denote the set of admissible potentials by $\mathcal{W}_q$.

(5.35) Theorem [Kirsch and Martinelli 1982; Englisch and Kursten 1983]. *The spectrum of* $-\Delta + q$ *is the closure of the union* $\bigcup_{w \in \mathcal{W}_q} \sigma(-\Delta + w)$.

Proof. We present a proof somewhat different from the original one. Let $w(x)$ be an admissible potential. For any cube $\Lambda \subset \mathbf{R}^d$ and any $\varepsilon > 0$, the event

$$\{\sup_{x \in \Lambda} |q(x) - w(x)| < \varepsilon\}$$

has positive probability, because of the independence of the random variables ξ_y. By Theorem 2.17 we find that $\sigma(-\Delta + q) \supset \sigma(-\Delta + w)$, so that

$$\sigma(-\Delta + q) \supset \overline{\bigcup_{w \in \mathcal{W}_q} \sigma(-\Delta + w)}.$$

For the opposite inclusion, it is enough to show that there exists a sequence of admissible potentials $w_n(x)$ such that $\lim_{n \to \infty} N(-\Delta + w_n, \lambda) = N(-\Delta + q, \lambda)$. Since $N(-\Delta + q, \lambda)$, being nonrandom, is an integrated density of states of almost all realizations of the random operator with potential (5.56), we fix a realization $\bar{q}(x)$ and, given a sequence of cubes $\Lambda_n \to \infty$ centered at the origin and having integer side lengths, we construct a sequence of admissible potentials $w_n(x)$ by periodic continuation to $\mathbf{Z}^d$ of the restriction of $\bar{q}(x)$ to Λ_n. If $\Lambda_n' \to \infty$ is another sequence of cubes such that $\Lambda_n' \subset \Lambda_n$ for all n and $\lim_{n \to \infty} \operatorname{dist}(\partial \Lambda_n, \partial \Lambda_n') = \infty$, we have

$$\lim_{n \to \infty} \sup_{x \in \Lambda_n'} \left| w_n(x) - q(x) \right| = 0. \tag{5.59}$$

Let $\Lambda_{n,\nu}$ be a cube composed of ν cubes congruent to Λ_n. By the periodicity of $w_n(x)$ and inequality (5.39a),

$$N_{\Lambda_{n,\nu}}^{(0)}(-\Delta + w_n, \lambda) \geqslant N_{\Lambda_n}^{(0)}(-\Delta + w_n, \lambda).$$

The right-hand side, by virtue of (5.40), (5.59) and Theorem 5.24(b), is bounded below by

$$\frac{V_n'}{V_n} N_{\Lambda_n'}^{(0)}(-\Delta + w_n, \lambda) \geqslant \frac{V_n'}{V_n} N_{\Lambda_n'}^{(0)}(-\Delta + \bar{q}, \lambda - \delta_n),$$

where δ_n is the supremum in (5.59). If we choose the cubes Λ_n' so that $\lim_{n \to \infty} V_n'/V_n$, the last two inequalities, together with the existence of the limiting integrated density of states, show that

$$\liminf_{n \to \infty} N(-\Delta + w_n, \lambda) \geqslant N(-\Delta + \bar{q}, \lambda - \delta)$$

for any $\delta > 0$.

A similar argument, using (5.39b), (5.40) and Theorem 5.24(b), leads to the opposite inequality

$$\limsup_{n\to\infty} N(-\Delta + w_n, \lambda) \leqslant N(-\Delta + \bar{q}, \lambda + \delta)$$

for any $\delta > 0$. Therefore, for all continuity points,

$$\lim_{n\to\infty} N(-\Delta + w_n, \lambda) = N(-\Delta + \bar{q}, \lambda). \qquad \square$$

This theorem suggests that the spectrum of a Schrödinger operator with potential of the form (5.56) can have gaps, as in the case of a periodic potential. The following theorem provides conditions for the existence of such gaps:

(5.36) Theorem [Kirsch and Martinelli 1982b]. *Suppose the function $u(x)$ in* (5.56) *has a definite sign. If an open interval (α, β) does not intersect the spectra of any Schrödinger operator with potential $w_g(x) = g\sum_{y\in\mathbf{Z}^d} u(x-y)$, for $g \in \operatorname{supp} F$, the same interval does not intersect the spectrum of the Schrödinger operator with random potential* (5.56).

Proof. By the preceding theorem, it is enough to show that (α, β) does not intersect the spectrum of any operator $-\Delta + w$, for $w \in \mathcal{W}_q$. Assume, on the contrary, that there is a point $\lambda_0 \in (\alpha, \beta)$ in the spectrum of $\Delta + w$. Then $\lambda_0 = \lambda_{n_0}(w, k_0)$ for some $n_0 \in \mathbf{Z}_+$ and some $k_0 \in B$, where $\lambda_n(w, k)$ is the n-th eigenvalue of operator (5.58) with periodic potential $w(x)$, and B is the corresponding first Brillouin zone. Let $\lambda_- = \inf \operatorname{supp} F$ and $\lambda_+ = \sup \operatorname{supp} F$. By the minimax principle.

$$\lambda_n(w_-, k) \leqslant \lambda_n(w, k) \leqslant \lambda_n(w_+, k),$$

where $w_+(x)$ and $w_-(x)$ are either $\lambda_+ \sum_y u(x-y)$ or $\lambda_- \sum_y u(x-y)$. Thus we can find $\tilde{g} \in [\lambda_-, \lambda_+]$ such that

$$\lambda_{n_0}(w, k_0) = \lambda_{n_0}\left(\tilde{g}\sum_y u(x-y), k_0\right),$$

contradicting the hypothesis of the theorem. $\square$

Theorems 5.35 and 5.36 have the following corollary:

(5.37) Corollary [Kirsch and Martinelli 1982b]. *Suppose that the function $u(x)$ in* (5.56) *has a definite sign, and that* $\operatorname{supp} F$ *is an interval. An open interval (α, β) lies in a gap of the spectrum of the Schrödinger operator with potential* (5.56) *if and only if it lies in a gap of the spectra of all the operators*

$-\Delta+g\sum_y u(x-y)$, for $g \in \operatorname{supp} F$. In other words, the gaps in the spectrum of $-\Delta+\sum_y q_y u(x-y)$ are intersections of the gaps in the spectra of all the operators $-\Delta+g\sum_y u(x-y)$, for $g \in \operatorname{supp} F$. □

Problems

1. Let A be a random Hermitian matrix of order ν, with spectrum $\sigma(A)$. If $n(A,\Delta)$ is the number of eigenvalues of A in an interval Δ, show that

$$\mathsf{P}\{\operatorname{dist}\{\lambda,\sigma(A)\} \geqslant \varepsilon\} \leqslant \mathsf{E}\{n(A,[\lambda-\varepsilon,\lambda+\varepsilon])\}.$$

2. Let D be a diagonal operator on $\ell_2(\mathbf{Z}^d)$, that is, $(D\psi)(x) = d(x)\psi(x)$, and let its integrated density of states be $N(D,d\lambda)$. If U is a unitary operator on the same space, satisfying $|U(x,y)-\delta_{xy}| \leqslant Ce^{-\rho|x-y|}$ for some $C,\rho > 0$, let $A = U^{-1}DU$ be the conjugate of D by U.

For a sequence of cubes $\Lambda \subset \mathbf{Z}^d$ centered at the origin and having volume V, show that, for any bounded measurable function f,

$$\lim_{V\to\infty} \frac{1}{V} \operatorname{Tr} \chi_\Lambda f(A) = \lim_{V\to\infty} \frac{1}{V} \operatorname{Tr} \chi_\Lambda f(D),$$

where χ_Λ is the operator of multiplication by the characteristic function of Λ. In particular, $N(A,d\lambda) = N(D,d\lambda)$ and, if D is metrically transitive,

$$N(A,d\lambda) = \mathsf{P}\{d(0) \in d\lambda\}.$$

3. Consider a metrically transitive operator $H = -\Delta + q(x)$ with potential of the form $q(x) = \sum_{j=1}^d q_j(x_j)$, where $x = (x_1,\ldots,x_d)$ and the q_j are metrically transitive fields on $\mathbf{R}$ such that $\mathsf{E}\{q_j^2(0)\} < \infty$. Prove that

$$N(H,\lambda) = N(H_1,\lambda) * \cdots * N(H_d,\lambda),$$

where $H_j = -d^2/dx^2 + q_j(x)$, for $1 \leqslant j \leqslant d$, and $*$ represents convolution of measures: $(N_1 * N_2)(\lambda) = \int_{\mathbf{R}} N_1(\lambda-\mu)\, dN_2(\mu)$.

4. Consider a metrically transitive Schrödinger operator on $L_2(\mathbf{R}^d)$ or a matrix operator on $\ell_2(\mathbf{Z}^d)$ whose integrated density of states has the representation (5.21) or (4.23) (see Theorems 5.20 and 4.8). Show that, for any interval $\Delta \subset \mathbf{R}$,

(a) $N(\Delta) = V^{-1}\mathsf{E}\{\operatorname{Tr} \chi_\Lambda E(\Delta)\chi_\Lambda\}$, where Λ is a cube of volume V and χ_Λ is the operator of multiplication by the characteristic function of Λ;

(b) $\mathsf{E}\{\operatorname{Tr} f E(\Delta) f\} = \|f\|^2 N(\Delta)$, where f is a function in $L_2(\mathbf{R}^d)$ or $\ell_2(\mathbf{Z}^d)$ (or the operator of multiplication by it);

(c) $\mathsf{E}\{(f, E(\Delta)f)\} \leqslant \|f\|_1^2 N(\Delta)$, where f is a function in $L_1(\mathbf{R}^d)$ or $\ell_1(\mathbf{Z}^d)$.

5. Let A be a metrically transitive Schrödinger operator or matrix operator satisfying the conditions of Theorems 5.20 and 4.8 and set $A_0 = \mathsf{E}\{A\}$. Let F be a convex function on $\mathbf{R}$, such that $\int_{-\infty}^{\infty} |F(\lambda)| N(A_0, d\lambda) < \infty$. Show that

$$\int_{-\infty}^{\infty} F(\lambda) N(A, d\lambda) \geqslant \int_{-\infty}^{\infty} F(\lambda) N(A_0, d\lambda).$$

In particular, if $q(x)$ is such that $\mathsf{E}\{q(0)\} = 0$ and $\tilde{N}(t)$ and $\tilde{N}_0(t)$ are the Laplace transforms of the integrated densities of states of $-\Delta + q$ and $-\Delta$, respectively, $\tilde{N}(t) \geqslant \tilde{N}_0(t)$.

Hint. Use part (a) of Problem II.4 and the inequality

$$\mathsf{E}\{(\phi, F(A)\phi)\} \geqslant F\big((\phi, \mathsf{E}\{A\}\phi)\big).$$

6. Consider the discrete analogue of the Schrödinger operator, the operator $h = h_0 + Q$ on $\ell_2(\mathbf{Z}^d)$ given by

$$(h\psi)(x) = \sum_{|y-x|=1} (\psi(x) - \psi(y)) + q(x)\psi(x),$$

for $x \in \mathbf{Z}^d$, where q is a metrically transitive field. Consider also the operators L^{xy} and S^{xy} on $\ell_2(\mathbf{Z}^d)$ given by

$$(L^{xy}\psi)(z) = \begin{cases} 0 & \text{if } z \neq x, y, \\ \psi(x) - \psi(y) & \text{if } z = x, \\ \psi(y) - \psi(x) & \text{if } z = y; \end{cases}$$

$$(S^{xy}\psi)(z) = \begin{cases} 0 & \text{if } z \neq x, y, \\ \psi(x) + \psi(y) & \text{if } z = x \text{ or } z = y. \end{cases}$$

Pairs $(x, y) \in \mathbf{Z}^d \times \mathbf{Z}^d$ of nearest neighbors, that is, such that $|x - y| = 1$, are also called *bonds*. For an arbitrary set B of bonds, consider the operators

$$h^{B,0} = h_0 + \sum_{(x,y)\in B} S^{xy} + Q,$$

$$h^{B,1} = h_0 - \sum_{(x,y)\in B} L^{xy} + Q,$$

said to be subject to the Dirichlet and Neumann conditions, respectively. Check the following relations:

$$h_0 = \sum_{x,y} L^{xy}, \qquad L^{x,y} \geqslant 0, \qquad S^{x,y} \geqslant 0, \qquad h^{B,1} \leqslant h \leqslant h^{B,0}.$$

For an arbitrary $L \in \mathbf{Z}_+$, consider the partition of $\mathbf{Z}^d$ into cubes $S_\alpha(L)$, for $\alpha = (\alpha_1, \dots, \alpha_d) \in \mathbf{Z}^d$, given by

$$S_\alpha(L) = \{(n_1, \dots, n_j) \in \mathbf{Z}^d : \alpha_j L + 1 \leqslant n_j \leqslant (\alpha_j + 1)L \quad \text{for } 1 \leqslant j \leqslant d\}.$$

Clearly, $\ell_2(\mathbf{Z}^d) = \bigoplus_\alpha \ell_2(S_\alpha(L))$. Denote by $B(L)$ the set of bonds between the different cubes $S_\alpha(L)$. Show that:

(a) the spaces $\ell_2(S_\alpha(L))$ reduce the operators $h^{B(L),a}$, for $a = 0, 1$, and we have the representations

$$h^{B(L),a} = \bigoplus_\alpha h_\alpha^{B(L),a};$$

(b) the integrated density of states of $h^{B(L),a}$, for $a = 0, 1$, can be written as

$$N(h^{B(L),a}, d\lambda) = L^{-d}\mathsf{E}\{n_L^a(d\lambda)\},$$

where $n_L^a(d\lambda)$ is the number of eigenvalues of the L^d-dimensional matrix $h_\alpha^{B(L),a}$ in the interval $d\lambda$;

(c) The following discrete analogues of the variational Weyl–Courant inequalities (5.38) and of equality (5.10) hold:

$$N(h^{B(L),0}, d\lambda) \leqslant N(h, d\lambda) \leqslant N(h^{B(L),1}, d\lambda),$$
$$\lim_{L\to\infty} N(h^{B(L),a}, d\lambda) = N(h, d\lambda) \quad \text{for } a = 0, 1.$$

Further details may be found in [Simon 1985a].

7. Let the operator $h = h_0 + Q$ be as in problem II.6, and assume that the $q(x)$, for $x \in \mathbf{Z}^d$, are independent, identically distributed random variables. Prove that, for any $\varepsilon > 0$, there exists $C(\varepsilon) > 0$ such that

$$\mathsf{P}\{|N_\Lambda(h_0 + Q, \Delta) - N(h_0 + Q, \Delta)| \geqslant \varepsilon\} \leqslant e^{-C(\varepsilon)V}.$$

for any interval $\Delta \subset \mathbf{R}$.

Hint. Use the theorem on large deviations

$$\mathsf{P}\left\{\frac{1}{n}\left|\sum_{j=1}^{n} \xi_j\right| \geqslant \varepsilon\right\} \leqslant e^{-C(\varepsilon)n},$$

where the ξ_j are independent, identically distributed random variables. Then use Theorem 3.4 and Problem II.6.

8. Let A and B be self-adjoint operators on $\ell_2(\mathbf{Z}^d)$, with A symmetric and B compact. Show that, if A has an integrated density of states, so does $A + B$, and the two distributions coincide.

9. Let A and B be symmetric matrix operators on $\ell_2(\mathbf{Z}^d)$. Assume that A has an integrated density of states $N(A, d\lambda)$ and that B satisfies

$$\lim_{V\to\infty} \frac{1}{V} \sum_{x,y\in\Lambda} |B(x,y)| = 0,$$

where Λ stands for cubes centered at the origin and having volume V. Show that $A+B$ has an integrated density of states, and that it coincides with $N(A, d\lambda)$.

Hint. Consider the traces of the resolvents of A_Λ and $(A+B)_\Lambda$ and use Proposition 4.9.

10. Show that, if A is a matrix operator of the form (4.47) and $\mathsf{E}\{|q(0)|^n\} < \infty$ for some $n \in \mathbf{Z}_+$, we have $\int |\lambda|^n N(A, d\lambda) < \infty$. Generalize to the case where $|\lambda|^n$ is replaced by a continuous, nonnegative function $f(\lambda)$ such that, for any $b > 0$,

$$\sup_{\substack{\lambda\in\mathbf{R}\\ |a|\leqslant b}} \frac{f(\lambda+a)}{1+f(\lambda)} < \infty.$$

Hint. Use the fact that $N(A, d\lambda) = \lim_{V\to\infty} \mathsf{E}\{N(A_\Lambda, d\lambda)\}$ and

$$\int f(\lambda)\mathsf{E}\{N(A_\Lambda, d\lambda)\} \leqslant \text{const}\, \mathsf{E}\{f(q(0))\}.$$

11. Let A and B be symmetric metrically transitive operators, both lying in $\mathcal{L}_{1,1}$ or $\mathcal{L}_{2,2}$. Show that $\lim_{g\to\infty} N(gA+B, g\,d\lambda) = N(A, d\lambda)$.

Hint. Check that $N(A, g\,d\lambda) = N(g^{-1}A, d\lambda)$, then use Theorem 4.18.

12. Let h_0 be a discrete Laplace operator in $\ell_2(\mathbf{Z}^d)$, defined by

$$(h_0\psi)(x) = -\sum_{|x-y|=1} \psi(y) + 2d\psi(x)$$

for $x \in \mathbf{Z}^d$. We represent a point $x \in \mathbf{Z}^d$ as $x = (x_1, \bar{x})$, with $x_1 \in \mathbf{Z}$ and $\bar{x} \in \mathbf{Z}^{d-1}$. Take an arbitrary metrically transitive field $q(\bar{x})$, for $\bar{x} \in \mathbf{Z}^{d-1}$, and consider the operator A on $\ell_2(\mathbf{Z}_+ \times \mathbf{Z}^{d-1})$, specified by the relations $(A\psi)(x) = (h_0\psi)(x)$, for $x \in \mathbf{Z}_+ \times \mathbf{Z}^{d-1}$, and $\psi(0,\bar{x}) + q(\bar{x})\psi(1,\bar{x}) = 0$. The entries of the matrix of A are

$$A(x,y) = h_0(x-y) + \delta_{x_1,1}\delta_{\bar{x}\bar{y}}q(\bar{x}),$$

for $x, y \in \mathbf{Z}_+ \times \mathbf{Z}^{d-1}$. For $q = 0$, we denote A by A_0. For each $L \in \mathbf{Z}_+$, consider in $\mathbf{Z}_+ \times \mathbf{Z}^{d-1}$ the cube

$$\Lambda = \{(x_1, \bar{x}) : 1 \leqslant x_1 \leqslant 2L+1 \text{ and } |x_j| \leqslant L \text{ for } 2 \leqslant j \leqslant d\}$$

and let A_L be the matrix with entries $A(x,y)$, for $x,y \in \Lambda$. Let $n(A_L, d\lambda)$ be the number of eigenvalues of A_L in the interval $d\lambda$.

(a) Using (4.27) and the method of the proof of Theorem 4.11, show that if q is bounded there exists a nonrandom measure $\nu(d\lambda)$ such that, with probability 1,

$$\lim_{L\to\infty} L^{-(d-1)}\big(n(A_L, d\lambda) - n(A_{0,L}, d\lambda)\big) = \nu(d\lambda).$$

(b) Extend the result in part (a) to an arbitrary metrically transitive field $q(\bar{x})$ by the methods of the proof of Theorem 4.4.

(c) Using the results of Problem II.24, compute the measure $\nu(d\lambda)$ if $q(\bar{x})$ has a Cauchy distribution.

(d) Extend the statements above from the discrete to the continuous case.

13. Let A_0 be a difference operator in $\ell_2(\mathbf{Z}^d)$ as in Theorem 4.16. For a fixed $1 \leqslant k \leqslant d-1$, write a point $x \in \mathbf{Z}^d$ as $x = (x^k, \bar{x})$, where $x_k \in \mathbf{Z}^k$ and $\bar{x} \in \mathbf{Z}^{d-k}$. Let $q(\bar{x})$, for $\bar{x} \in \mathbf{Z}^{d-k}$, be an arbitrary metrically transitive field, let Q be the random diagonal operator given by $Q(x) = \delta_{0,x^k} q(\bar{x})$, for $x \in \mathbf{Z}^d$, and set $A = A_0 + Q$. If Λ is the cube $\{x \in \mathbf{Z}^d : |x_j| \leqslant L \text{ for } 1 \leqslant j \leqslant d\}$, denote by A_L the matrix with entries $A(x,y)$, for $x,y \in \Lambda$, and by $n(A_L, d\lambda)$ the number of eigenfunctions of A_L in the interval $d\lambda$. Prove that there exists a nonrandom measure $\nu(d\lambda)$ such that, with probability 1,

$$\lim_{L\to\infty} L^{-d+k}\big(n(A_L, d\lambda) - n((A_0)_L, d\lambda)\big) = \nu(d\lambda).$$

Hint. See Problem II.12.

14. Let A be a Hermitian matrix and P the orthogonal projection onto a fixed unit vector in $\mathbf{C}^n$. Show that, for any positive t, the eigenvalues $\lambda_1 \leqslant \lambda_2 \leqslant \cdots$ of A and the eigenvalues $\lambda_1' \leqslant \lambda_2' \leqslant \cdots$ of $A + tP$ alternate, that is,

$$\lambda_1 \leqslant \lambda_1' \leqslant \lambda_2 \leqslant \lambda_2' \leqslant \cdots$$

Hint. Use Lemma 4.20.

15. Let A and B be Hermitian matrices of order ν, and, for any interval Δ, let $n(A, \Delta)$ be the number of eigenvalues of the A that lie in Δ. Prove that

$$\big|n(A,\Delta) - n(B,\Delta)\big| \leqslant \operatorname{rank}(A - B).$$

Hint. Denote by $\mathcal{L}_\Delta$ the set of linear subspaces $L \subset \mathbf{C}^\nu$ invariant under A and such that $(Ax, x) \in \Delta$ for all unit vectors $x \in L$. Use the fact that

$$n(A,\Delta) = \max_{L \in \mathcal{L}_\Delta} \{\dim L\}.$$

16. Let A and $n(A,\Delta)$ be as in the previous problem, and set $n(A,\lambda) = n(A,(-\infty,\lambda])$ for $\lambda \in \mathbf{R}$. If P is an orthogonal projection in $\mathbf{Z}^d$, show that:

(a) There exists a real number q_0, depending on A and λ, such that

$$n(A+qP,\lambda) = n(A+qP,\lambda)\big|_{q=-\infty} - \chi_{[0,\infty)}(q-q_0)$$

for all real q, that is,

$$\frac{\partial}{\partial q} n(A+qP,\lambda) = -\delta\big(q - q_0(A,\lambda)\big),$$

where δ is the Dirac delta function.

(b) $n(A+qP,\lambda) - n(A+tP,\lambda) = \chi_{(q,t]}(q_0)$.

If Q is a diagonal operator on $\mathbf{C}^n$ with real entries $q_1,\ldots,q_n$, show also that

$$n(A+Q,\lambda) - n(A+Q,\mu) = \sum_{j=1}^{n} \chi_{[q_j-\lambda,q_j-\mu]}(q_0^{(j)}),$$

for all $\lambda > \mu$, where $q_0^{(j)}$ depends on j, λ, μ and all the q_m excluding q_j.
Hint. Use Problem II.15 for (a) and the identity $n(A,\lambda) = n(A-\lambda I,0)$ for (b).

17. Let A be a Hermitian matrix of finite order with resolution of the identity $E(A,d\lambda)$; let $n(A,\lambda)$ be the number of eigenvalues of A less than λ, and let B be a nonnegative matrix of the same order. Prove that, for any real t and any interval $\Delta \subset \mathbf{R}$,

$$\frac{\partial}{\partial t}\int_\Delta n(A+tB,\mu)d\mu = \operatorname{Tr} BE(A+tB,\Delta).$$

Use this to show that, for any nonnegative, measurable function $f(t)$,

$$\int_{\mathbf{R}} f(t)\operatorname{Tr} BE(A+tB,\Delta) \leqslant \sup f(t) \operatorname{rank} B \operatorname{meas} \Delta.$$

18. Consider a metrically transitive Hermitian matrix A in $\mathcal{L}_{1,1} \subset \ell_2(\mathbf{Z}^d)$, and let $F\big(da \mid a(x,y), x \neq 0, y \neq 0\}$ be the conditional distribution of the random variable $a(0,0)$ with the other elements fixed. Show that:

(a) If $F(da \mid \cdot)$ is absolutely continuous and $F(da \mid \cdot) \leqslant f_0\, da$, where f_0 is a constant, the integrated density of states $N(A,d\lambda)$ is absolutely continuous and $N(A,d\lambda) \leqslant f_0\, d\lambda$.

(b) If there exist positive constants C and α for which $F\big([a_1,a_2]\cdot\big) \leqslant C|a_1 - a_2|^\alpha$ for any a_1 and a_2 such that $|a_1 - a_2| \leqslant 1$, the same inequality is satisfied, and with the same constants, by $N(A,d\lambda)$.

Hint. Use Problem II.16.

19. Let $A = A_0 + Q$ be a metrically transitive random field as in Theorem 4.16, with $q(x)$, for $x \in \mathbf{R}^d$, a Gaussian random field. Assume that the Fourier transform $\hat{b}(k)$, for $k \in \mathbf{T}^d$, of the correlation function $b(x) = \mathsf{E}\{q(x)q(0)\}$ satisfies $\int_{\mathbf{T}^d} \hat{b}^{-1}(k)\,dk < \infty$. Show that the corresponding integrated density of states is absolutely continuous and has a bounded density.

Hint. Use Problem I.7.

20. Let A be a metrically transitive operator in $\ell_2(\mathbf{Z}^d)$ of the form

$$(A\psi)(x) = \sum_y J(x,y)\big(\psi(x) - \psi(y)\big),$$

with $J(x,y) = J(y,x)$. Consider a partition $\mathbf{Z}^d \times \mathbf{Z}^d = \mathcal{A} \cup \mathcal{A}^c$ such that $(x,x) \in \mathcal{A}$ for all $x \in \mathbf{Z}^d$ and $(x,y) \in \mathcal{A}$ if and only if $(y,x) \in \mathcal{A}^c$, if $x \neq y$. Assume that:

(i) there exist positive constants R and J_0 such that $J(x,y) = 0$ if $|x-y| > R$, and $0 \leqslant J(x,y) \leqslant J_0$ with probability 1 for all x, y;
(ii) The $J(x,y)$, for $(x,y) \in \mathcal{A}$, are independent random variables, and, for any z with $|z| \leqslant R$, the random variables $J(x, x+z)$, for $x \in \mathbf{Z}^d$, have the same distribution function $F_z(dJ) = f_z(J)\,dJ$, and that, moreover, there exists a constant C such that $0 \leqslant f_z(J) \leqslant C$.

Prove that the integrated density of states $N(A,\lambda)$ is absolutely continuous and that

$$\frac{\partial}{\partial\lambda} N(A,\lambda) \leqslant \frac{2C_0J_0}{\lambda},$$

for all $\lambda > 0$.

Hint. Taking an arbitrary cube $\Lambda \subset \mathbf{Z}^d$ of volume V and defining the matrix A_Λ by

$$(A_\Lambda\psi,\psi) = \sum_{(x,y)\in\mathcal{A}\cap\Lambda^2} J(x,y)\big|\psi(x) - \psi(y)\big|^2,$$

derive the following relations between the eigenvalues λ_α the eigenfunctions ψ_α and the integrated density of states $N_\Lambda(\lambda)$ of A_Λ:

$$0 \leqslant -\frac{\partial N_\Lambda(\lambda)}{\partial J(x,y)} = \frac{1}{V}\sum_\alpha \delta(\lambda - \lambda_\alpha)\frac{\partial\lambda_\alpha}{\partial J(x,y)}$$

and

$$\frac{\partial\lambda_\alpha}{\partial J(x,y)} = \left(\frac{\partial A_\Lambda}{\partial J(x,y)}\psi_\alpha, \psi_\alpha\right) = \big|\psi_\alpha(x) - \psi_\alpha(y)\big|^2$$

for $(x,y) \in \mathcal{A} \cap \Lambda^2$;

$$J_0 \sum_{(x,y)\in\mathcal{A}\cap\Lambda^2} \frac{\partial\lambda_\alpha}{\partial J(x,y)} \geqslant (A_\Lambda\psi_\alpha,\psi_\alpha) = \lambda_\alpha;$$

$$\sum_{(x,y)\in\mathcal{A}\cap\Lambda^2} -\frac{\partial N_\Lambda(\lambda)}{\partial J(x,y)} \geqslant \frac{1}{V}\sum_\alpha \delta(\lambda-\lambda_\alpha)\frac{\lambda_\alpha}{J_0} = \frac{\lambda}{J_0}\frac{\partial N_\Lambda(\lambda)}{\partial\lambda};$$

$$\frac{\lambda}{J_0}\frac{\partial}{\partial\lambda}\mathsf{E}\{N_\Lambda(\lambda)\} \leqslant \mathsf{E}\Big\{-\sum_{(x,y)\in\mathcal{A}\cap\Lambda^2}\frac{\partial N_\Lambda(\lambda)}{\partial J(x,y)}\Big\}$$

$$\leqslant CV\mathsf{E}\{N_\Lambda(\lambda)\big|^0_{J(x,y)=J_0}\} \leqslant 2C.$$

In deriving the latter inequality, note that the rank of $A_\Lambda\big|^0_{I(x,y)=J_0}$ is equal to two, and use Problem II.15.

21. Let $A = A_0 + Q$ be a metrically transitive matrix as in Theorem 4.16, with potential $q(x)$ of the form

$$q(x) = \sum_{y\in\mathbf{Z}^d} \xi_y u(x-y),$$

where u is a nonnegative function with compact support and the ξ_y, for $y \in \mathbf{Z}^d$, are independent, identically distributed random variables. Assuming the distribution function ξ_0 has bounded density $f(\xi)$, show that the integrated density of states $N(A,d\lambda)$ also has bounded density.

Hint. Recall that $N(A,d\lambda) = \lim_{V\to\infty}\mathsf{E}\{(E(A_\Lambda,d\lambda)e_0,e_0)\}$, where A_Λ is the restriction of A to a cube Λ of volume V. Using Problem II.17, show that

$$\sum_{x\in\mathbf{Z}^d} u(x)\int (E_{\xi_0}(d\lambda)e_x,e_x) f(\xi_0)\,d\xi_0 \leqslant \sup_\xi f(\xi)\nu\,d\lambda,$$

where ν is the number x for which $u(x) > 0$, and $\big(E_{\xi_0}(d\lambda)e_0,e_0\big)$ is a matrix entry of the resolution of the identity of A with an explicitly written dependence on the quantity ξ_0.

22. Let $A = A_0 + Q$ be a metrically transitive matrix as in Theorem 4.16. Assume that $\sum_x |a_0(x)| = a < \infty$, that the $q(x)$ are independent, identically distributed random variables, and that, for some positive constants C and α, we have $|\mathsf{E}\{e^{itq(0)}\}| \leqslant Ce^{-\alpha t}$ for all real t. Show that, if $\varepsilon = \alpha - Ca > 0$, the integrated density of states $N(A_0+Q,\lambda)$ is an analytic function of λ on the strip $|\operatorname{Im}\lambda| < \varepsilon$.

Hint. Use the operator identity

$$e^{(A+B)t} = e^{Bt}$$
$$+\sum_{n\geqslant 1}\int_0^t d\tau_1\int_0^{t_1} d\tau_2\ldots\int_0^{t_{n-1}} d\tau_n e^{B\tau_n} A e^{B(\tau_{n-1}-\tau_n)} A\ldots A e^{B(t-\tau_1)}$$

with iA_0 for A and iQ for B, and show that

$$\left|\mathsf{E}\{(e_0, e^{it(A_0+Q)}e_0)\}\right| \leqslant C(e_0, e^{(CA_0^+ - \alpha)t}e_0)$$

for $t > 0$, where A_0^+ is the difference operator with kernel $|a_0(x-y)|$.

23. Let A_0 be a self-adjoint operator.

(a) Show that the absolutely continuous spectrum of A_0 does not change under perturbations of rank one.

(b) Show that, if ϕ is a cyclic vector of A_0, P_ψ is orthogonal projection on ψ, and $A_\tau = A_0 + \tau P_\psi$, where $\tau \in \mathbf{R}$, the singular parts of the spectral measures of A_{τ_1} and A_{τ_2} are mutually singular for $\tau_1 \neq \tau_2$.

Hint. Use Lemma 4.20 and Theorem 4.10; for part (b), see also [Reed and Simon IV].

24. Let A be a bounded operator in $\ell_2(\mathbf{Z}^d)$ and

$$a = \inf_{\|\psi\|=1} \left\{ \left(\psi, \frac{A - A^*}{2i} \psi \right) \right\}.$$

Pick an arbitrary positive number ε and consider the following family $\mathcal{E}$ of diagonal operators D:

$$(D\psi)(x) = d(x)\psi(x), \quad \operatorname{Im} d(x) \geqslant a + \varepsilon$$

for $x \in \mathbf{Z}^d$. It is easy to check that $\|(A+D)^{-1}\| \leqslant \varepsilon^{-1}$ for $D \in \mathcal{E}$. Let P_x be the orthogonal projection onto the basis vector e_x, for $x \in \mathbf{Z}^d$.

A random variable q is said to have a *Cauchy distribution* if, for some $\alpha > 0$ and $\beta \in \mathbf{R}$,

$$P\{q \in \Delta\} = \frac{1}{\pi} \int_\Delta \frac{\alpha}{(t-\beta)^2 + \alpha^2} dt;$$

we abbreviate the integrand as $p_C(t)$, and associate with the random variable q the complex parameter $\zeta(q) = \alpha + i\beta$.

(a) If q, q_1 and q_2 are independent random variables with Cauchy distribution, the random variables $-q$, q^{-1} and $q_1 + q_2$ also have Cauchy distribution, with $\zeta(-q) = \zeta^*(q)$, $\zeta(q^{-1}) = (\zeta^*(q))^{-1}$, and $\zeta(q_1 + q_2) = \zeta(q_1) + \zeta(q_2)$.

(b) For any real q and $D \in \mathcal{E}$ we have

$$(A + D + qP_x)^{-1} - (A+D)^{-1} = -\frac{q}{q + (e_x, (A+D)^{-1}e_x)} \times (A+D)^{-1}P_x(A+D)^{-1}.$$

(c) If $D \in \mathcal{E}$ and q is a random variable with Cauchy distribution,

$$\mathsf{E}\{(A+D+qP_x)^{-1}\} = \big(A+D+i\zeta(q)P_x\big)^{-1}.$$

(d) If $D \in \mathcal{E}$ and $q(x)$, for $x \in \mathbf{Z}^d$, are independent random variables with Cauchy distribution,

$$\mathsf{E}\{(A+D+\sum_x q(x)P_x)^{-1}\} = \big(A+D+\sum_x i\zeta(q(x))P_x\big)^{-1}.$$

(e) If A_0 is a bounded real Toeplitz operator in $\ell_2(\mathbf{Z}^d)$, the field $q(x)$ is as in part (d), with $\zeta(q(x))$ the same for all x, and $Q = \sum_x q(x)P_x$, we have

$$N(A_0+Q,\lambda) = \int_{\mathbf{R}} N(A_0, \lambda - q)p_C(q)\,dq.$$

25. Let $q(x)$, for $x \in \mathbf{Z}^d$, be a field of independent, identically distributed random variables with Cauchy distribution with parameter $\zeta(q) = \alpha + i\beta$ (see previous problem). Let $\phi(x)$, for $x \in \mathbf{Z}^d$, be a nonrandom sequence of nonnegative numbers such that $\sum_x \phi(x) = \hat{\phi} < \infty$. Consider the metrically transitive field $\tilde{q}(x) = \sum_y \phi(x-y)q(y)$, for $x \in \mathbf{Z}^d$, and the operator $\tilde{Q}$ of multiplication by it. Prove that if A_0 is a bounded real Toeplitz operator in $\ell_2(\mathbf{Z}^d)$,

$$N(A_0+\tilde{Q},\lambda) = \int_{\mathbf{R}} N(A_0, \lambda - q)\tilde{p}_C(q)\,dq,$$

where $\tilde{p}_C(q)$ is the Cauchy density with parameter $\zeta(\tilde{q}) = \hat{\phi}(\alpha + i\beta)$.

Hint. Use the operator identity from the hint to Problem II.21.

26. Let $q_1(x)$ and $q_2(x)$, for $x \in \mathbf{R}^d$, be nonnegative metrically transitive fields such that $\mathsf{E}\{|q_1(0)|^2\} < \infty$, that $q_2 \in C^d(\mathbf{R}^d)$ with probability 1, and that, for some $\alpha > d/2$,

$$\mathsf{E}\{|q_2(0)|^\alpha\}, \mathsf{E}\Big\{\Big|\frac{\partial}{\partial 1}q_2(0)\Big|^\alpha\Big\}, \ldots, \mathsf{E}\Big\{\Big|\frac{\partial}{\partial 1}\cdots\frac{\partial}{\partial d}q_2(0)\Big|^\alpha\Big\} < \infty.$$

Prove that the symmetric Schrödinger operator $-\Delta + q_1 - q_2$ is essentially self-adjoint on $C_0^\infty(\mathbf{R}^d)$.

Hint. Use the Titchmarsh–Sears criterion of essential self-adjointness (see [Titchmarsh 1958, Theorem 12.11]; for details, see [Figotin and Pastur 1983].)

27. Let $q(x)$, for $x \in \mathbf{R}^d$, be a metrically transitive field satisfying condition (b) of Theorem 5.8, and take $v \in L_{p+2}(\mathbf{R}^d)$. Show that, with probability 1, the integrated density of states $N(-\Delta + q + v, d\lambda)$ exists, and that it equals $N(-\Delta + q, d\lambda)$.

Hint. Use Lemma 5.7.

28. Let $q(x)$, for $x \in \mathbf{R}^d$, be a metrically transitive field, and let the Schrödinger operator $H = -\Delta + q$ have the integrated density of states $N(H, d\lambda)$. For a fixed n, consider the operator

$$H_n = \sum_{j=1}^{n} H(j) + \sum_{1 \leqslant l < m \leqslant n} u(x_l - x_m)$$

on $L_2(\mathbf{R}^{nd})$, where $H(j) = -\Delta + q(x_j)$, for $x_j \in \mathbf{Z}^d$, and $u(x)$ is a bounded real function that approaches 0 as $|x|$ increases. Prove that H_n has an integrated density of states $N(H_n, d\lambda)$ equal to $N(H, d\lambda) * \cdots * N(H, d\lambda)$, the n-fold convolution of $N(H, d\lambda)$.

29. Let $q(x)$, for $x \in \mathbf{R}^d$, be a smooth periodic field: this means there is a basis $\{a_j : 1 \leqslant j \leqslant d\}$ of $\mathbf{R}^d$ such that $q(x + a_j) = q(x)$. Let Γ denote the unit cell of the corresponding lattice:

$$\Gamma = \Big\{ x = \sum_{j=1}^{d} t_j a_j : 0 \leqslant t_j < 1 \text{ for } 1 \leqslant j \leqslant d \Big\}.$$

Let $\{b_j : 1 \leqslant j \leqslant d\}$ be the basis dual to $\{a_j : 1 \leqslant j \leqslant d\}$, that is, $(a_i, b_j) = \delta_{ij}$, and Γ the corresponding unit cell. Prove that $N(-\Delta + q, d\lambda)$ exists and that:

(a) $N(-\Delta + q, d\lambda) = (\operatorname{meas} \Gamma)^{-1} \operatorname{Tr} \chi_\Gamma E(-\Delta + q, d\lambda) \chi_\Gamma$, where χ_Γ is the operator of multiplication by the characteristic function of Γ;

(b) If $\lambda_n(k)$, for $n = 1, 2, \ldots$ and $k \in B$, is the set of eigenvalues of the component of the direct integral decomposition of the periodic operator $-\Delta + q$ into a direct integral (see [Reed and Simon IV]), we have

$$N(-\Delta + q, d\lambda) = \frac{1}{\operatorname{meas} B} \int_B \sum_n \chi_{d\lambda}(\lambda_n(k))\, dk.$$

30. For a real function $w(x) \in L_1(\mathbf{R}^d)$, consider the family of periodic potentials

$$q_L(x) = \sum_{n \in \mathbf{Z}^d} w(x - Ln).$$

Prove that $\lim_{L \to \infty} N(-\Delta + q_L, d\lambda) = N(-\Delta, d\lambda)$.

31. Show that if $q(x)$, for $x \in \mathbf{R}$, is a periodic function of period a and $q \in L_2^{\mathrm{loc}}(\mathbf{R})$, the integrated density of states $N(\lambda)$ of the operator $-d^2/dx^2 + q$ is absolutely continuous and

$$N'(\lambda) = \frac{A'(\lambda, a)}{a\sqrt{1 - A^2(\lambda, a)}},$$

where $A(\lambda, a) = \frac{1}{2}\big(c(\lambda, a) + s'(\lambda, a)\big)$ is the *Hill discriminant*, and c and s are solutions of the corresponding Schrödinger equation with initial conditions $s(\lambda, 0) = c'(\lambda, 0) = 0$ and $c(\lambda, 0) = s'(\lambda, 0) = 1$.

32. Let $q(x)$, for $x \in \mathbf{R}^d$, be a metrically transitive field such that $\mathsf{E}\{|q(0)|^{p+1}\} < \infty$, where p is the smallest even number strictly larger than $d/2$. For any $n \in \mathbf{Z}_+$, consider the periodic potential $q_n(x)$ having the cube $\{|x_j| \leqslant n \text{ for } 1 \leqslant j \leqslant d\}$ as its periodicity cell and coinciding with $q(x)$ on it. Prove that, with probability 1,

$$N(-\Delta + q, d\lambda) = \lim_{n \to \infty} N(-\Delta + q_n, d\lambda).$$

33. Let Q be a smooth periodic function on $\mathbf{R}^n$: this means there is an orthonormal basis $\{a_j : 1 \leqslant j \leqslant n\}$ of $\mathbf{R}^n$ such that $Q(x + a_j) = Q(x)$. Consider the quasiperiodic potential

$$q(x) = Q\left(\left(\sum_{j=1}^{n} \alpha_j a_j\right) x\right)$$

for $x \in \mathbf{R}$, where $\alpha = (\alpha_1, \ldots, \alpha_n)$ is a frequency vector. Prove that the integrated density of states $N(-d^2/dx^2 + q, d\lambda)$ is a continuous function of α on the set of α such that $\alpha_1, \ldots, \alpha_n$ are rationally independent.

34. Let $H_c = -\Delta + q_c(x)$, for $x \in \mathbf{R}^d$, be a Schrödinger operator with a Poisson random potential q_c, where c is the concentration of Poisson points (see Example 1.15(e)). Prove that

$$\lim_{c \to 0} N(H_c, d\lambda) = N(-\Delta, d\lambda).$$

35. Prove Proposition 5.19.

36. (a) Prove (5.16), starting from the probabilistic interpretation of the kernels of the operators $e^{-tH_{\Lambda,a}}$.
(b) Prove inequalities (5.19).
(c) Prove that $0 \leqslant \mathsf{E}\{K_{\Lambda,1} - K_{\Lambda,0}\} \leqslant (k_{\Lambda,1} - k_{\Lambda,0})\mathsf{E}\{e^{-tq(0)}\}$.

Hint. In part (c), use Trotter's formula and Hölder's inequality.

37. Let $Q(x)$, for $x \in \mathbf{R}^d$, be a metrically transitive, bounded below with probability 1 and such that $\mathsf{E}\{|Q(0)|^{p+1}\} < \infty$, where p is the smallest even number strictly larger than $d/2$. Consider the family of Schrödinger operators $H_R = -\Delta + Q(x/R)$, for $R > 0$. Prove that

$$\lim_{R \to \infty} N(H_R, \lambda) = \int_{\mathbf{R}} N_0(\lambda - q) F(dq),$$

where $N_0(\lambda) = N(-\Delta, \lambda)$ and $F(dq)$ is the distribution of the random variable $Q(0)$.

Hint. Show that $\tilde{N}(H_R, t)$ converges to $\tilde{N}_0(t)\tilde{F}(t)$ by using (5.17), Problem I.5 and the estimate

$$\mathsf{P}\left\{\int_0^t |Q(x(s)/R) - Q(0)|\, ds > \varepsilon\right\} \leqslant \frac{1}{\varepsilon}\int_0^t \mathsf{E}\{|Q(x(s)/R) - Q(0)|\}\, ds.$$

38. If $Q(x)$, N_0 and F are as in the preceding problem, show that

$$\lim_{\varepsilon\downarrow 0} \varepsilon^{d/2} N(-\varepsilon\Delta + Q, \lambda) = \int_{\mathbf{R}} N_0(\lambda - q) F(dq).$$

Hint. Reduce to the preceding problem by a linear change of coordinates.

39. Let $q(x)$, for $x \in \mathbf{R}^d$, be a metrically transitive such that $\mathsf{E}\{|q(0)|^{p+1}\} < \infty$, where p is the smallest even number strictly larger than $d/2$, and that $\mathsf{E}\{q(0)\} = 0$. Prove that if $\tilde{N}(t) = \int_{\mathbf{R}} e^{-t\lambda} N(d\lambda)$ we have

$$\tilde{N}(-\Delta + q, t) \geqslant \tilde{N}(-\Delta, t).$$

Show that, if $d = 1$, the inequality $N(-\Delta + q, \lambda) \geqslant N(-\Delta, \lambda)$ cannot be true for all $\lambda \in \mathbf{R}$.

Hint. Assume the converse and derive a contradiction using the inequality

$$\begin{aligned}\int_a^b (N_1(\lambda) - N_0(\lambda))\, d\lambda &\leqslant e^{tb} \int_{-\infty}^{\infty} e^{-\lambda t} (N_1(\lambda) - N_0(\lambda))\, d\lambda \\ &\leqslant \frac{e^{tb}}{t} (\tilde{N}_1(\lambda) - \tilde{N}_0(\lambda)),\end{aligned}$$

where N_1 and N_0 are the operators $-\partial^2/\partial x^2 + q$ and $-\partial^2/\partial x^2$, and the right-hand side approaches 0 as $t \downarrow 0$.

40. Let A and B be the metrically transitive operators in $L_2(\mathbf{R}^d)$ or $\ell_2(\mathbf{Z}^d)$, such that A has an integrated density of states $N(A, \lambda)$ and B is bounded with probability 1, that is, the nonrandom number $\|B\|$ is finite. Consider the family of operators $A_g = A + gB$, for $g \in \mathbf{R}$, and assume $N(A_g, \lambda)$ is absolutely continuous, with density $n(A_g, \lambda)$. Prove that

$$\left|\frac{\partial}{\partial g} N(A_g, \lambda)\right| \leqslant \sup_{|\mu-\lambda|\leqslant \|B\|} n(A_g, \mu).$$

Hint. Use Theorems 4.14 and 5.24.

41. For a given $0 < a < 1$ and independent, identically distributed random variables ξ_n, $n \in \mathbf{Z}$, consider the field $q(x) = \sum_n \xi_n \chi_{[0,a]}(x - n)$. Prove that every point $\left(\pi m/(1-a)\right)^2$, for $m \in \mathbf{Z}_+$, lies in the interior of a gap in the spectrum of the Schrödinger operator $-d^2/dx^2 + q$.

Hint. Use Theorem 5.35.

42. Consider a one-dimensional Schrödinger operator with potential

$$\sum_n \xi_n \delta(x-n),$$

where the ξ_n, for $n \in \mathbf{Z}$, are independent, identically distributed random variables whose distribution function $F(d\xi) = f(\xi)d\xi$ has bounded density, $f(\xi) \leqslant f_0 < \infty$. Show that the integrated density of states of this operator is absolutely continuous: $N(d\lambda) = n(\lambda)\, d\lambda$. (Concerning the meaning of the Dirac delta function as a potential, see (6.26) and the text below.)

Hint. Prove a formula similar to that of Lemma 4.20: if $q(x) = q_0(x) + \xi\delta(x)$,

$$G(x,y) = G_0(x,y) - \frac{\xi G_0(x,0)G_0(0,y)}{1+\xi G_0(0,0)},$$

where $G(x,y)$ is the Green's function for the Schrödinger operator with potential $q(x)$, and $G_0(x,y) = G(x,y)\big|_{\xi=0}$.

43. Prove that the bottom of the spectrum of a metrically transitive Schrödinger operator, or its discrete analogue, cannot be an eigenvalue, with probability 1.

Hint. Use Theorems 2.10 and 2.12 and the fact that for such operators and eigenvalue at the bottom must be simple [Reed and Simon IV].

Chapter III

Integrated Density of States in One-Dimensional Problems of Second Order

The spectral theory of one-dimensional differential and finite-difference operators of second order is the most developed area of spectral theory, in particular due to the existence of certain objects specific to dimension one, in particular the phase variables—amplitude and phase—of the solutions. In traditional spectral theory these variables are explicitly used almost exclusively in the analysis of boundary problems on a finite interval [Hartman 1964], although Weyl's theory of boundary problems on an infinite interval makes at least implicit use of the phase formalism at several central points.

In the case of metrically transitive coefficients, whose behavior is in a certain sense homogeneous with respect to the coordinate, the phase variables turn out to have fairly regular asymptotic probabilistic properties for large values of the argument. This important distinction from the case of arbitrary potential permits their use as an efficient tool for the spectral analysis of operators of second order, as will become clear in the next two chapters.

In this chapter we introduce one of these two variables, the phase, and show how it yields information on the integrated density of states. In subsequent chapters, after developing the phase formalism in more detail, we will apply it to the study of the spectrum of one-dimensional operators of order two.

6 The Oscillation Theorem and the Integrated Density of States

6.A The Phase and the Existence of the Integrated Density of States

Let (a, b) be an interval on the real axis. As in Section 5, $N_{(a,b)}(\lambda)$ will denote $(b-a)^{-1}$ times the number of eigenvalues not exceeding λ of the self-adjoint boundary problem

$$-\psi''(x) + q(x)\psi(x) = \lambda\psi(x), \tag{6.1}$$

$$\psi(a)\cos\alpha - \psi'(a)\sin\alpha = 0, \qquad \psi(b)\cos\beta - \psi'(b)\sin\beta = 0. \tag{6.2}$$

We now introduce the phase function $\theta(x)$, which plays a fundamental role in spectral problems in dimension one. Let $y(x)$ be the solution of the Cauchy problem for equation (6.1), that is, y satisfies (6.1) and the conditions

$$y(a) = \sin\alpha, \qquad y'(a) = \cos\alpha. \tag{6.3}$$

If the metrically transitive potential $q(x)$ is summable, that is, $\mathsf{E}\{|q(x)|\} < \infty$, it is easy to show by successive approximations that $y(x)$ exists and is unique, with probability 1, for all finite x. The *phase* $\theta(x)$ of this solution is defined by

$$\cot\theta(x) = \frac{y'(x)}{y(x)} \tag{6.4}$$

and the requirement that it be continuous. One can easily check that $\theta(x)$ satisfies the equation

$$\theta'(x) = \Phi\big(q(x), \theta(x)\big) \qquad \theta(a) = \alpha, \tag{6.5}$$

where $\Phi(q,\theta) = \cos^2\theta + (\lambda - q)\sin^2\theta$.

The importance of θ is immediately apparent:

(6.1) Proposition [Hartman 1964]. *If the potential $q(x)$ is square-summable on (a, b), we have*

$$N_{(a,b)}(\lambda) = \frac{1}{b-a}\left[\frac{\theta(b) - \theta(a)}{\pi}\right], \tag{6.6}$$

where $[t]$ denotes the integer part of t.

Proof. We will just sketch the origin of (6.6). By the definition of $\theta(x, \lambda)$ (where we make the dependence of λ explicit), the eigenvalues of (6.1)–(6.2) satisfy $\theta(b, \lambda_k) = \beta + \pi n$, where n is an integer depending on k. But $\theta(b, \lambda)$ increases monotonically with λ, because so does the function $\Phi(q, \theta)$ of (6.5). Therefore $(b - a)N_{(a,b)}(\lambda)$ is the number of points of the form $\beta + \pi n$, for $n \in \mathbf{Z}$, that fall within the interval $\big(\theta(b, -\infty), \theta(b, \lambda)\big]$. Now it only remains to check that $\theta(b, -\infty) = 0$. □

If $q(x)$ is such that

$$\mathsf{E}\{|q(0)|\} < \infty \quad \text{and} \quad q(x) \in L_2^{\text{loc}}(\mathbf{R}) \text{ with probability 1,} \tag{6.7}$$

$N_{(a,b)}(\lambda)$ can be analyzed using Proposition 6.1. Here is a simple illustration:

(6.2) Lemma. *If a one-dimensional Schrödinger operator has a metrically transitive potential satisfying* (6.7), *we have*

$$\mathsf{E}\{N_{(a,b)}(\lambda)\} \leqslant c(\lambda), \tag{6.8}$$

where $c(\lambda)$ *does not depend on* $b-a$ *when* $b-a \geqslant L_0 > 0$.

Proof. Standard methods [Hartman 1964] show that equation (6.1) has a unique solution if $q(x)$ is locally summable. We therefore have

$$\bigl|\theta(x)\bigr| \leqslant \alpha + (x-a) + \int_a^x \bigl|q(t)-\lambda\bigr|\,dt \tag{6.9}$$

which, after substitution in (6.6), leads to (6.8). □

The following result will be useful below. We say that a random process $\xi(t,s)$, for $s,t \in \mathbf{R}$, is *subadditive* if the following conditions are satisfied:

(a) for any $t < u < s$ we have $\xi(t,s) \leqslant \xi(t,u) + \xi(u,s)$;
(b) the joint distribution of any finite family of quantities $\xi(t_j, s_j)$, for $j = 1, \ldots, n$, coincides with the joint distribution of the family $\xi(t_j + a, s_j + a)$, for any $a \in \mathbf{R}$;
(c) the mathematical expectation $\mathsf{E}\{\xi(0,s)\} = m(s)$ exists, and it satisfies $m(s) \geqslant -cs$ for some constant $c > 0$ and all $s \geqslant 1$.

(6.3) Proposition (the subadditive ergodic theorem). *If* $\xi(t,s)$, *for* $s,t \in \mathbf{R}$, *is a subadditive random process, the limit* $\xi = \lim_{s\to\infty} \xi(0,s)/s$ *exists with probability* 1. *Therefore* $\mathsf{E}\{\xi\} = \inf_{s \geqslant 1} m(s)/s$, *and* ξ *is measurable with respect to the* σ*-algebra of events invariant under shifts* T_a, *for* $a \in \mathbf{R}$. *In particular, if the* T_a *are metrically transitive,* ξ *is a nonrandom quantity.*

Proof. See, for example, [Hall and Heyde 1980]. □

(6.4) Theorem [Benderskii and Pastur 1970]. *Consider a one-dimensional Schrödinger operator with a metrically transitive potential satisfying* (6.7), *the angles* α *and* β *in the boundary conditions being arbitrary random variables with values in* $[0,\pi]$. *There exists a nonrandom continuous function* $N(\lambda)$, *independent of* α *and* β *and such that*

$$\lim_{L\to\infty} N_{(0,L)}(\lambda) = \lim_{L\to\infty} N_{(-L,0)}(\lambda) = \lim_{L\to\infty} N_{(-L,L)}(\lambda) = N(\lambda) \tag{6.10}$$

with probability 1, *for any* $\lambda \in \mathbf{R}$, *and*

$$N(\lambda) = \lim_{n\to\infty} \frac{1}{\pi L}\mathsf{E}\{\theta(L)\}. \tag{6.11}$$

Proof. Equation (6.10) could certainly be derived from Theorem 5.20, by approximating with bounded potentials and using estimate (6.8), along the lines of the proofs in Sections 4 and 5. But we will use another method, which is technically simpler in this case, based on the variational inequalities (5.38) and (5.39). In the multidimensional case, this method was used to prove similar theorems by Slivnyak [1966] and Pastur [1971b].

Denote by $N^{(0)}_{(a,b)}(\lambda)$ and $N^{(1)}_{(a,b)}(\lambda)$ the functions $N_{(a,b)}(\lambda)$ corresponding to $\alpha = \beta = 0$ and $\alpha = \beta = \pi/2$, respectively, in (6.2); these are the Dirichlet and the Neumann problems. By the one-dimensional version of (5.39), the quantities $(b-a)N^{(0)}_{(a,b)}(\lambda)$ and $-(b-a)N^{(1)}_{(a,b)}(\lambda)$, as functions of a and b, represent subadditive random processes, and in each case the mathematical expectation is finite for all finite a and b with $a < b$. Using the subadditive ergodic theorem (Proposition 6.3), we conclude that for any $\lambda \in \mathbf{R}$ the limits

$$\text{(6.12)} \qquad \lim_{L\to\infty} N^{(0)}_{(0,L)}(\lambda) = N^{(0)}(\lambda), \qquad \lim_{L\to\infty} N^{(1)}_{(0,L)}(\lambda) = N^{(1)}(\lambda)$$

exist with probability 1. But it is known [Hartman 1964] that for a boundary value problem involving second-order equations on a finite interval, an arbitrary variation of the boundary condition changes the number of eigenvalues that fall within a fixed interval by no more than 2; this is because such a change in boundary conditions represents a perturbation of the resolvent of the original operator by an operator of rank at most 2. This means that, for any a, b and λ, we have

$$0 \leqslant N^{(1)}_{(a,b)}(\lambda) - N^{(0)}_{(a,b)}(\lambda) \leqslant 2(b-a)^{-1}.$$

Now it follows from (5.38) that $\lim_{L\to\infty} N_{(0,L)}(\lambda) = N(\lambda)$ with probability 1 for every λ, and

$$\text{(6.13)} \qquad N(\lambda) = \lim_{L\to\infty} \mathsf{E}\{N_{(0,L)}(\lambda)\}.$$

Since the shifts T_a, for $a \in \mathbf{R}$, acting on the space of realizations, form a group, the families of random variables $(b-a)N^{(0)}_{(-b,-a)}(\lambda)$ and $-(b-a)N^{(1)}_{(-b,-a)}(\lambda)$ are also subadditive processes. Thus, for any λ, there exists with probability 1 a nonrandom limit $\lim_{L\to\infty} N_{(-L,0)}(\lambda)$, which is almost sure to coincide with $\lim_{L\to\infty} \mathsf{E}\{N_{(-L,0)}(\lambda)\}$. The latter limit, by the invariance of the probability measure under the T_a, equals $\lim_{L\to\infty} \mathsf{E}\{N_{(0,L)}(\lambda)\}$, so it equals $N(\lambda)$ by (6.13). Finally, taking into account the bound in the drop of eigenvalues under changes of boundary conditions, we get

$$\left|N_{(-L,L)}(\lambda) - \tfrac{1}{2}(N_{(0,L)}(\lambda) + N_{(-L,0)}(\lambda))\right| \leqslant \text{const}\, L^{-1},$$

so for every λ the function $N_{(-L,L)}(\lambda)$ tends, with probability 1, to the same limit (6.13) as $L \to \infty$.

That the limit is a continuous function follows from Theorem 3.2. Finally, (6.11) follows from (6.10) and (6.6). □

Since the zeros of the solution $y(x)$ of the Cauchy problem—equations (6.1) and (6.3)—correspond, according to (6.4), to phase values that are multiples of π, and since each value is taken only once—equation (6.5) implies that $\theta'\big|_{\theta=\pi n} = 1$—we conclude from Proposition 6.1 that the number $(b-a)N_{(b,a)}(\lambda)$ of eigenvalues of (6.1)–(6.2) not exceeding λ differs by no more than one from the number of zeros of $y(x)$ on (a, b). This statement is often called the *oscillation theorem* for the Schrödinger equation; it also holds for the more general Sturm–Liouville equation of Example 1.4(b). It allows the derivation of expressions for $N_{(a,b)}(\lambda)$ other than (6.11).

For example, if we restrict ourselves to positive values of λ, the phase can be introduced in a somewhat different way, as the function $\chi(x)$ such that

$$\frac{y'(x)}{y(x)} = \sqrt{\lambda}\cot\chi(x). \tag{6.14}$$

From (6.1) it follows that χ satisfies

$$\chi' = \sqrt{\lambda} - \frac{q}{\sqrt{\lambda}}\sin^2\chi, \tag{6.15}$$

whence $\chi'\big|_{\chi=\pi n} = \sqrt{\lambda} > 0$. Therefore, $\chi(b)$ differs by no more than π from the number of eigenvalues of problem (6.1)–(6.2) not exceeding λ, and together with (6.11) we can write, for $\lambda > 0$,

$$N(\lambda) = \lim_{L\to\infty}\frac{1}{\pi L}\mathsf{E}\{\chi(L)\}. \tag{6.16}$$

6.B Simplest Asymptotics of the Integrated Density of States at the Edges of the Spectrum

Since in equation (6.15) the denominator of the term with the potential contains λ, the phase $\chi(x)$ is a convenient variable for λ large. By integrating (6.15) between 0 and L and substituting the result into (6.16), we verify that the asymptotic formula (5.36), obtained in Section 5 for a bounded potential, is true in the one-dimensional case even under the weaker condition (6.7). This formula can be sharpened further:

(6.5) Theorem. *If the metrically transitive potential in* (6.1) *has a finite second moment* $\mathsf{E}\{q^2(x)\} < \infty$,

$$N(\lambda) = \frac{\sqrt{\lambda}}{\pi}\left(1 - \frac{1}{2\lambda}\mathsf{E}\{q(0)\}\right) + o(\lambda^{-1})$$

as $\lambda \to \infty$.

Proof. Equations (6.15) and (6.16) immediately imply that this formula is true if we replace $o(\lambda^{-1})$ by

$$\frac{1}{2\lambda}\lim_{L\to\infty}\frac{1}{L}\int_0^L \mathsf{E}\{q(x)\cos 2\chi(x)\}\,dx.$$

Thus, we have to show that this expression is $o(\lambda^{-1})$ as $\lambda \to +\infty$. This is a result of fast oscillations of the integrand as $\lambda \to +\infty$. More precisely, we write the integral as a sum of integrals over intervals $[x_j, x_{j+1}]$, for $j = 0, \ldots, n-1$, where $x_j = jL/n$. We set $h = L/n$, $k = \pi/h$, $\lambda = k^2$, $\chi_j = \chi(x_j)$, and $\chi_j(x) = \chi_j + k(x - x_j)$, for $x \in [x_j, x_{j+1}]$. Using (6.15), we find that the error caused by replacing $\chi(x)$ by $\chi_j(x)$ in each integral is no greater than

$$O\big(k^{-2}h\mathsf{E}\{q^2(0)\}\big) = O(k^{-3}).$$

After the replacement we use the equation $\int_{x_j}^{x_{j+1}} \cos 2\chi_j(x) = 0$ and the fact that $\mathsf{E}\{q(x+y) - q(y)\}$ is independent of y—an immediate consequence of metrical transitivity—to get

$$\frac{1}{2\lambda}\lim_{L\to\infty}\sum_{l=0}^{n-1}\mathsf{E}\left\{\int_{x_l}^{x_{l+1}}(q(x) - q(x_l))\cos 2\chi_l(x)\,dx\right\}$$
$$\leqslant \frac{1}{2\lambda h}\int_0^h \mathsf{E}\{|q(x) - q(0)|\}\,dx.$$

Now $\lim_{x\to 0}\mathsf{E}\{|q(x) - q(0)|\} = 0$, because the definition of a metrically transitive process includes the condition of stochastic continuity (1.19), and the second moment of q is finite by assumption. This shows our integral is $o(\lambda^{-1})$. □

Remark. The formulation and the method of proof of this theorem go back to [Hartman and Putnam 1950] (see also [Glazman 1965]), who used similar arguments to show that if a (not necessarily metrically transitive) potential $q(x)$ is a uniformly continuous function on the whole axis, the asymptotic estimate

$$\frac{\chi(x)}{x} = \sqrt{\lambda}\left(1 - \frac{1}{2\lambda x}\int_0^x q(t)\,dt + o(\lambda^{-1})\right) \quad \text{as } \lambda \to \infty$$

holds uniformly for $x \to \infty$. Using this, the authors found that the gaps in the spectrum of a one-dimensional Schrödinger operator in $L_2(\mathbf{R}^d)$ with such a potential tend to zero as $\lambda \to \infty$. This question, in the metrically transitive case, is associated with the asymptotic behavior of $N(\lambda)$ as $\lambda \to \infty$, as we saw in Section 5. In this case, instead of uniform continuity, one uses the equation $\lim_{x\to 0} \mathsf{E}\{|q(x+y)-q(y)|^2\} = 0$—which holds uniformly in y, by metric transitivity—in order to provide a "fixed scale" for the change in potential in comparison with $\cos \chi(x)$, which oscillates increasingly rapidly and uniformly as $\lambda \to \infty$.

We now discuss some improvements of the formula of Theorem 6.5. But first notice that the formula cannot be improved for arbitrary metrically transitive potentials. This follows from the results in [Marchenko 1986]: for any sequence $\{h_k : k \geqslant 1\}$ of nonnegative numbers such that $\sum_{k\geqslant 1}(kh_k)^2 < \infty$, there exists a periodic potential $q(x)$ of period π, square-integrable over a period—and therefore satisfying the condition of Theorem 6.5—and satisfying the following property: the difference

$$N(\lambda) - \frac{\sqrt{\lambda}}{\pi}\left(1 - \frac{\mathsf{E}\{q\}}{2\lambda}\right)$$

has upper and lower bounds of the form const nh_n/λ, for λ in the allowed band $[\nu_{n-1}^+, \nu_n^-]$, and $\nu_n^\pm = n^2(1+o(1))$ as $n \to \infty$.

Clearly, then, we must impose additional conditions on the potential. There are two directions we can go in, which are in a sense opposite. One is to assume that the realizations of the potential get progressively smoother. If, for example, $q(x)$ is bounded with probability 1 and $\mathsf{E}\{|q'(x)|^2\} < \infty$, we can integrate by parts in (6.15) and then argue as in the proof of Theorem 6.5, to find that the difference $N(\lambda) - \pi^{-1}\sqrt{\lambda}(1 - \mathsf{E}\{q\}/2\pi)$ is of order $o(\lambda^{-1})$, rather than $o(\lambda^{-1/2})$, as in the proof of Theorem 6.5, where the only assumption was that $\mathsf{E}\{q^2\} < \infty$. If we further assume that $\mathsf{E}\{|q''|^2\} < \infty$ and integrate by parts once more, we obtain the following asymptotic expansion for $N(\lambda)$ as $\lambda \to \infty$:

$$N(\lambda) = \frac{\sqrt{\lambda}}{\pi}\left(1 - \frac{1}{2\lambda}\mathsf{E}\{q\} - \frac{1}{8\lambda^2}\mathsf{E}\{q^2\} + o(\lambda^{-2})\right). \tag{6.17}$$

We can continue in this way, assuming derivatives of higher and higher order of the potential to have finite moment. This process makes very little use metric transitivity; it is essentially a version of the process of improvement of quasiclassical approximations, which, as is well-known, require the potential to be sufficiently smooth.

The other direction we can go is to assume the metrically transitive potential is sufficiently random: for example, we can consider a Markov process with sufficiently good ergodic properties. Assuming for simplicity that

$\mathsf{E}\{q(0)\} = 0$, we can formally solve equation (6.15) down to terms of order q/λ, then substitute the result into (6.16) to get the following formula:

$$N(\lambda) \sim \frac{\sqrt{\lambda}}{\pi}\left(1 - \frac{1}{4\lambda^{3/2}}\int_0^\infty b(x)\sin 2\sqrt{\lambda}x\,dx\right), \tag{6.18}$$

where $b(x) = \mathsf{E}\{q(x)q(y+x)\}$ is the correlation function of $q(x)$.

After integrating by parts we again arrive at (6.17); but this new (formal) derivation assumes not that the potential is smooth, but rather that the correlation function $b(x)$ drops fast enough at infinity, and that the amplitude of the potential is small compared with λ. Therefore it is reasonable to think that (6.17) and (6.18) hold for random potentials whose realizations are not necessarily smooth. Indeed, we will see in Section 7 (cf. Problem III.27) that (6.18) holds for the potential representing the random telegraph signal of Example 1.15(b). The same formula agrees with the results of Section 7 on potentials representing generalized random processes with independent values, such as Gaussian white noise of the Poisson potential (6.25), for which $b(x) = \mathcal{D}\delta(x)$.

The arguments used in the derivation of (6.18) suggest that this formula should also describe the asymptotic behavior of $N(\lambda)$ for fixed $\lambda > 0$ and a small random potential, that is, one of the form $q(x) = gQ(x)$, where $Q(x)$ is a fixed Markov process and $g \to 0$. Using the explicit formulas (7.3) one can check that this is indeed the case if $Q(x)$ is a random telegraph signal.

Another application of the phase formalism to the investigation of the integrated density of states results if we consider the metrically transitive differential operator

$$(H_1\psi)(x) = -\frac{d}{dx}\left(s(x)\frac{d\psi}{dx}\right)(x), \tag{6.19}$$

where $s(x)$ is a metrically transitive process whose realizations are continuously differentiable and bounded functions

$$0 < s_0 \leqslant s(x) \leqslant s_1 < \infty. \tag{6.20}$$

In this case the boundary problem in the finite interval (a, b) is specified by the conditions

$$\psi(a)\cos\alpha - s(a)\psi'(a)\sin\alpha = 0,$$
$$\psi(b)\cos\beta - s(b)\psi'(b)\sin\beta = 0,$$

and the corresponding Cauchy problem by the conditions

$$y(a) = \sin\alpha, \qquad s(a)y'(a) = \cos\alpha.$$

The phase $\theta(x)$ of this solution is defined by

$$\cot\theta(x) = \frac{s(x)y'(x)}{cy(x)},$$

where c is a positive quantity, independent of x, whose value will be given below. This phase satisfies the equation

$$\theta' = \frac{c}{s}\cos^2\theta + \frac{\lambda}{c}\sin^2\theta. \tag{6.21}$$

The function $N_{(a,b)}(\lambda)$ for H_1 is defined as for the Schrödinger operator, and one can show that H_1 has an integrated density of states $N(H_1,\lambda)$ using arguments similar to those in the proof of (6.10). Moreover, $N(H_1,\lambda)$ is continuous, so the limits in (6.10) are attained with probability 1 for every λ. (It turns out that $N_{(a,b)}(\lambda)$ satisfies (5.38) and (5.39), so that the existence of $N(H_1,\lambda)$ for H_1 can be proved in the same way we proved Theorem 6.4. In fact, it is enough to assume that $\mathsf{E}\{1/s(x)\}$ is finite, because this implies, by (6.21), that $\mathsf{E}\{\theta(L)/L\}$ is finite.)

Here, too, the oscillatory properties of the solution of the Cauchy problem are related to the number of eigenvalues of the problem on a finite interval, that is, a result similar to Proposition 6.1 holds [Hartman 1964]. Thus, starting from (6.21), we get the following analogue for (6.11):

$$N(H_1,\lambda) = \lim_{L\to\infty} \mathsf{E}\left\{\frac{1}{\pi L}\int_0^L \left(\frac{c}{s}\cos^2\theta + \frac{\lambda}{c}\sin^2\theta\right)dx\right\}. \tag{6.22}$$

The operator H_1 is nonnegative and the bottom of its spectrum is always 0, as can be seen by considering the corresponding quadratic form on smooth functions with compact support that remain constant sufficiently far away. The asymptotic behavior of $N(H_1,\lambda)$ in the neighborhood of 0 is given by the following result:

(6.6) Theorem. *If, in* (6.19), $s(x)$ *is a metrically transitive process with continuously differentiable realizations satisfying* (6.20), *we have*

$$N(H_1,\lambda) = \frac{1}{\pi}\sqrt{\frac{\lambda}{\kappa}}(1+o(1)) \tag{6.23}$$

as $\lambda \downarrow 0$, *where* $\kappa^{-1} = \mathsf{E}\{1/s(x)\}$.

Proof. Set $\lambda = k^2 > 0$, $c = k\sqrt{\kappa}$ and $\delta(x) = \kappa(s^{-1} - \kappa^{-1})$. If $\chi(x)$ denotes the phase obtained with this choice of parameters, (6.22) and (6.21) become

$$N(H_1,\lambda) = \frac{k}{\pi\sqrt{\kappa}}\left(1 + \lim_{L\to\infty}\mathsf{E}\left\{\frac{1}{L}\int_0^L \delta(x)\cos^2\chi(x)\,dx\right\}\right), \tag{6.24}$$

$$\chi' = \frac{k}{\sqrt{\kappa}} + \frac{k}{\sqrt{\kappa}}\delta(x)\cos^2\chi. \tag{6.24a}$$

Now the limit in (6.24) is still valid if we restrict our attention to a subsequence, so we choose $L = \nu h$, where h is fixed and ν is an integer that tends to infinity. We set $x_n = nh$, for $n = 0, 1, \ldots, \nu$, $\Delta_n = (x_n, x_{n+1})$ and $\chi_n = \chi(x_n)$. By assumption, $\delta(x)$ is bounded, so from the equation for χ we get $\cos^2 \chi(x) - \cos^2 \chi_n = O(kh)$ for $x \in \Delta_n$. Thus, writing the integral in (6.24) as a sum of integrals over the intervals Δ_n, we have

$$\begin{aligned}
&\frac{1}{L}\int_0^L \delta(x)\cos^2\chi\,dx \\
&\quad = \frac{1}{L}\sum_{n=0}^{\nu-1}\left(\cos^2\chi_n \int_{\Delta_n}\delta(x)\,dx + \int_{\Delta_n}\delta(x)(\cos^2\chi - \cos^2\chi_n)\,dx\right) \\
&\quad \leqslant \frac{1}{L}\sum_{n=0}^{\nu-1}\left|\int_{\Delta_n}\delta(x)\,dx\right| + O(kh).
\end{aligned}$$

Thus, the second term in (6.24) is bounded above by $\mathsf{E}\{\left|\frac{1}{h}\int_0^h \delta(x)\,dx\right|\} + O(kh)$. Choosing $h = \kappa^{-\alpha}$, for $\alpha > 0$, we can prove that the first term vanishes as k approaches 0, based on the ergodic theorem and on the equality $\mathsf{E}\{\delta(x)\} = 0$. If $\alpha < 1$, the second term will go to zero as k approaches 0. The theorem is proved. □

We note in passing that asymptotic formulas for the discrete analogue (1.4) of H_1 are similar to (6.23), and can be found in [Lifshitz et al. 1982b].

Theorems 6.5 and 6.6 have in common the fact that the asymptotic behavior they describe corresponds to that of a differential operator with constant coefficients, namely, $-d^2/dx^2$ and $-\kappa^{-1}d^2/dx^2$, respectively (cf. Subsection 5.C). The reason in the first case is that the spectrum near $\lambda = \infty$ is determined chiefly by the $-d^2/dx^2$ term; in the second case it is the "divergent" form of the operator that gives the spectrum immediately to the right of the point $\lambda = 0$ for any function $s(x)$. In each case the spectrum near the boundary of interest arises from regions in the space of of realizations consisting of realizations of arbitrary form, so it is not surprising that the asymptotics of the integrated density of states depends only weakly on the random function that goes into the operator. Boundaries like this, near which the integrated density of states asymptotically coincides with that of an operator with constant coefficients, we call *stable*.

But for random operators a different situation can arise. For example, in the proof of Theorem 5.34 we saw that for a Schrödinger operator whose potential is a Poisson field with $u(x)$ nonnegative, the spectrum immediately to the right of 0 owes its existence to the fact that such a random field can be arbitrarily close to zero in an arbitrarily large cube. Since the spectrum near the boundary stems from regions in the space of realizations with very atypical behavior, straying, as they do, very far from the mathematical expectation, it seems natural to refer to such boundaries as *fluctuation boundaries*.

(In contrast, for an alloy-type potential (5.56), Theorem 5.35 says that the bottom of the spectrum coincides with the bottom of the spectrum of the Schrödinger operator with a periodic potential obtained from (5.56) when $\xi_y = \inf \operatorname{supp} F$; it is therefore strictly positive if $u(x) \geqslant 0$.)

The probability of large fluctuations of a random potential is typically very small—for the Poisson field, it decreases exponentially with increasing fluctuation volume—so the integrated density of states near fluctuation boundaries must also be small. Its exact asymptotic behavior depends strongly on the statistical properties of the potential, on the amount of correlation between values at points far apart.

Chapter IV will present a systematic study of this asymptotic behavior for multidimensional Schrödinger operators at fluctuation boundaries, for several classes of random potentials. Here we derive just one asymptotic formula near a fluctuation boundary, for the one-dimensional Poisson random potential of Example 1.15(d). We do this using the phase formalism, which is applicable in dimension one only.

In our example we take $u(x) = k_0\delta(x)$, where $k_0 > 0$ is constant and $\delta(x)$ is the Dirac delta function, modeling the interaction of a one-dimensional quantum particle with a point impurity. Thus the potential has the form

$$q(x) = k_0 \sum_j \delta(x - x_j). \tag{6.25}$$

(This potential and its multidimensional analogue have been considered in [Kirsch and Martinelli 1982b] and [Grossman et al. 1980]; other approaches are discussed in [Krein and Kac 1968] and [Zhikov 1967].)

Since the realizations of the potential are generalized functions, it is not obvious that the theory developed in Chapters I and II is applicable; but in fact all basic general results on metrically transitive operators still hold. This can be shown, for example, by considering the quadratic form

$$k_0 \sum_j \psi_1^*(x_j)\psi_2(x_j)$$

formally corresponding to the potential (6.25), on functions in the domain of the operator $-d^2/dx^2$.

The spectrum of the operator with potential $q(x)$ is the semiaxis $[0,\infty)$. But we will not pursue this question in detail, because for the purposes of finding an asymptotic formula for the integrated density of states it is enough to consider boundary problems on finite intervals. Here, since every finite interval contains with probability 1 only a finite number of points x_j, we can replace the delta functions by the following pairs of conditions:

$$\lim_{x \downarrow x_j} \psi'(x) = \lim_{x \uparrow x_j} \psi(x), \qquad \lim_{x \downarrow x_j} \psi'(x) - \lim_{x \uparrow x_j} \psi'(x) = k_0 \psi(x_j) \tag{6.26}$$

This is justified by taking the limit from smooth functions $u(x)$ with compact support, or by construction the self-adjoint extension of $-d^2/dx^2$ to a suitable set of functions: see [Berezin and Fadeev 1961], [Kotani 1976] and [Kirsch and Martinelli 1982b].

(6.7) Theorem [Gredeskul and Pastur 1975]. *For a one-dimensional Schrödinger operator with potential* (6.25), *where* $k_0 > 0$ *and the points* x_j *are distributed according to the Poisson law with density* n, *we have*

$$\ln N(\lambda) = -\frac{\pi n}{\sqrt{\lambda}}(1 + O(\lambda^{1/2})) \tag{6.27}$$

as $\lambda \downarrow 0$.

Proof. We use the phase $\chi(x)$ defined by (6.14). Set $\lambda = k^2 > 0$, $x_{j+1} - x_j = l_j$, $\lim_{x \downarrow x_j} \chi(x) = \chi_j$. From (6.15) and (6.26) we find that for $x_j < x < x_{j+1}$ we have

$$\cot \chi(x) = \cot(\chi_j + kx)$$

and

$$\cot \chi_{j+1} = \cot(\chi_j + kl_j) + k_0/k.$$

From this it follows readily that if $0 < l_j < l_{\text{cr}} = \pi/k - 4/k_0$ and $0 < \chi_j < \chi_{\text{cr}} = \tan^{-1} k_0/4k$, we have $\chi(x) \leqslant \pi$ for $x \in (x_j, x_{j+1})$, and furthermore $0 < x_{j+1} < \chi_{\text{cr}}$. In other words, if immediately to the right of x_j the phase is small enough and the distance to the next point x_{j+1} is not too large, the phase remains small throughout the interval (x_j, x_{j+1}).

By induction, we find that between two zeros of the solution $y(x)$ of the Cauchy problem there must be at least one interval of length not less than l_{cr} and containing no point x_j. On the other hand, $y(x)$ must have a zero on every interval with length greater than π/k, so denoting by $\mathcal{N}_L(l)$ the maximum number of disjoint intervals of length l not containing any point x_j that fit into the interval $(0, L)$, we get

$$\frac{1}{L}\mathcal{N}_L\left(\frac{\pi}{k}\right) + O(L^{-1}) \leqslant N_{(0,L)}(\lambda) \leqslant \frac{1}{L}\mathcal{N}_L(l_{\text{cr}}) + O(L^{-1}).$$

After taking expectations and passing to the limit $L \to \infty$, this yields

$$n \sum_{m=1}^{\infty} G\left(\frac{\pi m}{k}\right) \leqslant N(\lambda) \leqslant n \sum_{m=1}^{\infty} G\left(m\left(\frac{\pi}{k} - \frac{4}{k_0}\right)\right), \tag{6.28}$$

where $G(l) = e^{-nl}$. This is equivalent to (6.27). □

Remark. One can easily see that in this proof the Poisson nature of $q(x)$ comes in only in the form of the distribution function $F(l) = 1 - e^{-nl}$ of the random distances $l = x_{j+1} - x_j$ between the delta functions. Thus the same arguments apply to the more general case where these distances are arbitrary nonnegative independent random variables with finite mathematical expectation, whose distribution function $F(l)$ is continuous and strictly less than 1 for any finite l. More precisely, assume given a point x_0 and a sequence $l_1, l_2, \dots$ of independent, identically distributed random variables with a distribution function satisfying these conditions, and set $x_1 = x_0 + l_1$, $x_2 = x_1 + l_2$, and so on. The corresponding random potential is defined only for $x > x_0$, so it's not metrically transitive; but it may be turned into a metrically transitive one by the technique of Problem I.10. In this situation inequality (6.28) still holds, with $1 - F(l)$ for $G(l)$ and $1/\mathsf{E}\{l\}$ for n. Therefore if, for example, there is $a > 1$ such that $G(l_1)/G(l_2) \leqslant c(l_1/l_2)^{-a}$ for all $l_1 > l_2$, it follows from (6.28) that

$$\ln N(\lambda) = \ln G(\pi/\sqrt{\lambda})(1 + o(1)) \tag{6.29}$$

as $\lambda \downarrow 0$. If there is $a > 2$ such that $G(l)$ is asymptotically equal to $Al^{-a}(1 + o(1))$ as $l \to \infty$, (6.29) can be sharpened, also based on (6.28):

$$N(\lambda) = G(\pi/\sqrt{\lambda}) \sum_{n=1}^{\infty} n^{-a}(1 + o(1)) \tag{6.29a}$$

as $\lambda \downarrow 0$.

Other random potentials can be treated similarly, as well as certain one-dimensional finite-difference operators of order two: see [Gredeskul and Pastur 1975] and [Pastur 1982].

Formulas (6.27) and (6.29), as well as (7.4), (7.8) and (7.11) in the next section, are typical for fluctuation boundaries. This type of asymptotic behavior was discovered by [Lifshitz 1964] using the following heuristic arguments: it is clear that the existence of the spectrum immediately to the right of 0 for potential (6.25) with $k_0 > 0$ is due to the occurrence of arbitrarily wide potential wells, that is, arbitrarily large distances $l_j = x_{j+1} - x_j$. But the wider such wells are, the rarer their occurrence, so they are as a rule separated by very large distances—formula (6.27) says the distances are at least of the order of $\exp(l^{1+\varepsilon})$, for $\varepsilon > 0$. Thus, in quantum-mechanical terms, the motion of a particle in each such well, which seems infinitely deep to the particle, is quantized independently, and this part of the spectrum can be obtained by superposition of the spectra of the Dirichlet problems in each well, as expressed by (6.28). As to formulas (6.27) and (6.29), they are explained by the fact that the energy λ of a particle will coincide, with overwhelming probability, with the smallest eigenvalue $(\pi/l)^2$ of the spectrum of an infinitely deep well of length l; this follows essentially from the quantum-mechanical principle of uncertainty. Since the relative number of wells with length not less than l is $G(l)$, setting $l = \pi/\sqrt{\lambda}$ immediately gives (6.27) and (6.29). See also Problem III.31.

These heuristic arguments, of course, are very general, so we should expect similar phenomena to arise for many other problems, including multidimensional ones, where the methods of proof used above no longer apply. We will discuss multidimensional asymptotic formulas, both for stable and fluctuation boundaries, in Chapter IV; here we note only that the arguments used to explain the fluctuation asymptotics also give a picture of the structure of the spectrum near the fluctuation boundaries. One should expect the spectrum to be pure point in this region, with each eigenfunction mainly concentrated within a single fluctuation of the potential, dropping exponentially with distance from it. In Chapters V and VI we will see that these conclusions are indeed true in the one-dimensional case. The most recent rigorous proofs of existence of the point spectrum in the multidimensional Anderson model are based on the same ideas: see, for example, [Fröhlich et al. 1985] and Section 15. Moreover, this "fluctuation ideology" of Lifshitz [1964, 1967] is the basis for an efficient method in the theoretical physics of disordered systems which permits the calculation of a variety of important characteristics: see [Lifshitz et al. 1982b, Shklovskii and Efros 1984].

6.C Schrödinger Operator with Markov Potential

The oscillation theorem offers the possibility of computing $N(\lambda)$, for all λ. The formalism developed in this computation will be used in subsequent chapters, after suitable modifications, to study the nature of the spectrum of one-dimensional operators.

Let $a = 0$ in equation (6.5), integrate between 0 and L, and take the mathematical expectation of the resulting equation. Taking into account that $\Phi(q,\theta)$ is periodic in θ, with period π, we have

$$\text{(6.30)} \qquad \mathsf{E}\{\theta(L)\} = \mathsf{E}\{\alpha\} + \int_0^L dx \int_E \Phi(q,\phi) F(x; dq, d\phi),$$

where $F(x; dq, d\phi) = \mathsf{P}\{q(x) \in dq, \phi(x) \in d\phi\}$, the function $\phi(x) = \theta(x) - \pi[\theta(x)/\pi]$ is the so-called *reduced phase*, and E is the phase space of the process $(q(x), \phi(x))$, forming a cylinder $\mathbf{R} \times S^1$, where S^1 is a circle of length π. After changing the order of integration in this equation, we get

$$\text{(6.31)} \qquad \mathsf{E}\{\theta(L)\} = \mathsf{E}\{\alpha\} + L \int_W \Phi(q,\phi) F_L(dq, d\phi),$$

where the measures

$$F_L(dq, d\phi) = \frac{1}{L}\int_0^L F(x; dq, d\phi)\, dx \tag{6.32}$$

form a precompact family of probability measures on E (in the sense of weak convergence). That this is so follows from the obvious estimate

$$\int_{|q|\leqslant N} |q| F_L(dq, d\phi) \leqslant \int \bigl|q(0,\omega)\bigr| P(d\omega) \tag{6.33}$$

which holds uniformly in L, because of (6.7) and of the metric transitivity of $q(x)$.

Now choose a sequence L_k so that the sequence $F_{L_k}(dq, d\phi)$ converges to some probability measure $F(dq, d\phi)$. After dividing both sides of (6.31) by L, take the limit $L_k \to \infty$. The left-hand side, by (6.11) in Theorem 6.4, will go to $\pi N(\lambda)$, while the right-hand side will become $\int_E \Phi(q,\theta) F(dq, d\phi)$ because $\left|\Phi(q,\theta) \leqslant |q-\lambda| + 1\right.$. We thus get:

(6.8) Theorem [Benderskii and Pastur 1970]. *Let $q(x)$ be a metrically transitive process with $\mathsf{E}\bigl\{|q(0)|\bigr\} < \infty$. Then*

$$N(\lambda) = \frac{1}{\pi}\int_E \Phi(q, \phi) F(dq, d\phi), \tag{6.34}$$

where $F(dq, d\phi)$ is the limit of any weakly converging subsequence $\{F_{L_k}\}$ of the family (6.32) as $L_k \to \infty$. All such limits represent probability measures. □

Theorem 6.8 reduces the calculation of the integrated density of states of a one-dimensional operator to finding a stationary one-dimensional distribution of the two-component process $\bigl(q(x), \phi(x)\bigr)$, that is, the limits of measures of the form (6.32). For a general metrically transitive potential, there is no analytic procedure to solve this problem. But since equation (6.5), relation the potential to the phase, is "local," it is reasonable to think that such a procedure might exist if values of the potential at various points are as weakly dependent as possible. In the continuous case, which we examine now, the case of weakest correlation is that of a Markov process; cf. Example 1.15(b). For the remainder of this section we will concentrate on this class of potentials.

(6.9) Theorem [Benderskii and Pastur 1970]. *If $q(x)$ is a metrically transitive Markov process satisfying condition (6.7), the pair $\bigl(q(x), \phi(x)\bigr)$ is also a Markov process. If $\mathsf{P}_{a,q}$ is the family of Markov measures associated with $q(x)$, the transition probability of the two-component process has the form*

(6.35) $$P(x;q',\phi';dq,d\phi)=\mathsf{P}_{a,q'}\{q(x+a)\in dq \text{ and } \phi(x+a)\in d\phi\},$$

where $\phi(x)=\theta(x)-\pi[\theta(x)/\pi]$ and $\theta(x)$ is the solution of equation (6.5) *with the condition $\theta(a)=\phi'$.*

If $\mathcal{M}$ is the set of probability measures $F(dq,d\phi)$ on E such that

(6.36) $$\int_0^{\pi} F(dq,d\phi)=\mathsf{P}\{q(a)\in dq\}$$

and representing invariant distributions of the Markov process $(q(x),\phi(x))$, that is, satisfying the integral equation

(6.37) $$F(dq,d\phi)=\int_E \mathsf{P}(s;q',\phi';dq,d\phi)F(dq',d\phi'),$$

we can represent $N(\lambda)$ as

(6.38) $$N(\lambda)=\frac{1}{\pi}\int_E \Phi(q,\phi)F(dq,d\phi),$$

where $\Phi(q,\phi)=\cos^2\phi+(\lambda-q)\sin^2\phi$ and $F(dq,d\phi)\in\mathcal{M}$ is arbitrary.

(6.10) Lemma. *$(q(x),\phi(x))$ is a homogeneous Markov process.*

Proof. It is enough to show that $(q(x),\theta(x))$ is a Markov process, by verifying conditions (i)–(iv) of Example 1.15(b). We introduce the following notations: $\theta(x,a,\alpha)$ is the solution of (6.5) with initial condition $\theta(a,a,\alpha)=\alpha$;

$$\tilde{\Omega}=\Omega\times\mathbf{R}^{\mathbf{R}},\quad \tilde{\mathcal{F}}=\mathcal{F}\times\mathcal{B}^{\mathbf{R}}(\mathbf{R})$$
$$(q(x;\tilde{\omega}),\theta(x,\tilde{\omega}))=(q(x;\omega),\theta(x,\tilde{\omega}))\quad\text{for all } \tilde{\omega}\in\tilde{\Omega} \text{ and } x\geqslant 0;$$

$\tilde{\mathcal{F}}_y^x$, for $x\geqslant y$, is the σ-algebra on $\tilde{\Omega}$ generated by the σ-algebra $\mathcal{F}_y^x$ and the random variables $\theta(z)$, for $y\leqslant z\leqslant x$; and $\mathsf{P}_{x,q,\alpha}=\mathsf{P}_{x,q}\times\delta_{\theta(\,\cdot\,,x,\alpha)}$, where δ_ω is the probability measure on $\mathbf{R}^{\mathbf{R}}$ concentrated at point ω.

Notice first that

(6.39) $$\theta(x,y,\alpha)=\theta(x,z,\theta(z,y,\alpha))$$

for all $x\geqslant y\geqslant z$. Next, property (i) of Example 1.15(b) is obviously fulfilled by the definition of $\mathsf{P}_{x,q,\alpha}$. We also have, for $x\geqslant y$,

(6.40) $$\begin{aligned}P(y;q_1,\theta_1;x;dq_2,d\theta_2)&=\mathsf{P}_{y,q_1,\theta_1}\{q(x)\in dq_2,\theta(x)\in d\theta_2\}\\&=\mathsf{P}_{y,q_1}\{q(x)\in dq_2,\theta(x,y,\theta_1)\in d\theta_2\}.\end{aligned}$$

From (6.40), the properties of θ, the q-measurability of $\mathsf{P}_{x,q}$, and Theorem 0.6 of [Dynkin 1965], we conclude that $\mathsf{P}_{x,q}$ is measurable with respect to q_1 and θ_1, which proves property (ii).

To prove (iii), we use the following equalities, which are based on equation (6.39), the properties of Markov processes [Gihman and Skorohod II], the properties of conditional expectations [Gihman and Skorohod I] and the fact that, for fixed α, the function $\theta(x,y,\alpha)$ is an $\mathcal{F}_y^x$-measurable random variable. For $y \leqslant x \leqslant z$ and $Q, \Gamma \in \mathcal{B}(\mathbf{R})$ we have

$$\begin{aligned}
&\mathsf{P}_{y,q,\theta}\{q(z) \in Q, \theta(z) \in \Gamma \mid \tilde{\mathcal{F}}_y^x\} \\
&\qquad = \mathsf{E}_{y,q,\theta}\{\chi_Q(q(z))\chi_\Gamma(\theta(z,x,\theta(x))) \mid \tilde{\mathcal{F}}_y^x\} \\
&\qquad = \mathsf{E}_{y,q,\theta}\{\chi_Q(q(z))\chi_\Gamma(\theta(z,x,\alpha)) \mid \tilde{\mathcal{F}}_y^x\}\big|_{\alpha=\theta(x)} \\
&\qquad = \mathsf{E}_{y,q}\{\chi_Q(q(z))\chi_\Gamma(\theta(z,x,\alpha)) \mid \mathcal{F}_y^x\}\big|_{\alpha=\theta(x)} \\
&\qquad = \mathsf{E}_{x,q(x)}\{\chi_Q(q(z))\chi_\Gamma(\theta(z,x,\alpha)) \mid \mathcal{F}_y^x\}\big|_{\alpha=\theta(x)} \\
&\qquad = \mathsf{P}_{x,q(x),\theta(x)}\{q(z) \in Q, \theta(z) \in \Gamma\}.
\end{aligned}$$

Finally, the homogeneity of the Markov process $\big(q(x), \theta(x)\big)$ for $x \geqslant 0$ follows from the equalities below, which are based on the metric transitivity of the process $q(x)$. If $z \geqslant 0$, $0 \leqslant y < x$ and $Q, \Gamma \in \mathcal{B}(\mathbf{R})$,

$$\begin{aligned}
&P_{y+z,q,\theta}\{q(x+z) \in Q, \theta(x+z) \in \Gamma\} \\
&\qquad = P_{y+z,q}\{q(x+z) \in Q, \theta(x+z, y+z, \theta) \in \Gamma\} \\
&\qquad = P_{y,q}\{q(x+z) \in Q, \theta(x,y,\theta) \in \Gamma\} \\
&\qquad = P_{y,q,\theta}\{q(x) \in Q, \theta(x) \in \Lambda\}. \qquad \square
\end{aligned}$$

Proof of Theorem 6.9. By Theorem 6.8, we can plug into (6.34) any probability measure that is a limit point of the measures

$$\frac{1}{L}\int_0^L dx \int_E \mathsf{P}(x; q_1, \phi_1; dq, d\phi)\tilde{F}(dq_1, d\phi_1), \tag{6.41}$$

as $L \to \infty$, where $\tilde{F}(dq_1, d\phi_1)$ is an arbitrary probability measure on E satisfying (6.36). This follows from the fact, proved in Theorem 6.4, that the integrated density of states is independent of the quantity α in boundary condition (6.2): this independence allows us to define the random variable $\phi(0)$ arbitrarily, in particular in such a way that the joint distribution of the random variables $\big(q(0), \phi(0)\big)$ is equal to $\tilde{F}(dq, d\phi)$.

If we replace $\tilde{F}$ in (6.41) by any invariant distribution $F(dq, d\phi)$ of the process $\big(q(x), \phi(x)\big)$ satisfying (6.36), the result is evidently $F(dq, d\phi)$. Thus, any $F \in \mathcal{M}$ can be plugged into (6.38), which concludes the proof. It is easy to see that the limit of a convergent sequence of the form (6.41) is some invariant measure satisfying (6.36) and (6.37). $\square$

Thus, to find the integrated density of states of a one-dimensional Schrödinger operator with a Markov potential, one must find a probability measure on E satisfying (6.36) and (6.37). Now under fairly general conditions (see, for example, [Doob 1952]), the one-dimensional distribution $F(dq) = \mathsf{P}\{q(0) \in dq\}$ of $q(x)$ is the unique solution of the integral equation (1.26); in such cases condition (6.36) need not be verified, because it follows automatically from equation (6.37), given (6.35).

However, the problem of finding the explicit form, or a sufficiently rich set of properties, of the transition probability (6.35), which serves as the kernel of (6.37), is, in general, quite difficult. Indeed, for an arbitrary Markov process the only way to find it is the rather complicated *Kolmogorov–Chapman–Smoluchowski* equation:

$$(6.42)\quad P(x+y; q_1, \phi_1; dq_2, d\phi_2) = \int_E P(x; q_1, \phi_1; dq', d\phi') P(y; q', \phi'; dq_2 d, \phi_2).$$

(This equation implies, in particular, that if the transition probability (6.35) is known, the invariant distribution of the process may be found in another and, generally speaking, simpler way, than by solving (6.37). It is enough to apply the operation $L^{-1}\int_0^L \cdots\, dx$ to the transition function. Indeed, applying this operation to (6.42) shows that any limit point of a family of measures of this form is a solution of (6.37) satisfying (6.36), provided that the invariant potential distribution is unique.)

However, the theory of Markov processes provides an alternative, infinitesimal approach under certain conditions, which obviates the need to solve (6.42). Assume that the metrically transitive Markov potential $q(x)$ is a stochastically continuous, regular Feller process. What this means is explained by the following definitions [Gihman and Skorohod II]:

A homogeneous Markov process is *stochastically continuous* if, for any neighborhood U_q of a point q, we have $\lim_{x\downarrow 0} P(x; q, U_q) = 1$. It is called a *Feller process* if the semigroup corresponding to the associated transition probability maps into itself the set C of continuous, bounded functions on S, where S is a finite or infinite interval on $\mathbf{R}$ such that $q(x) \in S$ (the phase space of the process). It is *regular* if this same semigroup maps into itself the set $\tilde{C}$ of continuous, bounded functions $f(q)$ on S such that the set $\{q : |f(q)| > \varepsilon\}$ is compact for all $\varepsilon > 0$. If $S = \mathbf{R}$ this condition is obviously equivalent to $\lim_{|q|\to\infty} f(q) = 0$; if the process $q(x)$ is bounded, that is, if S is a finite interval, regularity reduces to the Feller property.

For processes $q(x)$ satisfying all three properties above, the domain $\mathcal{D}(\mathfrak{A}_q)$ of the infinitesimal operator $\mathfrak{A}_q$ is dense with respect to uniform convergence in the set of functions f +const, where $f \in \tilde{C}$ [Gihman and Skorohod II]. If, in addition, $q(x)$ is a bounded process, that is, $|q(x)| \leqslant q_0 < \infty$, the set $\mathcal{D}(\mathfrak{A}_q)$ is dense in $C([-q_0, q_0])$.

For such a process $q(x)$, the *infinitesimal operator* $\mathfrak{A}$ of the process $(q(x), \phi(x))$ is

$$\mathfrak{A} = \Phi(q, \phi)\frac{\partial}{\partial \phi} + \mathfrak{A}_q. \tag{6.43}$$

Such a form of the infinitesimal operator $\mathfrak{A}$ is easy to obtain from (6.5), by calculating the limit

$$\lim_{x \downarrow 0} \frac{1}{x} \int_E (f(q_2, \phi_2) - f(q_1, \phi_1)) P(x; q_1, \phi_1; dq_2, d\phi_2) = (\mathfrak{A}f)(q_1, \phi_1)$$

in the same way used in [Gihman and Skorohod III] for random processes that are solutions of stochastic differential equations. In this case, the domain $\mathcal{D}(\mathfrak{A})$ of $\mathfrak{A}$ contains the set $\tilde{\mathcal{D}}(\mathfrak{A})$ of linear combinations of functions $u(q)v(\phi)$, where $u \in \mathcal{D}(\mathfrak{A}_q)$ and v is a smooth function on the circle S^1. Because $\mathcal{D}(\mathfrak{A}_q)$ is dense in the above set of continuous functions, $\tilde{\mathcal{D}}(\mathfrak{A}_q)$ is a separating set for probability measures on E, that is, $\int_E f\, d\mu = 0$ for all $f \in \tilde{\mathcal{D}}(\mathfrak{A}_q)$ implies $\mu = 0$. Hence, for these processes $q(x)$, condition (6.37) of invariance of $F(dq, d\phi)$ on E is equivalent to

$$\mathfrak{A}^* F = \left(-\frac{\partial}{\partial \phi}\Phi(q, \phi) + \mathfrak{A}_q^*\right) F = 0, \tag{6.44}$$

where $\mathfrak{A}_q^*$ is the adjoint of the operator $\mathfrak{A}_q$ and acts on the space of measures of bounded variation on E.

In conclusion, then, one can find the transition probability (6.35) by solving the stationary *Fokker–Planck–Kolmogorov equation* (6.44), instead of the integral equation (6.37).

(6.11) Theorem. *Suppose $q(x)$, for $x \in \mathbf{R}$, is a metrically transitive Markov process that is stochastically continuous (or Fellerian, if $q(x)$ is bounded), and that equation* (6.37) *has as its solution a probability measure $F(dq, d\phi)$ with the following properties*:

(a) *$F(dq, d\phi) = \pi(q, \phi)m(dq)\, d\phi$, where $m(dq)$ is some probability measure on $\mathbf{R}$ and $d\phi$ is the Lebesgue measure on the circle S^1*;

(b) *The function $\pi(q, \phi)$ is differentiable with respect to ϕ and*

$$\int_E \left|\frac{\partial \pi(q, \phi)}{\partial q}\right| m(dq)\, d\phi < \infty.$$

Then, if we set $p(\phi) = \int \pi(q, \phi)m(dq)$, we have

$$N(\lambda) = p(0). \tag{6.45}$$

Proof. We show that $\int \Phi(q,\phi)\pi(q,\phi)m(dq)$ does not depend on ϕ. Indeed, let $f(\phi)$ be an arbitrary smooth function on S^1. Then

$$\begin{aligned}\int_E \frac{\partial}{\partial\phi} f(\phi)\Phi(q,\phi)\pi(q,\phi)m(dq)\,d\phi &= \int_E (\mathfrak{A}f)(q,\phi)F(dq,d\phi) \\ &= \int_E f(\phi)(\mathfrak{A}^*F)(dq,d\phi) = 0.\end{aligned}$$

Since f is arbitrary, it follows that

$$\int \Phi(q,\phi)\pi(q,\phi)m(dq) = \int \Phi(q,0)\pi(q,0)m(dq) = p(0). \tag{6.46}$$

In this derivation we used continuity of the left-hand side, which follows from condition (b) of the theorem, and the identity $\Phi(q,0) = 1$, which follows from (6.5). From (6.46) and Theorem 6.9 clearly imply (6.45). □

Remark. Formula (6.45), expressing the integrated density of states in terms of the stationary probability density $p(\phi)$ of the reduced phase $\phi(x)$, can also be obtained in a slightly different way [Benderskii and Pastur 1969]. This other derivation does not use the property that $q(x)$ is a Markov process, and requires only that its realizations be piecewise continuous with probability 1, that for x large enough there be a probability density $p(x;\theta)$ for the phase $\theta(x,\lambda)$, and that the probability density of the reduced phase $p(x;\phi) = \sum_n p_1(x;\phi + n\pi)$ stabilize as $x \to \infty$, in the sense that for every $\phi \in [0,\pi)$ the limit

$$p(\phi) = \lim_{L\to\infty} \frac{1}{L}\int_0^L p(x;\phi)\,dx$$

exists. The derivation as follows:

Denote by x_j, for $j = 0, 1, \ldots$, the zeros of the solution $y(x,\lambda)$ of the Cauchy problem (6.1) with boundary condition (6.3) for $\alpha = 0$; we have $x_0 = 0$. If ν_L is the number of zeros in the interval $[0,L]$, the oscillation theorem gives

$$N(\lambda) = \lim_{L\to\infty} \frac{\mathsf{E}\{\nu_L\}}{L}.$$

Since each of the points x_j is the only solution of the equation $\theta(x,\lambda) = \pi j$,

$$\nu_L = \int_0^L \sum_k \delta(x - x_k)\,dx = \int_0^L \sum_k \delta\big(\theta(x;\lambda) - \pi k\big)|\theta'|_{\theta=\pi k}dx.$$

Since, by (6.5), $\theta'|\big|_{\theta=\pi k} = 1$, we get

$$\mathsf{E}\{\nu_L\} = \int_0^L \sum_k p_1(x;\pi k)\,dx = \int_0^L p(x;0)\,dx.$$

Given our assumptions, this indeed implies formula (6.45).

This derivation reveals the connection between (6.45) and the *Rice–Kac formula* [Kramer and Leadbetter 1967] of the theory of random processes, according to which the mathematical expectation of the number of zeros of a continuously differentiable random function $f(x)$ that fall within the interval $(0, L)$ is

$$\int_{-\infty}^{\infty} \int_0^L p(x, 0, f')|f'|\,df'\,dx,$$

where $p(x, f, f')$ is the joint probability density of the values of the function and its derivative at x.

6.D The Brownian Motion Model

In this subsection we consider an important class of metrically transitive processes arising from homogeneous Markov processes that are sufficiently smooth and have sufficiently good ergodic properties. These properties enable one to derive a number of interesting and important expressions for various spectral characteristics of this class of one-dimensional Schrödinger operators in the simplest and clearest form possible: by means of the joint probability densities of the phases and related quantities. We have already seen the simplest example of such relations, namely, (6.45). Here and in the problems of this chapter other relations will be obtained for the integrated density of states and related quantities, while in Chapter VI we will apply the formalism developed here to the study of the nature of the spectrum. Among other things we will prove that the Lyapunov exponent is positive, that the spectrum is pure point and that the eigenfunctions decay exponentially.

This class of potentials is considered in [Goldsheidt et al. 1977].

Let K be a compact, ν-dimensional Riemannian manifold with metric ds^2 and volume element dX, expressed in local coordinates $x_1, \dots, x_n$ as

$$ds^2 = g_{ij}\,dx_i\,dx_j, \qquad dX = \sqrt{\det g}\,dx_1 \dots dx_\nu. \tag{6.47}$$

Here and below, repeated subscripts indicate summation.

Let $X(x)$, for $x \in \mathbf{R}$, be a homogeneous Markov process on K with infinitesimal operator $\frac{1}{2}\Delta$, where

$$\Delta = \frac{1}{\sqrt{\det g}} \frac{\partial}{\partial x_i} g_{ij} \sqrt{\det g} \frac{\partial}{\partial x_j} \tag{6.48}$$

is the *Laplace–Beltrami* operator on K (see [McKean 1969], for example). The process $X(x)$ has a symmetric transition density $p(x; X_1; X_2) = p(x; X_2; X_1)$ with respect to the measure dX, satisfying the parabolic equation

$$\frac{\partial p}{\partial x} = \tfrac{1}{2}\Delta_{X_1} p = \tfrac{1}{2}\Delta_{X_2} p, \qquad p(0, X_1, X_2) = \delta_{X_1}(X_2), \tag{6.49}$$

where $\delta_X(\,\cdot\,)$ is the delta function concentrated at $X \in K$. By virtue of the ergodic theorem [Doob 1952], there are positive constants C, β and x_0 such that, for all $X_1, X_2 \in K$ and $x > x_0$ we have

$$\bigl|p(x, X_1, X_2) - 1/V(K)\bigr| \leqslant Ce^{-\beta x}, \tag{6.50}$$

where $V(K)$ is the volume of K. Thus $X(x)$ satisfies the strongest mixing conditions. We assume that the process $X(x)$, for every $x \in \mathbf{R}$, has as a one-point distribution the distribution with constant density $V^{-1}(K)$ with respect to the measure dX, so that $X(x)$, for $x \in \mathbf{R}$, is a metrically transitive Markov process with values in K.

Now take a smooth, *nonflat* function $Q : K \to \mathbf{R}$: this means there exists n_0 such that for any $X \in K$ there is $k \leqslant n_0$ for which $d^k Q(K) \neq 0$. For example, *Morse functions* satisfy this condition with $n_0 = 2$. Assume, for definiteness, that

$$\min_{X\in K} Q(X) = 0, \qquad \max_{X\in K} Q(X) = 1. \tag{6.51}$$

Consider the metrically transitive random field

$$q(x) = Q\bigl(X(x)\bigr), \tag{6.52}$$

for $x \in \mathbf{R}$; we call this the *Brownian motion model.* For $\lambda > 0$ arbitrary, introduce the phase $\theta(x)$ satisfying equation (6.5) (see Subsection 6.A) and the reduced phase $\phi = \theta - \pi[\theta/\pi]$, which takes values on the circle S^1 of circumference π. We will study the spectral characteristics of the Schrödinger operator $-d^2/dx^2 + q(x)$, via the properties of the random process $\bigl(X(x), \phi(x)\bigr)$.

(6.12) Theorem [Goldsheidt et al. 1977]. *The process $\bigl(X(x), \phi(x)\bigr)$ is a homogeneous Markov process on $K \times S^1$, with infinitesimal operator $\mathfrak{A}$ given by*

$$\mathfrak{A} = \tfrac{1}{2}\Delta + \bigl(\cos^2\phi + (\lambda - Q(X))\sin^2\phi\bigr)\frac{\partial}{\partial\phi}. \tag{6.53}$$

The transition probability of $\bigl(X(x), \phi(x)\bigr)$ has density $p(x; X_0, \phi_0; X, \phi)$ with respect to the measure $dX\,d\phi$, which is continuous in the variables $x > 0$, (X_0, ϕ_0) and (X, ϕ), and also in the parameter λ.

Proof. Showing that $(X(x), \phi(x))$ is a homogeneous Markov process is done just as in Lemma 6.10 and is left to the reader. Expression (6.53) follows directly from the definition of the diffusion process $X(x)$ and equation (6.5) defining the phase: cf. (6.43). Since the transition density p is the fundamental solution of the parabolic equation

$$\frac{\partial p}{\partial x} = \mathfrak{A}^* p, \qquad p(0; X_0, \phi_0; X, \phi) = \delta_{X_0,\phi_0}(X, \phi), \tag{6.54}$$

where δ_{X_0,ϕ_0} is the Dirac delta function, proving that p is continuous amounts to proving the smoothness of the solutions of (6.54). This is done using the following result:

(6.13) Proposition [Hörmander 1967; Ichihara and Kunita 1974]. *Let M^n be a smooth manifold of dimension n, and $\mathfrak{A}$ the second-order differential operator*

$$\mathfrak{A} = \frac{1}{2} \sum_{i=1}^{r} A_j^2 + B, \tag{6.55}$$

where $1 \leqslant r \leqslant n$ and the A_j and B are smooth vector fields on M^n. Denote by $\mathcal{A}$ the Lie algebra generated by the fields A_i, $[B, A_j]$, $\big[[B, A_j], A_j\big]$, $\big[[B, A_j], B\big]$, and so on. If $\mathcal{A}$ has maximal dimension n, the diffusion process with infinitesimal operator L has transition probability $P(x; X, dY) = p(x; X; Y)\, dY$, where the density $p(x; X, Y)$ is a smooth function on $(0, \infty) \times M^n \times M^n$ satisfying

$$\left(\mathfrak{A}_X - \frac{\partial}{\partial x}\right) p(x; X, Y) = 0, \qquad \left(\mathfrak{A}_Y^* - \frac{\partial}{\partial x}\right) p(x; X, Y) = 0. \tag{6.56}$$ □

It is straightforward to show that $\mathfrak{A}$ can be put into the form (6.55), by a change of variables on K, so in order to apply the proposition we just have to check the dimension of the Lie algebra generated by the commutators. Consider on K the local coordinates $x_1, \ldots, x_\nu$ and the $\nu + 1$ vector fields

$$\begin{aligned} A_j &= \frac{\partial}{\partial x_j}, \qquad \text{for } j = 1, \ldots, \nu, \\ B &= \big(\cos^2 \phi + (\lambda - Q(x_1, \ldots, x_\nu)) \sin^2 \phi\big) \frac{\partial}{\partial \phi}. \end{aligned} \tag{6.57}$$

By forming commutators of the form $\big[A_j[A_k[\cdots[A_l, B]\cdots]]\big]$ we easily see that the fields

$$C_{i_1,\ldots,i_\nu} = \frac{\partial^k Q}{\partial x_1^{i_1} \ldots \partial x_\nu^{i_\nu}} \sin^2 \phi \frac{\partial}{\partial \phi},$$

for $k = i_1 + \cdots + i_n$, lie in $\mathcal{A}$. By the nonflatness of Q, the dimension of $\mathcal{A}$ is $\nu + 1$ everywhere, except perhaps on $K \times \phi^{-1}(0)$. But

$$[[C_{i_1,\ldots,i_\nu}, B], B] = 2\frac{\partial^k Q}{\partial x_1^{i_1} \ldots \partial x_\nu^{i_\nu}}(\cos^2\phi + (\lambda - Q)\sin^2\phi)\frac{\partial}{\partial \phi},$$

and the coefficient of $\frac{\partial}{\partial \phi}$ for $\phi = 0$ and the proper choice of $i_1, \ldots, i_\nu$ is again nonzero. Thus, the dimension of the algebra $\mathcal{A}$ is $\nu + 1$, so Proposition 6.13 applies, and the transition density $p(x; X_0, \phi_0; X, \phi)$ is smooth.

The continuous dependence of p on λ can be shown by the techniques used in proving Proposition 6.13. □

(6.14) Theorem. *For any $\lambda > 0$, there exists a positive number $x_0(\lambda)$ such that for any open set $U \subset K \times S^1$ and any $x \geqslant x_0(\lambda)$,*

$$P_{x_0,\phi_0}(x, U) = \int_U p(x; X_0, \phi_0; X, \phi)\, dX\, d\phi > 0, \tag{6.58}$$

that is, the process $\big(X(x), \phi(x)\big)$ has a positive probability of reaching any open set in $K \times S^1$. This process has a unique invariant measure, with density $\pi(X, \phi)$ smooth with respect to $dX\, d\phi$ on $K \times S^1$, and there are positive constants $\delta(\lambda)$ and $C(\lambda)$ such that, for $x > x_0(\lambda)$,

$$\big|p(x; X_0, \phi_0; X, \phi) - \pi(X, \phi)\big| \leqslant C(\lambda)e^{-\delta(\lambda)x}. \tag{6.59}$$

(6.15) Lemma. *Let $\tilde{\theta}(x, \mu)$, for $x \geqslant 0$, be the solution of the Cauchy problem*

$$\tilde{\theta}' = \cos^2\tilde{\theta} + \mu\sin^2\tilde{\theta}, \qquad \tilde{\theta}(0, \mu) = \phi_0 \in [0, \pi]. \tag{6.60}$$

Given $0 < \mu_1 < \mu_2 < \infty$, there exists a positive $T = T(\mu_1, \mu_2)$ such that, for any $\phi_0, \phi_1 \in [0, \pi)$ there exists $\mu \in [\mu_1, \mu_2]$ satisfying $\tilde{\theta}(T, \mu) = \phi_1 \pmod{\pi}$.

Proof. From equation (6.60) we get

$$\sqrt{\mu}\tan\tilde{\theta}(x, \mu) = \tan\big(\sqrt{\mu} + \phi(\mu, \phi_0)\big), \tag{6.61}$$

where $\phi(\mu, \phi_0)$ is the solution of the equation

$$\sqrt{\mu}\tan\phi_0 = \tan\phi(\mu, \phi_0). \tag{6.62}$$

From (6.62) we see that for fixed ϕ_0 we can define $\phi(\mu, \phi_0)$ so as to depend continuously on $\mu \in (0, \infty)$. Set $T = T(\mu_1, \mu_2) = 2\pi/(\sqrt{\mu_2} - \sqrt{\mu_1})$. Then, the argument of the tangent on the right-hand side of (6.61), for $x = T$ and $\mu \in [\mu_1, \mu_2]$, takes values on an interval with length greater than or equal to $(\sqrt{\mu_2} - \sqrt{\mu_1})T - \pi = \pi$. The lemma now follows immediately from equation (6.61), given that the arguments of the tangents are continuous. □

Proof of Theorem 6.14. To prove (6.58), note first that because of the semigroup property in x of the transition densities $p(x;\,\cdot\,;\,\cdot\,)$, it is sufficient to verify the inequality for $x = x_0(\lambda)$. Take $(X_0, \phi_0) \in K \times S^1$. Now $\phi(x)$ satisfies (6.5) by definition, and that equation, because of (6.52), can be written

$$(6.63)\qquad \phi' = \cos^2\phi + (\lambda - Q(X(x)))\sin^2\phi, \quad \phi(0) = \phi_0, \quad X(0) = X_0.$$

Choose positive numbers $\mu_1 < \mu_2$ in the interval $(\lambda - 1, \lambda)$. By (6.51), for δ small enough, the interval $[\mu_1 - \delta, \mu_2 + \delta]$ lies in the domain of $\lambda - Q(x)$. Set $x_0(\lambda) = T$, where T is to be determined in terms of μ_1 and μ_2 according to Lemma 6.15. Take an arbitrary open set $U \in K \times S^1$ and a point $(X_1, \phi_1) \in U$, then determine $\mu \in [\mu_1, \mu_2]$ from ϕ_0 and ϕ_1 in accordance with Lemma 6.15. Given $\varepsilon > 0$, which so far is arbitrary, construct the following set open in K:

$$K_\varepsilon = \{X \in K : |\lambda - Q(X) - \mu| < \varepsilon\}.$$

By the construction of the interval $[\mu_1, \mu_2]$, this set is nonempty. Choose open sets $B \subset K$ and $\Psi \subset S^1$ such that $X_1 \in B$, $\phi_1 \in \Psi$ and $U_1 = B \times \Psi \subset U$. Since the solution ϕ of (6.63) depends continuous on its right-hand side [Hartman 1964], it can be made arbitrarily close to the solution $\tilde{\theta}$ of equation (6.60), so long as

$$(6.64)\qquad X([\varepsilon, T - \varepsilon]) \subset K_\varepsilon \quad \text{and} \quad X(T) \in B$$

and ε is chosen small enough. Choose ε so small that $\phi(T) \in \Psi$ when (6.64) is satisfied. Then

$$P_{X_0,\phi_0}(T; U) \geqslant P_{X_0,\phi_0}(T; B \times \Psi) \geqslant \mathsf{P}_{X_0}\big(X([\varepsilon, T - \varepsilon]) \subset \text{ and } X(T) \in B\big).$$

The right-hand side of the last inequality is strictly positive, by well-known properties of the diffusion process $X(x)$ [McKean 1969]. This concludes the proof of (6.58).

We now prove (6.59). By (6.58), Theorem 6.12 and the continuity of p, there exist open sets U and V such that, for some $\delta_1 > 0$,

$$p\big(x_0(\lambda); X_0, \phi_0; X, \phi\big) \geqslant \delta_1$$

for $(X_0, \phi_0) \in U$ and $(X, \phi) \in V$. Therefore from inequality (6.58) and the semigroup property of p we get that

$$p\big(2x_0(\lambda); X_0, \phi_0; X, \phi\big) \geqslant P_{x_0,\phi_0}\big(x_0(\lambda); U\big)\delta_1 > 0$$

for $(X_0, \phi_0) \in K \times S^1$ and $(X, \phi) \in V$. Since p is continuous and $K \times S^1$ is compact, we conclude that there exists $\delta > 0$ such that

(6.65) $$p(2x_0(\lambda); X_0, \phi_0; X, \phi) \geqslant \delta$$

for $(X_0, \phi_0) \in K \times S^1$ and $(X, \phi) \in V$. But (6.65) is a stronger variant of the Doeblin condition, which yields the desired estimate (6.59) [Doob 1952].

The smoothness of π follows readily from the invariance of the measure $\pi(X, \phi)\, dX\, d\phi$ and the smoothness of p, that is, from the identity

$$\pi(x, \phi) = \int \pi(X', \phi')p(x; X', \phi'; X, \phi)\, dX'\, d\phi'. \qquad \square$$

For the process $q(x)$ defined by (6.52), an analogue of Theorem 6.11 holds, with a very similar proof.

(6.16) Theorem. *Let $q(x)$, for $x \in \mathbf{R}$, be the random process defined by (6.52). If $p(\phi) = \int_K \pi(X, \phi)\, dX$, where $\pi(X, \phi)$ is the invariant density of the Markov process $(X(x), \phi(x))$, the integrated density of states $N(\lambda)$ of the Schrödinger operator $-d^2/dx^2 + q(x)$ satisfies*

(6.66) $$N(\lambda) = p(0). \qquad \square$$

6.E Jacobi Matrices with Independent and Markov Coefficients

Consider the metrically transitive matrix J given by

(6.67) $$(J\psi)(x) = -s(x+1)\psi(x+1) + q(x)\psi(x) - s(x)\psi(x-1),$$

where $s(x, \omega)$ and $q(x, \omega)$, for $x \in \mathbf{Z}$, are metrically transitive sequences defined on some probability space $(\Omega, \mathcal{F}, \mathsf{P})$ with a metrically transitive group of shifts T_a, $a \in \mathbf{Z}$. Everywhere below we will assume that

(6.68) $$s(x) \geqslant s > 0.$$

The study of the spectral properties of J, which acts on $\ell_2(\mathbf{Z})$, is closely related to that of the second-order difference equation

(6.69) $$-s(x+1)\psi(x+1) + q(x)\psi(x) - s(x)\psi(x-1) = \lambda\psi(x)$$

considered for all $x \in \mathbf{Z}$. Along with this equation, we will consider, as we did for the one-dimensional differential operators of the preceding section, the same equation on a finite interval (a, b), for $a, b \in \mathbf{Z}$, $a < b$, with boundary conditions

(6.70) $$\begin{aligned} \psi(a)\cos\alpha - s(a+1)\psi(a+1)\sin\alpha &= 0, \\ \psi(b)\cos\beta - s(b+1)\psi(b+1)\sin\beta &= 0 \end{aligned}.$$

This boundary problem for equation (6.69) is equivalent to the eigenvalue problem for the following matrix $J_{a,b}$ of order $b-a$:

(6.71)

$$\begin{pmatrix} q(a+1)-s^2(a+1)\tan\alpha & -s(a+2) & 0 & & \\ -s(a+2) & q(a+2) & -s(a+3) & \cdots & \\ 0 & -s(a+3) & q(a+3) & & \\ & & & \ddots & -s(b) \\ & \vdots & & -s(b) & q(b)-\cot\beta \end{pmatrix}$$

where for $\alpha=\pi/2$ the first row and first column are omitted, while for $\beta=0$ the last row and column are omitted.

As in the continuous case, we let $(b-a)N_{(a,b)}(\lambda)$ be the number of eigenvalues of $J_{a,b}$ not exceeding λ. The relationship between the eigenvalues of $J_{a,b}$—the boundary problem defined by (6.69) and (6.70)—and the oscillatory properties of solutions of the finite-difference equation (6.69) is summarized by the following result:

(6.17) Proposition [Atkinson 1964]. *The number of eigenvalues of the boundary-value problem* (6.69)–(6.70) *not exceeding* λ, *when* $\alpha=\beta=0$, *equals the number of sign changes in the sequence* $y(a+1,\lambda),\dots,y(b+1,\lambda)$, *where* $y(x,\lambda)$ *is the solution of the Cauchy problem*

$$(6.72)\qquad y(a)=\sin\alpha,\qquad s(a+1)y(a+1)=\cos\alpha$$

of equation (6.69), *for* $\alpha=0$. □

We now introduce the reduced phase $\phi(x)$, for $x\geqslant 0$, of the solution $y(x)$, with values on the circle $S^1=[0,\pi)$:

$$(6.73)\qquad \cot\phi(x)=\frac{s(x)y(x)}{y(x-1)},$$

where $y(x)$ is the solution of the Cauchy problem defined by (6.69) and (6.73). If $y(x-1)=0$, we set $\phi(x)=0$. We have the following result (compare Theorems 6.4 and 6.11):

(6.18) Theorem. *Suppose that in* (6.69) *the metrically transitive sequence* $q(x)$ *is arbitrary and* $s(x)$ *is bounded from below, that is,* $s(x)\geqslant s>0$; *and let the angles* α *and* β *in* (6.70) *be arbitrary random variables with values in* $[0,\pi)$. *There exists a nonrandom continuous function* $N(J,\lambda)$, *independent of* α *and* β, *such that, with probability* 1,

$$(6.74)\quad N(J,\lambda)=\lim_{n\to\infty}N_{(0,n)}(\lambda)=\lim_{n\to\infty}N_{(-n,0)}(\lambda)=\lim_{n\to\infty}N_{(-n,n)}(\lambda),$$

$$(6.75)\quad N(J,\lambda)=F\big([\pi/2,\pi)\big)$$

for any $\lambda \in \mathbf{R}$, where $F(d\phi)$ is the weak limit of any subsequence of the sequence of probability measures

$$F_n(d\phi) = \frac{1}{n}\sum_{x=1}^{n} \mathsf{P}\{\phi(x) \in d\phi\}. \tag{6.76}$$

Proof. Equation (6.74) follows from Theorems 4.4 and 3.2. To prove (6.76), note that, since the integrated density of states is independent of α and β, we can consider just the case $a = 0$, $b = n$, $\alpha = \beta = 0$. From the definition of the phase (6.73) and from Proposition 6.17, we have

$$N_n(\lambda) = \frac{1}{n}\sum_{x=1}^{n} \chi_{[\pi/2,\pi)}\big(\phi(x)\big).$$

This, together with the definition of $F(d\phi)$ and relation (6.74), proves (6.76). □

If we assume that $\big(s(x), q(x)\big)$ is an ergodic Markov chain, and in particular that it consists of independent, identically distributed random variables, we can prove more, just as in the continuous case:

(6.19) Theorem. *Suppose that $\big(s(x), q(x)\big)$, for $x \in \mathbf{Z}$, is a homogeneous, metrically transitive Markov process, and that $s(0) > 0$ with probability 1. Then $\big(s(x), q(x), \phi(x)\big)$ is also a homogeneous Markov process, whose transition probability we denote by $p(s_0, q_0, \phi_0; ds, dq, d\phi)$. If $\mathcal{M}$ denotes the set of probability measures F invariant with respect to p, that is, such that*

$$F(ds, dq, d\phi) = \int_{\mathbf{R}^2\times S^1} F(ds', dq', d\phi')p(s', q', \phi'; ds, dq, d\phi), \tag{6.77}$$

and satisfying

$$\int_0^{\pi} F(ds, dq, d\phi) = \mathsf{P}\big\{s(0) \in ds \text{ and } q(0) \in dq\big\}, \tag{6.78}$$

the integrated density of states $N(J, \lambda)$ satisfies the relation

$$N(J, \lambda) = F\big([\pi/2, \pi)\big), \tag{6.79}$$

where $F(d\phi) = \int_{\mathbf{R}^2} F(ds, dq, d\phi)$ and $F(ds, dq, d\phi)$ is an arbitrary measure in $\mathcal{M}$.

Proof. By the definition of $\phi(x)$ (6.73) and equation (6.69) we can write the recurrence formula

$$\cot\phi(x+1) = \big(q(x) - \lambda\big) - \frac{s^2(x)}{\cot\phi(x)}. \tag{6.80}$$

This clearly implies that $\big(s(x), q(x), \phi(x)\big)$ is a homogeneous Markov process. Now take an arbitrary $F \in \mathcal{M}$ and define the random variable $\phi(0)$ so that the joint distribution of $s(x)$, $q(x)$ and $\phi(x)$ coincides with $F(ds, dq, d\phi)$. In view of (6.77), we get

$$\mathsf{P}\big\{\phi(x) \in d\phi\big\} = F(d\phi).$$

From this and from Theorem 6.18, equation (6.79) follows easily, because $F_n(d\phi) = F(d\phi)$. □

The following is an immediate consequence of Theorem 6.19.

(6.20) Theorem. *Let* $\big(s(x), q(x)\big)$, *for* $x \in \mathbf{Z}$, *be a sequence of independent, identically distributed random variables on* $\mathbf{R}^2$, *and assume that* $s(0) > 0$ *with probability* 1. *Denote by* $\mathcal{M}_1$ *the set of probability measures* $F(d\phi)$ *on* S^1 *having the property that, if a random variable* $\phi(0)$ *has distribution* $F(d\phi)$ *and is independent of* $\big(s(0), q(0)\big)$, *the random variable* $\big(q(0) - \lambda\big) - s^2(0) \tan \phi(0)$ *has the same distribution as* $\cot \phi(0)$. *Then* $N(\lambda)$ *satisfies the representation*

$$N(\lambda) = F\big([\pi/2, \pi)\big), \tag{6.81}$$

where $F(d\phi)$ *is an arbitrary probability measure in* $\mathcal{M}_1$.

Proof. From (6.80) it follows that $\phi(x)$ and $\big(s(x), q(x)\big)$, for $x \geqslant 0$, are independent random variables. If $F \in \mathcal{M}_1$, the conditions of the theorem imply that the measure $F(ds, dq, d\phi) = F(d\phi)\mathsf{P}\big\{s(0) \in ds, q(0) \in dq\big\}$ satisfies (6.77) and (6.78), that is, it belongs to $\mathcal{M}$. The theorem now follows directly from Theorem 6.19. □

Based on this discrete variant of the phase formalism, one can in particular study the asymptotic behavior of the integrated density of states of metrically transitive Jacobi operators (6.67), as was done for Schrödinger operators in Subsection 6.B. The resulting asymptotic formulas are by and large similar to (6.17), (6.23), (6.29), etc.; examples of them can be found, for example, in [Lifshitz et al. 1982b] and [Pastur 1982].

6.F Smoothness of $N(\lambda)$; Special Energies

In this subsection we will present some results on the smoothness of the integrated density of states of one-dimensional and finite-difference operators of the second order. Since these results were obtained by a variety of methods, based on various auxiliary arguments, we will restrict ourselves to stating them and discussing briefly the ideas of their proofs.

The smoothness of the integrated density of states and of closely related functions, such as the mathematical expectation of the prelimit integrated densities of state (6.6) and the conditional mathematical expectation of the spectral measure (as in Theorem 13.7, for example), is both interesting in itself and relevant to a variety of physical applications, and plays a significant part in the spectral analysis of metrically transitive operators, in particular in the proof that their spectra are pure point (Sections 13 and 15). But smoothness of the integrated density of states alone is insufficient to guarantee that a metrically transitive operator has a pure point spectrum. By Problem II.24(e) and Theorem 18.1, the Jacobi matrices (6.67) corresponding to $s(x) = 1$ and two different potentials, one quasiperiodic and one consisting of independent Cauchy distributions, have the same integrated density of states, analytic on a strip of the spectral plane. But the spectrum of the first of these operators, for irrational numbers α that are well approximated by rational numbers, is singular continuous (see (18.81)), while the spectrum of the second is pure point (Example 13.8(a)).

It was shown in Section 3 that under minimal conditions on the coefficients of a one-dimensional metrically transitive operator of the second order—in fact of any finite order—the integrated density of states is a continuous function of λ (Theorem 3.2). In Section 11 (Theorem 11.11) we will see that, essentially under the same conditions, the integrated density of states is also a log-Hölder function, that is,

$$|N(\lambda_1) - N(\lambda_2)| \leqslant \frac{C}{\ln|\lambda_1 - \lambda_2|} \tag{6.82}$$

where C depends only on the position of λ_1 and λ_2 and on the moments of the coefficients: $\mathsf{E}\{\ln(1 + |q(0)|)\}$ and $\mathsf{E}\{\ln s(0)\}$ in the discrete case and $\mathsf{E}\{q^2(0)\}$ in the continuous one.

If no restrictions are imposed on the coefficients, other than their metric transitivity and the above conditions on the moments, (6.82) is essentially the best possible estimate for the continuity modulus. This can be shown for a Jacobi matrix (6.67) with $q(x) = 0$, an example studied by Dyson [1953] and considered in the next section (7.27). For the Schrödinger operator, both discrete and continuous, examples were presented by Schmidt [1957] and Lax and Phillips [1958]. Schmidt considered the operator (6.67) with $s(x) = 1$ and $q(x)$ equal to a Bernoulli potential with distribution function

$$F(dq) = \big((1-c)\delta(q) + c\delta(q - k_0)\big)\, dq, \tag{6.83}$$

where $0 < c < 1$. Lax and Phillips considered a Schrödinger operator with Poisson potential (6.25), in which $k_0 < 0$. An approximate calculation of the invariant phase distribution, from equations similar to (6.37), (6.44) and (6.77) (see also Problem III.13) allows one to prove that in both cases there exists a point λ_0 (equal to $-k_0^2/4$ in the second case) such that, for c small enough,

$$|N(\lambda) - N(\lambda_0)| \sim |\lambda - \lambda_0|^{2c/\sqrt{|\lambda_0|}} \tag{6.84}$$

(in the continuous case, c is the Poisson point density).

Lifshitz [1964] arrived at formula (6.84) by an approximation method he developed to calculate the integrated density of states of a Poisson potential, which is also applicable to the multidimensional case and is based on the analysis of some determinants of large order. Halperin [1967] (see also [Simon and Taylor 1985]) obtained (6.84) as a lower bound (that is, with $\geqslant$ instead of $\sim$) using the variational principle (discussed, for example, in [Glazman 1967]).

A rigorous version of (6.84) was obtained by Morrison [1962], who showed by analyzing the equation for the invariant density of the quantity $z = \cot\phi$ (see Problem III.13) that for the Schrödinger operator with potential (6.25), when $k_0 < 0$ and $c \downarrow 0$,

$$N(\lambda) = \big(1 + o(1)\big)\frac{4c}{3 + |\varepsilon|^{a/(1+\varepsilon)}}\begin{cases} \varepsilon^{2c/(1+\varepsilon)} & \text{for } \varepsilon > 0, \\ 0 & \text{for } \varepsilon < 0. \end{cases} \tag{6.85}$$

Here $\varepsilon = \sqrt{\lambda/\lambda_0 - 1}$ and $a = 2c/\sqrt{|\lambda_0|}$, and this asymptotic formula is true for all $|\varepsilon| \leqslant \frac{1}{2}$.

It follows from this formula that near the point λ_0 ($\varepsilon = 0$), the integrated density of states satisfies a *Hölder condition*

$$|N(\lambda_1) - N(\lambda_2)| \leqslant C|\lambda_1 - \lambda_2|^\alpha \tag{6.86}$$

with an exponent α that can be made arbitrarily small by an appropriate choice of k_0 and c (although the constant C can also become small).

This has an important consequence. If we consider a potential $q_0(x) = k_0\delta(x)$, for $k_0 < 0$, that is, having a single impurity at the origin, the operator has a unique eigenvalue λ_0, equal to $-k_0^2/4$ in the continuous case and to $k_0 - 2e^{-k}$, for $k = \operatorname{arcsinh}|k_0/2|$, in the discrete case. Furthermore, the corresponding eigenfunction $\psi_0(x)$ decreases exponentially as $|x| \to \infty$, and the greater $|k_0|$ is, the faster the decay.

Therefore, by taking a potential with a sufficiently large $|k_0|$ at points far enough apart—that is, by making c small enough—one should expect, for the operator considered in the interval $(0, L)$, cL eigenvalues in a small neighborhood of λ_0. But this means that $N(\lambda)$ will change strongly in this neighborhood, because it should be expected from the above discussion that in the continuous case, for example,

$$\lim_{c\downarrow 0} \frac{N(\lambda)}{c} = \begin{cases} 0 & \text{for } \lambda \leqslant \lambda_0, \\ 1 & \text{for } \lambda > \lambda_0. \end{cases}$$

This does not exclude Hölder estimates of the form (6.86), with constants depending on the interval under consideration and on the form of the random potential. For example, [Le Page 1984] gives such an estimate in the discrete case for a random potential $q(x)$, for $x \in \mathbf{Z}$, independent at different points and satisfying $\mathsf{E}\{|q(0)|^a\} < \infty$ for some $a > 0$. See also Theorem 11.14 and Subsection 15.B.

Thus, by passing from general metrically transitive potentials to independent ones, the smoothness of the integrated density of states increases: instead of a log-Hölder estimates of the form (6.82), we arrive at locally Hölder estimates such as (6.86).

This discussion suggests that for a discrete metrically transitive operator of the form (6.67), with $s(x) = 1$ and $q(x)$ independent for different values of x, the smoothness of $N(\lambda)$ should not be expected to increase without some assumption on the regularity of the potential. We have already obtained such results in Section 4 (see Theorem 4.21 and Problems II.18 to II.22), where in particular it was shown that if $F([a,b]) \leqslant \text{const}\,|a-b|^\alpha$, (6.86) holds. However, since the proof of this fact performed an explicit average only over the values of the potential at a single point, it is natural to expect that, if we average over an increasingly large set of random variables $q(x)$, we obtain a smoother integrated density of states even for less regular distributions $F(dq)$.

In the multidimensional case, such sufficiently explicit averaging is rather difficult to perform (but see [Spencer 1984] and [Maier 1986]). But as we have seen more than once in this section, in dimension one there is a special method based on recurrence relations similar to (6.80) in the discrete case, and on differential equations (6.5) and (6.21) in the continuous case. By using this method, Simon and Taylor [1985] proved infinite differentiability of $N(\lambda)$ of the operator (6.67), with $s(x) = 1$ and a potential independent at different points and having distribution density $f(q)$ with compact support, such that for some $g(q) \in L_1(\mathbf{R})$ and $\alpha > 0$, the Fourier transforms $\hat{f}(k)$ and $\hat{g}(k)$ are related by $\hat{g}(k) = (1+k^2)^{\alpha/2}\hat{f}(k)$.

The starting point of the proof is formula (6.81), which shows that the smoothness of $N(\lambda)$ depends on the smoothness of the invariant measure $F(d\phi)$ of the Markov chain $\phi(x)$ with respect to λ and z. Such a measure can be found as a result of an infinite sequence of iterations of the equation

$$p(x+1,z) = \int_{\mathbf{R}} f\left(z + \lambda + \frac{1}{z'}\right) p(x,z')dz' \tag{6.87}$$

which follows from (6.77) and (6.80) for $s(x) \equiv 1$, where $p(x,z)$ is the distribution density of the random variable $z(x) = \cot\phi(x)$ and $f(q)$ is the density of the potential, $F(dq) = f(q)\,dq$.

But this equation, a sort of "skew convolution," is not very helpful in the study of smoothness, because of the complicated form of its argument: it is easy to see that, after several iterations, the presence of the term $1/z'$ in the integrand prevents further smoothing. Therefore, it is convenient to write the characteristically one-dimensional fact of the existence of iterative procedures with respect to the spatial variable x ($x \in \mathbf{R}$ or $\mathbf{Z}$) in a different form. This form results if the recurrence relation (6.69) for $s(x) = 1$ is written as a system of two first-order equations. It will be explained in detailed in Sections 11 and 14: see (11.14), for example.

Then, instead of (6.87), we obtain equation (14.12) containing the ordinary convolution (14.11) of measures on the group $\mathrm{SL}(2, \mathbf{R})$ and on the projective line $\mathbf{P}^1$. However, with this approach, because the matrix (11.15) has only one random element, the kernel of the operator of the convolution in the right-hand side of (14.12) is singular with respect to the Haar measure on $\mathrm{SL}(2, \mathbf{R})$, even if the potential has an absolutely continuous distribution. This difficulty can be overcome by considering the third power of the convolution operator because, roughly, the product of three successive matrices (11.15) contains three independent parameters, like the group $\mathrm{SL}(2, \mathbf{R})$. Successive convolutions after that increase smoothness. (For $f(q)$ bounded, the smoothness of the triple convolution was proved by Figotin [1983] in the proof of positiveness of the Lyapunov exponent (Section 14). In this case, one can even consider a potential that is a reversible Markov chain.)

The invariant measure is the eigenfunction of the operator of the l-tuple convolution for any l, corresponding to the eigenvalue 1. In the case of $f(q)$ with compact support, this operator is compact and 1 is an isolated and (geometrically and algebraically) simple eigenvalue. Therefore, it is easy to prove by elementary perturbation theory [Reed and Simon IV] that the invariant measure is an infinitely differentiable function of ϕ and λ, and then, by (6.81), that $N(\lambda) \in C^\infty(\mathbf{R})$.

This result cannot be improved if the distribution of the potential has compact support. Indeed, if it does, by Corollary 4.23, the spectrum is a bounded set and its edges are fluctuation boundaries (see the discussion at the end of Section 6.1). Therefore, at these edges the asymptotic behavior $N(\lambda)$ is given by (6.27) and (6.29) (for the discrete analogue, see Problems III.19–20). If, instead of the Jacobi matrices (6.67) with $s(x) = 1$ (the discrete analogue of the Schrödinger operator) we consider Jacobi matrices in which $q(x) = s(x+1) - s(x)$ (the discrete analogue of the divergence operator (6.19)), the bottom of the spectrum is always $\lambda = 0$ and is a stable boundary, an in its neighborhood $N(\lambda)$ has the behavior (6.23) (in fact, its discrete analogue: see Problem II.18). Therefore, $N(\lambda)$ for such an operator is a C^∞-function everywhere on the spectrum, except at $\lambda = 0$, where it only satisfies the Hölder condition (6.86) with $\alpha = \frac{1}{2}$.

For the potential (6.67) with periodic coefficients, where the spectrum consists of bands and $N(\lambda)$ is real analytic in the interior of the each band, $N(\lambda)$ can also satisfy only the Hölder condition with $\alpha = \frac{1}{2}$ at the edges of the bands [Marchenko 1986; Toda 1981]. Therefore the result of Simon and Taylor [1985] is a characteristic property of random operators (6.67) with $s(x) = 1$ and sufficiently regular distribution of the potential $q(x)$.

The smoothness of $N(\lambda)$ can also be studied using a formula slightly different from (6.81), that relates the density of states $n(\lambda) = N'(\lambda)$ of the operator (6.67) to the density $p(\phi)$ of the invariant distribution of the phase $\phi(x)$ from (6.73):

$$n(\lambda) = \int_{\mathbf{R}^2} p(\zeta)\, p\left(\frac{t^2}{\zeta}\right) f_s(t)\, d\zeta\, dt. \tag{6.88}$$

Here $p(\zeta)$ is the density of the invariant distribution of the random variable $\zeta(x) = s(x)y(x)/y(x-1)$ (6.73) and $f_s(t)$ is the density of the distribution of the random variables $s(x)$, which, like $q(x)$, are assumed to be independent and identically distributed and to have a bounded density. This formula is an example of "bilinear" formulas for mathematical expectations of a wide variety of spectral characteristics of one-dimensional random operators of the second order. Such formulas were first obtained by Halperin [1965] for the Schrödinger operator, though by somewhat complicated arguments. A simple and natural way to derive and use them in various problems of spectral analysis and applications can be found in [Goldsheidt et al. 1977], [Antsygina et al. 1981] and in Subsection 15.1 (see in particular Lemma 15.1, Theorem 15.3 and Problems VI.25–28).

The derivation of (6.88), including the proof of the existence of $p(\zeta)$ and its smooth dependence on λ, is based on the study of equation (6.87) for $x \to \infty$, and can be done in various ways. Figotin [1980] did it for the case where the distributions of the random variables $s(x)$ and $q(x)$ have continuous densities with compact support; Bougerol and Lacroix [1986] gave a proof for $s(x) = 1$ and potential having a distribution with finite second moment and continuous density. In all these cases $N(\lambda)$ is a continuously differentiable function of λ; the fact that its derivative is bounded follows from the general Theorem 4.21 (see also Problems II.18–22).

Till now we have obtained recurrence relations with respect to spatial variables for $N(\lambda)$ and related quantities by using, in some way or another, the oscillation theorem and the phase of the Cauchy solution. Campanino and Klein [1986] proposed a new representation for the mathematical expectation of the *Green's function* (the resolvent kernel) of the operator (6.71) for $s(x) = 1$ and independent $q(x)$, leading to somewhat different recurrence relations.

This representation is obtained by using integration over anticommuting (Grassmann) "variables" introduced by Berezin [1965] and widely used now in theoretical and mathematical physics (see, for example, the review by Efetov [1983], where the application of this technique to various problems of the theory of disordered systems is described).

Denote by h_L the operator (6.71) for $s(x) \equiv 1$, $-a = b = L$ and $\alpha = \beta = 0$; by Theorem 6.18), the study of $N(\lambda)$ is reduced to the study of h_L for increasing L. Campanino and Klein showed that

$$(6.89)\quad r_L(z) = \frac{1}{L}\mathsf{E}\{\mathrm{Tr}(h_L - z)^{-1}\} = i\int_0^\infty \{((TB)_L 1)(r^2)\}^2 \beta(r^2, z)\, dr^2$$

where $z = \lambda + i\varepsilon$, B is the operator of multiplication by $\beta(r^2, z) = \hat{f}(r^2)e^{izr^2}$ (treated as a function of the variable r^2), $\hat{f}(k)$ is, as above, the Fourier transform of the distribution function of the potential, and T is the Hankel transformation operator, given by

$$(T\phi)(r^2) = -\int_0^\infty J_0(rs)\psi(s^2)\, ds^2,$$

where J_0 is the Bessel function of order zero.

From this representation it follows that if $\sup_{t\in\mathbf{R}} e^{\alpha|t|}|\hat{f}(t)| < \infty$ for some $\alpha > 0$, the integrated density of states can be analytically continued into a strip $|\varepsilon| \leqslant \alpha_1(\alpha)$. This is an improvement on the result of Problem II.22. Now, proceeding from (6.89) and using a method similar to that employed in [Simon and Taylor 1985] (harmonic analysis for the Hankel transformation instead of harmonic analysis on the group $\mathrm{SL}(2, \mathbf{R})$, and the nondegeneracy of the maximal eigenvalue of TB), it is possible to prove that if $\hat{f}(k)$ has $n-1$ derivatives $\hat{f}^{(j)}(k)$, is absolutely continuous, and

$$\sup_{\substack{k\in\mathbf{R}\\ 0\leqslant j\leqslant n}} (1+k^2)^{\alpha/2}\hat{f}^{(j)}(k) < \infty$$

for some $\alpha > 0$, then $N(\lambda) \in C^\infty(\mathbf{R})$.

A convenient sufficient condition for the above requirement to hold is that $\mathsf{E}\{|q(0)|^{n+\delta}\} < \infty$, $\sup_{k\in\mathbf{R}}(1+k^2)^{\alpha/2}\hat{f}(k) < \infty$ for some δ, $\alpha > 0$. In particular, $N(\lambda) \in C^\infty(\mathbf{R})$ if the random potential has all moments (e.g, if the support of the distribution is compact) and $\sup_{k\in\mathbf{R}}(1+k^2)^{\alpha/2}\hat{f}(k) < \infty$ (this excludes distributions that are too singular, in particular Bernoulli distributions).

Thus, even a low degree of smoothness of the distribution of the potential leads to a fairly smooth integrated density of states. On the other hand, [Carmona et al. 1986] showed that for a Bernoulli potential (6.83) with k_0 large enough, $N(\lambda)$ has a singular component (singular continuous, by Theorem 3.2). In this case, the spectrum contains an infinite set, whose accumulation points include fluctuation boundaries at which $N'(\lambda) = 0$, as well as points at which $N'(\lambda) = \infty$. Such points, in the case of a discrete distribution (6.83) of the random potential, were found rather long ago [Dean 1960] by numerical calculation Their existence was explained by Matsuda [1964] (see also [Mattis and Lieb 1966]) as a result of the existence, of a sufficiently large set of realizations of the potential representing "islands" consisting of a prescribed sequence of zeros and k_0's surrounded by a "sea" of zeros (or k_0's) alone.

A rigorous study of the behavior of $N(\lambda)$ in the neighborhood of these points, known as *special energies* or *special frequencies*, was carried out by English and Endrullis [1987] by the method developed above to prove formulas (6.27) and (6.29) corresponding to the fluctuation boundaries. These authors define the special energies as follows:

Let Ω_{per} be the set of all periodic realizations of the Bernoulli potential (6.83). A point λ_s is called a special energy if there exists a periodic realization $\omega \in \Omega_{\text{per}}$ such that λ_s is contained in the spectrum $\sigma(\omega)$ of the respective periodic operator, and if for all $\omega' \in \Omega_{\text{per}} \setminus \{\omega\}$, the point λ_s is not in the interior of the spectrum $\sigma(\omega')$.

It turns out that if k_0 is sufficiently large,

$$\ln\left|N(\lambda) - N(\lambda_s)\right| = -\operatorname{const}|\lambda - \lambda_s|^{\alpha}\left(1 + o(1)\right) \tag{6.90}$$

as $\lambda \to \lambda_s$, where the exponent α may be either 1 or $\frac{1}{2}$. In the latter case, the asymptotics turn out to be the same as at the edge of the spectrum (6.27). However, under special conditions on the parameters, $N(\lambda)$ may have power-law behavior:

$$C_2|\lambda - \lambda_s|^{\alpha_2} \leqslant \left|N(\lambda) - N(\lambda_s)\right| \leqslant C_1|\lambda - \lambda_s|^{\alpha_1}, \tag{6.91}$$

for some $0 < \alpha_2 < \alpha_1 < 1$.

A more detailed, but not quite rigorous, analysis [Nieuwenhuizen et al. 1985] shows that the preexponential factor in (6.90) (in particular, at the fluctuation boundary, in (6.27), for example), is a periodic functions of $\ln\lambda$, with period depending on λ_s. The coefficient C in the formula

$$\left|N(\lambda) - N(\lambda_s)\right| = C|\lambda - \lambda_s|^{\alpha}\left(1 + o(1)\right) \qquad \text{for } \lambda \to \lambda_s$$

(which is more precise that (6.91)) enjoys the same property.

7 Examples of Calculation of the Integrated Density of States

7.A The Kronig–Penny Stochastic Model

Consider the one-dimensional Schrödinger operator

$$-\frac{d^2}{dx^2} + r(x),$$

for $x \in \mathbf{R}$, where $r(x)$ is the metrically transitive Markov process of Example 1.15(b). The infinitesimal operator $\mathfrak{A}$ of $r(x)$ is, according to (1.27),

$$(\mathfrak{A}f)(r) = n_r\big(-f(r) + f(1-r)\big),$$

for $r = 0, 1$. The phase space E consists of two circles of length π, and functions on E are conveniently written in the form $f(r, \phi)$, where f is periodic in ϕ, with period π, and $r = 0, 1$. Equation (6.44) becomes

$$-\frac{\partial}{\partial\phi}(\Phi(r,\phi)p(r,\phi)) = n_r p(r,\phi) + n_{1-r}p(r,\phi) = 0. \tag{7.1}$$

Using (6.46), which takes the form

$$\sum_{r=0,1} \Phi(r,\phi)p(r,\phi) = p(0,0) + p(0,1),$$

the system (7.1) of two ordinary differential equations can easily be reduced to a single equation integrable by quadratures. This is how both functions $p(r, \phi)$ are found. We will not present in full the resulting formulas, since they are cumbersome [Benderskii and Pastur 1969, 1970]. We will only remark that they clearly imply continuous differentiability of $p(r, \phi)$, which permits us to use Theorem 6.11. Equation (6.24) takes the form

$$N(\lambda) = p(1,0) + p(0,0). \tag{7.2}$$

After calculation, we obtain

$$N(\lambda) = \frac{kn_1}{(n_0 + n_1)B} \tag{7.3}$$

where $\lambda = k^2$ and, for $\lambda \geqslant 1$,

$$B = \pi - \frac{4n_0}{\sinh f(\pi/2)} \int_0^{\pi/2} \cosh\big(f(\pi/2) - f(\pi)\big)\, d\phi \int_0^{\phi} \cosh f(t)\, dt,$$

$$f(t) = n_0 t - \frac{n_1}{\sqrt{1-\lambda^{-1}}} \arctan \frac{\tan t}{\sqrt{1-\lambda^{-1}}},$$

and for $\lambda \in [0,1]$,

$$B = \pi - n_0 \int_{-\pi+\alpha}^{-\alpha} d\phi \int_{\phi}^{\alpha} R(\phi,t)\,dt + n_0 \int_{-\alpha}^{\alpha} d\phi \int_{-\alpha}^{\phi} R(\phi,t)\,dt,$$

where

$$R(\phi,t) = e^{n_0(\phi-t)} \left| \frac{\sin(\phi-\alpha)\sin(t+\alpha)}{\sin(\phi+\alpha)\sin(t-\alpha)} \right|^{\frac{n_1}{2\sqrt{\lambda^{-1}-1}}}$$

and α is the root of the equation $k^2 - \cos^2\phi = 0$. This general expression leads to the following asymptotics of $N(\lambda)$ for $\lambda \downarrow 0$:

$$N(\lambda) + \frac{n_0 n_1}{n_0+n_1} H^{-2}(n_0,n_1) e^{-\pi n_0/\sqrt{\lambda}} (1+o(1)), \tag{7.4}$$

where

$$H(x,y) = \int_0^\infty \left(\frac{t}{t+2x}\right)^{\frac{y}{2}} e^{-x+t}\,dt.$$

Let's consider the special case where $n_0 = n_1 = n$, and assume the potential takes on the values 0 and $q_0 > 0$, that is, $q(x) = q_0 r(x)$. The formula for $N(\lambda)$ is obtained by introducing the multiplier $\sqrt{q_0}$ and replacing the parameters λ and n by $q_0\lambda$ and $\sqrt{q_0}n$ on the right-hand side of (5.3). If we shift the origin in the resulting expression to the point $q_0/2 = \mathsf{E}\{q(x)\}$ and take the limit as $q_0 \to \infty$ and $n \to \infty$, in such a way that $q_0^2/4n = D$ remains fixed, we get

$$\frac{1}{N(\lambda)} = \pi^{1/2} D^{1/3} \int_{-\infty}^{\infty} dx \int_{-\infty}^{x} dy\, e^{\Phi(y)-\Phi(x)}, \tag{7.5}$$

where $\Phi(x) = x^3/3 + \lambda x D^{-2/3}$. With this normalization the correlation function of $q(x)$ is $\mathsf{E}\{q(0)q(x)\} = (q_0^2/4)e^{-2n|x|}$, so taking this limit corresponds to replacing the initial potential $q(x) = q_0 r(x)$ by *white noise*, a Gaussian process with zero mathematical expectation and correlation function $b(x) = D\delta(x)$. However, since this is a generalized process, it cannot be used as the potential in the Schrödinger equation without special justification; in principle, formula (7.5) may be regarded only as asymptotic for the more general expression (7.3). But the justification for actually passing to the limit can be found in [Fukushima and Nakao 1977], where the Schrödinger operator on a finite interval (a,b) is defined in terms of a symmetric bilinear form on the Sobolev space $W_2^1(a,b)$ by the relation

$$\mathcal{E}(u,v) = \int_a^b u'v'^*\,dx - \int_a^b (u'v^* + uv'^*) w(x)\,dx,$$

where $w(x)$ is the Wiener process, of which the Gaussian white noise is the formal derivative. In equation (6.3) for the phase, the potential $q(x)$

should be understood as a symmetric stochastic differential of $w(x)$ in the sense of Stratonovich. The Fokker–Planck–Kolmogorov stationary equation for the density of the invariant measure of the phase $\phi(x)$ becomes, after the introduction of the variable $z = -\cos\phi(x)$,

$$\frac{d}{dz}((z^2+\lambda)p) + D\frac{d^2p}{dx^2} = 0, \tag{7.6}$$

and the counterpart of (6.45) and (7.2), relating the probability density $p(z)$ to $N(\lambda)$ is

$$N(\lambda) = \lim_{|z|\to\infty} z^2 p(z).$$

It is easy to see that the $N(\lambda)$ obtained as described above coincides with (7.5). It was in this form that this function was found and analyzed by Frisch and Lloyd [1960] and Halperin [1965].

Formula (7.5) can be written as

$$\frac{1}{N(\lambda)} = \pi^{1/2}D^{-1/3}\int_0^\infty \frac{1}{\sqrt{x}}\exp\left(-\frac{x^3}{n} - \frac{\lambda x}{D^{2/3}}\right)dx.$$

Thus $N(\lambda)$ is analytic for all real λ, and has the following asymptotic behavior:

$$N(\lambda) = \frac{\sqrt{\lambda}}{\pi}\left(1 + \frac{5}{32}\frac{D^2}{\lambda^3} + O(\lambda^{-6})\right) \tag{7.7}$$

as $\lambda \to \infty$ and

$$N(\lambda) = \frac{\sqrt{\lambda}}{\pi}\exp\left(-\frac{4|\lambda|^{3/2}}{3D}\right)(1+o(1)) \tag{7.8}$$

as $\lambda \to -\infty$. It is interesting that for so singular a random potential the term in (7.7) following the universal one has order $O(\lambda^{-3})$, while for smooth potentials, and even potentials with finite variance, it has order $O(\lambda^{-2})$: cf. Theorem 6.5 and the subsequent discussion, including formulas (6.17) and (6.18).

Another example of a generalized process as a potential is obtained from the same potential $q_0 r(x)$, for $q_0 > 0$, by taking the limit

$$q_0 \to \infty, \quad n_1 \to \infty, \quad \frac{q_0}{n_1} \equiv \nu_1, \tag{7.9}$$

where ν_1 is constant. Since $1/\nu_1$ is the mathematical expectation of the lengths of barriers, that is, intervals where the potential equals q_0, this limiting transition corresponds to "contracting" the barriers, replacing them by delta functions with random amplitudes $k_j \geqslant 0$ equal to the areas of the original barriers. These amplitudes are independent, identically distributed random variables, with probability density $\nu_1 e^{-\nu_1 k}$. Thus, the resulting potential is

$$q(x) = \sum_j k_j \delta(x - x_j), \tag{7.10}$$

where the x_j are random points governed by a Poisson process with density n_0: cf. (6.25).

We will not present here the complete expression of $N(\lambda)$ for this potential. Together with many other quantitative results about $N(\lambda)$ for the one-dimensional Schrödinger operator, they can be found in [Lifshitz et al. 1982b]. We give only the asymptotic formula

$$N(\lambda) = \frac{n_0}{\int_0^\infty e^{-t-n_0/(\nu_1 t)} dt} e^{-\pi n_0 \sqrt{\lambda}} (1 + o(1)), \tag{7.11}$$

valid for $\lambda \downarrow 0$: cf. (6.27) and (7.4).

The general formula (7.3) yields an expression for $N(\lambda)$ for potential (7.10) with negative k_j and same distribution. We get it by making the following substitution:

$$N \to \frac{N}{\sqrt{|q_0|}}, \qquad \lambda \to |q_0|\lambda, \qquad n_r \to n_r\sqrt{|q_0|}; \tag{7.12}$$

then shifting the origin of λ to q_0 and taking a limit similar to (7.9):

$$|q_0| \to \infty, \qquad n_0 \to \infty, \qquad \frac{|q_0|}{n_0} \equiv \nu_0, \tag{7.13}$$

a constant. In this case, as in the case of the Gaussian white noise potential, the resulting operator is not bounded from below, and the spectrum occupies the entire axis; but the asymptotics of $N(\lambda)$ for $\lambda \to -\infty$ is different:

$$N(\lambda) = n_0 e^{-2\sqrt{|\lambda|}/\nu_0} (1 + o(1)). \tag{7.14}$$

Such random operators, whose potentials constructed from delta functions with equal, nonrandom amplitudes, were first analyzed by Frisch and Lloyd [1960], whose paper was an important contribution to the this theory. The various mathematical problems arising in the case of arbitrary identically distributed amplitudes k_j in (7.10) were considered in detail by Kotani [1976] and Luttinger and Sy [1973], who give general asymptotic formulas for $N(\lambda)$. These formulas can also be found in [Lifshitz et al. 1982b.]

7.B Random Jacobi Matrices

Let h be a Jacobi matrix of the form (6.67), describing the one-dimensional Anderson model (4.47): that is, $s(x) = 1$ and the $q(x)$, for $x \in \mathbf{Z}$, are independent, identically distributed random variables. Assume that the $q(x)$ have a Cauchy distribution (Problem II.24):

$$\mathsf{P}\{q(x) \in dq\} = \frac{1}{\pi} \frac{\alpha}{\alpha^2 + (q-\beta)^2} \, dq, \tag{7.15}$$

where $\alpha > 0$ and β is an arbitrary real number. We will compute $N(\lambda)$ using Theorem 6.15, whose conditions are obviously fulfilled. This will simply serve as an illustration of the phase-variable approach explained in this chapter, given that in Chapter II we found $N(\lambda)$ for the multidimensional Anderson model with a Cauchy-distributed potential, of which this is but a special case. See also Problems II.24–25 and Chapter VII.

If we introduce the random variable $t = \cot \phi$, Theorem 6.11 says that we just have to find a distribution for t such that, assuming that t and q_0 are independent, t and $(q_0 - \lambda) - t^{-1}$ are identically distributed. We look for t in the class of Cauchy distributions, using the properties listed in Problem II.24: Parametrizing Cauchy distributions by points $u = \alpha + i\beta$ in the right half-plane (for $\operatorname{Re} u = 0$ we assume, by convention, that the distribution is concentrated at $\operatorname{Im} u$), we know that for independent, Cauchy-distributed random variables ξ and ξ' with parameters u and u', the distributions $\xi + \xi'$, $-\xi$ and ξ^{-1} are also Cauchy-distributed, with parameters $u + u'$, $\bar{u}$ and $\bar{u}^{-1}$.

If $u = \alpha + i\beta$ and z denotes the parameter corresponding to the distribution of t, the condition that t and $q_0 - \lambda - t^{-1}$ be identically distributed is expressed as $z = -i\lambda + u - z^{-1}$, or

$$z = \tfrac{1}{2}\left(-i\lambda + u + \sqrt{(-i\lambda + u)^2 + 4}\right), \tag{7.16}$$

the square root having a positive real part. Using Theorem 6.15 we find

$$N(\lambda) = \int_{-\infty}^{0} \frac{\operatorname{Re} z}{\pi} \frac{dt}{(t - \operatorname{Im} z)^2 + (\operatorname{Re} z)^2} = \frac{1}{2} - \frac{\arg z}{\pi}. \tag{7.17}$$

We now consider, following [Dyson 1953], another example where the integrated density of states can be found explicitly. Here the specific character of the one-dimensional finite-difference operator of the second order is essential.

The operator is defined on $\ell_2(\mathbf{Z})$ by (6.67), with $q(x) = 0$ and the $s(x)$, for $x \in \mathbf{Z}$, being independent, identically distributed random variables with distribution

$$F(ds) = \frac{\nu^\nu}{\Gamma(\nu)} e^{-\nu s^2} s^{2(\nu-1)} \, ds^2. \tag{7.18}$$

That is, $s^2(x)$ has the so-called Γ-distribution: see (1.34).

According to the general results of Sections 4 and 6, $N(\lambda)$ exists and is the limit, as $L \to \infty$, of the mathematical expectations of

$$N_L(\lambda) = (2L+1) \sum_{\lambda_{n,L} \leqslant \lambda} 1,$$

where $\lambda_{n,L}$, for $|n| \leqslant L$, are the eigenvalues of the Dirichlet problem defined by equation (6.47) on the interval $|x| \leqslant L$ and by the boundary conditions $\psi(-L-1) = \psi(L+1) = 0$. In this case, it is more convenient to find $N(\lambda)$ not from the oscillation theorem, but more directly, by using the function

$$w(z) = -\lim_{L\to\infty} \frac{1}{2L+1} \sum_{n=-L}^{L} \ln(\lambda_{n,L} - z) = -\int_{\mathbf{R}} \ln(\mu - z) N(d\mu), \tag{7.19}$$

where $\operatorname{Im} z \neq 0$ and the branch of the logarithm is analytic in the complement of the negative semiaxis, so that, in particular,

$$\ln(\lambda + i0) = \ln|\lambda| + i\pi\chi_{(-\infty,0)}(\lambda) \tag{7.20}$$

for $\lambda \in \mathbf{R}$. By applying the unitary transformation $(U\psi)(x) = (-1)^x \psi(x)$ we see that the eigenvalue distribution is symmetric with respect to the origin, so $w(z)$ can be written as

$$w(z) = -\int_0^\infty \ln(z^2 - \mu^2) N(d\mu),$$

which we abbreviate as $\omega(z^2)$; and $N(\lambda)$ can be expressed in the following way:

$$\frac{1}{\pi}\omega(\lambda + i0) = -\int_{\sqrt{\lambda}}^\infty N(d\mu) = -1 + N(\sqrt{\lambda}). \tag{7.21}$$

On the other hand, let's consider the solution $u(x,\lambda)$ of the Cauchy problem for equation (6.47) satisfying the conditions $u(-L-1,\lambda) = 0$ and $u(-L,\lambda) = 1$. Obviously, $u(x,\lambda)$ is a polynomial of degree $x+L$ with respect to λ and the eigenvalues $\lambda_{n,L}$ will be zeros of the polynomial $u(L+1,\lambda)$. Since

$$\lim_{z\to\infty} (-z)^{L+x} u(x,z) = \left(\prod_{y=-L+1}^{x} s(y) \right)^{-1},$$

we have

$$w(z) = -\lim_{L\to\infty} \frac{1}{2L+1} \ln u(L+1,z) \prod_{y=-L+1}^{L+1} s(y).$$

Assume (cf. (6.53)) that $\zeta(x) = s(x+1)u(x+1)/u(x)$. Then

$$u(L+1,z)\prod_{x=-L+1}^{L+1} s(x)=\prod_{x=L}^{-L}\zeta(x),$$

and therefore

$$w(z)=-\int_{\mathbf{C}}\ln\zeta P(d\zeta) \tag{7.22}$$

where, as in Theorem 6.13, $P(d\zeta)$ is a weak limit point of the family of probability measures

$$P_L(d\zeta)=\frac{1}{2L+1}\sum_{x=-L}^{L}P\{\zeta(x)\in d\zeta\}.$$

The quantities $\zeta(x)$ satisfy a recurrence relation of the type (6.60):

$$\zeta(x)=z-\frac{s^2(x)}{\zeta(x-1)},\qquad \zeta(-L)=z. \tag{7.23}$$

Consider purely imaginary values of z, with imaginary part $\mu>0$, and introduce new variables $\xi(x)=\zeta(x)z^{-1}-1$, in terms of which (7.23) becomes

$$\xi(x)=\frac{s^2(x)}{\mu^2}\big(\xi(x-1)+1\big).$$

It follows that $\xi(x)$, for all $x\geqslant -L$, is nonnegative, and the distribution functions $P_x(d\xi)=\mathsf{P}\{\xi(x)\in d\xi\}$ are related by

$$P_x(d\xi)=\int_0^\infty P_{x-1}(d\xi')f\big(\mu^2\xi(\xi'+1)\big)\mu^2(1+\xi'^2),$$

where $f(t)$ is the density of the Γ-distribution (7.18).

Thus, every limit point of the family of measures

$$\frac{1}{2L+1}\sum_{x=-L}^{L}P_x(d\xi)$$

has density $p(\xi)$ satisfying the integral equation

$$p(\xi)=\int_0^\infty f\big(\mu^2\xi(\xi'+1)\big)\mu^2(1+\xi'^2)p(\xi')\,d\xi'.$$

One can easily verify that if the density $f(t)$ has the form (7.18), the solution to this equation is the function

$$p(\xi)=C_\nu\xi^{\nu-1}(1+\xi)^{-\nu}e^{-\mu^2\xi}, \tag{7.24}$$

where

$$C_\nu^{-1} = \int_0^\infty \xi^{\nu-1}(1+\xi)^{-\nu} e^{-\mu^2\xi} d\xi. \tag{7.25}$$

From this equation and (7.22), we get

$$w(i\mu) = -\ln i\mu - \int_0^\infty \ln(1+\xi) p(\xi)\, d\xi. \tag{7.26}$$

We now have to perform analytic continuation from imaginary to real values of the spectral parameter, that is, from positive to negative values of μ^2 in formulas (7.24) to (7.26). We refer the reader to [Dyson 1953] for the case of arbitrary ν, and consider here only the simplest case $\nu = 1$, corresponding to exponential distribution of off-diagonal entries $s^2(x)$. In this case, the integrals in (7.25) and (7.26) can be integrated by parts to yield, in both cases,

$$\int_0^\infty \left(\ln \frac{\mu^2+\xi}{\xi}\right)^\alpha e^{-\xi} d\xi$$

for $\alpha = 1, 2$, and the analytic continuation can be performed straightforwardly using (7.20). The explicit form of $N(\lambda)$ in terms of quadratures follows then from (7.20). In particular, it implies that $N(\lambda)$ is real analytic for $\lambda \neq 0$, and leads to the following asymptotic formula for the density of states $n(\lambda) = N'(\lambda)$ in the vicinity of 0:

$$n(\lambda) = \frac{C_2 - C_1^2}{\lambda\, |\ln \lambda|^3}(1 + o(1)) \tag{7.27}$$

as $\lambda \downarrow 0$, where $C_\alpha = \int_0^\infty \ln^\alpha t e^{-t}\, dt$ (C_1 is the Euler constant).

Thus, when $\lambda = 0$, the density of states has a singularity (a Dyson singularity) due to the possibility of infinitesimal values for the off-diagonal entries $s(x)$. The relation between these two facts is especially simple when the random variables $s(x)$ take the values 0 and $s > 0$ with probability c and $1 - c$, for $1 > c > 0$. Then the associated Jacobi matrix is subdivided into blocks whose sizes depend on the lengths of the series of nonzero values of $s(x)$, each block of odd order having eigenvalue zero. Therefore, the density of states $n(\lambda)$ corresponding to this matrix will have at zero a singularity of the form $A\delta(\lambda)$, for $A > c$. This singularity is certainly smoothed when the $s(x)$ have continuous distribution, and (7.27) is an example of such smoothing.

Problems

1. Show that the Schrödinger operator $-d^2/dx^2 + q(x)$ on the interval $[0, L]$ with zero boundary conditions at 0 and L is bounded from below by $-\frac{L}{4}\int_0^L q^2(x)\,dx$.

Hint. Use the representation $y(x) = -\int_x^L y'\,dt$, valid if $y(L) = 0$, and the resulting inequality $|y(x)| \leqslant \sqrt{L}\|y'\|_2$.

2. Show that the point spectrum of the operator $-d^2/dx^2 + k_0\delta(x)$, where $k_0 > 0$ and δ is the Dirac delta function, has a single eigenvalue $\lambda_0 = -k_0^2/4$.

3. Let $m(x)$, for $x \in \mathbf{R}$, be a metrically transitive field bounded from above and below by positive constants. Consider the eigenvalue problem for the equation $-\psi'' = \lambda m\psi$ on the interval $[-L, L]$ with specified boundary conditions, and let $\hat{N}_L(m, \lambda)$ be $(2L)^{-1}$ times the number of eigenvalues not exceeding λ. Prove that:

(a) the nonrandom limit $\lim_{L\to\infty} \hat{N}_L(m, d\lambda) = \hat{N}(m, d\lambda)$ exists with probability 1, and is independent of the boundary conditions;

(b) $\hat{N}(m, d\lambda) = N\left(-\frac{d}{dx}\frac{1}{m(x)}\frac{d}{dx}, d\lambda\right).$

Hint. Reduce to the equation $-(m^{-1}u')' = \lambda u$ by the substitution $m^{-1}u' = \psi$, and use the fact that the integrated density of states of the operator

$$-\frac{d}{dx}\frac{1}{m(x)}\frac{d}{dx}$$

is independent of the boundary conditions.

4. Show, using the phase formalism, that

$$N\left(-\frac{d^2}{dx^2} + q, \lambda\right) \leqslant \frac{2}{\pi}\mathsf{E}^{1/2}\{|q(0) - \lambda|\}.$$

Hint. Introduce the phase θ by the relation $\cot\theta = cy'/y$, where c is a constant with value to be determined.

5. Assuming that $\mathsf{E}\{|q(0)|^2\} < \infty$, show that

$$\left|N\left(-\frac{d^2}{dx^2} + q, \lambda\right) - N\left(-\frac{d^2}{dx^2}, \lambda\right)\right| \leqslant \text{const}\,|\lambda|^{-1/2}.$$

Hint. For $\lambda > 0$, use (6.15) and (6.16). For $\lambda < 0$, use (12.25) and Lemma 5.21.

6. Let $q(x)$, for $x \in \mathbf{R}$, be a metrically transitive Markov process, and assume that $-\infty < b \leqslant q(x) \leqslant B < \infty$ with probability 1. Show that, if $\lambda > B$ and $C = \max\{\lambda - b, 1\}/\min\{\lambda - B, 1\}$, any weak limit point $P(d\phi)$, $\phi \in S^1$, of the distributions

$$\frac{1}{x}\int_0^x P\{\phi(t) \in d\phi\}\, dt$$

for $x \to \infty$ satisfies

$$P(d\phi) = \frac{p(\phi)}{\pi} d\phi,$$

and $C^{-1} \leqslant p(\phi) \leqslant C$.

Hint. Use (6.5) and the obvious estimate $\min\{\lambda - B, 1\} \leqslant \theta'(x) \leqslant \max\{\lambda - b, 1\}$, and show that, with probability 1,

$$1 + \frac{k_1}{x} C^{-1} \leqslant \frac{1}{\pi(d\phi) x} \operatorname{meas}\{t \in [0, x] : \theta(t) \in d\phi \pmod{\pi}\} \leqslant 1 + \frac{k_2}{x} C,$$

where k_1 and k_2 are constants depending only on λ, b and B.

7. Estimate the lengths of possible gaps in the spectrum of the Schrödinger operator based on asymptotic formulas (6.17).

Hint. Use arguments like those in the proof of Theorem 5.32.

8. Show that, if the correlation function of the process of Theorem 6.6 is bounded and summable, the term $o(1)$ in (6.23) can be replaced by $O(\lambda^{-1/2})$.

9. Let $s(x)$ and $m(x)$, for $x \in \mathbf{R}$, be metrically transitive fields bounded from above and below by positive constants, with probability 1; assume $s \in C^1(\mathbf{R})$, and let $N(\lambda)$ be the integrated density of states for the boundary problem

$$-\frac{d}{dx} s(x) \frac{d}{dx} \psi = \lambda m(x) \psi, \qquad \psi(-L) = \psi(L) = 0.$$

Obtain an asymptotic formula for $N(\lambda)$ similar to (6.23):

$$N(\lambda) = \frac{1}{\pi}\sqrt{\frac{\lambda\mu}{\kappa}}(1 + o(1))$$

for $\lambda \downarrow 0$, where $\mu = \mathsf{E}\{m(x)\}$ and $\kappa = \mathsf{E}\{s^{-1}(x)\}$.

10. Let $s(x)$, for $x \in \mathbf{R}$, be a metrically transitive field bounded from above and below by positive constants, and assume there exists a positive number s such that, for all ε and $L > 0$,

$$\mathsf{P}\{|s(x) - s|\} \leqslant \varepsilon, |x| \leqslant L\} > 0.$$

Show that the spectrum of the operator $\dfrac{d}{dx} s(x) \dfrac{d}{dx}$ is $[0, \infty)$.

11. Find $N(\lambda)$ for the operator (6.19) with $s(x) = s_1 + s_2 r(x)$, where s_1 and s_2 are positive and $r(x)$ is the process from Example 1.15(b).

12. Find $N(\lambda)$ for the Jacobi matrix

$$(J\psi)(x) = s(x+1)\psi(x+1) + s(x)\psi(x-1),$$

where the $s(x)$, for $x \in \mathbf{Z}$, are independent, identically distributed random variables and $s(0)$ takes on only two values, 0 and 1, with probabilities c and $1-c$.

13. Show that for the Poisson potential (6.25) with concentration c and $k_0 > 0$, the Fokker–Planck equation for the stationary probability density $p(z)$ of the quantity $z = \cot\phi$, where ϕ is defined by (6.5), is

$$\frac{d}{dz}((z^2+\lambda)p) - c(p(z) - p(z-k_0)) = 0,$$

and that $N(\lambda) = \lim_{|z|\to\infty} z^2 p(z)$.

Hint. Derive the equation $dz/dx = -(z^2+\lambda)+q(x)$ from (6.5) and determine the infinitesimal operator of the process $z(x)$, noting that $\lim_{x\downarrow x_j} z(x) - \lim_{x\uparrow x_j} z(x) = k_0$ and that the probability that a given interval of small length l will contain one Poisson point is cl. Then derive the formula for the integrated density of states $N(\lambda)$ from (6.45).

14. Show that, under the conditions of the preceding problem, $N(\lambda)$ is a real analytic function of λ, when $\lambda > 0$.

Hint. Solve the Fokker–Planck equation by successive approximations, and represent $N(\lambda)$ as

$$N(\lambda) = \frac{k}{A(k)}, \qquad k = \sqrt{\lambda}, \qquad A(k) = \sum_{l=0}^{\infty} \left(\frac{c}{k}\right)^l A_l,$$

where

$$A_l = \int_{-\infty}^{\infty} f_l(\zeta)\,d\zeta, \quad f_{l+1}(\zeta) = \frac{1}{\zeta^2+1}\int_{\zeta-k_0/k} f_l(\zeta')\,d\zeta', \quad f_0(\zeta) = \frac{1}{\zeta^2+1}.$$

Using these relations, show that $0 \leqslant A_l \leqslant \pi^{l+1}/(l+1)!$.

15. Check formula (6.17), using equality (7.3).

16. Check that for the Gaussian white noise potential (with correlation function $b(x) = D\delta(x)$, formula (6.18) does not contradict the asymptotics (7.7).

17. The considerations used in the heuristic derivation of (6.18), the analogy of (6.23) and (6.17), and the formal perturbation theory in λ enable us to assume that the improved version of the asymptotic formula (6.23) is, when $\lambda \to 0$, as follows:

$$N(\lambda) = \frac{1}{\pi}\frac{\lambda}{\kappa}\left(1 + \frac{1}{4}\sqrt{\frac{\lambda}{\kappa}}\int_0^\infty b_\delta(x)\sin 2\sqrt{\frac{\lambda}{\kappa}}x\,dx\right)(1+o(1))$$
$$= \frac{1}{\pi}\frac{\lambda}{\kappa}\left(1 + \frac{\lambda}{2\kappa}\int_0^\infty b_\delta(x)x\,dx + o(\lambda)\right),$$

where $b_\delta(x) = \mathsf{E}\{\delta(x)\delta(0)\}$ is the correlation function of the random process $\delta(x) = xs(x) - 1$, and $\kappa^{-1} = \mathsf{E}\{s^{-1}(x)\}$. Verify the above asymptotic formula for $s(x) = s_0 + (s_1 - s_0)r(x)$, where $0 < s_0 < s_1$ and $r(x)$ is the random process of Example 1.15(b).

18. Derive an asymptotic formula similar to (6.23) for the Jacobi matrices

$$(J\psi)(x) = -J(x+1)\psi(x+1) + (J(x) + J(x+1))\psi(x) + J(x)\psi(x-1),$$

where $J(x)$, for $x \in \mathbf{Z}$, is a metrically transitive field such that, with probability 1, $0 < J_0 \leqslant J(x) \leqslant J_1 < \infty$.

19. Consider the Jacobi matrix

$$(J\psi)(x) = J(x+1)\psi(x+1) + J(x)\psi(x-1),$$

where $J(x)$, for $x \in \mathbf{Z}$, are independent, identically distributed random variables and $J(0)$ takes on only two values $J_2 > J_1 > 0$, with probabilities p and $1-p$, respectively. Show that $N(J,\lambda)$ satisfies the asymptotic formula

$$\ln N(J,\lambda) = \ln p\,\pi\sqrt{J_2|\lambda - a|}(1 + o(1))$$

where $a = \pm J_2$ and λ approaches J_2 from below or $-J_2$ from above.

Obtain a similar formula for the integrated density of states of the Jacobi matrix

$$(J_1\psi)(x) = -\psi(x+1) - \psi(x-1) + q(x)\psi(x),$$

where $q(x)$, for $x \in \mathbf{Z}$, is a random variable of the same type as $J(x)$ in the previous paragraph.

Hint. Use Theorem 6.18 and the discrete version of Theorem 6.7.

20. Consider a Schrödinger operator with a Markov potential on the interval $[0, L]$, with zero boundary conditions at the endpoints. Parametrize the eigenvalues $\lambda_n(L)$ of this operator by the relation $\theta(L, \lambda_n) = \pi n$, for $n \in \mathbf{Z}$, where $\theta(x, \lambda)$ is the phase (6.5). Introduce the distribution functions $w_n(L, \lambda) = \mathsf{P}\{\lambda_n(L) \leqslant \lambda\}$ and $F(L, \lambda, \eta) = \mathsf{P}\{\theta(L, \lambda) < \eta\}$. Assume the latter has a smooth density $f(L, \lambda, \eta)$. Show that

$$w_n(L, \lambda) = 1 - F(L, \lambda, \pi n) = \int_0^L f(l, \lambda, \pi n)\, dl.$$

Hint. The density of the joint distribution $p(x, q, \theta)$ of the random variables $q(x)$, $\theta(x)$ satisfies the equation

$$\frac{\partial p}{\partial x} = -\frac{\partial}{\partial \theta}(\Phi p) + \mathfrak{A}_q^* p,$$

which, after integration over θ in the interval $[-\infty, \pi n]$ and over q, and taking into account that $\Phi(q, \pi n) = 1$, gives

$$\frac{\partial}{\partial L} F(L, \lambda, \eta) = -f(L, \lambda, \eta).$$

21. Show that if λ_k, for $k = 1, 2, \ldots$, are the eigenvalues of the boundary-value problems for the Schrödinger operator (6.1)–(6.2) or for the Jacobi matrices (6.69)–(6.70), we have

$$\frac{\partial \lambda_k}{\partial \alpha} = -\left(\int_a^b y^2(x, \lambda_k)\, dx \right)^{-1},$$

where $y(x, \lambda)$ is the solution of the Cauchy problem for equation (6.1) or (6.69), with initial condition (6.3) or (6.72), and where the integral is understood as a sum in the discrete case. Obtain similar formulas for $\partial \lambda_k / \partial \beta$. Show that if the potential is $q(x) = q_0(x) + t q_1(x)$, we have

$$\frac{\partial \lambda_k}{\partial t} = \int_a^b q_1(x) |\psi_k(x)|^2\, dx,$$

where $\psi_k(x) = y(x, \lambda_k) \left(\int_a^b |y(x', \lambda_k)|^2\, dx' \right)^{-\frac{1}{2}}$.

Show that, when $\beta = 0$ in problem (6.1)–(6.2),

$$\frac{\partial \lambda_k}{\partial b} = \frac{\partial \lambda_k}{\partial \alpha} r^2(b),$$

where $r^2(b) = y^2(b, \lambda_k) + y'^2(b, \lambda_k)$.

22. Consider the Jacobi matrices

$$(J\psi)(x) = -s(x+1)\psi(x+1) + q(x)\psi(x) - s(x)\psi(x-1),$$
$$(J_+\psi)(x) = -|s(x+1)|\,\psi(x+1) + q(x)\psi(x) - |s(x)|\,\psi(x-1),$$

where $s(x) \neq 0$ for $x \in \mathbf{Z}$. Prove that there exists a unitary, diagonal matrix U, with entries ± 1, such that $J_+ = U^{-1}JU$. Show that the integrated densities of states for J and J_+ either both exist or both fail to exist, and if they exist they coincide.

In particular, this shows that the assumption that $s(x) \geqslant s > 0$ or $s(x) > 0$ in Theorems 6.18–6.20 may be replaced by the weaker conditions $|s(x)| \geqslant s > 0$ or $s(x) \neq 0$.

23. Show that the density of states for the Brownian motion model—the Schrödinger operator with potential (6.52)—is given by

$$\begin{aligned} n(\lambda) = N'(\lambda) &= \int_0^{\pi} d\phi \int_K dX\, V(K)\pi(X,\phi)\pi(X,-\phi)\sin^2\phi \\ &= \int_{\mathbf{R}} dz \int_K dX\, V(K)\tilde{\pi}(X,z)\tilde{\pi}(X,-z), \end{aligned}$$

where $\pi(X,\phi)$ is the invariant density of the Markov process $(X(x),\phi(x))$ (see Subsection 6.D), $\tilde{\pi}(X,z)$ is the invariant density for $(X(x),z(x))$, where $z = \cot\phi$, and $V(K)$ is the volume of the compact set K.

Hint. Let $\lambda_k(L)$, for $k = 1,2,\ldots$, be the eigenvalues of the corresponding Dirichlet problem. The density of states on $(0,L)$ can be formally written as

$$n_L(\lambda) = \frac{\partial}{\partial\lambda} N_{(0,L)}(\lambda) = \frac{1}{L}\sum_{k\geqslant 1}\delta(\lambda-\lambda_k),$$

where δ is the Dirac delta function. If $\theta(x,\lambda)$ is the phase corresponding to H, we have $\theta(L,\lambda_k) = \pi k$. Hence, in view of $\partial\theta/\partial\lambda \geqslant 0$, we find

$$(1)\qquad n_L(\lambda) = \frac{1}{L}\delta_p(\theta(L,\lambda))\frac{\partial\theta}{\partial\lambda}(L,\lambda), \qquad \delta_p(\mu) = \sum_{k\geqslant 1}\delta(\mu-\pi k).$$

Since the density satisfies the relation

$$(2)\qquad n(\lambda) = \lim_{L\to\infty} \mathsf{E}\{n_L(\lambda)\},$$

equation (1) indicates that we should consider, along with the process $\theta(x,\lambda)$, for $x \geqslant 0$, the process $\Theta(x,\lambda) = \partial\theta/\partial\lambda(x,\lambda)$ as well. Here θ and Θ satisfy

$$\begin{aligned} \theta' &= \cos^2\theta + (\lambda - Q(x))\sin^2\theta = \Phi(X,\theta), \\ \Theta' &= (\lambda - Q(X) - 1)\Theta\sin 2\theta + \sin^2\theta = \frac{\partial\Phi}{\partial\theta}\Theta + \sin^2\theta. \end{aligned}$$

Thus, the triple $(X(x), \theta(x), \Theta(x))$ is a Markov process, whose transition density $p_3(x, X_0, \theta_0, \Theta_0; X, \theta, \Theta) = p_3(x, X, \theta, \Theta)$ satisfies the equation

$$\text{(3)} \qquad \frac{\partial p_3}{\partial x} = \left(\tfrac{1}{2}\Delta_X - \frac{\partial}{\partial \theta}\Phi - \frac{\partial}{\partial \Theta}\left(\frac{\partial \Phi}{\partial \theta}\Theta + \sin^2\theta\right)\right) p_3$$

In view of (2), we introduce $\Gamma(x, X, \theta) = \int p_3(x, X, \theta, \Theta)\Theta\, d\Theta$. Then

$$\text{(4)} \qquad n(\lambda) = \lim_{L\to\infty} \frac{1}{L}\int_K \Gamma(L, X, 0)dX.$$

To determine Γ, consider the differential equation for it, which follows from (3):

$$\frac{\partial \Gamma}{\partial x} = \tfrac{1}{2}\Delta\Gamma - \Phi\frac{\partial}{\partial \theta}\Gamma + \sin^2\theta\, p_2(x, X_0, \theta_0; X, \theta),$$

where p_2 is the transition density of the process $(X(x), \theta(x))$. Recalling that $p_2(x, X_0, \theta_0; X, \theta)$ tends to $\pi(X, \theta)$ as $x \to \infty$ (see (6.59)), we replace the last equation by the asymptotic equation

$$\text{(5)} \qquad \frac{\partial \Gamma}{\partial x} = \tfrac{1}{2}\Delta\Gamma - \Phi\frac{\partial}{\partial \theta}\Gamma + \sin^2\theta\,\pi$$

as $x \to \infty$. The solution of (5) will be sought in the form

$$\text{(6)} \qquad \Gamma(x, X, \theta) = \frac{1}{V(K)}n(\lambda)x + f(X, \theta),$$

where the x-independent part of the first term on the right-hand side was determined from (4). By substituting (6) in (5) we obtain

$$\text{(7)} \qquad \tfrac{1}{2}\Delta f - \Phi\frac{\partial}{\partial \theta}f + \sin^2\theta\,\pi = \frac{n}{V(K)},$$

where π is such that

$$\text{(8)} \qquad \frac{1}{2} - \frac{\partial}{\partial \theta}\Phi\pi = 0.$$

We now multiply (7) by $\pi(X, -\theta)$ and (8) by $f(X, -\theta)$, and integrate the resulting equalities with respect to X and θ. Subtracting the results from one another gives the desired formula for $n(\lambda)$.

24. Let h be the discrete Schrödinger operator of (6.67), with $s(x) = 1$ for $x \in \mathbf{Z}$ and $q(x)$, for $x \in \mathbf{Z}$, consisting of independent, identically distributed random variables such that $\mathsf{P}\{q(0) \in dq\} = f(q)\,dq$. Show that the density of states $n(\lambda)$ of h can be represented as

$$n(\lambda) = \int_0^\pi \pi(\phi)\pi\left(\frac{\pi}{2} - \phi\right)\cos^2\phi\, d\phi = \int_{-\infty}^{\infty} p(z)p(z^{-1})\, dz,$$

where $\pi(\phi)$ is the density of the invariant distribution of the phase $\phi(x)$ given by (6.73), and $p(z)$ is the analogous density for $z = \cot\phi$.

Hint. Adapt the arguments in Problem 23 to the discrete case.

25. Consider a metrically transitive Schrödinger operator on the whole axis, whose potential has a finite second moment and is a strictly Markov process [Gihman and Skorohod II]. Let $\theta(x,\lambda)$ be the corresponding phase function satisfying (6.5), with boundary condition $\theta(0,\lambda) = 0$, and let L_k, for $k = 1, 2, \dots$ be defined by $\theta(L_k, \lambda) = \pi k$, in nondecreasing order. Show that, for any λ, we have $L_1 < L_2 < \cdots$ with probability 1, and that the random variables $l_k = L_k - L_{k-1}$ (with $L_0 = 0$) are independent and identically distributed, with

$$N(\lambda) = \mathsf{E}^{-1}\{l_1(\lambda)\}.$$

Hint. Use (6.6), noting that $\theta'(L_k, \lambda) = \Phi(q,\theta)\big|_{\theta=pk} = 1$.

26. Consider the Schrödinger operator H on the whole axis whose random potential q is defined as follows: Let l_k, for $k \in \mathbf{Z}$, be nonnegative, independent, identically distributed random variables with $\mathsf{E}\{l_0\} = \infty$; let q_k, for $k \in \mathbf{Z}$ also be independent and identically distributed, and also independent of l_k; and assume that $|q_k| \leqslant C < \infty$ with probability 1, where C is constant.

Consider the partition of $\mathbf{R}_+$ into adjacent intervals Δ_k of length l_k, with $\Delta_0 = (0, l_0)$, and define q by the relation $q(x) = q_k$ for $x \in \Delta_k$. Such a potential is not, in general, metrically transitive.

Prove that, with probability 1,

$$N(H,\lambda) = \int_{\mathbf{R}} N\left(-\frac{d^2}{dx^2}, \lambda - q\right) \mathsf{P}\{q_0 \in dq\}.$$

Compare this with the formula of Problem II.37.

Hint. Use (6.6) and the fact that, if $n(L) = \max\{k > 0 : \sum_{j=1}^{k} l_j \leqslant L\}$, we have $n(L)/L \to 0$ with probability 1.

27. Give a rigorous proof for formula (6.18), and estimate the error. Find similar formulas for the operator (6.19).

Hint. Use relations (6.14) to (6.16) and the scheme of the proof of Theorems 14.5 and 14.6.

28. Calculate further terms in (6.17), under the assumption that

$$\mathsf{E}\left\{\left|\frac{\partial^n}{\partial x^n} q(0)\right|\right\} < \infty.$$

29. In (6.67), let $\{s(x), q(x)\}$ be a sequence of independent, identically distributed random variables, the distribution of $(s(0), q(0))$ being nontrivial, that is, not concentrated at a single point (s,q). Show that the invariant distribution of the phase $F(d\phi)$ is (a) continuous, (b) unique, and (c) either absolutely continuous or singular continuous.

Hint. Consider first, for simplicity, the case of $s(x) \equiv 1$. Introduce the transformation T_q, for $q \in \operatorname{supp} P$, of the circle $S^1 = [0, \pi)$, given by $\cot(T_q\phi) = q - \lambda - \cot\phi$. Write the condition that F is invariant as

$$F(d\phi) = \int P(dq) F(T_q^{-1} d\phi).$$

For (a), assume that F has atoms and arrive at a contradiction by considering the powers $(T_{q_1} T_{q_2}^{-1})^n$, for $n = 1, 2, \ldots$, which satisfy

$$\cot(T_{q_1} T_{q_2}^{-1})^n \phi = \cot\phi + (q_1 - q_2)n.$$

For (b), assume there exist two continuous invariant measures F_1 and F_2 and arrive at a contradiction by applying the powers above to closed intervals $\Delta \subset S^1$ such that $(F_1 - F_2)(\Delta) = \sup(F_1 - F_2)(\Delta')$, for Δ' ranging over all intervals in S^1.

For (c), show that $F_{\text{ac}}(X) \geqslant \int P(dq) F_{\text{ac}}(T_q^{-1} X)$, where $F = F_{\text{ac}} + F_{\text{sc}}$.

30. In the preceding problem, let $s(x) \equiv 1$ and assume that the $q(x)$ are independent and take on two values q_1 and q_2 with probability p_1 and p_2. Show that there exist values of these parameters for which the distribution $F(d\phi)$ does not have a continuous density.

Hint. Consider the case of $|q_1 - q_2| > 4$ where, according to Corollary 4.3, the spectrum consists of intervals $I_j = \{\lambda : |\lambda - q_j| \leqslant 2\}$. If $\lambda \in I_1$, the transformation $T_{q_2}^{-1}$ has two fixed points ϕ_1 and ϕ_2. Set $z_1 = \cot\phi_1$ and $z_2 = \cot\phi_2$, arranging so that $z_1 > 0$; we have $z_2 = z_1^{-1}$, and, if $p_2 z_1^{-2} > 1$, the assumption that $F(d\phi) = f(\phi)\, d\phi$ implies that $f(T_1^{-n}\phi_1) = 0$ for all n or $f(T_1^{-n}\phi_1) = \infty$ for all n. But for a set of $\lambda \in I_1$ having full Lebesgue measure, $\{T_1^{-n}\phi_1 : n \in \mathbf{Z}\}$ is dense in $[0, \pi)$, so this is incompatible with the continuity of $f(\phi)$.

31. Consider the one-dimensional Schrödinger operator with potential (6.25) in which $k_0 = \infty$, and the points x_j are determined as in Problem I.11, with $f_0(l) = f_1(l)$. Show that the spectrum of this operator is, with probability 1, pure point and dense, and its density of states is

$$n(\lambda) = \frac{\pi}{2n\sqrt{\lambda}} \sum_{m=1}^{\infty} f\left(\frac{\pi m}{\sqrt{\lambda}}\right).$$

Prove that the asymptotic formulas (6.29) are valid.

32. Prove that $\mathsf{P}\{\lim_{L\to\infty} \sup_{\lambda\in\mathbf{R}} |N_{(0,L)}(\lambda) - N(\lambda)| = 0\} = 1$.

Hint. Use Theorem 6.4 or 6.18 and the scheme of the proof of the Glivenko–Cantelli theorem [Loeve 1962].

33. If, in the one-dimensional case, $N_{(0,L)}(\lambda)$ or $\mathsf{E}\{N_{(0,L)}(\lambda)\}$ depend continuously on a parameter ξ that ranges over a compact set $K \subset \mathbf{R}^n$, the convergence of $\mathsf{E}\{N_{(0,L)}(\lambda)\}$ to $N(\lambda)$ for fixed λ is uniform in $\xi \in K$.

Hint. Use Theorems 6.4 and 6.18, the inequalities of Problem II.6 (that is, (5.39) in the continuous case), Theorem 3.2, and the following analogue of Dini's theorem: if $l_k(\xi)$ is a sequence of continuous functions on K and $l_{k+j}(\xi) \leqslant l_k(\xi) + l_j(\xi)$, the limit $l(\xi) = \lim_{k\to\infty} k^{-1} l_k(\xi)$ exists and the convergence to it is uniform if l is a continuous function.

34. Prove that the Markov process considered in Subsection 6.D is reversible (see Problem I.20 for the definition of reversibility).

Chapter IV

Asymptotic Behavior of the Integrated Density of States at Spectral Boundaries in Multidimensional Problems

We saw in Chapter III that, in one dimension, $N(\lambda)$ can display various types of asymptotic behavior near the boundaries of the spectrum: formulas (6.17), (6.23), (6.27), (7.4), and so on. Those results were obtained by specifically one-dimensional methods. In this chapter we show, using different methods, that similar formulas are true in a wide range of multidimensional problems, and that the distinction between stable and fluctuation boundaries, made in Subsection 6.B, still applies. For each class we will find representative formulas of a universal character.

8 Stable Boundaries

In Subsection 6.B we called a spectrum boundary *stable* if the spectrum in its neighborhood exists owing to arbitrary regions of the space of realizations of the coefficients. The location of a stable boundary does not depend on the parameters involved in the probabilistic description of the coefficients, and the functional form of the asymptotic behavior of $N(\lambda)$ in its vicinity does not depend on the statistical properties, and in particular on the mixing properties, of the coefficients—it coincides with the behavior for some operator with constant coefficients (see formulas (6.17) and (6.23), for example).

The simplest example of a stable boundary is the point $\lambda = \infty$ for a Schrödinger operator. The asymptotic behavior of $N(\lambda)$ at this boundary is given, in the multidimensional case, by Theorem 5.29. Theorem 5.31 generalizes this to elliptic differential operators of arbitrary order. These results are particular cases of the semiclassical approximation of quantum mechanics, where nonleading terms of the differential operator are negligible, and for leading terms the contributions of all infinitesimal volumes to $N(\lambda)$ are additive [Titchmarsh 1958]. Asymptotic formulas of this type are very general and are not restricted to metrically transitive, and in particular to random, operators.

We now consider a spectral boundary that is less trivial from the point of view of the spectral theory of metrically transitive operators, namely, the point $\lambda = 0$ in the spectrum of the divergence operator of the second order

(8.1) $$A = -\frac{\partial}{\partial x_i} a_{ij}(x) \frac{\partial}{\partial x_j},$$

for $x \in \mathbf{R}^d$, where we use the convention that repeated indexes are to be summed from 1 to d, and where there are number $0 < a_- \leqslant a_+ < \infty$ such that

(8.2) $$a_-|\xi|^2 \leqslant a_{ij}(x)\xi_i\xi_j \leqslant a_+|\xi|^2$$

for all $\xi \in \mathbf{R}^d$. By Theorem 5.24, we have

(8.3) $$N_0\left(\frac{\lambda}{a_+}\right) \leqslant N(A, \lambda) \leqslant N_0\left(\frac{\lambda}{a_-}\right),$$

where $N_0(\lambda) = c_d \lambda^{d/2}$ is the integrated density of states of the operator $-\Delta$, according to (5.35). Thus as $\lambda \downarrow 0$, the function $N(A, \lambda)$ cannot decrease faster than $\lambda^{d/2}$, in accordance with formula (6.23) for the one-dimensional analogue (6.19) of the operator (8.1). We have the following rather general theorem:

(8.1) Theorem [Kozlov 1979a]. *If A is an operator of the form* (8.1), *with metrically transitive coefficients satisfying* (8.2), *we have*

(8.4) $$N(A, \lambda) = N(A_0, \lambda)(1 + o(1))$$

as $\lambda \downarrow 0$, where

(8.5) $$N(A_0, \lambda) = c_d(\det \hat{a})^{-1/2}\lambda^{d/2}$$

is, by (3.12) *and* (5.35), *the integrated density of states of a differential operator A_0 of the form* (8.1) *with nonrandom matrix $\hat{a} = \{\hat{a}_{ij} : 1 \leqslant i, j \leqslant d\}$ given by*

(8.6) $$\hat{a}_{ij} = \mathsf{E}\{\psi_{li}(x)a_{lk}(x)\psi_{kj}(x)\} = \mathsf{E}\{a_{ik}(x)\psi_{kj}(x)\}$$

and $\psi_{kj}(x)$, for $x \in \mathbf{R}^d$, are metrically transitive fields representing the unique solution of the following variational problem:

(8.7*a*) $$\mathsf{E}\{\psi_{li}(x)a_{lk}(x)\psi_{ki}(x)\} \to \infty,$$

(8.7*b*) $$\operatorname{rot}\psi_i := \frac{\partial\psi_{ij}}{\psi_k} - \frac{\partial\psi_{ki}}{\psi_j} = 0,$$

(8.7*c*) $$\mathsf{E}\{\psi_{ji}(x)\} = \delta_{ji}, \qquad \mathsf{E}\{\psi_{ji}(x)\psi_{ji}(x)\} < \infty.$$

Proof. The same argument used in the proof of Theorems 5.18 and 5.23 shows that the Laplace transform $\tilde{N}(t)$ of $N(A,\lambda)$ has the following representation, analogous to (5.17):

$$\tilde{N}(t) = \int_0^\infty e^{-\lambda t} N(A, d\lambda) = \mathsf{E}\{K(t,0,0)\}, \tag{8.8}$$

where $K(t,x,y)$ is the kernel of the semigroup e^{-tA}, which is a fundamental solution of the parabolic equation

$$\frac{\partial K}{\partial t} = -AK, \qquad K(0,x,y) = \delta(x-y). \tag{8.9}$$

If we make the substitutions $x = \sqrt{\rho}\xi$, $y\sqrt{\rho}\eta$ and $t = \rho\tau$, the function $\kappa_\rho(\tau,\xi,\eta) = \rho^{d/2}K(\rho\tau,\sqrt{\rho}\xi,\sqrt{\rho}\eta)$ will again satisfy an equation of the form (8.9), but with coefficients $a_{ij}(\sqrt{\rho}\xi)$. Since the asymptotic behavior of $N(A,\lambda)$ near $\lambda = 0$ is determined, according to (8.8), by the behavior of $K(t,x,x)$ for $x = 0$ and $t \to \infty$, it is sufficient to know the behavior of $\kappa_\rho(t,x,x)$ as $\rho \to \infty$. It turns out to be as follows (see Proposition 8.2):

$$\kappa_\rho(1,0,0) = (4\pi\rho)^{-d/2}(\det \hat{a})^{-1/2}(1+o(1)) \tag{8.10}$$

as $\rho \to \infty$. It follows then from (8.8) that

$$\tilde{N}(t) = (4\pi t)^{-d/2}(\det \hat{a})^{-1/2}(1+o(1))$$

as $t \to \infty$. We now apply the classical Tauberian theorem [Feller 1966], which says that if the Laplace transform has the asymptotic behavior

$$\tilde{N}(t) = ct^{-a}(1+o(1)) \tag{8.11}$$

as $t \to \infty$, the inverse Laplace transform has the behavior

$$N(\lambda) = \frac{c}{\Gamma(a+1)}\lambda^a(1+o(1)) \tag{8.12}$$

as $\lambda \downarrow 0$. □

We now discuss formula (8.10), the basis of the proof of Theorem 8.1.

(8.2) Proposition [Kozlov 1979a; Yurinskii 1982; Papanicolaou and Varadhan 1982]. *Let A_ρ, for $\rho > 0$, be self-adjoint operators on $L^2(\mathbf{R}^d)$ of the form*

$$A_\rho = -\frac{\partial}{\partial x_i} a_{ij}(\sqrt{\rho}x)\frac{\partial}{\partial x_j},$$

where $a_{ij}(x)$ is a metrically transitive field satisfying (8.2), and let $\kappa_\rho(\tau,\xi,\eta)$ be the kernel of the semigroup $e^{-\tau A_\rho}$. Then, as $\rho \to \infty$ uniformly with respect *to ξ and η on a compact set in $\mathbf{R}^d$, the following asymptotic formula is valid with probability* 1:

$$\kappa_\rho(\tau,\xi,\eta) = \rho^{-d/2}(4\pi\tau)^{-d/2}(\det \hat{a})^{-1/2}e^{-(\hat{a}(\xi-\eta),\xi-\eta)/4\tau} + o(\rho^{-d/2}) \tag{8.13}$$

as $\rho \to \infty$.

Proof. We outline the main steps in the proof, following mainly [Kozlov 1979a]. Based on the Trotter–Kato theorem [Kato 1966], we can consider the elliptic equation

$$A_\rho u_\rho = f, \tag{8.14}$$

instead of the parabolic equation

$$\frac{\partial \kappa_\rho}{\partial \tau} = -A_\rho \kappa_\rho.$$

Assume we have constructed metrically transitive fields satisfying conditions (8.7a,b,c), and in particular the equations

$$\frac{\partial}{\partial x_i}\big(a_{ij}(x)\psi_{jk}(x)\big) = 0, \tag{8.15}$$

which are the Euler equations of the variational problem (8.7a). Using these equations, integration by parts and the ergodic theorem, we can obtain another expression for the coefficients $\hat{a}_{ij}$:

$$\hat{a}_{ij} = \mathsf{E}\{a_{ik}\psi_{kj}\}. \tag{8.16}$$

Now the functions

$$z_j(x) = \int_0^1 \psi_{ij}(tx)x_i\,dt \tag{8.17}$$

satisfy the equation

$$A_1 z_j = 0; \tag{8.18}$$

furthermore, by (8.7c) and the ergodic theorem, and because the fields $\psi_{ij}(x)$ are metrically transitive, we also have

$$\lim_{\rho\to\infty} \frac{1}{\sqrt{\rho}} z_j(\sqrt{\rho}x) = x_j. \tag{8.19}$$

If we use the functions $z_j^\rho(x) \equiv \rho^{-1/2}z_j(\sqrt{\rho}x)$ as a new set of coordinates—called harmonic, in view of (8.18)—the entries of the change-of-variables matrix are $\psi_{jk}(\sqrt{\rho}x)$, so A_ρ takes the form

$$A_\rho = \psi_{ik}(\sqrt{\rho}x)a_{ij}(\sqrt{\rho}x)\psi_{jl}(\sqrt{\rho}x)\frac{\partial^2}{\partial z_k \partial z_l}.$$

Thus, according to (8.6), the coefficients of the operator obtained by making ρ approach ∞ can be calculated by merely averaging the coefficients of the initial operator A_ρ written in the new coordinates.

To prove the existence and uniqueness of the fields ψ_{ij} and of the harmonic coordinates z_k, one systematically uses techniques from the modern theory of functional spaces and differential equations. Then the operator A_ρ must be applied to the difference $u_\rho(x) - u_0(z^\rho(x))$, where $u_0(x)$ is the solution of the limit equation

$$\hat{a}_{ij}\frac{\partial^2 u_0}{\partial x_i \partial x_j} = f, \tag{8.20}$$

which leads to the following equality, to within terms that converge weakly to zero (that is, when integrated with suitable functions) as $\rho \to \infty$:

$$\begin{aligned} &A_\rho(u_\rho - u_0) \\ &\quad = A_\rho\big(u_\rho - u_0 - (z^\rho - x)\nabla u_0\big) \\ &\quad = \big(\hat{a}_{ij} - (a\psi)_{ij}(\sqrt{\rho}x)\big)\frac{\partial^2 u_0(x)}{\partial x_i \partial x_j} + \frac{\partial}{\partial x_i}(z_k^\rho - x_k)a_{ij}(\sqrt{\rho}x)\frac{\partial^2 u_0}{\partial x_i \partial x_k}. \end{aligned}$$

By (8.16), (8.17) and the ergodic theorem, both terms in the right-hand side converge weakly to zero as $\rho \to \infty$. The proposition follows readily. □

The variables in the proof above admit a natural physical interpretation, if we look at (8.14) as an equation of stationary diffusion or heat conduction. In a medium with diffusion coefficient a_{ij}, under the action of sources with density f, diffusion is governed by the equations

$$\frac{\partial c}{\partial t} + \frac{\partial q_i}{\partial x_i} = f, \qquad q_i = -a_{ij}\frac{\partial c}{\partial x_j}, \tag{8.21}$$

where $c(t,x)$ is the density of diffusing particles and q is their flux. Therefore, each quantity $w_k = z_k - x_k$, which, in view of (8.18), satisfies

$$\frac{\partial}{\partial x_i}\left(a_{ij}\frac{\partial w_k}{\partial x_j}\right) - \frac{\partial a_{ik}}{\partial x_i} = 0,$$

can be considered as the stationary concentration of particles under the action of a homogeneous gradient parallel to the k-axis, and having unit magnitude. Each $\psi_{jk}(x)$ is the concentration gradient in the direction of the j-axis induced by a unit gradient directed along the k-axis; and (8.16) means that $\hat{a}_{ij}$ is the mean flux caused by this gradient, that is, by definition, the mean diffusion coefficient of the medium.

If the solution of the initial equation (8.1) becomes smooth at "microscopic" distances of the order of $1/\sqrt{\rho}$, the true concentration gradient in the vicinity of a point x will be $\partial u_\rho/\partial x_j$, and therefore the vector $\hat{a}_{ij}\partial u_\rho/\partial x_j$ should be expected to be the effective flux of particles, and equation (8.20) the effective "macroscopic" equation of stationary diffusion. For this reason, the derivation of asymptotic formulas like (8.7) is one of the central problems of the so-called theory of averaging or homogeneization, whose mathematical aspects are treated in a number of papers: see the surveys [Zhikov et al. 1979, 1982] and the books [Bakhvalov and Panasenko 1984; Sanchez-Palencia 1980]. In physical terms, the goal of this theory is to give an effective, macroscopic description of a medium whose microstructure produces large variations of the characteristics of the medium within microscopic distances.

"Automodel" asymptotic formulas like (8.13) also occur when we consider problems connected with the central limit theorem, or, more generally, with the asymptotic behavior of random walks in random media, in the limit $t \to \infty$. In this context it is natural to consider the metrically transitive matrix operator A on $\ell_2(\mathbf{Z}^d)$ corresponding to an infinitesimal operator of random walk in the lattice $\mathbf{Z}^d$:

$$(A\psi)(x) = \sum_{y\in\mathbf{Z}^d} a(x,y)\big(\psi(x) - \psi(y)\big), \tag{8.22}$$

where the random coefficients $a(x,y)$ are nonnegative, equal to zero when $|x-y| > R$ for some $R < \infty$, and satisfy relation (1.45). As was shown by Anshelevich et al. [1982], Figari et al. [1982], Kozlov and Molchanov [1984] (see also the survey [Kozlov 1985]), in the discrete case an asymptotic formula similar to (8.10) is valid, so the integrated density of states of operators of the form (8.22) satisfy asymptotic formulas (8.4) and (8.5).

This, too, has a physical interpretation: we can consider A in (8.22) as an operator describing the harmonic oscillation of atoms at sites of the lattice $\mathbf{Z}^d$ and having equal masses M and random elastic constants $a(x,y)$ (see [Lifshitz et al. 1982], for example). Then $N(A,\lambda)$ is the number of oscillatory modes with frequency less than or equal to $\sqrt{\lambda/M}$ per unit volume. Small values of λ correspond to low frequencies and thus long wavelengths, for which the difference between a lattice and a continuous medium is negligible.

In one dimension the effective coefficient $\hat{a}$ is easy to find from (8.6) and (8.7): $\hat{a} = \mathsf{E}^{-1}\{a^{-1}(x)\}$. Combined with (8.4), this gives the same result as (6.23), as should be expected. But in higher dimension, $\hat{a}_{ij}$ is rather difficult to find, since one has to solve the differential equations

$$\frac{\partial}{\partial x_i} a_{ij}\psi_{jk} = 0, \qquad \frac{\partial \psi_{ji}}{\partial x_k} - \frac{\partial \psi_{ki}}{\partial x_j} = 0,$$

which are equivalent to the variational problem (8.7), but are no less complicated, in general, than the initial equation, since they include the same random coefficients.

The effective coefficients can be found approximately under certain assumptions: for example, if each $a_{ij}(x)$ fluctuates weakly, that is, $a_{ij} = \bar{a}_{ij} + \varepsilon b_{ij}(x)$, for $\bar{a}_{ij} \gg |\varepsilon b_{ij}|$, one can take expansions in powers of ε [Anshelevich et al. 1982].

For $d = 2$ the coefficients can sometimes be found explicitly. This was discovered by Dykhne [1970] for the case of a matrix operator $a_{ij}(x) = \alpha(x)\delta_{ij}$, where $\alpha(x)$ takes two equiprobable values (two-component structure). More generally, Kozlov [1979b] proved (see Problems IV.4 and IV.5) that, if the metrically transitive matrix fields $a(x)$ and $a_1(x) = a(x)/\det a$ are distribution-equivalent,

(8.23) $$N(A,\lambda) = N(-\Delta,\lambda)\big(1+o(1)\big)$$

as $\lambda \downarrow 0$, and if the matrix of coefficients of A in equation (8.1) is $a_{ij}(x) = \alpha(x)\delta_{ij}$, for $i = 1, 2$, and the fields $\pm\big(\ln\alpha(x) - \mathsf{E}\{\ln\alpha(x)\}\big)$ are distribution-equivalent and isotropic (that is, their distribution is invariant under rotations in $\mathbf{R}^2$),

(8.24) $$N(A,\lambda) = \frac{1}{q} N(-\Delta,\lambda)\big(1+o(1)\big)$$

as $\lambda \downarrow 0$, where $q = e^{\mathsf{E}\{\ln\alpha(x)\}}$.

In particular, if $\alpha(x)$ takes on two equiprobable values α_1, α_2 and is isotropic, we have $q = \sqrt{\alpha_1\alpha_2}$. The same result is valid for the chessboard structure, where $\alpha(x)$ takes on two equiprobable values, independently in different squares. In this case the field is not homogeneous with respect to shifts in $\mathbf{R}^2$, so Proposition 4.1 and Theorem 4.2 are not, rigorously speaking, applicable. But homogeneity under $\mathbf{Z}^2$ can easily be reduced to homogeneity under $\mathbf{R}^2$, as mentioned in Section 1: see (1.31).

In conclusion, notice that the asymptotic behavior of integrated densities of states of the type considered in this section occurs for operators other than those of the form (8.1). Kozlov [1982, 1983] shows it occurs for divergence operators of arbitrary order with general metrically transitive coefficients, and for nondivergence operators with almost-periodic coefficients as long as these coefficients are smooth enough, their frequencies satisfy a sufficiently strong incommensurability condition like (18.21) below, and the nonleading terms are sufficiently small. In particular, a multidimensional

Schrödinger operator with a sufficiently small quasiperiodic potential satisfies these conditions, and therefore (8.4).

9 Fluctuation Boundaries: General Discussion and Classical Asymptotics

9.A Introduction and Heuristic Discussion

In Subsection 6.B we defined a *fluctuation boundary* as one near which the spectrum is entirely due to sufficiently long or large fluctuations of the random potential. A typical example is the point $\lambda = 0$ for the Schrödinger operator with Poisson potential (1.29), for a nonnegative function $u(x)$. Near this point the spectrum is due to regions in the realizations of the potential that contain no points x_j, and the closer λ is to 0, the bigger such regions are.

If the points x_j in the potential (1.29) are arranged periodically, rather than randomly, the lower spectral boundary ν_0 is strictly positive for any fixed nonnegative $u(x)$ and lattice period. The asymptotic behavior of the integrated density of states is then of the from

$$N(\lambda) = c_d(\lambda - \nu_0)^{d/2}(1 + o(1)) \tag{9.1}$$

for $\lambda \downarrow \nu_0$, where the constant c_d depends on the periodic potential. This is not a fluctuation boundary, for in the vicinity of a fluctuation boundary $N(\lambda)$ decays exponentially, due to the (usually) exponentially small probability of large deviations of the potential: equations (6.27), (6.29), (7.4), (7.8), etc. Thus fluctuation boundaries are a particular feature of random metrically transitive operators.

Fluctuation boundaries were first identified by I. Lifshitz, who discovered the asymptotic behavior of the integrated density of states in their vicinity [Lifshitz 1964], gave arguments for the point structure of the spectrum, and developed an efficient method to compute various physical quantities [Lifshitz 1967]. See also [Lifshitz et al. 1982b] and [Efros and Shklovskii 1984].

Lifshitz's heuristic arguments for a Schrödinger operator with Poisson potential (1.29), where $u(x)$ is a nonnegative function with compact support, are given below. Essentially the same arguments were used in the proof of Theorem 6.7, in one-dimensional cases. We will use the symbol $\sim$ to denote asymptotic equivalence, without writing the order of magnitude of the error.

According to Theorem 5.20, $\lim_{\Lambda\to\infty} \mathsf{E}\{N_\Lambda(\lambda)\} = N(\lambda)$, with

(9.2) $$N_\Lambda(\lambda) = \frac{1}{V}\sum_i \chi_{(0,\infty)}(\lambda - \lambda_i),$$

where λ_i, for $i = 0, 1, \ldots$ are the eigenvalues, in nondecreasing order, of the approximating operator H_Λ. Therefore

(9.3) $$N(\lambda) \sim P_\Lambda(\lambda).$$

Here $P_\Lambda(\lambda)$ is the probability that an eigenvalue of H_Λ, taken "at random," does not exceed λ. For the random potential (1.29), as we discussed above (see also Theorem 5.33), the left edge of the spectrum is $\lambda = 0$, and in its vicinity the spectrum exists only because each realization of the potential has arbitrarily large regions where the potential is zero. Therefore it is these realizations that contribute most to $N(\lambda)$. The properties of the Poisson potential, (1.28) and (1.29), imply that the probability $P(V)$ of a realization vanishing over a region of volume V is

(9.4) $$P(V) \sim e^{-\operatorname{const} cV}.$$

In quantum-mechanical terms, a low-energy particle, for which λ is small, sees a potential well of any nonzero depth as infinitely deep. By the uncertainty principle, the lowest eigenvalue λ_0 in such a well is of order

(9.5) $$\lambda_0 \sim L^{-2},$$

where L is the linear dimension of the well, so $V \sim L^d$. Since larger wells, corresponding to smaller values of λ, become exponentially less probable, we can combine (9.3) and (9.4) to get

(9.6) $$N(\lambda) \sim \exp(-\operatorname{const} c\lambda^{-d/2}).$$

The heuristic arguments leading to (9.6) are sufficiently universal that we can expect the asymptotic formula

(9.7) $$N(\lambda) \sim P(V(\lambda))$$

to hold as $\lambda \downarrow 0$, where $V(\lambda)$ is the volume of the smallest region Λ in which a given λ is the lowest eigenvalue for the Dirichlet problem with uniform potential $\sim \inf_{y\in\mathbf{R}^d} q(y)$, and $P(V)$ is the probability that the random potential takes on its minimum value (or close enough values) in a region of this volume.

In particular, (9.6) should be valid not only for the Poisson potential (1.29) with nonnegative and sharply decreasing $u(x)$, as proved in Theorem 10.2, but also in the more general case of a nonnegative random potential whose correlation at far away points drops quickly enough, so long as the probability of small values of the potential is not too small (so that (9.4) holds).

Both of these conditions are essential, as can be seen from formula (6.29a) and Theorem 10.1. If the correlation decays slowly, say polynomially, (6.29a) shows that the asymptotics of $N(\lambda)$ as $\lambda \downarrow 0$ are also polynomial, though still described by (9.7). This is so even if the one-dimensional distribution of the potential has an atom at zero, as in the case, discussed above, of a Poisson potential with function $u(x)$ having compact support.

If, on the other hand, the potential is statistically independent for points far away, as in equation (10.6), but the one-dimensional distribution of the potential $F(q)$ behaves as $\exp(-\operatorname{const} q^{-a})$, for some $a > 0$, as $q \downarrow 0$, we have $N(\lambda) \sim \exp(-\operatorname{const} \lambda^{-(d/2+a)})$. See also (10.93) and (10.99).

Now consider the Poisson potential (1.29) with a smooth, nonpositive function $u(x)$ that decays rapidly enough as $|x| \to \infty$ and has its minimum at zero. Then, according to Theorem 5.33, the spectrum is unbounded from below. Choose $\varepsilon > 0$ such that $u(x) < u(0) + \delta$ for $|x| < 2\varepsilon$ and some $\delta > 0$. If ν is the number of Poisson points x_j within the sphere of radius ε around the origin, we have $u(x - x_j) < u(0) + \delta$ for $|x| < \varepsilon$, and the potential $q(x)$ is at most of order

(9.8) $$\nu\big(u(0) + \delta\big)$$

in the ball $B_\varepsilon = \{|x| \leqslant \varepsilon\}$. Using the variational principle, for example, we easily see that the lowest eigenvalue λ_0 in a smooth and very deep potential well $\nu u(x)$, as $\nu \to \infty$, is asymptotically

(9.9) $$\lambda_0 \sim \nu u(0).$$

Here, therefore, what is essential is not the width of the well, as in formula (9.5), but its depth $\nu u(0)$. Since the probability $P(\nu)$ of having $q(x) < \nu\big(u(0) + \delta\big)$ in B_ε is

(9.10) $$P(\nu) \sim \frac{\big(cV(B_\varepsilon)\big)^\nu}{\nu!} e^{-cV(B_\varepsilon)},$$

we conclude from (9.3) and (9.9) that

(9.11) $$N(\lambda) \sim \exp\left(\frac{-\lambda}{u(0)} \ln \frac{\lambda}{u(0)}\right),$$

thereby arriving at a version of formula (9.7):

(9.12) $$N(\lambda) \sim P\big(\nu(\lambda)\big).$$

Formulas (9.6) and (9.11) were obtained by similar probabilistic arguments, but the relevant expressions, (9.5) and (9.9), for the lowest eigenvalue of the potential well were quite different, connecting as they do the spectral parameter with one or another of the dimensions of the potential well: the width in the first case, the depth in the second. Equation (9.5) is essentially quantum-mechanical, while (9.9) corresponds to the quasi-classical approach; therefore we refer to the corresponding asymptotics (9.6) and (9.11) as quantum fluctuation and classical fluctuation asymptotics, respectively. Quantum asymptotic formulas, the subject of Section 10, are characteristic of potentials bounded from below, while the classical formulas, which we will prove rigorously in 9.B and 9.C, arise when the potential is not bounded from below, and has certain smoothness properties.

For example, consider the one-dimensional Poisson potential (1.29) with $u(x) = -k_0\delta(x)$, for $k_0 > 0$. As discussed in Section 6, we have $\lambda_0 = -k_0^2/4$, and therefore for a fluctuation of the potential having ν centers x_j concentrated in a small vicinity of the origin, the eigenvalue must be of order $-(\nu k_0)^2/4$. Using (9.12), we have the following counterpart for (9.11):

$$N(\lambda) \sim \exp(-\sqrt{|\lambda|}/k_0 \ln|\lambda|) \tag{9.13}$$

as $\lambda \to -\infty$. The proof of this formula can be found in [Lifshitz et al. 1982b]; see also formula (7.14).

Like formula (7.5) for the Gaussian white noise potential, formula (9.13) shows that quantum asymptotic formulas for potentials unbounded from below arise when realizations of these potentials are fairly irregular (here, in fact, generalized functions). By contrast, for random potentials bounded from below classical formulas apply if their realizations are functions that vary sufficiently smoothly, for example, if $u(x)$ in (1.29) decays more slowly than $|x|^{-(d+2)}$: see Theorem 9.13.

As we mentioned, formula (9.5), establishing an important relation between the size of the potential well and the lowest eigenvalue of the Dirichlet problem in the well, is the result of a minimization procedure—either the variational principle or the uncertainty principle of quantum mechanics, which can also be seen as the result of an optimization. (Similar optimization arguments suggest that the constant in (9.6) should be an eigenvalue of the Dirichlet problem in a sphere of unit radius; this will be proved rigorously in Section 10.) Therefore the same heuristic arguments of Lifshitz's that lead to (9.6) show that for every $\lambda \downarrow 0$ there exists a potential well of optimum size, that is, the optimum fluctuation of the potential essentially determining the form of the integrated density of states for such λ. A refinement of these arguments has led to the development of an efficient approximate method in the theoretical physics of disordered systems [Lifshitz et al. 1982b; Efros and Shklovskii 1984].

However, attempts to derive formulas (9.6), (9.11) and similar ones rigorously by following Lifshitz's arguments as far as possible, have so far resulted only in the relation

$$\frac{\ln(-\ln N(\lambda))}{-\ln\lambda} = \frac{d}{2}(1+o(1)) \tag{9.14}$$

for $\lambda \downarrow 0$ [Kirsch and Martinelli 1983a; Simon 1985a], which is the iterated logarithmic version of (9.6). We will discuss the derivation of (9.14) in Section 10.

In Sections 6 and 7, we applied the characteristically one-dimensional phase formalism to a number of random potentials, to derive formula (9.6) in its sharpest form, including the constant in the exponent and the preexponential factor. In this chapter we will describe a method to obtain, for a variety of random potentials in the multidimensional case, the leading term in the asymptotics of $\ln N(\lambda)$, that is, the log version of formulas (9.6) and (9.11).

We conclude this subsection with some remarks. As we have emphasized, the important information that we obtain from the above heuristic derivation of the fluctuation asymptotics (9.6) and (9.11) is the structure of the regions of realizations of the random potential that are essentially responsible for the spectrum of the Schrödinger operator in the vicinity of the fluctuation boundary. These regions represent potential wells whose dimensions (width or depth, in the quantum and classic cases, respectively) increase without bounds as λ approaches the fluctuation boundary. Such large deviations from the typical realizations are extremely unlikely, as attested by (9.4) and (9.10), are separated from one another by very large distances, of the order of $1/P(V)$. For example, in the quantum case (9.4)–(9.5), the dimension of the relevant potential wells, by (9.5), is of the order of $1/\sqrt{\lambda}$, while the distance between wells has order at least $e^{\text{const}\, c/\sqrt{\lambda}}$, and therefore is much greater than the dimension of the wells. In quantum mechanical terms, this says that there is practically no tunneling between such wells; therefore the spectrum of the Schrödinger operator in this fluctuation region should be very similar to the union of spectra of the Dirichlet problem in each of the fluctuation wells. In other words, the movement of a particle in each well is quantized practically independently, as in the simple example considered in Problem III.31. This means that the fluctuation spectrum of the Schrödinger operator should be pure point, and its eigenfunctions should be essentially concentrated within a single fluctuation well, decaying exponentially outside it. This will be proved rigorously in Subsection 15.C.

9.B Simplest Bounds. Gaussian and Negative Poisson Potentials

We have mentioned that all the asymptotic formulas for $N(\lambda)$ derived in this chapter result from lower and upper bounds of $N(\lambda)$ and related functions. One such upper bound for $N(\lambda)$, very useful in obtaining classical asymptotics, is as follows:

(9.1) Theorem [Pastur 1972b]. *Consider a Schrödinger operator with metrically transitive potential satisfying*

$$\mathsf{E}\{e^{-tq(0)}\} < \infty \tag{9.15}$$

for all $t > 0$. *The integrated density of states* $N(\lambda)$ *and its Laplace transform* $\tilde{N}(t) = \int_{-\infty}^{\infty} e^{-\lambda t} N(d\lambda)$ *satisfy*

$$\tilde{N}(t) \leqslant (4\pi t)^{-d/2} \mathsf{E}\{e^{-tq(0)}\}, \tag{9.16}$$

$$N(\lambda) \leqslant (4\pi t)^{-d/2} e^{\lambda t} \mathsf{E}\{e^{-tq(0)}\} \tag{9.17}$$

for all $t > 0$.

Proof. We will use the representation (5.17) of $\tilde{N}(t)$ by means of the Wiener integral; it is easy to check that we are allowed do that under condition (9.15). Since e^x is convex we can write

$$e^{-\int_0^t q(x(s))\, ds} \leqslant \frac{1}{t} \int_0^t ds\, e^{-tq(x(s))}.$$

Plugging this into (5.17) and taking into account that $q(x)$ is homogeneous, we get the bound (9.16). We also have

$$\tilde{N}(t) \geqslant \int_{-\infty}^{\lambda} e^{-\mu t} N(d\mu) \geqslant e^{-\lambda t} N(\lambda),$$

that is,

$$N(\lambda) \leqslant e^{\lambda t} \tilde{N}(t). \tag{9.18}$$

Now (9.17) follows immediately from (9.16) and (9.18). □

Remarks. 1. Since the first multiplier in the right-hand side of (9.16) is the Laplace transform $\tilde{N}_0(t)$ of the integrated density of states $N_0(\lambda)$ of the operator $-\Delta$ of (5.35) and the second is the Laplace transform $\tilde{F}(t)$ of the distribution function $F(q)$ of the random variable $q(0)$, the bound (9.16) can be written as

$$\tilde{N}(t) \leqslant \tilde{N}_0(t) \tilde{F}(t). \tag{9.19}$$

But the product of the transforms $\tilde{\mu}_1(t)$ and $\tilde{\mu}_2(t)$ of two measures $\mu_1(\lambda)$ and $\mu_2(\lambda)$ is the transform of the convolution $\mu_1 * \mu_2$, so we can also write

$$\int_{-\infty}^{\infty} e^{-\lambda t} N(d\lambda) \leqslant \int_{-\infty}^{\infty} e^{-\lambda t}(N_0 * F)(d\lambda). \tag{9.20}$$

2. Inequality (9.19) can also be proved without the Wiener integral, and for a wider class of metrically transitive operators: see Problems IV.8 and IV.9.

In many cases, the following theorem yields sufficiently sharp lower bounds on $N(\lambda)$.

(9.2) Theorem [Pastur 1976]. *Let Λ be a bounded domain in $\mathbf{R}^d$ with piecewise smooth boundary and volume V, and $\psi(x)$ a smooth function with support in Λ such that $\|\psi\|_2 = 1$. If H is the Schrödinger operator with potential q and we set $Q(\psi) = \int_{\mathbf{R}^d} q(x)\,|\psi(x)|^2\,dx$, we have*

$$N(\lambda) \geqslant \frac{1}{V} F_\psi(\lambda - J(\psi)), \tag{9.21}$$

where

$$F_\psi(q) = \mathsf{P}\{Q(\psi) \leqslant q\}, \tag{9.22}$$

$$J(\psi) = \int_{\mathbf{R}^d} |(\Delta\psi)(x)|^2\,dx. \tag{9.23}$$

Proof. By Theorem 5.25,

$$N(\lambda) \geqslant \mathsf{E}\{N_\Lambda^{(0)}(\lambda)\}, \tag{9.24}$$

where

$$N_\Lambda^{(0)} = \frac{1}{V} \sum_{i=0}^{\infty} \chi_{(0,\infty)}(\lambda - \lambda_i) \tag{9.25}$$

and $\lambda_0 \leqslant \lambda_1 \leqslant \cdots$ are the eigenvalues of the Dirichlet problem for the Schrödinger equation on Λ. By the minimax principle [Reed and Simon IV], any smooth function ψ with compact support in Λ and such that $\|\psi\|_2 = 1$ satisfies $\lambda_0 \leqslant (H\psi, \psi)$. The theorem now follows from (9.24) and (9.25). □

Theorems 9.1 and 9.2 readily lead to the asymptotics of $N(\lambda)$ at the left edge of the spectrum for a Schrödinger operator with a metrically transitive Gaussian potential. Recall that in this case the left edge is $-\infty$, by Theorem 5.33.

(9.3) Theorem [Pastur 1972b, 1977]. *Consider a d-dimensional Schrödinger operator with a random potential $q(x)$, for $x \in \mathbf{R}^d$, equal to a Gaussian metrically transitive field with zero mathematical expectation and correlation function $b(x)$ continuous at the origin. For such an operator,*

$$\ln N(\lambda) = -\frac{\lambda^2}{2b(0)}(1 + o(1)) \tag{9.26}$$

as $\lambda \to -\infty$.

Proof. Upper bound: Applying Theorem 9.1 to a Gaussian potential, we get

$$N(\lambda) \leqslant (4\pi t)^{-d/2} \exp(\lambda t + \tfrac{1}{2} b(0) t^2).$$

Choosing t so as to minimize the exponential, we get

$$\limsup_{\lambda \to -\infty} \frac{\ln N(\lambda)}{\lambda^2} \leqslant -\frac{1}{2b(0)}. \tag{9.27}$$

Lower bound: we use Theorem 9.2. Since any linear combination of Gaussian random variables is one, the distribution function (9.22) for a Gaussian potential is

$$F_\psi(dq) = \frac{1}{\sqrt{2\pi\sigma_\psi}} \exp\left(-\frac{q^2}{2\sigma_\psi}\right) dq, \tag{9.28}$$

where

$$\sigma_\psi = \int b(x-y) |\psi(x)|^2 |\psi(y)|^2 \, dx \, dy. \tag{9.29}$$

We substitute (9.28) into (9.21), making Λ the ball of radius R around the origin. We also choose for $\psi(x)$ the function

$$\psi(x) = R^{-d/2} \phi(x/R), \tag{9.30}$$

where $\phi(x)$ is a smooth function whose support is the ball of unit radius around the origin and such that $\|\phi\|_2 = 1$, and R is given by

$$R = |\lambda|^{-1/2+\beta} \tag{9.31}$$

for $0 < \beta < \frac{1}{2}$, and $\lambda \to -\infty$. We obtain

$$J(\psi) = o(\lambda^2), \qquad \sigma_\psi = b(0)(1 + o(1)) \tag{9.32}$$

as $\lambda \to -\infty$. From (9.32), (9.28) and (9.21) we get

$$\liminf_{\lambda \to -\infty} \frac{\ln N(\lambda)}{\lambda^2} \geqslant -\frac{1}{2b(0)}. \tag{9.33}$$

From this and (9.27) we get (9.26), concluding the proof. □

Problem IV.7 presents an improvement on formula (9.26).

The asymptotic behavior of $N(\lambda)$ is slightly more complicated to determine for the Poisson potential (1.29) with a nonpositive function $u(x)$, for which, by Theorem 5.33, the spectrum is the whole axis.

(9.4) Theorem [Pastur 1974b, 1977]. *Consider a Schrödinger operator with a Poisson potential* (1.29) *with* $u(x)$ *summable, nonpositive, continuous at zero and having a global minimum there. If there is a sphere centered at the origin outside of which* $u(x) \geqslant (1-\delta)u(0)$, *for some* $\delta > 0$, *we have*

$$\ln N(\lambda) = -\frac{\lambda}{u(0)} \ln|\lambda|\big(1+o(1)\big) \tag{9.34}$$

as $\lambda \to -\infty$.

Proof. Upper bound: Using expression (1.40) for the Laplace characteristic functional of the Poisson random field for $\phi(x) = t\delta(x)$, where $\delta(x)$ is the delta function, we find that

$$\mathsf{E}\{e^{-tq(0)}\} = \exp\left(c \int_{\mathbf{R}^d} (e^{-tu(x)} - 1)\, dx\right).$$

Given the assumptions on u, this integral is bounded from above by

$$\frac{e^{-tu(0)} - 1}{|u(0)|} \|u\|_1,$$

so by Theorem 9.1 we find that, as $\lambda \to -\infty$,

$$\ln N(\lambda) \leqslant -\frac{\lambda}{u(0)} \ln|\lambda|\big(1+o(1)\big). \tag{9.35}$$

Lower bound: We again use Theorem 9.2, estimating the distribution function $F_\psi(q)$ of (9.22) by the arguments used in the heuristic derivation of formula (9.11). Namely, we fix $\delta > 0$ and find $\varepsilon > 0$ such that $u(x) \leqslant u(0)+\delta$ for $|x| \leqslant 2\varepsilon$. Then, if the support of the function $\psi(x)$ in (9.21) is contained in the ball B_ε of radius ε, we have $Q_\psi \leqslant N_\varepsilon\big(u(0)+\delta\big)$, where N_ε is the number of Poisson points within B_ε. If we write $\nu = \big[q/u(0)+\delta\big]$ and let V_ε be the volume of B_ε, we get

$$F_\psi(q) \geqslant \mathsf{P}\{N_\varepsilon \leqslant \nu\} \geqslant e^{-cV_\varepsilon} \frac{(cV_\varepsilon)^\nu}{\nu!}. \tag{9.36}$$

Now, if $\psi(x)$ has the form (9.30), it follows from (9.36) and (9.21) that

(9.37)
$$\ln N(\lambda) \geqslant -\frac{\lambda}{u(0)} \ln|\lambda|(1+o(1))$$

as $\lambda \to \infty$. This, combined with (9.35), is equivalent to (9.34), as we wished to prove. □

Estimates (9.17) and (9.21) can, in principle, be used to find classical asymptotic formulas for other random potentials. However, it is generally far from easy to compute a straightforward and sufficiently sharp lower bound for the distribution function $F_\psi(q)$. Therefore, in order to investigate the asymptotic behavior of $F_\psi(q)$ as q tends to the spectral boundary and ψ changes simultaneously, it is convenient to use the Laplace transform of $F_\psi(q)$, that is, the Laplace characteristic functional (1.39). As a result, we have from (9.21) the following important inequality, which gives sufficiently sharp logarithmic lower bounds in many interesting cases, both classical and quantum:

(9.38)
$$\tilde{N}(t) \geqslant \frac{1}{V_\psi} e^{-tJ(\psi)-\Phi(t|\psi|^2)},$$

where V_ψ is the volume of the set $\{x : |\psi(x)| > 0\}$ and

(9.39)
$$\Phi(\phi) = -\ln \mathsf{E}\{e^{-\int_{\mathbf{R}^d} q(x)\phi(x)\,dx}\}.$$

We won't justify the transition from (9.21) to (9.38); instead, we will derive (9.38) independently, as a particular case of the following fairly simple and universal bound, suitable for the derivation of both classical and quantum asymptotics.

(9.5) Theorem [Pastur 1974b]. *Let A be a self-adjoint metrically transitive operator on $L_2(\mathbf{R}^d)$ and $f(\lambda)$, for $\lambda \in \mathbf{R}$, a nonnegative, convex function such that $f(A)$ is the integral operator with kernel $f(x,y)$. If $\mathcal{F}$ is a set of real-valued functions ψ with compact support and unit $\|\cdot\|_2$-norm, we have*

(9.40)
$$\mathsf{E}\{f(x,x)\} \geqslant \frac{1}{V_\psi} \mathsf{E}\{f((A\psi,\psi))\}$$

for all $\psi \in \mathcal{F}$, where V_ψ is the volume of the set $\{x : |\psi(x)| > 0\}$.

Proof. If Λ is the support of $\psi \in \mathcal{F}$ and π_Λ is the orthogonal projection onto the space of functions that vanish outside Λ, we can write

$$\int_\Lambda f(x,x)\,dx = \operatorname{Tr} \pi_\Lambda f(A) \geqslant (\psi, f(A)\psi) \geqslant f((\psi, A\psi))$$

by the convexity of f. This immediately implies the theorem. □

Inequality (9.40) is one of the forms of the variational inequalities widely used in mathematical physics. It clearly implies (9.38), with $f(\lambda) = e^{-t\lambda}$.

It is now convenient to combine (9.16) and (9.38).

(9.6) Theorem [Pastur 1974b]. *If the conditions of Theorems* 9.1 *and* 9.2 *are satisfied, the following bounds hold for the Laplace transform* $\tilde{N}(t)$ *of* $N(\lambda)$:

$$\frac{1}{V_\psi} e^{-tJ(\psi)-\Phi(t|\psi|^2)} \leqslant \tilde{N}(t) \leqslant (4\pi t)^{-d/2} \mathsf{E}\{e^{-tq(0)}\}. \tag{9.41}$$ □

As we will see later, $\tilde{N}(t)$ can often be efficiently estimated, so asymptotic formulas for $\ln \tilde{N}(t)$ for $t \to \infty$ can be obtained. This done, the asymptotics (both classical and quantum) for $N(\lambda)$ at the low edge of the spectrum can be found, in the form of logarithmic analogues for (9.6) and (9.11), including explicit constants. This approach, first used by Pastur [1972b] and later developed by Pastur [1974b, 1977] and Figotin [1979, 1981], is based on lower and upper bounds for $\ln \tilde{N}(t)$ that can be shown to coincide asymptotically as $t \to \infty$.

Another important element of this approach consists in passing from the asymptotic formula for $\ln \tilde{N}(t)$, as $t \to \infty$, to an asymptotic formula for $\ln N(\lambda)$, as $\lambda \to 0$ or $\lambda \to -\infty$. This essentially reduces the problem to an application the Tauberian theorem of [Minlos and Povzner 1967]. To formulate this theorem, we need the notion of regularly varying functions, which we now introduce.

We say that a real-valued function $f(t)$ of a real variable t varies regularly at the point 0 (or ∞) if for any c the limit

$$\lim_{t\to 0 \ (\text{or } \infty)} \frac{f(ct)}{f(t)} = \beta(c)$$

exists and is finite. If, moreover, $f(t)$ is convex or monotone, it is easy to show (see (10.23)–(10.25) below, or [Feller 1966]) that $\beta(c)$ is of the from c^κ for all $c > 0$, where κ is a constant.

(9.7) Theorem. *Let* $N(\lambda)$ *be a nonnegative, nondecreasing function on* $\mathbf{R}$ *which vanishes at* $\lambda = 0$ *(or* $-\infty$*) and is strictly positive for* $\lambda > 0$ *(or* $-\infty$*). Assume that the Laplace transform*

$$\tilde{N}(t) = \int_{0 \ (\text{or } -\infty)}^{\infty} e^{-\lambda t} dN(\lambda)$$

exists for positive t, *and that* $\ln \tilde{N}(t)$ *is a regularly varying function at* $t = \infty$, *that is*,

$$\lim_{t\to\infty} \frac{\ln \tilde{N}(ct)}{\ln \tilde{N}(t)} = c^{\kappa}$$

for $c > 0$, where $\kappa > 0$ and $\kappa \neq 1$ (whether $\kappa < 1$ or $\kappa > 1$ depends on whether the lower edge of the spectrum is 0 or $-\infty$).

Then, setting $\tilde{n}(t) = \ln \tilde{N}(t)$, we have

$$\ln N(\lambda) = \inf\{\lambda t + \ln \tilde{N}(t)\}(1 + o(1)) \tag{9.42a}$$

as $\lambda \to 0$ (or $-\infty$), or

$$\ln N(\lambda) = (\kappa^{-1} - 1)\kappa^{1/(1-\kappa)}\tilde{n}(t)\big|_{\tilde{n}(t)/t=-\lambda}(1 + o(1)) \tag{9.42b}$$

as $\lambda \to 0$ (or $-\infty$). If the equation $\tilde{n}(t)/t = -\lambda$ has more than one solution, any of them can be used in (9.42b).

Proof. For definiteness, we take the case that 0 is the left edge of the spectrum; the case of $-\infty$ is treated analogously. We show (9.42b) first. Setting $n(\lambda) = \ln N(\lambda)$, we get from the definition of $\tilde{N}(t)$, after some simple transformations, that

$$e^{\tilde{n}(t)\Phi_t(\beta)} = \int_0^\infty e^{\tilde{n}(t)(\lambda\beta - \phi_t(\lambda))}\, d\lambda$$

for any positive t and β, where $\Phi_t(\beta) = \tilde{n}(\beta t)/\tilde{n}(t)$ and

$$\phi_t(\lambda) = -\frac{n(-\tilde{n}(t)\lambda/t) + \ln(-\tilde{n}(t))}{\tilde{n}(t)}.$$

Clearly, then, $\tilde{n}(t) < 0$ and $\lim_{t\to\infty}|\tilde{n}(t)| = \infty$. Since $\tilde{n}(t) = \ln \tilde{N}(t)$ is a regularly varying function with exponent $\kappa \in (0,1)$, the conditions of the Tauberian theorem of [Minlos and Povzner 1967] are satisfied. This theorem says that for positive λ there exists the limit

$$\lim_{t\to\infty} \frac{\phi_t(\lambda)}{-\tilde{n}(t)} = \inf_{\beta>0}\{\lambda\beta - \beta^{\kappa}\},$$

the convergence being uniform on any bounded interval in $(0,\infty)$. Equation (9.42b) follows easily from this and from the definition of ϕ_t.

We now show that the right-hand sides of (9.42a) and (9.42b) coincide. If t_0 is an arbitrary positive number and $t = t_0\beta$, we obtain, in the notation above,

$$\inf_{t>0}\{\lambda t + \ln \tilde{N}(t)\} = \inf_{\beta>0}\{\lambda\beta t_0 + \tilde{n}(t_0\beta)\}.$$

We now choose t_0 so that $\lambda t_0 = -\tilde{n}(t_0)$. The last equality then becomes

$$\inf_{t>0}\{\lambda t + \ln \tilde{N}(t)\} = -\tilde{n}(t_0) \inf_{\beta>0}\left\{\lambda\beta - \frac{\tilde{n}(t_0\beta)}{\tilde{n}(t_0)}\right\}.$$

Since $\lambda \to 0$ as $t_0 \to \infty$ and $\tilde{n}(t)$ is a regularly varying function with $\kappa \in (0,1)$, the right-hand side is asymptotically equal to the right-hand side of (9.42b), as we wished to prove. □

(9.8) Lemma. *Let* $g_1(t)$ *and* $g_2(t)$ *be continuous, real-valued functions on* $(0, \infty)$, *and set*

$$n_j(\lambda) = \inf_{t>0} \{\lambda t + g_j(t)\}, \tag{9.43}$$

for $j = 1, 2$ *and* $\lambda > 0$. *Assume that:*

(a) $\lim_{t\to\infty} g_1(t) = -\infty$ *and* $\lim_{t\to\infty} g_1(t)/t = 0$;
(b) $\lim_{t\to\infty} g_1(t)/g_2(t) = 1$;
(c) *there exists* $x \in \mathbf{R}$ *such that* $\lim_{\lambda\downarrow 0} n_1(\lambda c)/n_1(\lambda) = c^\kappa$ *for* $c > 0$.

Then $\lim_{\lambda\downarrow 0} n_1(\lambda)/n_2(\lambda) = 1$.

Proof. Conditions (a) and (b) imply that $n_1(\lambda)$ and $n_2(\lambda)$ take on finite negative values for sufficiently small $\lambda > 0$, and the value of $t_j(\lambda)$ that minimizes $\lambda t + g_j(t)$ in (9.43) is finite and tends to infinity as $\lambda \downarrow 0$. In view of (b) and of (9.43) we have, for any fixed $\varepsilon > 0$ and sufficiently small $\lambda > 0$:

$$\begin{aligned} n_1(\lambda) = \lambda t_1(\lambda) + g_1(t_1(\lambda)) &\geqslant \lambda t_1(\lambda) + (1-\varepsilon) g_2(t_1(\lambda)) \\ &\geqslant (1-\varepsilon) n_2(\lambda(1-\varepsilon)^{-1}). \end{aligned}$$

Applying this inequality as it is and in reverse, after interchanging the subscripts 1 and 2, and taking into account condition (c), the lemma follows. □

The following statement is proved in a similar way.

(9.9) Lemma. *Let* $g_j(t)$ *and* $n_j(\lambda)$, *for* $\lambda < 0$, *be as in Lemma* 9.8, *with conditions (a)–(c) replaced by:*

(a) $\lim_{t\to\infty} g_1(t)/t = \infty$ *and* $\lim_{t\to\infty} g_1(t)/g_2(t) = 1$;
(b) *there exists* $\kappa \in \mathbf{R}$ *such that* $\lim_{\lambda\to-\infty} n_1(\lambda c)/n_1(\lambda) = c^\kappa$ *for* $c > 0$.

Then $\lim_{\lambda\to-\infty} n_1(\lambda)/n_2(\lambda) = 1$. □

All the asymptotic formulas obtained in this chapter for $\ln \tilde{N}(t)$ for a metrically transitive Schrödinger operator $H = -\Delta + q$, and many other formulas obtained at the level of rigor of theoretical physics (see [Lifshitz et al. 1982b], for example), can be written in the following unified form:

$$\ln \tilde{N}(t) = \sup_{\psi\in\mathcal{F}} \ln \mathsf{E}\{e^{-t(H\psi,\psi)}\}(1 + o(1)) \tag{9.44a}$$

for $t \to \infty$, or

(9.44b) $$\ln \tilde{N}(t) = -\inf_{\psi\in\mathcal{F}} \{tJ(\psi) + \Phi(t|\psi|^2)\}(1+o(1))$$

as $t \to \infty$, where $\mathcal{F}$ is the set of real-valued functions on $\mathbf{R}^2$ with compact support and unit $\|\cdot\|_2$-norm, and the functionals J and Φ are given by (9.23) and (9.39).

These relations can be regarded as generalizations of the variational principle for the lowest eigenvalue of a self-adjoint operator governing the asymptotic behavior for large times of the kernel of the semigroup generated by the operator. We can also consider (9.44b) from the viewpoint of statistical mechanics: $J(\psi)$, being the logarithm of the probability measure, is analogous to the entropy, while $\Phi(\psi)$ is analogous to the mean energy. Therefore, (9.44b) is an analogue of the Gibbs variational principle relating entropy to the thermodynamic potential, here represented by $\ln \tilde{N}(t)$. This approach is discussed in [Donsker and Varadhan 1983] and [Freidlin and Wentzell 1984].

In fact, the right-hand side of (9.44b) has already appeared, as a lower bound for $\tilde{N}(t)$, in (9.38), and we will see that this simple estimate is universal—it gives an asymptotically exact expression for $\ln \tilde{N}(t)$ as $t \to \infty$ is both he classical and quantum cases. Thus, we are left with the problem of finding a similarly precise upper bound. In many cases, this can be obtained by the technique of Wiener integrals, which will occupy most of Section 10. Here we will use this technique to propose simple heuristic arguments to explain the important and universal variational formula (9.44).

We proceed from expression (5.17) for $\tilde{N}(t)$. First, since we're dealing with large t, we can neglect the condition $x(t) = 0$ in this representation and just write

(9.45) $$\tilde{N}(t) = (4\pi t)^{-d/2}\mathsf{E}\{\mathsf{W}\{e^{-\int_0^t q(x(s))\,ds}\}\}.$$

We now introduce, for the Wiener process, the density $e^{-J_t(x(\cdot))}$ with respect to a certain hypothetical measure $d(\,\cdot\,)$ by the equality

$$J_t(x(\,\cdot\,)) = \int_0^\infty \dot{x}^2(s)\,ds.$$

If we also set

(9.46) $$\Phi_t(x(\,\cdot\,)) = -\ln \mathsf{E}\{e^{-\int_0^t q(x(s))\,ds}\},$$

expression (9.45) becomes

(9.47) $$\tilde{N}(t) = (4\pi t)^{-d/2}\int e^{-J_t(x(\cdot))-\Phi_t(x(\cdot))}\,d(\,\cdot\,).$$

Let's change over from the classical trajectories $x(s)$ to wave functions $\psi(x)$ by the well-known relation of the quasiclassical approximation $|\psi(x)|^2\,dx \sim t^{-1}\,\mathrm{meas}\{0 \leqslant s \leqslant t : x(s) \in dx\}$. The functions $t^{-1}J_t(\,\cdot\,)$ and $\Phi_t(t^{-1}(\,\cdot\,))$ are replaced by the functionals $J(\psi)$ and $\Phi(t|\psi|^2)$, respectively, specified by (9.23) and (9.39). After this transformation, equation (9.45) becomes

$$\tilde{N}(t) = (4\pi t)^{-d/2} \int e^{-tJ(\psi)-\Phi(t|\psi|^2)}\, d(\,\cdot\,), \tag{9.47a}$$

from which, neglecting the multiplier $(4\pi t)^{-d/2}$, we get the desired formula (9.44). This formula shows that the problem of finding the asymptotic behavior of $\ln \tilde{N}(t)$ for $t \to \infty$ boils down to justifying the analogue of the Laplace method of asymptotic calculation of integrals in the infinite-dimensional case (that is, in a certain functional space).

Formula (9.44) serves as a convenient tool for finding whether the asymptotic behavior of $N(\lambda)$ in the neighborhood of the lower edge of the spectrum of the given metrically transitive Schrödinger operator is classical or quantum. Classical asymptotics are characterized by the fact that the term $tJ(\psi)$ can be neglected, that is, by the condition that, as $t \to \infty$,

$$\inf_{\|\psi\|_2=1} \{tJ(\psi) + \Phi(t|\psi|^2)\} \sim \inf_{\|\psi\|_2=1} \{\Phi(t|\psi|^2)\}.$$

Formulas (9.26) and (9.34) for the integrated density of states of a Schrödinger operator with a Gaussian or Poisson potential with $u(x) \leqslant 0$ have the following counterparts in the time domain:

$$\ln \tilde{N}(t) = \tfrac{1}{2} b(0) t^2 (1 + o(1)), \tag{9.48}$$

$$\ln \tilde{N}(t) = \mathrm{const}\, t^{-d/2} e^{-tu(0)} (1 + o(1)), \tag{9.49}$$

as $t \to \infty$; the constant in (9.49) is calculated in Problem IV.8. These formulas are readily obtained from (9.41) if we choose the trial function $\psi(x)$ to be of the form (9.30), with $R = t^{-1/2+\beta}$, for $0 < \beta < \frac{1}{2}$: compare (9.31).

The features of the classical asymptotics, and in particular their independence from the dimension, become obvious if we rewrite (9.48) and (9.49) as

$$\ln \tilde{N}(t) = \ln \mathsf{E}\{e^{-tq(0)}\}(1 + o(1)) \tag{9.50}$$

as $t \to \infty$. Thus, in view of Lemma 9.9, we get

$$\ln N(\lambda) = \ln F(\lambda)(1 + o(1)) \tag{9.51}$$

as $\lambda \to -\infty$, where $F(\lambda) = \mathsf{P}\{q(0) \leqslant \lambda\}$ is the one-dimensional distribution of the random potential. This form of classical asymptotics is useful also because is shows the similarity between Theorems 9.3 and 9.4 and the following simple theorem about the discrete analogue of the Schrödinger operator (1.8).

(9.10) Theorem. *Let A be a metrically transitive operator on $\ell_2(\mathbf{Z}^d)$ of a the form $A = A_0 + Q$, where A_0 is a nonrandom, bounded finite-difference (Toeplitz) operator on $\ell_2(\mathbf{Z}^d)$ and Q is the operator of multiplication by the metrically transitive field $q(x)$, for $x \in \mathbf{Z}^d$. Assume that $q(x)$ is not bounded from above (or from below) and that the tails of its one-dimensional distribution $F(\lambda) = \mathsf{P}\{q(x) > \lambda\}$ (or $F(\lambda) = \mathsf{P}\{q(x) < \lambda\}$) vary sufficiently regularly, that is, for any fixed a either*

(a) $$\lim_{\lambda \to +\infty \ (\text{or } \infty)} F(\lambda + a)/F(\lambda) = 1$$

or

(b) $$\lim_{\lambda \to +\infty \ (\text{or } \infty)} \ln F(\lambda + a)/\ln F(\lambda) = 1.$$

Then we have, in cases (a) and (b), respectively,

(9.52) $$N(\lambda) = F(\lambda)(1 + o(1))$$

(9.53) $$\ln N(\lambda) = \ln F(\lambda)(1 + o(1))$$

as $\lambda \to +\infty$ (or $-\infty$).

Proof. This is a simple corollary of Theorem 4.15. □

[Pastur 1982] contains similar formulas for the discrete analogue of a general metrically transitive differential operator in which not only the potential, but also the leading terms, are random.

9.C Generalized Poisson Potential

To conclude this section, we will find that asymptotic behavior of $N(\lambda)$ for the Schrödinger operator with Poisson potential (1.29), with $u(x)$ nonnegative but slowly decreasing. This behavior is also classic.

We will study a somewhat more general form of the generalized Poisson field

$$q(x) = \int_{\mathbf{R}^d} u(x - y) m(dy)$$

for $x \in \mathbf{R}^d$, considered in example 1.5(e). In the calculation of the asymptotics of $N(\lambda)$, both classical and quantum, for such a potential, an important role is played by the function

$$f(t) = -\ln \mathsf{E}\{e^{-tm(C)}\}, \tag{9.54}$$

where C is the unit cube and the functional $\Phi(\phi)$ from (9.39) is defined on the set of summable real functions $\phi \in L_1^+(\mathbf{R}^d)$. We now list the properties of f and give a convenient representation for Φ.

(9.11) Lemma. *The function $f(t)$ defined by (9.54) takes on finite values for $t \geqslant 0$ and has the following properties:*

(i) *$f(t) \geqslant 0$ for $t \geqslant 0$, and $f(0) = 0$;*
(ii) *f is concave and monotonically increasing, and $f(t)/t$ is monotonically decreasing, for $t \geqslant 0$;*
(iii) $\lim_{t\to 0} f(t)/t = \mathsf{E}\{m(C)\}$;
(iv) *$f(a+b) \leqslant f(a) + f(b)$ for $a, b \geqslant 0$.*

If, moreover, $f(t)$ is defined by (9.54) also for $t < 0$, property (ii) is also valid for $t < 0$.

Proof. This is easy to verify. □

(9.12) Lemma. *Let $L_{1,a}$, for $a \leqslant 0$, denote the set of integrable functions ϕ on $\mathbf{R}^d$ such that $\phi(x) \geqslant a$, and set $L_1^+ = L_{1,0}$. The functional Φ given by (9.39) assumes finite values on L_1^+ and has the form*

$$\Phi(\phi) = \int_{\mathbf{R}^d} f(\phi(x))\, dx. \tag{9.55}$$

Further, Φ is continuous on L_1^+ and subadditive, that is,

$$\Phi(\phi_1 + \phi_2) \leqslant \Phi(\phi_1) + \Phi(\phi_2). \tag{9.56}$$

If, in addition, $f(t)$ is defined for negative t, the functional Φ takes on finite values on $L_{1,a}$ for any $a < 0$, is continuous on $L_{1,a}$, and satisfies (9.55).

Proof. The continuity of Φ on L_1^+ follows immediately from its definition (9.39). Now consider the functional $\nu(\,\cdot\,) = -\ln \mathsf{E}\{e^{-tm(\cdot)}\}$ defined on Borel sets in $\mathbf{R}^d$. From the properties of the generalized Poisson measure, Example 1.15(e), it follows that $\nu(\,\cdot\,) = f(t)\operatorname{meas}(\,\cdot\,)$. From this and from (9.54), we readily get (9.55) for step functions $\phi \in L_1^+$. Since, by Lemma 9.11,

$$|f(t_1) - f(t_2)| \leqslant f'(0)|t_1 - t_2| \tag{9.57}$$

for $t_1, t_2 \geqslant 0$, the functional Φ defined by (9.55) is continuous on L_1^+, and therefore coincides with the one defined by (9.39) on all of L_1^+, because step functions span a dense set in L_1^+. Finally, (9.56) follows from (9.55) and Lemma 9.11(d).

Exactly the same arguments apply in the case of the sets $L_{1,a}$, with the exception that $f'(0)$ in inequality (9.57) should be replaced by $f'(a)$, and $t_1, t_2 \geqslant 0$ by $t_1, t_2 \geqslant a$. □

In the case of the Poisson random measure from (1.5), we have

$$f(t) = c(1 - e^{-t}), \tag{9.58}$$

and for the generalized Poisson measure (1.6),

$$f(t) = \ln(1 + tc^{-1}). \tag{9.59}$$

(9.13) Theorem [Figotin 1980b]. *Let $q(x)$ be a generalized Poisson field as in Example* 1.15(*e*), *in which $u(x)$ is nonnegative and bounded, and*

$$u(x) = c_0|x|^{-\alpha}(1 + o(1)) \tag{9.60}$$

as $|x| \to \infty$, for some $d < \alpha < d+2$. Assume that the random measure $m(dy)$ is such that $\lim_{t\to\infty} f(t)t^{-\gamma} = 0$ for a certain constant $0 < \gamma < 1$ satisfying

$$\gamma + \frac{d(1-\gamma)^2}{d(1-\gamma)+2} < \frac{d}{\alpha}. \tag{9.61}$$

Then, the spectrum of the Schrödinger operator $H = -\Delta + q$ with potential $q(x)$ is the semiaxis $[0, \infty)$, and

$$\ln N(\lambda) = -(ab)^{1/(1-b)}(b^{-1} - 1)\lambda^{-b/(1-b)}(1 + o(1)) \tag{9.62}$$

as $\lambda \downarrow 0$, where

$$b = \frac{d}{\alpha}, \qquad a = \frac{c_0^b}{\alpha} S_d \int_0^\infty f(\tau)\tau^{-b-1}\, d\tau, \tag{9.63}$$

and S_d is the area of the sphere of unit radius in $\mathbf{R}^d$.

Proof. The equality $\sigma(H) = [0, \infty)$ was already proved in Theorem 5.33. To obtain (9.62), we follow the technique described in the previous subsection: the upper estimate (9.16) for $\tilde{N}(t)$, together with (9.54) and (9.55), gives

$$\ln \tilde{N}(t) \leqslant -\frac{d}{2}\ln(4\pi t) - \int_{\mathbf{R}^d} f\big(tu(x)\big)\, dx. \tag{9.64}$$

In order to find the behavior of this integral for large t, we take an arbitrary $\varepsilon > 0$ and pick R_0 so that $|x| > R_0$ implies

$$c_0(1-\varepsilon)|x|^{-\alpha} \leqslant u(x) \leqslant c_0(1+\varepsilon)|x|^{-\alpha}.$$

By dividing the region of integration into the two subregions $|x| \leqslant R_0$ and $|x| > R_0$, we get

$$\text{(9.65)} \quad \int_{|x|\geqslant R_0} f\big(tc_0(1-\varepsilon)|x|^\alpha\big)\, dx \leqslant \int_{\mathbf{R}^d} f\big(tu(x)\big)\, dx$$
$$\leqslant R_0^d V_d f(tc) + \int_{|x|\geqslant R_0} f\big(tc_0(1+\varepsilon)|x|^\alpha\big)\, dx,$$

where V_d is the volume of the d-dimensional ball of unit radius and $C = \sup_{x\in\mathbf{R}^d} u(x)$. Now by the theorem's assumptions, $\lim_{t\to\infty} f(t)t^{-\gamma} = 0$, so making the obvious change of variables in the integrals over $|x| \geqslant R_0$ we obtain, after taking limits,

$$\lim_{t\to\infty} t^b \int_{\mathbf{R}^d} f\big(tu(x)\big)\, dx = \frac{c_0^b}{\alpha} S_d \int_0^\infty f(\tau)\tau^{-b-1} d\tau.$$

Recalling (9.64) and the definition of a (9.63), we can write

$$\text{(9.66)} \qquad \limsup_{t\to\infty} t^{-b} \ln \tilde{N}(t) \leqslant -a.$$

We now turn to the lower bound for $\tilde{N}(t)$. From (9.38), (9.54) and (9.55), we see that for any continuous, real-valued function ψ of unit $\|\cdot\|_2$-norm,

$$\text{(9.67)} \qquad \ln \tilde{N}(t) \geqslant -\ln V_\psi - J(\psi) - \int_{\mathbf{R}^d} f\big(t(u*\psi^2)(x)\big)\, dx.$$

Substituting ψ of the form (9.30) in (9.67), we obtain, after elementary transformations,

$$\text{(9.68)} \quad \ln \tilde{N}(t) \geqslant -d\ln R - \ln V_\phi - tR^{-2}J(\phi) - R^d \int f\big(t(u_R*\phi^2)(x)\big)\, dx$$

for all $R > 0$, where $u_R(x) = u(Rx)$. We estimate this last integral using the second inequality in (9.65), where u is replaced by $u_r * \phi^2$ and we use the fact that ϕ is bounded, has compact support and satisfies $\|\phi\|_2 = 1$. Then, given $\varepsilon > 0$, one can choose $R_0 = R_0(\varepsilon)$ so that

$$R^d \int_{\mathbf{R}^d} f\big(t(u_R*\phi^2)(x)\big)\, dx$$
$$\leqslant R^d R_0^d V_d f(tc_1 R^{-d}) + R^d \int_{|x|\geqslant R_0} f\big(tc_0(1+\varepsilon)R^{-\alpha}|x|^{-\alpha}\big)\, dx$$

for $R > 1$, where $c_1 = \|u\|_1 \sup \phi^2(x)$. Since $\lim_{t\to\infty} f(t)t^{-\gamma} = 0$, it follows that

(9.69)
$$R^d \int_{\mathbf{R}^d} f\big(t(u_R * \phi^2)(x)\big)\, dx \leqslant c_2 t^\gamma R^{d(1-\gamma)} + \frac{t^b\big(c_0(1+\varepsilon)\big)^b}{\alpha} S_d \int_0^{tc_0(1+\varepsilon)(RR_0)^{-\alpha}} f(\tau)\tau^{-b-1} d\tau,$$

where the constant c_2 does not depend on t or R.

Now we choose the dependence of R on t so that

(9.70)
$$t^\gamma R^{d(1-\gamma)} = tR^2.$$

Expressing R in terms of t we see that both sides of (9.70) equal t^β, where β is the left-hand side of (9.61), so $\beta < d/\alpha = b$. In view of this and of the condition $d < \alpha < d+2$, we get $\lim_{t\to\infty} tR^{-\alpha}(t) = \infty$. From this and from (9.68) and (9.69), we get

$$\liminf_{t\to\infty} t^{-b} \ln \tilde{N}(t) \geqslant -\frac{\big(c_0(1+\varepsilon)\big)^b}{\alpha} S_d \int_0^\infty f(\tau)\tau^{-b-1} d\tau.$$

Since ε is arbitrary, it follows from (9.66) that

(9.71)
$$\lim_{t\to\infty} t^{-b} \ln \tilde{N}(t) = -a.$$

Substituting this expression for $\ln \tilde{N}(t)$ into (9.42) and using Lemma 9.8 we readily obtain (9.62). □

(9.14) Corollary [Pastur 1977]. *If $q(x)$ is the Poisson random field* (1.29) *with a bounded nonnegative function $u(x)$ satisfying* (9.60), *we have*

(9.72)
$$\ln N(\lambda) = -(ab)^{1/(b-1)}(b^{-1}-1)\lambda^{d/(d-\alpha)}\big(1+o(1)\big)$$

as $\lambda \downarrow 0$, where a and b are as in Theorem 9.13. □

10 Fluctuation Boundaries: Quantum Asymptotics

In this section we consider the quantum asymptotic formulas for the integrated density of states of a multidimensional Schrödinger operator with a nonnegative random potential, whose spectrum is bounded from below. In 10.A, coarse results like (9.14) will be obtained for $\ln\big(-\ln N(\lambda)\big)$ as $\lambda \downarrow 0$. Their proof is conceptually and technically simple, essentially a rigorous version of Lifshitz's arguments presented in 9.A.

In the rest of the section we derive sharper formulas giving the asymptotics of $\ln N(\lambda)$ as $\lambda \to \infty$ for a variety of random potentials: a nonnegative generalized Poisson potential, the smoothed square of the Gaussian potential, and so on. To obtain sufficiently sharp lower bounds for $\ln \tilde{N}(t)$, where $\tilde{N}(t)$ is the Laplace transform of $N(\lambda)$, we use the variational inequality (9.38), as we did in the proof of the classical asymptotic formulas in Section 9. The upper bounds are obtained using the Wiener integral technique, applying results of Donsker and Varadhan [1975a,b,c, 1976, 1983] and their generalizations.

10.A The Lifshitz Exponent

Given a Schrödinger operator (or its discrete analogue) with fluctuation boundary λ_f, we define the correspondent *Lifshitz exponent* as

(10.1) $$L = -\lim_{\lambda \to \lambda_f} \frac{\ln(-\ln|N(\lambda) - N(\lambda_f)|)}{\ln|\lambda - \lambda_f|}.$$

In this subsection we will assume $\lambda_f = 0$, which is possible without loss of generality.

It follows from the results of the preceding section that $L = -2$ for a Gaussian potential (Theorem 9.3) and $L = -1$ for a Poisson potential with a nonpositive function $u(x)$ (Theorem 9.4). For a Poisson potential with a nonnegative $u(x)$, by Corollary 9.14 and the arguments leading to (9.6), we have

(10.2) $$L = \begin{cases} \frac{d}{2} & \text{if } u(x) = o(|x|^{-d-2}) \text{ as } |x| \to \infty, \\ \frac{d}{\alpha - d} & \text{if } \lim\limits_{|x| \to \infty} u(x)|x|^{\alpha} = \text{const, with } d < \alpha < d + 2. \end{cases}$$

(For this situation, of course, more exact asymptotic formulas are known: see (9.72) and (10.84), and [Pastur 1977; Figotin 1979]. These formulas immediately imply (10.2), so (10.2) should be regarded as an illustration of an approach to the study of the asymptotics of $N(\lambda)$ distinct from that of the authors.)

The first equality in (10.2) was rigorously obtained by Kirsch and Martinelli [1983a] as a corollary of general estimates for $N(\lambda)$, obtained under the assumption that the random potential satisfies a sufficiently strong mixing condition, namely,

(10.3) $$|\mathsf{P}(X \cap Y) - \mathsf{P}(X)\mathsf{P}(Y)| \leqslant \mathsf{P}(X)\phi(d(\Lambda_1, \Lambda_2))$$

for any $X \in \mathcal{F}_{\Lambda_1}$ and $Y \in \mathcal{F}_{\Lambda_2}$, where $\mathcal{F}_{\Lambda}$, for $\Lambda \subset \mathbf{R}^d$, is the σ-algebra generated by the values of $q(x)$ for $x \in \Lambda$; $d(\Lambda_1, \Lambda_2)$ is the distance between Λ_1 and Λ_2; and the nonnegative function $\phi(t)$ on $[0, \infty)$ satisfies

$$\phi(t) \leqslant c_1 \exp(-c_2 t^{d(1+\varepsilon)}) \tag{10.4}$$

as $t \to \infty$, for fixed $c_1, c_2, \varepsilon > 0$.

Simon [1985a] extended and improved the arguments of Kirsch and Martinelli, and proved that $L = d/2$ for the discrete analogue of the Schrödinger operator (Anderson model) under the condition that the potential $q(x) \geqslant 0$, for $x \in \mathbf{Z}^d$, forms a family of independent, identically distributed random variables whose distribution function does not go to 0 too fast as $q \downarrow 0$, that is, satisfies

$$F(q) \geqslant Cq^l \tag{10.5}$$

for some $C, l > 0$. All these authors proceed from the variational inequalities (5.38) and their discrete analogues (see [Simon 1985a] and Problem II.6), and the main novelty was that they obtained an upper bound on $N(\lambda)$ (which led to a relation similar to (9.5)) by using certain inequalities due to Thirring [1981] and Temple [1928] (see also [Reed and Simon IV]) for the minimum eigenvalue of the Neumann problem with a nonnegative potential in a domain Λ. In Theorem 10.1 below we will demonstrate the conceptual simplicity of this approach by presenting the proof of a result of this type that is, in a sense, the simplest possible. The proof is an extension of Simon's arguments for the Schrödinger operator with potential

$$q(x) = \sum_{y \in \mathbf{Z}^d} \xi_y \chi_C(x - y), \tag{10.6}$$

where ξ_y, for $y \in \mathbf{Z}^d$, are nonnegative, independent, identically distributed random variables, and C is the unit cube centered at the origin (cf. (1.31)) with edges parallel to the coordinate axes.

(10.1) Theorem. *Consider the Schrödinger operator with random potential (10.6). If the random variables ξ_y, for $y \in \mathbf{Z}^d$, have a distribution function $F(\xi)$ satisfying*

$$\lim_{\xi \downarrow 0} \frac{\ln(-\ln F(\xi))}{-\ln \xi} = a > 0, \tag{10.7}$$

the integrated density of states satisfies

$$\lim_{\lambda \downarrow 0} \frac{\ln(-\ln N(\lambda))}{-\ln \lambda} = \frac{d}{2} + a, \tag{10.8}$$

that is, the Lifshitz exponent is $L = d/2 + a$.

Proof. We start from (5.41), which follows from the variational principle. Let Λ be a cube centered at the origin, with edges of integer length R and parallel to the coordinate axes.

Lower bound: As in Section 9, we estimate the left-hand side $\mathsf{E}\{N_\Lambda^{(0)}(\lambda)\}$ of inequality (5.41) using Theorem 9.2. For the function $F_\psi(q)$ from (9.22), we obviously have

$$F_\psi(q) = \mathsf{P}\left\{\int_\Lambda q(x)\psi^2(x)\,dx \leqslant q\right\} \leqslant \mathsf{P}\{\xi_y \leqslant q, y \in \Lambda\} = \big(F(q)\big)^V,$$

where $V = R^d$ is the volume of Λ. Set $R = \sqrt{2J(\phi)/\lambda}$ (compare (9.5)). Then, as $\lambda \downarrow 0$, we have

$$J(\psi) = \frac{J(\phi)}{R^2} = \frac{\lambda}{2}, \qquad \lambda - J(\psi) = \frac{\lambda}{2}, \qquad V = O(\lambda^{-d/2}).$$

These relations, together with (9.21) and (10.7), give

$$\ln N(\lambda) \geqslant O(\ln \lambda) + \text{const}\, \lambda^{-d/2} \ln F(\lambda/2),$$

which certainly implies the $\geqslant$ part of (10.8).

Upper bound: By the second inequality in (5.41), we have

$$N(\lambda) \leqslant \mathsf{E}\{N_\Lambda^{(1)}(\lambda)\} \leqslant \mathsf{E}\{N_\Lambda^{(1)}(\lambda)\chi_{[0,\infty)}(\lambda - \lambda_0^{(1)})\}. \tag{10.9}$$

Since $q(x) \geqslant 0$, Theorem 5.24(i) implies that $N_\Lambda^{(1)}(-\Delta+q, \lambda) \leqslant N_\Lambda^{(1)}(-\Delta, \lambda)$, so that

$$N(\lambda) \leqslant N_\Lambda^{(1)}(-\Delta, \lambda)\mathsf{P}\{\lambda_0^{(1)} \leqslant \lambda\}, \tag{10.10}$$

where $\lambda_0^{(1)}$ is the lowest eigenvalue for the Neumann problem for $-\Delta + q$ on Λ.

We now use Temple's inequality [1928], following Simon [1985a]. (Kirsch and Martinelli use instead Thirring's inequality [1981]; see also Problem IV.13.) The inequality says that for a self-adjoint operator A with discrete spectrum $\lambda_0 < \lambda_1 < \cdots$ and for a vector $\phi \in \mathcal{D}(A)$ such that $\|\phi\| = 1$ and $(A\phi, \phi) \leqslant \lambda_1$, we have

$$\lambda_0 \geqslant (\phi, A\phi) - \frac{(A^2\phi, \phi) - (A\phi, \phi)^2}{\lambda_1 - (A\phi, \phi)}. \tag{10.11}$$

By our assumptions on $q(x)$,

$$0 < \mathsf{P}\{\xi_y \leqslant \delta\} = F(\delta) \equiv p < 1 \tag{10.12}$$

for $\delta > 0$ small enough. Thus, setting $\xi_y^* = \delta\chi_{(0,\infty)}(\xi_y - \delta)$, and denoting by $q^*(x)$ the potential obtained from (10.6) by replacing ξ_y by ξ_y^*, we have $q^* \leqslant q$.

Now set $A = -\Delta_\Lambda^{(1)} + q$ in (10.11) and take ϕ to be the constant function $\sqrt{V}$, for which $-\Delta_\Lambda^{(1)}\phi = 0$. If n_Λ is the number of cubes $C_y \subset \Lambda$ such that $\xi_y > \delta$, we have

$$(A\phi, \phi) = \frac{\delta n_\Lambda}{V}, \qquad (A^2\phi, \phi) = \frac{\delta^2 n_\Lambda}{V}. \tag{10.13}$$

Taking $\delta = \pi^2 R^{-2}\alpha$, for $\alpha < \frac{1}{2}$, and recalling that $\lambda_1^{(1)} \geqslant \pi^2 R^{-2}$ because $q(x)$ is nonnegative, we get

$$\lambda_1^{(1)} - (A\phi, \phi) \geqslant \pi^2 R^{-2}(1 - \alpha). \tag{10.14}$$

Substituting (10.13) and (10.14) in (10.11) gives

$$\lambda_0^{(1)} \geqslant \frac{n_\Lambda}{VR^2C},$$

where C is a positive constant depending only on α. Therefore,

$$\mathsf{P}\{\lambda_0^{(1)} \leqslant \lambda\} \leqslant \mathsf{P}\{n_\Lambda \leqslant cVR^2\lambda\}. \tag{10.15}$$

Since $n_\Lambda = \delta^{-1}\sum_{y\in\Lambda}\xi_y^*$, we conclude as in (9.18) that, for any positive ν and t,

$$\mathsf{P}\{n_\Lambda \leqslant \nu V\} \leqslant e^{\nu V t}\mathsf{E}\{e^{-tn_\Lambda}\} = e^{V(\nu t - \kappa(t))}, \tag{10.16}$$

where

$$\kappa(t) = -\ln \mathsf{E}\{e^{-(t/\delta)\xi_y^*}\} = -\ln\bigl(p + (1-p)e^{-t}\bigr). \tag{10.17}$$

Now choose t so that $e^{-t} = p^2$; our reasoning here will be an example of estimation of the probability of large deviations [Loeve 1960]. Combining (10.16) and (10.17) we get

$$\mathsf{P}\{n_\Lambda \leqslant \nu V\} \leqslant e^{V((1-2\nu)\ln p + \ln(1+p))}. \tag{10.18}$$

Setting $\nu = CR^2\lambda = \frac{1}{3}$ and combining (10.10), (10.15) and (10.18), we get

$$\ln N(\lambda) \leqslant c_1\lambda^{-d/2}\ln F(c_2\lambda)$$

for $c_1, c_2 > 0$. Finally, using (10.7) and (10.12), we get the $\leqslant$ part of (10.8). □

One can also consider the asymptotic behavior of $N(\lambda)$ for a more general potential given by the sum of a periodic function of a sufficiently general form and a random potential. In (10.6), for example, one can replace $\chi_C(x)$ by a nonnegative function $u(x)$ such that $u(x) = O(|x|^{-d-2})$ as $|x| \to \infty$. If the spectrum has gaps, one can consider the asymptotic behavior not only near the bottom of the spectrum, but also near the edges of the gaps. These questions were taken up by Kirsch and Simon [1986] and Mezincescu [1986]. In all cases, the asymptotic behavior is given by a formula similar to (10.8).

10.B Generalized Poisson Potential with a Nonnegative, Rapidly Decreasing Function

Consider, in Example 1.15(e), a potential of the form

$$q(x) = \int_{\mathbf{R}^d} u(x-y)\, m(dy) \tag{10.19}$$

for $x \in \mathbf{R}^d$, where $m(dy)$ is a positive, metrically transitive random measure, and $u(x)$ satisfies

(10.20)
$$u(x) \geqslant 0, \quad \int u(x)\, dx = \|u\|_1 = 1, \quad u(x) = O(|x|^{-d-2}) \text{as } |x| \to \infty.$$

The condition $\|u\|_1 = 1$ is for normalization only, and is postulated in order to simplify subsequent formulas.

By Theorem 5.34, the Schrödinger operator with such a potential has its lowest boundary at the point $\lambda = 0$.

(10.2) Theorem [Pastur 1977; Figotin 1979]. *Consider the Schrödinger operator $-\Delta + q$, where $q(x)$, for $x \in \mathbf{R}^d$, is a random field of the form* (10.19), *with $u(x)$ satisfying conditions* (10.20). *Let γ_d be the lowest eigenvalue of the operator $-\Delta$ in a ball of unit volume in $\mathbf{R}^d$. As $\lambda \downarrow 0$, the bottom of the spectrum, $N(\lambda)$ has the following asymptotic behavior:*

(i) *for the Poisson potential,*

$$\ln N(\lambda) = -c\gamma_d^{d/2}\lambda^{-d/2}(1 + o(1)); \tag{10.21}$$

(ii) *for a potential of the form* (1.6),

$$\ln N(\lambda) = -\gamma_d^{d/2}\lambda^{-d/2}\ln\lambda^{-1}(1 + o(1)). \tag{10.22}$$

Other examples of asymptotic behavior for $\ln N(\lambda)$ as $\lambda \downarrow 0$ are given in (10.93) and (10.95).

As we saw in Section 9, to derive (10.21) and (10.22) one must find the asymptotic behavior of $\ln \tilde{N}(t)$ as $t \to \infty$, where $\tilde{N}(t)$ is the Laplace transform of $N(\lambda)$, and then use the Tauberian theorem (see Theorem 9.7).

The corresponding asymptotic formulas for $\ln \tilde{N}(t)$ are given by Theorem 10.4; after its proof, as corollaries, we derive (10.21) and (10.22).

To prepare the ground for Theorem 10.4 we introduce some auxiliary functions of t and study their properties.

Consider the function $f(t)$ associated in Section 9 with the random field $q(x)$ of the form (10.19):

$$f(t) = -\ln \mathsf{E}\{e^{-tm(C)}\},$$

where C is the unit cube in $\mathbf{R}^d$. Assume that, for any $c > 0$, the limit

$$\lim_{t\to\infty} \frac{f(ct)}{f(t)} = \beta(c) \tag{10.23}$$

exists. This clearly implies

$$\beta(c_1 c_2) = \beta(c_1)\beta(c_2) \tag{10.24}$$

for $c_1, c_2 > 0$. Since, by Lemma 9.11, f is convex on the semiaxis $\mathbf{R}_+$ and $f(0) = 0$, the function β is also convex on $\mathbf{R}_+$, and therefore continuous on $(0, \infty)$, and $\beta(0) = 0$. Therefore there must exist $\kappa \in [0, 1]$ such that

$$\beta(c) = c^\kappa \tag{10.25}$$

for $c \geqslant 0$, where, by convention, $0^0 = 0$.

Now let $\eta(t)$, for $t > 0$, be the solution of

$$\eta^{d/2}(t) f\left(\frac{t}{\eta^{d/2}(t)}\right) = \frac{t}{\eta(t)}. \tag{10.26}$$

(10.3) Lemma. *There exists a positive function $\eta(t)$ such that* (10.26) *is satisfied identically. Moreover,*

(i) *the functions $\eta(t)$, $t/\eta(t)$ and $t/\eta^{d/2}(t)$ are monotonically increasing and tend to ∞ as $t \to \infty$;*

(ii) $$\lim_{t\to\infty} \frac{t}{\eta^{d/2+1}(t)} = \lim_{t\to\infty} f(t) > 0;$$

(iii) $$\lim_{t\to\infty} \frac{\eta(t-\varepsilon)}{\eta(t)} = 1 \qquad \textit{for all } \varepsilon > 0;$$

(iv) $$\lim_{t\to\infty} \frac{\eta(t)}{t} \ln \eta(t) = 0.$$

Proof. Left to the reader. □

We now investigate the asymptotic behavior of $\ln \tilde{N}(t)$ as $t \to \infty$.

(10.4) Theorem [Donsker and Varadhan 1975c; Figotin 1979, 1981]. *Consider the Schrödinger operator* $-\Delta + q$, *where* $q(x)$, *for* $x \in \mathbf{R}^d$, *is a random field of the form* (10.19), *with* $u(x)$ *satisfying conditions* (10.20). *Assume that the limit* (10.23) *exists for all* $c > 0$. *If* $\eta(t)$ *is defined by equation* (10.26), *we have*

$$\lim_{t\to\infty} \frac{\eta(t)}{t} \ln \tilde{N}(t) = - \inf_{\|\psi\|_2=1} \left\{ J(\psi) + \int |\psi(x)|^{2\kappa} \, dx \right\}, \tag{10.27}$$

where

$$J(\psi) = \int |\nabla\psi(x)|^2 \, dx \tag{10.28}$$

and κ *is the constant of equation* (10.25) (*with* $0^0 = 0$).

A rigorous proof of this theorem is rather long and involved; therefore we precede it with a heuristic discussion outlining its main points.

As we saw in Section 9, we have a universal lower asymptotic bound for $\ln \tilde{N}(t)$, namely (9.38), so the problem is to get a sharp upper bound. In this case a simple upper bound similar to (9.17) turns out not to be sharp enough; this suggests we're dealing with quantum, not classical, asymptotics. We must then make rigorous the heuristic arguments of Section 9 about quantum asymptotics.

It is easy to see that the universal lower bound (9.38) with trial functions ψ of the form (9.30) implies that

$$\ln \tilde{N}(t) \geqslant - \text{const}\, t^{-d/(d+2)} \tag{10.29}$$

as $t \to \infty$, where the radius R of the support of ψ should grow as

$$R \sim t^{1/(d+2)}; \tag{10.30}$$

for more details, see Problem IV.16 and relation (10.78). If we formally substitute the right-hand side of (10.29) into (9.42), we get

$$\ln N(\lambda) \sim - \text{const}\, \lambda^{-d/2} \tag{10.31}$$

as $\lambda \downarrow 0$, coinciding with Lifshitz's formula (9.6). This indicates that the right-hand side of (10.29) is not only a lower bound for $\ln \tilde{N}(t)$, but actually describes the asymptotic behavior of this function. Therefore, in view of the formula

$$|\psi(x)|^2 \, dx \sim \frac{1}{t} \operatorname{meas}\{0 \leqslant s \leqslant t : x(s) \in dx\},$$

known from the semi-classical approximation to quantum mechanics, it is natural to assume that as $t \to \infty$ the main contribution to the Wiener integral (9.47) representing $\tilde{N}(t)$ is due to trajectories of the Wiener process which, within time t, move away from the origin a distance of the order of, at most, $t^{1/(d+2)}$. But typical Wiener trajectories behave in a different way: they move away a much greater distance, of the order of $\sqrt{t}$, so we're clearly dealing with a version of the large deviation problem, which consists in estimating the probability of such atypical events.

Thus, the asymptotic formulas for $\ln \tilde{N}(t)$ are obtained by a method based on a nontrivial generalization of the classical theorem on large deviations [Loeve 1960], developed in a series of papers by Donsker and Varadhan [1975a,b,c, 1976, 1983]; see also [Freidlin and Wentzell 1984]. We recall briefly the essence of this theorem by examining the example of a sequence ξ_n, for $n \in \mathbf{Z}_+$, or independent, identically distributed real random variables. By the law of large numbers, $\Sigma_n = \frac{1}{n}\sum_{k=1}^n \xi_k$ tends to $\langle\xi\rangle = \mathsf{E}\{\xi\}$ with probability 1. If we denote by P_{Σ_n} the distribution of the random variables Σ_n, that is,

$$P_{\Sigma_n}(d\sigma) = \mathsf{P}\{\Sigma_n \in d\sigma\},$$

it follows from the law of large numbers that

$$\lim_{n\to\infty} P_{\Sigma_n}(d\nu) = \delta_{\langle\xi\rangle}(d\nu), \tag{10.32}$$

where $\delta_a(d\sigma)$ is the probability measure concentrated at a. (Here and below, unless we say otherwise, convergence of measures means weak convergence [Gihman and Skorohod I].) The theorem on large deviations states that the convergence rate in (10.32) is exponential: there exists a convex function $h(s)$ such that, for any interval $\Delta \subset \mathbf{R}$,

$$\frac{1}{n}\ln P_{\Sigma_n}(\Delta) = -\inf_{s\in\Delta} h(s)\big(1 + o(1)\big) \tag{10.33}$$

as $n \to \infty$, where $h\big(\langle\xi\rangle\big) = 0$ and $h(s) > 0$ for $s \neq \langle\xi\rangle$. The function $h(s)$ is an analogue of the entropy in statistical mechanics: see [Ruelle 1969b], for example.

Thus, the classic theorem on large deviations describes the asymptotic behavior of the distribution functions of normalized sums Σ_n of independent random variables ξ_n. We can also consider more complicated objects associated with the ξ_n, for example, the sequence μ_n of sample distributions of the ξ_n, defined as

$$\mu_n(ds) = \frac{1}{n}\sum_{k=1}^{n} \delta_{\xi_k}(ds) = \frac{1}{n}\#\{k : 1 \leqslant k \leqslant n, \xi_k \in ds\}. \tag{10.34}$$

Clearly,

(10.35)
$$\lim_{n\to\infty} \mu_n(d\nu) = \pi_\xi(d\nu)$$

with probability 1, where $\pi_\xi(d\nu)$ is the distribution of the ξ_n.

In the current situation, we're dealing with random distributions (measures) μ_n, or, equivalently, with a probability measure $P_{\mu_n}(d\mu)$ on the set $\mathcal{M}$ of probability measures μ on $\mathbf{R}$, where, by definition,

(10.36)
$$P_{\mu_n}(A) = \mathsf{P}\{\mu_n \in A\}$$

and $A \subset \mathcal{M}$ is an event in $\mathcal{M}$. Therefore, (10.35) can be rewritten as

(10.37)
$$\lim_{n\to\infty} P_{\mu_n}(d\mu) = \delta_{\pi_\xi}(d\mu),$$

where $\delta_\alpha(d\mu)$ is the probability measure on $\mathcal{M}$ concentrated at the point $\alpha \in \mathcal{M}$. As was the case with Σ_n, we're interested in learning more about the convergence of (10.37). The situation turns out to be similar to the classical situation of large deviations (10.33): with probability exponentially close to 1, as $n \to \infty$, the distribution μ_n is concentrated in any given neighborhood of π_ξ. More precisely, Donsker and Varadhan proved that there exists a nonnegative, convex functional $I(\mu)$, for $\mu \in \mathcal{M}$, such that

(10.38)
$$\frac{1}{n} \ln P_{\mu_n}(A) \simeq - \inf_{\mu\in A} I(\mu)(1+o(1))$$

as $n \to \infty$. This functional play a role similar to that of $h(s)$ in the classical theorem on large deviations (10.33).

Formula (10.38), with n replaced by a continuous time t, is also valid in the case that interests us here, the Wiener process $x(t)$, for $t \geqslant 0$, on $\mathbf{R}^d$. If we introduce a random measure $\mu_t(dx)$ on $\mathbf{R}^d$, called *local time* and defined by

$$\mu_t(B) = \frac{1}{t} \operatorname{meas}\{s : 0 \leqslant s \leqslant t, x(s) \in B\},$$

we have

(10.39)
$$\frac{1}{t} \ln P_{\mu_t}(A) \simeq - \inf_{\mu\in A} I(\mu)(1+o(1))$$

as $t \to \infty$, where the functional $I(\mu)$ is given by

(10.40)
$$I(\mu) = J(\psi) = \int \left|\nabla\psi(x)\right|^2 dx$$

if $\mu(dx) = \left|\psi(x)\right|^2 dx$ with $\|\psi\|_2 = 1$, and $I(\mu) = \infty$ if $\mu(dx)$ is not absolutely continuous with respect to the Lebesgue measure dx.

We can now state the general plan for determining the asymptotic behavior of $\ln \tilde{N}(t)$, based on the theorem on large deviations (10.39). Using the Feynman–Kac formula and the notation introduced above, we get the following representation:

$$\tilde{N}(t) = \int_{\mathcal{M}} e^{-\Phi_t(\mu)} P_{\mu_t}(d\mu), \tag{10.41}$$

where

$$\Phi_t(\mu) = -\ln \mathsf{E}\{e^{-t\int q(x)\mu(dx)}\}. \tag{10.42}$$

However, (10.39) cannot be applied directly to (10.41) because Φ_t and $\ln P_{\mu_t}$ are of different orders as $t \to \infty$, that is,

$$\lim_{t\to\infty} \Phi(t)/t = \operatorname*{ess\,sup}_{\omega\in\Omega} q(0) = 0.$$

To make them of the same order, we use the invariance of the Wiener process under the transformations $x(t) \to \sqrt{\eta}\, x(\eta t)$. This scale invariance, applied to (10.41), gives

$$\tilde{N}(t) = \int e^{-\Phi_t(\eta;\mu)} P_{\mu_{t/\eta}}(d\mu). \tag{10.43}$$

To equate the orders of Φ_t and $\ln P_{\mu_t}$ it is sufficient to choose $\eta(t)$ so that the limit

$$\lim_{t\to\infty} \frac{\eta(t)}{t} \Phi(t)\big(\eta(t);\mu\big) = \Phi(\mu) \tag{10.44}$$

is finite and nonzero for $\mu(dx) = \left|\psi(x)\right|^2 dx$, with $\|\psi\|_2 = 1$. Setting $\eta = \eta(t)$ in (10.42) and using (10.41) and (10.44), we arrive at the desired asymptotic formula (10.27).

Proof of Theorem 10.4. Upper bound: Consider the representation (5.17) of $\tilde{N}(t)$ in terms of the Wiener integral and define the function

$$\tilde{N}_1(t) = \mathsf{E}\{\mathsf{W}\{e^{-\int_0^t q(x(s))\,ds}\}\}. \tag{10.45}$$

This function differs from $\tilde{N}(t)$ by the absence of the condition $x(t) = 0$ and of the multiplier $(4\pi t)^{-d/2}$. We will later see (in the upper-bound part of the proof) that this difference becomes insignificant as $t \to \infty$.

We must write $\tilde{N}_1(t)$ in a form appropriate to the application of the large deviation technique. Thus we introduce, as before, the following functional on $L_1^+(\mathbf{R}^d)$:

$$\Phi(\phi) = -\ln \mathsf{E}\{e^{-\int q(x)\phi(x)\,dx}\}. \tag{10.46}$$

Let $\mathcal{M}$ be the set of probability measures on $\mathbf{R}^d$, and consider the map on Wiener trajectories $x(\,\cdot\,)$ in $\mathbf{R}^d$ such that $x(0) = 0$ given by

$$x(\,\cdot\,) \mapsto \mu_{x(\cdot)}(A) = \frac{1}{t}\,\mathrm{meas}\{s : 0 \leqslant s \leqslant t, x(s) \in A\}, \tag{10.47}$$

where A is a Borel set in $\mathbf{R}^d$. This mapping, together with the Wiener measure on the set of trajectories $x(\,\cdot\,)$, induces in a standard way the probability measure $P_t(d\mu)$ on $\mathcal{M}$. If we introduce the notations

$$(u * \mu)(x) = \int u(y - x)\mu(dy), \qquad W_t\{\,\cdot\,\} = \int \cdot\, dP_t,$$

equation (10.45) can be rewritten

$$\tilde{N}_1(t) = W_t\{e^{-\Phi(tu*\mu)}\}. \tag{10.48}$$

We now formulate rigorously the results of Donsker and Varadhan's discussed at the beginning of this subsection. It is technically more convenient to consider the Wiener process on the d-dimensional torus $\mathbf{T}^d$, also denoted

$$\mathbf{T} = \mathbf{T}_r = \{x \in \mathbf{R}^d : |x_j| \leqslant r/2, 1 \leqslant j \leqslant d\}, \tag{10.49}$$

and to take the limit $r \to \infty$ at the suitable moment. Consider on $\mathbf{T}$ the Wiener process $x(t)$, for $t \geqslant 0$, with initial condition $x(0) = 0$, that is, the diffusion process with infinitesimal operator $-\Delta$ on $\mathbf{T}$. Denote by $\mathcal{M}_{\mathbf{T}}$ the set of probability measures on $\mathbf{T}$, endowed with the weak convergence of measures; the mapping (10.47) and the Wiener measure on the trajectories $x(\cdot)$ with values in $\mathbf{T}$ induce the corresponding probability measure $P_t^{\mathbf{T}}(d\mu)$ on $\mathcal{M}_{\mathbf{T}}$; we denote the integral with respect to this measure by $W_t^{\mathbf{T}}\{\,\cdot\,\}$.

(10.5) Proposition [Donsker and Varadhan 1975a]. *If Φ is a continuous functional on $\mathcal{M}_{\mathbf{T}}$,*

$$\lim_{t\to\infty} \frac{1}{t} \ln W_t^{\mathbf{T}}\{e^{-t\Phi}\} = -\inf_{\|\psi\|_{2,\mathbf{T}}=1} \{J(\psi) + \Phi(|\psi|^2\,dx)\},$$

where

$$J(\psi) = \int_{\mathbf{T}} \left|\nabla\psi(x)\right|^2 dx. \qquad \square$$

As stated, this proposition isn't quite sufficient for our purposes, since it talks about functionals on $\mathcal{M}_{\mathbf{T}}$, while the functional Φ of (10.46) is defined on functions $L_1^+(\mathbf{R}^d)$. To fix this, we consider a family of nonnegative, measurable functions $k_\varepsilon(x, y)$ on $\mathbf{T}^2$, for $\varepsilon > 0$, satisfying the following properties:

(a) $\int_{\mathbf{T}} k_\varepsilon(x,y)\,dy = 1$ for all $x \in \mathbf{T}$ and $\varepsilon > 0$;
(b) the mapping $x \mapsto k_\varepsilon(x,\cdot)$ from $\mathbf{T}$ into $L_1(\mathbf{T})$ is continuous for all $\varepsilon > 0$;
(c) $\lim_{\varepsilon\downarrow 0} \int_{\mathbf{T}} k_\varepsilon(x,y)\phi(y)\,dy = \phi(x)$ uniformly in $x \in \mathbf{T}$, for all $\phi \in C(\mathbf{T})$;
(d) $\lim_{\varepsilon\downarrow 0} \int_{\mathbf{T}} k_\varepsilon(x,y)\phi(y)\,dy = \phi(x)$ in the metric of $L_1(\mathbf{T})$, for all $\phi \in L_1(\mathbf{T})$.

We use the functions $k_\varepsilon(x,y)$ to smooth functionals in $\mathcal{M}_{\mathbf{T}}$, setting

$$\phi_\varepsilon(x) = (K_\varepsilon\mu)(x) = \int_{\mathbf{T}} k_\varepsilon(x,y)\mu(dy). \tag{10.50}$$

By our assumptions on k_ε, equation (10.50) defines a continuous mapping from $\mathcal{M}_{\mathbf{T}}$ into $L_1^+(\mathbf{T})$, since for any $\mu \in \mathcal{M}_{\mathbf{T}}$ we have $\mu(\mathbf{T}) = 1$ and therefore $\|K_\varepsilon\mu\|_1 = 1$.

(10.6) Proposition [Donsker and Varadhan 1975b]. *Let Φ be a continuous functional on $L_1^+(\mathbf{T})$ (or on the subset of $L_1^+(\mathbf{T})$ of functions of unit $\|\cdot\|_1$-norm), bounded from below, and let $\varepsilon(t)$ be a positive function tending to zero as $t \to \infty$, in such a way that*

$$\liminf_{t\to\infty} \varepsilon(t)t^{1/d} > 0. \tag{10.51}$$

Then

$$\lim_{t\to\infty} \frac{1}{t} \ln W_t^{\mathbf{T}}\{e^{-t\Phi(K_{\varepsilon(t)}\mu)}\} = - \inf_{\|\psi\|_{2,\mathbf{T}}=1} \{J(\psi) + \Phi(|\psi|^2)\}, \tag{10.52}$$

where

$$J(\psi) = \int_{\mathbf{T}} |\nabla\psi(x)|^2\,dx. \tag{10.53}$$ □

The version of this proposition that we will actually use admits a "weak" dependence of the functional Φ on t:

(10.7) Proposition [Donsker and Varadhan 1975b]. *Let Φ_t, for $t > 0$, be a family of nonnegative, continuous functionals on $L_1^+(\mathbf{T})$ (or on the subset of $L_1^+(\mathbf{T})$ of functions of unit $\|\cdot\|_1$-norm), having the following property: for any family of nonnegative functions ϕ_t of unit $\|\cdot\|_1$-norm, converging as $t \to \infty$ in the metric of $L_1(\mathbf{T})$ to some function ϕ such that $J(\sqrt{\phi}) < \infty$, there exists a weakly lower semicontinuous functional Φ on the subset of $L_1^+(\mathbf{T})$ of functions of unit $\|\cdot\|_1$-norm, such that*

$$\liminf_{t\to\infty} \Phi_t(\phi_t) \geqslant \Phi(\phi). \tag{10.54}$$

(In particular, $\liminf_{t\to\infty} \Phi(\phi_t) \geqslant \Phi(\phi)$.) Then, if $\varepsilon(t) \to 0$ as $t \to \infty$ and (10.51) is satisfied, we have

$$(10.55) \qquad \limsup_{t\to\infty} \frac{1}{t} \ln W_t^{\mathbf{T}}\{e^{-t\Phi_t(K_{\varepsilon(t)}\mu)}\} \leqslant - \inf_{\|\psi\|_{2,\mathbf{T}}=1} \{J(\psi) + \Phi(|\psi|^2)\},$$

where

$$(10.56) \qquad J(\psi) = \int_{\mathbf{T}} |\nabla\psi(x)|^2 \, dx. \qquad \square$$

We return to the study of the function $\tilde{N}_1(t)$ defined by (10.45) and (10.48). According to our plan for the proof, we will be able to apply Proposition 10.7 if we can scale the Wiener process in $\mathbf{R}^d$ so that (10.44) is satisfied. For we have, for any $\eta > 0$,

$$(10.57) \qquad \tilde{N}_1(t) = \mathsf{E}\{\mathsf{W}\{e^{-\int_0^t q(x(s))\,ds}\}\} = \mathsf{E}\{\mathsf{W}\{e^{-\eta\int_0^{t/\eta} q(\sqrt{\eta}x(s))\,ds}\}\}.$$

If we now introduce the function $u_\eta(x) = \eta^{d/2}u(\eta^{1/2}x)$ and the functional

$$(10.58) \qquad \Phi(\phi;\eta) = -\ln \mathsf{E}\{e^{-\eta^{-d/2}\int \phi(x)m(\sqrt{\eta}\,dx)}\},$$

defined for any $\phi \in L_1^+$, we conclude from (10.45), (10.48) and (10.57) that

$$(10.59) \qquad \tilde{N}_1(t) = W_{t/\eta}\{e^{-\Phi(tu_\eta*\mu;\eta)}\}.$$

Now set $\eta = \eta(t)$, where $\eta(t)$ is the solution to (10.26), and perform the change of variables $\tau = t/\eta(t)$. By Lemma 10.3(i), passing to the limit $t \to \infty$ is the same as making $\tau \to \infty$. In terms of the functional

$$(10.60) \qquad \Phi_\tau(\phi) = \frac{1}{\tau}\Phi(t\phi;\eta(t))\big|_{t=t(\tau)},$$

the representation (10.58) becomes

$$(10.61) \qquad \tilde{N}_1(t) = W_\tau\{e^{-\tau\Phi_\tau(u_\eta*\mu)}\}.$$

We're almost ready to apply (10.55): the only remaining obstacle is that (10.55) contains an integral over the Wiener process in the torus $\mathbf{T}_r$ of (10.49), while (10.61) uses an integral over the Wiener process in all of $\mathbf{R}^d$. It is reasonable to try to approximate (10.61) by something of the form $W_t^{\mathbf{T}}\{\,\cdot\,\}$, then apply (10.55) to the result, and finally, after taking the limit $\tau \to \infty$, go back to $\mathbf{R}^d$ by making $r \to \infty$. To do this we introduce yet another bit of notation:

$$(10.62) \qquad \begin{aligned} \phi_{\mathbf{T}_r}(x) &= \sum_{\ell\in\mathbf{Z}^d} \phi(x+rl) \qquad \text{for } \phi \in L_1^+, \\ \mu_{\mathbf{T}_r}(A) &= \sum_{\ell\in\mathbf{Z}^d} \mu\big((A\cap\tilde{\mathbf{T}}_r)+rl\big) \qquad \text{for } \mu \in \mathcal{M}, \end{aligned}$$

where $\tilde{\mathbf{T}}_r$ is the cube corresponding to the tours $\mathbf{T}_r$.

(10.8) Lemma. *For any $\phi \in L_1^+(\mathbf{R}^d)$, $\mu \in \mathcal{M}$ and $\eta > 0$, we have*

$$\Phi(\phi;\eta) \geqslant \Phi(\chi_\Lambda \phi_{\mathbf{T}};\eta,) \tag{10.63}$$

$$(u_\eta * \mu)_{\mathbf{T}} = (u_\eta)_{\mathbf{T}} * \mu_{\mathbf{T}}. \tag{10.64}$$

Proof. (10.64) follows immediately from (10.62); we must show (10.63). From (10.58) and (9.55) we have

$$\Phi(\phi;\eta) = \eta^{d/2} \int_{\mathbf{R}^d} f\left(\eta^{-d/2}\phi(x)\right) dx. \tag{10.65}$$

From Lemma 9.11(iv), we have

$$\begin{aligned}
\Phi(\phi;\eta) &= \sum_{\ell \in \mathbf{Z}^d} \eta^{d/2} \int_{\Lambda - rl} dx\, f\left(\eta^{-d/2}\phi(x)\right) \\
&= \eta^{d/2} \int_\Lambda dx \sum_\ell f\left(\eta^{-d/2}\phi(x+rl)\right) \\
&= \eta^{d/2} \int_\Lambda f\left(\eta^{-d/2} \sum_\ell \phi(x+rl)\right) dx \\
&= \eta^{d/2} \int_\Lambda f\left(\eta^{-d/2}\phi_{\mathbf{T}}(x)\right) dx = \Phi(\chi_\Lambda \phi_{\mathbf{T}};\eta).
\end{aligned}$$
□

Thus, from (10.60), (10.61), (10.63) and (10.64), we conclude that, for any $\mathbf{T}$,

$$\tilde{N}_1(t) \leqslant W_\tau\{\exp(-\tau\Phi_\tau(\chi_\Lambda((u_\eta)_{\mathbf{T}} * \mu_{\mathbf{T}})))\}. \tag{10.66}$$

But this is clearly equal to

$$\tilde{N}_1(t) \leqslant W_\tau^{\mathbf{T}}\{\exp(-\tau\Phi_\tau(\chi_\Lambda((u_\eta)_{\mathbf{T}} * \mu)))\}, \tag{10.67}$$

Now recall from (10.60), (10.65) and the definition of τ that

$$\Phi_\tau(\phi) = \frac{\eta}{t}\eta^{d/2} \int_{\mathbf{R}^d} f\left(t\eta^{-d/2}\phi(x)\right) dx \bigg|_{t/\eta(t)=\tau}. \tag{10.68}$$

If we use the fact that $f(0) = 0$ and introduce the functional on $L_1^+(\mathbf{T})$ given by

$$\Phi_\tau^{\mathbf{T}}(\phi) = \frac{\eta}{t}\eta^{d/2} \int_\Lambda f\left(t\eta^{-d/2}\phi(x)\right) dx \bigg|_{t/\eta(t)=\tau}, \tag{10.69}$$

we get

$$\tilde{N}_1(t) \leqslant W_\tau^{\mathbf{T}}\{\exp(-\tau\Phi_\tau^{\mathbf{T}}(\chi_\Lambda((u_\eta)_{\mathbf{T}} * \mu)))\}. \tag{10.70}$$

(10.9) Lemma. *Let $\Phi_\tau^{\mathbf{T}}$ be the functionals on $L_1^+(\mathbf{T})$ defined by* (10.69). *If ϕ_τ and ϕ are functions in $L_1^+(\mathbf{T})$ such that $\lim_{\tau\to\infty}\phi_\tau = \phi$ in the metric of $L_1(\mathbf{T})$ and κ is the constant of equation* (10.25),

$$\liminf_{\tau\to\infty} \Phi_\tau^{\mathbf{T}}(\phi_\tau) \geqslant \int_{\mathbf{T}} dx\, \phi^\kappa(x). \tag{10.71}$$

Moreover, the functional $\tilde{\Phi}^{\mathbf{T}}$ defined by the right-hand side of (10.71) *is lower semicontinuous on $L_1^+(\mathbf{T})$.*

Proof. It follows from (10.23), (10.25), (10.26) and Lemma 10.3(a) that

$$\lim_{t\to\infty} \frac{\eta(t)}{t}\eta^{d/2}(t) f\big(c t \eta^{-d/2}(t)\big) = c^\kappa \tag{10.72}$$

for all $c > 0$, still with the convention that $0^0 = 0$. But, as we have seen, $t \to \infty$ is equivalent to $\tau \to \infty$, so (10.72) and Fatou's lemma imply (10.71). Checking that $\tilde{\Phi}^{\mathbf{T}}$ is semicontinuous is straightforward. □

Given Lemma 10.9, the only thing that remains to be done before we can apply Proposition 10.7 is to check condition (10.51). Here the role of $k_\varepsilon(x,y)$ is played by $\chi_\Lambda(x)(u_\eta)_{\mathbf{T}}(y-x)$, so (10.51) reduces to

$$\liminf_{\tau\to\infty} \tau^{1/d}\eta^{-1/2}(t)\big|_{t/\eta(t)=\tau} > 0, \tag{10.73}$$

which is clearly a consequence of Lemma 10.3(ii)—just plug in $\tau = t/\eta(t)$.

Now apply Proposition 10.7 to the right-hand side of (10.70) to get

$$\limsup_{t\to\infty} \frac{\eta(t)}{t} \ln \tilde{N}_1(t) \leqslant -\inf_{\|\psi\|_{2,\mathbf{T}}=1}\left\{J(\psi) + \int_{\mathbf{T}} |\psi(x)|^{2\kappa}\, dx\right\}. \tag{10.74}$$

Replacing $\mathbf{T}$ by $\mathbf{T}_r$ and letting $r \to \infty$,

$$\limsup_{t\to\infty} \frac{\eta(t)}{t} \ln \tilde{N}_1(t) \leqslant -\inf_{\|\psi\|_2=1}\left\{J(\psi) + \int_{\mathbf{R}^d} |\psi(x)|^{2\kappa}\, dx\right\}. \tag{10.75}$$

(One must show that taking this limit is allowed. For $\kappa = 0$ this is not at all hard, and is done in [Donsker and Varadhan 1975c]; in the general case essentially the same proof given there also works.)

We must now show that $\tilde{N}_1$ can be replaced by $\tilde{N}$ in (10.75). If we set $p_t(x) = (4\pi t)^{-d/2}e^{-x^2/4t}$ and recall that $q(x) > 0$, we can write, for any $\varepsilon > 0$,

$$\begin{aligned}
&\mathsf{W}\{e^{-\int_0^t q(x(s))\,ds} \mid x(t)=0\}\\
&\qquad \leqslant \mathsf{W}\{e^{-\int_0^{t-\varepsilon} q(x(s))\,ds} \mid x(t)=0\}\\
&\qquad = \int dy\, \mathsf{W}\{e^{-\int_0^{t-\varepsilon} q(x(s))\,ds} \mid x(t-\varepsilon)=y\} \frac{p_{t-\varepsilon}(y)p_\varepsilon(-y)}{p_t(0)}\\
&\qquad \leqslant \frac{p_\varepsilon(0)}{p_t(0)} \int dy\, \mathsf{W}\{e^{-\int_0^{t-\varepsilon} q(x(s))\,ds} \mid x(t-\varepsilon)=y\} p_{t-\varepsilon}(y)\\
&\qquad = \frac{p_\varepsilon(0)}{p_t(0)} \mathsf{W}\{e^{-\int_0^{t-\varepsilon} q(x(s))\,ds} \mid x(t-\varepsilon)=y\}.
\end{aligned}$$

Using (5.17) and (10.45), we conclude that $\tilde{N}(t) \leqslant p_\varepsilon(0)\tilde{N}_1(t-\varepsilon)$; finally, using Lemma 10.3 and inequality (10.75), we arrive at the desired upper bound:

$$\limsup_{t\to\infty} \frac{\eta(t)}{t} \ln \tilde{N}(t) \leqslant - \inf_{\|\psi\|_2=1} \left\{ J(\psi) + \int_{\mathbf{R}^d} |\psi(x)|^{2\kappa}\, dx \right\}. \tag{10.76}$$

Lower bound: Starting from the universal lower bound (9.38) and equation (9.55), we easily get

$$\ln \tilde{N}(t) \geqslant -\ln V_\phi - tJ(\sqrt{\phi}) - \int dx\, f\big(t(u*\phi)(x)\big),$$

where $*$, as usual, denotes convolution and $J(\psi)$ is defined by (10.28). This inequality is valid for any nonnegative $\phi \in C_0^\infty(\mathbf{R}^d)$ such that $\|\phi\|_1 = 1$. If we replace $\phi(x)$ by

$$\phi_\eta(x) = \eta^{-d/2}\phi(\eta^{-1/2}x) \tag{10.77}$$

simple manipulations yield

(10.78)
$$\ln \tilde{N}(T) \geqslant -\frac{d}{2}\ln\eta - \ln V_\phi - \frac{t}{\eta}J(\sqrt{\phi}) - \eta^{-d/2}\int dx\, f\big(t\eta^{-d/2}(u_\eta * \phi)(x)\big).$$

As before, we take $\eta = \eta(t)$ to be the solution of (10.26), and divide both sides by $t/\eta(t)$. By Lemma 10.3(iv), we get

(10.79)
$$\liminf_{t\to\infty} \frac{\eta(t)}{t}\ln \tilde{N}(t) \geqslant -J(\sqrt{\phi}) - \lim_{t\to\infty}\frac{\eta}{t}\eta^{-d/2}\int dx\, f\big(t\eta^{-d/2}(u_\eta * \phi)(x)\big).$$

Again by Lemma 10.3, $\eta(t) \to \infty$ as $t \to \infty$, and since $\phi \in C_0^\infty(\mathbf{R}^d)$, we get $u_\eta * \phi \to \phi$ as $t \to \infty$. Therefore (10.72) gives

$$\lim_{t\to\infty}\frac{\eta}{t}\eta^{-d/2}\int dx\, f\big(t\eta^{-d/2}(u_\eta * \phi)(x)\big) = \int \phi^\kappa(x)\, dx, \tag{10.80}$$

for $\phi \in C_0^\infty(\mathbf{R}^d)$, with the additional assumption, if $\kappa = 0$, that the boundary of the set $\{x : \phi(x) > 0\}$ has zero Lebesgue measure. Details of the proof of (10.80) are left to the reader.

From (10.79) and (10.80) we get

$$\liminf_{t\to\infty} \frac{\eta(t)}{t} \ln \tilde{N}(t) \geqslant -J(\sqrt{\phi}) - \int \phi^\kappa(x)\,dx; \tag{10.81}$$

since ϕ is arbitrary, we get

$$\liminf_{t\to\infty} \frac{\eta(t)}{t} \ln \tilde{N}(t) \geqslant -\inf_{\|\psi\|_2=1}\left\{J(\psi) + \int dx\,|\psi(x)|^{2\kappa}\right\}. \tag{10.82}$$

In passing from (10.81) to (10.82) we replaced the infimum over the set of positive, normalized functions $\phi \in C_0^\infty(\mathbf{R}^d)$ by the infimum over a broader set of normalized functions in $L_1^+(\mathbf{R}^d)$. The justification for this is straightforward and we omit it: in the case $\kappa = 0$ it can be found in [Donsker and Varadhan 1975c]. □

Proof of Theorem 10.2. We just have to apply (10.27). For the Poisson random potential, (9.58) and (10.23) give $\kappa = 0$, then solving (10.26) we get

$$\eta(t) = \left(\frac{t}{c}\right)^{\frac{2}{(d+2)}}(1 + o(1))$$

as $t \to \infty$. Plugging this into (10.27) and (9.42) gives

$$\ln \tilde{N}(t) = -\frac{d+2}{2}c^{2/(d+2)}\left(\frac{2\gamma_d}{d}\right)^{\frac{d}{(d+2)}} t^{d/(d+2)}(1 + o(1)) \tag{10.83}$$

$$\ln N(\lambda) = -c\gamma_d^{d/2}\lambda^{-d/2}(1 + o(1)) \tag{10.84}$$

as $t \to \infty$ and $\lambda \downarrow 0$, respectively. Here γ_d is the smallest eigenvalue of the Dirichlet problem for the operator $-\Delta$ in a ball of unit volume, and we have made use of the following equation from [Donsker and Varadhan 1975c]:

(10.85)

$$\inf_{\|\psi\|_2=1}\left\{-\int \psi(x)\Delta\psi(x)\,dx + \text{meas}\{x : |\psi(x)| > 0\}\right\} = \frac{d+2}{2}\left(\frac{2\gamma_d}{d}\right)^{\frac{d}{(d+2)}}.$$

This equality is proved as follows. Let $\gamma(G)$ be the lowest eigenvalue of $-\Delta$ on the domain G with Dirichlet conditions. Since any function with compact support $\psi \in L_2$ can be represented as

$$\psi(x) = R^{-d/2}\tilde{\psi}(x/R) \tag{10.86}$$

with $\text{meas}\,\text{supp}\,\tilde{\psi} = 1$, the left-hand side of (10.85) equals

$$\text{(10.87)} \qquad \inf_{\|\psi\|_2=1}\Big\{-\int \psi(x)\Delta\psi(x)\,dx + \operatorname{meas}\{x : |\psi(x)| > 0\}\Big\} = \inf_{R>0}\{R^{-2}\inf_{\operatorname{meas} G=1}\gamma(G) + R^d\}.$$

But it is well-known (see [Courant and Hilbert 1968], for example) that $\gamma(G)$ achieves its infimum on a ball. Taking the minimum of (10.87) with respect to R gives (10.85).

For the generalized Poisson potential with a random measure (1.6) we get $\kappa = 0$ from (9.59). Solving (10.26) and using (10.85) gives

$$\text{(10.88)} \qquad \eta(t) = \left(\frac{d+2}{2}\frac{t}{\ln t}\right)^{\frac{2}{(d+2)}}(1+o(1)),$$

$$\text{(10.89)} \qquad \ln \tilde{N}(t) = \left(\frac{d+2}{d}\gamma_d\right)^{\frac{2}{(d+2)}} t^{d/(d+2)}\ln^{2/(d+2)} t(1+o(1)).$$

By using (9.42) we finally obtain (10.22). □

Comparison of (10.21) and (10.22) shows that, for the generalized Poisson field, there is an additional logarithmic factor in the asymptotics for $\ln N(\lambda)$. A simple analysis of (10.26), which is essentially responsible for the form of the asymptotics of $\ln \tilde{N}(t)$ and $N(\lambda)$, shows that this factor is due to the absence of an atom at zero in the distribution of the random variable $m(C)$, where m is the random measure in (10.19) and C is a cube in $\mathbf{R}^d$. Thus, among generalized Poisson potentials, a "genuine" Poisson potential has the greatest integrated density of states near the bottom of the spectrum.

The proof of the next theorem is an almost literal repetition of that of Theorem 10.4, and will not be given.

(10.10) Theorem. *Consider the Schrödinger operator* $-\Delta + q$, *where* $q(x)$, *for* $x \in \mathbf{R}^d$, *is a random field of the form*

$$\text{(10.90)} \qquad q(x) = \sum_{y\in\mathbf{Z}^d} u(x-y)\xi_y,$$

where $u(x)$ *satisfies conditions* (10.20) *and the* ξ_y, *for* $y \in \mathbf{Z}^d$, *are nonnegative, independent, identically distributed random variables with* $\mathsf{E}\{\xi_0\} < \infty$ *and* $\operatorname{ess\,sup}\xi_0 = 0$. *If* $f(t) = -\ln \mathsf{E}\{e^{-t\xi_0}\}$ *satisfies* (10.23) *and* $\eta(t)$ *is the solution of* (10.26), *we have*

$$\text{(10.91)} \qquad \lim_{t\to\infty}\frac{\eta(t)}{t}\ln \tilde{N}(t) = -\inf_{\|\psi\|_2=1}\Big\{J(\psi) + \int |\psi(x)|^{2\kappa}\,dx\Big\},$$

where

$$J(\psi) = \int \left|\nabla\psi(x)\right|^2 dx$$

and κ *is the constant of equation* (10.25) (*with* $0^0 = 0$). *The asymptotic behavior of* $N(\lambda)$ *is obtained from* (10.91) *by means of* (9.42). □

This theorem has simple corollaries describing the asymptotic behavior of $\ln N(\lambda)$ near $\lambda = 0$:

(10.11) Corollary. *Consider a Schrödinger operator* $-\Delta + q$ *as in the statement of Theorem* 10.10, *with the additional condition that the distribution function* $F(\xi) = \mathsf{P}\{\xi_y \leqslant \xi\}$ *of* ξ_y *satisfies*

$$\ln F(\xi) = -\kappa\xi^{-a} \ln^b \xi^{-1}\left(1 + o(1)\right) \tag{10.92}$$

as $\xi \downarrow 0$, *where* $k > 0$, $a \geqslant 0$ *and* b *are constants. Then*

$$\ln N(\lambda) = -k_1\lambda^{-d/2+a} \ln^b \lambda^{-1}\left(1 + o(1)\right) \tag{10.93}$$

as $\lambda \downarrow 0$, *where* k_1 *depends only on* k, a *and* b.

Proof. From the definition of $f(t)$ and from (10.92) we have

$$f(t) = \inf_{\xi>0}\left\{\xi t - \ln F(\xi)\right\}\left(1 + o(1)\right)$$

as $t \to \infty$. This expression, together with (10.92), gives

$$f(t) = \text{const}\, t^{a/(a+1)} \ln^{b/(a+1)} t \left(1 + o(1)\right)$$

as $t \to \infty$, where the constant, like the ones to follow, depends only on a, b and k. Solving equation (10.26) we get

$$\eta(t) = \text{const}\, t^{(d/2+a+1)^{-1}} \ln^{-b} t \left(1 + o(1)\right)$$

as $t \to \infty$, and comparing this with (10.91) gives

$$\ln \tilde{N}(t) = -\,\text{const}\, t^{1-(d/2+a+1)^{-1}} \ln^b t \left(1 + o(1)\right)$$

as $t \to \infty$. Substituting this into (9.42) and using Lemma 9.9 gives the desired asymptotic formula. □

It is instructive to compare (10.92) and (10.93) with the corresponding relations (10.7) and (10.8) of Theorem 10.1. Theorems 10.4 and 10.10 and Corollary 10.11 are clearly more precise than Theorem 10.1, since they give the asymptotic behavior of $\ln N(\lambda)$, while Theorem 10.1 only talks about $\ln\left(-\ln N(\lambda)\right)$. This is because Theorems 10.4 and 10.10 were proved by the more powerful technique developed by Donsker and Varadhan. Corollary

10.11, in particular, means that a condition like (10.7) cannot yield a more exact statement on the asymptotic behavior of $N(\lambda)$ than formula (10.8). It also indicates that $N(\lambda)$ is sensitive at least to such coarse characteristics of the random potential as $F(q) = \mathsf{P}\{q(x) \leqslant q\}$, if not to the potential itself.

(10.12) Corollary. *Consider a Schrödinger operator* $-\Delta + q$ *as in the statement of Theorem* 10.10, *with the additional condition that the distribution function* $F(\xi) = \mathsf{P}\{\xi_y \leqslant \xi\}$ *of* ξ_y *satisfies*

$$\mathsf{P}\{\xi_y = 0\} = 1 - p, \qquad \mathsf{P}\{\xi_y = 1\} = p, \tag{10.94}$$

where $0 < p < 1$. *Then*

$$\ln N(\lambda) = -\ln(1-p)\gamma_d^{d/2}\lambda^{-d/2}(1 + o(1)) \tag{10.95}$$

as $\lambda \downarrow 0$, *which coincides with the corresponding formula for the Poisson potential with* $c = -\ln(1-p)$.

Proof. We have $f(t) = -\ln\big(pe^{-t} + (1-p)\big)$, which is asymptotically equal to $-\ln(1-p)$. Formula (10.95) now follows easily from Theorem 10.10. □

10.C Smoothed Square of a Gaussian Random Field

In this subsection we study random potentials $q(x)$ of the form

$$q(x) = \frac{1}{2}\int_{\mathbf{R}^d} u(x-y)g^2(y)\,dy, \tag{10.96}$$

for $x \in \mathbf{R}^d$, where $u(x)$ satisfies (10.20) and $g(x)$ is a uniform Gaussian random field with zero mathematical expectation, and whose correlation function $b(x) = \mathsf{E}\{g(0)g(x)\}$ satisfies the following conditions:

(a) there exist positive constants c and γ so that the Fourier transform $\hat{b}(k)$, for $k = (k_1, \ldots, k_d) \in \mathbf{R}^d$, satisfies

$$\hat{b}(k) \leqslant c\prod_{j=1}^{d}\frac{1}{1+|k_j|^{1+\gamma}}; \tag{10.97}$$

(b) there exists $m > d + 2(\gamma^{-1} + 2)$ such that, as $|x| \to \infty$,

$$b(x) = O\big(|x|^{-m}\big). \tag{10.98}$$

The factor $\frac{1}{2}$ in the definition of $q(x)$ is meant to simplify some of the subsequent formulas.

For fields $q(x)$ of the form (10.96), the lower boundary of the spectrum of $-\Delta + q(x)$ is $\lambda = 0$; this is proved in the same way as for generalized Poisson fields (Theorem 5.34).

(10.13) Theorem [Figotin 1981]. *Consider a Schrödinger operator with potential $q(x)$ of the form* (10.96), *with $u(x)$ satisfying* (10.20). *The following asymptotic formulas hold for $N(\lambda)$:*

(i) *If $b(x) = \exp\left(-\sum_{j=1}^{d} |x_j|\right)$, where $x = (x_1, \ldots, x_d)$, we have*

$$\ln N(\lambda) = -c_d \lambda^{-d/2+1} \ln^{2d-2} \lambda^{-1} \left(1 + o(1)\right) \tag{10.99}$$

as $\lambda \downarrow 0$, where c_d is a known positive constant;

(ii) *if $\hat{b}(k) \in C^\infty(\mathbf{R}^d)$ and $\hat{b}(k) = b|k|^{-\alpha}\left(1 + o(1)\right)$ as $|k| \to \infty$, where b is a positive constant and $\alpha > d$, we have*

$$\ln N(\lambda) = -c\lambda^L \left(1 + o(1)\right) \tag{10.100}$$

as $\lambda \downarrow 0$, where $L = \dfrac{\alpha - d + 2}{2(\alpha d^{-1} - 1)}$ and c is a known positive constant.

These formulas are obtained by determining the asymptotic behavior, as $t \to \infty$, of $\ln \tilde{N}(t)$ and then using the Tauberian theorem (9.42). The formulas for $\ln \tilde{N}(t)$ are derived in Theorem 10.14; after its proof (10.99) and (10.100) will follow as corollaries.

The asymptotic behavior of $\ln \tilde{N}(t)$ for the potentials considered here is similar to that for generalized Poisson potentials. The function $f(t)$ in (9.54) is replaced here by

$$f(t) = \int_{\mathbf{R}^d} \ln\left(1 + t\hat{b}(k)\right) dk. \tag{10.101}$$

(10.14) Theorem [Figotin 1981]. *Consider a Schrödinger operator with potential $q(x)$ of the form* (10.96), *with $u(x)$ satisfying* (10.20), *and $g(x)$ a Gaussian random field with zero mathematical expectation and correlation function satisfying* (10.97) *and* (10.98). *Suppose the function $f(t)$ of* (10.101) *is such that the limit* (10.23) *exists, with the constant κ of* (10.25) *strictly positive.*

Then, if $\eta(t)$ is the solution of (10.26), *we have*

$$\lim_{t\to\infty} \frac{\eta(t)}{t} \ln \tilde{N}(t) = -\inf_{\|\psi\|_2=1} \left\{ J(\psi) + \frac{1}{2} \int dx \, |\psi(x)|^{2\kappa} \right\}, \tag{10.102}$$

where $J(\psi) = \int |\nabla\psi(x)|^2 \, dx$. The asymptotic behavior of $N(\lambda)$ is determined by (10.102) *and* (9.42).

The proof is similar to that of Theorem 10.4: we find an asymptotic upper bound for $\ln \tilde{N}(t)$ using Proposition 10.7, and an asymptotic lower bound using (9.38). We will tackle it after proving some preliminary results.

(10.15) Lemma. *Let $f(t)$ be the function defined by* (10.101).

(i) *$f(t)$ satisfies the properties listed in Lemma* 9.11, *the only difference being that here* $\lim_{t\downarrow 0} f(t)/t = b(0)$;

(ii) *for any $\varepsilon > 0$ there exists $C(\varepsilon) > 0$ such that*

$$f(t) \leqslant C(\varepsilon)t^{1/(1+\gamma)+\varepsilon}; \tag{10.103}$$

(iii) *the solution $\eta(t)$ of* (10.26) *has the properties listed in Lemma* 10.3;

(iv) *if $\zeta(a) = \sup_{|x|\geqslant a}|b(x)|$, there exists a constant $l > 0$ such that*

$$\lim_{t\to\infty} t\eta(t)\zeta\big(\sqrt{\eta(t)}\,t^{-l}\big) = 0. \tag{10.104}$$

Proof. Parts (i) and (iii) are obvious. Part (ii) is easily proved by induction on the dimension d, using condition (10.97). For part (iv), set $p = 1/(1+\gamma)$; from (10.26) and (10.103) we have

$$\frac{t}{\eta(t)} \leqslant \eta^{d/2}(t)C(\varepsilon)\big(t\eta^{-d/2}(t)\big)^{p+\varepsilon},$$

which can be written $\eta(t) \geqslant C_1(\varepsilon)t^{l_1}$, where

$$l_1 = \frac{1-(p+\varepsilon)}{1+d(1-(p+\varepsilon))/2}.$$

Hence, from (10.98), we have

$$t\eta(t)\zeta\big(\sqrt{\eta(t)}t^{-\alpha}\big) \leqslant C_2(\varepsilon)t^{-l_2},$$

where

$$l_2 = \frac{(1-(p+\varepsilon))(m-2)}{2+d(1-(p+\varepsilon))} - 1 - ml.$$

Since $m \geqslant d + (\gamma^{-1}+2)$ (see (10.98)) we have $l_2 > 0$ for ε and l small enough, concluding the proof. □

(10.16) Lemma. *Let B_n, for $n \in \mathbf{Z}_+$, be nonnegative operators on a Hilbert space such that $\operatorname{Tr} B_n < \infty$, and let B_∞ be an operator on the same space. Then:*

(i) *if* $\lim_{n\to\infty} \|B_n - B_\infty\| = 0$ *and* $\lim_{n\to\infty} \operatorname{Tr} B_n = \operatorname{Tr} B_\infty < \infty$,

$$\lim_{n\to\infty} \operatorname{Tr}\ln(I+B_n) = \operatorname{Tr}\ln(I+B_\infty);$$

(ii) $\operatorname{Tr}\ln(I+B_1+B_2) \leqslant \operatorname{Tr}\ln(I+B_1) + \operatorname{Tr}\ln(I+B_2)$;

(iii) $\left|\operatorname{Tr}\ln(I+B_1) - \operatorname{Tr}\ln(I+B_2)\right| \leqslant \|B_1 - B_2\| \left(\operatorname{Tr}\ln(I+B_1) + \operatorname{Tr}\ln(I+B_2)\right)$ *if* $\operatorname{Tr} B_1 = \operatorname{Tr} B_2$.

Proof. Left to the reader. □

Now consider the functional Φ on $L_1^+(\mathbf{R}^d)$ given by

$$\Phi(\phi) = -\ln \mathsf{E}\{e^{-\frac{1}{2}\int_{\mathbf{R}^d} g^2(x)\phi(x)\,dx}\} \tag{10.105}$$

and the self-adjoint operator B on $L_2(\mathbf{R}^d)$ given by

$$(B\psi)(x) = \int_{\mathbf{R}^d} b(x-y)\psi(y)\,dy \tag{10.106}$$

For any $\psi \in L_2(\mathbf{R}^d)$, the operator of multiplication by ψ will be denoted by the same letter.

(10.17) Lemma. *The functional* Φ *of* (10.105) *is continuous in the metric of* $L_1(\mathbf{R}^d)$ *and satisfies*

$$\Phi(\phi) = \frac{1}{2}\operatorname{Tr}\ln\left(I + \sqrt{\phi}B\sqrt{\phi}\right). \tag{10.107}$$

Furthermore, it is monotonic, that is, $\Phi(\phi_1) \leqslant \Phi(\phi_2)$ *if* $\phi_1(\,\cdot\,) \leqslant \phi_2(\,\cdot\,)$; *it is invariant under translations, that is,* $\Phi(\phi(\,\cdot\,+a)) = \Phi(\phi(\,\cdot\,))$; *and it is subadditive, that is,*

$$\Phi(\phi_1+\phi_2) \leqslant \Phi(\phi_1) + \Phi(\phi_2). \tag{10.108}$$

Proof. Continuity, monotonicity and invariance under translations follow immediately from the definition (10.105). The representation (10.107) is proved first for step functions and then extended to an arbitrary $\phi \in L_1^+$; this is done by using the fact that the realizations of the Gaussian field $g(x)$ are continuous with probability 1, thanks to (10.97) and (10.98) [Fernique 1964]. Subadditivity is first proved for step functions, using Lemma 10.16(ii), then extended to L_1^+ by continuity. The details are left to the reader. □

Let Λ_R be the cube with center at the origin and edges of length R parallel to the coordinate axes, and define functions $f_R(t)$, for $t \geqslant 0$, by

$$f_R(t) = R^{-d}\operatorname{Tr}\ln(I + t\chi_{\Lambda_R}B\chi_{\Lambda_R}). \tag{10.109}$$

(10.18) Lemma. *With the notations above,*

(10.110) $$\lim_{R\to\infty} f_R(t) = f(t) = \int_{\mathbf{R}^d} \ln\bigl(1+t\hat{b}(k)\bigr)\,dk$$

Proof. For any $l \in \mathbf{Z}_+$,

(10.111) $$\lim_{R\to\infty} R^{-d}\,\mathrm{Tr}(\chi_{\Lambda_R} B \chi_{\Lambda_R})^l = \int_{\mathbf{R}^d} \hat{b}^l(k)\,dk,$$

because $(\chi_{\Lambda_R} B \chi_{\Lambda_R})^l = \chi_{\Lambda_R} B^l \chi_{\Lambda_R}$ by (10.97) and (10.98). The lemma follows readily. □

Proof of Theorem 10.14. Upper bound: As in the proof of Theorem 10.4, we introduce a function $\tilde{N}_1(t)$ by (10.45). Clearly (10.48) is still valid, and by (10.57) we have the following counterpart for (10.59):

(10.112) $$\tilde{N}_1(t) = W_{t/\eta}\{e^{-\Phi(tu_\eta * \mu;\eta)}\}$$

for any $\eta > 0$, where $\Phi(\,\cdot\,;\eta)$ is the functional on L_1^+ defined by (10.105), but for the Gaussian field whose covariance function is $b(\sqrt{\eta}\,x)$ rather than $b(x)$.

To obtain an asymptotic upper bound for (10.112) using Proposition 10.7, we must approximate the integral $W_{t/\eta}\{\,\cdot\,\}$ over the Wiener trajectory in the whole space $\mathbf{R}^d$ by the integral over the Wiener process on the torus $\mathbf{T}_r$ (10.49). We use the following lemma.

(10.19) Lemma. *Let $\zeta(a)$ be the function of Lemma* 10.15(*iv*)*. Then, for any $\phi \in L_1^+$,*

(10.113) $$\Phi(\phi;n) \geqslant \Phi(\chi_{\Lambda_{r-2\delta}}\phi_{\mathbf{T}};\eta) - \zeta(\sqrt{\eta}\delta)b(0)\|\phi\|_1^2,$$

where $\phi_{\mathbf{T}}$ is defined by (10.62).

Proof. We first prove (10.113) for $\eta = 1$, that is, for $\Phi(\,\cdot\,;1) = \Phi(\,\cdot\,)$, and for $\phi \in L_1^+$ with compact support. Notice first that if $\phi_j \in L_1^+$ for $1 \leqslant j \leqslant k$ and $\phi_i\phi_j = 0$ for all $i \neq j$, we have

(10.114) $$\mathrm{Tr}\ln\Bigl(I + \sum_{j=1}^{k} \sqrt{\phi_j}B\sqrt{\phi_j}\Bigr) = \sum_{j=1}^{k} \mathrm{Tr}\ln(I + \sqrt{\phi_j}B\sqrt{\phi_j}).$$

Now set

(10.115) $$\phi_l(x) = \chi_{\Lambda_{r-2\delta}}(x - rl)\phi(x)$$

for $l \in \mathbf{Z}^d$. Since $\phi(x) \geqslant \sum_l \phi_l(x)$, it follows from Lemma 10.17 that

(10.116) $$\Phi(\phi) \geqslant \Phi\Bigl(\sum_l \phi_l(x)\Bigr)$$

Since ϕ has compact support, the number of terms in the sum $\sum_l \phi_l$ is finite. Using (10.107) we get

$$\Phi\Big(\sum_l \phi_l\Big) = \tfrac{1}{2}\,\mathrm{Tr}\ln\Big(I + \sum_l \sqrt{\phi_l}B\sqrt{\phi_l} + \sum_{l\neq k}\sqrt{\phi_l}B\sqrt{\phi_k}\Big). \tag{10.117}$$

If Σ and Σ' denote the first and second sums on the right-hand side, $\mathrm{Tr}\,\Sigma \leqslant b(0)\|\phi\|_1$ and $\mathrm{Tr}\,\Sigma' = 0$. By Lemma 10.16(iii) and the obvious fact that $\mathrm{Tr}\ln(I+A) \leqslant \mathrm{Tr}\,A$, we get

$$\big|\mathrm{Tr}\ln(I+\Sigma+\Sigma') - \mathrm{Tr}\ln(I+\Sigma)\big| \leqslant 2b(0)\|\phi\|_1\|\Sigma\|_1. \tag{10.118}$$

On the other hand,

$$\begin{aligned}\|\Sigma'\|^2 \leqslant \mathrm{Tr}\,\Sigma'^2 &= \sum_{l\neq k,p\neq q}\mathrm{Tr}\sqrt{\phi_l}B\sqrt{\phi_k}\sqrt{\phi_p}B\sqrt{\phi_q}\\ &= \sum_{l\neq k}\mathrm{Tr}\sqrt{\phi_l}B\sqrt{\phi_k}B\\ &= \sum_{l\neq k}\int dx\,dy\,\phi_l(x)b(x-y)\phi_k(y)b(y-x)\\ &\leqslant \sum_{l\neq k}\zeta^2(\delta)\int dx\,dy\,\phi_l(x)\phi_k(y) \leqslant \zeta^2(\delta)\|\phi\|_1^2.\end{aligned} \tag{10.119}$$

Putting together formulas (10.115) through (10.119), we get

$$\Phi(\phi) \geqslant \sum_l \Phi(\phi_l) - \zeta(\delta)b(0)\|\phi\|_1^2,$$

which, in view of Lemma 10.17, gives

$$\begin{aligned}\Phi(\phi) &\geqslant \sum_l \Phi\big(\phi_l(x+rl)\big) - \zeta(\delta)b(0)\|\phi\|_1^2\\ &\geqslant \Phi\Big(\sum_l \phi_l(x+rl)\Big) - \zeta(\delta)b(0)\|\phi\|_1^2 = \Phi(\chi_{\Lambda_{r-2\delta}}\phi_{\mathbf{T}}) - \zeta(\delta)b(0)\|\phi\|_1^2.\end{aligned}$$

This proves (10.113) for $\eta = 1$ and ϕ with compact support. The extension to arbitrary $\phi \in L_1^+$ is justified by the continuity of Φ, and the case of arbitrary $\eta > 0$ follows by the definition of $\Phi(\,\cdot\,;\eta)$. □

We now set $\eta = \eta(t)$ in (10.113), where $\eta(t)$ is the solution of (10.26), and $\delta = t^{-l}$, where l is the constant of Lemma 10.15(iv). By Lemma 10.15(iv) and inequality (10.112), we get

(10.120)
$$\limsup_{t\to\infty}\frac{\eta(t)}{t}\ln\tilde N_1(t)\leqslant\limsup_{t\to\infty}W_{t/\eta}\{\exp(-\Phi(t\chi_{\tilde{\mathbf{T}}_{r-2t-\alpha}}(u_\eta*\mu)_{\mathbf{T}};\eta))\}.$$

Again as in the proof of Theorem 10.4, we make the substitution $\tau=t/\eta(t)$ and set

$$\Phi_\tau(\phi)=\frac{1}{\tau}\Phi\big(t\chi_{\Lambda_{r-2t-\alpha}}\phi;\eta(t)\big)\big|_{t/\eta(t)=\tau} \tag{10.121}$$

Then (10.120) can be rewritten

$$\limsup_{t\to\infty}\frac{\eta(t)}{t}\ln\tilde N_1(t)\leqslant\limsup_{\tau\to\infty}\frac{1}{t}\ln W_\tau^{\mathbf{T}}\{\exp(-\tau\Phi_\tau((u_\eta)_\tau*\mu))\}. \tag{10.122}$$

Here we used Lemma 10.15(iii) to change the limit variable from t to τ; the substitution of W_t by $W_t^{\mathbf{T}}$ is justified just as in the proof of Theorem 10.4, based on equality (10.64).

We now have to check that the right-hand side of (10.122) satisfies (10.54), in order to apply Proposition 10.7. This is done in the next two lemmas. We need to introduce a variation of the functional Φ_τ of (10.121):

$$\tilde\Phi_\tau(\phi)=\frac{1}{\tau}\Phi\big(t\phi;\eta(t)\big)\big|_{t/\eta(t)=\tau}. \tag{10.123}$$

(10.20) Lemma. *For any* $c>0$,

$$\lim_{\tau\to\infty}\tilde\Phi_\tau(c\chi_{\Lambda_\rho})=\tfrac12 c^\kappa\rho^d. \tag{10.124}$$

Proof. Let B_η be the integral operator on $L_2(\mathbf{R}^d)$ with kernel $b\big(\sqrt{\eta}(x-y)\big)$. Clearly $B_1=B$. From the definition of the functional $\Phi(\,\cdot\,;\eta)$ and equations (10.107) and (10.123), we get

$$\lim_{\tau\to\infty}\tilde\Phi_\tau(c\chi_{\Lambda_\rho})=\lim_{t\to\infty}\frac{\eta(t)}{2t}\operatorname{Tr}\ln(I+ct\chi_{\Lambda_\rho}B\chi_{\Lambda_\rho}). \tag{10.125}$$

From the definition of $\eta(t)$, we get

$$\begin{aligned}\frac{\eta(t)}{t}\operatorname{Tr}\ln(I+ct\chi_{\Lambda_\rho}B\chi_{\Lambda_\rho})&=\frac{\operatorname{Tr}\ln(I+ct\chi_{\Lambda_\rho}B_\eta\chi_{\Lambda_\rho})}{\eta^{d/2}f(t\eta^{-d/2})}\\&=\frac{\operatorname{Tr}\ln(I+ct\chi_{\Lambda_\rho}B\chi_{\Lambda_\rho})}{\eta^{d/2}f(ct\eta^{-d/2})}\,\frac{f(ct\eta^{-d/2})}{f(t\eta^{-d/2})}.\end{aligned} \tag{10.126}$$

The first factor on the right can be estimated as follows. Set $R=n\rho$ for $n\in\mathbf{Z}_+$. By (10.107) and Lemma 10.17,

$$\operatorname{Tr}\ln(I+ct\chi_{\Lambda_R}B\chi_{\Lambda_R})\leqslant n^d\operatorname{Tr}\ln(I+ct\chi_{\Lambda_\rho}B\chi_{\Lambda_\rho}).$$

If we go from B_η to B, this can be rewritten

$$\eta^{d/2}(R\sqrt{\eta})^{-d}\operatorname{Tr}\ln(I+ct\eta^{-d/2}\chi_{\Lambda_{R\sqrt{\eta}}}B\chi_{\Lambda_{R\sqrt{\eta}}}) \leqslant \rho^{-d}\operatorname{Tr}\ln(I+ct\chi_{\Lambda_\rho}B_\eta\chi_{\Lambda_\rho}).$$

Using Lemma 10.18 and making R go to infinity we get

$$\eta^{d/2}f(ct\eta^{-d/2}) \leqslant \rho^{-d}\operatorname{Tr}\ln(I+ct\chi_{\Lambda_\rho}B_\eta\chi_{\Lambda_\rho}).$$

Using (10.23), (10.25) and (10.126), we get

$$\liminf_{t\to\infty}\frac{\eta(t)}{t}\operatorname{Tr}\ln(I+ct\chi_{\Lambda_\rho}B_\eta\chi_{\Lambda_\rho}) \geqslant c^\kappa\rho^d. \tag{10.127}$$

We now find an upper bound for the left-hand side of (10.126). For an arbitrary $\varepsilon > 0$, we set $R = n(\rho+\varepsilon)$. Then

$$\chi_{\Lambda_R}(x) \geqslant \sum_l \chi_{\rho,l}(x), \tag{10.128}$$

where $\chi_{\rho,l}(x) = \chi_{\Lambda_\rho}\big(x-(\rho+\varepsilon)l\big)$ and the sum is taken over all l such that $|l|_0 \leqslant n$. Using Lemma 10.17 we have

(10.129)
$$\begin{aligned}\operatorname{Tr}\ln(I+ct\chi_{\Lambda_R}B_\eta\chi_{\Lambda_R}) &\geqslant \operatorname{Tr}\ln\Big(I+ct\sum_l\chi_{\rho,l}B_\eta\sum_l\chi_{\rho,l}\Big)\\ &= \operatorname{Tr}\ln\Big(I+ct\sum_l\chi_{\rho,l}B_\eta\chi_{\rho,l}+ct\sum_{l\neq m}\chi_{\rho,l}B_\eta\chi_{\rho,m}\Big).\end{aligned}$$

Denoting by Σ and Σ' the first and second sums on the right, we have

$$\begin{aligned}\operatorname{Tr}\ln(I+ct\chi_{\Lambda_R}B_\eta\chi_{\Lambda_R}) &\geqslant \operatorname{Tr}\ln(I+ct\Sigma)\\ &\quad - ct\|\Sigma'\|\big(\operatorname{Tr}\ln(I+ct\chi_{\tilde{\mathbf{T}}_R}B_\eta\chi_{\Lambda_R})+\operatorname{Tr}\ln(I+ct\Sigma)\big),\end{aligned} \tag{10.130}$$

since the right-hand side of (10.129) is greater than or equal to the right-hand side of (10.130), by Lemma 10.16(iii) and the equalities

$$\operatorname{Tr}\Sigma = n^d\rho^d b(0), \qquad \operatorname{Tr}\Sigma' = 0.$$

To estimate $\|\Sigma'\|$, let $\psi \in L_2(\mathbf{R}^d)$. Then

$$\begin{aligned}\big|(\psi,\Sigma'\psi)\big| &\leqslant \sum_{l\neq m}\big|(\chi_{\rho,l}\psi, B_\eta\chi_{\rho,m}\psi)\big|\\ &\leqslant \zeta(\sqrt{\eta}\varepsilon)\sum_{l\neq m}\big(\chi_{\rho,l}|\psi|,\chi_{\rho,m}|\psi|\big) \leqslant \zeta(\sqrt{\eta}\varepsilon)\|\psi\|_2^2.\end{aligned}$$

Since Σ' is obviously self-adjoint, the latter inequality implies $\|\Sigma'\| \leqslant \zeta(\sqrt{\eta}\varepsilon)$. This, combined with (10.130), gives

(10.131)

$$\operatorname{Tr}\ln(I+ct\chi_{\Lambda_R}B_\eta\chi_{\Lambda_R}) \geqslant \frac{1-ct\zeta(\sqrt{\eta}\varepsilon)}{1+ct\zeta(\sqrt{\eta}\varepsilon)}\operatorname{Tr}\ln\left(I+ct\sum_l \chi_{\rho,l}B_\eta\chi_{\rho,l}\right).$$

By applying (10.114) to the right-hand side, switching from B_η to B, and using Lemma 10.17, we arrive at

$$\eta^{d/2}(R\sqrt{\eta})^{-d}\operatorname{Tr}\ln(I+ct\eta^{-d/2}\chi_{\Lambda_{R\sqrt{\eta}}}B\chi_{\Lambda_{R\sqrt{\eta}}}) \geqslant (\rho+\varepsilon)^{-d}\frac{1-ct\zeta(\sqrt{\eta}\varepsilon)}{1+ct\zeta(\sqrt{\eta}\varepsilon)}\operatorname{Tr}\ln(I+ct\chi_{\Lambda_\rho}B_\eta\chi_{\Lambda_\rho}).$$

Taking the limit $R\to\infty$, and taking into account Lemma 10.18, gives

$$\eta^{d/2}f(ct\eta^{-d/2}) \geqslant (\rho+\varepsilon)^{-d}\frac{1-ct\zeta(\sqrt{\eta}\varepsilon)}{1+ct\zeta(\sqrt{\eta}\varepsilon)}\operatorname{Tr}\ln(I+ct\chi_{\Lambda_\rho}B_\eta\chi_{\Lambda_\rho}).$$

Using Lemma 10.15(v) and choosing $\varepsilon=t^{-\alpha}$ we get

$$\limsup_{t\to\infty}\frac{\operatorname{Tr}\ln(I+ct\chi_{\Lambda_\rho}B_\eta\chi_{\Lambda_\rho})}{\eta^{d/2}f(ct\eta^{-d/2})} \leqslant \rho^d.$$

Combining this with (10.126), (10.23) and (10.25), and applying Lemma 10.15(iii), we get

$$\limsup_{t\to\infty}\frac{\eta(t)}{t}\operatorname{Tr}\ln(I+ct\chi_{\Lambda_\rho}B_\eta\chi_{\Lambda_\rho}) \leqslant c^\kappa\rho^d. \tag{10.132}$$

Now (10.127), (10.58) and (10.125) yield (10.124), as desired. □

(10.21) Lemma. *Let $\phi(x)=\sum_{i=1}^n c_i\chi_{\Lambda_{\rho_i}}$, where $c_i\geqslant 0$ and K_{ρ_i} is a closed cube of edge ρ_i. If* meas $K_{\rho_i}\cap K_{\rho_j}=0$ *for all $i\neq j$, we have*

$$\limsup_{t\to\infty}\tilde{\Phi}_\tau(\phi) \leqslant \frac{1}{2}\int\phi^\kappa(x)\,dx. \tag{10.133}$$

If, in fact, $K_{\rho_i}\cap K_{\rho_j}=\varnothing$ for all $i\neq j$, we have

$$\limsup_{t\to\infty}\tilde{\Phi}_\tau(\phi) = \frac{1}{2}\int\phi^\kappa(x)\,dx. \tag{10.134}$$

Proof. Inequality (10.133) follows easily from the definition of $\tilde{\Phi}_\tau$ (10.123), (10.108) and (10.124). To get (10.134), denote by ε the minimum distance between any two distinct cubes. Then

$$\operatorname{Tr}\ln(I+t\sqrt{\phi}B_\eta\sqrt{\phi}) = \operatorname{Tr}\ln\left(I+t\sum_{i=1}^n\sqrt{c_i}\chi_{\Lambda_{\rho_i}}B_\eta\sqrt{c_i}\chi_{\Lambda_{\rho_i}}+\sum_{i\neq j}\sqrt{c_i}\chi_{\Lambda_{\rho_i}}B_\eta\sqrt{c_j}\chi_{\Lambda_{\rho_i}}\right).$$

The right-hand side is estimated just like the right-hand side of (10.129): by repeating the arguments above, we obtain the following counterpart for (10.131):

$$\operatorname{Tr}\ln(I + t\sqrt{\phi}B_\eta\sqrt{\phi}) \geqslant \frac{1 - ct\zeta(\sqrt{\eta}\varepsilon)}{1 + ct\zeta(\sqrt{\eta}\varepsilon)} \operatorname{Tr}\ln\biggl(I + t\sum_{i=1}^{n} \sqrt{c_i}\chi_{\Lambda_{\rho_i}} B_\eta \sqrt{c_i}\chi_{\Lambda_{\rho_i}}\biggr),$$

where $c = \max c_i$. Using (10.114), (10.123), (10.124) and Lemma 10.5 we get

$$\liminf_{t\to\infty} \frac{\eta(t)}{t} \operatorname{Tr}\ln(I + t\sqrt{\phi}B\sqrt{\phi}) \geqslant \sum_{i=1}^{n} c_i^\kappa \rho_i^d = \int \phi^\kappa(x)\,dx.$$

This, together with (10.133), gives (10.134). □

(10.22) Lemma. *If ϕ is a function in L_1^+ with compact support,*

(10.135) $$\lim_{T\to\infty} \tilde{\Phi}_t(\phi) = \frac{1}{2}\int \phi^\kappa(x)\,dx.$$

Proof. First take a nonnegative $\phi \in C_0^\infty(\mathbf{R}^d)$. We can approximate ϕ by step functions, which satisfy (10.133) and (10.134); also, since $\kappa \in (0,1)$, the right-hand side of (10.135) is continuous on $L_1^+(K)$, for K compact. Since $\tilde{\Phi}_T$ is monotonic by Lemma 10.17, Lemma 10.21 shows that (10.135) hold for nonnegative $\phi \in C_0^\infty(\mathbf{R}^d)$.

To extend this to $\phi \in L_1^+$ with compact support, consider a positive function $h \in C_0^\infty(\mathbf{R}^d)$ such that $\|h\|_1 = 1$. We will show that

(10.136) $$\tilde{\Phi}_T(\phi) \leqslant \tilde{\Phi}_T(h * \phi).$$

Indeed, using Jensen's inequality and the homogeneity of $g(x)$, we have

$$\begin{aligned}\mathsf{E}\bigl\{e^{-\frac{1}{2}\int g^2(x)(h*\phi)(x)\,dx}\bigr\} &= \mathsf{E}\bigl\{e^{-\frac{1}{2}\int h(x)\,dx\int g^2(x+y)\phi(y)\,dy}\bigr\}\\ &= \mathsf{E}\biggl\{\int dx\, h(x)e^{-\frac{1}{2}\int g^2(x+y)\phi(y)\,dy}\biggr\}\\ &= \mathsf{E}\bigl\{e^{-\frac{1}{2}\int g^2(y)\phi(y)\,dy}\bigr\}\end{aligned}$$

This implies (10.136), in view of the definitions (10.105) and (10.121) of Φ and Φ_T. We already know that equality (10.135) holds for nonnegative $\phi \in C_0^\infty(\mathbf{R}^d)$, so we can extend one half of it to any $\phi \in L_1^+$ with compact support:

(10.137) $$\limsup_{T\to\infty} \tilde{\Phi}_T(\phi) \leqslant \frac{1}{2}\int \phi^\kappa(x)\,dx,$$

because h is arbitrary and $\kappa \in (0,1)$. For the converse, set $\psi_+(x) = \max\{\psi(x), 0\}$. By Lemma 10.17,

$$\tilde{\Phi}_\tau(\phi) + \tilde{\Phi}_\tau\big((\psi-\phi)_+\big) \geqslant \tilde{\Phi}_\tau\big(\phi + (\psi-\phi)_+\big) \geqslant \tilde{\Phi}_\tau(\psi).$$

Using (10.137) and (10.135) for nonnegative $\psi \in C_0^\infty(\mathbf{R}^d)$, we get

$$\liminf_{\tau\to\infty} \tilde{\Phi}_\tau(\phi) \geqslant \frac{1}{2}\int \psi^\kappa(x)\,dx - \frac{1}{2}\int (\psi-\phi)_+^\kappa(x)\,dx.$$

If we now substitute $h_n * \phi$ for ψ, where $h_n(x) = n^d h(nx)$ and h is as defined above, and then let $n \to \infty$, we get

$$\lim_{\tau\to\infty} \tilde{\Phi}_\tau(\phi) \geqslant \frac{1}{2}\int \phi^\kappa(x)\,dx$$

because $\kappa \in (0,1)$. This, together with (10.137), proves the lemma. □

(10.23) Lemma. *Consider a family of functions $\phi_\tau \in L_1^+$ such that:*

(a) *there exists a compact $K \subset \mathbf{R}^d$ such that* $\operatorname{supp}\phi_\tau \subset K$ *for all τ;*
(b) *there exists $\phi \in L_1^+$ such that* $\lim_{t\to\infty}\|\phi_\tau - \phi\|_1 = 0$.

Then

$$\liminf_{\tau\to\infty} \tilde{\Phi}_\tau(\phi_\tau) \geqslant \frac{1}{2}\int \phi^\kappa(x)\,dx. \tag{10.138}$$

If, in addition, there exists $C < \infty$ such that $\phi_\tau(x) \leqslant C$ for all $x \in \mathbf{R}^d$ and $\tau > 0$, we also have

$$\lim_{\tau\to\infty} \tilde{\Phi}_\tau(\phi_\tau) = \frac{1}{2}\int \phi^\kappa(x)\,dx. \tag{10.139}$$

Proof. Let τ_k, for $k \in \mathbf{Z}_+$, be a sequence achieving the lim inf in (10.138). By condition (b) we can assume, without loss of generality, that

$$\lim_{k\to\infty} \phi_{\tau_k}(x) = \phi(x) \tag{10.140}$$

almost everywhere. Since $\tilde{\Phi}_t$ is monotonic (Lemma 10.17 and Lemma 10.22), we have, for $n \in \mathbf{Z}_+$,

$$\begin{aligned}\liminf_{\tau\to\infty} \tilde{\Phi}_\tau(\phi_\tau) &= \lim_{k\to\infty} \tilde{\Phi}_{\tau_k}(\phi_{\tau_k}) \\ &\geqslant \lim_{k\to\infty} \tilde{\Phi}_{\tau_k}\big(\inf_{m\geqslant n} \phi_{\tau_m}\big) = \tfrac{1}{2}\inf_{m\geqslant n} \phi_{\tau_m}^\kappa(x)\,dx.\end{aligned} \tag{10.141}$$

Since n is arbitrary we have, by (10.140),

(10.142) $$\liminf_{\tau\to\infty} \tilde{\Phi}_\tau(\phi_\tau) \geqslant \tfrac{1}{2} \sup_{n\geqslant 1} \int dx \inf_{m\geqslant n} \phi^\kappa_{\tau_m}(x)\, dx = \frac{1}{2}\int \phi^\kappa(x)\, dx,$$

concluding the proof of (10.138).

For (10.139), let τ_k, for $k \in \mathbf{Z}_+$, be a sequence achieving the upper limit of $\tilde{\Phi}_\tau(\phi_\tau)$ as $\tau \to \infty$. Again we can assume that (10.140) holds. Instead of (10.141), we can write

$$\limsup_{\tau\to\infty} \tilde{\Phi}_\tau(\phi_\tau) \leqslant \frac{1}{2}\int \sup_{m\geqslant n} \phi^\kappa_{\tau_m}(x)\, dx,$$

and instead of (10.142) we have

$$\limsup_{\tau\to\infty} \tilde{\Phi}_\tau(\phi_\tau) \leqslant \tfrac{1}{2} \inf_{n\geqslant 1} \int dx \sup_{m\geqslant n} \phi^\kappa_{\tau_m}(x)\, dx = \frac{1}{2}\int \phi^\kappa(x)\, dx,$$

by the last condition of the lemma and equality (10.140). Together with (10.138), this proves (10.139). □

We now estimate $\tilde{N}_1(t)$ using (10.122) and Proposition 10.7. Verification of (10.54) reduces to checking that

(10.143) $$\liminf_{\tau\to\infty} \Phi_t(\phi_t) \geqslant \frac{1}{2}\int \phi^\kappa(x)\, dx.$$

But in view of the definitions (10.121) and (10.123),

(10.144) $$\Phi_\tau(\phi_\tau) = \tilde{\Phi}_\tau(\chi_{\Lambda_{r-2t-\alpha}} \phi_t)\big|_{t/\eta(t)=\tau}.$$

Since, obviously,

$$\lim_{\tau\to\infty} \|\chi_{\Lambda_{r-2t(\tau)-\alpha}} \phi_\tau - \phi\|_1 = 0,$$

we conclude from (10.144) and (10.138) that (10.143) holds, so that condition (10.54) of Proposition 10.7 is fulfilled. Condition (10.51) is also fulfilled; the proof uses the same arguments as in Subsection 10.B, but relies on Lemma 10.15.

Thus, from (10.122) and Proposition 10.7, we have, for any $\mathbf{T} = \mathbf{T}_r$:

(10.145) $$\limsup_{t\to\infty} \frac{\eta(t)}{t} \tilde{N}_1(t) \leqslant - \inf_{\|\psi\|_{2,\mathbf{T}}=1} \left\{ J(\psi) + \frac{1}{2}\int_{\mathbf{T}} |\psi(x)|^{2\kappa}\, dx \right\},$$

whence, for $r \to \infty$, we obtain

(10.146) $$\limsup_{t\to\infty} \frac{\eta(t)}{t} \ln \tilde{N}(t) \leqslant - \inf_{\|\psi\|_2=1} \left\{ J(\psi) + \frac{1}{2}\int_{\mathbf{T}} |\psi(x)|^{2\kappa}\, dx \right\}.$$

In passing from (10.145) to (10.146) the same arguments are used as in going from (10.74) to (10.75), thanks to Lemma 10.17.

Lower bound: As in the proof of Theorem 10.4, we use the universal lower bound (9.38) and a similar argument. We use the following counterpart for (10.78):

$$\ln \tilde{N}(t) \geqslant -\frac{d}{2}\ln\eta - \ln V_\phi - \frac{t}{\eta}J(\sqrt{\phi}) - \Phi(tu_\eta * \phi;\eta), \tag{10.147}$$

where the functional $\Phi(\,\cdot\,,\eta)$ is defined by (10.105), but for the Gaussian field whose covariance function is $b(\sqrt{\eta}\,x)$ rather than $b(x)$. Using Lemmas 10.3(iv) and 10.17 and definition (10.123), we conclude that, for nonnegative ϕ with unit norm,

$$\liminf_{t\to\infty}\frac{\eta(t)}{t}\ln\tilde{N}(t) \geqslant -J(\sqrt{\phi}) - \limsup_{\tau\to\infty}\tilde{\Phi}_\tau(u_\eta * \phi), \tag{10.148}$$

where u_η stands for $u_{\eta(t)}\big|_{t/\eta(t)=\tau}$.

Now choose $R>0$ such that $\operatorname{supp}\phi \subset \Lambda_{R/2}$ (see (10.49)). Using definition (10.123) and (10.108), we have

(10.149)

$$\limsup_{\tau\to\infty}\tilde{\Phi}_\tau(u_\eta*\phi) \leqslant \limsup_{\tau\to\infty}\tilde{\Phi}_\tau(\chi_{\Lambda_R}(u_\eta*\phi)) + \limsup_{\tau\to\infty}\tilde{\Phi}_\tau((1-\chi_{\Lambda_R})(u_\eta*\phi)).$$

But clearly $\lim_{\tau\to\infty}\chi_{\Lambda_R}(u_\eta*\phi)=\phi$ and $\chi_{\Lambda_R}(x)(u_\eta*\phi)(x)\leqslant\|\phi\|_\infty$, so we can apply Lemma 10.23. By (10.139), we can write

$$\lim_{\tau\to\infty}\tilde{\Phi}_\tau(\chi_{\Lambda_R}(u_\eta*\phi)) = \frac{1}{2}\int\phi^\kappa(x)\,dx. \tag{10.150}$$

To estimate the second term on the right-hand side of (10.149), write $\phi_\tau = (1-\chi_{\Lambda_R})(u_\eta*\phi)$ and use (10.123) and (10.107):

$$\begin{aligned}\tilde{\Phi}_\tau(\phi_\tau) &= \frac{\eta(t)}{2t}\operatorname{Tr}\ln(I + t\sqrt{\phi_t}B_\eta\sqrt{\phi_t})\\ &\leqslant \frac{\eta(t)}{2t}\operatorname{Tr}(t\sqrt{\phi_t}B_\eta\sqrt{\phi_t})\tfrac{1}{2}b(0)\|\phi_\tau\|_1 \leqslant O(\eta^{-1}(t)),\end{aligned}$$

where the last inequality comes from (10.20). In view of (10.150) and (10.149), we conclude that

$$\limsup_{\tau\to\infty}\tilde{\Phi}_\tau(u_\eta*\phi) \leqslant \tfrac{1}{2}\phi^\kappa(x)\,dx.$$

Together with (10.148), this gives, for any nonnegative $\phi\in C_0^\infty(\mathbf{R}^d)$ of unit norm:

$$\liminf_{t\to\infty}\frac{\eta(t)}{t}\ln\tilde{N}(t) \geqslant -J(\sqrt{\phi}) - \frac{1}{2}\int\phi^\kappa(x)\,dx,$$

or, in other words,

$$(10.151)\qquad \liminf_{t\to\infty}\frac{\eta(t)}{t}\ln\tilde{N}(t) \geqslant -\inf_{\|\psi\|_2=1}\left\{J(\sqrt{\psi}) - \frac{1}{2}\int |\psi(x)|^{2\kappa}\,dx\right\}.$$

In view of (10.146), this implies the desired formula (10.102). □

Proof of Theorem 10.13. The correlation function $b(x)$ in the hypothesis of the theorem satisfies (10.97) and (10.98), so we can use Theorem 10.14 to study the asymptotic behavior of $\ln\tilde{N}(t)$.

In case (i), a simple analysis shows that the functions $f(t)$ and $\eta(t)$ of (9.54) and (10.26) satisfy

$$f(t) = \sigma\sqrt{t}\ln^{d-1}t\,(1+o(1))$$
$$\eta(t) = \alpha^{(d-1)/\alpha}\sigma^{-1/\alpha}t^{1/(2\alpha)}\ln^{-(d-1)/\alpha}t\,(1+o(1))$$

as $t\to\infty$, where $\alpha = d/4+1$ and

$$\sigma = \frac{2^{d-1}}{\pi^{d/2-1}(d-1)!}.$$

Thus $\kappa = \frac{1}{2}$, which justifies the application of Theorem 10.14.

Hence, using (10.102), we define the asymptotic behavior of $\ln\tilde{N}(t)$, which, being substituted into (9.42), will provide the desired formula (10.99) for $\ln N(\lambda)$ as $\lambda\downarrow 0$.

Case (ii) is handled similarly. We have

$$f(t) = \text{const}\,t^{d/\alpha}(1+o(1)),$$
$$\ln\tilde{N}(t) = -\text{const}\,t^{\omega}(1+o(1))$$

as $t\to\infty$, where $\omega = (\alpha-d+2)/\big(\alpha d^{-1}(d+2)-d\big)$. Using (9.42), we get formula (10.100). □

Equation (10.100) implies that the Lifshitz exponent L can take any value in the interval $(d/2,\infty)$, as α ranges through (d,∞): cf. (10.8) and (10.93). In particular, if we formally set $\alpha=\infty$ in (10.100), the asymptotic behavior of $\ln N(\lambda)$ becomes identical to what it would be for a Poisson random field. But this behavior is in fact excluded for potentials of the form (10.96), because it would require that $f(t)$ in (10.101) tend to a finite constant as $t\to\infty$, which would mean that the one-dimensional distribution of the potential has an atom at zero.

Thus, for potential (10.96), $N(\lambda)$ always goes to zero as $\lambda\downarrow 0$ faster than it does for the Poisson potential (that is, $L > d/2$). This is natural, because the probability of the potential being close to zero in any region is smaller for the square of the Gaussian field than for the Poisson field. This is especially easy to see when $u(x)$ has compact support. Then the distribution function of $q(x)$ in the Poisson case has an atom at zero, which is not the case for potential (10.96).

It is natural to search for counterparts for the results of this chapter in the discrete case, that is, for operator (1.8). Theorem 9.10 gives a counterpart in the case of classical asymptotic behavior; see also [Pastur 1982]. In the quantum case, one could try to use the approach of 10.B and 10.C, based on the representation of $\tilde{N}(t)$ as the mathematical expectation over trajectories of the Markov process. This was done by Fukushima [1974] (see also [Fukushima et al. 1975]), who used, instead of (5.17), a similar representation over trajectories of the standard random walk in $\mathbf{Z}^d$, for which the discrete Laplacian is an infinitesimal operator. However, Fukushima obtained upper and lower asymptotic bounds for $\ln N(\lambda)$ with different degrees in λ. This estimate was sharpened by Romerio and Wreszinski [1979], who showed, using the same technique, that if $p = \mathsf{P}\{q(0) = 0\} > 0$, we have

$$-c_2 \leqslant \liminf_{\lambda \downarrow 0} \lambda^{d/2} \ln N(\lambda) \leqslant \limsup_{\lambda \downarrow 0} \lambda^{d/2} \ln N(\lambda) \leqslant -c_1,$$

where $c_1, c_2 > 0$ and $c_2 \to \infty$ as $p \to 0$. One can also obtain an asymptotic formula of the type (10.93) based on the corresponding result of Donsker and Varadhan [1975a,b] for random walks in $\mathbf{Z}^d$.

Analogues for the coarser formula (10.8), which holds for the edges of all spectral gaps, were obtained by Mezincescu [1985, 1986], Kirsch and Simon [1986] and Simon[1986].

Problems

1. Give a formal proof of Proposition 8.2 by the so-called two-scale method, which is the extension of one of the versions of the averaging method in nonlinear mechanics to equations involving partial derivatives. The method consists in assuming that, for $\rho \to \infty$, the solutions of (8.14) are

$$u_\rho(x) = u^{(0)}(x) + \rho^{-1} u^{(1)}(x, y) + \rho^{-2} u^{(2)}(x, y) + \cdots,$$

where $y = \rho x$ and $u^{(k)}(x, y)$ are sufficiently smooth functions that are metrically transitive fields in y, for each x.

2. Show that Proposition 8.2 is implied by the following two facts:

(i) the possibility of reducing the divergence operator (8.1) to the form

$$\tilde{a}_{ij}(\rho x) \frac{\partial}{\partial z_i} \frac{\partial}{\partial z_j}$$

by passing to harmonic coordinates (8.17)–(8.19);

(ii) the general result from averaging theory [Papanicolaou and Varadhan 1982; Yurinskii 1982; Zhikov and Sirazhudinov 1981], according to which the solutions of the boundary-value and initial boundary-value problems for the operator

$$B_\rho = -a_{ij}(\rho x)\frac{\partial^2}{\partial x_i \partial x_j} + b_j(\rho x)\frac{\partial}{\partial x_j} + c(\rho x),$$

where the coefficients $k(x) = \{a_{ij}(x), b_j(x), c(x)\}$ are metrically transitive fields, converge strongly, as $\rho \to \infty$, to solutions of the same problems for the same operator, with coefficients

$$\langle k \rangle = \lim_{V \to \infty} \frac{\int_\Lambda k(x)\mu(x)\,dx}{\int_\Lambda \mu(x)\,dx},$$

where $\mu(x) \geqslant 0$ is a metrically transitive field such that

$$\frac{\partial^2}{\partial x_i \partial x_j} a_{ij}\mu = 0.$$

In other words, $\mu(x)$ is the density of the invariant distribution of the diffusion process in $\mathbf{R}^d$ with infinitesimal operator $a_{ij}\partial^2/\partial x_i \partial x_j$: a random walk in a random medium.

3. Prove that the matrix $\mathsf{E}\{a\} - \hat{a}$ of (8.16) is nonnegative.

Hint. From Jensen's inequalities and Proposition 8.2 we can write

$$\mathsf{E}\{(e^{-tA_\rho}\phi, \phi)\} \geqslant e^{-t\int_{\mathbf{R}^d}(E\{a\}\nabla\phi, \nabla\phi)\,dx}.$$

4. Prove that in dimension $d = 2$, the matrices $\hat{a}$ and $\hat{a}'$ of (8.6), corresponding to operators (8.1) with coefficients a_{ij} and $a'_{ij} = a_{ij}/\det a$, are related by the equation $\det \hat{a}\hat{a}' = 1$. In particular, if a and a' are identically distributed, $\det \hat{a} = \det \hat{a}' = 1$.

Hint. Show that, for a two-dimensional vector field $b = (b_1, b_2)$,

$$\operatorname{div} pb = -\operatorname{rot} b, \qquad \operatorname{rot} pb = \operatorname{div} b,$$

where $p = \left(\begin{smallmatrix} 0 & -1 \\ 1 & 0 \end{smallmatrix}\right)$ and div, rot are the divergence and rotation (or curl) operators. By (8.6) and (8.7), $\hat{a} = \mathsf{E}\{j\}\mathsf{E}^{-1}\{\psi\}$, where $j = a\psi$. Set $j' = -p\psi$ and $\psi' = pj$, and use (8.7b), (8.15) and the duality between div and rot applied to the columns of the matrices j and ψ.

5. Prove formulas (8.23) and (8.24).

Hint. Use the previous problem and (8.5).

6. Based on Theorem 9.6, prove that, for a Gaussian random potential,

$$\ln \tilde{N}(t) = \frac{b(0)t^2}{2}(1 + o(1))$$

as $t \to \infty$.

7. Show that if the correlation function is smooth at the origin, the asymptotic formula of Theorem 9.3 can be sharpened as follows:

$$\ln N(\lambda) = -\frac{\lambda^2}{2b(0)}(1 + O(|\lambda|^{-1/2}))$$

as $\lambda \to -\infty$.

8. Show that, for the generalized Poisson potential (1.5e), with a nonnegative function $u(x)$ satisfying the conditions of Theorem 9.4 and having continuous first and second derivatives at the origin, we have

$$\ln \tilde{N}(t) = \left(\frac{2\pi}{\alpha\gamma}\right)^{\frac{d}{2}} D^{-1/2} t^{-\alpha d/2} \exp(\gamma t^\alpha |u(0)|^\alpha)(1 + o(1)),$$

$$\ln N(\lambda) = \frac{-|\lambda|}{\gamma^{1/\alpha}|u(0)|} \ln^{1/\alpha} |\lambda|(1 + o(1)),$$

where $t \to \infty$, $\lambda \to -\infty$, the positive constants α and γ are defined by

$$\ln(-f(-t)) = \gamma t^\alpha + O(1)$$

as $t \to \infty$, and D is the Hessian

$$D = \det\left\{ \frac{\partial^2 u(x)}{\partial x_i \partial x_j} \bigg|_{x=0} \right\}.$$

Hint. Use the scheme of the proof of Theorem 9.13.

9. Show that, for the discrete analogue of the Schrödinger operator (Problem II.6), there is an estimate similar to (9.16) and (9.19), which can be interpreted as saying that the Laplace transform of $N(\lambda)$ is bounded above by the Laplace transform of the distribution function of an independent superposition of the eigenvalues of the discrete Laplacian with those of the operator of multiplication by the potential (cf. Problem II.28).

10. Consider on $\ell_2(\mathbf{Z}^d)$ the operator

$$(\hat{J}\psi)(x) = \sum_{m=1}^{d} \left(\hat{J}_m(x)\psi(x+\delta_m) + \hat{J}_m(x-\delta_m)\psi(x-\delta_m)\right),$$

where $\{\delta_m\}_1^d$ is a basis for $\mathbf{Z}^d$ and the $\hat{J}_m(x)$, for $x \in \mathbf{Z}^d$ and $m = 1, \ldots, d$ are independent, identically distributed random variables such that $0 < b \leqslant \hat{J}_m(0) \leqslant B \leqslant \infty$ and, for all $\varepsilon > 0$,

$$0 < \mathsf{P}\{\hat{J}_m(0) > B - \varepsilon\} \leqslant C\varepsilon^\alpha$$

for some $\alpha > d^{-1}$. If $J = \|\hat{J}\| - \hat{J}$, let $N(\lambda)$ be the integrated density of states of J and $N_1(\lambda)$ the function

$$N_1(\lambda) = \lim_{V\to\infty} \frac{1}{V} \sum_{\lambda_k<\lambda} \left|\sum_{x\in\Lambda} \psi_k(x)\right|^2,$$

where Λ is a cube of volume V centered at the origin of $\mathbf{Z}^d$, and λ_k and ψ_k are the eigenvalues and eigenfunctions of J_Λ. The function $N_1(\lambda)$, like $N(\lambda)$, is nonrandom and nondecreasing in λ. Prove that

$$\int_0^\infty \frac{dN(\lambda)}{\lambda} < \infty \quad \text{and} \quad \int_0^\infty \frac{dN_1(\lambda)}{\lambda^2} < \infty.$$

Hint. This is equivalent to

$$\int_0^\infty \tilde{N}(t)\, dt < \infty \quad \text{and} \quad \int_0^\infty t\tilde{N}(t)\, dt < \infty,$$

where $\tilde{N}$ and $\tilde{N}_1$ are the Laplace transforms of N and N_1. Show that

$$\tilde{N}(t) = \mathsf{E}\{K(t;0,0)\} \quad \text{and} \quad \tilde{N}_1(t) = \mathsf{E}\Big\{\sum_x K(t;0,x)\Big\},$$

where $K(t;x,y)$ is the kernel of the operator e^{-tJ}. Write J as $J(x,y) = I(x,y) + q(x)\delta_{xy}$, where

$$q(x) = \|J\| - \sum_{m=1}^{d} \left(\hat{J}_m(x) + \hat{J}_m(x-\delta_m)\right),$$

and show that (1) $\|J\| = 2dB$ and $q(x) \geqslant 0$; (2) I is the infinitesimal operator of a process $r(s)$, for $s \geqslant 0$, on the lattice $\mathbf{Z}^d$, whose transition probability we denote by $p(t;x,y)$. Then, by the Feynman-Kac formula [Reed and Simon II],

$$K(t;x,y) = p(t;x,y)\mathsf{E}_x\{e^{-\int_0^t q(r(s))\,ds} \mid x(t) = y\},$$

where E_x is the mathematical expectation with respect to the process $r(s)$ with initial condition $r(0) = x$. From this and the representations above for $\tilde{N}(t)$ and $\tilde{N}_1(t)$, conclude that

$$0 < \tilde{N}(t) \leqslant \tilde{N}_1(t) \leqslant \mathsf{E}\{e^{-tq(0)}\} = \mathsf{E}^{2d}\{e^{-t(B-\hat{J}_m(0))}\}.$$

11. Show that $N(0^+) = 0$ under the conditions of the previous problem. *Hint.* It is enough to check that $K(t;x) = \sum_y K(t;x,y)$ goes to 0 with probability 1 as $t \to \infty$. Show that, with probability 1, there exists the limit $K(x) = \lim_{t\to\infty} K(t;x)$, and that $(I+q)K = 0$. Then show that $\mathsf{E}\{q(x)K(x)\} = 0$, and hence that $q(x)K(x) = 0$, since $q(x)$ and $K(x)$ are nonnegative. Therefore $IK = 0$; conclude that $K(x) = \text{const}$ with probability 1, that is, $K(x) \equiv 0$.

12. Based on Theorems 9.1 and 9.2, show that for the potential (10.6), where the ξ_y are independent, identically distributed random variables unbounded from below, the Lifshitz exponent for $\lambda \to -\infty$ can be any positive number.

13. [Thirring 1981] Let A be a nonnegative operator on a Hilbert space and P_ψ the orthogonal projection on a vector ψ of unit norm. Show that

$$A \geqslant \frac{1}{(A^{-1}\psi, \psi)} P_\psi.$$

14. Consider the Schrödinger operator with potential (10.6), with $\mathsf{P}\{\xi_0 = 0\} = p_0 > 0$. Show that if $\lambda_0 > 0$ is small enough, $\ln N(\lambda) \leqslant -C(p_0)\lambda^{-d/2}$ for $0 \leqslant \lambda \leqslant \lambda_0$.

Hint. Use (10.10) and the previous problem, A being the potential and ψ an eigenvector with eigenvalue 0 for the Neumann problem for $-\Delta$ on the cube Λ. Then use arguments similar to those used for the proof of the upper bound in Theorem 10.1.

15. Using Problem II.6, prove the discrete version of Theorem 10.1.

16. Prove that, for a nonnegative function $u(x)$ with support in the ball of radius R_1, the functional Φ satisfies

$$\Phi(\psi) \leqslant (R + R_1)^d \omega_d,$$

where ψ has support in a sphere of radius R. Obtain from this the lower estimate for $\tilde{N}(t)$, using (10.83).

17. Prove the analogue of Theorem 9.3 for potential (10.90), where the ξ_t are Gaussian random variables.

Chapter V

Lyapunov Exponents and the Spectrum in One Dimension

We saw in Chapter III that, in the study of second-order, metrically transitive differential or finite-difference operators in dimension one, the phase (6.4) or (6.73) of the solutions $y(x)$ of the Cauchy problem can be used to derive information on the integrated density of states $N(\lambda)$. To learn more about the spectral properties of such operators, we must study the behavior of the corresponding equations for large values of the argument. In this chapter we do this in as much generality as possible—that is, for arbitrary one-dimensional metrically transitive operators of the second order, without making additional assumptions on the coefficients unless absolutely essential.

Even in such a general setting the solutions behave quite regularly: we show in Section 11 that the envelope of $y(x)$, for large values of x, behaves as $e^{\pm\gamma(\lambda)|x|(1+o(1))}$, where the nonnegative and nonrandom quantity $\gamma(\lambda)$ is called the Lyapunov exponent. Thus the Lyapunov exponent is important in spectral theory: it governs the asymptotic behavior of the solutions, and therefore also the character of the spectrum. Section 11 also explains the relationship between the Lyapunov exponent and $N(\lambda)$: it turns out that $\gamma(\lambda) + iN(\lambda)$, seen as a complex function on $\mathbf{R}$, can be extended to an analytic function on the upper half-plane.

In Section 12, we show that the absolutely continuous spectrum σ_{ac} coincides, within a set of Lebesgue measure zero, with the set $\{\lambda \in \mathbf{R} : \gamma(\lambda) = 0\}$; recall that for an operator with periodic coefficients, $\sigma = \sigma_{\mathrm{ac}}$. We also give conditions for the Lyapunov exponent to be positive, and therefore for the absolutely continuous component to be absent.

Section 13 discusses some general mathematical techniques for obtaining the point spectrum. In particular, we investigate criteria which, allied to the positiveness of the Lyapunov exponent, ensure that the spectrum is pure point and that the eigenfunctions decrease exponentially.

We will mainly deal with the simplest, but very interesting, case of a one-dimensional Schrödinger equation (6.1) and its discrete analogue (6.67) for $s(x) \equiv 1$. The changes needed to deal with more general operators of the form (1.4) and (1.6) and multicomponent operators of the form (11.86) and (12.41) will be briefly discussed toward the end of the respective subsections.

11 Existence and Properties of Lyapunov Exponents

11.A The Multiplicative Ergodic Theorem and the Existence of Lyapunov Exponents

We consider the phase $\theta(x)$ of the solution of the Cauchy problem (6.3) for the Schrödinger equation (6.1) as the angular coordinate of the point (y', y) in the "phase plane." The other polar coordinate of this point,

$$r(x) = \sqrt{|y(x)|^2 + |y'(x)|^2}, \tag{11.1}$$

is called the *amplitude*, and it clearly satisfies

$$y(x) = r(x)\sin\theta(x), \qquad y'(x) = r(x)\cos\theta(x). \tag{11.2}$$

While $\theta(x)$ is responsible for the oscillatory properties of $y(x)$, the amplitude plays the role of the envelope of $y(x)$, governing its growth or decay as $|x| \to \infty$.

Likewise, in the discrete case the envelope of the solution $y(x)$ of the Cauchy problem

$$y(0) = \sin\alpha, \qquad y(1) = \cos\alpha, \qquad \alpha \in [0, \pi) \tag{11.3}$$

for the equation

$$-\psi(x+1) - \psi(x-1) + q(x)\psi(x) = \lambda\psi(x) \tag{11.4}$$

is the quantity

$$r(x) = \sqrt{|y(x)|^2 + |y(x+1)|^2}. \tag{11.5}$$

Given the standard results about the solutions of differential and finite-difference equations with constant and periodic coefficients—the simplest examples of metrically transitive functions—it seems reasonable to expect exponentials to provide a useful scale of measurement for the growth of the solutions of (6.1) and (11.4). Therefore the following definition is natural:

Definition. *Let the solution to the Cauchy problem defined by* (6.1) *and* (6.3), *or* (11.3)–(11.4), *with metrically transitive potential* $q(x)$, *have amplitude* $r(x)$ *given by* (11.1) *or* (11.5). *Suppose the limits*

$$\gamma_{\pm}(\lambda, \omega, \alpha) = \lim_{\pm\infty} \frac{1}{|x|} \ln r(x) \tag{11.6}$$

exist for every $\lambda \in \mathbf{R}$ *and* $\alpha \in [0, \pi)$, *and every* ω *in a set of realizations of full measure, possibly depending on* λ. *The numbers* $\gamma_{\pm}(\lambda, \omega, \alpha)$ *are called the Lyapunov characteristic exponents of the solution (in the positive and negative direction).*

Lyapunov characteristic exponents are very important in the theory of ordinary differential and finite-difference equations, and in particular in stability theory [Bylov et al. 1966]. In the general case of arbitrary coefficients one must define upper and lower exponents, replacing the lim in (11.6) by lim sup and lim inf; then, if the upper and lower exponents coincide, we say the Lyapunov characteristic exponents exist. For metrically transitive coefficients, which have a certain homogeneity, the upper and lower exponents always coincide.

A simple calculation shows that for a constant potential $q(x) \equiv q$ the Lyapunov exponents $\gamma_{\pm}(\lambda, \alpha)$ exist and take on the two values $\pm\gamma_0(\lambda)$, where

$$\gamma_0(\lambda) = \sup_{\alpha\in[0,\pi)} \gamma(\lambda, \alpha) \geqslant 0. \tag{11.7}$$

In the continuous case, we have

$$\gamma_0(\lambda) = \sqrt{\max\{q - \lambda, 0\}}, \tag{11.8}$$

while in the discrete case,

$$\gamma_0(\lambda) = \operatorname{arccosh} \max\{\tfrac{1}{2}|q - \lambda| - 1, 0\}. \tag{11.9}$$

Furthermore, if $\gamma_0(\lambda) > 0$,

$$\gamma_{\pm}(\lambda, \alpha) = \gamma_0(\lambda) \qquad \text{for } \alpha \neq \alpha_0(\lambda), \tag{11.10}$$

$$\gamma_{\pm}(\lambda, \alpha) = -\gamma_0(\lambda) \qquad \text{for } \alpha = \alpha_0(\lambda), \tag{11.11}$$

where the exceptional value $\alpha_0(\lambda)$ corresponds to a solution whose increasing exponential has coefficient zero: for example, in the continuous case, $\alpha_0(\lambda) = -\arctan\sqrt{\max\{q - \lambda, 0\}}$.

For a periodic potential, one can prove similar results using Floquet theory: see [Titchmarsh 1961] and [Marchenko 1986], which treat the Schrödinger equation. The case of finite-difference equations can be considered similarly [Toda 1981].

Even these well-studied cases illustrate the role of Lyapunov characteristic exponents in spectral theory. Indeed, the spectrum of a Schrödinger operator with a periodic potential can be defined as the set $\{\lambda \in \mathbf{R} : \gamma(\lambda) = 0\}$, and is purely absolutely continuous. The profound relationships between $\gamma(\lambda)$ and the spectral theory of periodic Schrödinger operators, especially the role of the analytic continuation of $\gamma(\lambda)$ to the upper half-plane in the inverse problem of spectral analysis, are discussed in [Marchenko 1986]; see also Sections 16 and 17.

Other applications of Lyapunov characteristic exponents arise under the assumption of randomness of the coefficients: if the statistical correlation of the coefficients decays fast enough, $\gamma(\lambda)$ is, as a rule, positive on the spectrum. For all these reasons, characteristic exponents are central to the spectral analysis of one-dimensional metrically transitive operators and deserve to be studied in detail.

We start by providing another definition for characteristic exponents, which also works for complex values of the spectral parameter $z = \lambda + i\varepsilon$. We rewrite the second-order equations (6.1) and (11.4) as systems of two first-order equations: in the continuous case,

$$Y'(x) = A(x)Y(x), \qquad Y(x) = \begin{pmatrix} y'(x) \\ y(x) \end{pmatrix} \tag{11.12}$$

with

$$A(x) = \begin{pmatrix} 0 & q(x) - z \\ 1 & 0 \end{pmatrix}; \tag{11.13}$$

and, in the discrete case,

$$Y(x+1) = A(x+1)Y(x), \qquad Y(x) = \begin{pmatrix} y(x+1) \\ y(x) \end{pmatrix} \tag{11.14}$$

with

$$A(x) = \begin{pmatrix} q(x) - z & -1 \\ 1 & 0 \end{pmatrix}. \tag{11.15}$$

In this section we assume the metrically transitive process $q(x)$ to satisfy

$$\mathsf{E}\{|q(x)|\} < \infty \tag{11.16a}$$

in the continuous case (6.1), and

$$\mathsf{E}\{\ln(1 + |q(x)|)\} < \infty \tag{11.16b}$$

in the discrete case (11.4).

We denote by $\mathcal{T}(x_2, x_1)$ the *fundamental matrix*, or *transfer matrix*, of these linear systems, that is, the second-order matrix satisfying

$$Y(x_2) = \mathcal{T}(x_2, x_1)Y(x_1). \tag{11.17}$$

As a consequence of (6.1) and (11.4), we have

$$\frac{\partial \mathcal{T}(x_2, x_1)}{\partial x_2} = A(x_2)\mathcal{T}(x_2, x_1), \qquad \mathcal{T}(x_1, x_1) = I \tag{11.18}$$

in the continuous case, and, in the discrete case,

$$\mathcal{T}(x_2+1,x_1)=A(x_2)\mathcal{T}(x_2,x_1),\qquad \mathcal{T}(x_1,x_1)=I. \tag{11.19}$$

Equations (11.1), (11.5) and (11.6) show that the characteristic exponents can be represented, for any $z \in \mathbf{C}$, by

$$\gamma_{\pm}(z,\omega,\alpha)=\lim_{x\to\pm\infty}\frac{1}{|x|}\ln\|\mathcal{T}(x,0)Y(0)\| \tag{11.20}$$

where $Y(0)=\binom{\cos\alpha}{\sin\alpha}$ and $\|\cdot\|$ denotes the Euclidean norm in $\mathbf{C}^2$.

We will write $\mathcal{T}(x_2,x_1,\omega)$ when we need to make explicit the dependence of the transfer matrix on the realizations of the metrically transitive potential.

Here are some important properties of the transfer matrix:

$$\mathcal{T}(x_3,x_1,\omega)=\mathcal{T}(x_3,x_2,\omega)\mathcal{T}(x_2,x_1,\omega),\qquad \mathcal{T}(x,x,\omega)=I, \tag{11.21}$$

and, in particular,

$$\mathcal{T}(x_2,x_1,\omega)\mathcal{T}(x_1,x_2,\omega)=I. \tag{11.21a}$$

Also,

$$\mathcal{T}(x_2+a,x_1+a,\omega)=\mathcal{T}(x_2,x_1,T_a\omega). \tag{11.22}$$

These relations follow from the fact that the Cauchy problem (6.1) and (6.3), or (11.3)–(11.4), admits a unique solution under conditions (11.16), and from the metric transitivity condition $q(x,T_a\omega)=q(x+a,\omega)$.

If we set

$$\mathcal{T}(x,\omega)=\mathcal{T}(x,0,\omega),$$

equations (11.21) and (11.22) give

$$\mathcal{T}(x_1+x_2,\omega)=\mathcal{T}(x_1,T_{x_2}\omega)\mathcal{T}(x_2,\omega). \tag{11.23}$$

A family $\mathcal{T}(x,\omega)$ of $m\times m$ matrices of nonzero determinant, defined on an abstract dynamical system $(\Omega,\mathcal{F},\mathsf{P},T_x)$, and such that $\mathcal{T}$ is a measurable function of (x,ω) with respect to the product σ-algebra of the Borel σ-algebra in $\mathbf{R}$ and $\mathcal{F}$, is called a *multiplicative cocycle* if it satisfies (11.23) [Oseledec 1968].

Every metrically transitive process $f(x,\omega)$ gives rise to an associated scalar (one-dimensional) multiplicative cocycle $\exp\bigl(\int_0^x f(t,\omega)\,dt\bigr)$. Thus we can consider multiplicative cocycles as a generalization of metrically transitive processes, arising naturally when we consider linear equations with metrically transitive coefficients.

From now on we will restrict our attention to unimodular matrices, that is, those of determinant 1. This assumption is not restrictive in considering linear systems, since any multiplicative cocycle becomes unimodular after being divided by the scalar cocycle $\det \mathcal{T}(x_2, x_1)$. In any case, the transfer matrices (11.12) and (11.14) are already unimodular.

We now reach the central point of this subsection: the study of asymptotic properties of transfer matrices as $|x_2 - x_1| \to \infty$. We start with the following statement, which goes back in essence to Kesten and Furstenberg [1960].

(11.1) Lemma. *For a metrically transitive potential given by equation* (6.1) *or* (11.4) *and satisfying* (11.16), *let* Ω_z, *for a given* $z \in \mathbf{C}$, *be the set of realizations such that the limits*

$$\lim_{x\to\pm\infty} \frac{1}{|x|} \ln\left\|\mathcal{T}(x,0,\omega)\right\| \tag{11.24}$$

exist and are finite, nonrandom and equal. Then Ω_z *has full measure, and the limits equal*

$$\gamma(z) = \lim_{|x|\to\infty} \frac{1}{|x|}\mathsf{E}\{\ln\|\mathcal{T}(x,0,\omega)\|\}. \tag{11.25}$$

In addition, we have

$$\gamma(z) \geqslant 0 \tag{11.26}$$

and

$$\gamma(z) \leqslant \mathsf{E}\{\hat{q}\}, \tag{11.27a}$$

$$\gamma(z)\tfrac{1}{2}\mathsf{E}\{\ln(2+\hat{q}^2)\}, \tag{11.27b}$$

in the continuous and discrete case, respectively, where $\hat{q} = \max\{|q(0) - z|, 1\}$.

Proof. From the general relations (11.21) and (11.22) and the inequality $\|AB\| \leqslant \|A\|\,\|B\|$ for matrix norms, it follows that the quantities $\ln\left\|\mathcal{T}(x_2,x_1,\omega)\right\|$ form a subadditive random process, similar to the one considered in Section 6.A for $N(\lambda)$. Thus the existence, finiteness and nonrandomness of the limits (11.24), as well as formula (11.25), will follow from (Proposition 6.3), the subadditive ergodic theorem, if we verify that

$$\mathsf{E}\{\ln\|\mathcal{T}(x,0,\omega)\|\} \geqslant C|x|$$

uniformly in x. In the discrete case, this follows from (11.16b), (11.21a) and the relations

$$(11.28) \qquad \mathcal{T}(x_2, x_1) = A(x_2 - 1) \ldots A(x_1 + 1) A(x_1) \qquad \text{if } x_2 > x_1,$$

$$(11.29) \qquad \|A^{-1}\| \leqslant \sqrt{2 + \hat{q}^2}, \qquad \|A\| \leqslant \sqrt{2 + \hat{q}^2}.$$

In the continuous case it follows from the Bellman–Gronwall estimate [Hartman 1964],

$$(11.30) \qquad \big\|\mathcal{T}(x_2, x_1)\big\| \leqslant \exp\left|\int_{x_1}^{x_2} \big\|A(t)\big\|\, dt\right|,$$

together with (11.6a) and the inequality $\|A\| \leqslant \hat{q}$ for the norm of matrices (11.13). The same inequalities also lead to the estimates (11.27).

Now notice that (11.21a), together with the metric transitivity of the potential, gives

$$(11.31) \qquad \mathsf{E}\{\ln \|\mathcal{T}(x, 0)\|\} = \mathsf{E}\{\ln \|\mathcal{T}^{-1}(-x, 0)\|\}$$

for $x < 0$. But it is easy to see that the norm of a 2×2 matrix of determinant 1 equals the norm of its inverse; therefore the limits (11.24) coincide for $x \to \pm\infty$. Finally, the fact that $\gamma(z)$ is nonnegative follows from the fact that the norm of a unimodular matrix cannot be less than 1. □

The main assertion of Lemma 11.1, the existence of the limits in (11.24), is a particular case of the following general result by Furstenberg and Kesten [1960]: if $\mathcal{T}(x, \omega)$, for $x \in \mathbf{Z}$, is a multiplicative cocycle satisfying

$$\mathsf{E}\{\ln(1 + \|\mathcal{T}(1, \omega)\|)\} < \infty,$$

the limit of the sequence $x^{-1} \ln\big\|\mathcal{T}(x, \omega)\big\|$ exists with probability 1. (Such a cocycle can always be written in the form

$$\mathcal{T}(x, \omega) = A(x - 1, \omega) A(x - 2, \omega) \ldots A(0, \omega),$$

where $A(x, \omega) = \mathcal{T}(1, T_x\omega)$ is a metrically transitive matrix-valued process on $\mathbf{Z}$.) The method of our proof, based on the subadditive ergodic theorem and equally applicable to the continuous and discrete cases, was proposed by Derriennic [1975].

In view of Lemma 11.1, it is natural to pose the following definition:

Definition. *The Lyapunov exponent of equations* (6.1) *and* (11.4), *or of the systems* (11.12) *and* (11.14), *is the nonnegative quantity*

$$\gamma(z) = \lim_{|x| \to \infty} \frac{1}{|x|} \mathsf{E}\{\ln \|\mathcal{T}(x, 0, \omega)\|\}.$$

Remark. Let $\mu(x,\omega)$ denote the maximum eigenvalue of the matrix $(\mathcal{T}^*\mathcal{T})^{1/2}$, where $\mathcal{T}^*$ is the Hermitian conjugate of $\mathcal{T}$. By Lemma 11.1, the Lyapunov exponent can be written

$$\gamma(z) = \lim_{|x|\to\infty} \frac{1}{|x|} \ln \mu(x,\omega) = \lim_{|x|\to\infty} \frac{1}{|x|} \mathsf{E}\{\ln \mu(x,\omega)\}. \tag{11.32}$$

The Lyapunov exponent indicates the rate of exponential growth of the norm of the fundamental matrix of equation (6.1) or (11.4), but it is not sufficient for characterizing the growth of its solutions, because the solutions depend on the vector of initial values, as shown by (11.17). The solutions are more fully characterized by the Lyapunov characteristic exponents of (11.6) and (11.20). That these quantities exist is a consequence of the following important result, called the *multiplicative ergodic theorem*:

(11.2) Proposition [Oseledec 1968]. *Let $\mathcal{N}(x,\omega)$ be a $\mathrm{GL}(m,\mathbf{R})$-valued multiplicative cocycle on a dynamical system $(\Omega,\mathcal{F},\mathsf{P},T_x)$, where $x\in\mathbf{Z}$ or $\mathbf{R}$. If*

$$\mathsf{E}\{\sup_{|x|\leqslant 1} \max\{\ln\|\mathcal{N}(x,\omega)\|,0\}\} < \infty, \tag{11.33}$$

there exists a T_x-invariant set $\Omega_1\subset\Omega$ of full P-measure and such that, for every $\omega\in\Omega_1$, the following statements hold:

(i) *If $\mu_1(x,\omega)\geqslant\cdots\geqslant\mu_m(x,\omega)$ are the eigenvalues of the matrix $(\mathcal{N}^T\mathcal{N})^{1/2}$ (cf. (11.32)), the limits*

$$\gamma_i = \lim_{x\to\pm\infty} \frac{1}{|x|} \ln \mu_i(x,\omega) \tag{11.34}$$

for $i=1,\ldots,m$, exist and are nonrandom.

(ii) *If the distinct values of the limits in (i) are denoted by $\chi_1>\chi_2>\cdots>\chi_r$, for $r\leqslant m$, and their multiplicities by k_i, we can decompose $\mathbf{R}^m$ into direct summands $L_\pm^{k_i}$ of dimension k_i, where for almost every $\omega\in\Omega$ we have*

$$\lim_{x\to\pm\infty} \frac{1}{|x|} \ln\big\|\mathcal{N}(x,\omega)e\big\| = \chi_i, \tag{11.34a}$$

for all $e\in L_\pm^{k_i}$, the convergence being uniform in e.

The set of pairs (χ_i,k_i) is called the *Lyapunov cocycle spectrum*. It can also be defined as the spectrum of the random symmetric matrix $\mathcal{M} = \lim_{x\to\infty}\big(\mathcal{N}^T(x,\omega)\mathcal{N}(x,\omega)\big)^{1/2}$ [Ruelle 1979].

Proof. We will only sketch the proof. First one deals with the case of triangular cocycles consisting of lower triangular matrices $\mathcal{N}$ with positive elements; using arguments that go back to Lyapunov, one can reduce to the ordinary (additive) ergodic theorem. Then one shows that a general multiplicative cocycle can be reduced to triangular form by transformations of the form

$$\mathcal{N}(x,\omega) \to C^{-1}(T_x\omega)\mathcal{N}(x,\omega)C(\omega),$$

where $C(\omega)$ is a measurable $\mathrm{GL}(m,\mathbf{R})$-valued function on Ω, in such a way that the limits are not changed.

Millionsčikov [1969] gave another proof of this proposition. Yet another proof, conceptually and technically transparent, is due to Raghunathan [1979]; see also [Ruelle 1979b]. It works by reducing Proposition 11.2 to the subadditive ergodic theorem (Proposition 6.3), and is based on a simple lemma, essentially about linear algebra (Problems V.6–7): if $\mathcal{N}(x)$ is a $\mathrm{GL}(m,\mathbf{R})$-valued function on $\mathbf{R}$ or $\mathbf{Z}$ such that

$$\lim_{x\to\infty} \frac{1}{x} \sup_{0\leqslant\,\delta\leqslant\,1} \left|\ln \|\mathcal{N}(x+\delta)\mathcal{N}^{-1}(x)\|\right| = 0, \tag{11.35}$$

and if the limits

$$\lim_{x\to\infty} \frac{1}{x} \ln \mu_i(x) = \gamma_i \tag{11.36}$$

exist for $i = 1,\ldots,m$, where $\mu_1(x) \geqslant \cdots \geqslant \mu_m(x)$ are the eigenvalues of $\left(\mathcal{N}^T(x)\mathcal{N}(x)\right)^{1/2}$, there exists a limit matrix

$$\mathcal{M} = \lim_{x\to\infty} \left(\mathcal{N}^T(x)\mathcal{N}(x)\right)^{1/(2x)}$$

whose eigenvalues equal e^{γ_i}, and there is a decomposition as in part (ii) of Proposition 11.2, each summand being the eigenspace of $\mathcal{M}$ corresponding to an eigenvalue e^{γ_i}.

The proof is completed by observing that the existence of the limits (11.34) is equivalent to the existence of the limits

$$\lim_{x\to\infty} \frac{1}{x} \ln\left\|\mathcal{N}(x)^{\wedge q}\right\|,$$

for $q = 1,\ldots,m$, where $\mathcal{N}(x)^{\wedge q}$ is the q-th exterior power of $\mathcal{N}$. Thus (11.35) and (11.36) follow from the subadditive ergodic theorem (Proposition 6.3). □

Remarks. 1. Proposition 11.2 also holds for complex matrices in $\mathrm{GL}(m,\mathbf{C})$, if instead of transposes we take Hermitian conjugates. In fact, the complex case can be reduced to the real one in dimension twice as high.

2. For the current state of knowledge on the multiplicative ergodic theorem, see [Arnold and Wihstutz 1986] and references therein.

(11.3) Theorem. *Consider a metrically transitive process* $q(x)$, *for* $x \in \mathbf{R}$ *or* $\mathbf{Z}$, *satisfying* (11.16). *For every value of the spectral parameter* $z \in \mathbf{C}$, *there exists a set of realizations* Ω_z *of full* P*-measure such that, for every* $\omega \in \Omega_z$, *both characteristic Lyapunov exponents* (11.20) *exist for all values of* $\alpha \in [0, \pi)$ *in the initial condition* (6.3) *or* (11.3). *Each of them takes on no more than two values* $\pm\gamma(z)$, *where the nonnegative* $\gamma(z)$ *is the Lyapunov exponent of equations* (6.1) *or* (11.4), *and is given by* (11.32). *The value* $-\gamma(z)$ *occurs only for a single value* $\alpha_\pm(z, \omega)$ *of the angle* α:

$$\gamma_\pm(z, \omega, \alpha_\pm(z, \omega)) = -\gamma(z). \tag{11.37}$$

Proof. By (11.24), the fundamental matrices $\mathcal{T}(x, 0)$ of (11.12) and (11.14) form a multiplicative cocycle, which satisfies (11.33) if (11.16) holds. The theorem follows. □

Thus, as far the existence of characteristic exponents is concerned, there is a fairly general class of metrically transitive potentials that behave like constant and periodic potentials.

11.B The Lyapunov Exponent and the Integrated Density of States

We now discuss the properties of the Lyapunov exponent that are used in the spectral theory of metrically transitive operators. One of these properties, whose importance was demonstrated in [Herman 1983] and [Craig and Simon 1983], is subharmonicity.

An upper semicontinuous function $f(\zeta)$ defined on a region $\mathcal{D} \subset \mathbf{C}$ is called *subharmonic* if, for any $\zeta \in \mathcal{D}$, there exist arbitrarily small numbers $\rho > 0$ such that

$$f(\zeta) \leqslant \frac{1}{2\pi} \int_0^{2\pi} f(\zeta + \rho e^{i\phi})\, d\phi. \tag{11.38}$$

We will need the following properties of subharmonic functions [Hayman and Kennedy 1980]:

(i) If $f(\zeta)$ is analytic on $\mathcal{D}$, the functions $\ln|f(\zeta)|$ and $|f(\zeta)|^\alpha$, for $\alpha > 0$, are subharmonic on $\mathcal{D}$.

(ii) A convex combination of subharmonic functions is subharmonic.

(iii) If $\{f_\nu(\zeta)\}$ is a family of subharmonic functions and $f(\zeta) = \sup_\nu f_\nu(\zeta)$ is upper semicontinuous, $f(\zeta)$ is also subharmonic; if $\{f_\nu(\zeta)\}$ is a finite family, $f(\zeta)$ is always subharmonic.

(iv) If $f_n(\zeta)$, for $n \in \mathbf{Z}_+$, is a nonincreasing sequence of subharmonic functions, the function $f(\zeta) = \lim_{n\to\infty} f_n(\zeta)$ is also subharmonic.

(11.4) Lemma [Craig and Simon 1983b; Herman 1983]. *Suppose the transfer matrix $\mathcal{T}(x,\omega)$ is, with probability* 1, *an analytic function of some parameter $\zeta \in \mathcal{D}$ for every x. Then the Lyapunov exponent $\gamma(z)$ of* (11.32) *is a subharmonic function of this parameter.*

Proof. By properties (i)–(iii), $\ln\|\mathcal{T}(x,\omega)\|$ is a subharmonic function of ζ. By (11.23), the quantities $\gamma_l = 2^{-l}\mathsf{E}\{\ln\|\mathcal{T}(2^l,\omega)\|\}$ form a monotonically decreasing sequence. The lemma follows by property (iv) and Lemma 11.1. □

(11.5) Theorem [Craig and Simon 1983b]. *The Lyapunov exponent* (11.32) *is a subharmonic function of the spectral parameter $z \in \mathbf{C}$.*

Proof. In the discrete case this is a consequence of Lemma 11.4, formula (11.28) and the form (11.15) of the matrices $A(x)$. In the continuous case, one shows that $\mathcal{T}(x,\omega)$ is analytic by solving (11.18) by successive approximations. □

Theorem 16.13 provides another important application of Lemma 11.4, to the estimation from below of the Lyapunov exponent of the almost-Mathieu equation.

The following example of a subharmonic function will be important in the sequel:

$$f(\zeta) = \int_{\mathbf{R}} \ln|\zeta - \mu| m(d\mu), \tag{11.39}$$

where $m(d\mu)$ is a nonnegative measure on $\mathbf{R}$ satisfying the condition

$$\int_{\mathbf{R}} \ln(1+|\mu|) m(d\mu) < \infty,$$

and where, by convention, $f(\zeta) = -\infty$ if the integral diverges to $-\infty$.

We now discuss the relationship, originally published in the physics literature [Herbert and Jones 1971; Thouless 1972], between the Lyapunov exponent of a second-order, one-dimensional equation with metrically transitive coefficients and the integrated density of states of the corresponding operator. This relationship is fairly simple: as we saw in Sections 6 and 7, the zeros λ_i of the solution $y(x,\lambda)$, for x fixed, of the Cauchy problems for equations of order two, are eigenvalues of the self-adjoint boundary problems (6.1)–(6.2) or (6.67) and (6.70), for $a = 0$, $b = x$ and $\beta = 0$. Therefore, if $N_x(\lambda)$ is x^{-1} times the number of eigenvalues $\lambda_i \leqslant \lambda$, we can write

$$\frac{1}{x}\ln\bigl|y(x,\lambda)\bigr| = \frac{1}{x}\sum_i \ln|\lambda-\lambda_i| = \int_{\mathbf{R}} \ln|\lambda-\mu|\, N_x(d\mu). \tag{11.40}$$

If we take the formal limit $x \to \infty$ in this expression, using Theorems 6.4 and 6.18 on the existence, with probability 1, of the weak limit of the measures N_x and Theorem 11.3 on the existence, with probability 1, of the limit of the quantity $x^{-1}\ln r(x)$ in (11.1) and (11.5), we formally obtain

$$\gamma(\lambda,\omega,\alpha) = \int_{\mathbf{R}} \ln|\lambda-\mu|\, N(d\mu). \tag{11.41}$$

Since the right-hand side is independent of α and ω with probability 1 (Theorems 6.1 and 6.18), $\gamma(\lambda,\omega,\alpha)$ must coincide, on a sufficiently big set of values of these variables, with the Lyapunov exponent $\gamma(\lambda)$.

Making these arguments rigorous is nontrivial for a number of reasons. First, since $\ln|\lambda-\mu|$ is not continuous, we cannot take the limit on the right-hand side of (11.41) based just on the weak convergence of the measures N_x. Second, by Theorem 6.5, the integrated density of states of the Schrödinger operator behaves as $\sqrt{\lambda}/\pi$ as $\lambda \to \infty$; therefore (11.41) cannot be true in the continuous case, because the integral on the right-hand side requires some regularization. Finally, unlike the amplitude $r(x)$ that is used in the definition (11.6) of characteristic exponents, the solution of the Cauchy problem may quite often be zero (see Proposition 6.1, for example), so the replacement of $x^{-1}\ln\bigl|y(x,\lambda)\bigr|$ by $x^{-1}\ln r(x)$ also requires justification.

A derivation of (11.41) following the outline above is given in [Pastur and Figotin 1984a]; a similar result was obtained by Avron and Simon [1983]. The theorem below essentially follows the paper of Pastur and Figotin.

(11.6) Theorem. *Let $\gamma_\pm(z,\omega,\alpha)$ be the characteristic Lyapunov exponent of the Schrödinger operator* (6.1) *or* (11.4) *with a metrically transitive potential satisfying* (11.16). *Assume that, in the continuous case,*

$$\mathsf{E}\{q^2(x)\} < \infty. \tag{11.42}$$

Let $N(\lambda)$ be the integrated density of states of the corresponding metrically transitive operator, and set $\Omega_{\mathbf{C}} = \mathbf{C}\times\Omega\times[0,\pi)$ and $\Omega_{\mathbf{R}} = \mathbf{R}\times\Omega\times[0,\pi)$, where the first two factors are given the Lebesgue measure and the third the probability measure P.

In the continuous case (6.1), *there exists for any z_0 with $\operatorname{Im} z_0 \neq 0$ a set $\Omega_{\mathbf{C},\mathbf{R}}(z_0)$ of full measure such that on it we have*

$$\gamma_\pm(z,\omega,\alpha) = \gamma_\pm(z_0,\omega,\alpha) + \int_{\mathbf{R}} \ln\left|\frac{z-\mu}{z_0-\mu}\right| N(d\mu). \tag{11.43a}$$

In the discrete case (11.4), *for almost all points of $\Omega_{\mathbf{C}}$ and $\Omega_{\mathbf{R}}$,*

$$\gamma_{\pm}(z,\omega,\alpha) = \int_{\mathbf{R}} \ln|z-\mu|\, N(d\mu). \tag{11.43b}$$

Proof. We start with the continuous case, which is more complicated, then see how the arguments used can be applied to the discrete case. We consider only the limit $x \to \infty$, since the argument for $x \to -\infty$ is identical.

Let $y(x,z)$ be the solution of the Cauchy problem (6.1) and (6.3), for $a = 0$ and $\lambda \in \mathbf{R}$ replaced by $z \in \mathbf{C}$. Using the local summability of the potential, which is a consequence of (11.42), one easily sees that $y(x,z)$ and $y'(x,z) = \partial y(x,z)/\partial x$, for each fixed x, are entire functions of z, whose zeros $\lambda_n^{(r)}$, for $n \geqslant 1$ and $r = 0, 1$, are real, simple and coincide with the eigenvalues of the boundary problems (6.1)–(6.2) for $\beta = 0, \pi/2$, $a = 0$, $b = x$ and $r = 0, 1$, respectively. If $N_x^{(r)}(\lambda)$ denotes x^{-1} times the number of eigenvalues not exceeding λ, the oscillation theorem (Proposition 6.1) and the phase $\chi(x)$ of (6.14) can be used to give, for $\lambda \geqslant 0$ and $r = 0, 1$,

$$N_x^{(r)}(\lambda) \leqslant \frac{\sqrt{\lambda}}{\pi} + \frac{2}{x} + \frac{1}{\pi\sqrt{\lambda}x}\int_0^x |q(t)|\, dt, \tag{11.44a}$$

$$N(\lambda) \leqslant \frac{\sqrt{\lambda}}{\pi} + \frac{1}{\pi\sqrt{\lambda}}\mathsf{E}\{|q(0)|\}, \tag{11.44b}$$

where $N(\lambda)$ is the integrated density of states of the Schrödinger operator $-d^2/dx^2 + q$. From (11.44a), by standard theorems of function theory [Marchenko 1986], we find, setting $y^{(0)} = y$ and $y^{(1)} = y'$, that

$$\frac{1}{y^{(r)}(x,z)}\frac{\partial}{\partial z}y^{(r)}(x,z) = \sum_{n\geqslant 1}\frac{1}{z-\lambda_n^{(r)}}, \tag{11.45a}$$

$$\frac{1}{y^{(r)}(x,z)}\frac{\partial}{\partial z}y^{(r)}(x,z) = \int \frac{xN_x^{(r)}(d\mu)}{z-\mu}, \tag{11.45b}$$

where the integral or sum converges uniformly on compact subsets of $\mathbf{C}$ that do not contain points $\lambda_n^{(r)}$.

Integrating (11.45) along a path connecting the points z and z_0 we obtain, for the real part of the resulting expression (in the continuous case):

$$\delta_x^{(r)}(z) = \delta_x^{(r)}(z_0) + \int_{\mathbf{R}} \ln\left|\frac{z-\mu}{z_0-\mu}\right| N_x^{(r)}(d\mu), \tag{11.46}$$

where

$$\delta_x^{(r)}(x) = \frac{1}{x}\ln\left|y^{(r)}(x,z)\right|. \tag{11.47}$$

Introduce the quantities

(11.48) $$\gamma_x(z) = \frac{1}{2x} \ln\big(|y(x,z)|^2 + |y'(x,z)|^2\big),$$

and notice that

(11.49) $$\big|\gamma_x - \max_{r=0,1} \delta_x^{(r)}\big| \leqslant \frac{\ln\sqrt{2}}{x}.$$

Combining (11.46) and (11.49) and using the obvious identities $\max\{a_1 + b, a_2 + b\} = \max\{a_1, a_2\} + b$ and $\big|\max\{a_1 + b_1, a_2 + b_2\} - \max\{a_1, a_2\}\big| \leqslant \max\{|b_1|, |b_2|\}$, we obtain

(11.50) $$\left|\gamma_x(z) - \gamma_x(z_0) - \int_{\mathbf{R}} \ln\left|\frac{z-\mu}{z_0-\mu}\right| N(d\mu)\right| \leqslant \max_{r=0,1}\left\{\left|\int_{\mathbf{R}} \ln\left|\frac{z-\mu}{z_0-\mu}\right| \big(N(d\mu) - N_x^{(r)}(d\mu)\big)\right|\right\} + \frac{\ln 2}{x}.$$

(11.7) Lemma. *Let $\mathcal{M}_c$, for $0 < c < \infty$, be the set of positive measures $m(d\lambda)$ on $\mathbf{R}$, such that*

(11.51) $$m\big((-\infty, \lambda]\big) \leqslant c\sqrt{\lambda}$$

for $\lambda \geqslant 1$. Fix a complex number z_0 with $\operatorname{Im} z_0 \neq 0$, and consider, for every $m \in \mathcal{M}_c$, the function

(11.52) $$\tau(z,m) = \tau(z) = \int_{\mathbf{R}} \ln\left|\frac{z-\mu}{z_0-\mu}\right| m(d\mu)$$

for $z \in \mathbf{C}$.

(i) *The family $\tau(z,m)$, for $m \in \mathcal{M}_c$, is uniformly bounded and equicontinuous on any compact K in the region $\operatorname{Im} z \neq 0$.*

(ii) *For every compact K of $\mathbf{R}$ or $\mathbf{C}$ there are positive constants c_K depending only on c and K and such that*

(11.53) $$-\infty \leqslant \tau(z,m) \leqslant c_K$$

for $z \in K$, and

(11.54) $$\int_K \big|\tau(\lambda,m)\big|\, d\lambda \leqslant c_K, \qquad \int_K \big|\tau(z,m)\big|\, d^2z \leqslant c_K,$$

where d^2z denotes the Lebesgue measure on the plane.

(iii) *If $\lim_{x\to\infty} m_x = m$, where $m_x \in \mathcal{M}_c$ and the limit is with respect to weak convergence of measures (this implies $m \in \mathcal{M}_c$ as well), and if $\tau_x(z) = \tau(z, m_x)$, we have*

(11.55) $$\lim_{x\to\infty} \tau_x(z) = \tau(z)$$

if $\operatorname{Im} z \neq 0$ *and*

$$\lim_{x\to\infty} \int_K \left|\tau_x(\lambda) - \tau(\lambda)\right| d\lambda = 0, \tag{11.56}$$

$$\lim_{x\to\infty} \int_K \left|\tau_x(z) - \tau(z)\right| d^2 z = 0 \tag{11.57}$$

for any compact K *in* $\mathbf{R}$ *or* $\mathbf{C}$.

Proof. All statements up to and including (11.55) follow immediately from condition (11.51) and from the inequalities $\ln t \leqslant \ln_+ t$,

$$\ln\left|\frac{z-\mu}{z_0-\mu}\right| \leqslant \frac{\text{const}}{|\mu|}$$

as $|\mu| \to \infty$, and

$$\int_{\mathbf{R}} m(d\mu) \int_K \left|\ln\left|\frac{\lambda-\mu}{z_0-\mu}\right|\right| d\lambda < \infty \tag{11.58}$$

(and a similar relation with λ and $d\lambda$ replaced by z and $d^2 z$).

For any bounded Borel set B in $\mathbf{R}$ or $\mathbf{C}$, consider the function

$$t_B(\mu) = \int_B \ln\left|\frac{\lambda-\mu}{z_0-\mu}\right| d\lambda \tag{11.59}$$

for $\mu \in \mathbf{R}$ (or a similar relation with λ and $d\lambda$ replaced by z and $d^2 z$). It is easy to see that for any compact K in $\mathbf{R}$ or $\mathbf{C}$ the family of functions $\{t_B(\mu) : B \in \mathcal{B}, B \subset K\}$, where $\mathcal{B}$ is the family of Borel sets, is uniformly bounded and equicontinuous on any bounded interval. Furthermore,

$$\sup_{\substack{B\in\mathcal{B}\\ B\subset K}} \left|t_B(\mu)\right| \leqslant \frac{\text{const}}{1+|\mu|}. \tag{11.60}$$

But also, from (11.58), we have

$$\int_B \tau(\lambda, m)\, d\lambda = \int_{\mathbf{R}} t_B(\mu) m(d\mu) \tag{11.61}$$

(and a similar relation in the case of $B \subset \mathbf{C}$).

Since the family $t_B(\mu)$ is uniformly bounded and equicontinuous and the measures $m_x \in \mathcal{M}_c$ converge weakly to m, relations (11.60) and (11.61) and condition (11.51) imply that

$$\lim_{x\to\infty} \sup_{\substack{B\in\mathcal{B}\\ B\subset K}} \left|\int_K (\tau_x(\lambda) - \tau(\lambda))\, d\lambda\right| = 0, \tag{11.62}$$

for $K \subset \mathbf{R}$ compact. Let β_x be the expression of which the limit is taken in (11.62). We have

$$\int_K |\tau_x(\lambda) - \tau(\lambda)|\, d\lambda = \int_{K\cap\{\tau_x > \tau\}} (\tau_x - \tau)\, d\lambda - \int_{K\cap\{\tau_x < \tau\}} (\tau_x - \tau)\, d\lambda \leqslant 2\beta_x.$$

Since the right-hand side goes to 0 as $x \to \infty$, the lemma is proved (the case $K \subset \mathbf{C}$ being analogous). □

We continue the proof of theorem 11.6. Using the notation of Lemma 11.7, we can rewrite (11.50) as

$$(11.63)\quad |\gamma_x(z) - \gamma_x(z_0) - \tau(z, N)| \leqslant \frac{\ln 2}{x} + \max_{r=0,1}\{|\tau(z, N) - \tau(z, N_x^{(r)})|\}.$$

Theorem 6.4 implies that, with probability 1, the measures $N_x^{(r)}$ converge weakly to the measure N, and because of (11.44) we can use Lemma 11.7. Indeed, if we make $x \to \infty$ in (11.63), where $\operatorname{Im} z \neq 0$, we get from (11.55) and Theorem 11.3 that (11.43a) holds for almost every point in $\Omega_{\mathbf{C}}$.

For $\lambda \in \mathbf{R}$ we do instead the following. Denote by Ω_1 the set of realizations ω for which $N_x^{(r)}$ converges weakly to N; we have $\mathsf{P}(\Omega_1) = 1$. Taking an arbitrary $\omega \in \Omega_1$, we can choose, by (11.56), a sequence of positive numbers x_n, for $n \geqslant 1$, tending to ∞ and such that, for Lebesgue-almost all λ,

$$(11.64)\qquad \lim_{n\to\infty} \max_{r=0,1} |\tau(\lambda, N) - \tau(\lambda, N_{x_n}^{(r)})| = 0.$$

As the measures $N_x^{(r)}$ depend on α, so does the set of λ where (11.64) is satisfied, and the sequence x_n. Now set $z = \lambda$ and $x = x_n$ in (11.63), and take the limit $n \to \infty$. By (11.64) and Theorem 11.3, relation (11.43a) is again true for almost any point in $\Omega_{\mathbf{R}}$. This concludes the proof of the continuous case.

We now consider the discrete case. We can assume $\alpha \neq \pi/2$, since (11.43a) only has to be proved for almost all points in $\Omega_{\mathbf{R}}$ and $\Omega_{\mathbf{C}}$. Let $h_{0,x}^{(\alpha)}$ be the matrix $J_{a,b}$ of (6.71) with $s(x) \equiv 1$, $a = 0$, $b = x$ and $\beta = 0$; the order of this matrix is clearly $x - 1$. Denote by $y(x, z)$ the solution of the Cauchy problem (11.3)–(11.4). Clearly, $y(x, z)$ is a polynomial in z of degree $x - 1$, and its zeros coincide with the eigenvalues of $h_{0,x-1}^{(\alpha)}$. Hence,

$$(11.65)\qquad \frac{1}{x-1} \ln|y(x, z)| = \int_{\mathbf{R}} \ln|z - \lambda|\, N_{x-1}(d\lambda),$$

where $xN_x(\lambda)$ is the number of eigenvalues of the matrix $h_{0,x}^{(\alpha)}$ not exceeding λ. If we set

(11.66) $$\gamma_x(z) = \frac{1}{2x}\ln\big(|y(x,z)|^2 + |y(x+1,z)|^2\big),$$

(11.67) $$\delta_x^{(0)}(z) = \frac{1}{x}\ln\big|y(x,z)\big|, \qquad \delta_x^{(1)}(z) = \frac{1}{x}\ln\big|y(x+1,z)\big|,$$

(11.68) $$N_x^{(0)}(\mu) = \frac{x-1}{x}N_{x-1}(\mu), \qquad N_x^{(1)}(\mu) = N_x(\mu),$$

we have

(11.69) $$\delta_x^{(r)}(z) = \int_{\mathbf{R}} \ln|z-\mu| N_x^{(r)}(d\mu),$$

for $r = 0, 1$. In addition, γ_x and $\delta_x^{(r)}$ satisfy (11.49). The analogue of (11.50) here takes the form

(11.70) $$\left|\gamma_x(z) - \int_{\mathbf{R}} \ln|z-\mu|\, N(d\mu)\right| \leqslant \frac{\ln 2}{x} + \max_{r=0,1}\left\{\int_{\mathbf{R}} \ln|z-\mu|\big(N(d\mu) - N_x^{(r)}(d\mu)\big)\right\},$$

where $N(d\mu)$ is the integrated density of states of the metrically transitive operator h defined on $\ell_2(\mathbf{Z})$ by (11.4). By the variational principle (cf. (4.42)), we have

(11.71)
$$\frac{1}{x}\#\{1 \leqslant n \leqslant x : q(n) + 2 < \mu\} \leqslant N_x(\mu) \leqslant \frac{1}{x}\#\{1 \leqslant n \leqslant x : q(n) - 2 < \mu\},$$

(11.72) $$F(\mu - 2) \leqslant N(\mu) \leqslant F(\mu + 2).$$

Let μ be the set of positive measures m on $\mathbf{R}$ such that $m(\mathbf{R}) \leqslant 1$ and

(11.73) $$\int_{\mathbf{R}} \ln(1 + |\mu|)m(d\mu) < \infty,$$

and $\tilde{\mathcal{M}}$ an arbitrary subset $\mathcal{M}$ having the property that

(11.74) $$\lim_{R\to\infty} \sup_{m\in\tilde{\mathcal{M}}} \int_{|\mu|\geqslant R} \ln|\mu| m(d\mu) = 0.$$

Each measure m determines a function

(11.75) $$\tau(z, m) = \tau(z) = \int_{\mathbf{R}} \ln|z-\mu|\, m(d\mu).$$

If we replace the set $\mathcal{M}_c$ in Lemma 11.7 by $\tilde{\mathcal{M}}$ and consider the functions $\tau(z)$ defined by (11.75), the conclusions of the lemma remain valid. By Theorem 6.18, with probability 1, the measures $N_x^{(r)}(d\mu)$ converge weakly to $N(d\mu)$, and, by virtue of (11.71), (11.72), (11.16b) and the ergodic theorem, the measures $N_x^{(r)}$ and N will also belong to $\tilde{\mathcal{M}}$. Applying Lemma 11.7 with the changes above, we get the following counterpart for (11.63):

$$|\gamma_x(z) - \tau(z, N)| \leqslant \frac{\ln\sqrt{2}}{x} + \max_{r=0,1}\{|\tau(z,N) - \tau(z, N_x^{(r)})|\}. \tag{11.76}$$

Using this and repeating the arguments after (11.63), we prove (11.43b) for almost all points in $\Omega_{\mathbf{C}}$ and $\Omega_{\mathbf{R}}$. □

From now on we use ζ to denote $z \in \mathbf{C}$ or $\lambda \in \mathbf{R}$, as the case may be.

(11.8) Theorem [Pastur and Figotin 1984a]. *Assume the conditions of Theorem* 11.6 *hold.*

(i) *For almost all points in* $\Omega_{\mathbf{C}}$ *or* $\Omega_{\mathbf{R}}$,

$$\gamma_+(\zeta, \omega, \alpha) = \gamma_-(\zeta, \omega, \alpha) = \gamma(\zeta). \tag{11.77}$$

(ii) *For almost all* $z \in \Omega_{\mathbf{C}}$ *or* $\lambda \in \Omega_{\mathbf{R}}$, *with respect to Lebesgue measure,*

$$\gamma(\zeta) = \gamma(z_0) + \int_{\mathbf{R}} \ln\left|\frac{\zeta - \mu}{z_0 - \mu}\right| N(d\mu) \tag{11.78a}$$

in the continuous case, where $\operatorname{Im} z_0 \neq 0$, *and*

$$\gamma(\zeta) = \int_{\mathbf{R}} \ln|\zeta - \mu|\, N(d\mu) \tag{11.78b}$$

in the discrete case.

Similar results are proved in [Avron and Simon 1983] and [Kotani 1984].

Proof. By Theorem 11.3, for ζ and ω in a set of full P-measure, we have $\gamma_\pm(\zeta, \omega, \alpha) = \gamma(\zeta)$ except for one value of α. Therefore the exceptional set has measure 0 in $\Omega_{\mathbf{C}}$ or $\Omega_{\mathbf{R}}$, which, together with (11.43b), implies (11.78b). In the continuous case, there are two more values of α to be excluded, those for which $\gamma_\pm(z_0, \omega, \alpha) = -\gamma(z_0)$, but otherwise the same argument holds. □

It turns out, however, that part (ii) of this theorem can be strengthened:

(11.9) Theorem [Craig and Simon 1983b]. *Under the conditions of Theorem* 11.6, *equations* (11.78) *hold for all values of* ζ *and* $z_0 \in \mathbf{C}$ *with* $\operatorname{Im} z_0 \neq 0$.

Proof. By Theorem 11.4, $\gamma(\zeta)$ is a subharmonic function. The right-hand side of (11.78b) has the form (11.39) and thus is also subharmonic. But two subharmonic functions that coincide almost everywhere in $\mathbf{C}$ are actually identical [Hayman and Kennedy 1980], so the result follows in the discrete case.

To prove the continuous case, fix $R > 0$ and z_0 with $\operatorname{Im} z_0 \neq 0$. We show that the integral in (11.78a) is also a subharmonic function in the region $|\zeta| < R$:

$$\int_{\mathbf{R}} \ln\left|\frac{\zeta-\mu}{z_0-\mu}\right| N(d\mu) =$$

$$-\int_{|\mu|\leqslant 2R} \ln|z_0-\mu|\, N(d\mu) + \int_{|\mu|\leqslant 2R} \ln|\zeta-\mu|\, N(d\mu) + \int_{|\mu|>2R} \ln\left|\frac{\zeta-\mu}{z_0-\mu}\right| N(d\mu).$$

In view of (11.44b), the first integral on the right is independent of ζ, the second is a subharmonic function, and the third a harmonic function, when $|\zeta| < R$. Since R is arbitrary, the integral in (11.78a) is actually subharmonic everywhere, which implies the theorem as in the discrete case. □

Formula (11.78a) for the Schrödinger equation is more complicated than its discrete version (11.78b) because $N(\lambda)$ in the continuous case does not decay at infinity (see formula (6.17), for example), and therefore we must make a subtraction in order to prove the convergence of the integral in (11.78a). The subtraction method used in the proof of (11.78a) was first employed by [Kotani 1982] in a similar situation. It is not the only possible one:

(11.10) Corollary [Craig and Simon 1983b]. *Under condition* (11.42), *we have*

(11.79)
$$\gamma(z) = \gamma_0(z) + \int_{\mathbf{R}} \ln|z-\mu|\big(N(d\mu) - N_0(d\mu)\big),$$

where $\gamma_0(z) = \left|\operatorname{Re}\sqrt{-z}\right|$ *and* $N_0(\lambda) = \sqrt{\lambda}/\pi$ *are the Lyapunov exponent and the integrated density of states of the operator* $-d^2/dx^2$.

Proof. Write, along with (11.78a), a similar formula for $-d^2/dx^2$, and subtract the two. After separating the terms that depend on z and those that depend on z_0 on each side of the resulting equation, we get

(11.80)
$$\gamma(z) = \gamma_0(z) + \int_{\mathbf{R}} \ln|z-\mu|\big(N(d\mu) - N_0(d\mu)\big) + C,$$

where C does not depend on $z \in \mathbf{C}$. The integral converges at $\pm\infty$ because of the following estimate, which holds under condition (11.40) (see Problem III.5):

(11.81)
$$\left|N(\lambda) - N_0(\lambda)\right| \leqslant \operatorname{const}\big(1+|\lambda|\big)^{-1/2}.$$

We now show that $C = 0$ in (11.80). Setting $z = \lambda + i\varepsilon$ and taking into account (11.81) and the fact that $\gamma_0(z) = |\mathrm{Re}\sqrt{-z}|$, we verify that the first two terms on the right in (11.80) go to 0 as $\lambda \to \infty$ if $\varepsilon = \lambda^{1/3}$. Since $\gamma(z) \geqslant 0$ by definition, $C \geqslant 0$.

On the other hand, by Problem V.2, $\gamma(z) \leqslant \gamma_0(z) + \mathsf{E}\{|q(0)|\}|z|^{-1/2}$, which, together with (11.80), means that, for all $z \in \mathbf{C}$,

$$\int_{\mathbf{R}} \ln|z-\mu|\big(N(d\mu) - N_0(d\mu)\big) + C \leqslant \mathrm{const}\,|z|^{-1/2}.$$

If we set $z = i\varepsilon$ and take the limit $\varepsilon \to \infty$, the left-hand side of this expression goes to 0 by (11.81), and we conclude that $C \leqslant 0$. Thus $C = 0$, and the corollary is proved. □

Remark. We have just considered the case of "diagonal disorder," where it is only the coefficients of the diagonal parts of the Jacobi matrices (1.4) and the Sturm–Liouville operators (1.6) that are metrically transitive processes. Using essentially the same arguments, analogous results could be proved for the *Herbert–Jones–Thouless formulas* (11.78) and (11.79) in the general case of a one-dimensional metrically transitive operator of the form (1.4) or (1.6). For the discrete case the analogue is

$$\gamma(\lambda) = -\mathsf{E}\{\ln s_0\} + \int_{\mathbf{R}} \ln|\lambda-\mu| N(d\mu). \tag{11.82}$$

One of the simple consequences of these formulas is the following statement strengthening Theorem 3.2.

(11.11) Theorem [Kotani 1984; Craig and Simon 1983a]. *Under the hypotheses of Theorem* 11.6, *there exists a nonrandom constant* C_R *such that, for all* $|\lambda|, |\lambda'| \leqslant R < \infty$ *the integrated density of states* $N(\lambda)$ *satisfies the log-Hölder condition*

$$\big|N(\lambda) - N(\lambda')\big| \leqslant C_R\big|\ln|\lambda-\lambda'|\big|^{-1}. \tag{11.83}$$

Proof. We give the proof only in the discrete case (11.2), under the assumption that the potential is bounded; the other cases differ only in simple, if somewhat cumbersome, detail.

It is enough to prove (11.83) for $|\lambda - \lambda'| \leqslant 1$ and, say, $\lambda \leqslant \lambda'$. By (11.78b), we have

$$0 \leqslant \int_{\lambda}^{\lambda'} \ln(\mu-\lambda)\, N(d\mu) + \int_{|\lambda-\mu|\geqslant 1} \ln|\lambda-\mu|\, N(d\mu), \tag{11.84}$$

because the Lyapunov exponent is nonnegative. If the potential is bounded by $C < \infty$ with probability 1, Theorem 4.14 says that the support of the measure is concentrated in the interval $|\lambda| \leqslant 2 + C$, so the second integral on the right in (11.84) is approximated by $\ln\bigl(|\lambda| + 2 + C\bigr)$. Thus (11.84) becomes

$$-\ln|\lambda - \lambda'| \int_{\lambda}^{\lambda'} N(d\mu) \leqslant \ln(R + 2 + C),$$

which gives (11.83) for small $|\lambda - \lambda'|$. □

Remarks. 1. It is obvious from the proof that to get (11.83) it is enough to have (11.78b) for any z such that $\operatorname{Im} z \neq 0$. The proof of this relation, which contains a nonsingular integral, is much simpler than that of Theorem 11.6, and follows almost immediately from Theorem 11.4 and the results in Sections 4–6 on the existence of $N(\lambda)$.

2. In general, one cannot improve on the estimate (11.84) for the continuity modulus of $N(\lambda)$. This can be seen even from formula (7.27). (This formula applies to the purely "nondiagonal" form of operator (1.4). But (11.83) also holds for this operator, as can be seen by applying (11.82) instead of (11.78b).) This conclusion corroborates the results of Pöschel [1983] and Craig [1983], who construct Jacobi matrices (11.4) with an almost-periodic potential and such that

$$\sup_{|\lambda-\lambda'|\leqslant \varepsilon} \bigl|N(\lambda) - N(\lambda')\bigr| \geqslant \frac{C}{\ln \varepsilon^{-1}} (\ln\ln \varepsilon^{-1})^{-1-\delta} \tag{11.85}$$

for some $\delta > 0$ and $\varepsilon \downarrow 0$.

We discuss now the question of analogues for the Herbert–Jones–Thouless formulas (11.78) and (11.79) for systems of differential and finite-difference equations with metrically transitive coefficients. Craig and Simon [1983b] studied this question for the operator

$$(h_n\psi)(x) = -\psi(x+1) - \psi(x-1) + q_n\psi(x) \tag{11.86}$$

for $x \in \mathbf{Z}$, acting on the space of n-component functions $\psi(x)$ such that

$$\sum_{x\in\mathbf{Z}} \|\psi(x)\|^2_{\mathbf{C}^n} < \infty,$$

where $\|\cdot\|_{\mathbf{C}^n}$ is the Euclidean norm in $\mathbf{C}^n$, under the assumption that $q_n(x,\omega)$ is a metrically transitive sequence of Hermitian $n \times n$ matrices satisfying

$$\mathsf{E}\bigl\{\ln(1 + \|q_n(x)\|_{\mathbf{C}^n})\bigr\} < \infty. \tag{11.87}$$

As in the case of a single equation, we pass to a first-order recurrence relation such as (11.12)–(11.13). Here we write $Y_n(x+1) = A_n(x)Y_n(x)$, where $Y_n(x) = \bigl(\psi(x+1), \psi(x)\bigr)$ has $2n$ components, and A_n is a $2n \times 2n$ block matrix of the form

(11.88)
$$A_n(x) = \begin{pmatrix} q_n(x) & I_n \\ -I_n & 0 \end{pmatrix},$$

where I_n is the identity matrix of order n. Based on (11.86), one can show that $A_n(x)$ is symplectic, that is,

(11.89)
$$A_n^T \varGamma_n A_n = \varGamma_n,$$

where
$$\varGamma_n = \begin{pmatrix} 0 & I_n \\ -I_n & 0 \end{pmatrix}, \qquad \varGamma_n^2 = -I_{2n}.$$

(Equation (11.89) is just the condition that the Wronskian of the system of equations $h_n\psi = \lambda\psi$ be constant.)

Also as in the scalar case (cf. (11.19)), we can define the fundamental matrix $\mathcal{T}_n(x_2, x_1, \omega)$ of the system, and (11.89) implies that $\mathcal{T}_n(x_2, x_1, \omega)$ is symplectic, and the eigenvalues μ_j of $(\mathcal{T}^*\mathcal{T})^{1/2}$, for $j = 1, \ldots, 2n$, satisfy $\mu_j = \mu_{2n-j+1}^{-1}$. The matrices $\mathcal{T}(x, 0, \omega)$ constitute a multiplicative cocycle (see (11.24)) because the potential is metrically transitive; therefore the limits γ_j in (11.34), under condition (11.87), exist by Proposition 11.2(a), and satisfy

(11.90)
$$\gamma_1 \geqslant \gamma_2 \geqslant \cdots \gamma_n \geqslant 0 \geqslant \gamma_{n+1} = -\gamma_n \geqslant \cdots \geqslant \gamma_{2n} = -\gamma_1.$$

Since, for this metrically transitive operator, $N(\lambda)$ exists and satisfies $N(\mathbf{R}) = 1$ even under the condition that the potential is finite with probability 1, we have all the ingredients that might be expected to take part in the generalization of formula (11.78b).

The generalization turns out to be

(11.91)
$$\frac{1}{n}\sum_{j=1}^{n} \gamma_j(z) = \int_{\mathbf{R}} \ln|z-\mu|\, N(d\mu),$$

that is, we must take the arithmetic mean of the nonnegative Lyapunov exponents of the systems of equations. We explain briefly the derivation of (11.91). We consider, as in the proof of (11.78b), the boundary problem on the interval $[1, x]$ for the equation associated with (11.86). We restrict ourselves, for simplicity, to the boundary condition $\psi(0) = 0$ (at the other boundary it is always convenient to set $\psi(x+1) = 0$ anyway: cf. (11.43)). The general solution of the system, expressed in terms of the fundamental matrix

(11.92)
$$\mathcal{T}_n(x, 0) = A_n(x) \ldots A_n(1) = \begin{pmatrix} \mathcal{T}_n^{11}(x) & \mathcal{T}_n^{12}(x) \\ \mathcal{T}_n^{21}(x) & \mathcal{T}_n^{22}(x) \end{pmatrix},$$

will take the form

$$\begin{pmatrix} \psi(x+1) \\ \psi(x) \end{pmatrix} = \mathcal{T}_n(x,0) \begin{pmatrix} \psi(1) \\ 0 \end{pmatrix}. \tag{11.93}$$

Therefore, the condition $\psi(x+1) = 0$, which defines the eigenvalues of our boundary problem $\psi(0) = \psi(x+1) = 0$, is $\det \mathcal{T}_n^{11}(x) = 0$. It is easy to see that this determinant is a polynomial of degree nx in λ, that is, the eigenvalues are the zeros of this polynomial. On the other hand, the multiplicative ergodic theorem (Proposition 11.2) implies that the maximum rate of exponential growth of minors of order r for $\mathcal{T}_n(x,0)$ is $\sum_{j=1}^r \gamma_j$. We obtain (11.91) by replacing λ by the variable $z \in \mathbf{C}$ (where $\operatorname{Im} z \neq 0$) in $\det \mathcal{T}_n^{11}$, taking the logarithm of this determinant, dividing by nx, and passing to the limit $x \to \infty$. Using more refined arguments (see the proof of Theorem 11.9), one can prove (11.91) for real $z = \lambda$ as well.

The fact that (11.91) gives no information on individual characteristic Lyapunov exponents, but only on their arithmetic mean, makes it, generally speaking, less useful than (11.78b). We can still draw some conclusions: for example, the trivial solution is unstable if and only if the maximum exponent γ_1 is positive, and this obviously follows from a positive average, so if the right-hand side of (11.91) can be proved to be positive (see Problem V.13, for example), we have learned something. But the absence of the absolutely continuous component in the spectrum of the metrically transitive operator h_n of (11.86) follows from the positiveness of the minimum exponent γ_n (Section 12.B), which is not implied if the average is positive.

However, (11.91) does provide an estimate of the form (11.83) for the continuity modulus of $N(\lambda)$ for the operator (11.86). Indeed, by Remark 1 after Theorem 11.11, (11.83) can be obtained from the inequality

$$0 \leqslant \int_{\mathbf{R}} \ln |z-\mu| \, N(d\mu)$$

for all $z = \lambda + i\varepsilon$, with $\varepsilon \neq 0$. This inequality clearly follows from (11.91).

Moreover, a similar inequality can be obtained for second-order finite-difference operators with bounded potential in arbitrary dimension. Namely, consider the operator (4.47)

$$(A\psi)(x) = -\sum_{|y-x|=1} \psi(y) + q(x)\psi(x) \tag{11.94}$$

for $x \in \mathbf{Z}^d$, acting on $\ell_2(\mathbf{Z}^d)$. It follows from the proof of Theorem 4.4 that, for any metrically transitive field $q(x)$ on $\mathbf{Z}^d$, the measure $N(A; d\lambda)$ corresponding to operator (11.94) is the limit, as $m \to \infty$, of the measures $\mathsf{E}\{N(A_m; d\lambda)\}$ corresponding to the operators A_m obtained by restricting to the slab

(11.95) $$S_m = \{x \in \mathbf{Z}^d : |x_j| \leqslant m \text{ for } j = 2, \dots, d\}$$

with cross-section

(11.96) $$C = \{(x_2, \dots, x_n) : |x_j| \leqslant m \text{ for } j = 2, \dots, d\} \subset \mathbf{Z}^{d-1}.$$

The operator A_m is clearly the same as (11.86), where $\psi(x) = \{\psi(x,y), y \in C_m\}$ for $x \in \mathbf{Z}$. Therefore equation (11.91) applies, hence also inequality (11.84), with $N(\lambda)$ replaced by $\mathsf{E}\{N(A_m;\lambda)\}$ and λ by z. Passing to the limit $m \to \infty$ in the latter inequality, always assuming $\operatorname{Im} z \neq 0$, we get (11.84) for $N(A; d\lambda)$, whence, by arguments similar to those used to prove Theorem 11.11, we obtain an estimate of the form (11.83) for the continuity modulus of $N(A, d\lambda)$, and thus a refinement of Theorem 3.4 [Craig and Simon 1983].

11.C Simplest Asymptotic Formulas and Estimates for Lyapunov Exponents

We now consider some asymptotic results for $\gamma(\lambda)$ in special regions in the spectrum—chiefly near spectral boundaries—and for special (large or small) values of various parameters affecting the coefficients: amplitudes, scales, and so on. Such results apply to all metrically transitive coefficients, but they are few and not very deep. In the next two chapters we will derive more complex quantitative results on Lyapunov exponents of random and almost-periodic operators.

We start with bounds for the Lyapunov exponent near stable boundaries. Recall from Sections 6 and 8 that we call a boundary stable if, in its vicinity, $N(\lambda)$ is asymptotically close to what it would be for some operator with constant coefficients. But the Lyapunov exponent of an operator with constant coefficients vanishes exactly on the spectrum, so it is reasonable to assume that as λ approaches a stable boundary the Lyapunov exponent tends to zero. We will verify this by considering two examples: the point $\lambda = \infty$ for the Schrödinger operator and the point $\lambda = 0$ for (6.19). Other examples can be found in [Lifshitz et al. 1982b].

The bounds and other quantitative results for Lyapunov exponents are based on the following representations, which are similar to the representations (6.11), (6.16) and (6.22) for $N(\lambda)$.

(11.12) Theorem. *For a Schrödinger operator with metrically transitive potential satisfying* (11.16a), *we have*

(11.97a) $$\gamma(\lambda) = \lim_{L\to\infty} \frac{1}{L} \int_0^L dx \int_0^\pi d\alpha\, \mathsf{E}\{(q(x) - \lambda + 1) \sin 2\theta(x)\}$$

for any real λ, *where the phase* $\theta(x)$ *is given by* (6.4) *and* (6.5), *and*

$$\gamma(\lambda) = \lim_{L\to\infty} \frac{1}{2L\sqrt{\lambda}} \int_0^L dx \int_0^\pi d\alpha\, \mathsf{E}\{q(x)\sin 2\chi(x)\} \tag{11.97b}$$

for $\lambda > 0$, *where* $\chi(x)$ *is given by* (6.14) *and* (6.15).

For the operator (6.19) *with* $s(x)$ *satisfying* (6.20), *we have*

$$\gamma(\lambda) = \lim_{L\to\infty} \frac{1}{2L}\sqrt{\frac{\lambda}{\kappa}} \int_0^L dx \int_0^\pi d\alpha\, \mathsf{E}\{\delta(x)\sin 2\chi(x)\}, \tag{11.98}$$

for $\lambda > 0$, *where* $\kappa = \mathsf{E}^{-1}\{s^{-1}(x)\}$, $\delta(x) = \kappa s^{-1}(x) - 1$, *and the phase* $\chi(x)$ *is given by* (6.24*a*).

Proof. From (11.2) and the Schrödinger equation, we get

$$r'(x) = \tfrac{1}{2} r(x)\big(q(x) - \lambda + 1\big)\sin 2\theta(x). \tag{11.99}$$

which, by (6.3) and (11.2), gives

$$\ln r(x) = \tfrac{1}{2}\int_0^x dx'(q - \lambda + 1)\sin 2\theta. \tag{11.100}$$

From (11.17), with $x_1 = 0$, $x_2 = x$ and $Y(0) = (\cos\alpha, \sin\alpha)$, and recalling that $r^2(x) = \|Y(x)\|^2$, we get, after elementary integration,

$$\frac{1}{\pi}\int_0^\pi \ln r^2(x)\, d\alpha = \ln\big(\tfrac{1}{2}\operatorname{Tr}(\mathcal{T}^*(x)\mathcal{T}(x))^{1/2}\big)^2.$$

Thus, by integrating (11.100) over α, dividing by π and taking the limit using Theorems 11.3 and 11.8, we get (11.97a).

Formulas (11.97b) and (11.98) are obtained similarly, using definitions for $r(x)$ other than (11.1) and (11.2), but leading to the same values of the Lyapunov exponent: for (11.97b) we use

$$r^2(x) = y^2(x) + \frac{1}{\lambda}y'^2(x), \qquad y = r\sin\chi, \qquad y' = \sqrt{\lambda}\, r\cos\chi,$$

and for (11.98) we use

$$r^2(x) = y^2(x) + \frac{1}{c^2}(sy')^2(x), \qquad y = r\sin\chi, \qquad y' = cr\cos\chi,$$

where $c = \sqrt{\lambda\kappa}$. □

Equations (11.97) and (11.98) can be readily recast into a form similar to (6.34), representing the Lyapunov exponent as the mathematical expectation, with respect to the limiting distribution of the phase and of the coefficient, of the functions on the right-hand side of (11.97) and (11.98).

Equation (11.97b) immediately shows that, if the mathematical expectation of the potential is finite, $\gamma(\lambda) = O(\lambda^{-1/2})$ as $\lambda \to \infty$ for the Schrödinger equation. Equation (11.98) shows that $\gamma(\lambda) = O(\lambda^{1/2})$ as $\lambda \downarrow 0$ for the divergence operator (6.19). More careful arguments, similar to those used in the proof of Theorems 6.5 and 6.6, leads to a refinement of these estimates:

(11.13) Theorem. *For a Schrödinger operator with metrically transitive potential satisfying* (11.42), *we have*

$$\gamma(\lambda) = o(\lambda^{-1/2}) \qquad \textit{as } \lambda \to \infty. \tag{11.101}$$

For the operator (6.19) *with* $s(x)$ *satisfying* (6.20), *we have*

$$\gamma(\lambda) = o(\lambda^{1/2}) \qquad \textit{as } \lambda \downarrow 0. \tag{11.102}$$

□

Thus the Lyapunov exponent is 0 in these cases, in accordance with the heuristic description of stable boundaries given in Sections 6 and 8, by which the spectrum in the neighborhood of such boundaries is similar to what it is for equations with constant coefficients. Near fluctuation boundaries this is normally not so: fluctuation boundaries are a characteristic feature of random operators. We postpone the discussion of the properties of the Lyapunov exponent in their vicinity to Section 14.

We now discuss the question of improving the estimates of Theorem 11.13. If the potential in the Schrödinger equation is bounded with probability 1 and its derivatives of all orders $\leqslant n$ have finite absolute moments, we can integrate by parts n times in (11.97b) and use (6.15) to find

$$\gamma(\lambda) = O(\lambda^{-(n+1)/2}) \tag{11.103}$$

as $\lambda \to \infty$. (Delyon and Foulon [1986] obtained the same result by a different method, based on a version of the adiabatic theorem of nonlinear mechanics. This technique can also be used to prove that, if the metrically transitive potential is analytic in a strip containing the real axis, $\gamma(\lambda) = O(e^{-\text{const}\sqrt{\lambda}})$ as $\lambda \to \infty$.)

This estimate can only be improved for arbitrary metrically transitive potentials up to the replacement of O by o, as shown by the following example from [Marchenko 1986]: given any sequence h_k, for $k \geqslant 1$, such that $\sum_{k=1}^{\infty}(k^{n+1}h_k)^2 < \infty$, there exists a potential $q(x)$, periodic of period π, whose first n derivatives are absolutely continuous, with the n-th derivative square-integrable, and such that $h_k = \sup_{\lambda\in\Delta'_k}\gamma(\lambda)$, where Δ'_k is the k-th gap in the spectrum of the Schrödinger operator on $L_2(\mathbf{R})$ with potential $q(x)$. This gap lies in the vicinity of the point $\nu_k = k^2$, and has size of order $\delta_k k^{-n}$, where $\sum_{k\geqslant 1}\delta_k^2 < \infty$. Therefore, for $\lambda \in \Delta'_k$, we have

$\gamma(\lambda) = O(h_k) = O(\alpha_k \lambda^{-(n+1)/2})$, where α_k, for $k \geqslant 1$, is a sequence such that $\sum_{k\geqslant 1} \alpha_k^2 < \infty$.

Since the point $\lambda = \infty$ is not a stable boundary for the discrete analogue (11.4) of the Schrödinger operator, there is no sense in talking about a discrete analogue for formulas (11.101) and (11.103). This is also clear from formula (11.103) itself, because the order of the Lyapunov exponent when $\lambda \to \infty$ is controlled by the smoothness of the potential, a property that has no counterpart in the discrete case. Equation (11.102) does have a discrete version, which can be obtained in the same way and has the same form (see Theorem 14.6).

Thus, for arbitrary metrically transitive potentials, (11.103) is the best possible estimate. But if we only consider sufficiently random potentials, the Lyapunov exponent as $\lambda \to \infty$, for the same degree of smoothness, becomes smaller. We will consider this question in detail in the next chapter; here we give just some heuristic arguments.

If we formally solve equation (6.15) for the phase χ to within terms of order q/λ and substitute the result into (11.97b), we get, under the assumption that $\mathsf{E}\{q(x)\} = 0$:

$$\gamma \simeq \frac{1}{4\pi}\int_0^\infty b(x)\cos 2\sqrt{\lambda}x\,dx = \frac{\hat{b}(2\sqrt{\lambda})}{8\lambda}, \tag{11.104}$$

where $b(x) = \mathsf{E}\{q(0)q(x)\}$ is the correlation function of the potential and $\hat{b}(k)$, for $k \in \mathbf{R}$, is its Fourier transform.

Even a formal derivation of this expression requires the correlation between potentials at points far away to decay quickly. Thus, it is natural to regard this expression as the asymptotic formula for the Lyapunov exponent of the Schrödinger equation for $\lambda \to \infty$ with a random potential—for example, a Markov process. The two potentials of this sort described in Section 1 (see Example 11.5(b)), namely, the Ornstein–Uhlenbeck process and the random telegraph signal, have correlation function

$$b(x) = b_0 e^{-ax} \tag{11.105}$$

for $a > 0$. Substituting this into (11.104), we get

$$\gamma(\lambda) = -\frac{b'(0)}{16\lambda^2}(1 + o(1)) \tag{11.106}$$

as $\lambda \to \infty$. This formula will be proved in Section 14, but for now what is important is that, because both of these random processes are nondifferentiable with probability 1, they correspond to $n = 0$ in (11.103), yet that formula predicts the quadratic decay of $\gamma(\lambda)$ as $\lambda \to \infty$. (The realizations of a random telegraph signal are piecewise constant functions, and those of

Ornstein–Uhlenbeck process are continuous functions for which, with probability 1, $\limsup_{\varepsilon\downarrow 0}|q(x_1)-q(x_2)|\sqrt{2\varepsilon\ln\varepsilon^{-1}}=1$ for $0\leqslant x_1<x_2\leqslant 1$ and $x_2-x_1=\varepsilon$.)

Formula (11.104) also demonstrates another feature of random potential, even more important for spectral theory. Since, for such potentials, $\hat{b}(k)$ is generally strictly positive on a big set—for example, $\hat{b}(k)=2b_0a/(a^2+k^2)$ in the case of (11.105)—if we assume that (11.104) gives the right asymptotic behavior for $\gamma(\lambda)$ as $\lambda\to\infty$, we must conclude that $\gamma(\lambda)$ is strictly positive on a big set of sufficiently large values of λ—in the case of (11.105), for all sufficiently large λ. This will indeed be proved in Sections 12 and 14.

The above arguments are a particular case of a fairly general heuristic rule, which says that any formula or result that implies that the Lyapunov exponent is positive (and so, by Theorem 12.1, that the spectrum has no absolutely continuous component) can only be true for sufficiently random potentials, and generally requires some sort of special assumption. (By Theorem 12.9, such a potential should be, in general, a nondeterministic random process, as defined in Section 1.) On the other hand, an upper bound on the Lyapunov exponent or a statement that the Lyapunov exponent tends to zero under some limit, can be very general. We will demonstrate this heuristic rule with several examples, which will also illustrate the use of the Herbert–Jones–Thouless formulas (11.78b) and (11.79) in the study of the properties of the Lyapunov exponent.

(11.14) Examples

(a) Consider a metrically transitive Jacobi matrix, that is, a metrically transitive operator on $\ell_2(\mathbf{Z})$ corresponding to equation (11.4), and assume that the spectrum contains neighborhoods of ∞ and $-\infty$. In view of formulas (11.15), (11.20) and (11.28) for $\gamma(\lambda)$, or of formula (11.78b), it is natural to assume that

(11.107) $$\gamma(\lambda)=\ln|\lambda|\bigl(1+o(1)\bigr)$$

as $|\lambda|\to\infty$. However, attempts to prove this asymptotic formula by using only (11.78b) run into trouble, caused by the singularity of the log function, which, without additional assumptions on the smoothness of $N(\lambda)$, can change formula (10.107). If we assume that $N(\lambda)$ satisfies a Hölder condition, that is,

(11.108) $$\bigl|N(\lambda)-N(\lambda')\bigr|\leqslant C|\lambda-\lambda'|^{\alpha},$$

at least for $|\lambda|$ large enough, the contribution of a neighborhood of the point λ to the integral in (11.78b) becomes finite and, after some simple estimates, we find that, under the addition assumption (11.16b), formula (11.107) is indeed true.

Similar considerations in the continuous case lead to the formula

$$\gamma(\lambda) = \sqrt{|\lambda|}\bigl(1 + o(1)\bigr) \tag{11.109}$$

as $\lambda \to -\infty$, which is true when (11.40) and (11.108) hold.

And, as we saw in Sections 4–7, estimates like (11.108) can be proved for various types of random potentials: in the discrete case, independent, identically distributed or Gaussian potentials, satisfying the condition of Problem II.19; in the continuous case, Markov potentials like those of Sections 6 and 7.

(b) We next consider the asymptotic behavior of the Lyapunov exponent of a metrically transitive Jacobi matrix for large amplitudes of the potential. Let the potential $q(x)$ in (11.4) have the form

$$q(x) = gQ(x), \tag{11.110}$$

where $Q(x)$ is a fixed metrically transitive sequence, which we shall, for simplicity, assume bounded with probability 1:

$$\bigl|Q(x)\bigr| \leqslant 1. \tag{11.111}$$

From (11.14), (11.15) and (11.25), we can assume that the asymptotic behavior of the Lyapunov exponent $\gamma_g(\lambda)$, for $g \to \infty$ and fixed $\lambda \in \mathbf{R}$, is

$$\gamma_g(\lambda) = \ln|g|\bigl(1 + o(1)\bigr) \tag{11.112}$$

This is fairly easy to prove if we assume that the one-dimensional distribution $F(Q)$ satisfies a Hölder condition

$$\bigl|F(Q) - F(Q')\bigr| \leqslant \text{const}\,|Q - Q'|^{\alpha},$$

and the corresponding integrated density of states $N_g(\lambda)$ satisfies (11.108) with $C_g = O(g^{-\alpha})$ and $\alpha_g = \alpha$. (All these conditions are fulfilled, for example, in the situation of Theorem 4.21 or of Problems II.18–21.) Under such conditions, the contribution of the neighborhood $|\lambda - \mu| \leqslant 1$ of the point λ to the integral in (11.78b) has order $O\bigl(|g|^{-\alpha}\bigr)$, and the contribution of its complement is estimated by using the inequalities

$$F\bigl(g^{-1}(\lambda - 2)\bigr) \leqslant N_g(\lambda) \leqslant F\bigl(g^{-1}(\lambda + 2)\bigr) \tag{11.113}$$

for $g > 0$, which follow from Theorem 4.21. As a result,

$$\gamma_g(\lambda) = \ln|g| + \int_{\mathbf{R}} \ln|q|\, F(dq) + O\left(\frac{\ln|g|}{|g|^{\alpha}}\right) \tag{11.114}$$

as $g \to \infty$.

The formulation of the asymptotic problem for $\gamma_g(\lambda)$ as $g \to \infty$ can be slightly changed by considering $\gamma_g(\lambda g)$ for fixed λ and $g \to \infty$. In this case, (11.78b) yields

(11.115) $$\gamma_g(\lambda g) = \ln |g| + \int_{\mathbf{R}} \ln |\lambda - \mu| \, \hat{N}_g(d\mu),$$

where $\hat{N}_g(\lambda) = N_g(\lambda g)$. It follows from (11.113) that, as $|g| \to \infty$, $\hat{N}_g$ converges to $F(\lambda)$ for every λ. Thus, under the same assumptions made in deriving (11.114), we get from (11.115) the following generalization of (11.114):

(11.116) $$\gamma_g(\lambda g) = \ln |g| + \int_{\mathbf{R}} \ln |\lambda - \mu| \, F(d\mu) + O\left(\frac{\ln |g|}{|g|^{\alpha}}\right).$$

A discussion of the remainder in this formula and related problems can be found in [Avron et al. 1983].

(c) In the previous example, the integrated density of states converged weakly, within the asymptotic region, to a fixed limit, and the asymptotics of the Lyapunov exponent were also determined by that limit. A similar situation arises with a *slowly varying potential* like the metrically transitive process

(11.117) $$q_R(x) = Q(x/R)$$

as $R \to \infty$, where $Q(x)$ is a fixed metrically transitive process. By Problem II.37, the integrated density of states $N_R(\lambda)$ of the Schrödinger operator or Jacobi matrix with such a potential converges weakly, as $R \to \infty$, to the limit $N_\infty(\lambda)$ given by

(11.118) $$N_\infty(\lambda) = \lim_{R \to \infty} N_R(\lambda) = \int_{\mathbf{R}} N_0(\lambda - Q) F(dQ),$$

where $N_0(\mu) = N(\mu)\big|_{Q \equiv 0}$. Thus, based on (11.78), we can imagine that as $R \to \infty$ the Lyapunov exponent $\gamma_R(\lambda)$ goes to

(11.119) $$\gamma_\infty(\lambda) = \int_{\mathbf{R}} \gamma_0(\lambda - Q) F(dQ),$$

where $\gamma_0(\lambda) = \gamma(\lambda)\big|_{Q \equiv 0}$ (see (11.8) and (11.9)). This is, in fact, easily provable if, as in the previous example, we assume the Hölder condition (11.108) to hold uniformly as $R \to \infty$, which is the case, for example, in the situations of Theorem 4.21, Subsection 6.6 or Problems II.18–21. This condition yields an estimate of the contribution of a small neighborhood of the point $\mu = \lambda$ in (11.78b) and (11.79).

From (11.118), (11.119) and the equations $N_0(\lambda) = \sqrt{\max\{\lambda, 0\}}/\pi$ for the continuous case and $N_0(\lambda) = \arcsin(\lambda/2)/\pi + \frac{1}{2}$ for the discrete case, we conclude that $\gamma_\infty(\lambda)$ (and hence also $\gamma_R(\lambda)$ for R big enough) is positive on the closure of the set $(-\infty, 0] + \operatorname{supp} F$ in the continuous case and on the set $\{|\lambda| \geqslant 2\} + \operatorname{supp} F$ in the discrete case, where $+$ indicates the arithmetic sum of sets (see before Theorem 4.22). Conversely, $\gamma_\infty(\lambda)$ is zero in the complement of the same sets. In particular, in the continuous case, if we have

$$\sup_x q_R(x) = \sup_x Q(x) = c < \infty \tag{11.120}$$

with probability 1, we get

$$\lim_{R\to\infty} \gamma_R(\lambda) = 0 \quad \text{for } \lambda > \sup_x Q(x). \tag{11.121}$$

According to the heuristic rule formulated above, the limit of the Lyapunov exponent in (11.121) is 0 as $R \to \infty$ for any metrically transitive potential of the form (11.117) satisfying (11.16b) and (11.42), that is, without the restrictive condition (11.108). To verify this, notice that, since $N_0(\lambda) = \sqrt{\max\{\lambda, 0\}}/\pi$, formula (11.79) implies that, for $0 < \varepsilon < 1$,

$$\gamma_R(\lambda) \leqslant \int_{|\lambda-\mu|\geqslant\varepsilon} \ln|\lambda - \mu|\, \nu_R(d\mu) + \text{const}\,|\varepsilon \ln \varepsilon|, \tag{11.122}$$

where $\nu_R(\mu) = N_R(\mu) - N_0(\mu)$ and the constant depends only on λ and ε, not on R. By taking the limit $R \to \infty$ and using (11.118), the fact that

$$|\nu_\infty(\lambda)| \leqslant \frac{1}{\pi\sqrt{\lambda}} \mathsf{E}\{|Q(0)|\}$$

for $\lambda > 0$ (a consequence of (6.15) and (6.16)), the absolute continuity of $N_\infty(\lambda)$, and the fact that $\varepsilon > 0$ is arbitrary, we conclude that

$$0 \leqslant \limsup_{R\to\infty} \gamma_R(\lambda) \leqslant \gamma_\infty(\lambda).$$

From this and from equation (11.119), which implies that $\gamma_\infty(\lambda) = 0$ for $\lambda \geqslant \sup Q(x)$, we obtain (11.121).

By the definition of $\gamma(\lambda)$ and equation (11.78), for a general metrically transitive potential, the Lyapunov exponent is just an upper semicontinuous function, while $N(\lambda)$ is continuous and even satisfies (11.83). But if $N(\lambda)$ satisfies the stronger inequality (11.108), even with constants C and α that depend on the interval under consideration, $\gamma(\lambda)$ also satisfies a similar inequality with same exponent α. For simplicity, let's take the discrete case, dividing the integral in (11.78b) into two, for the regions $|\mu| \geqslant 2a$ and

$|\mu| < 2a$, with $a > 0$. Integrating the second integral by parts, we have, for $|\lambda| \leqslant a$,

(11.123)

$$\gamma(\lambda) - \int_{|\mu|\geqslant 2a} \ln|\lambda-\mu|\, N(d\mu) - N(\mu)\ln|\lambda-\mu| \Big|_{\mu=\pm 2a} = \int_{|\mu|<2a} \frac{N(\mu)}{\mu-\lambda} d\mu.$$

This, by well-known estimates from the theory of singular integrals [Muskhelishvili 1962], implies that, under conditions (11.16b) and (11.108), for any interval $\Delta \subset \mathbf{R}$, there exists $C > 0$ such that

$$|\gamma(\lambda_1) - \gamma(\lambda_2)| \leqslant C|\lambda_1 - \lambda_2|^\alpha \tag{11.124}$$

for $\lambda_1, \lambda_2 \in \Delta$, where the constant α is the same as in (11.108) if $\alpha < 1$, and $\alpha = 1 - \varepsilon$, for any $\varepsilon > 0$, if $\alpha = 1$ in (11.108). (In fact, in this case the right-hand side can be $C|\lambda_1 - \lambda_2|\big|\ln|\lambda_1 - \lambda_2|\big|$.)

We now demonstrate the use of the Herbert–Jones–Thouless formulas (11.78) to obtain information on $N(\lambda)$, starting with known information on $\gamma(\lambda)$; until now we have been proceeding in the opposite direction. For similar results, see [Carmona et al. 1986].

(11.15) Theorem [Le Page 1984; Bougerol and Lacroix 1986]. *Consider the second-order equations* (6.1) *and* (11.4), *and assume that conditions* (11.42) *and* (11.16b) *are satisfied. If* $\gamma(\lambda)$ *satisfies* (11.124), *perhaps with* C *and* α *depending on* $\Delta \subset \mathbf{R}$, *the integrated density of states* $N(\lambda)$ *for the corresponding metrically transitive operator also satisfies the Hölder condition* (11.108), *with* α *the same as in* (11.108) *if* $\alpha < 1$, *and* $\alpha = 1 - \varepsilon$, *for any* $\varepsilon > 0$, *if* $\alpha = 1$ *in* (11.108). (*In fact, in this case the right-hand side can be* $C|\lambda_1 - \lambda_2|\,\big|\ln|\lambda_1 - \lambda_2|\big|$.)

Proof. The right-hand side of (11.123), multiplied by π, is the Hilbert transform [Koosis 1980] $\tilde{N}_a(\lambda)$ of the function $N(\lambda)\chi_{[-2a,2a]}(\lambda)$. Since, for any function $f(x) \in L_2(\mathbf{R})$, we have $\tilde{\tilde{f}} = -f$, equation (11.123) implies that, for $|\lambda| \leqslant 2a$,

$$-\tilde{\tilde{N}}_a(\lambda) = N(\lambda) = \frac{1}{\pi}\int_{\mathbf{R}} \frac{\tilde{N}_a(\mu)d\mu}{\lambda-\mu}.$$

Splitting this into two integrals, over the regions $|\lambda| \geqslant a/2$ and $|\lambda| \leqslant a/2$, we find that the former, for $|\lambda| \leqslant a/2$, is a continuously differentiable (and even real analytic) function, and the latter, for $|\lambda| \leqslant a/2$, includes only the values of $\tilde{N}_a(\mu)$ from the interval $|\mu| \leqslant a$. But in this interval, the function $\tilde{N}_a(\mu)$, according to (11.123) and (11.124), satisfies a Hölder condition with exponent α. Therefore, by using again the above-mentioned estimates from the theory of singular integrals [Muskhelishvili 1962], we arrive at the statement of the theorem. □

Remarks. 1. We could have used, instead of the Hilbert transform, the analogous integral representation of $N(\lambda)$ via $\gamma(\lambda)$ obtained in Problem V.23. An important case in which the condition of Theorem 11.15 is met is the discrete equation (11.4), where the potential $q(x)$ consists of a sequence of independent, identically distributed random variables satisfying (11.16b). Then it can be shown [Bougerol and Lacroix 1985] (see also Section 14) that $\gamma(\lambda)$ is a locally Hölder function, and therefore so is $N(\lambda)$.

2. The asymptotic formulas obtained in Examples 11.14(a)–(c) under the assumption (11.108) are merely upper estimates without that assumption, because the contribution of an infinitesimal neighborhood of the point $\mu = \lambda$ in (11.78b) and (11.79) is nonpositive.

As will be shown in Section 16 for a variety of discrete, quasiperiodic metrically transitive potentials, a typical example being $q(x) = 2g\cos 2\pi(\alpha x + \omega)$ for α irrational, the lower bound $\gamma(\lambda) \geqslant \ln|g|$ is also valid for all λ. In view of the preceding paragraph, it follows that (11.108) is not necessary for the validity of (11.112).

3. If we consider the average of the Lyapunov exponent $\gamma(\lambda)$ over an arbitrarily small interval,

$$\gamma_\varepsilon(\lambda) = \frac{1}{2\varepsilon}\int_{|\mu-\lambda|\leqslant\varepsilon} \gamma(\mu)\, d\mu,$$

all the asymptotic formulas above hold without the assumption (11.108). Indeed, we can write

$$\gamma_\varepsilon(\lambda) = \int_{\mathbf{R}} l_\varepsilon(\lambda-\mu) N(d\mu),$$

where the function

$$l_\varepsilon(\lambda) = \frac{1}{2\varepsilon}\int_{|t|\leqslant\varepsilon} \ln|\lambda - t|\, dt$$

is obviously continuous and has the same asymptotic behavior as $\ln|\lambda|$ as $|\lambda| \to \infty$.

Another way of reaching the same conclusion is by observing that, for a general metrically transitive potential, the Lyapunov exponent is an upper semicontinuous function of λ and of any other parameters on which $N(\lambda)$ may depend continuously (in the weak topology).

4. Since the derivation of the heuristic formula (11.104) uses only the fact that the term containing the potential in (6.15) and (11.97b) is small, we should expect, using the same formal arguments, to get the asymptotic behavior of the Lyapunov exponent for fixed $\lambda > 0$ and small random potential, that is, a potential of the form (11.110) for $|g| \to 0$ (see Theorems 14.5 and 14.6).

Now, since the equation (6.24a) for the phase $\chi(x)$, which enters into formula (11.98), is similar to equation (6.15), the Lyapunov exponent of equation (6.19) should satisfy the asymptotic formula

$$\gamma(\lambda) = \frac{\lambda}{4\kappa} \int_0^\infty b_\delta(x)\, dx \tag{11.125}$$

as $\lambda \downarrow 0$, where $b_\delta(x)$ is the correlation function of the process $\delta(x) = \kappa s^{-1}(x) - 1$, with $\kappa = \mathsf{E}^{-1}\{s^{-1}(x)\}$. A similar formula should be true also for the discrete analogue of (6.19), namely,

$$-s(x+1)\Delta(x) - s(x)\Delta(x-1) = \lambda y(x), \tag{11.126}$$

where

$$\Delta(x) = y(x+1) - y(x). \tag{11.127}$$

(Naturally, the integral in (11.125) should be replaced by a sum.) These formulas are derived in Section 14 for a variety of potentials: see Theorems 14.5 and 14.6.

12 Lyapunov Exponents and the Absolutely Continuous Spectrum

In this section we discuss the relationship between the Lyapunov exponent and the absolutely continuous spectrum. Recall that for a periodic potential the spectrum can be characterized not only as the support of $N(\lambda)$, but also as the complement of the support of $\gamma(\lambda)$. In the more general case of a metrically transitive potential, we will show that the complement of $\operatorname{supp}\gamma(\lambda)$ coincides with the absolutely continuous spectrum. In particular, if the Lyapunov exponent is positive on the spectrum—and we'll see in Chapter VI that this is a property of all random operators—it has no absolutely continuous component (Theorem 12.1). A similar result, Theorem 12.10, says that if the Lyapunov exponent is zero on a set of positive Lebesgue measure, the potential is a deterministic metrically transitive process.

We also discuss methods to prove that the absolutely continuous spectrum is empty for some types of deterministic potentials, and some simple inverse problems, in which information on the Lyapunov exponent or closely related quantities yields information on the metrically transitive potential.

12.A Basic Facts About the Spectrum of One-Dimensional Operators of the Second Order

We start by presenting, in convenient form, some facts from the spectral theory of one-dimensional differential and finite-difference operators of the second order, which will be needed in this chapter and the next. Most of this material can be found in [Levitan and Sargsyan 1975; Titchmarsh 1946; Berezanskii 1968; Marchenko 1986].

Consider the operators $H_{a,b}^{(\alpha,\beta)}$ and $h_{a,b}^{(\alpha,\beta)}$ on the spaces $L_2(a,b)$ and $\ell_2(a+1,b)$, respectively, determined by equations (6.1) and (11.4) and the boundary conditions (6.2) and (6.70) with $s(x) \equiv 1$ (so that $h_{a,b}^{(\alpha,\beta)} = J_{a,b}\big|_{s=1}$, the matrix $J_{a,b}$ being specified by (6.71)). We assume that the potential $q(x)$ is a metrically transitive process on $\mathbf{R}$ or $\mathbf{Z}$ such that

$$\mathsf{E}\{q^2(x)\} < \infty \tag{12.1a}$$

in the continuous case, or

$$\mathsf{E}\{\ln(1+|q(x)|)\} < \infty \tag{12.1b}$$

in the discrete case: see (11.16b) and (11.42).

In particular, in the continuous case the potential is, with probability 1, locally square-integrable, and in the discrete case it is, with probability 1, finite for all $x \in \mathbf{Z}$. Then $H_{a,b}^{(\alpha,\beta)}$ and $h_{a,b}^{(\alpha,\beta)}$ are, with probability 1, essentially self-adjoint on the set of functions with compact support and smooth (in the continuous case) on (a,b).

Denote by $y_a^{(\alpha)}(x,\lambda)$ the solution of the Cauchy problem for equation (6.1) or (11.4) with initial conditions (6.3) and (6.70) at the point a, where $-\infty < a < \infty$. It can readily be shown by successive approximations that, under the conditions above, this solution exists, is unique, is measurable with respect to the σ-algebras $\mathcal{F}_a^x$, and for x fixed can be continued into the complex plane $\mathbf{C}$ with spectral parameter $z = \lambda + i\varepsilon$ as an entire function.

Given these solutions, we will write a representation for the *Green's function* $G_{a,b}^{(\alpha,\beta)}(x,x',z)$ of the operators $H_{a,b}^{(\alpha,\beta)}$ and $h_{a,b}^{(\alpha,\beta)}$, that is, of the kernel of the operators $(H_{a,b}^{(\alpha,\beta)} - z)^{-1}$ and $(h_{a,b}^{(\alpha,\beta)} - z)^{-1}$. We start by introducing some notation: if $f_1(x)$ and $f_2(x)$ are functions on the real line, we set

$$f_1(x) \wedge f_2(x') = \begin{cases} f_1(x)f_2(x') & \text{for } x \leqslant x', \\ f_1(x')f_2(x) & \text{for } x \geqslant x', \end{cases} \tag{12.2}$$

and

$$W(f_1,f_2) = \begin{cases} f_1'(x)f_2(x) - f_1(x)f_2'(x) \\ f_1(x+1)f_2(x) - f_1(x)f_2(x+1) \end{cases} \tag{12.3}$$

in the continuous and discrete case, respectively. Then we can write

$$G^{(\alpha,\beta)}_{a,b}(x,x',z) = W^{-1} y^{(\alpha)}_a \wedge y^{(\beta)}_b(x'), \tag{12.4}$$

where $W = W(y^{(\alpha)}_a, y^{(\beta)}_b)$ does not depend on x and is an entire function. This function has only simple real zeros, which are the eigenvalues λ_k, for $k = 1, 2, \ldots$, of the corresponding operator.

Let $\Delta \subset \mathbf{R}$ be an interval whose endpoints do not coincide with an eigenvalue. By the spectral theorem, the Green's function of a self-adjoint operator is related to the kernel $E(x,x',d\lambda)$ of its resolution of the identity by

$$G(x,x',z) = \int_{\mathbf{R}} (\lambda - z)^{-1} E(x,x',d\lambda). \tag{12.5}$$

Applying this to the kernel $E^{(\alpha,\beta)}_{a,b}(x,x',d\lambda)$ of the resolution of the identity of $H^{(\alpha,\beta)}_{a,b}$ and $h^{(\alpha,\beta)}_{a,b}$, and using formula A.3 from the Appendix, we get

$$E^{(\alpha,\beta)}_{a,b}(x,x',\Delta) = \sum_{\lambda_k \in \Delta} y^{(\alpha)}_a(x,\lambda_k) y^{(\alpha)}_a(x',\lambda_k) \mathcal{N}^{-1}_{(a,b)}(\lambda_k),$$

where

$$\mathcal{N}_{a,b}(\lambda) = \int_a^b \left(y^{(\alpha)}_a(x,\lambda)\right)^2 dx \tag{12.6a}$$

in the continuous case and

$$\mathcal{N}_{a,b}(\lambda) = \sum_{a+1}^{b} \left(y^{(\alpha)}_a(x,\lambda)\right)^2 \tag{12.6b}$$

in the discrete case.

The formula for $E^{(\alpha,\beta)}_{a,b}(x,x',\Delta)$ can be rewritten as

$$E^{(\alpha,\beta)}_{a,b}(x,x',\Delta) = \int_{\Delta} y^{(\alpha)}_a(x,\lambda) y^{(\alpha)}_a(x',\lambda) \rho^{(\alpha,\beta)}_{a,b}(d\lambda), \tag{12.7}$$

where

$$\rho^{(\alpha,\beta)}_{a,b}(\Delta) = \sum_{\lambda_k \in \Delta} \mathcal{N}^{-1}_{a,b}(\lambda_k) \tag{12.8}$$

is the so-called *spectral measure* of the operator.

We will be mostly interested in the case when the interval (a,b) is semi-infinite, that is, $a = -\infty$ and $b = 0$ or $a = 0$ and $b = \infty$. We call the corresponding operators $H^{(\alpha)}_{\pm}$ and $h^{(\alpha)}_{\pm}$, that is, $H^{(\alpha)}_{\pm}$ and $h^{(\alpha)}_{\pm}$ are the operators defined on $L_2(\mathbf{R}_\pm)$ and $\ell_2(\mathbf{Z}_\pm)$ by equations (6.1) and (11.4) with boundary conditions

(12.9a) $$\psi(0)\cos\alpha - \psi'(0)\sin\alpha = 0$$

and

(12.9b) $$\psi(0)\cos\alpha - \psi(1)\sin\alpha = 0,$$

respectively. If these operators are essentially self-adjoint on the set of functions with compact support and (in the continuous case) smooth, Weyl and Hamburger showed that, for every nonreal $z = \lambda + i\varepsilon$, the Cauchy problem has solutions $\phi_\pm(x,z)$, square-integrable near the points $\pm\infty$, and unique within a multiplicative constant. (One way to prove existence is to apply the resolvent of the operator to a function with compact support, concentrated on the semiaxis (a,b), and then extend the resulting function to the other semiaxis as the solution of the respective equation.) These solutions play the role of y_a and y_b in formulas (12.4) and (12.7). Equations (6.1) and (11.4) lead to the relations

(12.10a) $$\pm\operatorname{Im}\phi_\pm'(0,z)\overline{\phi_\pm(0,z)} = \varepsilon\int_{\mathbf{R}_\pm}|\phi_\pm(x,z)|^2\,dx,$$

(12.10b) $$\pm\operatorname{Im}\phi_\pm(1,z)\overline{\phi_\pm(0,z)} = \varepsilon\sum_{\mathbf{Z}_\pm}|\phi_\pm(x,z)|^2.$$

Therefore $\phi_\pm(0,z)\neq 0$ for $\operatorname{Im} z\neq 0$, and these solutions can be normalized by the condition $\phi_\pm(0,z) = 1$, which we assume from now on to be true. Using the fundamental systems of solutions $s(x,z)$ and $c(x,z)$, specified by the conditions

(12.11a) $$c(0,z) = 1,\quad c'(0,z) = 0,\qquad s(0,z) = 0,\quad s'(0,z) = 1$$

or

(12.11b) $$c(0,z) = 1,\quad c(1,z) = 0,\qquad s(0,z) = 0,\quad s(1,z) = 1,$$

the functions $\phi_\pm(x,z)$ can be represented as

(12.12) $$\phi_\pm(x,z) = c(x,z) \pm m_\pm(z)s(x,z),$$

where the coefficients $m_\pm(z)$ are usually called the *Weyl functions* on the respective semiaxes. By (12.11), $m_\pm(z)$ are *Nevanlinna functions*, that is, satisfy the property

(12.13) $$\operatorname{Im} m_\pm(z)\operatorname{Im} z > 0 \quad \text{for } \operatorname{Im} z\neq 0.$$

(Such functions are also called *Herglotz functions*: see [Duren 1970], for example. We follow here the tradition current in Russian.)

In the situation we're interested in—where, by Corollary 2.6, essential self-adjointness holds—the functions $m_\pm$, for every nonreal z, are the only limit points, as $L \to \pm\infty$, of the sequence of Weyl circles

$$\Delta_L^\pm(z) = \left\{ m \in \mathbf{C} : \pm \int_0^L \left|c(x,z) + ms(s,z)\right|^2 dx \leqslant \frac{\operatorname{Im} m}{\operatorname{Im} z} \right\} \tag{12.14}$$

(in the discrete case, naturally, the integral should be replaced by a sum). These circles have an important monotonicity property: $\Delta_{L_1}^\pm \supset \Delta_{L_2}^\pm$ for $|L_1| < |L_2|$.

The quantities

$$m_{0,L}^{(0,\beta)}(z) = -\frac{c(L,z)\cot\beta + c'(L,z)}{s(L,z)\cot\beta + s'(L,z)}, \tag{12.15}$$

which give, for a fixed nonreal z and $\beta \in [0,\pi)$, the boundaries of the circles as functions of z, are Nevanlinna functions, and

$$\operatorname{Im} m_{0,L}^{(0,\beta)}(z) = \varepsilon \int_{\mathbf{R}} |\mu - z|^{-2} \rho_{0,L}^{(0,\beta)}(d\mu). \tag{12.16}$$

From this and from the fact that the circles (12.14) converge, as $L \to \infty$, to a unique limit point, we conclude that the measures $\rho_{0,L}^{(0,\beta)}$ converge weakly to a limit measure $\rho_+^{(0)}$ independently of the boundary conditions at the point L (see Lemma 5.22, for example). Similar arguments for the case of an arbitrary boundary condition (6.2) lead to the relation

$$\lim_{L\to\infty} \rho_{0,L}^{(\alpha,\beta)}(d\lambda) = \rho_+^{(\alpha)}(d\lambda), \tag{12.17}$$

and the same relation for the measures $\rho_{-L,0}^{(\alpha,\beta)}$.

The *Green's functions* of the self-adjoint operators $H_\pm(\alpha)$ and $h_\pm(\alpha)$, by formulas (12.2)–(12.4), have the form

$$\begin{aligned} G_+^{(\alpha)}(x,x',z) &= \frac{y_0^{(\alpha)}(x,z) \wedge \phi_+(x',z)}{\cos\alpha - m_+(z)\sin\alpha}, \\ G_-^{(\alpha)}(x,x',z) &= \frac{\phi_-(x,z) \wedge y_0^{(\alpha)}(x',z)}{-\cos\alpha - m_-(z)\sin\alpha}. \end{aligned} \tag{12.18}$$

From (12.5) and from the equation obtained from (12.16) by letting $L \to \infty$, and using the inversion formula A.4 of the Appendix, we conclude that, for any interval $\Delta \subset \mathbf{R}$ whose endpoints are not atoms of $E_\pm^{(\alpha)}(d\lambda)$,

(12.19) $$E_{\pm}^{(\alpha)}(x,x',\Delta) = \int_{\Delta} y_0^{(\alpha)}(x,\lambda) y_0^{(\alpha)}(x',\lambda)\rho_{\pm}^{(\alpha)}(d\lambda).$$

It is thus clear that the *spectral measures* $\rho_{\pm}^{(\alpha)}(d\lambda)$, which are, by (12.17), weak limits of the measures (12.8) from (12.7), are sufficient to determine the spectral type of the operators $H_{\pm}^{(\alpha)}$ and $h_{\pm}^{(\alpha)}$. Furthermore, their Stieltjes transforms are the Nevanlinna functions

(12.20) $$m_{\pm}^{(\alpha)}(z) = \frac{\sin\alpha \pm m_{\pm}(z)\cos\alpha}{\pm\cos\alpha - m_{\pm}(z)\sin\alpha},$$

which are also limits of the functions $m_{0,L}^{(\alpha,\beta)}$ and $m_{-L,0}^{(\alpha,\beta)}$ as $L \to \infty$, with $\operatorname{Im} z \neq 0$.

The spectral measures $\rho_{\pm}^{(\alpha)}(d\lambda)$ also satisfy the conditions

(12.21a) $$\int_{\mathbf{R}} \frac{1}{1+|\mu|^{\delta}} \rho_{\pm}^{(\alpha)}(d\mu) < \infty$$

in the continuous case, where $\delta = 1$ if $\alpha \neq 0$ and $\delta = 2$ if $\alpha = 0$, and

(12.21b) $$\rho_{-}^{(\alpha)}(\mathbf{R}) = \sin^{-2}\alpha, \qquad \rho_{+}^{(\alpha)}(\mathbf{R}) = \cos^{-2}\alpha$$

in the discrete case.

It follows from (12.19) that the spectrum of a one-dimensional operator of the second order on a semiaxis always has multiplicity 1, and the operator itself is unitarily equivalent to the operator of multiplication by an independent variable in the space $L_2\big(\mathbf{R}, \rho_{\pm}^{(\alpha)}(d\lambda)\big)$.

When $\alpha = 0$, equation (12.20) gives

$$m_{\pm}^{(0)}(z) = m_{\pm}(z),$$

so that the Weyl functions $m_{\pm}(z)$ of (12.12) determine the spectral properties of the Dirichlet problems on the semiaxes $\mathbf{R}_{\pm}$, $\mathbf{Z}_{\pm}$, and

(12.22) $$m_{\pm}(z) = \left.\frac{\partial^2 G(x,x',z)}{\partial x\,\partial x'}\right|_{x=x'=0}$$

in the continuous case. In the discrete case, we can consider two Dirichlet problems, corresponding to zero boundary conditions at the points $x = 0$ or $x = 1$. In the first case, the Green's functions on the left and right semiaxes are

$$G_{1,+}(x,x') = s(x) \wedge \phi_{+}(x'),$$
$$G_{1,-}(x,x') = -\phi_{-}(x) \wedge s(x'),$$

while in the second,

$$G_{2,+}(x,x') = -m_+^{-1} c(x) \wedge \phi_+(x'),$$
$$G_{2,-}(x,x') = -m_-^{-1} \phi_+(x) \wedge c(x').$$

Hence we have

$$\begin{aligned} G_{1,+}(1,1) &= m_+, \qquad & G_{1,-}(-1,-1) &= m_- - z + q(0), \\ G_{2,+}(2,2) &= -m_+^{-1} - z + q(1), \qquad & G_{2,-}(0,0) &= -m_-^{-1}. \end{aligned} \tag{12.23}$$

These relations enable use to find the bounds that we'll need below for the Weyl functions. The simplest case is the discrete one, where from (12.23) and the inequality

$$|((A-z)^{-1}\psi,\psi))| \leqslant \frac{1}{\varepsilon}(\psi,\psi)$$

for $z = \lambda + i\varepsilon$, valid for any self-adjoint operator, we obtain

$$\begin{aligned} |m_+(z)| &\leqslant |\varepsilon|^{-1}, \qquad & \operatorname{Im} m_+(z) &\geqslant |\varepsilon|\big(|q(1)| + |z| + |\varepsilon|^{-1}\big)^{-2}, \\ |m_-(z)| &\leqslant |\varepsilon|^{-1} + |z| + |q(0)|, & \operatorname{Im} m_-(z) &\geqslant |\varepsilon|. \end{aligned} \tag{12.25}$$

In the continuous case, we set $R = (H_\pm^{(0)} - z)^{-1}$ and $R_0 = R\big|_{q=0}$. By the resolvent identity, we have

$$R = R_0 - R_0 q R_0 + R_0 q R q R_0. \tag{12.25}$$

Form this and the explicit form of the kernel $g_\pm(x,x',z)$ of R_0, namely

$$g_\pm(x,x',z) = \frac{e^{i\sqrt{z}|x-x'|}}{2i\sqrt{z}} - \frac{e^{i\sqrt{z}|x+x'|}}{2i\sqrt{z}}, \tag{12.25a}$$

we find, based on (12.22) and (12.25) (cf. the proof of Theorem 5.27):

$$\left| m_\pm(z) - i\sqrt{z} + \int_{\mathbf{R}_\pm} q(x) e^{2i\sqrt{z}|x|}\,dx \right| \leqslant \frac{1}{|\operatorname{Im} z|} \int_{\mathbf{R}_\pm} q^2(x) e^{-2\operatorname{Im} z|x|}\,dx. \tag{12.26}$$

We will also need the fact that the Weyl functions, for every fixed nonreal z, depends continuously on the potential, in the topology of pointwise convergence in the discrete case, and in the topology of L_2^{loc} in the continuous case. In the discrete case this follows from the fact that pointwise convergence of the potential implies strong convergence of resolvents of the corresponding operators. In both the discrete and the continuous cases, for an essentially self-adjoint operator (which ours are by Corollary 2.6), the same result can also be proved as follows: The value of the Weyl function is the unique limit point of the Weyl circles (12.14), which shrink monotonically as $L \to \infty$. To prove the continuity of the Weyl functions, then, it is enough to check that for any nonreal z the solutions $s(x,z)$ and $c(x,z)$ of (12.11) are continuous in the relevant topology, uniformly on any finite interval of **R** or **Z**. This, in turn, follows from the fact that these solutions satisfy the Volterra integral equations (or triangular systems of equations, in the discrete case).

The Weyl functions are important in spectral theory particularly because they uniquely determine the potential on the corresponding semiaxis. The simplest example occurs in the discrete case: by (12.16), $m_+(z)$ uniquely determines the spectral measure $\rho_+^{(0)}$, and then (12.19) shows that the solutions $s(x,z)$ are, as functions of λ, orthogonal polynomials with respect to $\rho_+^{(0)}$, and that, for $x \geqslant 1$,

$$q(x) = h_+^{(0)}(x,x) = \int_{\mathbf{R}} \lambda s^2(x,\lambda)\rho_+^{(0)}(d\lambda). \tag{12.27}$$

Therefore, to solve the inverse problem on the semiaxis $\mathbf{Z}_+$, that is, to find $q(x)$ for $x \geqslant 1$, it is enough to construct orthogonal polynomials with respect to $\rho_+^{(0)}$ and then use (12.27).

This method for constructing the potential has some drawbacks. First, it is difficult to study thoroughly the resulting solutions, and in particular to describe the class of measures that are spectral measures of operators of the second order. Second, the method cannot be generalized directly to the continuous case. A different method for solving the inverse problem, based on the technique of transformation operators and largely free of these drawbacks, was described by Marchenko [1986].

An important consequence of formulas like (12.5) or (12.19), containing eigenfunction expansions, is the derivation of polynomial bounds for generalized eigenfunctions. We start with the inequality

$$\mathsf{E}\{\operatorname{Im} G_\pm^{(\alpha)}(x,x,z)\} < \infty,$$

which holds for any nonreal z. This follows, in the discrete case, from the inequality $|G(x,x',z)| \leqslant |\varepsilon|^{-1}$ for the kernel of the resolvent of any self-adjoint operator in $\ell_2(\mathbf{Z}_\pm)$. In the continuous case it, together with relation (12.1a), follows from (12.25).

This inequality shows that, for any $\delta > 0$,

$$\int_{\mathbf{R}_\pm} \mathsf{E}\{\operatorname{Im} G_\pm^{(\alpha)}(x,x,z)\}\frac{dx}{1+|x|^{2\delta+1}} < \infty;$$

taking into account the spectral representations (12.5) and (12.9), we find that, with probability 1, for $\rho_\pm^{(\alpha)}$-almost all $\lambda \in \mathbf{R}$,

$$\int_{\mathbf{R}_\pm} \frac{\left|y_0^{(\alpha)}(x,\lambda)\right|^2 dx}{1+|x|^{1+2\delta}} < \infty. \tag{12.28a}$$

(In the discrete case, the integral should be interpreted as a sum.) From this follows, in the discrete case, the local bound

$$|y_0^{(\alpha)}(x,\lambda)| \leqslant C(\lambda,\delta,\omega)\big(1+|x|\big)^{\delta+1/2}. \tag{12.28b}$$

The existence of such local bounds in the continuous case requires additional conditions on the potential, on which we will not dwell now, because for our purposes it is sufficient, and more convenient, to have integral bounds like (12.28a).

Polynomial bounds like (12.28) are proved in essentially the same way for generalized eigenfunctions of operators on the whole axis.

It is important to note that such bounds do not apply only to metrically transitive operators, although for such operators they are proved very simply, as we have seen. Using techniques from the modern theory of differential and finite-difference operators, these bounds can be proved for multidimensional finite-difference and differential elliptic operators of a fairly general form [Schnol' 1954; Berezanskii 1968; Simon 1982]. Such bounds imply, in particular, that the spectrum of the operator is the closure of the set of values of λ for which the corresponding equation has polynomially bound solutions. This characterization of the spectrum turns out to be very useful in dealing with a variety of problems in spectral theory: for example, it provides an efficient technique for investigating the nature of the spectrum when the coefficients are of general form, that is, do not increase, decrease, or oscillate systematically enough.

We now consider operators on the whole axis **R** or **Z**. Here the Green's function is

$$G(x,x',z) = -\frac{\phi_-(x)\wedge\phi_+(x')}{\big(m_+(z)+m_-(z)\big)}, \tag{12.29}$$

and the kernel of the resolution of the identity may be written as

$$E(x,x',\Delta) = \sum_{r,r'=0,1}\int_\Delta \psi_r(x,\lambda)\psi_{r'}(x',\lambda)\rho_{r,r'}(d\lambda), \tag{12.30}$$

where

$$\psi_0(x,\lambda) = c(x,\lambda), \qquad \psi_1(x,\lambda) = s(x,\lambda), \tag{12.31}$$

and the measures $\rho_{r,r'}$ are given by

(12.32a)

$$\rho_{00}(d\lambda) = E(0,0,d\lambda), \qquad \rho_{01}(d\lambda) = \frac{\partial}{\partial x'}E(0,x',d\lambda)\bigg|_{x'=0},$$

$$\rho_{10}(d\lambda) = \frac{\partial}{\partial x}E(x,0,d\lambda)\bigg|_{x=0}, \qquad \rho_{11}(d\lambda) = \frac{\partial^2}{\partial x\,\partial x'}E(x,x',d\lambda)\bigg|_{x=x'=0}$$

in the continuous case, and

$$\rho_{r,r'}(d\lambda) = E(r, r', d\lambda) \tag{12.32b}$$

in the discrete case. This measure is specified by the inversion formula

$$\rho_{r,r'}(\varDelta) = \lim_{\varepsilon \downarrow 0} \frac{1}{\pi} \int_{\varDelta} \operatorname{Im} m_{r,r'}(\lambda + i\varepsilon)\, d\lambda, \tag{12.33}$$

where the $m_{r,r'}$ are the entries of the matrix

$$-\frac{1}{m_+ + m_-} \begin{pmatrix} 1 & m_+ \\ -m_- & -m_+ m_- \end{pmatrix}. \tag{12.34}$$

The matrix of measures $\rho_{r,r'}$, for $r, r' = 0, 1$, is Hermitian-positive, and therefore each $\rho_{r,r'}$ is absolutely continuous with respect to the measure

$$\rho(d\lambda) = \tfrac{1}{2}\big(\rho_{00}(d\lambda) + \rho_{11}(d\lambda)\big), \tag{12.35}$$

called the spectral measure of the corresponding operator H or h. This measure determines the spectral type of the operator.

12.B Lyapunov Exponents and the Absolutely Continuous Spectrum

We start by proving that if the Lyapunov exponent is positive in the neighborhood of some point of the spectrum, the spectrum has no absolutely continuous component. This type of result was first established by Casher and Lebowitz [1970] for a particular situation, and later by Ishii [1973] in more generality.

(12.1) Theorem [Pastur 1974a, 1980]. *Assume the Lyapunov exponent* $\gamma(\lambda)$ *of an equation of the form* (6.1) *or* (11.4), *with metrically transitive potential satisfying* (12.1), *is positive for Lebesgue-almost all* λ *in some Borel set* X *of positive Lebesgue measure. Then, with probability* 1, *the corresponding operators have no absolutely continuous spectrum on* X, *that is,* $\operatorname{meas}(\sigma_{\mathrm{ac}} \cap X) = 0$.

We present two proofs here; additional ones can be found in [Deift and Simon 1983; Kotani 1986a].

First proof. We show first that the theorem is true for the operators $H_\pm^{(\alpha)}$ and $h_\pm^{(\alpha)}$ given by the boundary problem (12.9) on a semiaxis. To do so, we have to show that $\rho_\pm^{(\alpha)}(X) = 0$ with probability 1. According to (12.17), the spectral measures $\rho_\pm^{(\alpha)}$ are weak limits of the measures $\rho_{0,L}^{(\alpha,\beta)}$ and $\rho_{-L,0}^{(\beta,\alpha)}$ of

(12.8), corresponding to the boundary problem on the intervals $(0, L)$ and $(-L, 0)$, and these measures, in turn, are related to $N_{0,L}(d\lambda)$ and $N_{-L,0}(d\lambda)$, the prelimit normalized distribution function of the eigenvalues, in the following way:

$$N_{0,L}(x) = \int_X Y_L(\lambda)\rho_{0,L}^{(\alpha,\beta)}(d\lambda), \tag{12.36}$$

where the function $Y_L(\lambda)$, by (12.9), is given by

$$Y_L(\lambda) = \frac{1}{L}\int_0^L \left(y_0^{(\alpha)}(x,\lambda)\right)^2 dx \tag{12.37a}$$

in the continuous case, and

$$Y_L(\lambda) = \frac{1}{L}\sum_{x=1}^{L}\left(y_0^{(\alpha)}(x,\lambda)\right)^2 \tag{12.37b}$$

in the discrete case (the expression for $N_{-L,0}$ is similar). By (11.5), we have

$$2LY_L(\lambda) \geqslant r^2(1) + \cdots + r^2(L') \tag{12.38}$$

in the discrete case, where L' is the largest even number strictly smaller than L. Therefore it follows from Theorem 11.18 that, for every α in a set of full Lebesgue measure in $[0, \pi)$, we have $\lim_{L\to\infty} Y_L(\lambda) = \infty$ with probability 1 for Lebesgue-almost $\lambda \in \mathbf{R}$. On the other hand, according to Theorem 6.18, the measures $N_{0,L}$ have a finite limit as $L \to \infty$, with probability 1. These facts are only consistent with formula (12.36) when the measure $\rho_+(\alpha)$ has no absolutely continuous component on X for almost every $\alpha \in [0, \pi)$.

The measure $\rho_-^{(\alpha)}$ clearly possesses the same property. This means that the spectra of $h_+^{(\alpha)}$ and $h_-^{(\alpha)}$, for almost every $\omega \in \Omega$ and almost every $\alpha \in [0, \pi)$, have no absolutely continuous component in X. Hence, the operator $h_0 = h_+^{(\alpha)} \oplus h_-^{(\alpha)}$ acting on $\ell_2(\mathbf{Z})$ has the same property. But the difference $h_1 = h - h_0$ is a matrix of the form

$$h_1 = \begin{pmatrix} \ddots & \vdots & \vdots \\ \cdots & \cot\alpha & -1 \\ \cdots & -1 & \cot\alpha \end{pmatrix}, \tag{12.39}$$

where all the entries indicated by dots vanish. Since the absolutely continuous spectrum is stable with respect to finite-dimensional perturbations (see Problem II.23(a) and [Kato 1966]), we conclude that h has no absolutely continuous component in X, as we wished to show.

The proof in the continuous case is similar, and based on the following facts. First, the resolvents of the operators H and $H_0 = H_+^{(\alpha)} \oplus H_-^{(\alpha)}$ differ by a one-dimensional operator. Second, the absolutely continuous component is stable with respect to finite-dimensional, and even trace-class, perturbations of the resolvent [Kato 1966]. Finally, for almost all $\lambda \in X$, the functions $Y_L(\lambda)$ from (12.37b) tend to infinity as $L \to \infty$, with probability 1. This last fact is proved as follows:

By integrating equation (6.5) for the phase $\theta(x)$ from δL to L, where $0 < \delta < 1$, and using some elementary estimates, we see that, for any $A > 0$,

$$\left|L^{-1}\theta(L) - L^{-1}\theta(\delta L) + (1-\delta)\right| - \frac{1}{L}\int_{\delta L}^{L} Q(t)\chi_{[0,\infty)}\big(A - Q(t)\big)\,dt \leqslant \frac{A}{L}\int_{\delta L}^{L} \sin^2\theta(t)\,dt,$$

where $Q(t) = \left|q(t) - \lambda + 1\right|$. As $L \to \infty$, the first term on the left goes to $(1-\delta)\left|\pi N(\lambda) - 1\right|$ with probability 1, by Theorem 6.5, and the second term goes to $(1-\delta)\mathsf{E}\left\{Q\chi_{[0,\infty)}(A - Q)\right\}$, also with probability 1, by the ergodic theorem. But this latter expression can be made arbitrarily small for large A and fixed λ, by condition (12.1a). Therefore, with probability 1, we can guarantee that the integral $L^{-1}\int_{\delta L}^{L}\sin^2\theta(t)\,dt$ is strictly positive for almost all $\lambda \in X$. Since $\delta \in (0,1)$ is arbitrary, we conclude from (12.37a), (11.2) and Theorem 11.18 that

$$\lim_{L\to\infty}\frac{1}{L}\ln Y_L(\lambda) = 2\gamma(\lambda),$$

showing that, with probability 1, $Y_L(\lambda)$ grows exponentially as $L \to \infty$ for almost all $\lambda \in X$ and $\alpha \in [0,\pi)$. □

Second proof. This proof is based on polynomial bounds like (12.28a). It follows from Theorem 2.12 that, for the operators we're considering, any fixed λ has probability 1 of not being an eigenvalue. From Theorem 11.3 it follows then that for any $\lambda \in X$, with probability 1,

$$\max\{\gamma_+(\lambda,\alpha,\omega), \gamma_-(\lambda,\alpha,\omega)\} = \gamma(\lambda) > 0$$

for $\alpha \in [0,\pi)$; otherwise, for some α we would have $\gamma_+ = \gamma_- = -\gamma$, that is, λ would be an eigenvalue with an exponentially decreasing eigenfunction. Thus, with probability 1, for Lebesgue-almost $\lambda \in X$, the amplitude $r^2(x)$ of the solution $y_0^{(\alpha)}(x,\lambda)$, for any $\alpha \in [0,\pi)$, is exponentially increasing for at least one of the limits $x \to \infty$ or $x \to -\infty$.

By the same argument as in the first proof, we conclude that, with probability 1 and for Lebesgue-almost $\lambda \in X$, the integral $Y_L(\lambda)$ of $y_0^{(\alpha)}(x,\lambda)$, given by (12.37a), increases exponentially as $L \to \infty$. In particular, this is the case for all generalized eigenfunctions that are solutions of the corresponding second-order equation and satisfy an initial condition of the form (6.3) or (6.72) for some α (which generally depends on λ and ω).

On the other hand, one can write an inequality like (12.28a) for operators on the whole axis, and that shows that for ρ-almost all $\lambda \in X$, where ρ is the spectral measure (12.35), the integral (12.37a) for generalized eigenfunctions cannot grow more rapidly than $O(L^{1+2\delta})$. Consequently, with probability 1, the measure ρ has no absolutely continuous component on X, as we wished to show. □

Theorem 12.1 can be interpreted as saying that, with probability 1, the spectral measure $\rho(d\lambda)$ is singular with respect to the Lebesgue measure. Thus the next result can be seen as a generalization of Theorem 12.1:

(12.2) Theorem. *Under the same conditions as Theorem* 12.1, *the spectral measure* $\rho(d\lambda)$ *is, with probability* 1, *singular with respect to any fixed nonrandom positive measure* $\nu(d\lambda)$ *on the set* X, *that is, the Radon-Nikodym derivative* $d\rho/d\nu$ *vanishes* ν*-almost everywhere on* X.

Proof. The second proof of Theorem 12.1 still works, if we notice that, because Theorem 11.3 is true for every fixed λ, we could have considered any positive nonrandom measure instead of the Lebesgue measure on the spectral axis $\mathbf{R}$. □

Theorems 12.1 and 12.2 are still valid for multi-component finite-difference or differential metrically transitive operators (or systems of one-component operators) of the second order, that is, for operator (11.86) and for its continuous analogue

$$(H_n\Psi)(x) = -\Psi''(x) + q_n(x)\Psi(x) \tag{12.41}$$

for $x \in \mathbf{R}$, where $q_n(x)$ is a metrically transitive field whose values are $n \times n$ Hermitian matrices and $\Psi(x)$ is an n-component vector function such that

$$\int_{\mathbf{R}} \|\Psi(x)\|^2_{\mathbf{C}^n}\, dx < \infty.$$

Under condition (11.87), in the discrete case, and

$$\mathsf{E}\{\|q_n(x)\|^2_{\mathbf{C}^n}\} < \infty,$$

in the continuous case, such operators admit nonrandom characteristic Lyapunov exponents $\gamma_j(\lambda)$, ordered according to (11.90). A sufficient condition for the absence of the absolutely continuous component in this case is that the smallest nonnegative Lyapunov exponent $\gamma_n(\lambda)$ be positive. The second proof of Theorem 12.1 can be repeated almost verbatim, because polynomial bounds like (12.28) and (12.40) apply equally well to the generalized eigenfunctions of systems of operators [Berezanskii 1968].

The first proof can also be applied with very little change, at least in the discrete case. We just have to rewrite formulas (12.36)–(12.38) with $\|y(x,\lambda)\|^2_{\mathbf{C}^n}$ instead of $y^2(x,\lambda)$, $\|r(x)\|^2_{\mathbf{C}^n}$ instead of $r^2(x)$, and so on. Obviously (12.39) is to be interpreted as a block matrix, so its rank is at most $2n$ and we're still applying a perturbation of finite rank.

In the continuous case, the first proof of Theorem 12.1 requires some modification, for then the phase $\theta(x)$ is replaced by a vector in projective space $\mathbf{P}^{n-1}$: see Problem V.16. We will not discuss this modification.

We now present an elegant converse of Theorem 12.1, due to Kotani [1984]. The technique used in the proof will be applied later on to establish a number of results on the spectrum: see Theorems 12.9 and 12.10. All these results were first obtained by Kotani for Schrödinger operators with a bounded potential. They were extended to the discrete case, still with a bounded potential, by Simon [1983]. More general operators of the form (1.6) are considered in [Minami 1986], and unbounded potentials in [Kirsch et al. 1985].

(12.3) Lemma. *Let* $\gamma(z)$, *for nonreal* z, *be the Lyapunov exponent* (11.25) *of equation* (6.1) *or* (11.4), *and let* $m_\pm(z)$ *be the Weyl functions* (12.12) *of the same equation. We have*

$$\mathsf{E}\{\operatorname{Re} m_\pm(z)\} = -\gamma(z), \tag{12.43a}$$

$$\mathsf{E}\left\{\frac{1}{\operatorname{Im} m_\pm(z)}\right\} = \frac{2\gamma(z)}{\operatorname{Im} z} \tag{12.44a}$$

in the continuous case, and

$$\mathsf{E}\{\ln|m_\pm(z)|\} = -\gamma(z), \tag{12.43b}$$

$$\pm\mathsf{E}\left\{\ln\left(1 \pm \frac{\operatorname{Im} z}{\operatorname{Im} m_\pm(z)}\right)\right\} = 2\gamma(z) \tag{12.44b}$$

in the discrete case.

Proof. Consider the Weyl solutions $\phi_\pm(x,z,\omega)$ from (12.12), which are square-integrable for $\operatorname{Im} z \neq 0$ on the right and left semiaxes, respectively (we explicitly indicate here the dependence of these solutions on the potential). These solutions are uniquely defined by this integrability condition at infinity and by the condition $\phi_\pm(0,z,\omega) = 1$ at the origin. This implies that

$$\phi_\pm(x+y,z,\omega) = \phi_\pm(x,z,T_y\omega)\phi_+(y,z,\omega). \tag{12.45}$$

Therefore, if we introduce the metrically transitive process

(12.46) $$\zeta_\pm(x,\omega,z) = m_\pm(z, T_x\omega),$$

bounded for $\operatorname{Im} z \neq 0$ by virtue of (12.26), we will have, in the continuous case,

(12.47a) $$\zeta_\pm = \pm\frac{d\phi_\pm/dx}{\phi_\pm}.$$

Using (12.26) and the ergodic theorem (1.15), we conclude that, with probability 1,

(12.48a)
$$\mathsf{E}\{\operatorname{Re} m_\pm(z)\} = \lim_{|x|\to\infty}\frac{1}{|x|}\int_0^{|x|}\operatorname{Re}\zeta_\pm(\pm|t|)\,dt = \lim_{|x|\to\infty}\frac{1}{|x|}\ln|\phi_\pm(x)|.$$

Now, from the identity

$$|\phi_\pm(x)|^2 + |\phi'_\pm(x)|^2 = |\phi_\pm(x)|^2(1+|\zeta_\pm(x)|^2)$$

and from (12.26), it follows that the right-hand side of (12.48a) equals $\lim_{|x|\to\infty}|x|^{-1}\ln r(x)$. But this limit, by Theorem 11.3 and the fact that $\phi_\pm$ is square-integrable, equals $-\gamma(z)$, which proves (12.43a).

To prove (12.44a), we notice that (12.46) and the Schrödinger equation imply that the process $\zeta_\pm(x,\omega)$ satisfies the Ricatti equation

(12.49a) $$\pm\zeta'_\pm = -\zeta_\pm^2 - z + q(x).$$

Writing $\zeta = \xi + i\eta$ and $z = \lambda + i\varepsilon$, this implies that

(12.50a) $$\pm\frac{\eta'_\pm}{\eta_\pm} = -2\xi_\pm - \frac{\varepsilon}{\eta_\pm}.$$

We now take the mathematical expectation of both sides of this relation. Using (12.43a) and the fact that the mathematical expectation of the derivative of a differentiable metrically transitive process is zero, we reach (12.44a).

Formulas (12.43b) and (12.44b) are proved in a similar way, using the relations

(12.47b) $$\zeta_\pm(x,\omega) = \pm\frac{\phi_\pm(x)}{\phi_\pm(x-1)},$$

(12.48b)
$$\mathsf{E}\{\ln|m_\pm|\} = \lim_{|x|\to\infty}\frac{1}{|x|}\sum_1^x \ln\left|\frac{\phi_\pm(x)}{\phi_\pm(x-1)}\right| = \lim_{|x|\to\infty}\frac{1}{|x|}\ln|\phi_\pm(x)| = -\gamma(z),$$

(12.49b) $$\mp(\zeta_\pm(x+1) + \zeta_\pm^{-1}(x)) + q(x) - z = 0,$$

(12.50b) $$\eta_\pm(x-1)|\zeta_\pm(x-1)|^{-2} = \eta_\pm(x) \pm \varepsilon,$$

and inequalities (12.24). □

(12.4) Lemma. *Assume that $\gamma(\lambda) = 0$ for Lebesgue-almost all $\lambda \in X$, where X is a Borel set of positive Lebesgue measure. With probability 1, for almost all $\lambda \in X$,*

$$\operatorname{Im} m_{\pm}(\lambda + i0) > 0, \tag{12.51}$$

$$m_{+}(\lambda + i0) + \overline{m_{-}(\lambda + i0)} = 0. \tag{12.52}$$

Furthermore, for any fixed $x \in \mathbf{R}$ or $\mathbf{Z}$,

$$\operatorname{Re} G(x, x, \lambda + i0) = 0, \tag{12.53}$$

where $G(x, x', z)$ is the Green's function of the corresponding operator.

Proof. Consider the function $w(z)$ of complex variable z defined by

$$w(z) = -\sqrt{-z} - \int_{\mathbf{R}} \ln(\mu - z)\big(N(d\mu) - N_0(d\mu)\big) \tag{12.54a}$$

in the continuous case and

$$w(z) = -\int_{\mathbf{R}} \ln(\mu - z) N(d\mu) \tag{12.54b}$$

in the discrete case, where the branches for the log and square root functions are chosen in the standard way: $\ln 1 = 0$ and $\sqrt{1} = 1$. It is clear that $w(z)$ is analytic outside the real axis, and, by Theorem 11.9,

$$-\operatorname{Re} w(z) = \gamma(z). \tag{12.55}$$

By the condition of the lemma and the analyticity of $w(z)$, we conclude that, for Lebesgue-almost $\lambda \in X$,

$$\begin{aligned} \lim_{\varepsilon \downarrow 0} \frac{1}{\varepsilon} \gamma(\lambda + i\varepsilon) &= \lim_{\varepsilon \downarrow 0} \frac{\partial}{\partial \varepsilon} \gamma(\lambda + i\varepsilon) \\ &= -\lim_{\varepsilon \downarrow 0} \frac{\partial}{\partial \varepsilon} \operatorname{Re} w(\lambda + i\varepsilon) = \operatorname{Im} w'(\lambda + i0). \end{aligned} \tag{12.56}$$

But (12.54) implies that

$$w'(z) = \int_{\mathbf{R}} \frac{1}{\mu - z} N(d\mu), \tag{12.57}$$

that is, $w'(z)$ is the Stieltjes transform of the nonnegative measure $N(d\lambda)$, so we can apply Theorem A.4 of the Appendix to conclude that $\operatorname{Im} w'$ has finite limit values for Lebesgue-almost all $\lambda \in \mathbf{R}$. Thus the limit in (12.56) is finite.

The same reasoning applied to (12.44a), together with Fatou's theorem on the limit of a mathematical expectation, gives

$$\mathsf{E}\{(\operatorname{Im} m_\pm(\lambda + i0))^{-1}\} < \infty, \tag{12.58}$$

from which (12.51) follows in the continuous case. For the discrete case we apply a similar argument to (12.44b) and (12.56), keeping in mind the elementary inequalities $\ln(1 + a) \geqslant a/(1 + a)$ if $a \geqslant 0$, and $-\ln(1 - a) \geqslant a$ if $0 \leqslant a \leqslant 1$.

To prove (12.52), notice that, by (12.56), for almost all $\lambda \in X$,

$$\lim_{\varepsilon \downarrow 0}(-\varepsilon^{-1} \operatorname{Re} w(\lambda + i\varepsilon) - \operatorname{Im} w'(\lambda + i0)) = 0. \tag{12.59}$$

But, because of (12.12), (12.29) and (12.57),

$$w'(z) = \mathsf{E}\{G(0,0,z)\} = -\mathsf{E}\{(m_+(z) + m_-(z))^{-1}\}. \tag{12.60}$$

Thus, in the continuous case, because of (12.44a), we can rewrite (12.59) as

$$\begin{aligned}
&\lim_{\varepsilon \downarrow 0} \mathsf{E}\{\tfrac{1}{4}(\eta_+^{-1} + \eta_-^{-1}) + \operatorname{Im}(m_+ + m_-)^{-1}\} \\
&= \lim_{\varepsilon \downarrow 0} \tfrac{1}{4}\mathsf{E}\{(\eta_+^{-1} + \eta_-^{-1})((\xi_+ + \xi_-)^2 + (\eta_+ - \eta_-)^2)\,|m_+ + m_-|^{-2}\} = 0,
\end{aligned}$$

where $m_\pm = \xi_\pm + i\eta_\pm$. This shows (12.52) in the continuous case, again because the limit values of Nevanlinna functions on the real axis are almost everywhere finite. In the discrete case, the proof is similar, starting from (12.44b) instead of (12.44a).

Finally, to prove (12.53), we use (12.52) and the representation

$$G(0,0,z) = -\big(m_+(z) + m_-(z)\big)^{-1}, \tag{12.61}$$

which follows from (12.12) and (12.29). We conclude that, with probability 1, $\operatorname{Re} G(0,0,\lambda + i0) = 0$ for $\lambda \in X$, and then use the fact that

$$G(x,x,z,\omega) = G(0,0,z,T_x\omega),$$

which holds for the kernel of any metrically transitive operator, by Problem I.16. (But in our case this relation also follows from (12.29) and (12.45).) □

(12.5) Theorem [Kotani 1984; Simon 1983]. *For an equation of the form* (6.1) *or* (11.4) *with potential satisfying* (12.1), *suppose that* $\gamma(\lambda)$ *vanishes for Lebesgue-almost all* λ *in a Borel set* $X \subset \mathbf{R}$. *Then, with probability* 1,

(i) $\operatorname{meas}(\sigma_{\mathrm{ac}} \cap X) = \operatorname{meas} X$, *and*

(ii) *if* $X = \Delta$ *is an open interval, the spectrum of the corresponding operator on* Δ *is purely absolutely continuous and the spectral measure* $\rho(d\lambda)$ *has analytic density on* Δ.

Proof. By general principles (see (12.30)), the existence of the absolutely continuous spectrum on X is equivalent to the positiveness on X of the density $\rho'_{\text{ac}}(\lambda)$ of the absolutely continuous component ρ_{ac} of the spectral measure ρ (12.35). Since, by (12.34) and (12.35), ρ is the measure corresponding to the Nevanlinna function

$$g(z) = \tfrac{1}{2}\left(-(m_+ + m_-)^{-1} + (m_+^{-1} + m_-^{-1})^{-1}\right), \tag{12.62}$$

we can use Theorem A.10(i) of the Appendix to conclude that

$$\rho'_{\text{ac}}(\lambda) = \frac{1}{\pi} \operatorname{Im} g(\lambda + i0).$$

But, by (12.52), we have

$$\operatorname{Im} g(\lambda + i0) = \frac{1 + \left|m_+(\lambda + i0)\right|^2}{4 \operatorname{Im} m_+(\lambda + i0)},$$

so that, by (A.7), (12.51) is also finite and positive almost everywhere on X. Hence, by Theorem A.10, part (i) of the theorem is verified.

To prove (ii), we notice that, by (12.53), $\operatorname{Re} g(\lambda + i0) = 0$ almost everywhere on Δ. We then use Theorem A.11(iv) (the symmetry principle). □

Theorems 12.1 and 12.5 can be formulated in a compact and elegant form via the concept of the essential closure of a set:

Definition. *The essential closure $\bar{X}^{\text{e}}$ of a measurable set $X \subset \mathbf{R}$ is the set of points $x \in \mathbf{R}$ such that every neighborhood of x has an intersection of positive Lebesgue measure with X.*

Clearly, $\bar{X}^{\text{e}}$ is closed for any $X \subset \mathbf{R}$, and if σ_{ac} is the absolutely continuous spectrum of a self-adjoint operator, $\bar{\sigma}^{\text{e}}_{\text{ac}} = \sigma_{\text{ac}}$.

(12.6) Theorem [Kotani 1984]. *Under the conditions of Theorems* 12.1 *and* 12.5, *σ_{ac} is the essential closure of the set Γ of $\lambda \in \mathbf{R}$ such that* $\gamma(\lambda) = 0$.

(12.7) Theorem [Kotani 1984]. *Under the conditions of Theorems* 12.1 *and* 12.5, *with probability* 1, *σ_{ac} is the essential closure of the set Γ_G of $\lambda \in \mathbf{R}$ such that* $\operatorname{Re} G(0, 0, \lambda + i0) = 0$.

Proof of Theorems 12.6 and 12.7. From Theorems 12.1 and 12.5 and from equality (A.8) and Theorem A.10(i) of the Appendix, we have

$$\operatorname{meas}(\sigma_{\text{ac}} \setminus \Gamma) = \operatorname{meas}(\Gamma \setminus \Gamma_G) = \operatorname{meas}(\Gamma_G \setminus \sigma_{\text{ac}}) = 0.$$

But if X and Y are Borel sets and $\text{meas}(X \setminus Y) = 0$, it is easy to see that $\bar{X}^e = \bar{Y}^e$, so

$$\sigma_{ac} \subset \bar{\Gamma}^e \subset \bar{\Gamma}^e_G \subset \sigma_{ac},$$

which concludes the proof. □

12.C Multiplicity of the Spectrum

We investigate the multiplicity of the spectrum of a one-dimensional Schrödinger operator H or Jacobi matrix h. Because of (12.30) (see [Reed and Simon I], for example), such an operator is unitarily equivalent to the operator of multiplication by λ in the space $L_2(\mathbf{R}, \rho_2) \oplus L_2(\mathbf{R}, \rho_2) \oplus L_2(\mathbf{R}, \rho_1)$, where the nonnegative measures ρ_1 and ρ_2 are mutually singular and their sum is the spectral measure ρ of the operator, $\rho = \rho_1 \oplus \rho_2$. We represent the spectrum σ of the operator as $\sigma_1 \cup \sigma_2$, where $\rho_1(\sigma_2) = \rho_2(\sigma_1) = 0$; the sets σ_1 and σ_2 are called the *homogeneous* (or *maximally homogeneous*) *components* of multiplicity 1 and 2, respectively.

Since the Weyl functions $m_\pm$ uniquely determine the spectral measures $\rho_\pm^{(0)}$ of the operators $H_\pm^{(0)}$ or $h_\pm^{(0)}$, it follows from Lemma 12.4 that on the set $\Gamma = \{\lambda \in \mathbf{R} : \gamma(\lambda) = 0\}$ the spectra of these operators are absolutely continuous and, by a general theorem (see (12.19), for example), have multiplicity 1. But, as explained in the first proof of Theorem 12.1, the orthogonal sum of these operators differs from the original operator H or h by an operator of rank at most 2 (after taking resolvents, in the continuous case). Since the absolutely continuous parts of self-adjoint operators that differ by a finite-dimensional operator are unitarily equivalent [Kato 1966], the spectrum of H or h on Γ has multiplicity 2. In view of Theorem 12.6, this actually means that the absolutely continuous spectrum of a one-dimensional metrically transitive operator of second order has homogeneous multiplicity 2 [Deift and Simon 1983].

On the other hand, the Wronskian (12.3) of two solutions with same λ does not depend on x, and therefore is 0 for any pair of decreasing solution. Thus, the point spectrum of our operators always has multiplicity 1.

Finally, to determine the multiplicity of the singular continuous component of the spectrum (if it exists), we use the following general result:

(12.8) Proposition [Kac 1963]. *Let A be an essentially self-adjoint Sturm–Liouville operator* (1.6) *such $s^{-1}(x)$ and $q(x)$ are locally summable, or an essentially self-adjoint Jacobi matrix* (1.4). *Let $m_\pm(z)$ be the Weyl functions for A, and $S_\pm$ the set of points $\lambda \in \mathbf{R}$ such that the limit $m_\pm(\lambda + i0)$ exists and is finite. If the set $S = S_+ \cap S_-$ has zero Lebesgue measure, the operator A has a simple spectrum. If S has positive measure, it coincides, up to a set of zero spectral measure, with the homogeneous part of multiplicity* 2 *of* *the spectrum (in other words, $\rho_1(S) = 0$). Finally, the spectrum of A is absolutely continuous on S, and consequently, so is ρ.*

Sketch of proof. The proof of this important statement is based on the unitary equivalence of A with an operator of multiplication by an independent variable in the space of two-component vector functions, square-integrable with respect to the matrix measure $\rho_{r,r'}$ of (12.32). This equivalence is a consequence of (12.30). One studies the rank of the matrix $\delta_{r,r'}(\lambda)$ of derivatives of the measures $\rho_{r,r'}$ with respect to the spectral measure ρ of (12.35), as a function of the character of the boundary values of the function $m_\pm$, which determine, according to (12.33) and (12.34), the measures $\rho_{r,r'}$. It turns out that the rank of $\delta_{r,r'}(\lambda)$ is 2 at points in S, while on $\operatorname{supp}\rho \setminus S$ it is 1 ρ-almost everywhere. Then one shows that the ρ-measure of S is positive if and only if its Lebesgue measure is. □

It follows from Proposition 12.8 that the singular spectrum of one-dimensional operators of second order always has multiplicity 1. This leads to the following theorem, which completely characterizes the multiplicity of the spectrum:

(12.9) Theorem. *For a metrically transitive operator H or h of the form* (6.1) *or* (11.4), *with potential satisfying* (12.1), *the absolutely continuous spectrum has multiplicity* 2, *with probability* 1, *and the singular spectrum has multiplicity* 1, *again with probability* 1. *In other words, if $\rho = \rho_1 \oplus \rho_2$ is the decomposition of the spectral measure into components according to multiplicity, $\rho_2 = \rho_{\mathrm{ac}}$ and $\rho_1 = \rho_{\mathrm{sc}} \oplus \rho_p$.*

Proof. We use Proposition 12.8 and the notations introduced there. From Theorem 12.5 it follows that, with probability 1 and up to sets of zero Lebesgue measure, the sets $S_\pm$ contain the absolutely continuous spectrum σ_{ac}, that is, $\operatorname{meas}(\sigma_{\mathrm{ac}} \setminus S_\pm) = 0$. On the other hand, since the absolutely continuous spectra of the operators H and $H_+^{(0)} \oplus H_-^{(0)}$ coincide (and similarly for h) by the argument in the first proof of Theorem 12.1, we have $\sigma_{\mathrm{ac}} = S_+ \cup S_-$, again up to sets of zero Lebesgue measure. Therefore $S_+ \cup S_- \,\triangle\, \sigma_{\mathrm{ac}}$ has zero Lebesgue measure, which, in view of Proposition 12.8, implies the theorem. □

12.D Deterministic Potentials

Till now we have made no assumptions on the metrically transitive potential, except for (12.1), which is necessary in order to prove the existence of various spectral objects. Thus our results so far apply to one-dimensional operators of order two having a wide range of statistical properties, from periodic to totally random (independent at different points, or Markovian). The theorem below is an example of a result of a different type: it requires the potential to be nondeterministic (see 1.C), that is, "random" in a sense that agrees at least somewhat with the intuitive concept of the word. As usual, we denote by H or h a one-dimensional, metrically transitive differential or finite-difference operators of the second order, of the form (6.1) or (11.4).

(12.10) Theorem [Kotani 1984 and Simon 1983]. *If the potential of H or h satisfies* (12.1) *and is a nondeterministic random process on* $\mathbf{R}$ *or* $\mathbf{Z}$,

(i) *the Lyapunov exponent $\gamma(\lambda)$ is positive for Lebesgue-almost all $\lambda \in \mathbf{R}$, and*

(ii) *with probability* 1, *the spectrum of H or h has no absolutely continuous component* (*that is, is singular*) *and has multiplicity* 1.

Proof. Part (ii) follows from (i) and Theorems 12.6 and 12.9. For part (i), let $\mathcal{F}^0_{-\infty}$ denote, as in Section 1, the σ-algebra generated by the values of the potential on the left semiaxis $\mathbf{R}_-$ or $\mathbf{Z}_-$. It is clear that the Weyl function $m_-(z)$ on this semiaxis is measurable with respect to this σ-algebra.

Assume that $\gamma(\lambda) = 0$ for almost all $\lambda \in X$, where X is a Borel set of positive Lebesgue measure. By Lemma 12.4, the limit values of the Weyl functions are related by $m_+(\lambda + i0) = -\overline{m_-(\lambda + i0)}$, for $\lambda \in X$. Now Theorem A.5 in the Appendix says that a Nevanlinna function is uniquely determined by its limit values on a set of positive measure, so $m_+(z)$ is uniquely determined by $q_- \in \mathcal{F}^0_{-\infty}$. But then $q_+ = \{q(x) : x \in \mathbf{R}_+ \text{ or } \mathbf{Z}_+\}$ is also uniquely determined, because the inverse problem of spectral analysis for the boundary problem on the semiaxis has a unique solution: see Section 12.A and [Marchenko 1986]. By stationarity, a whole realization of the potential $q(x)$ can be reconstructed from its restriction to $(-\infty, a]$ for any a. But this implies (see 1.C) that $\mathcal{F}_{-\infty} = \bigcap_a \mathcal{F}^a_{-\infty} = \mathcal{F}^{\infty}_{-\infty}$,that is, the random process is deterministic, contradicting the assumption of the theorem. □

This theorem provides a fairly general sufficient criterion for $\gamma(\lambda)$ to be positive for almost all λ, for operators of the form (6.1) or (11.4). For example, for the discrete counterpart of the Gaussian random potential of Example 1.15(c), the Lyapunov exponent is positive for almost all $\lambda \in \mathbf{R}$ if the spectral function $\hat{b}(k)$ of the potential satisfies $\int_{-\infty}^{\infty} \ln \hat{b}(k)\, dk > -\infty$.

We have mentioned that the concept of a nondeterministic process is one of several natural ways to formalize the notion of a "sufficiently random" function; thus Theorem 12.10 can be seen as saying that positive Lyapunov exponents for almost all $\lambda \in \mathbf{R}$ and an empty absolutely continuous spectral component go hand in hand with randomness of the potential. Conversely, for a periodic potential, which is highly nonrandom, the whole spectrum is absolutely continuous and the Lyapunov exponent vanishes on the spectrum, and only there.

The absolutely continuous component may be nonempty for almost periodic operators. For such operators and related ones, Theorem 12.10 is especially interesting, and we will see in Chapter VII several examples of almost periodic operators with a big absolutely continuous spectrum.

Theorem 12.10, although fairly general, is nonetheless inapplicable to many metrically transitive potentials that intuitively seem quite random. For example, the Poisson potential (1.29) and the potential (1.31) with independent, identically distributed random variables ξ_a are deterministic if the function $u(x)$ decays fast enough and can be analytically continued onto a strip. But intuitively they still seem very random, and one might expect their absolutely continuous spectrum to be empty.

And indeed, deterministic metrically transitive potentials can give rise to an empty absolutely continuous spectrum. (They can also give rise to an everywhere positive Lyapunov exponent: see Theorem 16.13.) This can be proved [Kotani 1985a; Kirsch et al. 1985] by means of topological arguments like those used in Theorem 2.17. We must make the space of realizations into a topological space: in the discrete case, where $\Omega = \mathbf{R}^{\mathbf{Z}}$, that is, the set of sequences $q(x)$ such that

$$|q(x)| < \infty \tag{12.63}$$

for all $x \in \mathbf{Z}$, we give Ω the topology of pointwise convergence; while in the continuous case, Ω is the set of measurable, locally square-integrable functions on $\mathbf{R}$ such that

$$\left(\int_n^{n+1} |q(x)|^2\, dx\right)^{\frac{1}{2}} \leqslant C(q)\left(|n|^{1+\delta} + 1\right) \tag{12.64}$$

for all $n \in \mathbf{Z}$, and some $\delta > 0$, we use the metric of L_2^{loc}-convergence. It is easy to see that this condition implies, with probability 1, that the second moment is finite, that is, (12.1) is satisfied.

As in Theorem 2.17, we denote by $\operatorname{supp} \mathsf{P}$ the *topological support* of a probability measure P in the relevant space.

(12.11) Theorem [Kotani 1985a]. *Let* P *be a metrically transitive measure on the space* Ω *defined above; let* $q(x)$*, for* $x \in \mathbf{R}$ *or* $\mathbf{Z}$*, be the corresponding metrically transitive potential; and let* $A = H$ *or* h *be the operator with this potential. If there exists* $q_\infty(x) \in \operatorname{supp} \mathsf{P}$ *such that*

$$\sup_{n\in\mathbf{Z}} \int_n^{n+1} |q_\infty(x)| + \mathsf{E}\{q^2(0)\} < \infty$$

in the continuous case, and

$$\sup_{x\in\mathbf{Z}}\bigl|q_\infty(x)\bigr| + \mathsf{E}\bigl\{\ln(1+|q(0)|)\bigr\} < \infty$$

in the discrete case, where the mathematical expectation is taken with respect to P, *we have*

$$\sigma_{\mathrm{ac}}(A) \subset \sigma_{\mathrm{ac}}\left(-\frac{d^2}{dx^2} + q_\infty(x)\right). \tag{12.65}$$

A similar result can be found in [Kirsch et al. 1985].

Proof. We will use the fact that A is essentially self-adjoint with probability 1 if q satisfies (12.64) [Kotani 1985a; Kirsch et al. 1985]. (Self-adjointness with probability 1 on $C_0^\infty(\mathbf{R})$ for metrically transitive operators with potential satisfying (11.16a) follows from Corollary 2.6; for a potential such that $\sup_{n\in\mathbf{Z}}\int_n^{n+1}\bigl|q^2(x)\bigr|\,dx$ is finite, see [Reed and Simon IV].) Also, $\sigma_{\mathrm{ac}}(a)$ is a nonrandom set by Theorem 2.16.

Since $q_\infty \in \operatorname{supp}\mathsf{P}$, we can apply Theorem 12.7 to find a sequence $q_n \in \operatorname{supp}\mathsf{P}$, for $n \in \mathbf{Z}_+$, such that $\lim_{n\to\infty} q_n = q_\infty$, and a nonrandom set $X \subset \mathbf{R}$ such that $\bar{X}^{\mathrm{e}} = \sigma_{\mathrm{ac}}(A)$ and $\operatorname{Re} G_n(0,0,\lambda+i0) = 0$ for $n \in \mathbf{Z}_+$ and $\lambda \in X$, where $G_n(x,x',z)$ is the Green's function of the operator with potential q_n. Using representation (12.61) for the functions $G_n(z) = G_n(0,0,z)$, and the fact that $m_\pm(z)$ depends continuously on the potential (see the end of 12.A), we conclude that $\lim_{n\to\infty} G_n(z) = G_\infty(z)$ for all nonreal z. By Theorem A.11(ii) of the Appendix it follows that $\operatorname{Re} G_\infty(\lambda+i0) = 0$ for almost all $\lambda \in X$.

Now if $G(x,x',z)$ is the Green's function of any self-adjoint operator in $L_2(\mathbf{R})$ or $\ell_2(\mathbf{Z})$, Theorem A.10(i) says that the essential closure of the set of λ such that $\operatorname{Re} G(0,0,\lambda+i0)$ vanishes is contained in the absolutely continuous spectrum of the operator, because on this set we have $\operatorname{Im} G(0,0,\lambda+i0) \neq 0$ for almost all λ (otherwise, by Theorem A.5, $G(0,0,z) = 0$ for all $z \in \mathbf{C}$, which is impossible). This clearly implies the theorem. □

Here is a simple corollary of Theorem 12.11:

(12.12) Theorem [Kotani 1985a]. *Let* P_1 *and* P_2 *be metrically transitive measures on the topological space* Ω *specified by* (12.63) *or* (12.64), *with the group of shifts* $(T_x q)(y) = q(x+y)$, *and let* $q_1(x,\omega)$ *and* $q_2(x,\omega)$ *be the corresponding metrically transitive potentials. If* A_1 *and* A_2 *are the corresponding operators, we have* $\sigma_{\mathrm{ac}}(A_1) \supset \sigma_{\mathrm{ac}}(A_2)$ *if* $\operatorname{supp}\mathsf{P}_1 \subset \operatorname{supp}\mathsf{P}_2$. □

This statement is analogous to Theorem 2.17. It formalizes the intuitive idea that the absolutely continuous component of the spectrum decreases when the randomness of the potential increases. Later in this chapter we will show that, much along the same lines, the spectrum of a second-order metrically transitive operator with a very random potential (Markovian in the continuous case, independent in the discrete) is pure point under mild conditions.

Since the spectrum of a periodic operator is absolutely continuous, Theorem 12.12 implies the following result:

(12.13) Corollary [Kirsch et al. 1985]. *If the potential* $q_\infty(x)$ *of Theorem* 12.11 *is periodic,*

$$\sigma_{\mathrm{ac}}(A) \subset \sigma_{\mathrm{ac}}\left(-\frac{d^2}{dx^2} + q_\infty(x)\right). \qquad \square$$

We now consider some examples of deterministic metrically transitive potentials, for which the absolutely continuous spectrum is nonetheless empty.

(1.14) Examples

(a) [Kirsch et al. 1985] Consider the Poisson random potential (1.29), where $u(x)$ is nonnegative and can be analytically continued onto the strip $\{x+iy : |y| \leqslant \beta\}$, with $\beta > 0$, and that the analytic continuation satisfies

$$|u(z)| \leqslant \mathrm{const}\bigl(1+|z|\bigr)^{2+\delta}, \tag{12.66}$$

where $\delta > 0$. This potential is deterministic, because its realizations, with probability 1, are analytic functions on the strip. On the other hand, the topological support of the random process contains periodic functions of the form $w(x) = \sum_{n\in\mathbf{Z}} u(x-na)$ for any $a > 0$. Since u is positive and not identically zero, we can assume that $u(x) \geqslant \varepsilon > 0$ for $|x| \leqslant \delta$. Thus, choosing $a = 2\delta/k$, we find that $w(x) \geqslant k\varepsilon$, so that

$$\sigma_{\mathrm{ac}}\left(-\frac{d^2}{dx^2} + w\right) \subset [k\varepsilon, \infty).$$

Since k is arbitrary, it follows from Corollary 12.13 that $\sigma_{\mathrm{ac}} = \varnothing$.

The same is true for the random alloy potential (1.31), if the random variables ξ_a are independent and identically distributed and $u(x)$ satisfies (12.66). Other similar examples can be found in [Kotani 1985a; Kirsch et al. 1985]. The latter paper, in particular, shows that if $\ell_1(L_2)$ is the Banach space of functions on $\mathbf{R}$ with norm $\|u\| = \sum_{n\in\mathbf{Z}}\bigl(\int_n^{n+1} |u(x)|^2\,dx\bigr)^{1/2}$, the set of $u \in \ell_1(L_2)$ that lead to deterministic potentials for any bounded sequence of independent, identically distributed random variables is a dense G_δ-set. Thus potentials with this property are typical.

(b) [Kotani 1985a] Let $X(x)$, for $x \in \mathbf{R}$, be a metrically transitive Gaussian process not identically equal to a constant. By Section 1 (see also [Gihman and Skorohod I]), this means that the quadratic form defined on the space $L_2(-2T, 2T)$ by the correlation function is strictly positive for any $T > 0$. Thus, the support of the probability measure induced in this space by $X(x)$ is the whole space. Let $F(t)$ be a nonconstant, continuous function on $\mathbf{R}$, and let a and b be its lower and upper bounds. The support of the probability measure induced by the process $F(X(x))$ contains the set of continuous functions $q(x)$ such that $a \leqslant q(x) \leqslant b$ for all x; but this set is the support of the measure generated by the random process $F_1(Q(x))$, where $Q(x)$ is the Brownian motion on a compact, smooth Riemannian manifold K, and $F_1(Q)$, for $Q \in K$, is a continuous function on the manifold with lower and upper bounds a and b. By Theorem 15.4, the spectrum of such an operator is pure point with probability 1; thus, by Theorem 12.12, the spectrum of the Schrödinger operator with potential $F(X(x))$ has no absolutely continuous component.

12.E Some Inverse Problems

We have seen that the Lyapunov exponent $\gamma(\lambda)$ allows a pretty complete study of the absolutely continuous spectrum of one-dimensional metrically transitive operators of order two. We will see now that in some cases $\gamma(\lambda)$ uniquely determines the potential:

(12.15) Theorem. *For a metrically transitive operator H or h with potential satisfying* (12.1), *let a and b be the lower and upper bounds of the spectrum σ. Suppose that $\gamma(\lambda) = 0$ for almost all $\lambda \in [a, b]$ with respect to the Lebesgue measure.*

(i) *In the discrete case, if $b - a = 4$, the potential is constant and equals $(a + b)/2$.*
(ii) *In the continuous case, if $q(x) \geqslant a$, the potential is constant and is equal to a.*

Proof. For (i), we can assume, without loss of generality, that $b = -a = 2$. It follows from (12.54b), Theorem A.11(iv) and the hypotheses of the theorem that the function $w(z)$ is analytic on $(-2, 2)$. Then $w'(z)$ is also analytic, and $\operatorname{Re} w'(\lambda + i0) = 0$ for $\lambda \in (a, b)$. Applying Theorem A.11(iv) again, we conclude from (12.57) that $N(\lambda)$ is analytic in $(-2, 2)$. Therefore, if $n(\lambda)$ is the derivative of $N(\lambda)$, we can use the Sokhotski–Plemelj formula (A.6) to rewrite $\operatorname{Re} w'(\lambda + i0) = 0$ as

$$\int_{-2}^{2} \frac{n(\mu)}{\mu - \lambda} d\mu = 0$$

for $\lambda \in (-2, 2)$. If we consider this relation as a singular integral equation for $n(\lambda)$, we find that

$$n(\lambda) = \begin{cases} (\pi\sqrt{4-\lambda^2})^{-1} & \text{for } |\lambda| \leqslant 2, \\ 0 & \text{for } |\lambda| \geqslant 2 \end{cases}$$

(see [Muskhelishvili 1962], for example). Now Theorem 4.16 implies that $q(x) = 0$.

For part (ii), we can take $a = 0$ (here the upper bound b is ∞). Using essentially the same argument as in part (i), we find that $n(\lambda) = (2\pi\sqrt{\lambda})^{-1}$. (Here $n(\lambda)$ is uniquely determined by taking into account the asymptotic formula of Theorem 6.5, rather than the normalization condition used in the discrete case.) But then $N(\lambda) = N_0(\lambda) = \sqrt{\lambda}/\pi$ for $\lambda \geqslant 0$, and then Theorem 5.26 implies that $q(x)$ vanishes identically. □

The proof of this theorem is based on two simple theorems, 4.16 and 5.26, which are valid for metrically transitive Schrödinger operators (discrete and continuous) in any dimension. More detailed arguments particular to dimension one allow one to prove Theorem 12.15 without the need for the assumptions $b - a = 4$ and $q \geqslant a$: see [Kotani 1984] and Problem V.36.

Theorems 12.15, 4.16 and 5.26 solve special cases of the inverse problem of metrically transitive spectral analysis, which in essence [Marchenko 1986] consists in obtaining information on the coefficients of a metrically transitive operator, based on spectral data. Several cases similar to that of Theorem 12.15 are solved in [Simon and Deift 1983]. This paper showed, in particular, that in the discrete case the Lebesgue measure of the set of λ where $\gamma(\lambda)$ vanishes is no greater than 4, and that it equals 4 if and only if the potential vanishes identically.

A fairly general inverse problem is discussed in [Kotani 1985b], where it is proved that for any function $w(z)$, analytic in the upper half-plane and such that w, iw and w are Nevanlinna functions and that

$$w(z) = iz^{1/2} + \frac{q_1}{2i} z^{-1/2} + \frac{q_2}{2i} z^{-3/2} + o(z^{-3/2})$$

as $\operatorname{Im} z \to \infty$, there exists a Schrödinger operator with a stationary potential $q(x)$ for which this w is the function of formula (12.54a). (The function of (12.54a) can also be defined as the mathematical expectation of $m_\pm$: see Problem V.26. That it must satisfy the Nevanlinna conditions and the asymptotic condition above is proved in [Kotani 1987b]; see also Problem V.22.) Moreover, the potential $q(x)$ satisfies, with probability 1:

$$\int_{\mathbf{R}} \frac{|q(x)|^2}{1 + |x|^3} dx < \infty$$

and $q_1 = \mathsf{E}\{q\}$, $q_2 = E\{q^2\}$.

If, in addition, $\gamma(\lambda) = \operatorname{Re} w(\lambda + i0)$ equals 0 for almost all λ with respect to the measure $N(d\lambda)$ associated with the nondecreasing function $N(\lambda) = \pi^{-1} \operatorname{Im} w(\lambda + i0)$, the potential $q(x)$ can be chosen as a metrically transitive process in $\mathbf{R}$.

Similar problems for the discrete Schrödinger operator are considered in [Carmona and Kotani 1987].

In Section 17 we will consider an inverse problem whose solution is in a sense complete: the potential not only exists, but is unique and can be described in considerable detail. It has to do with limit-periodic potentials that are approximated sufficiently well (superexponentially) by periodic ones: in which case it turns out that $\int_{\mathbf{R}} \gamma(\lambda) N(d\lambda) = 0$.

13 Lyapunov Exponents and the Point Spectrum

We saw in Section 12 that Lyapunov exponents determine many properties of the absolutely continuous spectrum of a one-dimensional metrically transitive operator of order two. In particular, Theorem 12.1 says that if the Lyapunov exponent is positive almost everywhere on the spectrum, the spectrum is singular with probability 1. In this section we give additional conditions that, together with the positiveness of $\gamma(\lambda)$, actually ensure that the spectrum is pure point.

We start with a mostly heuristic discussion of such conditions, in Subsection 13.A. In 13.B we give a formal proof that under these conditions the spectrum is pure point, and that the eigenfunctions decay exponentially at infinity, the rate of decay being determined by the Lyapunov exponents. Following the general trend of this chapter, we limit ourselves to results that are valid for as broad as possible a class of metrically transitive coefficients; but we also consider briefly important concrete examples of metrically transitive operators to illustrate the general concepts. A more detailed analysis is deferred to Chapters VI and VII, devoted to random and almost periodic operators, respectively.

13.A Heuristic Discussion

Theorem 2.11 says that any neighborhood of an eigenvalue of a one-dimensional metrically transitive operator of order two contains infinitely many eigenvalues. Thus, if the closure $\bar{\sigma}_{\mathrm{p}}$ of the point spectrum contains an interval Δ, the set $\sigma_{\mathrm{p}} \cap \Delta$ is dense in itself in this interval. The very possibility that one-dimensional operators of order two possess a dense in itself point spectrum is a consequence of the fact that the inverse problem of spectral analysis for such operators has a unique solution [Marchenko

1986; Berezanskii 1968], for this means there is an algorithm to associate a potential with any spectral measure satisfying some natural conditions.

However, from the standpoint of traditional spectral theory, which deals mainly with potentials that decay or increase at infinity, a dense point spectrum is a rarity, rather than the rule. In any case, in the inverse problem formalism it is very hard, if at all possible, to describe in detail and efficiently the class of potentials that give rise to dense point spectra: in particular, it is hardly possible within this framework to establish when a reconstructed potential is metrically transitive. (But in Section 17 we will use the inverse problem formalism to investigate the absolutely continuous spectrum of Schrödinger operators with limit-periodic potentials.) Therefore one needs to develop special methods to study the point spectrum of metrically transitive operators.

Theorem 12.1 implies that if $\gamma(\lambda)$ is positive on an interval Δ contained in the spectrum, $\sigma_{\mathrm{ac}} \cap \Delta = \varnothing$ with probability 1. It would seem, however, that in this case one can say more, namely, that the spectrum must be pure point in Δ. To see this, assume for definiteness that we are in the continuous case. Take a long enough interval (a, b), with $-\infty < a < 0 < b < \infty$, and consider the boundary problem (6.1)–(6.2) in this interval. The spectrum $\{\lambda_j\}$ is discrete and can usually be found by substituting the solution $y_a^{(\alpha)}(x, \lambda)$ of the Cauchy problem (6.1) and (6.3) into the second condition in (6.2) (the first is automatically satisfied).

Let us modify this traditional procedure [Goldsheidt et al. 1977] and consider the two Cauchy problems determined by the conditions at each end of the interval (a, b). Denoting their solutions by $y_a^{(\alpha)}(x, \lambda)$ and $y_b^{(\beta)}(x, \lambda)$, we can find the eigenvalues by specifying that the two solutions and their derivatives should coincide at the origin:

$$\left(y_a^{(\alpha)} \frac{dy_b^{(\beta)}}{dx} - y_b^{(\beta)} \frac{dy_a^{(\alpha)}}{dx} \right) \bigg|_{x=0} = 0. \tag{13.1}$$

From the results of Section 11 (Lemma 11.1, Theorems 11.3 and 11.8), it follows that if $\gamma(\lambda) > 0$ for $\lambda \in \Delta$, the envelopes (11.1) of both solutions grow exponentially as we move away from the endpoints of the interval (a, b) toward the middle. Thus the corresponding eigenfunctions must decay exponentially as we move from the middle of the interval to the endpoints. If we assume this behavior is preserved as a and b tend to $-\infty$ and ∞, we conclude that in the interval Δ the spectrum is pure point, with eigenfunctions decaying exponentially at infinity.

The conjecture that the spectrum of one-dimensional operators with random coefficients is pure point was first made by Mott and Twose [1961], based on heuristic arguments similar to the ones given above. These arguments were refined and developed in [Borland 1963] and [Halperin 1967]. But these authors implicitly accepted another important assumption besides the positiveness of the Lyapunov exponent, which turns out to be essential to the argument.

It will be more convenient to discuss this assumption by considering the whole axis to begin with, rather than a sequence of expanding finite intervals. By Theorem 11.3, for each ω in a set Ω_λ of full measure, there exist values $\alpha_+(\lambda, \omega)$ and $\alpha_-(\lambda, \omega)$ of the angle α in (6.3) that make $y_a^{(\alpha)}$ decay exponentially for $x \to \infty$ and $x \to -\infty$, respectively. Therefore, in order to get exponentially decaying solutions at both ends, which would of course be eigenfunctions, it is necessary that

$$\alpha_+(\lambda, \omega) = \alpha_-(\lambda, \omega), \tag{13.2}$$

an equation that plays here the role of (13.1). Theorem 11.3 allows us to determine both of these angles for fixed λ on a set of realizations of full measure, but the set generally depends on λ: conversely, for a fixed realization ω, the angles are determined on a set $U(\omega)$ of full Lebesgue measure (or some other fixed measure), but not for all λ.

Now for fixed ω, the set of values of λ for which (13.2) holds should form the set σ_{p} of eigenvalues of the operator. Hence, σ_{p} is countable, though it may be dense (see Theorem 2.11). Moreover, as Theorem 2.12 shows, σ_{p} depends nontrivially on the realization. Thus, if we use arguments based only on the fact that the Lyapunov exponent is positive and consider only individual solutions of the Cauchy problem (albeit ones that possess full probability measure for each λ), it is not clear how to establish that, with probability 1, σ_{p} lies outside the exceptional set (the complement of $U(\omega)$), which has zero Lebesgue measure but depends on the realization.

To make matters worse, there are in fact examples of almost periodic operators with positive Lyapunov exponent and purely singular continuous spectrum (see Theorem 16.17, Subsection 16.F and Theorem 18.14), showing that the spectral measure can be concentrated, with probability 1, on the exceptional set. To prevent this situation we will lay down nontrivial assumptions on the coefficients; as we mentioned above, these assumptions were originally implicit in the physics literature.

The discussion above hints at what these assumptions should be. The main one is that the potential have weak correlation at distant points, to guarantee that the solutions $y_a^{(\alpha)}$ and $y_b^{(\beta)}$ be "independent" enough that they can be matched together. Another way to ensure this matching, as we will see later, is to introduce one or more continuous parameters into the definition of the potential.

In terms of the angles $\alpha_\pm(\lambda,\omega)$, this means that the smoother and the more independent the potential distribution, the more uniform and independent are the angle distributions. This decreases the probability of correlation of transversal values for these angles; we will see in Section 15 that the spectral characteristics contain expressions including the generalized function $\delta\big(\alpha_+(\lambda,\omega) - \alpha_-(\lambda,\omega)\big)$, which is well-defined for continuously distributed random variables $\alpha_\pm(\lambda,\omega)$, and positive if the random variables are independent.

This weak correlation condition is important also from the quantum mechanical point of view, since it decreases the probability of *resonance tunneling* that can arise when two similar patterns are present in the graph of the potential. The tunneling effect consists in that, no matter how distant the patterns are from one another, there exist solutions such that the amplitudes within each pattern are comparable. Thus, in general, the solutions do not decay at infinity (see Subsection 15.C).

13.B Conditions for Positive Lyapunov Exponents to Imply a Pure Point Spectrum

To clarify the role of the exceptional sets of zero Lebesgue and probability measure discussed above, we will prove some simple propositions from the general spectral theory of one-dimensional operators of order two.

Definition. *A solution $y(x)$ of a second-order differential equation is*

(a) *polynomially bounded at ∞ (or $-\infty$) if, for some $C, a, x_0 > 0$,*

$$\left|\int_{x_0\,(\mathrm{or}\,-\infty)}^{\infty\,(\mathrm{or}\,-x_0)} \frac{|y(x)|^2}{1+|x|^{2a}}\,dx\right| \leqslant C < \infty \tag{13.3a}$$

in the continuous case and

$$|y(x)| \leqslant C(1+|x|^a) \tag{13.3b}$$

for $x \geqslant x_0$ (or $x \leqslant -x_0$) in the discrete case;

(b) *exponentially decaying at ∞ (or $-\infty$) with rate γ if*

$$\lim_{x\to\infty\,(\mathrm{or}\,-\infty)} \frac{1}{|x|}\ln r(x) = -\gamma, \tag{13.4}$$

where $r(x)$ is the envelope defined in (11.1) *and* (11.5).

The following results, implicitly contained in [Carmona 1982, 1983], state what properties of a second-order equation ensure that the corresponding operator has a point spectrum.

(13.1) Theorem. *Let $\rho_+^{(\alpha)}$ be the spectral measure of an operator $H_+^{(\alpha)}$ or $h_+^{(\alpha)}$ defined by a second-order equation and boundary conditions* (12.9) *on the right semiaxis. If, for $\rho_+^{(\alpha)}$-almost all λ in a Borel set X of positive $\rho_+^{(\alpha)}$-measure, every solution to the equation that is polynomially bounded at $+\infty$ decays exponentially at $+\infty$, with rate $\gamma(\lambda)$, it follows that for $\rho_+^{(\alpha)}$-almost all $\lambda \in X$ the generalized eigenfunctions $\psi_\lambda(x)$ of the corresponding operator decay exponentially at $+\infty$ with rate $\gamma(\lambda)$.*

The same is true for an operator defined on the left semiaxis, if we replace + by − throughout. It is also true for an operator H or h defined on the whole axis if we replace $\rho_+^{(\alpha)}$ by the operator's spectral measure ρ.

Proof. Almost all generalized eigenfunctions of $H_+^{(\alpha)}$ or $h_+^{(\alpha)}$ (with respect to the spectral measure) are polynomially bounded: see (12.28), [Berezanskii 1968; Simon 1982]. □

(13.2) Corollary. *Under the conditions of Theorem* 13.1, *the spectrum σ of $H_\pm^{(\alpha)}$ or $h_\pm^{(\alpha)}$ is pure point in X, that is, $\sigma \cap X = \bar{\sigma}_{\mathrm{p}} \cap X$ and $\sigma_{\mathrm{c}} \cap X = \varnothing$. Furthermore, all eigenfunctions decay exponentially with rate $\gamma_\pm(\lambda)$.*

For an operator H or h on the whole axis, satisfying the conditions of Theorem 13.1 *at both ends (that is, if every solution that is polynomially bounded at $\pm\infty$ for ρ-almost $\lambda \in X$ decays exponentially with rate $\gamma_\pm(\lambda)$), the spectrum is pure point and simple in X, and the corresponding eigenfunctions decay at $\pm\infty$ with the same rate.*

Proof. This follows immediately from Theorem 13.1, except for simplicity, which follows from Proposition 12.9. □

Remark. Theorem 13.1 and Corollary 13.2 are valid under somewhat weaker conditions: we can change the definition (13.4) of exponentially decaying functions to require only that there exist positive constants δ_+ and δ_- such that

$$\limsup_{x\to\pm\infty} \frac{1}{|x|} \ln r(x) \leqslant -\delta_\pm < 0. \tag{13.5}$$

As we will see in 15.C, one can establish an analogue of this property for the fluctuation spectrum of multidimensional operators.

The next theorem shows that the exponential decay property in the condition of Theorem 13.1 can be replaced by a property that is easier to check:

(13.3) Theorem. *Let $\mathcal{T}(x, 0)$ be the fundamental matrix* (11.7) *for a one-dimensional equation of order two, and let $\rho_\pm^{(\alpha)}$ be the spectral measure of* *the operator* $H_\pm^{(\alpha)}$ *or* $h_\pm^{(\alpha)}$. *If, for* $\rho_\pm^{(\alpha)}$*-almost all* $\lambda \in X$, *where* $\rho_\pm^{(\alpha)}(X) > 0$, *the exact Lyapunov exponents*

$$\gamma_\pm(\lambda) = \lim_{x\to\pm\infty} \frac{1}{|x|} \ln\|\mathcal{T}(x,0)\| \tag{13.6}$$

exists and is positive, the spectrum of the operator is pure point in X, *and all eigenfunctions decay exponentially at* $\pm\infty$ *with rate* $\gamma_\pm(\lambda)$.

If both limits (13.6) *exist and are positive for* ρ*-almost all* $\lambda \in X$, *where* ρ *is the spectral measure of the operator* H *or* h *and* $\rho(X) > 0$, *the spectrum is pure point and simple in* X, *and the eigenfunctions decay exponentially with rate* $\gamma_\pm(\lambda)$ *for* ρ*-almost* $\lambda \in X$.

Proof. We use the exponential dichotomy property for fundamental matrices, which says that for every λ for which limit (13.6) exists, there is a unique two-component vector $e_\pm = (\cos\alpha_\pm, \sin\alpha_\pm)$, for $\alpha_\pm \in [0,\pi)$, such that

$$\lim_{x\to\pm\infty} \frac{1}{|x|} \ln\|\mathcal{T}(x,0)e\| = \begin{cases} \gamma_\pm(\lambda) & \text{if } e \neq e_\pm, \\ -\gamma_\pm(\lambda) & \text{if } e = e_\pm, \end{cases} \tag{13.7}$$

where e is an arbitrary vector of the form $e = (\cos\alpha, \sin\alpha)$: see Problems V.6–7. (We have mentioned this property in our comments about the multiplicative ergodic theorem (Proposition 11.2), a proof of which is based on the counterpart of (13.7) for matrices of arbitrary order and on the subadditive ergodic theorem (Proposition 6.3). A similar dichotomy also arises in the qualitative theory of differential equations [Hartman 1964].)

It follows from (13.7) and the conditions of the theorem that for $\rho_\pm^{(\alpha)}$-almost all $\lambda \in X$ each solution of the equation either grows exponentially or decreases exponentially. Thus we can apply Theorem 13.1 and Corollary 13.2, which clearly imply the desired conclusion. □

Theorem 13.3 says that one can prove that the spectrum of a one-dimensional operator of the second order is pure point by proving (13.6) for almost all $\lambda \in \sigma$ with respect to the spectral measure, where σ is the spectrum. But, according to Lemma 11.1, the limits in (13.6), for any given λ, exist for a set of realizations Ω_λ of full P-measure. Using Fubini's theorem, we conclude that for every realization in a set of full P-measure these limits exist for Lebesgue-almost all $\lambda \in \sigma$, or for almost all λ with respect to any other nonrandom measure. Thus, one possible way to prove (13.6) would be as follows: first, to show that the Lyapunov exponent (11.25) is positive for all $\lambda \in \sigma$; then, to deduce from Lemma 11.1 the existence of the limits in the right-hand side of (13.6) not for Lebesgue-almost all λ, but for almost all λ with respect to the spectral measure.

That the Lyapunov exponent is positive for a broad class of random potentials and for many quasiperiodic potentials will be proved in Section 14 and Subsection 16.C. The second step in the scheme above— replacing "Lebesgue-almost all λ" with "almost all λ with respect to the spectral measure"—would be justified, for example, if

$$\mathsf{P}\Big(\bigcap_{\lambda\in\mathbf{R}}\Omega_\lambda\Big)=1. \tag{13.8}$$

Since the transition from Lebesgue-almost all λ to all $\lambda \in \mathbf{R}$ in (13.5) and (13.6) is related to the question of whether these equations are stable under slight changes in λ, it would seem that the proof of (13.8), and consequently the pure point character of the spectrum, can be reduced to the proof of this stability. It turns out, however, that (13.8) can never be satisfied: see [Carmona 1983] and Problem V.37. One of the reasons is that $\|\mathcal{T}(x,0,\omega)\|$ as a function of x does not have, in general, the exponential asymptotic behavior $e^{\gamma|x|(1+o(1))}$ for all λ and ω: this exponential merely provides an upper bound for $\|\mathcal{T}(x,0,\omega)\|$ in the sense that if we define

$$\bar{\gamma}_\pm(\lambda,\omega)=\limsup_{x\to\pm\infty}\frac{1}{|x|}\ln\|\mathcal{T}(x,0,\omega)\| \tag{13.9}$$

we can write

$$0\leqslant\bar{\gamma}_\pm(\lambda,\omega)\leqslant\gamma(\lambda) \tag{13.10}$$

for all λ, with probability 1 (see Problem V.5). This inequality can be strict, as can be seen from the existence of almost-periodic potentials for which the Lyapunov exponent is positive but the corresponding equation has no solutions that decay at infinity: an example is the potential $q(x) = 2g\cos 2\pi(\alpha x+\omega)$, for $x \in \mathbf{Z}$, where $g > 1$ and α is an irrational number that can be approximated well by rationals (see Corollary 16.9, Example 16.14 and Subsection 16.D).

This lack of stability for (13.5) and (13.6) for infinitesimal changes in λ stands in the way of proving that the spectrum is pure point. This is another form of the difficulty discussed above, namely, the existence of sets whose Lebesgue measure is zero but whose spectral measure is nonzero.

The simplest illustration of how to overcome this difficulty, and at the same time an example of the role of the technique of continuously varying parameters mentioned in the discussion at the end of 13.A, concerns boundary problems on the semiaxis.

(13.4) Theorem [Kotani 1986]. *Assume the Lyapunov exponent $\gamma(\lambda)$ of* (11.25) *for a second-order equation with metrically transitive coefficients is* *positive on a Borel set X of positive Lebesgue measure. Then, with probability 1 and for Lebesgue-almost all $\alpha \in [0, \pi)$, the operators $H_{\pm}^{(\alpha)}$ and $h_{\pm}^{(\alpha)}$ have a pure point, simple spectrum in X, and all the eigenfunctions decay exponentially at infinity with rate $\gamma(\lambda)$.*

Proof. It is enough to check the conditions of Theorem 13.3 for $\mathsf{P} \times \mathsf{L}$-almost all pairs $(\omega, \alpha) \in \Omega \times [0, \pi)$, where L is the Lebesgue measure. To this end, in addition to the upper Lyapunov exponent of (13.9), we introduce lower ones

$$\underline{\gamma}_{\pm}(\lambda, \omega) = \liminf_{x \to \pm\infty} \frac{1}{|x|} \ln \left\| \mathcal{T}(x, 0, \omega) \right\|, \tag{13.11}$$

and consider the sets

$$S_{\pm}(\omega) = \left\{ \lambda \in \mathbf{R} : \bar{\gamma}_{\pm}(\lambda, \omega) \neq \underline{\gamma}_{\pm}(\lambda, \omega) \right\} \tag{13.12}$$

consisting of those $\lambda \in \mathbf{R}$ that, in terms of stability theory, have no left or right exact Lyapunov exponent. From Lemma 11.1 and Fubini's theorem it follows that, with probability 1,

$$\operatorname{meas} S_{\pm}(\omega) = 0. \tag{13.13}$$

We now use the identity

$$\int_0^{\pi} \rho_{\pm}^{(\alpha)}(d\lambda)\, d\alpha = d\lambda, \tag{13.14}$$

which, in view of the inversion formula (A.4), follows from

$$\frac{1}{\pi} \int_0^{\pi} m_{\pm}^{(\alpha)}(z)\, d\alpha = i, \tag{13.15}$$

where $m_{\pm}^{(\alpha)}$ is the Weyl function (12.20) of the corresponding operator. Equation (13.15) is obtained from (12.20) by elementary integration (see also Problem V.17).

Combining (13.3) and (13.4), we arrive at the important equality

$$\rho_{\pm}^{(\alpha)}\left(S_{\pm}(\omega)\right) = 0, \tag{13.16}$$

which holds for almost all pairs $(\omega, \alpha) \in \Omega \times [0, \pi)$. But for all pairs satisfying (13.16), equation (13.6) is also satisfied for almost all $\lambda \in X$: hence, the conditions of Theorem 13.3 are met, and the theorem is proved. □

Remark. The probabilistic nature of H and h was used in the proof only via the assumption that the exact Lyapunov exponent exists. Thus Theorem 13.4 holds for every one-dimensional self-adjoint operator on a semiaxis that has a positive Lyapunov exponent (13.6) on the semiaxis for Lebesgue-almost all $\lambda \in X$.

A closer look at the proof of Theorem 13.4 easily shows that the dependence of $H_{\pm}^{(\alpha)}$ and $h_{\pm}^{(\alpha)}$ on the angle α comes in in only two ways: we use the absolute continuity of the spectral measure integrated over α, and the fact that the sets (13.12) are independent of α. Thus we can use some other parameter instead of α as long as these two properties remain true.

Consider an operator that depends on N parameters $(\xi_1, \dots, \xi_N) = \xi$, ranging over a set $\Xi \subset \mathbf{R}^N$ on which is given a measure $F(d\xi)$. We say that ξ is *randomizing* if, first, the spectral measure of the operator, integrated over Ξ, is absolutely continuous with respect to the Lebesgue measure (or perhaps some other fixed measure on $\mathbf{R}$), and, secondly, if the sets (13.12) outside of which the exponential growth or decay of solutions of the corresponding equation occurs, are independent of ξ. For an operator with randomizing parameters, we can repeat verbatim the proof of Theorem 13.4, to show that for F-almost all $\xi \in \Xi$ the spectrum is pure point and the eigenfunctions decay exponentially. Thus, the concept of randomizing parameters provides a way to formalize the heuristic arguments given in the preceding subsection.

From this more general viewpoint it is clear that Theorem 13.4 holds also if the angle α is randomized using any measure on $[0, \pi)$ that is absolutely continuous with respect to the Lebesgue measure. In this sense one can say that, for the boundary problem on a semiaxis, positive Lyapunov exponents lead to a pure point spectrum with exponentially decaying eigenfunctions for all typical boundary conditions.

An important example of randomizing parameters in the discrete case is the set of potential values at a fixed, finite, set of points. It is clear that the sets (13.12) defined by the asymptotic behavior of the solutions to the Cauchy problem are independent of these parameters. The other condition, absolute continuity of the spectral measure after integrating with a randomizing measure, can be proved in many cases from the following result, as we shall see later:

(13.5) Theorem [Pastur 1985; Simon et al. 1985]. *Let A be a self-adjoint operator on a Hilbert space $\mathcal{H}$, and P the operator of orthogonal projection on a unit vector $\psi \in \mathcal{D}(A)$. If $E_{A+\xi P}(d\lambda)$, for $\xi \in \mathbf{R}$, is the resolution of the identity of the operator $A + \xi P$ and $f(\xi)$ is a summable, bounded, nonnegative function, we have*

$$\int (E_{A+\xi P}(d\lambda)\psi, \psi) f(\xi) d\xi \leqslant \|f\|_{L_\infty(\mathbf{R})} d\lambda. \tag{13.17}$$

Proof. This is proved word-by-word like Theorem 4.21, based on Lemma 4.20. For variants of this result in other concrete situations, see [Kotani

1986; Delyon et al. 1985a,b]. The abstract operator content of this result is discussed in [Thomas and Wayne 1986] and [Howland 1987]. □

The importance of the potential values as randomizing parameters is due to the fact that the potential, unlike the angle α, can be used even if the operators are defined by a second-order equation on the entire axis—in particular, in the case of a metrically transitive operator. However, as we mentioned above, there are examples of finite-difference operators with quasiperiodic coefficients whose frequencies are well-approximable by rationals (Sections 16 and 18), whose corresponding operator has a purely singular continuous spectrum even though the Lyapunov exponent is positive. This shows that we cannot use the potential as a randomizing parameter in all cases.

The discussion in the previous subsection, the deterministic nature of the potential in the example just mentioned, and Theorem 13.5 indicate that the obstacle resides in a strong correlation of potential values at points far apart. However, by using the potential to carry out the randomization, one can prove that a singular continuous spectrum coexisting with positive Lyapunov exponents is unstable, and turns into a pure point spectrum under "typical" infinitesimal local perturbations.

Definition. *If h is the operator associated with a potential $q(x)$, for $x \in \mathbf{Z}$, by* (11.4), *the operator h_ξ associated with the $q+\xi$, where $\xi(x)$ is a function with finite support, is called a local perturbation of h.*

Clearly, the set of local perturbations by functions with support contained in some fixed finite set S is parametrized by $\mathbf{R}^S$.

(13.6) Theorem [Delyon et al. 1985a,b,c]. *Suppose that the Lyapunov exponent $\gamma(\lambda)$ of* (11.25), *for the operator h given by* (11.4), *is positive for Lebesgue-almost all λ in a Borel set $X \subset \mathbf{R}$ such that* $\operatorname{meas}(X) > 0$. *If we perturb h by functions $\xi(x)$ whose finite support S contains two adjacent points, the perturbed operator h_ξ, for Lebesgue-almost all $\xi \in \mathbf{R}^S$, has a pure point, simple spectrum in X with probability* 1, *and its eigenfunctions decay exponentially at $\pm\infty$ with rate $\gamma(\lambda)$.*

Proof. Let the consecutive points in S be x_0 and x_0+1, and introduce the measure

$$F_S(d\xi) = \prod_{x \in S} f\big(\xi(x)\big)\, d\xi(x) \tag{13.18}$$

on the set $\Xi = \mathbf{R}^S$, where $f(t)$ is any summable, bounded, positive function. We claim that the spectral measure $\rho_\xi(d\lambda)$ of h_ξ satisfies

$$\int_{\mathbf{R}^S} \rho_\xi(d\lambda) F_S(d\xi) \leqslant \|f\|_{L_\infty(\mathbf{R})} d\lambda; \tag{13.19}$$

in other words, the measure on the right is absolutely continuous with respect to the Lebesgue measure. To see this, notice that, by (12.32b) and (12.35) (where we can clearly replace 0 and 1 by x_0 and $x_0 + 1$),

$$\rho_\xi(d\lambda) = \tfrac{1}{2}\big(E_\xi(x_0, x_0, d\lambda) + E_\xi(x_0 + 1, x_0 + 1, d\lambda)\big), \tag{13.20}$$

where $E_\xi(x, y, \lambda)$ is the resolution of the identity of h_ξ. To get (13.19), we integrate (13.20) with respect to $F_S(d\xi)$ and estimate the result with the help of Theorem 13.5, the role of ξP being played by the operators $L_t = \xi(t)(\,\cdot\,, e_t)e_t$ with $e_t(x) = \delta_{xt}$ for $t = x_0, x_0 + 1$, and the role of A being played by $h_\xi - L_t$.

Since the sets (13.12) do not depend on ξ, we conclude that ξ, together with the measure F_S of (13.18), is a randomizing parameter. Repeating the proof of Theorem 13.4, we see that for F_S-almost $\xi \in \mathbf{R}^S$ relations (13.6) hold for ρ_ξ-almost $\lambda \in X$. Now Theorem 13.3 can be applied, concluding the proof. □

Theorems 13.4 and 13.6 make reasonable the conditions of the following result, a useful criterion for the verification that the spectrum of a metrically transitive operator is pure point.

(13.7) Theorem [Kotani 1986]. *Let h be a metrically transitive, one-dimensional, discrete Schrödinger operator h such that:*

(a) *The Lyapunov exponent $\gamma(\lambda)$ defined by* (11.25) *for the associated equation is positive for Lebesgue-almost all λ in a Borel set $X \subset \mathbf{R}$ of positive Lebesgue measure;*
(b) *if $\mathcal{F}(a, b)$, where a and b are integers such that $b \geqslant a+3$, is the σ-algebra generated by the values of the potential $q(x)$ for $x \in (-\infty, a] \cup [b, \infty)$, we have, with probability* 1,

$$\mathsf{E}\{(E(d\lambda)\psi, \psi) \mid \mathcal{F}(a, b)\} \leqslant C(\omega, \psi) d\lambda, \tag{13.21}$$

where $C(\omega, \psi)$ is finite, $E(d\lambda)$ is the resolution of the identity of h, and $\psi(x)$ is a function whose support is contained in (a, b).

Then the spectrum of h is pure point and simple in X, and its eigenfunctions decay exponentially as $|x| \to \infty$, with rate $\gamma(\lambda)$.

Proof. Exactly like the proofs of 13.4 and 13.6, but using (13.21) instead of (13.14) or (13.19). We already know that, for any finite x, the sets $S_\pm(\omega)$ of (13.12) are independent of the values of q on $(-\infty, x)$ and (x, ∞), respectively. Together with (13.12), this means we can use the values of q in the interval (a,b) as randomizing parameters, the randomizing measure being the conditional distribution of q with respect to $\mathcal{F}(a,b)$. □

Remarks. 1. One can use $\mathcal{F}_{-\infty}^{x}$ and $\mathcal{F}_{x}^{\infty}$ instead $\mathcal{F}(a,b)$ in (13.21), but then (13.21) must be satisfied for both σ-algebras. Another possibility is to consider $b=a+2$ and $\psi(x)=\delta_{x,a+1}$, and require $C(\omega,\psi)$ to be independent of ω. Then the mathematical expectation can be calculated under the condition that the potential is fixed except at $x=a+1$; but $E(a+1,a+1,d\lambda)$ appears in the expression whose expectation is being taken.

2. In Theorems 13.4, 13.6 and 13.7 one can assume something weaker than the positiveness of the Lyapunov exponent (11.25) and (13.6): namely, that there exists $\delta(\lambda)>0$ and some vector $e_{\pm}$ such that, for Lebesgue-almost all $\lambda \in X$,

$$\limsup_{x\to\pm\infty} \frac{1}{|x|} \ln\|\mathcal{T}(x,0)\tilde{e}_{\pm}\| \leqslant -\delta(\lambda). \tag{13.22}$$

The existence of these limits is independent of the values of the parameter affecting the potential (α in Theorem 13.4, ξ in 13.6 and 13.7). Thus, if we consider, instead of (13.12), the sets $\tilde{S}_{\pm}$ of $\lambda \in \mathbf{R}$ that do not satisfy (13.22), we can still apply (13.14), (13.19) or (13.21) to establish that almost all polynomially bound solutions (with respect to the spectral measure) decay exponentially in the sense of (13.5). By the remark following Corollary 13.2, this implies that the spectrum is pure point, and that the eigenfunctions decay exponentially.

(13.8) Examples

(a) Consider a discrete Schrödinger operator h with potential $q(x) = q_1(x)+q_2(x)$, where q_1 and q_2 are metrically transitive processes statistically independent of each other, and q_1 is a sequence of independent, identically distributed random variables whose distribution function has a bounded density:

$$F(dq)=f(q)\,dq, \qquad 0<\sup_{q\in\mathbf{R}} f(q)\leqslant f_0<\infty. \tag{13.23}$$

Since such a potential is nondeterministic, Theorem 12.10 says that its Lyapunov exponent $\gamma_1(\lambda)$ is positive for Lebesgue-almost all $\lambda \in X$ (actually, we will see in Section 14 that $\gamma_1(\lambda)>0$ for all $\lambda \in \mathbf{R}$). As for inequality (13.21), it should be considered (see Remark 1 to Theorem 13.7) at $b=a+2$. Then the conditional mathematical expectation is merely an integral over $q(a+1)$ with density $f\big(q(a+1)\big)$. Thus Theorem 13.5 applies, proving (13.21).

These results were obtained first by Kunz and Souillard [1980], with $q_2\equiv 0$, and by Delyon et al. [1983]. These authors used a different method, related to the ones discussed in Subsections 15.A and 15.B.

Theorem 13.6 showed the instability of the singular continuous spectrum of a one-dimensional Schrödinger operator with positive Lyapunov exponent under arbitrarily small, continuously distributed local perturbations, independent of the potential being perturbed. The example we have just seen shows that the spectrum of a one-dimensional metrically transitive operator of order two becomes pure point after we add to a metrically transitive potential q_2 a continuously distributed metrically transitive potential q_1 that is independent of q_2 and takes independent values at different points, even if its amplitude is arbitrarily small. In this sense, the typical behavior for a one-dimensional metrically transitive operator is to have a pure point, simple spectrum and exponentially decaying eigenfunctions. However, if the Lyapunov exponent of the unperturbed potential q_2 vanishes on the spectrum, as is the case, for example, for the absolutely continuous spectrum that is realized for periodic potentials or a number of almost-periodic potentials considered in 16.B and Section 17, then the results of 11.C, and in particular (11.104), show that the Lyapunov exponent for a small q_1 is of order $O(q_1)$, and therefore, if q_1 is small, the eigenfunctions start decaying only very far away.

One can weaken somewhat the conditions on q_1. It is enough to assume that the distribution function $F(dq)$ of the random variables $q_1(x)$ has a nonzero absolutely continuous component with a bounded density: $F = F_{\mathrm{ac}} + F_s$, with $F_{\mathrm{ac}}(dq) = f(q)\,dq$, where $0 < \sup_{q\in\mathbf{R}} f(q) \leqslant f_0 < \infty$. In this case, considering only those realizations where $q_1(a+1)$ and $q_2(a+2)$ belong to the support of F_{ac}, one establishes that the spectrum is pure point with probability $F_{\mathrm{ac}}(\mathbf{R}) > 0$. Further, using Theorem 2.16, we find that the above event occurs with probability 1. To verify that the eigenfunctions decay exponentially with rate equal to the Lyapunov exponent, also with probability 1, one uses the result of Problem V.40, according to which the probability of the spectral measure being concentrated outside the set (13.12) is either 0 or 1.

If the potential $q(x)$ is independent at distinct points but does not have a bounded density, we cannot apply Theorem 13.7 to deduce that the spectrum is pure point. However, the theorem does allow us to say that if we apply a continuously distributed and arbitrarily small local perturbation, the spectrum becomes pure point, since the Lyapunov exponent is positive for any metrically transitive potential independent at distinct points, provided that $\mathsf{E}\{\ln(1+|q(x)|)\} < \infty$: see Theorem 12.10 and Section 14.

That the spectrum is pure point for certain discretely distributed, independent potentials is proved in [Klein et al. 1986] and [Carmona et al. 1986], using other methods. In the latter paper it is shown that for the Bernoulli random potential, where $q(x)$ takes on only two values, the integrated density of states can have a nontrivial singular continuous component, which is inconsistent with (13.21).

(b) For a discrete Schrödinger operator, let $q(x)$ be a Gaussian metrically transitive process with zero mean value and correlation function $b(x)$ whose spectral density $\hat{b}(k)$, for $k \in [0,1)$, satisfies

$$\int_0^1 \hat{b}^{-1}(k)\,dk < \infty. \tag{13.24}$$

We know that a Gaussian process is nondeterministic if $\int_0^1 \ln \hat{b}(k)\,dk > -\infty$: see [Gihman and Skorohod I] and Section 1. This is certainly the case if (13.24) holds, so the Lyapunov exponent is positive Lebesgue-almost everywhere on $\mathbf{R}$. To check condition (13.21), one should, according to Remark 1 to Theorem 13.7, prove that

$$\mathsf{E}\{E(a,a;d\lambda) \mid \mathcal{F}(a-1,a+1)\} \leqslant C\,d\lambda. \tag{13.25}$$

But, according to Problem I.7, for a Gaussian process satisfying (13.24), the distribution

$$\mathsf{P}\{q(a) \in dq \mid \mathcal{F}(a-1,a+1)\} \tag{13.26}$$

has a bounded density. So, to obtain (13.25) one should use Theorem 13.5, as in the proof of Theorem 13.6. This approach is used in [Pastur 1985]; for a different one, see [Simon 1985c].

The degree of interdependence of the Gaussian variables $q(x)$, for $x \in \mathbf{Z}$, is determined by the rate of decay of the correlation function $b(x)$, that is, by the smoothness of the spectral density $\hat{b}(k)$. This property of $\hat{b}(k)$ is only tangentially related to the convergence of the integral in (13.24), which gives a sufficient condition for the existence of a bounded density for the conditional distribution (13.26), and hence to the property that we called smoothness in 13.A. Thus the two properties that, according with the heuristic reasoning of 13.A, increase the likelihood that the spectrum is pure point—smoothness of the metrically transitive potential and independence of its values—independent of one another.

Condition (13.24) is rather strong. Consider, for example, a Gaussian process

$$q(x) = \sum_{y \in \mathbf{Z}} u(x-y)\,\xi_y, \tag{13.27}$$

where the ξ_y, for $y \in \mathbf{Z}$, are independent, identically distributed Gaussian variables and $u(x)$ is a function with finite support. In this case $\hat{b}(k) = |\hat{u}(k)|^2$, where $\hat{u}(k)$ is a trigonometric polynomial that can be chosen in such a way that (13.24) does not hold: for example, if $u(x) = 0$ for $|x| \neq 1$ and $u(\pm 1) = \frac{1}{2}$, we have $\hat{u}(k) = \cos 2\pi k$. On the other hand, for any nonnegative $u(x)$ with finite support and independent variables ξ_y in (13.27), the spectrum of the operator is pure point with probability 1: see Problem V.44.

In Section 15 we will use Theorem 13.7 to prove that the Schrödinger operator has a pure point spectrum under various assumptions: see Theorem 15.8, Examples 14.9 and Subsection 15.A, and also Problems V.44–45, which deal with the discrete case. Here we will consider only one more aspect to the questions posed above. As we have mentioned, the main difficulty in proving that the spectrum is pure point, given that the Lyapunov exponent is positive, lies in excluding the possibility of a singular continuous spectrum caused by the existence of exceptional sets of zero Lebesgue and P-measure, on which the solutions to Cauchy problem do not have exponential asymptotic behavior (13.7) or (13.23).

The way we have seen to get around this difficulty is to find, by means of randomization, conditions for the instability of the singular continuous spectrum with respect to almost every finite-dimensional perturbation. In this abstract formulation, the results fall within the framework of traditional and important problems of spectral theory. For example, Kato [1966] showed that the essential spectrum of any self-adjoint operator is stable against compact perturbations, while the absolutely continuous spectrum is stable against trace-class perturbations. On the other hand, simple examples (see [Reed and Simon IV] and Problem II.23(b)) show that the singular spectrum is unstable even against perturbations of rank one, while Theorem 13.6 implies that the singular continuous spectrum of one-dimensional operators with a positive Lyapunov exponent is unstable against typical perturbations of rank one.

Abstract versions of Theorem 13.6 and 13.7 were proposed by Simon and Wolf [1986]. They are based on the results by Aronszajn [1957] and Donoghue [1965] on the variation of the spectrum of a self-adjoint operator under rank-one perturbations (see Lemma 4.20, for example), and on a version of the randomization method that we employed above. We will explain what form the positiveness condition for the Lyapunov exponent assumes in such an abstract approach. As was first pointed out by Ishii [1973] (see also [Delyon et al.1985b] and Problem V.31), if the Lyapunov exponent of a one-dimensional second-order equation (11.4) with a metrically transitive potential is positive on the spectrum, it is possible to determine the Green's function $G(x, y, \lambda + i0)$ of the corresponding operator for $\mathsf{P} \times \mathsf{L}$-almost all pairs $(\omega, \lambda) \in \Omega \times \mathbf{R}$, where L is the Lebesgue measure; besides, for all such pairs and for every $x \in \mathbf{Z}$,

$$\lim_{|y| \to \infty} \frac{1}{|y|} \ln \bigl| G(x, y, \lambda + i0) \bigr| = -\gamma(\lambda). \tag{13.28}$$

It follows that for all such pairs and all $x \in \mathbf{Z}$,

$$\sum_{y\in\mathbf{Z}} \left|G(x,y,\lambda+i0)\right|^2 < \infty. \tag{13.29}$$

This relation, because of the spectral representation (12.5), can be written

$$\int_{\mathbf{R}} (\lambda-\mu)^{-2} E(x,x,d\mu) < \infty. \tag{13.30}$$

That is why the abstract analogue of Theorem 13.6 proved by Simon and Wolf [1986] states that if A is a self-adjoint operator on a Hilbert space $\mathcal{H}$, $\psi \in \mathcal{H}$ is a cyclic vector of A and P is the operator of orthogonal projection onto this vector, the measure $\rho_\xi(d\lambda) = \big(E_{A+\xi P}(d\lambda)\psi, \psi\big)$ is pure point for Lebesgue-almost all $\xi \in \mathbf{R}$, provided the integral

$$\int_{\mathbf{R}} (\lambda-\mu)^{-2} \rho_0(d\mu) \tag{13.31}$$

is finite for Lebesgue-almost all $\lambda \in \mathbf{R}$.

Applications of this statement to the derivation of results similar to the ones in this subsection can be found in [Simon 1985], [Kotani and Simon 1987], and in Section 15.

Problems

1. Show that, for equation (6.1), the amplitude $r(x)$ of (11.1) satisfies $r(x) \leqslant Cr_0 e^{xQ(x)}$ for $x > 0$, where

$$Q(x) = \left(\frac{1}{x}\int_0^x \left|q(t)-\lambda\right| dt\right)^{\frac{1}{2}}$$

and $C = \sup_{x>0}\{Q(x) + Q^{-1}(x)\}$.

Further, if $r_1(x)$ and $r_2(x)$ are the functions corresponding to two potentials q_1 and q_2, and if $r_1(0) = r_2(0) = 1$, we have, for $x > 0$,

$$\left|r_1(x) - r_2(x)\right| \leqslant \max\{C_1, C_2\} e^{3x\max\{Q_1(x),Q_2(x)\}} \int_0^x \left|q_1(t) - q_2(t)\right| dt.$$

Hint. For the first inequality, set $\rho^2(x) = y^2(x) + y'^2(x)/Q^2(x)$ and, using a relation similar to (11.100), obtain

$$\ln\frac{\rho(x)}{\rho(0)} \leqslant \frac{1}{2}\int_0^x \big(Q(t) + Q^{-1}(t)|q(t)-\lambda|\big)\, dt.$$

For the second inequality, set $q_s(x) = q_1(x) + s(q_2(x) - q_1(x))$ and, using relations similar to (11.100), obtain an estimate for $\partial \ln\rho_s(x)/\partial s$. You can also use Gronwall's inequality: see [Hartman 1964].

2. Derive the following sharper versions for (11.27), in the continuous and discrete case, respectively:

$$\gamma(z) \leqslant \gamma_0(x) + \left(2\sqrt{|z|}\right)^{-1} \mathsf{E}\{|q(0)|\},$$
$$\gamma(z) \leqslant \gamma_0(x) + \left|\sin \arccos(z/2)\right|^{-1} \mathsf{E}\{|q(0)|\}.$$

Hint. For the continuous case, set $\rho^2(x) = |y(x)|^2 + |y'(x)|^2/|z|$, where $y(x)$ is the solution of the Cauchy problem (6.1), (6.3). Then show that

$$\frac{\rho'(x)}{\rho(x)} \leqslant \sqrt{|z|}\left|1 - \frac{z}{|z|}\right| + \frac{1}{2\sqrt{|z|}}|q(x)|;$$

the first term on the right is $\gamma_0(z)$. The discrete case is analogous.

3. If a Schrödinger operator on the semiaxis has an exponentially decreasing solution, all other solutions are exponentially increasing with an equal or higher rate.

Hint. The Wronskian of any pair of solutions is a constant.

4. The upper and lower characteristic exponents $\bar{\gamma}_\pm(z,\omega,\alpha)$ and $\underline{\gamma}_\pm(z,\omega,\alpha)$ and the upper and lower Lyapunov exponents $\bar{\gamma}(z,\omega)$ and $\underline{\gamma}(z,\omega)$ are defined by replacing the lim by lim sup and lim inf in equations (11.20) and (11.24). Show that:

(i) $\underline{\gamma}(z,\omega) + \bar{\gamma}_\pm(z,\omega,\alpha) \geqslant 0$ and $\bar{\gamma}(z,\omega) + \underline{\gamma}_\pm(z,\omega,\alpha) \geqslant 0$.

(ii) $\gamma_\pm(z,\omega,\alpha) \geqslant -\gamma(z)$;

(iii) if $\bar{\gamma}_\pm(z,\omega) \leqslant -\gamma(z)$ for some α, we have $\bar{\gamma}_\pm(z,\omega,\alpha) = \underline{\gamma}_\pm(z,\omega,\alpha) = -\gamma(z)$ and $\bar{\gamma}_\pm(z,\omega) = \underline{\gamma}(z,\omega) = \gamma(z)$.

5. Prove that $\bar{\gamma}_\pm(z,\omega) \leqslant \gamma(z)$ with probability 1.

Hint. Denoting the expression under the lim sup sign by $\gamma_x(z,\omega)$, show that γ_x is a subharmonic function, and thus

$$\gamma_x(z,\omega) \leqslant \frac{1}{\pi\rho^2}\int_{|\zeta - z| \leqslant \rho} \gamma_x(\zeta,\omega) d^2\zeta$$

(recall that $d^2\zeta$ stands for the two-dimensional Lebesgue measure on $\mathbf{C}$). Show that, for any compact $K \subset \mathbf{C}$, there exists $C_K(\omega) < \infty$ such that $|\gamma_x(z,\omega)| \leqslant C_K(\omega)$ with probability 1, for all $z \in K$ and $|x| \geqslant 1$. Use this to pass to the limit $x \to \pm\infty$ in the equation above. Then use the fact that, with probability 1, $\gamma_\pm(x,\omega) = \gamma(z)$ for almost all $z \in \mathbf{C}$, and the following equality, which is valid for any subharmonic function $f(z)$ [Hayman and Kennedy 1980]:

$$f(z) = \lim_{\rho \downarrow 0} \frac{1}{\pi\rho^2}\int_{|\zeta - z| \leqslant \rho} f(\zeta)\, d^2\zeta.$$

6. Let $A(x)$, for $x \in \mathbf{Z}$, be a sequence of unimodular 2×2 matrices such that

$$\limsup_{x\to\infty} \frac{1}{x} \ln\|A(x)\| = 0,$$

and that there exists the limit

$$\lim_{x\to\infty} \frac{1}{x} \lim\|\mathcal{T}(x)\| = \gamma > 0,$$

where $\mathcal{T}(x) = A(x-1)\dots A(0)$. If $e(\alpha) = (\cos\alpha, \sin\alpha)$ for $\alpha \in [0,\pi)$, show that there exists $\alpha_+ \in [0,\pi)$ such that

$$\lim_{x\to\infty} \frac{1}{x} \ln\|\mathcal{T}(x)e(\alpha)\| = \begin{cases} \gamma & \text{for } \alpha \neq \alpha_+, \\ -\gamma & \text{for } \alpha = \alpha_+. \end{cases}$$

Hint. Denote by $e_\pm(x)$ the normalized eigenvectors of the matrix $\mathcal{T}^*(x)\mathcal{T}(x)$ corresponding to the eigenvalues $\|\mathcal{T}^*(x)\mathcal{T}(x)\|^{\pm 1}$, and such that the angles $\alpha_\pm(x)$ in the relation $e_\pm(x) = e(\alpha_\pm(x))$ belong to $[0,\pi)$. Then

$$\begin{aligned} \|\mathcal{T}(x+1)e_+(x+1)\|\,|(e_+(x+1), e_-(x))| &\leqslant \|\mathcal{T}(x+1)e_-(x)\| \\ &\leqslant \|A(x+1)\|\,\|\mathcal{T}(x)e_-(x)\|, \end{aligned}$$

so that $|(e_+(x+1), e_-(x))| \leqslant e^{-2(\gamma-\delta(x))x}$, where $\lim_{x\to\infty}\delta(x) = 0$. From this prove that the vectors $e_\pm(x)$ converge to $e_\pm$. Now α_+ can be defined by $e_- = e(\alpha_+)$.

7. Let $A(x)$, for $x \in \mathbf{R}$, be a unimodular 2×2 matrix measurable in x, and let $\mathcal{T}(x)$ be the solution of the differential equation $\mathcal{T}' = A\mathcal{T}$, so that for A and $\mathcal{T}$ the conditions of the preceding problem are fulfilled. Show that the conclusion of the preceding problem holds.

Hint. Introduce the vectors $e_\pm(x)$ as in the preceding problem and, writing $\lambda(x) = \|\mathcal{T}^*\mathcal{T}(x)\|$, obtain the relation

$$\mathcal{T}^*(A^* + A)\mathcal{T}e_- + \mathcal{T}^*\mathcal{T}e_-' = -(\lambda^{-1})'e_- - \lambda^{-1}e_-',$$

from which it follows that

$$\lambda\|e_-'\| = \lambda\|(e_-', e_+)\| \leqslant 2\|A\|.$$

Then argue as in the preceding problem.

8. Any invertible matrix $\mathcal{N}$ of order n can be represented as $\mathcal{N} = U_1 \mathcal{D} U_2$, where U_1 and U_2 are unitary matrices and $\mathcal{D}$ is a diagonal matrix with elements $\delta_1 \geqslant \delta_2 \geqslant \cdots \geqslant \delta_n$. Show that the exponent γ_j of (11.34) can be represented as

$$\gamma_j = \lim_{x\to\infty} \frac{1}{|x|} \ln \mathsf{E}\{\delta_j(\mathcal{N}(x))\}.$$

Hint. Use the relation

$$\gamma_1 + \cdots + \gamma_p = \lim_{x\to\infty} \frac{1}{|x|} \ln \|\mathcal{N}^{\wedge p}(x)\|,$$

where $\mathcal{N}^{\wedge p}$ is the p-th exterior power of $\mathcal{N}$.

9. Let $\alpha_\pm(\lambda, \omega)$ be the angles defined by (11.37). Show that

$$\mathsf{P}\{\omega : \alpha_+(\lambda,\omega) = \alpha_-(\lambda,\omega)\} = 0$$
$$\operatorname{meas}\{\lambda : \alpha_+(\lambda,\omega) = \alpha_-(\lambda,\omega)\} = 0.$$

Hint. Use Theorem 2.12.

10. Let H be a one-dimensional differential operator of order two and $\gamma(\lambda)$ its Lyapunov exponent. Suppose that $\gamma(\lambda_0) = 0$ for some $\lambda_0 \in \mathbf{R}$. Show that $\gamma(\lambda)$ is continuous at λ_0.

Hint. $\gamma(\lambda)$ is nonnegative and upper semicontinuous.

11. Let H be a one-dimensional differential operator of order two satisfying the conditions of Theorem 11.6. Suppose that the open interval (a, b) lies outside the spectrum, but b is in the spectrum. If $\gamma(\lambda)$ is the Lyapunov exponent, show that

$$\gamma(b) = \lim_{\varepsilon\downarrow 0} \gamma(b - \varepsilon).$$

Hint. Use Theorem 3.1 and formula (11.79).

12. Let $\{J_\alpha\}$ be a family of Jacobi matrices with $s_\alpha(x) \equiv 1$ and $|q_\alpha(x)| \leqslant C < \infty$. Prove that, if $\lim_{\alpha\to\alpha_0} N(J_\alpha, \lambda) = N(J_{\alpha_0}, \lambda)$ for every λ, the convergence is uniform in λ on any compact set.

Hint. Argue as in the proof of Theorem 11.11.

13. Let h_n be an operator of the form (11.86), with matrix

$$q_n(x) = \begin{pmatrix} q(x,1) & -1 & \cdots & 0 & 0 \\ -1 & q(x,2) & \cdots & 0 & 0 \\ \vdots & \vdots & \ddots & \vdots & \vdots \\ 0 & 0 & \cdots & q(x,n-1) & -1 \\ 0 & 0 & \cdots & -1 & q(x,n) \end{pmatrix},$$

where the $q(x, k)$, for $k = 1, \ldots, n$, are independent, identically Cauchy-distributed random variables with parameter ζ (see Problem II.24). Denote by $\Gamma_n(\zeta)$ the arithmetic mean of the nonnegative characteristic Lyapunov exponents corresponding to h_n (cf. (11.91)). Consider also the operator $h_n^{(0)} = h_n \mid_{q(x,k)\equiv 0}$ and the corresponding function $\Gamma_n^{(0)}(z)$. Show that $\Gamma_n(z) = \Gamma_n^{(0)}(z + \zeta)$ and

$$\Gamma_n^{(0)}(z) = \frac{1}{n} \sum_{k=1}^{n} \gamma(z - \mu_k),$$

where $\gamma(z)$ is the Lyapunov exponent of the one-component operator $h_1^{(0)}$ and the μ_k are the eigenvalues of the matrix $q_n(x)$ in the case $q(x, 1) = \cdots = q(x, n) = 0$.

14. Show that if, in equation (11.4), the potential consists of independent, Cauchy-distributed random variables with parameter $\zeta = \alpha + i\beta$ (see Problem II.24), we have $\gamma(\lambda) = \operatorname{arcsinh} \delta$, where δ is the positive root of the equation

$$4\delta^2 + (4 - (\lambda - \beta)^2 - \alpha^2)\delta + \alpha^2 = 0.$$

15. Show that if, for the operator of Example 11.14(b) (see equations (11.110) and (11.111)), and if $\mathsf{P}\{Q(0) = 0\} = p > 0$, the Lyapunov exponent satisfies $\gamma_g(\lambda) \leqslant (1 - p) \ln |g| + O(1)$, as $g \to \infty$, showing that the asymptotic formula (11.114) is not always valid.

Hint. Use the inequality $\gamma_g(\lambda) \leqslant \mathsf{E}\{\ln \|A(x)\|\}$, which follows from (11.28).

16. For the multicomponent Schrödinger operator H_n of (12.41), develop a phase formalism similar to the one in Section 6.A.

Hint. To solve the equation $H_n \Psi = \lambda \Psi$, introduce the vector $Y = (\Psi', \Psi)$, and define the phase by $\phi = Y/\|Y\|$.

17. Verify the following representation for the spectral measure:

$$\rho_{a,b}^{(\alpha,\beta)}(d\lambda) = \delta_p\big(\theta_b^{(\beta)}(a, \lambda) - \alpha\big)d\lambda,$$

where $\delta_p(\theta) = \sum_{n \in \mathbf{Z}} \delta(\theta - \pi n)$ ($\delta(\cdot)$ being the Dirac delta function) and $\theta_b^{(\beta)}(x, \lambda)$ is the solution of the differential equation (6.5) satisfying the condition $\theta_b^{(\beta)}(b, \lambda) = \beta$. Use this representation to prove (13.14).

Hint. Use formula (12.8) and Problem III.21.

18. For an arbitrary $x \in \mathbf{R}$ or $\mathbf{Z}$, let $\rho_x(d\lambda)$ be the spectral measure of the order-two operator defined by (12.32)–(12.35), with x playing the role of the origin. For any x_1 and x_2, show that ρ_{x_1} is absolutely continuous with respect to ρ_{x_2}.

Hint. Use the fact that ρ_x is a weak limit of the measures $\rho_{x,a,b}$ of problems on the finite interval (a, b), as $a \to -\infty$ and $b \to \infty$ (cf. (12.8) and (12.17)), and the relation

$$\rho_{x,a,b}(\Delta) = \int_{\Delta} \frac{r^2(x)}{r^2(y)} \rho_{y,a,b}(d\lambda),$$

where $r^2(x)$ is defined by formulas (11.1) and (11.4) and by the condition $r^2(a) = 1$.

19. Prove that the Green's function $G(x, x', z)$ of an operator of order two, for x and x' fixed and $\operatorname{Im}(z) > 0$, depends continuously on the potential $q(x)$, in the topology of pointwise convergence in the discrete case and the topology determined by the norm

$$\left(\int_{-\infty}^{\infty} e^{-|\operatorname{Im} z|\,|x|/2} |q(x)|^2 \, dx \right)^{\frac{1}{2}}$$

in the continuous case.

Hint. Use a relation similar to (12.26).

20. Show that, if $w(z)$ is defined by (12.54), the function $w_0 = w \,|_{q \equiv 0}$ is equal to $-\sqrt{-z}$ in the continuous case and to $-\frac{1}{2} \arccos z/2$ in the discrete case.

21. Show that, under condition (11.42),

$$w(z) = -\sqrt{-z} + O(|z|^{-1/2})$$

as $\operatorname{Im} z \to \infty$.

Hint. Use (12.54) and Problem III.5.

22. Prove that the function $w(z)$ of (12.54) is a Nevanlinna function (see the Appendix), and so are $-iw$ and w' in the continuous case, and $-i \ln w$ and w' in the discrete case. Prove that $w(z)$ maps $\mathbf{C}_+$ conformally into itself.

23. Prove that the function $w(z)$ for the Schrödinger equation admits the representation

$$-iw(z) = \alpha + \frac{1}{\pi} \int_{\mathbf{R}} \frac{1 + \mu z}{\mu - z} \frac{\gamma(\mu)}{1 + \mu^2} \, d\mu,$$

where $\alpha \in \mathbf{R}$.

Hint. Use the fact that $-iw$ is a Nevanlinna function, together with Theorem A.12 and Problem V.21.

24. Prove that (a) and (b) are equivalent:

(i) In the continuous case,

(a) $w(z)$ and $w'(z)$ are Nevanlinna functions;

(b) there exist $\alpha \in \mathbf{R}$, $\beta \geqslant 0$ and a nondecreasing, nonnegative function $\nu(\mu)$ such that

$$\int \frac{\nu(\mu)}{1+\mu^2}\, d\mu < \infty$$

and

$$\begin{aligned} w(z) &= \alpha + \beta z + \int \left(\frac{1}{\mu - z} - \frac{\mu}{1+\mu^2} \right) \nu(\mu)\, d\mu \\ &= w(i) - \int \big(\ln(\mu - z) - \ln(\mu - i)\big)\nu(\mu)\, d\mu. \end{aligned}$$

(ii) In the discrete case,

(a) $w(z)$ and $w'(z)$ are Nevanlinna functions, and $w(iy) = -\ln(-iy) + O(1)$ as $y \to \infty$;

(b) $w(z) = -\int \ln(\mu - z)\nu(d\mu)$, where $\nu(\mathbf{R}) \leqslant 1$ and $\int \ln(1+|\mu|)\nu(d\mu) < \infty$.

25. Let $w(z)$ be a function such that w, $-iw$ and w' are Nevanlinna functions. Show that:

(a) $w(\lambda + i0) = -\gamma(\lambda) + i\pi\nu(\lambda)$, where $\gamma, \nu \geqslant 0$, the function $\nu(\lambda)$ is nondecreasing, and

$$\big|\nu(\lambda_1) - \nu(\lambda_2)\big| \leqslant \text{const}\, \big|\ln|\lambda_1 - \lambda_2|\big|^{-1}.$$

(b) There exist $\alpha_1 \in \mathbf{R}$ and $\beta_1 \geqslant 0$ such that

$$\begin{aligned} -iw(z)\sqrt{z} = \alpha_1 + \beta_1 z &+ \int_{\mathbf{R}_-} \left(\frac{1}{\mu - z} - \frac{\mu}{1+\mu^2} \right) \sqrt{-\mu}\, \nu(\mu)\, d\mu \\ &+ \frac{1}{\pi} + \int_{\mathbf{R}_+} \left(\frac{1}{\mu - z} - \frac{\mu}{1+\mu^2} \right) \sqrt{\mu}\gamma(\mu)\, d\mu. \end{aligned}$$

Hint. For part (b), show that if w_0 and $-iw_0$ are Nevanlinna functions, so is $-w_0 w$; then use (A.12).

26. Show that the function $w(z)$ of (12.54) is related to the Weyl functions as follows: $w = \mathsf{E}\{m_\pm\}$ in the continuous case, and $w = \pm\mathsf{E}\{\ln m_\pm\}$ in the discrete case.

Hint. First method: (12.43) and (12.55) show that the real parts of the analytic functions $w_1 = w$ and $w_2 = \mathsf{E}\{m_\pm\}$ coincide. Then Problem V.21 implies that w_1 and w_2 coincide.

Second method: Using (12.29), (12.47) and (12.49), show that

$$\frac{d}{dx}\ln G(x,x,z) = \zeta_+(x) - \zeta_-(x),$$
$$G(x+1,x+1,z)G^{-1}(x,x,z) = -\zeta_+(x+1)\zeta_-(x+1)$$

in the continuous and discrete case, respectively; that

$$\frac{d}{dx}A(x) = -\frac{2}{\zeta_+(x)+\zeta_-(x)} + \frac{d}{dz}(\zeta_+(x) - \zeta_-(x)),$$

where

$$A(x) = \frac{1}{\zeta_+(x)+\zeta_-(x)}\frac{\partial}{\partial z}(\zeta_+(x) - \zeta_-(x))$$

and the left-hand side is replaced by $A(x+1) - A(x)$ in the discrete case. From this, obtain that

$$2\mathsf{E}\{G(0,0,z)\} = \frac{\partial}{\partial z}\mathsf{E}\{m_\pm(z)\}$$

in the continuous case, because the derivative of a metrically transitive process is zero; in the discrete case the expectation on the right is $\mathsf{E}\{\ln m_\pm(x)\}$. From this and from (12.57) it follows that $w_1' = w_2'$; then Problem V.21 leads to $w_1 = w_2$.

27. Consider a one-dimensional Schrödinger operator whose potential is an infinitely differentiable metrically transitive process. Show that the function $w(z)$ of (12.54a) admits, for $|z| \to \infty$, the asymptotic expansion

$$w(z) = s + \sum_{k=1}^{\infty} \frac{w_k}{(2s)^k},$$

where $s = -\sqrt{-z} = i\sqrt{z}$.

Hint. Use Problem V.26 and equation (12.49a).

28. Show that, if (11.42) holds, we have

$$\left| w(z) - i\sqrt{z} - \frac{\mathsf{E}\{q(0)\}}{2i\sqrt{z}} \right| \leqslant \frac{\mathsf{E}\{q^2(0)\}}{2\,\mathrm{Im}\, z\, \mathrm{dist}(\sigma, z)}$$

for $\mathrm{Im}\, z \to \infty$, where σ is the spectrum.

Hint. Use (12.22), (12.26) and Problem V.26.

29. Show that under the conditions of Theorem 12.5, there exists, for $\lambda \in \Delta$, two linearly independent solutions $\chi_{\pm}$ to the Schrödinger equation:

$$\chi_{\pm}(x,\lambda,\omega) = \sqrt{\frac{\rho'_{\pm}(\lambda,\omega)}{\rho'_{\pm}(\lambda,T_x\omega)}}\, e^{\pm i\pi \int_0^x \rho'_{\pm}(\lambda,T_t\omega)\,dt},$$

where $\rho'_{\pm}$ is the density of the spectral measure $\rho_{\pm}(d\lambda,\omega)$ of the Schrödinger operator $H_{\pm}$ on the respective semiaxis.

Check that, for a periodic potential, these solutions coincide with the Floquet–Bloch solutions (16.1). Obtain similar formulas for the discrete case.

30. Show that, for any fixed $y \in \mathbf{R}$ and $z \in \mathbf{C}$ with $\operatorname{Im} z > 0$, we have, with probability 1:

$$G(x,y,z) = e^{w(z)|x-y|+o(|x-y|)}$$

as $|x| \to \infty$.

Hint. Use (12.29), (12.46), (12.47a) and Problem V.26.

31. Suppose the Lyapunov exponent is positive for Lebesgue-almost $\lambda \in X \subset \mathbf{R}$. Show that, for $\mathsf{P} \times \mathsf{L}$-almost all pairs $(\omega,\lambda) \in \Omega \times X$, where L is the Lebesgue measure, the Green's function $G(x,y,\lambda+i0)$ exists and, if it does,

$$\lim_{x\to\infty} \frac{1}{2|x|} \ln\bigl(|G(x,y,\lambda+i0)|^2 + |G(x+1,y,\lambda+i0)|^2\bigr) = -\gamma(\lambda)$$

for every y.

Hint. First method: Use Problems V.20 and V.30. Second method: Set $g(x) = G(x,0,\lambda+i0)$. Then

$$-g(x+1) - g(x-1) + (q(x)-\lambda)g(x) = \delta_{x,0},$$

which implies $\binom{g(x+1)}{g(x)} = \mathcal{T}(x,0)\binom{g(1)}{g(0)}$ and $\binom{g(x+1)}{g(x)} = \mathcal{T}(x,0)\binom{g(1)+1}{g(0)}$ for $x \geqslant 0$ and $x \leqslant 0$, respectively. Let $E_{\pm}$ be one-dimensional subspaces in $\mathbf{R}^2$ spanned by the vectors $e_{\pm} = (\cos\alpha_{\pm}, \sin\alpha_{\pm})$ of Theorem 11.3. Then, if $\binom{g(1)}{g(0)} \in E_+$ and $\binom{g(1)+1}{g(0)} \in E_-$, the function $g(x)$ decreases with rate $\gamma(\lambda)$ as $x \to \pm\infty$. These conditions uniquely determine $g(1)$ and $g(0)$ if $\alpha_+ \neq \alpha_-$; according to Problem V.9, this occurs for almost all pairs (ω,λ). Then consider $G(x,y)$ in a similar way and prove that $G(x,y) = G(y,x)$.

This method works in the multicomponent case (11.86) as well: see [Delyon et al. 1985b].

32. Verify that the spectrum of a Schrödinger operator with a periodic potential coincides with the set of zeros of the Lyapunov exponent.

Hint. See 16.A and references therein.

33. Let Γ be the set of $\lambda \in \mathbf{R}$ such that $\gamma(\lambda) = 0$, and $\rho'_{r,r'}(\lambda)$ the derivative of the spectral matrix (12.32)–(12.34). Show that $\det \rho' = \frac{1}{4}$ for Lebesgue-almost $\lambda \in \Gamma$.

Hint. Use (12.52).

34. Verify by direct calculation that Theorem 12.7 is not valid for the two-dimensional Laplacian acting on $\ell_2(\mathbf{Z}^2)$.

35. Prove the inequalities

$$\frac{\operatorname{Im} w}{\operatorname{Im} z} \geqslant -\frac{1}{2 \operatorname{Re} w}$$

in the continuous case, and

$$\frac{\operatorname{Im} \mathsf{E}\{m_\pm\}}{\operatorname{Im} z} \geqslant (e^{-2\mathsf{E}\{\ln |m_\pm|\}} \mp 1)^{-1}$$

in the discrete case, where equality holds if and only if, with probability 1, $m_\pm(z) = w(z)$ or $\pm \ln m_\pm = w(z)$, respectively.

Hint. Use (12.43) and (12.44), the convexity of the functions $1/t$ and $\pm \ln(1 \pm 1/t)$, and Problem V.26 (or else (12.10) and the convexity of e^t).

36. Prove Theorem 12.15 in the continuous case, without the assumption that $q(x) \geqslant a$.

Hint. The proof given in text shows that $w(z) = -\sqrt{-z+a}$. Hence we have $-2 \operatorname{Re} w \operatorname{Im} w = \operatorname{Im} z$ for $z \in \mathbf{C}$; then use the preceding problem.

37. Show that, if the Lyapunov exponent is positive on an interval Δ, a relation like (13.8) is, in general, impossible.

Hint. Let Δ be contained in $\sigma(H_+^{(\alpha)})$, for some $\alpha \in [0, \pi)$ (see Theorem 4.22), and let $\alpha_+(\lambda, \omega)$ be as in (11.37). Then, if (13.8) is fulfilled, Theorem 11.3 says that $\alpha_+(\lambda, \omega)$ is defined, with probability 1, for all $\lambda \in \Delta$. Then it follows from Theorem 13.4 that there exist α_1 and α_2 such that the sets $\Delta_j = \{\lambda \in \Delta : \alpha_+(\lambda, \omega) = \alpha_j\}$, for $j = 1, 2$, are dense and do not intersect. This contradicts the results of Problems V.6 and V.7 and Baire's theorem, according to which a pointwise limit (here $\alpha_+(\lambda, \omega)$) of a sequence of continuous points has a dense set of continuity points.

38. Prove Theorem 13.2 by using, instead of Lebesgue measure on $[0, \pi)$, any measure having a bounded density with respect to Lebesgue measure.

39. Show that, if $\rho_{0,\pm L}^{(\alpha,\beta)}(d\lambda)$ is the spectral measure (12.8), we have

$$\rho_{0,\pm L}^{(\alpha,\beta)}(d\lambda) = \frac{\partial \nu_L}{\partial \alpha} d\lambda,$$

where $\nu_L(\lambda)$ is the number of eigenvalues less that λ. Similarly, for the spectral measure of a boundary-value problem on the interval $[-L, L]$, show that

$$\rho_{-L,L}^{(\alpha,\beta)}(d\lambda) = \frac{1}{2}\left(\frac{\partial \nu_L}{\partial q(0)} + \frac{\partial \nu}{dq(1)}\right) d\lambda.$$

Hint. Use (12.6), (12.8) and Problem III.21.

40. Prove that the probability of the spectral measure ρ of a second-order metrically transitive operator being concentrated outside the set (13.12) is either 0 or 1.

Hint. Prove that $\rho(S_\pm(\omega))$ is measurable, and that the expression (13.11) is invariant with respect to shifts. Then use Problem V.17.

41. Prove (13.19), based on Problems II.15 and V.38.

42. Prove Theorem 12.1, with the help of Theorem 13.4.

Hint. As in the first proof of Theorem 12.1, use the stability of the absolutely continuous component of the spectrum under rank-one perturbations of the resolvent.

43. Suppose the conditions of Theorem 13.5 are fulfilled and set $\rho_\xi(d\lambda) = (E_{A+\xi P}(d\lambda)\psi, \psi)$. Set also $F_0(z) = \big((A - zI)^{-1}\psi, \psi\big)$ for $z \in \mathbf{C}$, and

$$F_0'(z) = \int_{\mathbf{R}} (\lambda - \mu)^{-2} \rho_0(d\mu)$$

for $\lambda \in \mathbf{R}$. Show that:

(i) the measure ρ_ξ has an atom at λ_0 if and only if $F_0(\lambda + i0) = -\xi^{-1}$ and $F_0'(\lambda_0) < \infty$;

(ii) the absolutely continuous, pure point and singular continuous components of the measure ρ_ξ, for $\xi \neq 0$, are concentrated on the sets

$$X = \{\lambda \in \mathbf{R} : \operatorname{Im} F_0(\lambda + i0) > 0\};$$
$$Y = \{\lambda \in \mathbf{R} : F_0'(\lambda) < \infty\};$$
$$Z = \mathbf{R} \setminus (X \cup Y),$$

and $X \cap Y = \varnothing$.

(iii) Given an interval Δ, $F_0'(\lambda) < \infty$ for Lebesgue-almost $\lambda \in \Delta$ if and only if ρ_ξ is pure point on Δ for Lebesgue-almost $\xi \in \mathbf{R}$.

Hint. Use Lemma 4.20, Theorem 13.5 and the Appendix.

44. If the potential of a one-dimensional Schrödinger operator h is as in Problem II.21 for $d = 1$, the spectrum of h is pure point.

Hint. Use the scheme of the proofs of Theorems 13.6 and 13.7, replacing $\mathcal{F}(a, b)$ in (13.21) by the σ-algebra generated by ξ_a, with $a \neq 0, 1$. You will need the results of Problems II.17 and II.21.

45. Suppose the potential of the operator h has the form (13.27), where $u(x)$ is a nonnegative function and $\lim_{x\to\infty} e^{b|x|}u(x) = 0$ for any $b > 0$; suppose also that the ξ_a are independent, identically distributed random variables whose distribution function has an absolutely continuous component with a bounded density. Assume also that $q(x)$ is a nondeterministic process (as is the case, for example, if u has compact support). Then, with probability 1, the spectrum is pure point.

Hint. Using the superexponential decrease in $u(x)$ and relation (11.15), check that the sets (13.12) do not depend on the values of ξ_a, when a runs over any bounded set. Then work as in the preceding problem.

46. Show that the conditions of Theorem 13.7 are fulfilled for a metrically transitive potential which is an ergodic Markov chain with a compact phase space, whose invariant measure has a bounded density, and whose transition probability has an everywhere positive and bounded density.

47. Let $\xi(x)$, for $x \in \mathbf{Z}$, be a sequence of independent, identically Cauchy-distributed random variables (see Problem II.24), and let $u(x)$ be a nonnegative function in $\ell_1(\mathbf{Z})$. Show that Theorem 13.7 is true for the potential in (13.27).

Hint. Prove, by arguing as in Problem II.22, that averaging over $\xi(0)$ replaces this quantity in the potential by $-i\alpha$. Then, if R is the resolvent to the resulting operator with a complex potential and U is the operator of multiplication by $\sqrt{u}$, show that $(1 - i\alpha r_0)r = r_0$, where $r = URU$ and $r_0 = r\,|_{\alpha=0}$. Prove that the operator $1 - i\alpha r_0$ is invertible, and that $\operatorname{Im}(r\psi, \psi) \leqslant \alpha^{-1}$ for $\psi \in \ell_2(\mathbf{Z})$ with $\|\psi\| = 1$. Then choose $\psi(x) = \delta_{x,x_0}$, where x_0 is a point where u is positive.

48. Prove that, if a metrically transitive potential assumes only finitely many values and is not periodic, the absolutely continuous spectrum of the corresponding Schrödinger operator is empty with probability 1.

Hint. Suppose the set Γ of Theorem 12.6 has positive Lebesgue measure. Then it suffices to prove that the space of realizations Ω is finite. By Theorem 12.10, if $\operatorname{meas}\Gamma > 0$, each $q_\pm(x) = \{q(x) : x \in \mathbf{R}_\pm \text{ or } x \in \mathbf{Z}_\pm\}$ determines $q(\cdot)$ uniquely. Give Ω a metric—for example, $d(q_1, q_2) = \sum_{x\in\mathbf{Z}} 2^{-|x|}|q_1(x) - q_2(x)|$ in the discrete case—and show that the maps $q_\pm \mapsto q_\mp$ are continuous. Then, for all $\varepsilon > 0$, there exists $\delta > 0$ and $M \geqslant 1$ such that $|q_1(x) - q_2(x)| \leqslant \delta$ for $x = -1, \ldots, -M$ implies $|q_1(0) - q_2(0)| < \varepsilon$. Since the set S of values of the potential is finite, if ε and δ are small enough, we conclude that $q_1(x) = q_2(x)$ for $x = -1, \ldots, -M$ implies $q_1(0) = q_2(0)$. Using the translational invariance of the probability measure, we conclude that the restriction of q to $[-M, -1]$ determines the restriction q_+ of q to $[0, \infty)$. Thus the cardinality of $\Omega_+ = \{q_+(x) : q(\cdot) \in \Omega\}$ is no greater than $M^{|S|}$. By Theorem 12.10, the same is true of Ω.

Chapter VI

Random Operators

Random operators, that is, differential and finite-difference operators whose coefficients exhibit weak correlation at far enough points, form one of the main and best studied classes of metrically transitive operators. Their basic spectral property, and one that has many physical applications, is that their spectrum has a point component—in the one-dimensional case, equal to the whole spectrum.

The discovery and investigation of this property required the development of methods new to spectral analysis—in particular, methods from differential and finite-difference equations with random coefficients. In one dimension, one important quantity associated with solutions of such equations is the Lyapunov exponent, whose relationship with the spectrum was studied in Chapter V in the general metrically transitive situation. The first section of this chapter also deals with the Lyapunov exponent: in it we prove that, for random one-dimensional equations, the Lyapunov exponent is positive for all $\lambda \in \mathbf{R}$ (and, in particular, for all λ in the spectrum of the corresponding operator), with the possible exception of a countable and discrete set. We give two proofs of this fact, one based on the well-known Furstenberg theorem, a powerful result applicable to the discrete case, and one based on the phase formalism of Sections 6 and 11, which works for one-dimensional Schrödinger operators whose potential is a Markov process with good enough ergodic properties, or the Markov-type process considered in 6.D.

In many problems involving random equations and random operators, it is important to have detailed quantitative information on the Lyapunov exponent, in particular its asymptotic behavior for various ranges of λ. Several such formulas are derived in 14.B.

The fact that the Lyapunov exponent is positive on the spectrum was, historically speaking, the first essential difference discovered between the behavior of the solutions of equations with random (independent, Markovian) coefficients as $|x| \to \infty$ and those with periodic coefficients. Because of Theorem 12.1, this fact rules out the possibility that the operators might have an absolutely continuous spectrum. Moreover, this result itself, and the methods developed to prove it, turn out to be important elements in one of the approaches to the proof that the spectrum is pure point. This method, proposed by Goldsheidt et al. [1977], and the results obtained thereby, are presented in 15.A. We also describe there other methods for analyzing pure point spectra, including a continuous variant of the randomization method of Section 13.

We then discuss briefly several other spectral properties of one-dimensional operators (Subsection 15.B) and the proof that the fluctuation spectrum of a multidimensional finite-difference operator is pure point (15.C). These results are all fairly recent [Fröhlich et al. 1985; Delyon et al. 1985; Simon et al. 1985; von Dreifus and Klein 1989], and are based on the fact [Fröhlich and Spencer 1983; von Dreifus 1987; Spencer 1988] that the Green's function of such operators decreases exponentially, with probability 1, for Lebesgue-almost values of λ in the fluctuation spectrum. This is a multidimensional analogue of the results on the positiveness of the Lyapunov exponent in one-dimensional random operators (and, by Problem V.30, follows from it in the one-dimensional case).

14 The Lyapunov Exponent of Random Operators in One Dimension

As we saw in Sections 12 and 13, the Lyapunov exponent is an important spectral quantity associated with one-dimensional metrically transitive operators. If it is positive the spectrum is different from that of periodic potentials: it has no absolutely continuous component, it has a point component, etc. By Theorem 12.10, the Lyapunov exponent of a (continuous or discrete) Schrödinger operator, whose potential is a nondeterministic random process, is positive for almost all values of λ on the operator's spectrum (the support of the integrated density of states).

In this section we show that if the coefficients of a second-order equation are random (independent, Markovian), the Lyapunov exponent $\gamma(\lambda)$ is positive for all $\lambda \in \mathbf{R}$, except perhaps for a countable number of isolated points that can, as a rule, be explicitly found, and moreover the behavior of $\gamma(\lambda)$ is understood in the vicinity of such points. To do this we use a number of special methods for estimating the growth rate of products of independent random matrices and their continuous analogues (multiplicative cocycles), and apply them to the fundamental matrices of the equations involved.

One of these methods is an important theorem of Furstenberg (Proposition 14.1), giving a general criterion for the exponential growth, with probability 1, of products of independent, identically distributed matrices. We will also use more specialized techniques, developed in the remainder of 14.A.

Subsection 14.B presents some asymptotic formulas for the Lyapunov exponent, more precise than those of 11.C and applicable to a number of typical cases: small random coefficients, λ near a spectral edge, λ near points with $\gamma(\lambda) = 0$, and so on.

14.A Positiveness of the Lyapunov Exponent

We start with a general theorem of Furstenberg [1963] that has played an important role in the development of the spectral theory of one-dimensional random operators. It can be used to describe well the set of λ for which the Lyapunov exponent is positive, for finite-difference equations with random coefficients and a number of differential equations.

Let $G = \mathrm{SL}(m, \mathbf{R})$ be the group of real, unimodular, $m \times m$ matrices. There is a natural action of G on the projective space $\mathbf{P}^{m-1}$ of one-dimensional subspaces of $\mathbf{R}^m$, which is a compact, connected manifold of dimension $m - 1$. The element of $\mathbf{P}^{m-1}$ corresponding to a vector $Y \in \mathbf{R}^m$ will be denoted by $\bar{Y}$, that is, $\mathbf{P}^{m-1} \ni \bar{Y} = \{tY : t \in \mathbf{R}\}$. The action is defined by

$$g\bar{Y} = \overline{gY}$$

for $g \in G$ and $Y \in \mathbf{R}^m$.

In the case $m = 2$—the one that interests us most at the moment—the projective line $\mathbf{P}^1$ can be parametrized by the angle $\phi \in [0, \pi)$, which we called the *reduced phase* in Sections 6, 11 and 13.

If F is a Borel measure on $\mathbf{P}^{m-1}$, we denote by gF the image of this measure under the transformation g, namely, $(gF)(\Gamma) = F(g^{-1}\Gamma)$ for any Borel set $\Gamma \subset \mathbf{P}^{m-1}$. A measure F is called G'-invariant if $gF = F$ for all g in the subgroup $G' \subset G$.

(14.1) Proposition [Furstenberg 1963]. *Let $A(x)$, for $x \in \mathbf{Z}$, be a family of independent random matrices in $\mathrm{SL}(m, \mathbf{R})$, with common distribution $P(dA)$ such that*

$$\int_G \ln \|A\| \, P(dA) < \infty,$$

and let G_P be the smallest closed subgroup containing the support of P. If G_P is not compact and there is no G_P-invariant probability measure on $\mathbf{P}^{m-1}$, the limit

$$\lim_{x \to \infty} \frac{1}{x} \mathsf{E}\{\ln \|A(x-1) \dots A(0)\|\} \tag{14.2}$$

is positive.

Proof. Denote by L the probability measure on $\mathbf{P}^{m-1}$ invariant with respect to the group $\mathrm{SO}(m,\mathbf{R})$ (the "Lebesgue measure" on $\mathbf{P}^{m-1}$) and consider, in $L_2(\mathbf{P}^{m-1},L)$, the following unitary representation of $\mathrm{SL}(m,\mathbf{R})$:

$$(V_g f)(e) = \sqrt{\frac{dgL}{dL}(e)}\,f(g^{-1}e), \quad \frac{dgL}{dL}e = \|g^{-1}e\|^{-m} \tag{14.3}$$

for $e \in \mathbf{P}^{m-1}$ and $g \in G$. Further, set

$$\mathcal{N}(x) = A(x-1)\dots A(0) \tag{14.4}$$

and $f_0(e) = 1$ for all $e \in \mathbf{P}^{m-1}$. Using the fact that the $A(x)$ are independent, we get

$$\mathsf{E}\{(V_{\mathcal{N}(x)}f_0, f_0)\} = (V^x f_0, f_0), \tag{14.5}$$

where

$$V = \int_G V_A P(dA). \tag{14.6}$$

Assume that the limit (14.2) vanishes. Equation (14.3) implies that

$$\|g\|^{-m} \leqslant \big|(V_g f_0, f_0)\big| \leqslant 1 \tag{14.7}$$

for $g \in G$, so we obtain

$$\begin{aligned} 1 &\geqslant \lim_{x\to\infty} \mathsf{E}\{(V_{\mathcal{N}(x)}f_0, f_0)^{1/x}\} \geqslant \lim_{x\to\infty} \mathsf{E}\{\|\mathcal{N}(x)\|^{-m/x}\} \\ &\geqslant \lim_{x\to\infty} \exp\Big(-\frac{m}{x}\mathsf{E}\{\ln\|\mathcal{N}(x)\|\}\Big) = 1, \end{aligned} \tag{14.8}$$

where we have used Jensen's inequality. But this implies, again by Jensen's inequality, that

$$\lim_{x\to\infty} \mathsf{E}^{1/x}\{(V_{\mathcal{N}(x)}f_0, f_0)\} = 1.$$

This, because of (14.5), means that the spectral radius of the operator V equals 1, that is, some points in the spectrum of V lie on the unit circle. Since any such point ζ is a limit point for points outside the circle that belong to the resolvent set of V, we conclude that for each such ζ there exists a sequence of unit vectors $f_n(e) \in L_2(\mathbf{P}^{m-1},L)$ such that $\lim_{n\to\infty}\big\|(V-\zeta)f_n\big\| = 0$. Then the identity

$$\int \big\|(V_A-\zeta)f_n\big\|^2 P(dA) = \big\|(V-\zeta)f_n\big\|^2 + (1-\|Vf_n\|^2)$$

implies that the limit of the expression on the left equals zero when $n\to\infty$. Hence, for some subsequence $n_j\to\infty$, we have $\big\|(V_A-\zeta)f_{n_j}\big\| \to 0$, almost everywhere with respect to P. Since $|V_A f| = V_A f$ and $\big||V_A f| - |f|\big| \leqslant |V_A f - f|$, we have

$$\lim_{j\to\infty} \big\|(V_A - 1)|f_{n_j}|\big\| = 0. \tag{14.9}$$

Denote by F_n the probability measure on $\mathbf{P}^{m-1}$ with density $dF_n/dL = |f_n(e)|^2$. It follows from (14.9) that

$$\lim_{j\to\infty} \int_{\mathbf{P}^{m-1}} \left| \frac{dAF_{n_j}}{dL} - \frac{dF_{n_j}}{dL} \right| L(de) = 0,$$

which means that the measures $AF_{n_j} - F_{n_j}$ converge weakly to zero. Because $\mathbf{P}^{m-1}$ is compact, there exists a measure F such that $AF = F$ almost everywhere with respect to P. Hence F is G_P-invariant, contradicting our hypotheses. □

Remark. We have stated only the part of Furstenberg's theorem that we will need later on. The proof we gave is due to Virtser [1979]. The full form of the theorem [Furstenberg 1963] includes the following results:

For any nonzero $Y \in \mathbf{R}^m$, with probability 1,

$$\lim_{x\to\infty} \frac{1}{x} \ln\|\mathcal{N}(x)Y\| = \gamma, \tag{14.10}$$

so that in the case of a product of independent matrices, expression in (11.34a), we have $\chi_1 = \gamma$ with probability 1, that is, for fixed (nonrandom) e, the probability of (11.37) being true is zero (see also (11.77)).

A measure F on $\mathbf{P}^{m-1}$ is called P-invariant (cf. (6.37) and (6.77)) if

$$\int_{\mathbf{P}^{m-1}} f(e)F(de) = \int_{G\times\mathbf{P}^{m-1}} f(Ae)P(dA)F(de). \tag{14.11}$$

for any Borel function $f(e)$ on $\mathbf{P}^{m-1}$. The right-hand side of this equation defines a natural convolution of the measures P and F. If we denote it by $P * F$, equation (14.11) can be rewritten as

$$F = P * F. \tag{14.12}$$

Thus, we arrive at the following formula for the limit in (14.2) and (14.10):

$$\gamma = \int_{G\times\mathbf{P}^{m-1}} \ln\frac{\|AY\|}{\|Y\|} P(dA)F(d\bar{Y}), \tag{14.13}$$

the right-hand side being the same for all solutions of (14.12) which are probability measures (cf. Theorems 6.8 and 6.9). These results are obtained by considering the Markov chain $\big(A(x), \mathcal{N}(x)\bar{Y}\big)$ on $G \times \mathbf{P}^{m-1}$. Using the identity

$$\ln \frac{\|\mathcal{N}(X)Y\|}{\|Y\|} = \sum_{y=1}^{x} f\big(A(y), \mathcal{N}(y)\bar{Y}\big),$$

where $f(A, \bar{Y}) = \ln\big(\|AY\|/\|Y\|\big)$, one reduces the proof of (14.10) and (14.13) to the study of the ergodic properties of this chain. This study takes up the bulk of Furstenberg's paper [1963]. For the current state of this question see [Bougerol and Lacroix 1986] or [Arnold and Wihstutz 1986].

Furstenberg's theorem has been used to prove that the Lyapunov exponent is positive for a number of finite-difference equations of order two with independent coefficients [Matsuda and Ishii 1970; Ishii 1973] and for the Schrödinger equation with piecewise constant, Markov potential [Benderskii and Pastur 1975]. The following proposition gives a similar result for Jacobi matrices.

(14.2) Theorem [Figotin 1980]. *Let J be a Jacobi matrix* (6.67) *whose coefficients $\big(s(x), q(x)\big)$ are independent at different points, independent of each other, identically distributed, satisfy condition* (6.68) *and*

$$\mathsf{E}\big\{\ln(1 + |s(0)| + |q(0)|)\big\} < \infty. \tag{14.14}$$

Then, if their joint distribution $\mathsf{P}\big\{s(0) \in ds, q(0) \in dq\big\}$ is not concentrated on a single line $q = \text{const}$, the Lyapunov exponent $\gamma(\lambda)$ of the equation

$$(Jy)(x) = \lambda y(x) \tag{14.15}$$

is positive for all λ. If the distribution is concentrated on a line $q = q_0$ (that is, $q(x) = q_0$ with probability 1), and is nondegenerate, $\gamma(\lambda)$ vanishes at the point $\lambda = q_0$ and nowhere else.

Proof. We write down equation (14.15) in the form (11.4), where (14.16)

$$Y(x) = \begin{pmatrix} s(x+1)y(x+1) \\ y(x) \end{pmatrix}, \quad A(x) = \frac{1}{s(x)} \begin{pmatrix} (q(x) - \lambda) & -s^2(x) \\ 1 & 0 \end{pmatrix}.$$

Then the Lyapunov exponent is given by (11.25) and (11.28), where the matrices (14.16) are unimodular, independent, identically distributed and satisfy (14.1), because of (14.14). Thus we are in the situation of Proposition 14.1, and must check if its conditions are met. Now, if

$$A_i = \frac{1}{s_i} \begin{pmatrix} q_i & -s_i^2 \\ 1 & 0 \end{pmatrix} \tag{14.17}$$

and $s_1 \geqslant s_2$, we have

$$(14.18)\qquad \begin{aligned} B_1 = A_1^{-1}A_2 &= \frac{1}{s_1 s_2}\begin{pmatrix} s_1^2 & 0 \\ q_1 - q_2 & s_2^2 \end{pmatrix}, \\ B_2 = A_1 A_2^{-1} &= \frac{1}{s_1 s_2}\begin{pmatrix} s_1^2 & q_1 s_2^2 - q_2 s_1^2 \\ 0 & s_2^2 \end{pmatrix}. \end{aligned}$$

Since

$$\begin{pmatrix} a & b \\ 0 & a^{-1} \end{pmatrix}^n = \begin{pmatrix} a^n & b_n \\ 0 & a^{-n} \end{pmatrix},$$

where $b_n = b(a^n - a^{-n})/(a - a^{-1})$, the group G_P is not compact.

Now associate with each nonzero vector $Y \in \mathbf{R}^2$ the point $\bar{Y} \in \mathbf{P}^1$, parametrized by $\phi \in [0, \pi)$ such that $\cot\phi = y_1/y_2$. A direct calculation shows that the set of fixed points of the transformations $\bar{Y} \to \overline{B_1 Y}$ and $\bar{Y} \to \overline{B_2 Y}$ are

$$\begin{aligned} Z_1 &= \{\pi/2, \operatorname{arccot}((s_1^2 - s_2^2)/(q_1 - q_2))\}, \\ Z_2 &= \{\operatorname{arccot}((q_1 s_2^2 - q_2 s_1^2)/(s_2^2 - s_1^2)), 0\}, \end{aligned}$$

the first point of each set being an attractor. (This means that if $\bar{Y}$ does not coincide with the second point, its iterates under B_1 or B_2 converge to the first point.)

If the joint distribution of $(s(x), q(x))$ is not concentrated on a single line $q(x) = q_0$, there exists in G_P a matrix (14.8) with $q_1 \neq q_2$. In this case $Z_1 \cap Z_2 = \varnothing$, and therefore there exists no G_P-invariant measure on $\mathbf{P}^1$, since such a measure would have to be concentrated on Z_1 and Z_2 at the same time.

A similar situation occurs if $q(x) \equiv q_0$ with $q_0 \neq \lambda$ and $s(x)$ nondegenerate, for in (14.18) we can put $q_1 = q_2 \neq 0$ and $s_1 \neq s_2$.

Finally, if $q(x) \equiv q_0 = \lambda$, we have, for every x,

$$\ln\|\mathcal{T}(x)\| = \left|\ln \prod_{k=0}^{x/2-1} \frac{s(2k+1)}{s(2k)}\right|.$$

Since the $s(x)$ are identically distributed and because of (14.14), we have $\gamma(q_0) = 0$, concluding the proof of the theorem. □

Remark. In an example of application of the heuristic rule formulated just before Example 11.14, we have not used the independence of the random variables $s(x)$ in the proof that $\gamma(q_0) = 0$. Thus this fact is still true for any metrically transitive $s(x)$. Similar results will be seen in Theorem 14.3 and Problem VI.15.

Thus, Furstenberg's theorem give us full information on the positiveness of the Lyapunov exponent of a finite-difference equation of order two with independent coefficients. The following proposition is an application of this knowledge to the Schrödinger equation.

(14.3) Theorem [Ishii 1973]. *Suppose the potential in the Schrödinger equation is of the form*

$$q(x) = \sum_{n \in \mathbf{Z}} \xi_n \delta(x - n) \tag{14.19}$$

(*a particular case of* (1.31)), *where the* ξ_n *are bounded, independent, identically distributed random variables. Then the Lyapunov exponent* $\gamma(\lambda)$ *vanishes if and only if* $\lambda = k_n^2$, *where* $k_n = \pi n$, *for* $n \in \mathbf{Z}$.

The meaning of the Dirac δ-function in the potential of the Schrödinger equation is explained by formula (6.26) and the subsequent text. The condition that ξ_n is bounded is given for simplicity; Theorem 14.3 remains true if $\mathsf{E}\{\ln(1 + |\xi_0|)\} < \infty$.

Proof. Suppose $\lambda = k_n^2$. Then the Schrödinger equation has two linearly independent solutions: $y_1(x) = k_n^{-1} \sin k_n x$ and $y_2(x) = \cos k_n x + k_n^{-1} \sin k_n x \sum_{l=1}^{[x]} \xi_l$. Both grow no faster than $|x|$, so $\gamma(\lambda_n) = 0$. If, on the other hand, $\lambda = k^2 \neq k_n^2$, we can write down the general solution of the Schrödinger equation with $n < x < n + 1$ as

$$y(x) = y(n) \cos k(x - n) + \eta_n \sin k(x - n). \tag{14.20}$$

Taking into account conditions (6.26), we find that the quantities $y(n)$, for $n \in \mathbf{Z}$, satisfy the equation

$$-y(n+1) - y(n-1) + \xi_n \frac{\sin k}{k} y = -2 \cos k y(n), \tag{14.21}$$

which coincides with the discrete analogue of the Schrödinger equation (11.4), with potential $q(n) = k^{-1} \xi_n \sin k$ and spectral parameter $-2 \cos k$. (A similar reduction is useful with other problems involving δ-functions: see [Khomskii 1966; Lifshitz et al. 1982b; Bellissard 1985].) Thus, denoting the Lyapunov exponent (11.4) by $\gamma_1(q, \lambda)$, we find from (14.21) that

$$\gamma(\lambda) = \gamma_1(k^{-1} \xi \sin k, -2 \cos k) \tag{14.22}$$

for $\lambda = k^2$. Combining this with Theorem 14.12, we conclude that $\gamma(\lambda) > 0$ when $\lambda \neq k_n^2$. □

Remark. Furstenberg's theorem allows us to find the set of λ such that $\gamma(\lambda) = 0$ in other cases as well. For example, if $\delta(x)$ in (14.19) is replaced by the characteristic function $\chi_{(0,1)}(x)$ and ξ_n is assumed to take on only two values, 0 and $\xi > 0$, a representation similar to (14.20) reduces the problem to finding the asymptotic behavior of the norm of the product of independent random matrices. One can then prove that the Lyapunov exponent vanishes only at points of the form $\lambda = \pi^2 n^2 + \xi$, for $n \in \mathbf{Z}$. This and many similar examples are considered in [Benderskii and Pastur 1975].

Before moving on to other random potentials in the Schrödinger equation, we discuss briefly other results concerning finite-difference equations. Proposition 14.1 enables us to study the case of independent coefficients, but more general situations are also important. In particular, the case of Markov coefficients is of substantial interest, both in itself and because in passing to the continuous case Markov coefficients are the best we can expect in terms of statistical independence. Indeed, potentials assumed to take on independent values on distinct intervals, or even at distinct points (in which case they must be generalized functions), abound in the literature (especially in theoretical physics: see [Lifshitz et al. 1982b]), but it is clear that Markov processes, both discrete and continuous, are a very natural and important class of random coefficients.

Pastur [1974a, 1980] proved that the Lyapunov exponent is positive for Jacobi matrices with Markov entries, if their ergodic properties satisfy certain conditions. We will present here an example of such conditions. Let $P(S,q;dS,dq)$ be the transition probability and $F(dS,dq)$ the stationary distribution of the Markov chain $\big(S(x),q(x)\big) = \big(s^2(x),q(x)\big)$. The conditions are:

(a) $P(S,q;dS',dq') = p(S,q;S',q')F(dS',dq')$ and $F(dS,dq) = \chi(q) \times \pi(S,q)\,dq\,m(dS)$, where $\chi(q) : \mathbf{R} \to [0,1]$ has finite support and $m(dS)$ is concentrated on the interval (α,β), for $0 < \alpha \leqslant \beta < \infty$;

(b) the densities $p(S,q;S',q')$ and $\pi(S,q)$ are bounded uniformly with respect to all variables by K^{-1} and K, respectively, for $1 < K < \infty$;

(c) $p(S,q;S',q') = p(S',q';S,q)$.

For the case of the discrete Schrödinger equation (11.4), where the potential forms a Markov chain with a finite number of states $q_1,\dots,q_L$, having transition matrix p_{jk} and stationary distribution $p_j > 0$, these conditions reduce to the relation

$$p_j p_{jk} = p_k p_{kj} \tag{14.23}$$

and to the requirement that the chain be indecomposable, that is, that the state space do not split into subsets such that for each j the entries p_{jk} are nonzero in only one of the subsets.

The advantage of the methods used in [Pastur 1974a, 1980] is that it applies, after slight modifications, to the continuous case: see Theorem 14.4. Virtser [1979] gave general conditions for the limit (14.2) to be positive for unimodular matrices of arbitrary dimension m and forming a Markov chain; our proof of Proposition 14.1 uses Virtser's method, adapted to the independent case. Other extensions of Furstenberg's theorem to the Markov case with arbitrary m were obtained by Royer [1980] and Guivarc'h [1984].

As we noted at the end of 11.B, however, for equations of order $m > 2$, it is not the positiveness of (14.2), the largest Lyapunov exponent, that matters, but that of the smallest nonnegative exponent. For a multicomponent equation of order two, corresponding to the operator (11.86), this is the quantity γ_n from (11.90). This quantity is positive when the Lyapunov spectrum (11.34) is simple: for example, when the inequalities in (11.90) are strict.

Simplicity of the Lyapunov spectrum is important because, for $m > 2$, Theorem 12.10 has no analogue. Furstenberg showed in the same paper [1963], although only implicitly, that the Lyapunov spectrum is simple if the measure $P(dA)$ has a bounded density with compact support with respect to Haar measure in $\mathrm{SL}(m, \mathbf{R})$. In [Sazonov and Tutubalin 1966] simplicity of the Lyapunov spectrum was proved explicitly under the assumption that $P(dA)$ has a density and that $\mathsf{E}\{\ln^2 \|A\|\} < \infty$.

A fairly general condition for the simplicity of the Lyapunov spectrum in the case of the product of the independent symplectic matrices (11.88) appearing in the study of the operator (11.86) was given by Guivarc'h and Raugi [1985]. The condition is that the minimal semigroup containing the support of $P(dA)$ should contain the open subset of $\mathrm{SL}(m, \mathbf{R})$. Very general algebraic conditions for the simplicity of the Lyapunov spectra were recently found by Goldsheidt and Margulis [1987].

For the case of Markov matrices $A(x)$ from (14.2), Virtser [1983] established that all Lyapunov exponents are different if, for some s, the s-fold convolution of $P(dA)$ is absolutely continuous with respect to the Haar measure on $\mathrm{SL}(m, \mathbf{R})$, and the transition probability $P(A, dA')$ is absolutely continuous with respect to $P(dA)$ and depends continuously on A in the topology of convergence in variation.

We now turn to the continuous case.

(14.4) Theorem [Pastur 1980]. *Assume the potential $q(x)$ in the Schrödinger equation is a Markov-type process defined by formulas* (6.47)–(6.52) (*the Brownian motion model*). *Then the Lyapunov exponent is positive for all $\lambda \in \mathbf{R}$.*

Proof. We use the notation and results from 6.D. Theorems 6.14 and 11.12 imply that

$$\gamma(\lambda) = \frac{1}{2} \int \big(Q(X) - \lambda + 1\big) \sin 2\phi \, \pi(X, \phi) \, dX \, d\phi, \tag{14.24}$$

where $\pi(X,\phi)$ is the density of the invariant measure of the Markov process $(X(x),\phi(x))$ satisfying the equation

$$(14.25)\qquad -\frac{\partial}{\partial\phi}((\cos^2\phi+(\lambda-Q(X))\sin^2\phi)\pi)+\tfrac{1}{2}\Delta\pi=0$$

(see (6.53)). Integrating this with respect to X, we get

$$\int(Q(x)-\lambda+1)\sin^2\phi\,\pi(X,\phi)\,dX=p(\phi)-p(0),$$

where

$$p(\phi)=\int_K\pi(X,\phi)\,dX.$$

Combining this with (14.24) gives

$$(14.26)\qquad \gamma(\lambda)=\frac{1}{2}\int\cot\phi(p(\phi)-p(0))\,d\phi.$$

We now pass to the variable $z=\cot\phi$, denoting the corresponding probability densities by $\tilde{\pi}(X,z)$ and $\tilde{p}(z)$. Set

$$(14.27)\qquad g(X,s)=\int_{\mathbf{R}}e^{isz}\tilde{\pi}(X,z)\,dz$$

for $s\geqslant 0$. From the results of 6.D and the equality

$$(14.28)\qquad \frac{\partial}{\partial z}((z^2+\lambda-Q(X))\tilde{\pi})+\tfrac{1}{2}\Delta\tilde{\pi}=0,$$

obtained by rewriting (14.25) for z, we easily conclude that there are positive constants C and $C_l>0$, for $l=1,2,\ldots$, such that

$$(14.29)\qquad \begin{gathered}\left|\frac{\partial^l}{\partial z^l}\tilde{\pi}\right|,\left|\frac{\partial^l}{\partial z^l}\Delta\tilde{\pi}\right|\leqslant C_l(1+|z|^{2+l})\\ |g(X,s)|+|\Delta g(X,s)|\leqslant C(1+s)^{-1}\end{gathered}$$

for $s>0$. We also conclude that

$$(14.30)\qquad \frac{\partial^l}{\partial z^l}g(X,s)=\lim_{R\to\infty}\int_{|z|\leqslant R}e^{isz}(iz)^l\tilde{\pi}(X,z)\,dz$$

for $l=0,1,2,\ldots$ and $s\geqslant 0$, and

$$g(X,0)=\frac{1}{V(K)},\qquad \left.\frac{\partial g(X,s)}{\partial s}\right|_{s=0+}=i\int_0^{\pi}\cot\phi(\pi(X,\phi)-\pi(X,0))\,d\phi.$$

Combining this with (14.28) and (14.30), we obtain, after elementary transformations,

$$(14.31) \qquad is\left(-\frac{\partial^2}{\partial s^2}g + (\lambda - Q(X))g\right) = \tfrac{1}{2}\Delta g$$

for $s > 0$. Multiplying this equation by $\bar{g}$, and integrating the imaginary part of the result over s from 0 to ∞, we get

$$(14.32) \qquad -\operatorname{Im}\left(\overline{g(X,s)}\frac{\partial g(X,s)}{\partial s}\Big|_{s=0+}\right) = \frac{1}{2}\int_0^\infty \frac{ds}{s}\operatorname{Re}(g\Delta\bar{g}),$$

where the convergence of the integral on the right is guaranteed by (14.29). Integrating (14.32) over X and taking (14.29) and (14.30) into account, we get the following representation for the Lyapunov exponent:

$$(14.33) \qquad \gamma(\lambda) = \frac{V(K)}{2}\int_0^\infty \frac{ds}{s}\big(g(\,\cdot\,,s), -\tfrac{1}{2}\Delta g(\,\cdot\,,s)\big)_{L_2(K)}.$$

Now assume that $\gamma(\lambda) = 0$. By (14.33), $g(X,s)$ is independent of X for almost all s. Because of (14.27), this would mean that the function $\tilde{\pi}(X,z)$ is also independent of X. It is easy to see that this contradicts (14.28), so $\gamma(\lambda)$ has to be positive. □

This method of proof is not based on Proposition 14.1 and can be applied to other random potentials. Indeed, it can be used to show that, for a potential that is a function of an ergodic Markov process,

$$(14.34) \qquad \gamma(\lambda) = -\int_0^\infty \frac{ds}{s}\operatorname{Re}\big(g(\,\cdot\,,s), \mathfrak{A}_X^* g(\,\cdot\,,s)\big)_X,$$

where $\mathfrak{A}_X$ is the infinitesimal operator of the process $X(x)$ and the inner product is taken in the Hilbert space of functions on the phase space of the process that are square-integrable with weight $\pi^{-1}(X)$, where $\pi(X)$ is the (possibly generalized) density of the invariant distribution of $X(x)$.

In particular, if the process is reversible, that is,

$$(14.35) \qquad \mathsf{P}\{X(0) \in dX_1, X(x) \in dX_2\} = \mathsf{P}\{X(0) \in dX_2, X(x) \in dX_1\},$$

the operator $\mathfrak{A}_X^*$ is self-adjoint and nonpositive definite on this same space. Thus, the question of whether $\gamma(\lambda)$ vanishes reduces to the study of the equation $\mathfrak{A}_X^* f = 0$, that is, in fact, the study of the ergodicity of the process $X(x)$. This is just the case with Theorem 14.4.

Another example is that of a random telegraph signal $r(x)$: see Example 1.15(b). By (1.27), this process is reversible; by the description for $\mathfrak{A}$ given at the beginning of 7.A, equation (14.34) takes the form

$$(14.36) \qquad \gamma = \frac{n_0 + n_1}{n_0 n_1}\int_0^\infty \frac{ds}{s}\big|n_0 g(0,s) - n_1 g(1,s)\big|^2.$$

It is not difficult to prove that this is true, since in this case the phase space of the process is $\{0,1\}$. Thus, for this random potential the Lyapunov exponent is everywhere positive.

By Problem I.9, the intervals where the process $r(x)$ are constant are independent variables with density $n_r e^{-n_r l}$. In the more general case of arbitrary densities, the process is not Markovian (we sometimes call such processes semi-Markovian). But, by Problem I.9, it can be regarded as a component of a Markov process, whose infinitesimal operator is self-adjoint and nonpositive. That is enough to show that the Lyapunov exponent is positive for all λ: see Problem VI.13.

Markov and semi-Markov jump processes can also be approached using Proposition 14.1.

Other examples of application of (14.34) will be found in [Pastur 1980], [Lifshitz et al. 1982b], and Problems VI.9 and VI.13.

Molchanov [1978] proposed a method to prove that the Lyapunov exponent is positive without using Furstenberg's theorem or the method we used to prove Theorem 14.4.

14.B Asymptotic Formulas for the Lyapunov Exponent

We discussed in 11.C the asymptotic behavior of the Lyapunov exponent for general metrically transitive operators in different spectral regions and for limits of a one-parameter potential. Here some of the formulas stated there will be proved or sharpened, under the assumption that the coefficients are random functions. We start with (11.104) and (11.125).

(14.5) Theorem.

(i) *For the Schrödinger equation with potential*

(14.37) $$q(x) = Q\big(X(x)\big),$$

where $Q\big(X(x)\big)$ *is a random Markov process of the type discussed in* 6.*D, let* $g = \sup_{X\in K}\big|Q(X)\big|$, *and assume that* $\mathsf{E}\{Q(X(0))\} = 0$. *Then, for any* $\lambda > 0$,

(14.38a) $$\gamma(\lambda) = \frac{1}{8\lambda}\hat{b}(2\sqrt{\lambda}) + O(g^3)$$

as $g \downarrow 0$, *and*

(14.38b) $$\gamma(\lambda) = -\frac{1}{16\lambda^2}b'(0) + O(\lambda^{-5/2})$$

as $\lambda \to \infty$, *where* $b(x) = \mathsf{E}\{Q(X(0))Q(X(x))\}$ *is the correlation function of the process* $Q\big(X(x)\big)$, *and* $\hat{b}(x)$ *is its Fourier transform.*

(ii) *For the operator* (6.19), *with coefficient* $s(x) = S(X(x))$ *a Markov process satisfying* (6.20), *we have*

$$\gamma(\lambda) = \frac{\lambda}{8\kappa}\hat{b}_\delta(0) + O(\lambda^{3/2}) \tag{14.39}$$

as $\lambda \downarrow 0$, *where* $b_\delta(x)$ *is the correlation function of the process*

$$\delta(x) = \frac{s^{-1}(x)}{\mathsf{E}\{s^{-1}(x)\}} - 1.$$

A formula similar to (14.39), for the more general case of Equation (1.6) with $q \equiv 0$, is treated in Problems VI.18–19.

Proof. Part (i). We set $\lambda = k^2$. Reasoning as in the proof of formula (14.24), but starting from (11.79b), we get

$$\gamma(\lambda) = \frac{1}{2k}\int_E Q(X)\sin 2\chi\,\pi(X,\chi)\,dX\,d\chi, \tag{14.40}$$

where $E = K \times [0,\pi)$ and $\pi(X,\chi)$ is the density of the invariant distribution of the Markov process $(X(x),\chi(x))$, which has the same properties as the density $\pi(X,\phi)$ from 6.D (to see this, recall that ϕ and χ are related by $k\cot\phi = \cot\chi$). In particular, $\pi(X,\chi)$ satisfies the stationary Fokker–Planck–Kolmogorov equation (cf. (14.25))

$$-\frac{\partial}{\partial\chi}\big((k - k^{-1}Q(X)\sin^2\chi)\pi\big) + \tfrac{1}{2}\Delta\pi = 0. \tag{14.41}$$

Set

$$\pi_n(X) = \int_0^\pi e^{2in\chi}\pi(X,\chi)\,d\chi \tag{14.42}$$

for $n \in \mathbf{Z}$; then

$$\pi_0(X) = \pi(X) = \frac{1}{V(K)}. \tag{14.43}$$

is the density of the invariant distribution of $X(x)$ (see (6.50)). Combining (14.40) and (14.42) and recalling from 6.D that $\pi(X,\chi)$ is smooth, we conclude that

$$\gamma(\lambda) = \frac{1}{2k}(Q, \operatorname{Im}\pi_1)_{L_2(K)}, \tag{14.44}$$

$$2in\Big(k\pi_n + \frac{Q}{4k}(\pi_{n+1} + \pi_{n-1} - 2\pi_n)\Big) + \tfrac{1}{2}\Delta\pi_n = 0. \tag{14.45}$$

The second relation, for $n \neq 0$, can be rewritten as

(14.46) $$\pi_n = \frac{in}{2k}(\Delta + 2ink)^{-1} Q(\pi_{n+1} + \pi_{n-1} - 2\pi_n).$$

By (14.42) and (14.43),

(14.48) $$\|\pi\|_{L_2(K)} \leqslant \|\pi_0\|_{L_2(K)} = \frac{1}{V(K)},$$

so that (14.45) implies

(14.48) $$\|\pi\|_{L_2} \leqslant \left(\frac{g}{k^2}\right)^n \frac{1}{V(K)}$$

Combining this with (14.41), (14.45) and (6.51), we get

(14.49) $$\gamma = -\frac{1}{4k^2}(Q, \mathrm{Re}(\tfrac{1}{2}\Delta + 2ik)^{-1} Q\pi_0) + O(g^3 k^{-5}).$$

Now, to obtain (14.38a), we recall the operator identity

(14.50) $$(\tfrac{1}{2}\Delta + 2ik)^{-1} = -\int_0^\infty e^{x(\frac{1}{2}\Delta + 2ik)}\, dx.$$

Since the kernel of the operator $e^{\frac{1}{2}x\Delta}$ is the density of the transition probability of the Markov process $X(x)$, the right-hand side of (14.49) can be rewritten as

(14.51) $$\gamma = \frac{1}{4k^2}\int_0^\infty b(x)\cos 2kx\, dx + O(g^3 k^{-5}),$$

which yields (14.38a). To get (14.38b), we must integrate (14.51) by parts, taking into account that

$$b'(0^+) = (Q, \tfrac{1}{2}\Delta Q)_{L_2(K)} < 0,$$

provided that $Q(X)$ is not constant.

Part (ii). To prove (14.39), we use (11.98), which implies, if we set $\delta(x) = \tilde{\delta}(X(x))$,

(14.52) $$\gamma = \frac{k_1}{2}(\tilde{\delta}, \mathrm{Im}\, \pi_1)_{L_2(K)},$$

where $k_1^2 = \lambda \mathsf{E}\{s^{-1}(x)\}$ and $\pi_n(X)$ is the Fourier coefficient (14.41) of the density of the invariant distribution of the process $(X(x), \chi(x))$ (see equation (6.24)). Using relations similar to those involved in the derivation of (14.46), we find the following expression for these coefficients:

(14.53) $$\pi_n = \tfrac{1}{2} ink_1(\tfrac{1}{2}\Delta + 2ink_1)^{-1}\tilde{\delta}(\pi_{n+1} + \pi_{n-1} + 2\pi_n).$$

Combining (14.52) and (14.53), we get a relation similar to (14.51), from which equation (14.39) follows as $k_1 \to 0$. □

Remarks. 1. Formulas (14.38) and (14.39) are variants of formulas well-known in solid-state theory for the inverse mean free path calculated under the Born approximation [Lifshitz et al. 1982b].

2. Based on (14.44) and (14.45), or on (14.52) and (14.53), we can see that the Lyapunov exponent can be represented as an asymptotic series, and that for small potentials ($g \to 0$), for examples, this series converges: its terms are estimated by expressions similar to the right-hand side of (14.48).

3. The technique used in the proof of Theorem 14.5 also works in deriving the asymptotic behavior of $\gamma(\lambda)$ for slowly varying Markov potentials $q_R(x) = Q\big(X(x/R)\big)$, as $R \to \infty$. We discussed this question, for general metrically transitive potentials, in Example 11.14(c). For Markov potentials, we can add the following. First, $N(\lambda)$ has a continuous density, by (6.45), Problem III.23(a) and Subsection 6.D. Therefore, condition (11.108) is fulfilled with $\alpha = 1$ and, as explained in 11.C, (11.79) and (11.118) imply (11.119), which gives the leading term in the asymptotic expansion of $\gamma_R(\lambda)$ for $R \to \infty$. Second, we can find further terms in this expansion: in the simplest case, where $\lambda > \sup_{x \in K} Q(X)$, the next term after (11.119) is

$$\gamma_R(\lambda) = \frac{1}{32R^2} \int_K \left(\frac{\nabla Q(X)}{\lambda - Q(X)} \right)^2 dX \big(1 + O(R^{-2})\big) \tag{14.54}$$

as $R \to \infty$ (see [Wihstutz 1985] and Problem VI.22).

4. Formulas (14.38) and (14.39) can be obtained for other random potentials as well. Indeed, the main ingredient in their derivation was equations (14.46) and (14.53) for the Fourier coefficients (14.42). But these coefficients can be regarded as conditional mathematical expectations of the random variables $\zeta_n(x) = e^{2in\phi(x)}$, calculated from the invariant distributions of the pair $\big(\zeta_n(x), X(x)\big)$, provided that $X(x)$ is fixed. For our purposes it is sufficient that the limit of

$$\frac{1}{x} \int_0^x \mathsf{E}\big\{\zeta_n(t) \mid X(t) = X\big\}\, dt$$

satisfy an equation like (14.46) for some subsequence $x_k \to \infty$.

The role of the operator $\frac{1}{2}\Delta$ in (14.41), for a general Markov process $X(x)$, should be played by its infinitesimal operator $\mathfrak{A}_X^*$. For example, such a scheme works out for a Markov or semi-Markov jump process $X(x)$ with a finite number of states, when $\mathfrak{A}_X^*$ is a finite-dimensional matrix, the simplest example being the random telegraph signal from 7.A, for which formula (14.38a) was obtained by Benderskii and Pastur [1975].

Formula (14.39) turns out to be valid also for generalized random potentials, for example, the white noise and the Poisson potentials: see [Lifshitz et al. 1982a] and Problems VI.10–12, where it is shown that $\gamma(\lambda)$ is a ratio of two entire functions of the parameters $\mathcal{D}/k^3$ and c/k, respectively. Thus $\gamma(\lambda)$ admits a Taylor expansion in these parameters, (14.38) giving its first term.

5. In 11.C we mentioned that, since the asymptotic behavior of $N(\lambda)$ in the neighborhood of a stable boundary is the same as the behavior for some operator with constant coefficients, and since the Lyapunov exponent for such operators is zero on the spectrum, we would expect $\gamma(\lambda)$ for an equation with random coefficients to approach zero as λ approaches a stable boundary. Formulas (14.38b) and (14.39) confirm this observation.

We will now establish a discrete version of Theorem 14.5. Note that no discrete analogue for (14.38b) can exist because $\lambda = \infty$ is not a stable boundary of the discrete Schrödinger operator h. But an analogue of (14.39) can be expected for the equation

$$-s(x+1)(y(x+1)-y(x)) + s(x)(y(x)-y(x-1)) = \lambda y(x), \tag{14.55}$$

which describes oscillations of a one-dimensional chain of particles with random force constants.

(14.6) Theorem. *Assume the coefficients $s(x)$ and $q(x)$ in* (11.4) *and* (14.55) *are independent and identically distributed random variables. Then*

(i) *for the discrete Schrödinger equation with $q(x) = gQ(x)$, where $g > 0$,*

$$|Q(x)| \leqslant 1 \quad \textit{and} \quad \mathsf{E}\{Q(x)\} = 0, \tag{14.56}$$

we have

$$\gamma(\lambda) = \frac{g^2}{2(4-\lambda^2)}\mathsf{E}\{Q^2(x)\}(1+O(g)) \tag{14.57}$$

as $g \downarrow 0$, if $|\lambda| < 2$ and $\lambda \neq 0$;

(ii) *for the equation* (14.55) *with*

$$0 < s_0 \leqslant s(x) \leqslant s_1 < \infty \tag{14.58}$$

we have

$$\gamma(\lambda) = \frac{\lambda\kappa}{8}\mathsf{E}\{(s^{-1}(x)-\kappa^{-1})^2\}(1+O(\lambda^{1/2})) \tag{14.59}$$

as $\lambda \downarrow 0$, where $\kappa = \mathsf{E}^{-1}\{s^{-1}(x)\}$ (cf. Theorem 6.6*).*

Notice that the condition $|\lambda| < 2$ in part (i), like the constraint $\lambda > 0$ in Theorem 14.5, leaves out only the edges of the spectrum, since the spectrum lies within the interval $|\lambda| \leqslant 2 + O(g)$ by Theorem 4.22.

For results similar to (14.59), see also [Ishii and Matsuda 1970] and [O'Connor and Lebowitz 1973].

Proof. Part (i). Let $y(x)$ be the solution of the Cauchy problem (11.3)–(11.4). Following Gredeskul and Pastur [1975] we set

$$z(x) = \frac{y(x)}{y(x-1)} = \frac{\sin(\phi(x)+k)}{\sin\phi(x)}, \tag{14.60}$$

where

$$\lambda = -2\cos k \tag{14.61}$$

for $k \in [0,\pi)$. From this and from (11.4) we get

$$\cot\phi(x+1) = \cos(\phi(x)+k) + v(x), \qquad \phi(0) = \alpha, \tag{14.62}$$

where

$$v(x) = \frac{q(x)}{\sin k}. \tag{14.63}$$

Further, set, as in (11.5),

$$\rho^2(x) = \big(y(x) - \cos k\, y(x-1)\big)^2 + \sin^2 k\, y^2(x-1). \tag{14.64}$$

Reasoning as in the proof of Theorem 11.2, we easily see that

$$\gamma(\lambda) = \lim_{x\to\infty} \frac{1}{\pi x}\int_0^{\pi} d\alpha\, \mathsf{E}\{\ln\rho(x)\}. \tag{14.65}$$

But relations (14.60)–(14.64) imply that

$$\frac{\rho^2(x+1)}{\rho^2(x)} = 1 + v(x)\sin 2\phi_k(x) + v^2(x)\sin^2\phi_k(x), \tag{14.66}$$

where $\phi_k(x) = \phi(x) + k$. Combining this with (14.65), we get

$$\gamma = \tfrac{1}{8}\mathsf{E}\{v^2(0)\} + \lim_{x\to\infty}\frac{1}{2\pi x}\int_0^{\pi} d\alpha\, \mathsf{E}\{f(x')\} + O(g^3), \tag{14.67}$$

where $f(x') = v(x')\sin 2\phi_k(x') - \frac{1}{2}v^2(x')\cos 2\phi_k(x') + \frac{1}{4}v^2\cos 4\phi_k(x')$. It follows from (14.62) that the random variables $v(x)$ and $\phi(x)$ are independent, and hence, taking (14.56) into account, that the mathematical expectation of the first term in the definition of $f(x')$ vanishes. Thus it is already clear that for independent potentials the order of smallness of the leading term in the asymptotic expansion of $\gamma(\lambda)$ is twice as high as in the case of a general metrically transitive potential, where (14.67) just gives $\gamma = O(g)$ (cf. (11.97b)).

As regards (14.57), notice that, by (14.61) and (14.63), its first term coincides with the first term in (14.67). Therefore it is enough to check that

$$\text{(14.68)} \qquad \mathsf{E}\{\cos 2\phi_k(x)\} = O(g), \quad \text{and} \quad \mathsf{E}\{\cos 4\phi_k(x)\} = O(g)$$

uniformly as $x \to \infty$. To do this, it is convenient to use the variable $\zeta(x) = e^{2i\phi(x)}$, and to rewrite (14.62) as

$$\text{(14.69)} \qquad \zeta(x+1) = \zeta_k(x) - \frac{v(x)}{2}\frac{\left(\zeta_k(x)-1\right)^2}{1+\frac{1}{2}v(x)\left(\zeta_k(x)-1\right)},$$

where $\zeta_k(x) = \zeta(x)e^{2ik}$. Hence it follows that, for $\lambda \neq \pm 2, 0$,

$$\text{(14.70)} \qquad \left|\langle\zeta(x)\rangle\right| \leqslant O\left(|v|/|\sin k|\right) \quad \text{and} \quad \left|\langle\zeta^2(x)\rangle\right| \leqslant O\left(|v|/|\sin 2k|\right),$$

where we've set

$$\langle\,\cdot\,\rangle = \lim_{x\to\infty}\frac{1}{x}\sum_{1}^{x}\mathsf{E}\{\,\cdot\,\}.$$

This proves (14.68) for $|\lambda| < 2$, $\lambda \neq 0$.

Part (ii). Instead of $y(x)$, it will be convenient to consider the sequence

$$\text{(14.71)} \qquad \Delta(x) = s(x)\left(y(x) - y(x-1)\right),$$

in terms of which (14.55) can be rewritten as

$$\text{(14.72)} \qquad -\Delta(x+1) - \Delta(x-1) + 2\Delta(x) = \lambda s^{-1}(x)\Delta(x).$$

Now this is identical with (11.4) after the substitutions

$$\text{(14.73)} \qquad \lambda \to \lambda\kappa^{-1} - 2, \qquad q(x) \to -\lambda\left(s^{-1}(x) - \kappa^{-1}\right),$$

where $\kappa(x) = \mathsf{E}\{s^{-1}(x)\}$. By arguments similar to those used in the derivation of (14.67), we find that $\gamma(\lambda) = \frac{1}{8}\mathsf{E}\{v^2(x)\} + O(v^3)$. But this equals the right-hand side of (14.59), by (14.61), (14.63) and (14.73). □

In the case of a general metrically transitive potential, the arguments used in the proof of part (ii) show that $\gamma = O(\lambda^{-1/2})$, which is similar to the behavior (11.98) for the continuous case.

Remarks. 1. Formula (14.57) is similar to (14.38a) if we formally set $b(x) = \text{const}\,\delta(x)$, which corresponds to independent values of the potential at different points (see Remark 4 to Theorem 14.5). To make the similarity complete in the discrete case, we must consider a potential representing an ergodic Markov chain as in the proof of Theorem 14.5.

2. The procedure used in the proof of Theorem 14.6 can be extended to give further terms in the asymptotic expansions. However, this procedure gives the singularities in terms of order $O(g^n)$ at the points $\lambda = 2\cos \pi m/n$, with integer m and n. These points were considered by Campanino and Klein [1990] by using a different method.

3. In the proof of Theorem 14.6, as in the proof of (14.38) (see Remark 4 to Theorem 14.5) we made no assumptions on the smoothness of the coefficients. However, other asymptotic formulas, discussed in 11.C, do require such assumptions. For example, (11.114) and (11.116), which give the behavior of $\gamma(\lambda)$ for a large potential ($q \to \infty$), do not work if the one-dimensional distribution of the potential has an atom at zero. This can be seen from Problem V.15 and the work of Martinelli and Micheli [1986], who showed that for a potential independent at distinct points and taking on the values 0 and g with probabilities p and $1-p$,

$$\lim_{g\to\infty} \frac{\gamma_g(\lambda)}{\ln g} = (1-p) - \frac{(1-p)p^2}{1-p^{L+1}}$$

for $\lambda = 2\cos\big(\pi k/(L+1)\big)$, where k and L are relatively prime integers with $0 < k \leqslant L$.

We conclude this section by discussing briefly the dependence of $\gamma(\lambda)$ on various parameters, starting with the spectral parameter λ. By Lemma 11.4, the Lyapunov exponent for a metrically transitive operator is a subharmonic, and therefore upper semicontinuous, function of any parameter on which the coefficients depend analytically. For random coefficients, using formulas of the type (14.13) and (14.65), or their continuous counterparts (14.40) and (14.52), one can show that $\gamma(\lambda)$ is a continuous function of any parameter on which the coefficients depend continuously. To get the exact form of this dependence, one can require the measure $P(dA)$ in (14.12) to be a weakly continuous function of the parameter, and the integral $\int \ln \|A\| P(dA)$ to converge uniformly with respect to the parameter. In this case, the fact that γ in (14.13) is independent of the solution of (14.12), together with standard arguments based on weak compactness and a passage to the limit under the integral sign, yield the desired result.

Such results constitute the simplest case of various theorems on perturbations of products of independent random matrices, treated in more generality in [Furstenberg and Kifer 1983].

For the spectral parameter λ in equations (11.4), (14.15) and (14.69), whose coefficients satisfy conditions (11.16b), (14.14) and (14.58), the requirement mentioned above is evidently fulfilled, so the Lyapunov exponent is a continuous function of λ.

A more detailed analysis of (14.12) shows [Le Page 1984; Bougerol and Lacroix 1986] that for the equation (11.4), under the condition that

$$\int |q|^{\alpha} P(dq) < \infty$$

for some $\alpha > 0$, the Lyapunov exponent is locally Hölder-continuous (that is, the exponent depends on the interval of values of λ under consideration). This follows from (14.13) and from the results in the papers mentioned, where it is established that the integrals $f(\bar{Y})F(d\bar{Y})$ of functions $f(\bar{Y})$ that meet the Hölder condition with exponent β, for $0 < \beta < \beta(\Delta)$, also satisfy a Hölder condition, with exponent $\beta/2$.

Finally, in a number of cases the Lyapunov exponent is even real analytic in λ. This is easily seen for the Schrödinger equation with a singular potential: a white noise potential, or the potential (6.25) (see [Lifshitz et al. 1982a] and Problems VI.10–12). Ruelle [1978] found fairly general conditions under which $\gamma(\lambda)$ depends analytically on λ.

15 The Point Spectrum of Random Operators

15.A The Pure Point Spectrum in One Dimension

In Section 13 we discussed the origin of the pure point spectrum of one-dimensional operators. We described one method to prove that the spectrum is pure point, and applied it to some important classes of one-dimensional finite-difference operators: see, in particular, Theorem 13.7, Example 13.8 and Problems V.44–45 and VI.24.

In this subsection we consider mainly one-dimensional differential operators. We first present a proof, due to Goldsheidt et al. [1977], that the spectrum of the Schrödinger operator is pure point for the Brownian motion model of 6.D (Theorem 15.4). This proof was originally designed for sufficiently smooth Markov potentials such as the ones discussed in 6.D, and was technically rather complicated, especially the part devoted to the exponential decrease of the eigenfunctions [Molchanov 1978]. But the same techniques turned out to be applicable to a variety of other problems involving random operators in one dimension, including physical applications [Lifshitz et al. 1982b], and to give more detailed information on spectral properties (see 15.B).

The original proof in [Goldsheidt et al. 1977] and [Molchanov 1978] that the spectrum is pure point was substantially improved by Carmona [1982]. Carmona's paper implicitly introduced the method, described in Section 13 in the context of finite-difference operators, of considering operators that depend on one or more continuously varying parameters. In [Kotani 1984] the same method appears in an explicit and fairly complete form, allowing a considerable extension of the class of operators under discussion, including multidimensional ones. We will apply this method to differential operators in the second part of this subsection.

As in Sections 5, 12 and 13, we denote by H and h the metrically transitive operator defined by the Schrödinger equation and its discrete analogue on the whole axis, and by $H_{a,b}^{(\alpha,\beta)}$ and $h_{a,b}^{(\alpha,\beta)}$ the same operators defined on a finite interval (see 12.A). If the potential is locally summable in the continuous case and finite at every point in the discrete one, the spectrum of these operators will consist of isolated eigenvalues $\lambda_n > -\infty$, with eigenfunctions $\psi_n(x)$, for $x \in (a,b)$, which we normalize so that $\|\psi_n\|_2 = 1$. The spectral measures of these operators are given by (12.8) and (12.6), but we will now denote them by $\tilde{\rho}_{a,b}^{(\alpha,\beta)}$. These spectral measures are suitable to the study of boundary-value problems on a semiaxis, that is, to the limit $b \to \infty$, because they are "tied" to the other end of the interval. For problems on the whole axis, that is, in the limit $a \to -\infty$ and $b \to \infty$, it is natural to consider measures tied to an interior point of the interval. For fixed a and b, all these measures are equivalent to each other: for instance, the measure tied to the point $x = 0$ is

$$\rho_{a,b}^{(\alpha,\beta)}(\Delta) = \sum_{\lambda_n \in \Delta} \tau_n, \tag{15.1}$$

where

$$\tau_n = \psi_n^2(0) + \psi_n'^2(0) \tag{15.2a}$$

in the continuous case and

$$\tau_n = \psi_n^2(0) + \psi_n^2(1) \tag{15.2b}$$

in the discrete case.

The following measure, somewhat more general than (15.1), will play an important role in the sequel:

$$\tau_{a,b}^{(\kappa)}(\Delta) = \sum_{\lambda_n \in \Delta} \tau_n^{\kappa} \tag{15.3}$$

We will express this measure in terms of the reduced phase $\phi_a^{(\alpha)}(x,\lambda)$ of Section 6. Recall that, in the continuous case,

$$\phi_a^{(\alpha)}(x,\lambda) = \pi\{\theta_a^{(\alpha)}(x,\lambda)/\pi\}, \tag{15.4}$$

where $\{t\}$ denotes the fractional part of a real number t, and the phase $\theta_a^{(\alpha)}(x,\lambda)$ is specified by (6.4) and (6.5) and the condition that it be continuous in x. In the discrete case, with $s(x) \equiv 1$, $\phi_a^{(\alpha)}(x,\lambda)$ is defined by (6.73) and by the condition $0 \leqslant \phi \leqslant \pi$. We recall also that $r_a^{(\alpha)}(x,\lambda)$ stands for the amplitude, or radial variable, of the solution $y_a^{(\alpha)}(x,\lambda)$ of the Cauchy problem (6.1) and (6.3) or (6.69) and (6.72): see (11.1) and (11.5).

(15.1) Lemma [Goldsheidt et al. 1977]. *Consider the Schrödinger equation or its discrete analogue* (11.4), *with a locally summable potential in the first case or an everywhere finite one in the second. The generalized density*

$$\dot{\tau}_{a,b}^{(\kappa)} = \sum_n \delta(\lambda - \lambda_n)\tau_n^{\kappa} \tag{15.5}$$

of the measure (15.3) *admits the representation*

$$\dot{\tau}_{a,b}^{(\kappa)} = \left(\nu_{a,b}(\lambda)\right)^{1-\kappa} \delta_p\left(\phi_a^{(\alpha)}(0,\lambda) - \phi_b^{(\beta)}(0,\lambda)\right), \tag{15.6}$$

where

$$\nu_{a,b}(\lambda) = \left(r_a^{(\alpha)}(0,\lambda)\right)^{-2} \int_a^0 \left(y_a^{(\alpha)}(x,\lambda)\right)^2 dx + \left(r_b^{(\beta)}(0,\lambda)\right)^{-2} \int_0^b \left(y_b^{(\beta)}(x,\lambda)\right)^2 dx, \tag{15.7}$$

δ_p *is the Dirac delta function on the circle* $[0,\pi)$, *and equalities* (15.5) *and* (15.6) *are to be understood in the sense that any continuous function with compact support gives the same result when integrated with the two measures.*

A similar representation, with the integrals in (15.7) *replaced by sums* (*cf.* (12.6)), *holds in the discrete case.*

Proof. We give the proof only in the continuous case. We construct the spectrum of $H_{a,b}^{(\alpha,\beta)}$ using the solutions of the Cauchy problems $y_a^{(\alpha)}(x,\lambda)$ and $y_b^{(\beta)}(x,\lambda)$ specified by the initial conditions at the ends a and b of the interval. Then

$$\psi_n(x) = \begin{cases} c_{n,+} y_a^{(\alpha)}(x,\lambda_n) & \text{for } a \leqslant x \leqslant 0, \\ c_{n,-} y_b^{(\beta)}(x,\lambda_n) & \text{for } 0 < x \leqslant b, \end{cases} \tag{15.8}$$

where the eigenvalues λ_n and the normalizing factors $c_{n,\pm}$ are determined by the continuity conditions on $\psi_n(x)$ and its derivatives at $x = 0$, and the normalization condition $\int_a^b |\psi_n(x)|^2\,dx = 1$. If we set $\theta_{a,b}^{(\alpha,\beta)}(\lambda_n) = \theta_a^{(\alpha)}(0,\lambda_n) - \theta_b^{(\beta)}(0,\lambda_n)$, this leads to the equation

$$\theta_{a,b}^{(\alpha,\beta)}(\lambda_n) = \pi n \tag{15.9}$$

for the eigenvalues, and to the following expression for the atoms (15.2) of the measure (15.1):

$$\tau_n = \nu_{a,b}^{-1}(\lambda_n). \tag{15.10}$$

We now show that

$$\nu_{a,b}(\lambda) = \frac{\partial}{\partial\lambda}\theta_{a,b}^{(\alpha,\beta)}(\lambda). \tag{15.11}$$

Setting

$$\Theta_a^{(\alpha)}(x,\lambda) = \frac{\partial}{\partial\lambda}\theta_a^{(\alpha)}(x,\lambda) \tag{15.12}$$

and differentiating (6.5) with respect to λ results in the following linear equation:

$$\frac{d}{dx}\Theta_a^{(\alpha)} = (\lambda - q(x) - 1)\sin 2\theta_a^{(\alpha)}\Theta_a^{(\alpha)} + \sin^2\theta_a^{(\alpha)}, \tag{15.13}$$

and $\Theta_a^{(\alpha)}(a,\lambda) = 0$. Using (11.1), (11.2) and (11.99), we find

$$\Theta_a^{(\alpha)}(x,\lambda) = \left(r_a^{(\alpha)}(x,\lambda)\right)^{-2}\int_a^x \left(y_a^{(\alpha)}(x',\lambda)\right)^2 dx', \tag{15.14}$$

from which (15.11) follows. From (15.11) and (15.7) it also follows that

$$\frac{\partial}{\partial\lambda}\theta_{a,b}^{(\alpha,\beta)} > 0,$$

and therefore the λ_n are simple roots of the entire function $\theta_{a,b}^{(\alpha,\beta)}(\lambda)$. Then, using the formula

$$\delta\big(\theta(\lambda)\big) = \sum_n \delta(\lambda - \lambda_n)\left|\frac{\partial\theta}{\partial\lambda}\right|^{-1}\Bigg|_{\lambda=\lambda_n}, \tag{15.15}$$

which is true for any smooth monotonic function $\theta(\lambda)$, we obtain from (15.15), (15.2) and (15.11) that

$$\dot{\tau}_{a,b}^{(\kappa)}(\lambda) = \sum_n \delta\big(\theta_{a,b}^{\alpha,\beta}(\lambda) - \pi n\big)\nu_{a,b}^{(1-\kappa)}(\lambda_n).$$

But this is the same as (15.6), because

$$\delta_p(\phi) = \delta_p\big(\pi\{\theta/\pi\}\big) = \sum_n \delta(\theta - \pi n). \qquad \square$$

(15.2) Corollary. *Let* $\tilde{\delta}_{a,b}^{\alpha,\beta}(\lambda)$ *and* $\delta_{a,b}^{\alpha,\beta}(\lambda)$ *be the generalized densities of the spectral measures* (12.8) *and* (15.1). *Then*

$$\delta_{a,b}^{\alpha,\beta}(\lambda) = \delta_p\big(\phi_a^{(\alpha)}(0,\lambda) - \phi_b^{(\beta)}(0,\lambda)\big), \tag{15.16a}$$

$$\tilde{\delta}_{a,b}^{\alpha,\beta}(\lambda) = \delta_p\big(\phi_b^{(\beta)}(a,\lambda) - \alpha\big), \tag{15.16b}$$

$$\delta_{a,b}^{\alpha,\beta}(\lambda) = \int_0^\pi \tilde{\delta}_{a,0}^{(\alpha,\phi)}(\lambda)\tilde{\delta}_{0,b}^{(\phi,\beta)}(\lambda)\, d\phi. \tag{15.16c}$$

Proof. The first formula follows from (15.6) for $\kappa = 1$; the second, by applying the arguments used to prove Lemma 15.1 to (12.6) and (12.8); and the third from the first two. $\square$

Remark. These formulas actually hold for any one-dimensional operator of order two, without the metric transitivity condition. But they, like the related formulas derived in Problems VI.25–28, are mostly useful in the case of random potentials $q(x)$ for which the joint distribution of $\big(\phi(x), q(x)\big)$ is sufficiently smooth with respect to ϕ. A typical example is that of potentials satisfying the conditions of Theorem 6.11: the potential (6.52), the random telegraph signal of 7.A, and so on.

In such cases one can analyze quite effectively the asymptotic behavior, for large values of x, of the phase functions that appear in formulas (15.16). It turns out that to understand this behavior it is sufficient to consider the mathematical expectation of these formulas, and the smoothness condition is essential if we want to meaningfully take the expectation of the δ-functions appearing in them. This procedure reduces the problem to the study of the ergodic properties of $\big(q(x), \phi(x)\big)$ and of other, related, random functions, like the ones defined in (15.12) and in Problems VI.26–28.

These two points—proving that the distributions involved are smooth enough and that the phase has a "stable" behavior for large x—form the technical side of the proofs below.

Before starting, we state and prove a simple criterion for the spectrum to be pure point. Together with Lemma 15.1, this criterion will constitute the essence of our proof that the spectrum of a Schrödinger operator with a Markov-type potential is pure point.

(15.3) Lemma [Goldsheidt et al. 1977]. *Let* $\rho(L, d\lambda)$, *for* $L > 0$, *be a family of nonnegative, atomic random measures on* $\mathbf{R}$, *such that:*

(a) *for any bounded interval* Δ *there exists a finite nonrandom constant* $C(\Delta)$ *such that* $\rho(L,\Delta) \leqslant C(\Delta)$ *if* $L \geqslant 1$;

(b) *with probability* 1, *the measures* $\rho(L,\,\cdot\,)$ *converge weakly to a measure* ρ, *as* $L \to \infty$.

Assume also that

$$\limsup_{\kappa\downarrow 1}\,\limsup_{L\to\infty} \mathsf{E}\Big\{\sum_n \alpha_n^\kappa(L)\Big\} \geqslant \mathsf{E}\{\rho(\Delta)\}, \tag{15.17}$$

where the $\alpha_n(L)$ *are the atoms of* $\rho(L,\,\cdot\,)$ *in the interval* Δ. *Then* ρ *is pure point (atomic) with probability* 1.

Proof. Let ρ_p be the pure point component of ρ and α_n its atoms in a prescribed bounded interval Δ. For any $\varepsilon > 0$, there exists a partition of the interval Δ into a finite number of small intervals Δ_j, whose endpoints do not coincide with any of the atoms of the measures $\rho(L,\,\cdot\,)$ and ρ, and such that, for any fixed $\kappa > 1$,

$$\sum_j \rho^\kappa(\Delta_j) \leqslant \sum_n \alpha_n^\kappa + \varepsilon.$$

Clearly, we also have

$$\sum_n \alpha_n^\kappa(L) \leqslant \sum_j \rho^\kappa(L,\Delta_j).$$

From these inequalities and condition (b) of the lemma we have

$$\limsup_{L\to\infty} \sum_n \alpha_n^\kappa(L) \leqslant \sum_n \alpha_n^\kappa \leqslant \Big(\sum_n \alpha_n\Big)^\kappa = \rho_p^\kappa(\Delta).$$

From this and from condition (a), we get

$$\sum_n \alpha_n^\kappa(L) \leqslant \Big(\sum_n \alpha_n(L)\Big)^\kappa \leqslant C^\kappa(\Delta)$$

and

$$\limsup_{\kappa\downarrow 1}\,\limsup_{L\to\infty} \sum_n \alpha_n^\kappa(L) \leqslant \rho_p(\Delta) \leqslant C(\Delta).$$

This, together with Fatou's theorem on limits of mathematical expectations, gives

$$\mathsf{E}\{\rho_p(\Delta)\} \geqslant \limsup_{\kappa\downarrow 1}\,\limsup_{L\to\infty} \mathsf{E}\Big\{\sum_n \alpha_n^\kappa(L)\Big\}.$$

Together with (15.17), this gives $\rho_p(\Delta) = \rho(\Delta)$ with probability 1. □

(15.4) Theorem [Goldsheidt et al. 1977]. *Consider a one-dimensional Schrödinger operator with a potential $q(x)$ of Brownian motion, that is,*

$$q(x) = Q\big(X(x)\big) \tag{15.18}$$

where $X(x)$, for $x \in \mathbf{R}$, is a homogeneous Markov process on a compact Riemannian manifold K with infinitesimal operator given by (6.48), and $Q(X)$, for $X \in K$, is an infinitely differentiable nonflat function (see 6.D) with $\min_{x\in K} Q(K) = 0$ and $\max_{x\in K} Q(K) = 1$. Then the spectrum of the metrically transitive operator $H = -d^2/dx^2 + q(x)$, with probability 1, coincides with $[0, \infty)$ and is pure point.

Proof. That $\sigma(H) = [0, \infty)$ follows from Theorem 5.33. To prove that the spectrum is pure point we will assume that certain technical statements are true, but we defer their proof to subsequent lemmas.

We start with the operator $H_{-L,L}^{(\phi_-,-\phi_+)}$, for $\phi_\pm \in [0, \pi)$. We set

$$X_\pm(x) = X\big(\pm(L-x)\big), \qquad q_\pm(x) = Q\big(X_\pm(x)\big); \tag{15.19a}$$

$$\theta_\pm(x) = \pm\theta_{\mp L}^{(\pm\phi_\mp)}\big(\pm(L-x)\big),\ \theta_\pm(0) = \phi_\pm; \tag{15.19b}$$

$$r_\pm(x) = r_{\mp L}^{(\pm\phi_\mp)}\big(\pm(L-x)\big), \quad r_\pm(0) = 1; \tag{15.19c}$$

$$\Theta_\pm(x) = \Theta_{\mp L}^{(\pm\phi_\mp)}\big(\pm(L-x)\big),\ \Theta_\pm(0) = 0. \tag{15.19d}$$

It follows from (6.5), (11.99), (15.12) and (15.13) that these quantities satisfy the following system of first order equations:

$$\theta_\pm' = \cos^2\theta_\pm + \big(\lambda - q_\pm(x)\big)\sin^2\theta_\pm, \qquad \theta_\pm(0) = \phi_\pm; \tag{15.20a}$$

$$\Theta_\pm' = -\big(q_\pm(x) - \lambda + 1\big)\sin 2\theta_\pm\,\Theta_\pm + \sin^2\theta_\pm, \Theta_\pm(0) = 0; \tag{15.20b}$$

$$r_\pm' = \tfrac{1}{2} r_\pm\big(q_\pm(x) - \lambda + 1\big)\sin 2\theta_\pm, \qquad r_\pm(0) = 1. \tag{15.20c}$$

Therefore, each of the triples $U_\pm(x) = \big(X_\pm(x), \phi_\pm(x), \Theta_\pm(x)\big)$, where $\phi_\pm = \pi\{\theta_\pm/\pi\}$, is a Markov process (see Lemma 6.10). Since equations (15.20) for both triples coincide, and the diffusion process $X(x)$ is reversible (Problem III.34), the processes $U_+(x)$ and $U_-(x)$ coincide in distribution, and we denote both by $U(x)$. (This coincidence is not essential for the subsequent arguments; it just makes them somewhat simpler.)

Let $P(x; U_0, dU)$ be the transition probability of $U(x)$. It follows from the properties of the potential (15.18) that this probability has a density $p(x; U_0, U)$ with respect to the measure $dX\,d\phi\,d\Theta$ on the phase space $K \times [0, \pi) \times [0, \infty)$, for $x \geqslant x_0(\lambda)$, where $\lambda > 0$. Let's write the mathematical expectation in the left-hand side of (15.17), using the functions introduced above, $\rho(L, \cdot\,)$ and ρ being the spectral measures of the operators H_L and H. By Lemma 15.1, we have

$$(15.21)\quad \mathsf{E}\Big\{\sum_n \alpha_n^\kappa(L)\Big\} = \int_\Delta d\lambda \int_K \frac{dX}{\pi(X)} \int_0^\pi d\phi \int_0^\infty d\Theta_1 \int_0^\infty d\Theta_2 \times p(L;\pi,\phi_+,0;X,\phi,\Theta_1)p(L;\pi,\phi_-,0;X,-\phi,\Theta_2)(\Theta_1+\Theta_2)^{1-\kappa},$$

where $p(x;\pi,\phi,\Theta;U_1)$ stands for the integral of the density $p(x,X,\phi,\Theta;U_1)$ over X, with invariant distribution $\pi(X) = V^{-1}(K)$ (see 6.D), and we have taken into account that the statistical dependence between the triples $U_\pm(x)$, because $X(x)$ is Markov, reduces to the condition $X_\pm(L) = X(0)$ (which results in the density $\pi(X)$ in the denominator).

First consider relation (15.21) for $\kappa = 1$. In this case, the left-hand side is $\mathsf{E}\{\rho(L,\cdot)\}$, where $\rho(L,\cdot)$ is the spectral measure of H_L. Now the measures $\rho_{a,b}^{(\alpha,\beta)}(\Delta)$, for Δ fixed, are uniformly bounded in all the other parameters, because the potential is locally summable (see [Levitan and Sargsyan 1976] and Subsection 12.A). Therefore, by the weak convergence of $\rho(L,\,\cdot\,)$ to ρ and the good ergodic properties of the pairs $(X_\pm,\phi_\pm)$ (see (6.59)) in (15.21) for $\kappa = 1$, we can take the limit $L \to \infty$, and obtain

$$(15.22)\qquad \mathsf{E}\{\rho(\Delta)\} = \int_\Delta d\lambda \int_K \frac{dX}{\pi(X)} \int_0^\pi \pi(X,\phi)\pi(X,-\phi)\,d\phi,$$

where $\pi(X,\phi)$ is the density of the invariant distribution of the pairs $(X_\pm,\phi_\pm)$. The existence of this density was proved in Theorem 6.12.

Now, we will see in Lemmas 15.5 and 15.7 that the density of the transition probability of $U(x)$ has some weak ergodic properties, expressed by the following formulas:

$$(15.23)\qquad \sup_{\substack{\lambda\in\Delta \\ L\geqslant L_0(\Delta),U_0,U}} p(L;U_0,U) \leqslant C(\Delta) < \infty,$$

$$(15.24)\qquad \lim_{L_j\to\infty} p(L_j;U_0,U) = \pi(U)$$

for some sequence $L_j \to \infty$ and every U_0 and $U \in K \times [0,\pi) \times [0,\infty)$, where $\pi(U)$ is a continuous and bounded function; and

$$(15.25)\qquad \int_0^\infty \pi(X,\phi,\Theta)\,d\Theta = \pi(X,\phi).$$

Given these facts and Fatou's theorem, we can take the limit $L_j \to \infty$ for $\kappa > 1$ in (15.21), to get

$$\limsup_{L\to\infty} \mathsf{E}\Big\{\sum_n \alpha_n^\kappa(L)\Big\} \geqslant \int_\Delta d\lambda \int_K \frac{dX}{\pi(X)} \int_0^\pi d\phi \int_0^\infty d\Theta_1 \int_0^\infty d\Theta_2 \times \pi(X,\phi,\Theta_1)\pi(X,-\phi,\Theta_2)(\Theta_1+\Theta_2)^{1-\kappa}.$$

Splitting this integral into two, over the regions $\Theta_1 + \Theta_2 > 1$ and $\Theta_1 + \Theta_2 < 1$, we can take in each of them the limit $\kappa \downarrow 1$. Taking (15.22) into account, we get (15.17), which is the condition for the spectral measure ρ of H to be pure point with probability 1. This proves the theorem. □

We now prove the auxiliary statements that have been used above.

(15.5) Lemma. *The transition probability of the Markov processes $U_\pm(x)$ for $\lambda > 0$ and $x \geqslant x_0(\lambda) > 0$ has a probability density $p(x; U_0, U)$ with respect to the measure $dX\, d\phi\, d\Theta$ on the phase space $K \times [0,\pi) \times [0,\infty)$. This density is bounded and continuous in all variables and in λ, for λ in any finite interval $\Delta \subset (0,\infty)$.*

Proof. In Theorem 6.12, a similar statement was proved for the pairs $(X_\pm(x), \phi_\pm(x))$. In our case, we use the same Proposition 6.13 that guarantees the existence of the classical fundamental solution of a parabolic equation with a degenerate elliptic (or hypoelliptic) operator. According to (15.20), the transition density $p(x; U_0, U)$ is a fundamental solution of the equation

$$\frac{\partial p}{\partial x} = \tilde{\mathfrak{A}}_{U_0} p, \tag{15.26}$$

where $\tilde{\mathfrak{A}}_{U_0} = \frac{1}{2}\Delta + C_1 + C_2$, with

$$C_1 = \left(\cos^2\phi + (\lambda - Q(X))\sin^2\phi\right)\frac{\partial}{\partial\phi},$$

$$C_2 = \left(\sin^2\phi + \Theta(\lambda - Q(X) - 1)\sin 2\phi\right)\frac{\partial}{\partial\Theta}$$

(cf. (6.43) and (6.53)). By Proposition 6.13, we must show that the Lie algebra generated by the vector fields C_1, C_2 and $A_j = \partial/\partial x_j$, where $j = 1, \dots, \nu = \dim K$ and $x_1, \dots, x_\nu$ are local coordinates on K, has dimension $\nu + 2$ (cf. (6.57)). The arguments are very similar to those used to prove Theorem 6.12, and we omit them. □

(15.6) Lemma. *For every $\lambda > 0$ there exist $\delta(\lambda) > 0$ and $x_0(\lambda) > 0$ such that*

$$\mathsf{E}\{\Theta^\delta(x) \mid \Theta(0) = 0\} < \infty \tag{15.27}$$

uniformly in $x \geqslant x_0(\delta)$.

Proof. By integrating (15.20b), or using (15.14) and (11.100), we conclude that, for any solution $\phi(x)$ of (6.5),

$$\Theta(x) = \int_0^x \left(r_t^{(\phi(t))}(x)\right)^2 \sin^2 \phi(t)\, dt, \tag{15.28}$$

where

$$r_t^{(\alpha)}(x) = \exp\left\{\frac{1}{2}\int_t^x B(s)\, ds\right\}, \tag{15.29}$$

where

$$B(s) = \big(q(s) - \lambda + 1\big) \sin 2\phi(s). \tag{15.30}$$

From this and from the elementary inequality

$$\left|\sum_j a_j\right|^\delta \leqslant \sum_j |a_j|^\delta,$$

for $0 \leqslant \delta \leqslant 1$, we find

$$\Theta^\delta(x) \leqslant \text{const} \sum_{s=0}^{[x]} \left(r_j^{(\phi(j))}([x]+1)\right)^{-2\delta}. \tag{15.31}$$

Now introduce in the space $C(E)$, where $E = K \times [0, \pi)$, the integral operator

$$(T_\delta f)(u) = \mathsf{E}\left\{\left(r_0^{(\phi)}(1)\right)^{-2\delta} f\big(u(1)\big) \mid u(0) = u\right\}, \tag{15.32}$$

where $u(x) = \big(X(x), \phi(x)\big)$ and $u = (X, \phi)$. This operator is compact, and from (6.58) and the preceding lemma it follows that, if $f \geqslant 0$ is not identically zero, $T_\delta f > 0$. Hence, its maximum eigenvalue $e^{\mu(\delta)}$ is simple, positive and real analytic, as a function of δ: see [Krasnosel'skii 1962]. In addition, for all $\delta > 0$ small enough,

$$\|T_\delta^n f\| \leqslant C(f, \delta) e^{n\mu}, \tag{15.33}$$

$\mu(0) = 0$, and, by perturbation theory [Kato 1966],

$$\mu'(0) = -\int_E \big(Q(X) - \lambda + 1\big) \sin 2\phi\, \pi(X, \phi)\, dX\, d\phi.$$

But, in view of (14.24), the right-hand side of this inequality is $-2\gamma(\lambda)$, and therefore strictly negative, by Theorem 14.4. Thus, for $\delta > 0$ small enough, $\mu(\delta)$ is strictly negative.

From (15.29)–(15.32) and the fact that $\big(X(x), \phi(x)\big)$ is a Markov process, it follows that the left-hand side of (15.27) does not exceed

$$\text{const} \sum_{j=0}^{[x]} \int_E (T_\delta^{[x]+1-j} \mathbf{1})(u) \pi(u)\, du,$$

where **1** is the function identically equal to 1. This expression, because of (15.33), remains bounded for $x \to \infty$. This proves the lemma. □

Remark. By (15.29), $\left(r_0^{(\phi)}(x)\right)^{-2\delta}$ is a multiplicative functional of the Markov process $u(x) = \left(X(x), \phi(x)\right)$. Therefore the quantity

$$Z_\delta(x,u) = \mathsf{E}\{(r_0^{(\phi)}(x))^{-2\delta} \mid u(0) = u\} \tag{15.34}$$

satisfies the equation [Dynkin 1965]

$$\frac{\partial}{\partial x} Z_\delta = \mathfrak{A} Z_\delta - \delta(Q(X) - \lambda + 1)\sin 2\phi\, Z_\delta \equiv \mathfrak{A}(\delta) Z_\delta \tag{15.35}$$

and

$$\mu(\delta) = \lim_{x\to\infty} \frac{1}{x} \ln \sup_u Z_\delta(x,u) \tag{15.36}$$

is the maximum eigenvalue of the operator $\mathfrak{A}(\delta)$ from the right-hand side of (15.35). Now, if $\eta(x) = \left(1 + \Theta(x)\right)^\delta$, we have

$$\eta' = -\delta B\eta + \delta(1+\Theta)^{\delta-1}(B + \sin^2\phi),$$

and therefore, integrating this equation and using (15.18), (15.30) and (15.34), we obtain

$$\mathsf{E}\{\Theta^\delta(x) \mid \Theta(0) = 0\} \leqslant \text{const}\left(Z_\delta(x,\pi) + \int_0^x Z_\delta(t,\pi)\,dt \right),$$

where $Z_\delta(x,\pi) = \int_E Z_\delta(x,u)\pi(u)\,du$ and we have taken into account that $u(x)$ is a homogeneous Markov process whose transition probability density is, by Theorems 6.12 and 6.14, uniformly bounded when $x \geqslant x_0(\lambda)$. Therefore, inequality (15.27) for small $\delta > 0$ could be obtained from (15.35), (15.36) and the fact that $\mu(\delta) < 0$, which can be derived by perturbation theory applied to the operator $\mathfrak{A}(\delta)$ [Pastur and Feldman 1975]. (In fact, (15.27) holds for all $0 < \delta < 1$: see Problem VI.29.)

However, here we used arguments based on the introduction of the operator $T_\delta = \exp \mathfrak{A}(\delta)$ because they are the simplest version of a fairly general method for the study of ergodic properties of Markov processes [Nagaev 1957]. In the spectral theory of random operators, a variety of important results have been obtained by various versions of this method: the point character of the spectrum of the one-dimensional discrete Schrödinger operator [Kunz and Souillard 1980], smoothness properties of $\gamma(\lambda)$ and $N(\lambda)$ (see 6.F and [Bougerol and Lacroix 1985]), the central limit theorem for the difference $\sqrt{L}\left(N_L(\lambda) - N(\lambda)\right)$ [Reznikova 1981], and so on.

(15.7) Lemma. *Under the conditions of Theorem* 15.3, *relations* (15.24) *and* (15.25) *hold.*

Proof. By Lemma 15.6, the family of transition probabilities $P(\,\cdot\,;\,\cdot\,,dU)$ is weakly compact in the space $E \times \mathbf{R}_+$. Therefore, writing the Chapman–Smoluchovski–Kolmogorov equation

$$p(L,\,\cdot\,;U) = \int_{E\times\mathbf{R}_+} p(1;U_1;U)P(L-1;\,\cdot\,;dU_1)$$

and noting that, by Lemma 15.5, the first factor in the integrand is a bounded and continuous function, we can, in view of Lemma 15.6, pass to the limit of a subsequence $L_j \to \infty$. This proves (15.24) and (15.25). □

Molchanov [1978] proved that in (15.24) all the limits coincide, that is, the Markov process $U(x)$ has a unique invariant distribution, having a continuous and bounded density.

Remarks. 1. The scheme of the proof of Theorem 15.4 works to show that other types of random operators have a pure point spectrum. Molchanov and Seidel [1982] considered Sturm–Liouville operators (1.6) on $L_2(\mathbf{R}_+)$ of the following type: all three coefficients are random processes of the form (15.18), with functions $Q(X)$, $R(X)$ and $S(X)$ playing the role of $Q(X)$; all three functions Q, R and S belong to $C^2(K)$, and at least one of the functions QR^{-1} and SR^{-1} is not constant. Under these conditions, the transition probability of the process $u(x)$ does not have, generally speaking, a continuous density with respect to the measure $dX\,d\phi$. But by taking into account the special form of the infinitesimal operator of the process $u(x)$ and using a version of the classical parametrix method to construct the fundamental solution of the corresponding parabolic equation, the authors could prove that the part of the process $U(x)$ on the manifold

$$E_0 = \{u \in E : \operatorname*{grad}_X(S(X)\cos^2\phi + (Q(X) - \lambda R(X))\sin^2\phi) \neq 0\},$$

that is, the Markov process on E_0 obtained from $U(x)$ by "killing" the trajectories as soon as they leave E_0, does have a continuous density. From this it follows, roughly speaking, that the process has a continuous transitional density after it has reached the set E_0, and this is enough to provide the ergodic properties of $u(x)$ needed to prove a version of Theorem 15.4.

2. The method used to prove Theorem 15.4 works also in the discrete case to yield an analogous result. This result, valid even for the multicomponent operator (11.86), was announced by Goldsheidt [1981]. However, the absence of an analogue for Hörmander theory (i.e., Proposition 6.13) makes the proof of the existence of the relevant densities more complicated than in the continuous case.

Subsection 15.B and Problems VI.27, VI.30 and VI.34 discuss further problems related with Theorem 15.4 and with the point spectrum of various classes of one-dimensional random operators. Here we just point out that Theorem 15.4 only states that the eigenfunctions of the Schrödinger operator with potential (15.18) lie in $L_2(\mathbf{R})$. Meanwhile, it is natural to expect, based on the heuristic arguments of 13.A and on general results from the spectral theory of differential operators (see [Reed and Simon IV] and [Glazman 1965]) that the eigenfunctions decrease exponentially at infinity. This was proved by [Molchanov 1978] by improving the techniques used in the proof of Theorem 15.4. The next important step was taken by Carmona [1982], who proved that the exponential rate of decrease does not exceed the Lyapunov exponent $\gamma(\lambda)$. Since, under pretty general conditions, this rate cannot be smaller than $\gamma(\lambda)$ (see [Craig and Simon 1983a] and Problem V.7), it must equal $\gamma(\lambda)$, which is to be expected in view of the arguments of 13.A.

An important contribution made by Carmona [1982], also using relations like (15.6) and (15.16), was a more profound understanding of the role of existence of the continuous and bounded density of the pair $(X(x), \phi(x))$ in the proof that the spectrum is pure point and that the eigenfunctions decay exponentially. More precisely, the author showed that the existence of this density allows one to go from statements on the exponential decay of solutions for Lebesgue-almost all values of λ (such as Propositions 11.2 and 14.1) to similar statements true for almost all λ with respect to the random and point spectral measure (see also [Carmona 1984]). In the language of Section 13, it allows one to use a randomization procedure with the random potential as a randomizing agent. In [Carmona 1983] this important observation was used to study the spectrum of a Schrödinger operator with a potential that has different behaviors at ∞ and $-\infty$ (and consequently is not a metrically transitive process). The author found, for example, that if the potential is the sum of a periodic function $q_1(x)$ and a random function identically equal to zero on $\mathbf{R}_-$ and equal to (15.18) on $\mathbf{R}_+$ (the *bicrystal model*), the spectrum is $\sigma = \cup_j [a_j, b_j + 1]$, where $\cup_j [a_j, b_j]$ is the spectrum σ_1 of the operator with potential $q_1(x)$. Further, σ is purely absolutely continuous on σ_1 and pure point on $\sigma \setminus \sigma_1$, and the eigenfunctions decay exponentially with rate $\gamma(\lambda)$.

The randomization technique, used only implicitly in Carmona's papers, was formulated explicitly and deftly used in the important paper by Kotani [1984], as we discussed in 13.B. Using this technique, the author established quite simply that the spectrum of a finite-difference operator whose coefficients have a smooth distribution is pure point. We now show

how similar arguments can be applied to the one-dimensional Schrödinger operator. These arguments, by their very nature, are highly efficient for cell-like potentials such as the alloy-type potential (1.31), whereas, when applied to Markov-type potentials of the type of (15.18), they yield essentially the same results obtained by the formalism used to prove Theorem 15.4. (Meanwhile, this formalism leads to complications when applied to (1.31): see, for example, [Bentosela et al. 1983].)

The following fairly general theorem is an analogue of Theorem 13.7.

(15.8) Theorem [Kotani and Simon 1987]. *Consider the Schrödinger operator H with potential $q(x)$, where $q(x)$ is a metrically transitive process with $\mathsf{E}\{q^2(0)\} < \infty$. Let Ω_0 be a measurable set in the space of realizations Ω; let $\mathcal{F}(c,0)$, for $c<0$, be the σ-algebra generated by $q(x)$ for $x \notin (c,0)$ and let Ω_1 be in $\mathcal{F}(c,0)$. Suppose that the conditional probability $\mathsf{P}\{\Omega_0 \mid \Omega_1\}$ is positive, and that*

$$\mathsf{P}\{\Omega_0 \cap \{\phi_0^{(\alpha)}(0,\lambda) \in d\phi\} \mid \mathcal{F}(c,0)\} \leqslant C\, d\phi \tag{15.37}$$

for $\omega \in \Omega_1$, where C is finite and does not depend on $\alpha \in [0,\pi)$ and $\lambda \in \Delta$, for Δ a finite interval, but can depend on Δ and on $\omega \in \Omega_1$. If, in addition,

$$\sup_{\substack{\omega\in\Omega_0\\ x\in[c,0]}} |q(x,\omega)| = q_0 < \infty,$$

the spectrum of H is pure point with probability 1 and its eigenfunctions decay exponentially with rate $\gamma(\lambda)$.

Proof. Our proof will be slightly different from the original one. We notice first that the conditions imply that the potential $q(x)$ is nondeterministic, so in view of Theorem 12.10, $\gamma(\lambda) > 0$ for Lebesgue-almost all λ.

Take an arbitrary bounded interval $\Delta \subset \mathbf{R}$ and assume $\lambda \in \Delta$ everywhere below. From the obvious equality

$$\phi_a^{(\alpha)}(0,\lambda) = \phi_c^{(\psi)}(0,\lambda)\big|_{\psi=\phi_a^{(\alpha)}(c,\lambda)}$$

and formula (15.16a) for the generalized density of the spectral measure $\rho_{a,b}^{(\alpha,\beta)}$, we get

$$\delta_{a,b}^{(\alpha,\beta)}(\lambda) = \delta(\Phi(\lambda)), \tag{15.38}$$

where

$$\Phi(\lambda) = \phi_c^{(\phi)}(0,\lambda)\big|_{\psi=\phi_a^{(\alpha)}(c,\lambda)} - \phi_b^{(\beta)}(0,\lambda).$$

Now the derivatives $\partial\phi_a^{(\alpha)}/\partial\lambda$, $-\partial\phi_b^{(\beta)}/\partial\lambda$ and $\partial\phi_a^{(\alpha)}/\partial\alpha$ are nonnegative and, by (15.20c),

$$0 < \frac{1}{C_0} \leqslant \frac{\partial \phi_c^{(\phi)}(0,\lambda)}{\partial \lambda} \leqslant C_0 < \infty,$$

where the constant C_0 depends only on Δ and on the constant q_0 from the theorem's statement. Therefore

$$0 < \frac{1}{C_1} \leqslant \left| \frac{\partial \Phi}{\partial \lambda}(\lambda) \right| \leqslant C_1 < \infty, \tag{15.39}$$

where C_1 depends on q_0, Δ and on the realization of the potential $q(x)$ for $x \notin (c, 0)$.

Hence, since $\phi = \pi\{\theta/\pi\}$ is the reduced phase, it follows that, on any interval $\tilde{\Delta} \subset \Delta$ of length $|\tilde{\Delta}| < \pi/C_1$, the function $\Phi(\lambda)$ has at most one zero. From (15.15), (15.38) and (15.39), it follows then that, for any interval, whose center we denote by $\tilde{\lambda}$,

$$\mathsf{E}\{\chi_{\Omega_0}\rho_{a,b}^{(\alpha,\beta)}(\tilde{\Delta}) \mid \mathcal{F}(c,0)\} \leqslant C_1 \mathsf{P}\{\omega \in \Omega_0 \text{ and } |\Phi(\tilde{\lambda})| \leqslant C_1|\tilde{\Delta}| \mid \mathcal{F}(c,0)\},$$

where we have used the measurability of the random variables $\phi_a^{(\alpha)}(c)$ and $\phi_b^{(\beta)}(0)$ with respect to $\mathcal{F}(c,0)$. From this and from (15.37) we get

$$\mathsf{E}\{\chi_{\Omega_0}\rho_{a,b}^{(\alpha,\beta)}(\tilde{\Delta}) \mid \mathcal{F}(c,0)\} \leqslant CC_1^2|\tilde{\Delta}|$$

for $\omega \in \Omega_1$. Since $\rho_{a,b}^{(\alpha,\beta)}$ converges weakly to ρ as $a \to -\infty$ and $b \to \infty$, this inequality implies that the measures $\mathsf{E}\{\chi_{\Omega_0}\rho(d\lambda) \mid \mathcal{F}(c,0)\}$, for $\omega \in \Omega_1$, are absolutely continuous with respect to Lebesgue measure. This fact is analogous to (13.21). Thus, using arguments like those in the proof of Theorem 13.7, we arrive at the conclusion of our theorem for $\omega \in \Omega_0 \cap \Omega_1$, which is a set of positive probability by assumption. Since the potential is metrically transitive, the conclusion of the theorem is true with probability 1 (see Theorem 2.16 and Problem V.40). □

(15.9) Examples

(a) Consider a Schrödinger operator with potential (15.18), but without the assumption that $Q(X)$ is nonflat: instead, we just assume that $Q(X) \in C^\infty(K)$ is not constant. It is clear that $\mathcal{F}(c,0)$ in the hypothesis of Theorem 15.8 can be replaced by the σ-algebra $\tilde{\mathcal{F}}(c,0)$ determined by the Markov process $X(x)$, for $x \notin (c,0)$. Because $(X(x), \phi(x))$ is a Markov process, the conditional distribution in (15.37) for $\Omega_0 = \Omega$ is just $p(c; X(c), \alpha; X(0), \phi)\, d\phi$. Therefore, under the assumptions of Theorem 15.4, if $Q(X)$ is nonflat, Theorems 15.8 and 6.12 imply that the spectrum is pure point.

With the weaker assumption that $Q(X)$ is not a constant, we let K_0 be the subset of K where the gradient of $Q(X)$ does not vanish, choose a nonflat function $Q_1(X) \in C^\infty(K)$ coinciding with $Q(X)$ when $X \in K_0$, and set $\Omega_0 = \{\omega \in \Omega : X(x) \in K_0 \text{ and } x \in [c,0]\}$. Now the conditional distribution in (15.37), multiplied by $\mathsf{P}(\Omega_0)$, is majorated by a similar distribution, with Ω_0 replaced by Ω and $\phi(x)$ by the reduced phase $\phi_1(x)$ of the solution of (6.63) with $Q_1(x)$ instead of $Q(X)$. Therefore the conclusion of Theorem 15.8 will still be true if we know that the transition probability of $(X(x), \phi(x))$ has a bounded density; but this is exactly what Theorem 6.12 says. So Theorem 15.8 stays true as long as $Q(X)$ is not a constant [Kotani and Simon 1987].

The technique used here is essentially the one used in [Molchanov and Seidel 1982], and is based on the intuitively obvious fact that it is not necessary to trace all the trajectories in the Brownian motion $X(x)$, because, if the trajectories visit the set where $\operatorname{grad} Q \neq 0$ sufficiently often, any initial distribution of the pair $(X(x), \phi(x))$ will become continuous.

(b) Consider the potential

(15.40) $$q(x) = \sum_{n \in \mathbf{Z}} \xi_n u(x-n),$$

for $x \in \mathbf{R}$, where the ξ_n are independent, identically distributed random variables whose distribution $F(d\xi)$ has absolutely continuous component $F_{\mathrm{ac}}(d\xi) = f(\xi)\, d\xi$, and $u(x)$ is a nonnegative, continuous function with support contained in the interval $[c_1, 0]$, where $-1 < c_1 < 0$. As in Example 15.9(a), we can impose conditions not on the potential, but on the random variables ξ_n. Thus we fix ξ_n, for $n \neq 0$, that is, we set $\xi_n = \xi_n^{(0)}$ for $n \neq 0$, where the $\xi_n^{(0)}$ are numbers to the support of F, and we consider

$$q(x,t) = (t + \xi_0^{(0)})u(x) + \sum_{\neq 0} \xi_n^{(0)} u(x-n),$$

where $\xi_0^{(0)} \in \operatorname{supp} F_{\mathrm{ac}}$, for $|t| \leqslant \varepsilon$, with $\varepsilon > 0$. It follows from (6.5) (or from (15.20a)) that, for $-1 < c < 0$,

$$\frac{\partial}{\partial t} \phi_c^{(\alpha)}(0,\lambda) = \int_c^0 u(x) \sin^2 \phi_c^{(\alpha)}(x,\lambda) e^{\int_c^x (q(s,t) - \lambda + 1) \sin 2\phi_c^{(\alpha)}(s)\, ds}\, dx$$

(cf. (15.14)). From this formula, the nonnegativity of $u(x)$ and our assumptions, which in particular imply that $q(x,t)$ is bounded for $x \in [c,0]$ and $|t| \leqslant \varepsilon$, we conclude that $\partial \phi_c^{(\alpha)}(0,\lambda)/\partial t$ is strictly positive and bounded when $\alpha \in [0,\pi)$, $|t| < \varepsilon$ and $\lambda \in \Delta$. Therefore, choosing for Ω_0 the set of realizations whose ξ_0 belongs to the set where the density $f(\xi)$ is bounded, and as $\mathcal{F}(c,0)$ the σ-algebra generated by ξ_n with $n \neq 0$, we fulfill the conditions of Theorem 15.8.

For a somewhat different analysis of this example, and in particular a different verification of (15.37), see [Delyon et al. 1987].

The arguments we used here are similar to the ones in the proof of Theorem 13.5. That theorem dealt essentially with the continuous dependence of the spectral function of an abstract self-adjoint operator on the amplitude of a perturbation of rank one. The present case is different in that here the local change of the potential (that is, ξ_0), is not one-dimensional, nor even a perturbation of finite rank (with the exception of the case where $u(x) = \delta(x)$, corresponding to the potential (14.19): see Problems II.42 and VI.24). However, as the above arguments show, the continuous analogue of Theorem 13.5 depends not on the one-dimensionality of the perturbation P, but on its positiveness: see Problem V.47. This observation allows other results, where Lemma 4.21 was used in the discrete version, to be extended to the continuous case. For example, one can prove a continuous version of Theorem 4.20 for the multidimensional Schrödinger operator with potential (10.6), in which the random variables are independent and have a bounded density [Kotani and Simon 1987].

15.B Other One-Dimensional Results

We saw in Section 13 and Subsection 15.A that one-dimensional operators of order two with random metrically transitive coefficients with certain smoothness properties have, with probability 1, a pure point spectrum and eigenfunctions that decay exponentially at a rate given by the Lyapunov exponent. In the physics literature this phenomenon is called *Anderson localization*, or *exponential localization*. In this subsection we will briefly study other classes of random operators for which Anderson localization takes place. We will also discuss some other results on the spectral properties of one-dimensional random operators.

We first describe the results obtained by the methods of of Section 13 and Subsection 15.A in more complex cases. The simplest of these is the Dirac operator, defined on the space of square-integrable, two-component functions $\Psi(x) = (\psi_1(x), \psi_2(x))$ on $\mathbf{R}$ by the equation

$$(15.41) \qquad -i\begin{pmatrix} 1-s & 0 \\ 0 & -1-s \end{pmatrix}\Psi' + \begin{pmatrix} q(x) & \Delta(x) \\ \Delta(x) & q(x) \end{pmatrix}\Psi = \lambda\Psi,$$

where $q(x)$ and $\Delta(x)$, called the potential and the gap, are metrically transitive processes, and s is a nonnegative number, representing the dimensionless velocity of motion of the reference system. This metrically transitive operator, for $s \equiv 0$ and coefficients $q(x)$ and $\Delta(x)$ of the form (15.8), that is, smooth, nonflat functions of Brownian motion on a compact manifold, was considered by Vardazaryan [1977], who obtained for (15.41) a result similar to Theorem 15.4. Gredeskul and Pastur [1978] subsequently found the explicit form for $N(\lambda)$ in the case when $q(x)$ and $\Delta(x)$ are functions of the random telegraph signal of Example 1.15(b).

Bratus' et al. [1986] showed that Anderson localization occurs for $0 < s < 1$ and the potentials considered by Vardazaryan, and also that if $\Delta(x)$ is constant and $q(x)$ satisfies $\mathsf{E}\{|q(0)|\} < \infty$, the spectrum of (15.41) is absolutely continuous for $s > 1$. Thus, the spectrum of this operator changes its character from pure point to absolutely continuous as s goes through the value 1.

We now consider the multicomponent discrete operator (11.86). We discuss the case where the matrices $q_n(x)$, for $x \in \mathbf{Z}$, are diagonal and their matrix elements $q_{n,\alpha}(x)$, for $x \in \mathbf{Z}$ and $\alpha = 1, \ldots, n$, are independent (for different n and α) and identically distributed random variables with a bounded density. Such a choice of q_n corresponds to the discrete Schrödinger operator on a slab with an n-point cross-section.

Goldsheidt [1981] announced that the spectrum of this operator is pure point with probability 1, based on a generalization of the technique of the proof of Theorem 15.4. Lacroix [1983b] gave a complete proof of this statement and of the exponential decrease of the eigenfunctions, using a different method (see also [Bougerol and Lacroix 1985]). Both of these proofs were technically far from simple. A simple proof of Anderson localization that applies to a fairly wide class of multicomponent operators was given in [Delyon et al. 1985b], based on a generalization of the randomization procedure introduced in [Kotani 1984] (see Theorems 13.4, 13.6 and 13.7). (It should be noted that Lacroix's proof also used, although in a somewhat different form, randomization of boundary conditions, which leads to a bounded density of the spectral measure even for the problem on a finite interval. This form of randomization is well-known: see, for example, [Atkinson 1964].)

Delyon et al. proved that, as in the one-component case, the spectrum is simple (in the case of Theorem 13.7 this follows from Proposition 12.8), and that the rate of exponential decrease coincides with the smallest positive Lyapunov exponent γ_n of (11.90): cf. Theorem 13.4 and Examples 13.8. A similar result was obtained by Simon [1985] by the somewhat different randomization procedure presented at the end of 13.B.

The assumption that $\gamma_n > 0$ is present, in some form or another, in all of the above proofs of the Anderson localization for the operator (11.86), as well as in the (simpler) proof of the absence of an absolutely continuous component to the spectrum [Lacroix 1983a]. For $n > 1$, this assumption is stronger, because the methods used to show that $\gamma(\lambda) > 0$ in the one-component case (Section 14), if extended naïvely, only show that the *largest* Lyapunov exponent, called γ_1 in (11.90), is positive. The currently known condition—which implies, in fact, that the Lyapunov spectrum (11.90) is simple—is that the distribution function of the potential should be absolutely continuous with respect to the Lebesgue measure. The proof of this fact, based on general results of Guivarc'h and Raugi [1985] and Virtser [1983], can be found in [Lacroix and Bougerol 1985].

For a different criterion for the simplicity of the Lyapunov spectrum (11.90), see [Goldsheidt and Margulis 1987]. This criterion does not require, generally speaking, that the distribution of the potential be absolutely continuous.

An interesting problem, also having physical applications, is to determine the nature of the spectrum of a one-dimensional operator whose coefficients are given by adding a random metrically transitive process to a fixed, generally nonrandom, function. The discrete Schrödinger operator with potential

$$q(x) = q_1(x) + q_2(x), \tag{15.42}$$

where the $q_1(x)$ are independent at distinct points and $q_2(x)$, for $x \in \mathbf{Z}$, is an arbitrary metrically transitive process independent of $q_1(x)$, was considered in Example 13.8(a), where it was shown that if the distribution of $q_1(x)$ has an absolutely continuous component, Anderson localization takes place. Delyon et al. [1983] proved that this result is true for any function $q_2(x)$. If $q_2(x)$ is an almost periodic function, this has the following physical interpretation: no matter how small the concentration of impurities in a one-dimensional incommensurate structure (e.g., a quasicrystal), localization takes place, whether or not it takes place in the absence of impurities. (It is believed now that for an ideal quasicrystal the spectrum may be singular continuous: see [Kalugin et al. 1986].) But the distance at which the exponential decay of the eigenfunction becomes appreciable—called the *localization radius*—can be very large if the perturbation potential is small, because in this case $\gamma(\lambda)$ should be small (see, for example, Theorems 14.5 and 14.6).

Another example of physical interest is that of a potential of the form (15.42) in the continuous case, with $q_2(x) = -\mathcal{E}x$ for $x \in \mathbf{R}$, where $\mathcal{E} > 0$. This can be seen as a model for the joint influence on a quantum particle of the field of a disordered medium $q_1(x)$ and a homogeneous electric field of strength $\mathcal{E}$. If $q_1(x) \equiv 0$, the spectrum is the whole axis, has multiplicity one and is absolutely continuous, and the generalized eigenfunctions have the following asymptotic behavior:

$$\psi_\lambda(x) = \frac{\mathcal{N}}{\xi^{1/4}} \begin{cases} e^{-2|\xi|^{3/2}/3}(1+o(1)) & \text{as } x \to -\infty, \\ \frac{1}{2}\sin(\frac{2}{3}\xi^{3/2} + \frac{1}{4}\pi)(1+o(1)) & \text{as } x \to \infty, \end{cases} \tag{15.43}$$

where $\xi = (x + \lambda/\mathcal{E})\mathcal{E}^{1/3}$ and $\mathcal{N} = \pi^{-1/2}\mathcal{E}^{-1/6}$ [Landau and Lifshitz 1963]. Bentosela et al. [1983] showed that the picture is qualitatively the same whenever $q_1(x)$ and its first derivative are bounded, uniformly continuous and absolutely continuous, and the second derivative is essentially bounded. The authors adapted the techniques of modern scattering theory to prove the non-existence of a singular continuous component, and standard arguments from ordinary differential equations to show the absence of eigenvalues. If q_1 has two continuous and bounded derivatives, the same results follow from purely ODE techniques, based on a special change of variables: see Sections 5.4–5.7 of [Titchmarsh 1946].

The same authors also proved, using the techniques in [Carmona 1982], that localization occurs for the potential (15.40) with $u(x)$ nonpositive and supported in $[0,1]$, and ξ_n having a continuous density with compact support.

Moreover, according to Example 15.9(b), the random potential (15.40) with nonnegative, twice differentiable $u(x)$ also gives rise to Anderson localization. Thus we conclude that the occurrence of localization is unstable under addition of the term $-\mathcal{E}x$, for $\mathcal{E} > 0$, to a random potential that is smooth in the above sense. The assumption of smooth $q_1(x)$ cannot be omitted completely. Indeed, as we saw in Section 14, a Schrödinger operator with potential (14.19) can be replaced, for the purposes of studying its spectral properties, by the discrete operator associated with equation (14.21) (see also Problems II.42 and VI.24). But if we consider a discrete potential of the form (15.42) with $q_2(x) = -\mathcal{E}x$, we get, for any $q_1(x)$, a spectrum consisting of isolated eigenvalues (the resolvent is easily seen to be a Hilbert–Schmidt operator).

This observation helps us understand the result by Delyon, Simon and Souillard [1985], who considered a Schrödinger operator with potential (15.42), in which $q_1(x)$, for $x \in \mathbf{R}$, is given by (14.19) and $q_2(x) = -\mathcal{E}x$. According to Problem VI.24, the spectrum for $\mathcal{E} = 0$ is pure point, with exponentially decreasing eigenfunctions. The authors showed that for $0 \leqslant \mathcal{E} \leqslant \mathcal{E}_1$, for some finite $\mathcal{E}_1 > 0$, the spectrum remains pure point and the eigenfunctions satisfy

$$(15.44a) \qquad C_2|x|^{-a_2/\mathcal{E}} \leqslant \bar{\psi}_\lambda(x) \leqslant C_1|x|^{-a_1/\mathcal{E}} \qquad \text{as } x \to \infty,$$

$$(15.44b) \qquad \bar{\psi}_\lambda(x) \leqslant C_3 \exp\bigl(-a_3\sqrt{\mathcal{E}}|x|^{3/2}\bigr) \qquad \text{as } x \to \infty,$$

where $\bar{\psi}_\lambda(x) = \bigl(\int_x^{x+1} |\psi_\lambda(t)|^2\, dt\bigr)^{1/2}$.

Thus, a weak electric field does not destroy the point spectrum due to the singular potential (14.19), it just alters the asymptotic behavior of the eigenfunctions. But a sufficiently high field $\mathcal{E} > \mathcal{E}_2$ does destroy the localization, since, by (15.44a), the generalized eigenfunction $\psi_\lambda(x)$ is no longer square-integrable.

The results above were proved by reducing the Schrödinger equation to its finite-difference analogue by explicit integration over the intervals $(n, n+$

1), for $n \in \mathbf{Z}$. We used such a reduction to prove Theorem 14.3. But since in this case the solutions in these intervals are Airy functions, rather than trigonometric ones as in (14.20), the role of the potential in the analogue of (14.21) is played by the sequence $g_n \xi_n$, where the nonrandom amplitudes g_n, up to inessential oscillations, behave as $g/\sqrt{|n|}$ as $|n| \to \infty$, where $g = \text{const}/\mathcal{E}$.

In this connection it is interesting to consider the more general finite-difference equation (11.4) with random potential

$$q(x) = ga(x)\xi(x) \tag{15.45}$$

for $x \in \mathbf{Z}$, where the function $a(x)$ is nonrandom and

$$C_2|x|^{-\alpha} \leqslant |a(x)| \leqslant C_1|x|^{-\alpha}$$

for $\alpha > 0$ and $\xi(x)$, for $x \in \mathbf{Z}$, is made up of independent, identically distributed random variables having a bounded and rapidly decreasing density (such that, for example, $\mathsf{E}\{|\xi(0)|^{2+\delta}\} < \infty$ for some $\delta > 0$). In this case, since the potential tends to zero at infinity, the essential spectrum is the interval $[-2, 2]$, outside of which there can be only isolated eigenvalues, possibly with ± 2 as limit points. The spectrum in the interior of this interval is described in [Delyon, Simon and Souillard 1985]:

(i) for $0 < \alpha < \frac{1}{2}$ the spectrum is pure point and the envelope $r(x)$ of the eigenfunctions $\psi_\lambda(x)$ (see (11.1)) satisfies

$$C_2 \exp(-\delta_2|x|^{1-2\alpha}) \leqslant r(x) \leqslant C_1 \exp(-\delta_2|x|^{1-2\alpha});$$

(ii) for $\alpha = \frac{1}{2}$ (corresponding to the Schrödinger equation with random potential (14.19) and an applied electric field) there exist constants g_1 and g_2 such that, if $g > g_1$, the spectrum is pure point, and if $g < g_2$ it is purely continuous in the interval $|\lambda| \leqslant \lambda_1 < 2$;

(iii) for $\alpha > \frac{1}{2}$ the spectrum is purely continuous (the standard methods of scattering theory give absolute continuity only for $\alpha > 1$).

An essential difference between these problems and those involving metrically transitive operators (corresponding to $a(x) \equiv \text{const}$ in (15.45)) is that here we cannot use the results of Sections 11 and 14 on the exponential (Lyapunov) behavior of the solutions of the Cauchy problem. This causes the need for additional (say to those of Examples 13.8) conditions to be imposed on the random variables $\xi(x)$, such as the existence of a bounded density and finite moments of order no less than 2 (for $a(x)$ constant, by Example 13.8(a), it is enough that the distribution function of $\xi(x)$ have an absolutely continuous component.) The existence of the moments, which is a form of boundedness condition on the potentials, cannot be omitted: if $\xi(x)$

has Cauchy distribution (7.15), the spectrum of both potentials (metrically transitive ones and those of the form (15.45)), for any $\alpha \in [0,1]$, is pure point, and the eigenfunctions decrease as $\exp\bigl(-\operatorname{const}|x|^{1-\alpha}\bigr)$.

As soon as a sufficiently regular (i.e., exponential or even polynomial) behavior has been established for the solutions of the Cauchy problem, one can use statements similar to Theorems 13.7 and 15.8. This was done in [Kotani and Ushiroya 1988], which will be discussed below. Delyon, Simon and Souillard [1895] used a slightly different method, which simultaneously proves both the growth of the solutions and the pure point character of the spectrum, and represents an extension of the method in [Delyon et al. 1983] to the case of random coefficients that are not identically distributed. This method, in turn, goes back to the method of Kunz and Souillard [1980], by which they gave the first proof of the occurrence of Anderson localization for the discrete Schrödinger operator with independent, identically distributed coefficients.

Delyon [1985] further development and improved these results, showing, through arguments that generalize those in the proof of Theorem 12.1, that for $\alpha = \frac{1}{2}$ the spectrum in the interval $|\lambda| \leqslant \lambda_1$ is singular continuous. Later, Kotani and Ushiroya [1988] treated the continuous version of this problem, that is, the Schrödinger equation with potential $q(x) = g(x)Q\bigl(X(x)\bigr)$, where the nonrandom function $g(x)$ is even and tends to zero monotonically as $|x| \to \infty$, and $Q\bigl(X(x)\bigr)$ is the Markov-type random process described in 6.D and Theorem 15.3. They showed, using martingale analysis, that for solutions of the Cauchy problem with such a potential,

$$\lim_{|x|\to\infty} G^{-1}(x)\ln\bigl\|\mathcal{T}(x,0)\bigr\| = \gamma_a(\lambda) \tag{15.46}$$

with probability 1, where

$$G(x) = \int_0^x g^2(t)\,dt, \tag{15.47}$$

$\mathcal{T}(x,0)$ is the fundamental matrix (11.17) of the Schrödinger equation, and $\gamma_a(\lambda)$ is the first term (14.38a) of the asymptotic expansion of the Lyapunov exponent as $g \to 0$.

(It seems intuitively natural that the behavior as $|x| \to \infty$ of the solutions for a random potential $g(x)Q\bigl(X(x)\bigr)$ with variable amplitude $g(x)$ that goes to 0 as $|x| \to \infty$ should be determined by the Lyapunov exponent for the potential $gQ\bigl(X(x)\bigr)$, with a small constant amplitude g. Since this asymptotic behavior, by (14.38a) and (14.57), involves the second moment of the potential, this fact explains the need for the condition on the existence of the moments $\xi(x)$ of order $2+\delta$, for $\delta > 0$, imposed in [Delyon, Simon and Souillard 1985].)

Based on (15.46) and arguing essentially as in the proof of Theorems 13.7 and 15.8, Kotani and Ushiroya showed that, with probability 1:

(i) if $G(\infty) = \infty$, the spectrum is pure point in the interval $[0, \lambda_0]$, where

$$\lambda_0 = \sup\left\{ \lambda > 0 : \int_{\mathbf{R}} e^{-2\gamma_a(\lambda)G(x)}\, dx < \infty \right\},$$

and purely singular continuous in the interval $[\lambda_0, \infty)$; further, the generalized eigenfunctions satisfy

$$\lim_{|x|\to\infty} G^{-1}(x) \ln(\psi_\lambda^2(x) + \psi_\lambda'^2(x)) = -\gamma_a(\lambda); \tag{15.48a}$$

(ii) if $G(\infty) < \infty$, $\mathsf{E}\{Q(X(0))\} = 0$ and $\int_{\mathbf{R}} g^2(x)|x|^\delta\, dx < \infty$ for some $\delta \in [0, 1]$, the spectrum is purely absolutely continuous in the interval $[0, \infty)$, and

$$\psi_\lambda(x) = e^{i\sqrt{\lambda}x} + o(1), \qquad \psi_\lambda'(x) = i\sqrt{\lambda}e^{i\sqrt{\lambda}x} + o(1) \tag{15.48b}$$

as $|x| \to \infty$.

These relations show, in particular, that the spectrum has multiplicity 1 in case (i) and 2 in case (ii) (cf. Theorem 12.8).

Thus, we see that for such a decreasing random potential, as for the problem with an electric field, the spectrum changes from pure point to absolutely continuous as a result of a change in the randomness of the potential: the randomness here being measured by the rate of decrease at infinity. For the critical rate $\alpha = \frac{1}{2}$ we get a mixed spectrum, pure point in one interval and singular continuous in another. Other examples of such transitions and coexistence, although of a somewhat different nature, can be found in [Carmona 1983; Kunz and Souillard 1983; Bratus' et al. 1986] and in Subsection 18.D.

As we mentioned, in the problems above the results are somewhat different than for metrically transitive operators; stronger conditions were required on the random potentials than in Theorems 13.7 and 15.8 and Examples 13.8 and 15.9. Stronger conditions must also be imposed if we go back to metrically transitive operators but desire more detailed information on the spectrum. For example, suppose we want to study the statistic properties of eigenvalues, seen as a point process on the spectral axis. From this point of view—which turns out to be useful also in multidimensional statistical analysis [Girko 1986]—the density of states, that is, the derivative of $N(\lambda)$, is just the density of the point process. An important characteristic of the process is the distribution of distances between neighboring points (eigenvalues). This problem has an interpretation in solid-state physics [Gor'kov and

Eliashberg 1965; Efetov 1983], but it first arose in nuclear physics [Wigner 1965], in the following context:

Let there be a sequence H_L of random self-adjoint operators with discrete spectrum $\{\lambda_{j,L}\}$, for which the integrated density of states $N(\lambda)$ exists. This means, roughly, that the distance between neighboring eigenvalues $\lambda_{n+1,L}$ and $\lambda_{n,L}$, with $n = O(L)$, is of order $O\big((N(\lambda)L)^{-1}\big)$, and thus it is natural to assume that the random variables $\Delta_{n,L} = (\lambda_{n+1,L} - \lambda_{n,L})N(\lambda)L$ for $\lambda_{n,L} = O(\lambda)$ have a limiting distribution $F_\lambda(x)$ as $L \to \infty$. Depending on whether $\lim_{x\downarrow 0} F_\lambda(x)/x$ is zero, finite, or infinite, we talk about adjacent eigenvalues exhibiting *repulsion*, *no interaction* or *attraction*.

Repulsion has been proved to occur for random matrices of order L whose entries are roughly of the same order of magnitude (for example, Hermitian matrices with independent, identically normally distributed entries, or unitary matrices with the Haar measure on the group $U(L)$ as the probability measure), and whose eigenvalues are widely believed to model the spectrum of excited states of heavy nuclei [Mehta 1967; Dyson 1962]. (Unlike the random operators considered above, whose entries decay as the distance from the main diagonal increases, in this case the corresponding Gaussian matrices require a special normalization factor $L^{-1/2}$ for the limit integrated density of states to exist; that is, its entries should be $L^{-1/2}a_{j,k}$, for $j, k = 1, \ldots, L$, with $\mathsf{E}\{a_{j,k}\} = 0$ and $\mathsf{E}\{a_{j,k}^2\} = a^2 > 0$. Clearly, in this case, the limiting operator for $L \to \infty$ does not exist and there is no formula similar to (3.9). Nonetheless this class of operators gives rise to many interesting problems: see [Wigner 1965; Mehta 1967; Dyson 1962; Brody et al. 1981; Marchenko and Pastur 1966; Pastur 1972; Girko 1986] and Appendix B.) This repulsion property of the spectrum is also regarded as the simplest criterion for the chaotic behavior of quantum systems with a finite (even small) number of degrees of freedom, whose classical analogues display dynamical chaos, that is, "random" behavior of trajectories due to high motion instability [Zaslavskii 1984; Casati 1985].

In contrast, for the one-dimensional Schrödinger operator it was proved by Molchanov [1981] that, under the conditions of Theorem 15.3, for any fixed positive numbers $a_1 < b_1 \leqslant \cdots \leqslant a_l < b_l$, integers $k_1, \ldots, k_l$ and $L \to \infty$, we have

(15.49)

$$\lim_{L\to\infty} \mathsf{P}\{LN_L(\Delta_1) = k_1, \ldots, LN(\Delta_l) = k_l\} = \prod_{j=1}^{l} \frac{\big(n(b_j - a_j)\big)^{k_j}}{k_j!} e^{-n(b_j - a_j)},$$

where $\Delta_j = (\lambda + a_j/2L, \lambda + b_j/2L)$ and $n(\lambda) = N'(\lambda)$. This shows that the random point process $N_L^*(t) = LN_L(\lambda, \lambda + t/2L)$ asymptotically approaches a Poisson process as $L \to \infty$ (in the sense of convergence of finite-dimensional distributions). Since the limit (15.49), when the eigenvalues $\lambda_{j,L}$ are independent, identically distributed random variables with density $n(\lambda)$, has the same form, we can regard (15.49) as evidence of the absence of statistical correlation (interaction) between the eigenvalues.

We emphasize that (15.49) is true for any finite interval of λ not containing 0, that is, at the internal points of the spectrum $[0, \infty)$. The case of λ asymptotic close to 0 (or, more accurately, of eigenvalues $\lambda_{i,L}$ with i bounded as $L \to \infty$) was considered in [Grenkova et al. 1983]. It turns out that if $\lambda = 0$ is a fluctuation boundary of the spectrum of an operator on the whole axis (as for the Schrödinger operator with potential (15.18)), the situation is qualitatively the same as that for the internal points of the spectrum. If, on the contrary, $\lambda = 0$ is a stable boundary (as for the operator (6.19) in which $s(x)$ is the Markov-type process (15.18)), there exists repulsion between eigenvalues near zero.

Since, in the vicinity of a stable boundary, localization is weak (by (14.39) and Problem VI.34, the closer λ is to zero, the slower the rate of decay of the eigenfunctions of (6.19)), it seems plausible that repulsion of levels and exponential localization of eigenfunctions arc mutually inconsistent.

These results require a detailed analysis of the spectral properties of the operator $H_{a,b}^{(\alpha,\beta)}$ of the boundary-value problem on the finite interval (a, b), as $a \to -\infty$ and $b \to \infty$, and are based on the phase formalism of Sections 6 and 14, further developed in the proof of Theorem 15.3. The use of this formalism in the study of other spectral properties of the one-dimensional Schrödinger operator and of various other quantities arising in physical applications is considered in [Antsygina et al. 1981; Lifshitz et al. 1982b; Gor'kov et al. 1983], and in the papers cited therein. Many of the results in these papers are derived with physical, rather than mathematical, rigor.

The preceding discussion has referred to one-dimensional random operators whose probabilistic characteristics are sufficiently smooth. This is, from the technical point of view, an important requirement of the methods covered in Section 13 and in this subsection, for proving that the spectrum of a one-dimensional random operator is pure point. Carmona et al. [1987] and Klein et al. [1986] used a different technique, basically a modification of the arguments of Fröhlich and Spencer [1983] in the multidimensional case (see 15.C) to prove that exponential localization takes place for a one-dimensional discrete Schrödinger operator with a singular distribution. In particular, the potential can be a sequence of Bernoulli random variables taking values k_0 and $-k_0$ with probability p and $1 - p$. It was also proved that for k_0 large enough in this example, $N(\lambda)$ has a singular continuous component; in view of results analogous to Theorems 4.20 and 13.7, this suggests that the mathematical mechanism of formation of the point spectrum in the case of discrete distributed potentials is somewhat different. We will discuss these results further in the next subsection.

15.C The Point Spectrum in Multidimensional Problems

We now discuss questions related to the proof that the operator h of the Anderson model—a discrete multidimensional Schrödinger operator of the form (11.94), with a metrically transitive potential independent at distinct points—has a pure point spectrum.

A very important contribution to the study of this class of problems was made by Fröhlich and Spencer [1983], who treated the operator h with metrically transitive potential $q(x)$, for $x \in \mathbf{Z}^d$, independent at distinct points, and having distribution function $F(dq)$ with bounded density: $F(dq) = f(q)\,dq$. It is natural to consider the quantity

$$g = \left(\sup_{q\in\mathbf{R}} f(q)\right)^{-1} \tag{15.50}$$

as a measure of the randomness of the potential.

(15.10) Theorem [Frölich and Spencer 1983]. *Consider the operator h of (11.94), subject to the conditions above, and let $G(x,y,z)$, for $x,y \in \mathbf{Z}^d$ and $z \in \mathbf{C}$ such that $\operatorname{Im} z \neq 0$, denote its Green's function (the kernel of the resolvent). For any $p > 1$ there exist $\lambda_0(p)$ and $g_0(p)$ such that, if either $|\lambda| > \lambda_0(p)$ or $g > g_0(p)$, there exists δ such that*

$$\mathsf{P}\big\{|G(x,y,\lambda+i\varepsilon)| \leqslant e^{\delta(K-|x-y|)} \text{ for all } y \in \mathbf{Z}^d\ \big\} \geqslant 1 - C(p)K^{-p}$$

for any $\varepsilon > 0$ and $x \in \mathbf{Z}^d$.

This conditions $|\lambda| > \lambda_0(p)$ and $g > g_0(p)$ single out the part of the spectrum $\sigma(h)$ adjacent to the boundaries $\inf\sigma(h)$ and $\sup\sigma(h)$, which are fluctuation boundaries and, by Corollary 4.23, equal to $\inf_{x\in\mathbf{Z}^d} q(x) - 2d$ and $\sup_{x\in\mathbf{Z}^d} q(x) + 2d$, respectively.

Theorem 15.10 follows, as we will see, from a similar estimate, under the same conditions, for the Green's function G_Λ of the operator h_Λ, the restriction of h to the cube Λ in $\mathbf{Z}^d$ centered at the origin:

$$\begin{aligned}\mathsf{P}\big\{|G_\Lambda(x,y,\lambda+i\varepsilon)| \leqslant e^{\delta|x-y|} &\text{ for all } x,y \in \mathbf{Z}^d \\ &\text{with } \operatorname{dist}(x,\partial\Lambda) \geqslant |x-y| \geqslant l\big\} \geqslant 1 - \operatorname{const} Vl^{-p},\end{aligned} \tag{15.51}$$

where V is the number of points in Λ.

We will sketch the proof of (15.51). We saw in Sections 6, 9 and 10 that the main contribution to $\ln N(\lambda)$ near a fluctuation boundary is made by parts of realizations of the potential that are large deviations from the mathematical expectation and form deep wells where $q(x) \gtrsim \inf q(x)$, separated by distances much larger than the well sizes. For the Schrödinger operator with a nonnegative potential, when $\nu_0 = 0$, the well sizes L, by (9.5), are of order $1/\sqrt{\lambda}$ as $\lambda \downarrow 0$, whereas the distance l between them, which should be of order $(N(\lambda))^{-1/d}$, are of order $l \sim \exp(const\lambda^{-d/2})$.

Thus, one idea to prove (15.51) is as follows: divide the cube Λ into blocks Λ_j having size of order l, and impose Dirichlet conditions at their boundaries, that is, replace h_Λ by $\bigoplus_j h_{\Lambda_j}$. By the preceding discussion on the structure of typical realizations of the potential, we should expect that each such block will, with probability close to 1, contain a deep well. As $N(\lambda)$ in this part of the spectrum is small (see (9.6), (9.11) and (9.51)), this value of λ, with high probability, should not be too close to the spectrum of h_{Λ_j}.

In this situation, the Green's function of h_{Λ_j} will decrease exponentially. Indeed, assume $|q(x)-\lambda| \geqslant 2d+\tau$, for some $\tau > 0$ and any $x \in \Lambda_j$. Expanding $(h_{\Lambda_j} - z)^{-1}$ in powers of the discrete Laplacian, whose norm is $2d$ and whose matrix entries are nonzero only for nearest neighbors, we find that

$$|G_{\Lambda_j}(x, y, \lambda + i\varepsilon)| \leqslant O(\tau^{-1})e^{-\delta_0|x-y|}$$

for $\delta_0 = \ln\left(1 + \frac{\tau}{2d}\right)$. (A similar estimate is true in the more general case where $\operatorname{dist}(\sigma(h_{\Lambda_j}), \lambda) = \tau > 0$: see Problem VI.36.)

After proving this inequality, one must show that for the overwhelming majority of realizations the exponential decay is not affected by switching on the interaction between blocks, that is, by adding to the orthogonal sum $\bigoplus_j h_{\Lambda_j}$ boundary operators that will complete the sum to h_Λ. Is natural to expect that the behavior will stay the same, because $L \ll l$.

A similar scheme of proof was formulated, albeit implicitly, by Lifshitz [1964], in his derivation of the asymptotic formula (9.6) for $N(\lambda)$; see also [Halperin and Lax 1967] and [Lifshitz 1967]. It forms the basis for an efficient approximate method in the theory of disordered systems: see [Lifshitz et al. 1982b] and [Efros and Shklovski 1984], where it is refereed to as the optimal fluctuation method.

As we saw in Sections 6, 9 and 10, this reasoning can yield asymptotic formulas for $\ln N(\lambda)$ near the fluctuation boundaries of the spectrum. But attempts to use it to prove (15.52) run into a fundamental difficulty; namely, in passing from $\bigoplus_j h_{\Lambda_j}$ to h_Λ using, say, perturbation theory, with the boundary operators being the perturbation, we must take into account the probability, however small, that the block operators have eigenvalues very close to the value of λ for which the Green's function is being calculated. This effect is known as resonance in quantum mechanics (see Problem VI.37), and has much in common with the problem of small denominators in classical mechanics; it will resurface in 16.B and 16.D in connection with the spectral analysis of almost-periodic operators. This difficulty did not appear in the estimation of $N(\lambda)$ in Section 9, because there it was enough to consider a single block if its size was chosen in an optimal way.

This difficulty was overcome by Fröhlich and Spencer [1983] (see also the reviews by Spencer [1986] and Martinelli and Scoppola [1987]), who developed a kind of multiscale perturbation theory, which can be considered as a probabilistic version of the Kolmogorov–Arnold–Moser method, supplemented with renormalization group ideas borrowed from modern statistical physics and solid state theory [Wilson and Kogut 1974; Altshuler et al. 1982]. Their proof involved a complex rescaling procedure and requires some hard probabilistic estimates.

Later von Dreifus [1987] and Spencer [1988] proposed a new proof of Fröhlich and Spencer's results, which uses the same basic ideas but is technically much simpler. The key new idea consists of rescaling arguments first developed in the context of percolation theory [Chayes and Chayes 1986]. We discuss this proof in some detail now.

It turns out that Theorem 15.10 can be strengthened. For L a natural number and $x \in \mathbf{Z}^d$, denote by $\Lambda_L(x)$ the cube centered at x with sides of length L. Let $|x| = \max\{|x_1|, \dots, |x_d|\}$, for $x = (x_1, \dots, x_d) \in \mathbf{Z}^d$, and denote by $\partial\Lambda_L(x)$ the boundary of $\Lambda_L(x)$, that is, the set of pairs (y, y') such that $y \in \Lambda_L(x)$, $y' \notin \Lambda_L(x)$, and $|y - y'| = 1$. We also write $y \in \partial\Lambda_l$ if there is y' such that $(y, y') \in \partial\Lambda_l$, and we let $\sum_{y\in\partial\Lambda}$ stand for $\sum_{(y,y')\in\partial\Lambda}$. Finally, we write $\Lambda = \Lambda_L = \Lambda_L(x)$.

The cardinality of Λ_L, denoted by V, does note exceed $(L+1)^d$, and that of $|\partial\Lambda_L|$ does not exceed $c_d L^{d-1}$, where c_d depends only on d. Let Γ_Λ be the operator defined by the matrix $\Gamma_\Lambda(x, y) = \chi_{\partial\Lambda}((x, y))$, set $h_{\Gamma_\Lambda} = h + \Gamma_\Lambda = h_\Lambda \oplus h_{\Lambda^c}$, and let R, R_{Γ_Λ}, R_Λ and R_{Λ^c} be the respective resolvents. Then $R_{\Gamma_\Lambda} = R_\Lambda \oplus R_{\Lambda^c}$ and, by the resolvent identity,

$$R = R_{\Gamma_\Lambda} + R_{\Gamma_\Lambda}\Gamma_\Lambda R. \tag{15.52}$$

The matrix form of this identity, written in terms of Green's functions (the matrix of the resolvent) is

$$\begin{aligned} G(x, y, \lambda + i\varepsilon) &= G_{\Gamma_\Lambda}(x, y, \lambda + i\varepsilon) \\ &\quad + \sum_{z\in\partial\Lambda} G_{\Gamma_\Lambda}(x, z, \lambda + i\varepsilon) G(z', y, \lambda + i\varepsilon). \end{aligned} \tag{15.53}$$

Observe that if $\partial\Lambda$ separates x and y, $G_{\Gamma_\Lambda}(x, y) = 0$.

Definition. *Let $\delta > 0$ and $\lambda \in \mathbf{R}$. A cube $\Lambda(x)$ is called (δ, λ)-regular (for a fixed potential) if, for all small enough $\varepsilon > 0$,*

$$\max_{y\in\partial\Lambda_L(x)}\left|G_{\Lambda_L(x)}(x,y,\lambda+i\varepsilon)\right| \leqslant e^{-\delta L/2}. \tag{15.54}$$

We will need the following important estimate, essentially due to Wegner [1981] (see also [Frölich and Spencer 1983]).

(15.11) Theorem. *Suppose the metrically transitive potential $q(x)$, for $x \in \mathbf{Z}^d$, consists of independent, identically distributed random variables with a bounded density: $F(dq) = f(q)\,dq$ with $\sup_q f(q) = f_0 < \infty$. Then, for all $\varepsilon > 0$,*

$$\mathsf{P}\{|G_\Lambda(x,y,\lambda+i\varepsilon)| \geqslant W\} \leqslant 2W^{-1}Vf_0. \tag{15.55}$$

Proof. We repeat almost literally the arguments used in the proof of Theorem 4.21 and Problem II.1, to find that $\mathsf{P}\{\operatorname{dist}(\lambda,\sigma(h_\Lambda)) \geqslant W\} \leqslant 2W^{-1}Vf_0$. Now (15.55) follows from the inequality $\left|(h-z)^{-1}(x,y)\right| \leqslant \left(\operatorname{dist}(z,\sigma(h))\right)^{-1}$. □

Here is the stronger version of Theorem 15.10.

(15.12) Theorem [Frölich et al. 1985; Spencer 1986; Martinelli and Scoppola 1987]. *Suppose that, for a given $p > 2d$ and $\delta_0 > 0$, there exists L_0 such that*

$$\mathsf{P}\{\Lambda_{L_0}(\cdot)\ \textit{is}\ (\delta_0,\lambda)\textit{-regular}\} \geqslant 1 - L_0^{-p}. \tag{15.56}$$

Then there exists $\delta \in (0,\delta_0)$ such that Theorem 15.10 *holds.*

Proof of Theorem 15.10 given Theorem 15.12. It suffices to prove that, for some $L_0 > 0$, there exist $g_0(L_0)$ and $\lambda_0(L_0)$ such that, if either $g > g_0$ or $|\lambda| > \lambda_0$, (15.56) holds. We first give a proof for $|\lambda| > \lambda_0$. Suppose that, for some λ_0,

$$|q(y)| \leqslant -2d + \lambda_1 \tag{15.57}$$

for all $y \in \Lambda_{L_0}(x)$, where $\lambda_1 = \lambda_0 - e^{\delta_0 L_0/2} \geqslant 2d$. Then $\sigma(h_{\Lambda_{L_0}(x)}) \subset [-\lambda_1,\lambda_1]$ and, for $|\lambda| > \lambda_0$, $D = \operatorname{dist}\left(\lambda,\sigma(h_{\Lambda_{L_0}(x)})\right) > e^{\delta_0 L_0/2}$. This implies that $\Lambda_{L_0}(x)$ is (δ_0,λ)-regular for $|\lambda| > \lambda_0$, since

$$\left|G_{\Lambda_{L_0}}(x,y,\lambda+i\varepsilon)\right| \leqslant D^{-1}. \tag{15.58}$$

But since the $q(y)$, for $y \in \mathbf{Z}^d$, are independent and identically distributed, the probability that (15.57) is valid equals $\left(1 - \mathsf{P}\{|q(0)| > -2d + \lambda_1\}\right)^{V_0}$, which is greater than $1 - L_0^{-p}$ if λ_0 is large enough.

For the case when the quantity (15.50) is large, we again use (15.58) with $D = e^{\delta_0 L_0/2}$ and λ satisfying $\left|q(y) - \lambda\right| \geqslant 2d + D$ for all $y \in \Lambda_{L_0}(x)$.

These inequalities hold with probability at least $1 - \left(g^{-1}(2d + e^{\delta_0 L_0/2})\right)^{V_0}$, so we can just take g_0 such that

$$\left(g_0^{-1}(2d + e^{\delta_0 L_0/2})\right)^{V_0} < L_0^{-p}. \qquad \square$$

Theorem 15.12, in turn, can be derived from its finite-volume version:

(15.13) Theorem [Frölich and Spencer 1983; Spencer 1986]. *Under the conditions of Theorem* 15.12 *there exists* $\delta \in (0, \delta_0)$ *such that, for all* $\varepsilon > 0$,

$$\mathsf{P}\{|G_\Lambda(x, y, \lambda + i\varepsilon)| \leqslant e^{-\delta|x-y|} \text{ for all } x, y \in \Lambda \text{ with } \operatorname{dist}(x, \partial\Lambda) \geqslant |x-y| \geqslant l\} \geqslant 1 - C_f V l^{-p},$$

where C_f *depends only on the distribution of the random potential.*

Proof of Theorem 15.12 given Theorem 15.13. Let $\{\Lambda_n\}$ be a sequence of cubes centered at the origin, with sides of length $L_n = 3^n(2|x| + K)$. Using the resolvent identity, we can write

$$R = R_{\Gamma_n} + R_{\Gamma_n}\Gamma_n R. \tag{15.59}$$

In particular, for $n = 0$ and $z = \lambda + i\varepsilon$,

$$G(0, x, z) = G_{\Lambda_0}(0, x, z) + \sum_{y_0 \in \partial\Lambda_0} G_{\Lambda_0}(0, y_0, z) G(y_0', x, z).$$

By repeatedly applying (15.59) to the right-hand side of this identity, for $n = 1, 2, \ldots$, we obtain the so-called *block resolvent expansion*:

$$G(0, x) = \left(R_{\Lambda_0} + R_{\Lambda_0}\Gamma_0 R_{\Lambda_1} + R_{\Lambda_0}\Gamma_0 R_{\Lambda_1}\Gamma_1 R_{\Lambda_2} + \cdots\right)(0, x). \tag{15.60}$$

Since $\operatorname{dist}(\partial\Lambda_n, \partial\Lambda_{n+1}) = L_n$, Theorem 15.13 implies that the entries of $\Gamma_n R_{n+1}\Gamma_{n+1}$ are smaller than $e^{-\delta L_n/2}$ with probability at least $1 - C_f V_{n+1} L_n^{-p}$. Further, by using Theorem 15.13 for $|x| > K/2$ and the Wegner estimate (15.55) for $|x| < K/2$, we find that $|G_{\Lambda_0}(0, x)| \leqslant e^{\delta(K-|x|)}$ with probability at least $1 - C_f K^{-p+d}$ for any $p > d$. Theorem 15.12 now follows. $\square$

We now come to the main technical result of this theory:

(15.14) Theorem [von Dreifus 1987; Spencer 1988]. *Suppose that, for a given* $p > 2d$ *and* $\delta_0 > 0$, *there exists* $\alpha = \alpha(p, d) \in (1, 2)$ *and* L_0 *such that*

$$\mathsf{P}\{\Lambda_{L_0}(\,\cdot\,) \text{ is } (\delta_0, \lambda)\text{-regular}\} \geqslant 1 - L_0^{-p}. \tag{15.61}$$

Then, if we set

$$L_{n+1} = L_n^\alpha \tag{15.62}$$

for $n = 0, 1, 2, \ldots$ *and pick* $0 < \delta < \delta_0$, *we can find* $Q = Q(p, d, \alpha, \delta_0, \delta)$ *such that, for any* $L_0 > Q$ *and any* $n = 0, 1, 2, \ldots$,

$$\mathsf{P}\{\Lambda_{L_n}(\,\cdot\,) \text{ is } (\delta, \lambda)\text{-regular}\} \geqslant 1 - L_n^{-p}. \tag{15.63}$$

Proof of Theorem 15.13 given Theorem 15.14. It is enough to prove the theorem with l equal to L_n, for all n. To this end, for any $x, y \in \Lambda$ such that $\operatorname{dist}(y, \partial\Lambda) \geqslant |x - y| \geqslant L_n$, starting from y, we apply iteratively the resolvent identity to a chain of at least $2|x - y|/L_n$ cubes $\Lambda_{L_n}(\,\cdot\,)$ in order to reach x. By Theorem 15.14, we conclude that the inequality

$$\left|G_\Lambda(x, y)\right| \leqslant (e^{-\delta_1 L_n/2})^{2|x-y|/L_n}\left|G_\Lambda(t, y)\right| \leqslant e^{-\delta_1|x-y|}\left|G_\Lambda(t, y)\right|,$$

where $\operatorname{dist}(t, x) \leqslant L_n/2$ and $\delta_1 < \delta_0$, holds with probability at least $1 - VL_n^{-p}$. By the Wegner estimate (15.55), we also have, for a certain $\bar{C}_f$,

$$\mathsf{P}\{|G_\Lambda(t, y)| \leqslant L_n^p\} \leqslant 1 - \bar{C}_f V L_n^{-p}.$$

Therefore (15.58) is valid with $l = L_n$ for all n. □

We now have to prove Theorem 15.14, which we do by induction on n. The proof has a deterministic and a probabilistic component: we start with the deterministic one.

(15.15) Lemma. *Let the numbers* L *and* l, *where* L/l *is sufficiently large, and a number* $\delta > 0$ *be given. Let* $x \in \Lambda_{2L}(0)$ *be such that* $\operatorname{dist}(x, \partial\Lambda_{2L}(0)) \geqslant l/2$, *and set* $A_{2L}(x) = \Lambda_{2L}(0) \setminus \Lambda_l(x)$. *Suppose that*

(a) $\left|G_{\Lambda_{2L}(0)}(r, s)\right| \leqslant W$ *for all* $r, s \in \Lambda_{2L}(0)$;
(b) *there exists* $x \in \Lambda_{2L}(0)$ *such that, for all* $u \in A_{2L}(x)$ *whose distance to* $\Lambda_{2L}(x)$ *is at least* $l/2$, *the cubes* $\Lambda_l(u)$ *are* (δ_0, λ)*-regular.*

Then

$$\max_{y \in \partial\Lambda_L(0)} \left|G_{\Lambda_{2L}(0)}(0, y)\right| \leqslant We^{-\delta(L/2 - 2l)}. \tag{15.64}$$

Proof. We use a block resolvent expansion similar to (15.60), but now for the chain of overlapping congruent cubes $\Lambda_l(\,\cdot\,)$. We start with $\Lambda_l(0)$. If $\Lambda_l(0) \cap \Lambda_l(x) = \varnothing$, we have

$$G_{\Lambda_{2L}(0)}(0, y) = \sum_{z_1 \in \partial\Lambda_l(0)} G_{\Lambda_l(0)}(0, z_1) G_{\Lambda_{2L}(0)}(z_1', y).$$

Thus

$$\left|G_{\Lambda_{2L}(0)}(0,y)\right| \leqslant \Bigl(\sum_{z_1\in\Lambda_l(0)} \left|G_{\Lambda_l(0)}(0,z_1)\right|\Bigr)\left|G_{\Lambda_{2L}(0)}(\bar z_1,y)\right|$$

for some $\bar z_1 \in \partial\Lambda_l(0)$. Now we use the resolvent identity with the cube $\Lambda_l(\bar z_1)$. We continue this iterative process until either we have accumulated at least $n = 2|y|/l - 4$ iterations or we have found that $\Lambda_l(\bar z_k) \cap \Lambda_l(x) \neq \varnothing$ for some $k < n$. It the first case, we arrive at (15.64) by using assumption (b) to estimate the sums

$$\sum_{z_i\in\partial\Lambda_l(\bar z_{i-1})} \left|G_{\Lambda_l(\bar z_{i-1})}(\bar z_{i-1},z_i)\right|,$$

and assumption (a) to estimate the factor $\left|G_{\Lambda_{2L}(0)}(\bar z_n,y)\right|$. In the second case we continue the iteration process by applying the resolvent identity to the cube $\Lambda_l(y)$. In other words, in this case we iterate starting from both endpoints 0 and y. (See Problems VI.37–38 for simpler applications of the same idea.) This way we again accumulate n iterations before the cube $\Lambda_l(x)$ is encountered, and by using again assumptions (a) and (b), we arrive at (15.64). □

Here is the probabilistic part of the argument:

(15.16) Lemma. *Suppose that*

(15.65) $$\mathsf{P}\{\Lambda_l(\,\cdot\,) \textit{ is } (\delta_0,\lambda)\textit{-regular}\} \geqslant 1 - l^{-p}.$$

Then, bound (15.64) *of Lemma* 15.15 *holds with probability at least*

(15.66) $$1 - (V_{2L}^2 l^{-2p} + 2V_{2L}\hat W^{-1}),$$

where $\hat W = W/\sup_q f(q)$.

Proof. It suffices to prove that (15.66) is a lower bound for the probability that conditions (a) and (b) of the preceding lemma be valid. By Theorem 15.11, condition (a) holds with probability at least

(15.67) $$1 - 2V_{2L}\hat W^{-1}.$$

If condition (b) fails, there exist $x \in \Lambda_{2L}(0)$ and $u \in \Lambda_{2L}(x)$ such that $\operatorname{dist}(u,\Lambda_{2L}(x)) \geqslant l/2$, $\max_{y\in\partial\Lambda_l(x)}\left|G_{\Lambda_l(x)}(y,x)\right| \geqslant e^{-\delta_0 l/2}$ and $\max_{y\in\partial\Lambda_l(u)}$ $\left|G_{\Lambda_l(u)}(y,u)\right| \geqslant e^{-\delta_0 l/2}$. Since $\Lambda_l(x) \cap \Lambda_l(u) = \varnothing$ by definition, and our random potential is independent at distinct points, the events determined by the last two inequalities are independent, and the probability that they occur simultaneously is bounded from above by $V_{2L}^2 l^{-2p}$, the factor V_{2L}^2 being an upper bound for the number of all possible choices of cubes $\Lambda_l(x)$ and

$\Lambda_l(u)$ (that is, the entropy factor). Combining this with bound (15.67), we find (15.66). □

Lemmas 15.15 and 15.16 give information on the Green's function with Dirichlet boundary conditions on $\partial\Lambda_{2L}(0)$ and with $y \in \Lambda_L(0)$. In the next lemma we sharpen this information, replacing $\Lambda_{2L}(0)$ in (15.64) by $\Lambda_L(0)$. As a result, we obtain the main auxiliary bound needed to carry out the induction step in Theorem 15.14.

(15.17) Lemma. *Suppose that*

$$\mathsf{P}\{\Lambda_l(\,\cdot\,) \text{ is } (\delta_l, \lambda)\text{-regular}\} \geqslant 1 - l^{-p},$$

where $p > 2d$ *and* l *is large enough. Then, if* $L = l^\alpha$ *for some* $\alpha \in (1, 2)$, *we have*

$$\mathsf{P}\{\Lambda_L(\,\cdot\,) \text{ is } (\delta_L, \lambda)\text{-regular}\} \geqslant 1 - L^{-p},$$

where

$$\delta_L \geqslant \delta_l - \big(4\delta_l l^{-\alpha+1} + (C_1 + C_2 \ln l) l^{-\alpha}\big), \tag{15.68}$$

with $C_1 = 2\ln(c_d 2^{2d+7})$ *and* $C_2 = 2\alpha(3d + 2p - 1)$.

Proof. We use the resolvent identity for the cubes $\Lambda_{2L}(0)$ and $\Lambda_L(0)$ and for $x \in \partial\Lambda_l$:

$$G_{\Lambda_{2L}(0)}(0, x) = G_{\Lambda_L(0)}(0, x) + \sum_{y \in \partial\Lambda_L(0)} G_{\Lambda_{2L}(0)}(0, y) G_{\Lambda_L(0)}(y', x).$$

Therefore, if

$$\big|G_{\Lambda_L(0)}(y', x)\big| \leqslant W \tag{15.69}$$

for all $y', x \in \Lambda_L(0)$, we have

$$\big|G_{\Lambda_L(0)}(0, x)\big| \leqslant \big(1 + |\partial\Lambda_L(0)|\, W\big)\big|G_{\Lambda_{2L}(0)}(0, x)\big|. \tag{15.70}$$

Combining Theorem 15.11 and Lemma 15.16, we find from (15.69) and (15.70) that

$$\max_{x \in \partial\Lambda_L(0)} \big|G_{\Lambda_L(0)}(0, x)\big| \leqslant \big(1 + |\partial\Lambda_L(0)|\, W\big) W e^{-\delta_l(L/2 - 2l)} \tag{15.71}$$

with probability at least $1 - (V_{2L}^2 l^{-2p} + 4V_{2L}\hat{W}^{-1})$.

If we set

$$\hat{W} = 2^{d+3} L^{p+d}, \tag{15.72}$$

we have $4V_{2L}\hat{W}^{-1} = \frac{1}{2}L^{-p}$, and for $L = l^{\alpha}$ the inequality $V_{2L}^2 l^{-2p} + 4V_{2L}\hat{W} \leqslant L^{-p}$ holds if

$$2^{2d+1} \leqslant l^{2p-\alpha(2d+p)}. \tag{15.73}$$

Let $\bar{\alpha}$ satisfy the equation $2p - \bar{\alpha}(p+2d) = 0$; then $\bar{\alpha} \in (1,2)$ if $p \in (2d, \infty)$. For $\alpha = \frac{1}{2}(\bar{\alpha}+1)$, we have $2p - \alpha(p+2d) = \frac{1}{2}(p-2d) > 0$, and (15.73) holds for

$$l \geqslant 2^{(2d+1)/(2p-\alpha(2d+p))}. \tag{15.74}$$

Now we conclude from (15.71) that

$$\delta_L = \delta_l - L^{-1}\big(4\delta_l l + 2\ln W(1 + |\partial \Lambda_L(0)|\, W)\big)$$

Combining this with (15.72), we obtain (15.68), which concludes the proof. □

Proof of Theorem 15.14. For a given $p > 2d$, let α be as in Lemma 15.17. Suppose that $\Lambda_{L_n}(0)$ is (δ_n, λ)-regular for some n. Using the preceding lemma with $l = L_n$, $L = L_{n+1}$, $L_{n+1} = L_n^{\alpha}$ and L_0 satisfying (15.74), we find that $\Lambda_{L_{n+1}}(0)$ is (δ_{n+1}, λ)-regular, where

$$\delta_{n+1} = \delta_n - \big(4\delta_n L_n^{-\alpha+1} + (C_1 + C_2\alpha^n)L_n^{-\alpha}\big). \tag{15.75}$$

Now pick $\delta \in (0, \delta_0)$. Then Theorem 15.14 will follow by induction if $\sum_{n=0}^{\infty}(\delta_n - \delta_{n+1}) \leqslant \delta_0 - \delta$. But, by (15.62) and the monotonicity of δ_n, we have $\sum_{n=0}^{\infty}(\delta_n - \delta_{n+1}) \leqslant C_3 \sum_{n=0}^{\infty} L_0^{-(\alpha-1)\alpha^n}$, where C_3 is independent of L_0. Thus the left-hand side of this inequality can be made smaller than $\delta_0 - \delta$ if L_0 is large enough. This proves Theorem 15.14. □

Remark. Theorems 15.12–14 remain valid if we replace $e^{-\delta L/2}$ in the definition of (δ, λ)-regularity by some $\rho_0 \in (0,1)$. It suffices to replace δ_0 in the statements and proofs by $-\frac{1}{2}\ln \rho_0 / L_0$: see [von Dreifus 1987].

We now consider some spectral facts that can be obtained from Theorem 15.10 Recall that for a self-adjoint operator h on an abstract Hilbert space $\mathcal{H}$, one can define a Borel measure ρ on the spectral axis by setting, for any basis $\{e_n\}$ of $\mathcal{H}$ and any summable sequence $\{a_n\}$ of positive numbers,

$$\rho(\Delta) = \sum_n \big(e_n, E_h(\Delta)e_n\big)a_n$$

Different choices of basis and of sequence lead to equivalent measures, and we call any of these equivalent measures a *spectral measure* of operator [Berezanskii 1968]. The spectral measure determines the spectral type of h: in particular, the spectrum of h is pure point on a Borel set $X \subset \mathbf{R}$

if and only if the restriction of the spectral measure to X is pure point. Using (12.30), it is easy to show that for one-dimensional operators of order two the measure (12.35) is a spectral measure in the sense defined here. For a finite-difference operator of order two in higher dimension (acting on $l_2(\mathbf{Z}^d)$), the type we're considering now, the sum (12.35) of diagonal entries of the resolution of the identity at the points 0 and 1 (or at any other two neighboring points) should be replaced by the sum of entries corresponding to points in the layer $S_{(0,1)} = \{0,1\} \times \mathbf{Z}^{d-1} \subset \mathbf{Z}^d$:

$$\rho(\Delta) = \sum_{x \in S_{(0,1)}} 2^{-|x|} E_h(x, x, \Delta). \tag{15.76}$$

We will need the following fact from the spectral theory of self-adjoint differential and finite-difference operators, due to Schnol' [1957]: a one-dimensional version of which was mentioned in Section 12 (see (12.28)). In its simplest form it goes as follows (for a more complete treatment, see [Berezanskii 1968] and [Simon 1982]):

(15.18) Proposition [Berezanskii 1968]. *Let $\psi_\lambda(x)$, for $x \in \mathbf{Z}^d$, be a solution of the finite-difference equation*

$$(h - \lambda)\psi_\lambda = 0 \tag{15.77}$$

corresponding to the operator (11.94), *and ρ its spectral measure. Then, for ρ-almost all $\lambda \in \mathbf{R}$ and $\varepsilon > 0$,*

$$|\psi_\lambda(x)| \leqslant C(\lambda, \varepsilon)(1 + |x|)^{d/2+\varepsilon}. \tag{15.78}$$

It is natural to call such polynomially bounded solutions *generalized eigenfunctions*, and the corresponding values of λ *generalized eigenvalues*; the spectrum of the operator coincides with the closure of the set of generalized eigenvalues.

(15.19) Theorem [Martinelli and Scoppola 1985]. *Suppose the potential $q(x)$ in the operator* (11.94) *of the Anderson model consists of independent, identically distributed random variables, whose distribution has density $f(q)$ bounded by the constant g^{-1} (see* (15.50)*). Then there exist constants λ_0 and g_0 such that $\sigma_{ac}(h) \cap [\lambda_0, \infty) = \varnothing$ and $\sigma_{ac}(h) = \varnothing$ for $g > g_0$.*

Proof. We consider λ in a finite interval Δ of the spectral axis for which Theorem 15.10 is true. Using this theorem and Fatou's theorem on limits under the integral sign, we find that for Lebesgue-almost $\lambda \in \Delta$ in that inequality, $\lambda + i\varepsilon$ can be replaced by $\lambda + i0$. Then, because the event defined by this new inequality increases monotonically with K, we conclude that, for a fixed $x \in \mathbf{Z}^d$,

$$|G(x, y, \lambda + i0)| \leqslant C(\lambda, \omega, x)e^{-\delta|x-y|} \tag{15.79}$$

for $y \in \mathbf{Z}^d$ and $\mathsf{L} \times \mathsf{P}$-almost all pairs $(\lambda, \omega) \in \Delta \times \Omega$, where L is the Lebesgue measure. Now let $\psi_\lambda(x)$ be a polynomially bounded solution (15.78) to equation (15.77). By applying the Green's function $G(x, y, \lambda+i0)$ to this equation and using (15.78) and (15.79), we found that $\psi_\lambda(x) \equiv 0$. From this and from Fubini's theorem, we have that, with probability 1, the spectral measure ρ has no absolutely continuous component. □

Remark. The reasoning at the end of this proof is essentially the same that we used for a similar purpose in the second proof of Theorem 12.1, in the one-dimensional case (see also [Pastur 1974a, 1980]). These arguments are general enough that they can be applied whenever we know that for almost all (λ, ω) the equation has no polynomially bounded solutions. In the one-dimensional case, this fact was obtained from Proposition 11.2 and the positiveness of the Lyapunov exponent on the spectrum (see Theorem 12.10 and Section 14), while here, in the multidimensional case, we used the exponential decay of the Green's function (15.51). In the one-dimensional case this fact follows also from the positiveness of the Lyapunov exponent, according to Problems V.30–31. Thus, the exponential decay of the Green's function can be considered as a multidimensional analogue of the positiveness of the Lyapunov exponent.

Actually, Theorem 15.10 implies not only that the fluctuation spectrum of the Anderson model has no absolutely continuous component, but even that it is pure point. This was proved simultaneously by Delyon et al. [1985a], Fröhlich et al. [1985] and Simon et al. [1985].

(15.20) Theorem. *Under the conditions of Theorem* 15.19, *for the same constants* λ_0 *and* g_0, *the spectrum is pure point and simple for* $|\lambda| \geqslant \lambda_0$. *If* $g > g_0$ *the spectrum is pure point and simple everywhere.*

Proof. We will present the proof given in [Delyon et al. 1985a] and then discuss briefly the other two.

We use the scheme applied in Section 13 to one-dimensional random operators. There we showed (Theorem 13.1, Corollary 13.2) that a sufficient condition for the spectrum to be pure point is for ρ-almost every polynomially bounded solution to decay exponentially. This is easily seen to be true also in several dimensions, the characterization of exponential decay being that (cf. (13.5))

$$\limsup_{|x|\to\infty} \frac{1}{x} \ln|\psi_\lambda(x)| \leqslant -\delta < 0. \tag{15.80}$$

The next step is to show that, if (15.80) is true for Lebesgue-almost all λ, it is true for ρ-almost all λ. In Section 13 this was done by the Carmona–Kotani randomization procedure, which uses as parameters the values of the independent potential at a finite set of point $x \in \mathbf{Z}^d$, and the verification of two conditions: the spectral measure averaged over the parameters should be absolutely continuous, and whether or not a solution decays exponentially should not depend on the parameters. The first condition, as in the one-dimensional case, follows from a general result, Theorem 13.5 (Delyon et al. [1985a], however, give a different proof).

To verify the second condition, consider, for $x \in \mathbf{Z}^d$ fixed, a polynomially bounded solution $\psi_{\lambda,\xi,x}(y)$ of the equation for the perturbed potential $q(y) + \xi\delta_{xy}$ (instead of $q(y)$). The equation is

$$((h-\lambda)\psi_{\lambda,\xi,x})(y) = \xi\psi_{\lambda,\xi,x}(x)\delta_{xy}, \tag{15.81}$$

where $\xi \in \mathbf{R}$ and $y \in \mathbf{Z}^d$. Apply the Green's function $G(x, y, \lambda + i0)$ to (15.81). Since the solution is polynomially bounded, (15.79) gives

$$|\psi_{\lambda,\xi,x}(y)| \leqslant |\xi\psi_{\lambda,\xi,x}(x)| C(\lambda, \omega, x) e^{-\delta|x-y|}.$$

Hence, for $\mathsf{L} \times \mathsf{P}$-almost all pairs $(\lambda, \omega) \in \Delta \times \Omega$, where L is the Lebesgue measure, any such solution is unique up to a multiplicative constant, and satisfies (15.80) (and even the similar equation for

$$\left(\sum_{y \in S_{(0,1)}} |\psi_{\lambda,\xi,x}(y)|^2\right)^{\frac{1}{2}},$$

if $|x| \to \infty$ in any direction other than parallel to the layer $S_{(0,1)}$).

Now let P_x be the restriction of P to the set Ω_x of values of the potential on $\mathbf{Z}^d \setminus \{x\}$. Since the potential is independent at different points, for $\mathsf{L} \times \mathsf{P}_x$-almost all pairs $(\lambda, \omega_x) \in \Delta \times \Omega_x$ and for any value of the potential at x, every polynomially bounded solution of (15.81) is unique and decays exponentially in the sense of (15.80). Using Proposition 15.18 and an argument like the one in the proof of Theorems 13.6 and 13.7, we conclude that, with probability 1, the measure $\rho_x(d\lambda) = E_h(x, x, d\lambda)$ is pure point, and for ρ_x-almost every $\lambda \in \Delta$ the eigenfunction is unique and decreases exponentially. By shift invariance the same is true for ρ_y with any $y \in \mathbf{Z}^d$, and by (15.76) also for the spectral measure ρ, which proves the theorem. □

The proof in [Simon et al. 1985], although similar to the one above in essence, differs from it in form. Instead of considering equation (15.81), it uses an abstract version of Theorem 13.6, proved also by using randomization and giving a criterion of stability for almost all $\xi \in \mathbf{R}$ for the pure point spectrum of an abstract self-adjoint operator of the form $h + \xi P$, where A is a fixed self-adjoint operator and P a one-dimensional orthogonal projection. The finiteness of integral (13.31) is such a criterion. For the finite-difference operator (11.94) of the Anderson model, this inequality, in view of (15.76) and of the spectral theorem, is equivalent to (13.30) and (13.29), which evidently follows from (15.79), that is, from Theorem 15.10. Therefore, choosing as P the operator of projection onto the function concentrated at zero, $(P\psi)(x) = \psi(0)\delta_0(x)$, we see that for $\mathsf{L} \times \mathsf{P}$-almost all (ξ, ω) the measure $\mathcal{E}_{A+\xi P}(0,0,d\lambda)$ is pure point. But the operator $A + \xi P$ has potential $q(x) + \xi\delta_{x,0}$, so the fact that $q(x)$ is independent and identically distributed at different points and has an absolutely continuous distribution implies that the spectrum of the initial operator A is pure point with probability 1.

The proof in [Fröhlich et al. 1985] is based on an essential sharpening of Theorem 15.10. The authors proved that, under the hypotheses of Theorem 15.10 (and even weaker ones: see Theorem 15.21 below), the Green's function of h admits an exponential bound not for a fixed λ as in Theorem 15.10, but for all the generalized eigenvalues $\lambda(\omega)$ with $|\lambda(\omega)| \geqslant \lambda_0$ in (15.77), which depend on ω in quite a nontrivial way.

An important ingredient in this result is a deep analysis of long-range tunneling carried out by Jona-Lasinio et al. [1985] for certain nonrandom, so-called hierarchical potentials. These potentials are functions assuming two values, with wells separated by barriers of rapidly growing width by the law (15.62), so that the potential is almost self-similar on all the scales (15.62). These potentials turn out to be a good deterministic model for the realizations of a random potential in the fluctuation region of the spectrum, and can be analyzed in detail. Thus they serve as an important intermediate step in the proof of localization of the random potential.

The original proof in [Fröhlich et al. 1985], being based, especially in its probabilistic part, on the method of [Fröhlich and Spencer 1983], was rather complicated. Some simplification and refinement was achieved in [Martinelli and Scoppola 1987]. But a much simpler and more transparent proof, which uses the approach of [von Dreifus 1987] and [Spencer 1988] was given by Klein and von Dreifus [1989]. We formulate and partly prove the relevant results:

(15.21) Theorem [Fröhlich et al. 1985; Martinelli and Scoppola 1987; Klein and von Dreifus 1989]. *Let $\lambda_0 \in \mathbf{R}$ and $l > 0$ be given, and assume that*

$$P\{\Lambda_l(0) \text{ is } (\delta_0, \lambda_0)\text{-regular}\} \geqslant 1 - l^{-p} \tag{15.82}$$

for some $p > d$ and $\delta_0 > 0$, and that

(15.83) $$P\{\text{dist}(\lambda, \sigma(h_{\Lambda_L(0)})) \leqslant e^{-L^\beta}\} \leqslant L^{-q}$$

for some $\beta \in (0,1)$ *and* $q > 4p+6d$, *all* $\lambda \in [\lambda_0 - \eta, \lambda_0 + \eta]$, *where* $\eta > 0$, *and all* $L \geqslant l$. *Then, given* $\delta \in (0, \delta_0)$, *there exists* $l_0 = l_0(p, d, \beta, q, \delta_0, \delta)$ *such that, if* $l > l_0$, *we can find* $\tau = \tau(l, \delta_0, \delta, \beta, \eta)$, *for which, with probability* 1, *the spectrum of* h *in* $(\lambda_0 - \tau, \lambda_0 + \tau)$ *is pure point and the corresponding eigenfunctions decay exponentially in the sense of* (15.86).

Proof. Using the resolvent identity for the pair $(h_{\Lambda_l(0)} - \lambda_0)^{-1}$ and $(h_{\Lambda_l(0)} - \lambda)^{-1}$ with $|\lambda - \lambda_0| \leqslant \frac{1}{2}e^{-l^\beta}$ and $\text{dist}(\lambda_0, \sigma(h_{\Lambda_l(0)})) \geqslant e^{-l^\beta}$, we find from (15.82) and (15.83) that, for a given $\delta_0' \in (0, \delta_0)$ and $p' \in (d, p)$, we have

$$P\{\Lambda_l(0) \text{ is } (\delta_0', \lambda)\text{-regular for all } \lambda \in (\lambda_0 - \tau, \lambda_0 + \tau)\} \geqslant 1 - l^{-p'}$$

if l is large enough. The theorem now follows from the next two results:

(15.22) Theorem [Klein and von Dreifus 1989]. *Let* $\Delta \subset \mathbf{R}$ *be an interval, and suppose that, for some* L_0, *we have*

(15.84)
$$P\{\text{either } \Lambda_{L_0}(x) \text{ or } \Lambda_{L_0}(y) \text{ is } (\delta_0, \lambda)\text{-regular for all } \lambda \in \Delta\} \geqslant 1 - L_0^{-2p}$$

for some $p > d$ *and* $\delta_0 > 0$, *and any* $x, y \in \mathbf{Z}^d$ *with* $|x - y| \geqslant L_0$. *Suppose also that*

(15.85) $$P\{\text{dist}(\lambda, h_{\Lambda_L(0)}) < e^{-L^\beta}\} \leqslant L^{-q}$$

for some $\beta \in (0,1)$ *and* $q > 4p + 6d$, *all* λ *with* $\text{dist}(\lambda, \Delta) \leqslant \frac{1}{2}e^{-L^{-\beta}}$, *and all* $L \geqslant L_0$. *Then there exists* $\alpha = \alpha(p, d) \in (1, 2)$ *such that, if* $L_{n+1} = L_n^\alpha$ *for* $n = 0, 1, 2, \ldots$ *and* $\delta \in (0, \delta_0)$, *we can find* $Q = Q(p, d, \beta, q, \delta_0, \delta, \alpha)$ *such that, for any* $L_0 > Q$ *and any* $n = 0, 1, \ldots$, *we have*

$$P\{\text{either } \Lambda_{L_n}(x) \text{ or } \Lambda_{L_n}(y) \text{ is } (\delta, \lambda)\text{-regular for all } \lambda \in \Delta\} \geqslant 1 - L_n^{-2p}$$

for all $x, y \in \mathbf{Z}^d$ *with* $|x - y| \geqslant L_n$.

This theorem is an analogue of Theorem 15.14. Its proof is based on a combination of the ideas from [Fröhlich et al. 1985], [von Dreifus 1987] and [Spencer 1988], and consists in an extension of the proof of Theorem 15.14.

(15.23) Theorem [Klein and von Dreifus]. *Let* $\Delta \subset \mathbf{R}$ *be an interval*, $p > d$, $L_0 > 0$, $\alpha \in (1, 2p/d)$, $\delta > 0$, *and* $L_{n+1} = L_n^\alpha$. *Suppose that, for* $n = 0, 1, \ldots$,

$$P\{\text{either } \Lambda_{L_n}(x) \text{ or } \Lambda_{L_n}(y) \text{ is } (\delta, \lambda)\text{-regular for all } \lambda \in \Delta\} \geqslant 1 - L_n^{-2p}$$

for any $x, y \in \mathbf{Z}^d$ *with* $|x - y| > L_n$. *Then, with probability* 1, *the spectrum of* h *in* Δ *is pure point, and the eigenfunctions decay exponentially in the sense of* (15.86).

Proof. By Theorem 15.18, if ρ is the spectral measure of h, then ρ-almost every $\lambda \in \sigma(h)$ is a generalized eigenvalue of (15.77). Thus, it suffices to show that each generalized eigenfunction with an eigenvalue in Δ decays exponentially with rate δ. Let b be a natural number and, for $x_0 \in \mathbf{Z}^d$, set $A_{n+1}(x_0) = \Lambda_{2bL_{n+1}} \setminus \Lambda_{2L_n}(x_0)$. Define the event

$$\mathcal{E}_n(x_0) = \begin{Bmatrix} \Lambda_{L_n}(x_0) \text{ and } \Lambda_{L_n}(x) \text{ are } (\delta, \lambda)\text{-singular} \\ \text{for some } \lambda \in \Delta \text{ and } x \in A_{n+1}(x_0) \end{Bmatrix}.$$

By assumption, $P\{\mathcal{E}_n(x_0)\} \leqslant (2b+1)^d L_n^{\alpha d - 2p}$, and, since $\alpha < 2p/d$, the sum $\sum_{n=0}^{\infty} P\{\mathcal{E}_n(x_0)\}$ is finite. Therefore, if

$$\Omega_0 = \{\mathcal{E}_n(x_0) \text{ occurs only finitely many times for each } x_0 \in \mathbf{Z}^d\},$$

we have $P\{\Omega_0\} = 1$.

Now let the independent, identically distributed random potential $q(x)$ belong to Ω_0, let $\lambda \in \Delta$ be a generalized eigenvalue of h, and $\psi_\lambda(x)$ the corresponding generalized eigenfunction, with $\psi_\lambda(x_0) \neq 0$. If $\lambda \notin \sigma(h_{L_n(x_0)})$, we can express $\psi_\lambda(x_0)$ in terms of the values of $\psi_\lambda(x)$ on $\partial\Lambda_{L_n}(x_0)$ by the formula

$$\psi_\lambda(x_0) = \sum_{u \in \partial\Lambda_{L_n}(x_0)} G_{L_n(x_0)}(x_0, u, \lambda)\psi_\lambda(u'). \tag{15.86}$$

If $\Lambda_{L_n}(x_0)$ is (δ, λ)-regular, we get from (15.60) and (15.84) the inequality

$$|\psi_\lambda(x_0)| \leqslant c_d L_n^{d-1} e^{-\delta L_n/2} c(\lambda, t)\big(L_n + |x_0|\big)^t.$$

Since the sequence $\{L_n\}$ is unbounded from above, this inequality is incompatible with our hypothesis that $\psi_\lambda(x_0) \neq 0$. Thus, $\Lambda_{L_n}(x_0)$ is (δ, λ)-singular for all $n \geqslant n_1(q, \lambda, x_0)$. On the other hand, since $q \in \Omega_0$, the event $\mathcal{E}_n(x_0)$ does not occur for any $n \geqslant n_2$. Taking $n = \max\{n_1, n_2\}$, we see that $\Lambda_{L_n}(x)$ is (δ, λ)-regular for all $x \in A_{n+1}(x_0)$. Now, for a given $\beta \in (0,1)$, choose $b \geqslant 1 + \beta + \beta^{-1}$, and define

$$\tilde{A}_{n+1}(x_0) = \Lambda_{[2b/(1+\beta)]L_{n+1}}(x_0) \setminus \Lambda_{[2/(1-\beta)]L_n}(x_0).$$

Then $\tilde{A}_{n+1}(x_0) \subset A_{n+1}(x_0)$, and, if $x \in \tilde{A}_{n+1}(x_0)$, we have

$$\operatorname{dist}\big(x, \partial A_{n+1}(x_0)\big) > \beta|x - x_0|.$$

Moreover, if $|x - x_0| > L_0/(1-\beta)$, we have $x \in \tilde{A}_{n_3+1}(x_0)$ for some n_3. For $n > n_3$, $\Lambda_{L_n}(y)$ is (δ, λ)-regular for all $y \in A_{n+1}(x_0)$. Using (15.85) with y instead of x_0, we find that

$$|\psi_\lambda(y)| \leqslant c_d L_n^{d-1} e^{-\delta L_n/2} \psi_\lambda(u')$$

for some $u' \in \partial \Lambda_{L_n}(y)$.

If $x \in \tilde{A}_{n+1}(x_0)$, we can repeat this procedure at least $\beta |x - x_0|/(1 + L_n/2)$ times and use the polynomial bound (15.84) to prove that $\psi_\lambda(x)$ decays exponentially with rate δ. This proves Theorem 15.23, and with it Theorem 15.21. □

Theorem 15.21 is very general. Its first hypothesis (15.82) expresses the fact that a point inside the cube feels the cube's boundary only weakly. In one dimension, this hypothesis follows from the positiveness of the Lyapunov exponent (Problem V.31), a fairly typical property of random operators (Section 14). The absence of a similar property in the multidimensional case is one of the reasons stronger conditions are needed here, such as a large potential, or proximity to a fluctuation boundary.

The second condition (15.83) of Theorem 15.21 means that long-range tunneling is highly unlikely. In the proof of Theorem 15.10, this condition follows from the Wegner estimate (15.55), which is valid for an independent, identically distributed random potential with a bounded probability density (see also Theorem 4.21 and Problem II.18). But, in fact, (15.83) is true in one dimension even if nothing is assumed about the metrically transitive potential's smoothness, so long as the potential is independent at distinct points and has a finite moment of a certain order (see [de Page 1984], and also 6.E and 11.C). Thus, Anderson localization can be proved in one dimension for all independent potentials whose distribution is not concentrated at a point and which have a finite moment. This class essentially coincides with the class of random potentials, for which, by Theorem 14.2, the Lyapunov exponent is positive. This result was proved by Carmona et al. [1987].

For the multidimensional Anderson model, the same authors proved that localization occurs for independent potentials whose distribution can be represented as a convex combination $F = pF_1 + (1-p)F_2$, where p is sufficiently close to 1, F_2 is an arbitrary probability measure (which can, in particular, be singular) and F_1 is such that, for some $\alpha > 0$, the quantity

$$g_\alpha = \Big(\inf_{\tau > 0} \sup_{|b-a| \leqslant \tau} |b-a|^{-\alpha} F_1([a,b])\Big)^{-1},$$

which is analogous to (15.50), is sufficiently large.

These facts demonstrate the advantages of the third proof of Theorem 15.20 [Fröhlich et al. 1985], which for potentials with absolutely continuous distribution seems more cumbersome than the other two. In particular, it goes beyond showing that the spectrum is pure point and that the eigenfunctions decay exponentially: it yields rich information on the structure of typical realizations of the potential and on the behavior of the eigenfunctions for the whole array of scales (15.62). It also answers other questions: the asymptotic properties of e^{ith} as $t \to \infty$, the behavior of some perturbations of the Anderson model, and so on. This approach, including interesting applications, generalizations and extensions, can be found in the survey by Martinelli and Scoppola [1987] and the references therein.

Here we mention only an analogue of Theorem 15.10 for the Schrödinger operator on $L_2(\mathbf{R}^d)$, for $d \geqslant 1$, studied in [Martinelli and Holden 1984]. The potential is (10.6), with independent, identically distributed random variables ξ_a, for $a \in \mathbf{Z}^d$, taking values in $[0,1]$, having a bounded density and such that $\alpha = \mathsf{P}\{\xi_0 \leqslant \frac{1}{2}\} < 1$. Under such conditions, an estimate similar to the one in Theorem 15.10 holds, if $0 = \nu_0 \leqslant \lambda \leqslant \lambda_0$, where λ_0 is a small enough number depending on α and on $\sup f(\xi)$. An analogue of Theorem 4.21 for this potential was proved by Kotani and Simon [1986]. Thus there is exponential localization for such an operator, in the interval $[\nu_0, \lambda_0]$.

It also turns out that with the proper choice of α and $\sup f(\xi)$, we can arrange for the set $\{x \in \mathbf{R}^d : q(x) \leqslant \lambda_0\}$ to contain, with probability 1, an infinite set of unit cubes $C(a)$, for $a \in \mathbf{Z}^d$, which are nearest neighbors (an infinite cluster, in the terminology of percolation theory: see [Kesten 1986]). In such a potential, a classical particle can, as $t \to \infty$, go off to infinity, while its quantum motion, because the spectrum is pure point for $\lambda < \lambda_0$, should be restricted to a finite region of space. The reason is roughly that here the role of points of the lattice is played, not by the unit cubes $C(a)$, for $a \in \mathbf{Z}^d$, that appear in the potential (10.6), but by much larger cubes $C_\lambda(a)$, for $a \in L(\lambda)\mathbf{Z}^d$, whose side length $L(\lambda)$ is of the order of the characteristic length $\lambda^{-1/2}$ of the fluctuation spectrum (see (9.5)).

Problems

1. Let $\tilde{G}$ be a noncompact subgroup of $\mathrm{SL}(m, \mathbf{R})$. Prove that either of the first two statements below implies the third:

(1) There exists no $\tilde{G}$-invariant measure on $\mathbf{P}^{m-1}$.
(2) $\tilde{G}$ has no invariant subgroups H of finite index.
(3) There exists no finite set $S \subset \mathbf{P}^{m-1}$ such that $\tilde{G}S = S$.

Hint. If $\tilde{G}S = S$, then $\tilde{G}$ acts on S as a subgroup $\tilde{H}$ of the group of permutations of elements of S. Now consider the kernel of the group homomorphism $\tilde{G} \to \tilde{H}$.

2. If F is the P-invariant distribution (14.12), the support of F is the smallest closed set in $\mathbf{P}^{m-1}$ that is invariant with respect to all $A \in G_p$.

3. Prove that, under the conditions of Proposition 14.1, any P-invariant measure F on $\mathbf{P}^m$ (that is, $P * F = F$) has no atoms.

Hint. Assuming F has atoms, consider the set $S \subset \mathbf{P}^{m-1}$ of all points at which there is an atom of maximal weight.

4. Prove that, if the conditions of Proposition 14.1 are satisfied and $m = 2$, equation (14.10) holds with probability 1.

Hint. By Proposition 11.2, there exists $e_0(\omega) \in \mathbf{P}^1$ such that, with probability 1,

$$\mathsf{P}\{\lim_{x\to\infty} x^{-1} \ln \|\mathcal{N}(x)e_0\| = \gamma\} = 1 - \mathsf{P}\{e_0(\omega) = e_0\} = 1 - p$$

for any $e_0 \in \mathbf{P}^1$ and $e_0 \neq e_0(\omega)$. Let $F_n(de) = \mathsf{P}\{\mathcal{N}(n)e_0 \in de\}$ and $\mu(de) = \mathsf{P}\{e_0(\omega) \in de\}$. Then, for all $n \in \mathbf{Z}_+$,

$$p = \int_{\mathbf{P}^1} \mu(de) F_n(\{e\}) = \int_{\mathbf{P}^1} \mu(de) \tilde{F}_n(\{e\}),$$

where $\tilde{F}_n(de) = n^{-1} \sum_{k=1}^{n} F_k(de)$. Passing to the limit (possibly after taking a subsequence), we obtain a similar representation for p, in which F is a P-invariant measure that is always continuous, by Problem VI.3.

5. Prove (14.13) for $m = 2$.

Hint. Use the results and notation of Problem VI.4 and show that, for any sequence $\{n\}$ tending to infinity and such that $\tilde{F}_n \to F$,

$$\begin{aligned}\gamma &= \lim_{n\to\infty} \frac{1}{n} \sum_{k=1}^{n} \mathsf{E}\left\{\ln \frac{\|\mathcal{N}(k+1)e_0\|}{\|\mathcal{N}(k)e_0\|}\right\} \\ &= \lim_{n\to\infty} \int \ln \frac{\|AY\|}{\|Y\|} P(dA)\tilde{F}_n(d\bar{Y}) = \int \ln \frac{\|AY\|}{\|Y\|} P(dA)(d\bar{Y}).\end{aligned}$$

6. Let the conditions of Proposition 14.1 be fulfilled and let F be a P-invariant measure on $\mathbf{P}^{m-1}$, absolutely continuous with respect to the unique probability measure invariant under $SO(m, \mathbf{R})$. Then

$$\gamma = -\frac{1}{m} \int_{G\times\mathbf{P}^{m-1}} \ln \frac{dA^{-1}F}{dF}(e) P(dA) F(de).$$

Hint. Use (14.11) and (14.12) and the equality $dA^{-1}L/dL(e) = \|Ae\|^{-m}$.

7. Show that, under the conditions of Problem VI.6, γ is strictly positive.

Hint. Use Jensen's inequality and the hypothesis that there is no G_P-invariant measure.

8. Show that equation (6.69), for $q(x) \equiv 0$, $\lambda = 0$ and $s(x)$ consisting of independent, identically distributed random variables such that $\infty > s_1 \geqslant s(x) \geqslant s_0 > 0$, has solutions that are unbounded as $|x| \to \infty$.

Hint. Use the arguments of Theorem 14.2 and the law of iterated logarithms.

9. Show that the Lyapunov exponent of the Schrödinger operators with potential (6.25) and Gaussian white noise potential (Section 7.1) can be represented as

$$\gamma = c \int_0^\infty (1 - \cos k_0 s) \frac{|g(s)|^2}{s} ds,$$

$$\gamma = D \int_0^\infty s\,|g(s)|^2\,ds,$$

respectively, where $g(s)$ is the Fourier transform of the invariant distribution of $\cot \phi(x)$.

10. For the Gaussian white noise potential (Section 7.1), show that the Lyapunov exponent is

$$\gamma(\lambda) = -\left(\frac{D}{16}\right)^{\frac{1}{3}} \frac{F'(\varepsilon)}{F(\varepsilon)},$$

where $\varepsilon = \lambda(D/2)^{-2/3}$ and

$$F(z) = \int_0^\infty \frac{dt}{\sqrt{t}} e^{-t^2/12 - tz}$$

is an entire function having no zeros on the real axis.

Hint. Use (7.6) and the relation $\gamma = \int_{-\infty}^{\infty} z p(z)\,dz$.

11. Prove that, as in the preceding problem, the Lyapunov exponent for the Poisson potential (6.25) with $k_0 > 0$ is the logarithmic derivative of an entire function of the parameter $\varepsilon = c/\sqrt{\lambda}$ that is strictly positive when $\varepsilon > 0$. Therefore $\gamma(\lambda)$ is real analytic when $\lambda > 0$.

12. Show that the Lyapunov exponent for the potential (1.30) with $u_i(x) = k_i \delta(x)$, where the k_i are independent positive random variables and have common density $\kappa e^{-k\kappa}$, for $\kappa > 0$, is given, for $\lambda = 0$, by the logarithmic derivative of the function

$$F(\kappa) = -\int_0^\infty \frac{e^{-(\kappa z + c/z)}}{z} dz.$$

From this obtain that $\gamma(0) = -c \ln c\kappa \big(1 + o(1)\big)$ as $c \downarrow 0$.

13. Show that, for the random potential of Problem 1.10, formula (14.34) becomes

$$\gamma = \left(\frac{1}{n_0} + \frac{1}{n_1}\right) \sum_{r=0,1} \int_0^\infty \frac{ds}{s} \int_0^\infty dl\, f_r(l)$$
$$\times \left| \frac{g(r,l,s)}{F_r(l)} - \int_0^\infty dl' \frac{f_r(l')}{F_r(l')} g(r,l',s) \right|^2 .$$

Obtain (14.36) from this formula, by taking $f_r(l) = n_r e^{-n_r l}$.

14. Consider the metrically transitive potential

$$q(x) = \sum_{n \in \mathbf{Z}} \xi_n u(x-n)$$

for $x \in \mathbf{R}$, there the ξ_n are independent, identically distributed random variables and the support of u is contained in $(0,1)$. Write the fundamental matrix $\mathcal{T}(x)$, for x an integer, as a product of independent, identically distributed random matrices.

15. Consider the potential of the preceding problem, with the random variables ξ_n being equal to either 0 or ξ and $u(x) = \chi_{(0,1)}$. Show that $\gamma(\lambda) > 0$ if ξ is not of the form $\xi = \pi(l_2^2 - l_1^2)$, where $l_2 > l_1$ are natural numbers. If, on the contrary, such numbers exist, $\gamma(\lambda) = 0$ if and only if $\lambda = \pi l_2$.

Hint. Use the preceding problem.

16. Consider the equation of harmonic oscillation of a one-dimensional chain of atoms:

$$-s(x+1)\big(u(x+1) - u(x)\big) - s(x)\big(u(x) - u(x-1)\big) = m(x)\nu^2 u(x).$$

Show that, if the pairs $\big(s(x), m(x)\big)$ are independent for different values of x and their distribution function is not concentrated at one point, the Lyapunov exponent of this equation is zero only at the point $\nu = 0$.

Hint. Introduce the vector-valued function $Y(x) = \big(s(x)(u(x) - u(x-1)),\, u(x-1)\big) \in \mathbf{R}^2$, and show that the fundamental matrix in terms of $Y(x)$ is (11.28), where

$$A(x) = \begin{pmatrix} -1 - m(x)\nu^2 s^{-1}(x) & -m(x)\nu^2 \\ s^{-1}(x) & 1 \end{pmatrix}.$$

Then argue as in the proof of Theorem 14.2.

17. Show that, for the equation

$$-y(x+1) - y(x-1) + q(x)y(x) = m(x)\lambda y(x)$$

with pairs $(q(x), m(x))$ that are independent for different values of x, the Lyapunov exponent is 0 only if the joint distribution $F(dq, dm)$ is concentrated on some straight line $q - \lambda_0 m = c$, for $|c| < 2$, and $\gamma(\lambda) = 0$ if and only if $\lambda = \lambda_0$.

Hint. Use the scheme of the proof of Theorem 14.2.

18. Show the following asymptotic formulas for the Lyapunov exponent of the one-dimensional wave equation

$$-\frac{d}{dx}\left(s(x)\frac{d}{dx}y(x)\right) = \lambda m(x)y(x),$$

when $s(x)$ and $m(x)$ are Markov-type process (Section 6.4):

$$\gamma(\lambda) = \frac{\lambda\mu}{4\kappa}\int_0^\infty dx\, \mathsf{E}\{(\delta_s - \delta_m)(0)(\delta_s - \delta_m)(x)\} + O(\lambda^{3/2})$$

as $\lambda \downarrow 0$, where $\kappa = \mathsf{E}^{-1}\{s^{-1}(x)\}$, $\delta_s(x) = \kappa s^{-1}(x) - 1$, $\mu = \mathsf{E}\{m(x)\}$ and $\delta_m(x) = \mu^{-1}m(x) - 1$.

Show that, if δ_m and δ_s are proportional to a parameter g, the asymptotic behavior of $\gamma(\lambda)$ as $g \downarrow 0$ for $\lambda > 0$ is given by

$$\gamma(\lambda) = \frac{\lambda\mu}{4\kappa}\int_0^\infty dx\, \mathsf{E}\{(\delta_s - \delta_m)(0)(\delta_s - \delta_m)(x)\}\cos\left(2\sqrt{\frac{\mu}{\kappa}}x\right) + O(g^3).$$

Hint. Generalize the arguments used to prove Theorem 14.5.

19. Show that the asymptotic formulas of the preceding problem remain valid even when $s(x)$ and $m(x)$ are functions of the random telegraph signal $r(x)$ of Example 1.15(b). Using (1.27), compute the corresponding integral and check that, for example, for $n_1 = n_2 = n$,

$$\gamma(\lambda) = \frac{\lambda}{16(\lambda n^{-2} + 1)}\left(\frac{s_1 - s_2}{s_1 + s_2} + \frac{m_1 - m_2}{m_1 + m_2}\right)^2 + O(g^3),$$

$$\gamma(\lambda) = \frac{\lambda}{16}\left(\frac{s_1 - s_2}{s_1 + s_2} + \frac{m_1 - m_2}{m_1 + m_2}\right)^2 + O(\lambda^{3/2}),$$

as $g \downarrow 0$ and $\lambda \downarrow 0$, respectively, where $s_1 > s_2$ and $m_1 > m_2$ are the values assumed by $s(x)$ and $m(x)$.

20. For the discrete "wave equation" of Problem 16, with independent, identically distributed pairs $(s(x), m(x))$, show asymptotic formulas similar to those of Problem VI.18 (low-frequency and quasihomogeneous).

Hint. Generalize the arguments used to prove Theorem 14.6.

21. Find the asymptotic behavior of the Lyapunov exponent with potential (14.19) near the points $\lambda_n = (\pi n)^2$.

Hint. Use (14.22) and Theorem 14.6.

22. **(a)** For a Schrödinger equation with potential (11.117), where $Q(x)$ is a bounded metrically transitive process and $\mathsf{E}\{|Q'(x)|\} < \infty$, show the following refinement of (11.121):

$$\gamma_R(\lambda) \leqslant \frac{1}{2R}\mathsf{E}\{|Q'(x)|(\lambda - Q(x))^{-1}\}.$$

(b) Derive formula (14.54) and thus find that the order of vanishing of the Lyapunov exponent is twice as high for operators with random coefficients as for general metrically transitive operators—for example, see formulas (14.38b) and (11.97b) for the Schrödinger equation as $\lambda \to \infty$, (14.38a) and (11.97b) for the same equation as $\sup_x Q(x) \to 0$, and (14.39) and (11.98) for the equation $-(s(x)y')' = \lambda y$ as $\lambda \downarrow 0$.

Hint. **(a)** Introduce phase variables through the formulas $\cot\phi = y'/\kappa y$ and $r^2 = y^2 + \kappa^{-2}y'^2$, where $\kappa^2 = \lambda - Q(x)$.

(b) Reduce the problem to equation (14.41), with $Q(x)$ replaced by $R^2Q(x)$ and k by kR, so that the small term is that with the operator $\frac{1}{2}\Delta$ (that is, multiplied by $(kR)^{-1}$). Conclude that

$$\pi_n\big|_{R=\infty} = (-1)^n \pi_0 e^{-2\kappa|n|},$$

where $\cosh^2\kappa = k^2/Q(X)$. Then carry out one iteration in (14.45), compute the integrals and estimate the error by using the smoothness of $Q(X)$.

23. Show that if a metrically transitive potential can be represented as a function of a multicomponent Markov ergodic process $X(x)$ whose infinitesimal operator $\mathfrak{A}$ is dissipative in the Hilbert space considered in the proof of Theorem 14.4, the eigenvalues μ_α of the infinitesimal operator of the Markov family $(X(x), \phi(x))$ have negative real parts.

Hint. Arguing as in Theorem 14.4, derive a representation for $-\operatorname{Re}\mu_\alpha$ similar to (14.34).

24. Prove that for the Schrödinger operator with potential described in Problem II.42, where the ξ_n are independent, identically distributed random variables having a distribution with an absolutely continuous component, the spectrum is pure point and the eigenfunctions decay exponentially.

Hint. Use Problem II.42, the scheme of the proof of Theorem 13.5 and 13.7, Example 13.8(a) and Theorem 14.3.

25. Set

$$\zeta_a^{(\alpha)} = \frac{1}{y_a^{(\alpha)}(x,\lambda)} \frac{\partial y_a^{(\alpha)}}{\partial x}(x,\lambda),$$

$$Z_{a,b}^{(\alpha,\beta)} = \zeta_a^{(\alpha)}(0,\lambda) - \zeta_b^{(\beta)}(0,\lambda),$$

$$\eta_x(t) = \frac{y_b^{(\beta)}(t,\lambda)}{y_a^{(\alpha)}(x,\lambda)},$$

where $y_a^{(\alpha)}(x,\lambda)$ is the solution of the Cauchy problem (6.1) and (6.3). If we denote by $e_{a,b}^{(\alpha,\beta)}(x,x',\lambda)$ the generalized density of the kernel $E_{a,b}^{(\alpha,\beta)}(x,x',\lambda)$ of the resolution of the identity of the operator $H_{a,b}^{(\alpha,\beta)}$, show that

$$e_{a,b}^{(\alpha,\beta)}(0,x,\lambda) = \delta(Z_{a,b}^{(\alpha,\beta)})\eta_x(0)$$

for $x > 0$.

26. Using the preceding problem, show that the mathematical expectation of the density $e(0,x,\lambda)$ of the kernel of the resolution of the identity of a metrically transitive Schrödinger operator with potential (15.18) can be written as

$$V^2(K)\int_{K^2\times\mathbf{R}^2} \tilde{\pi}(X_1,-z_1)\tilde{\pi}(X_2,z_2)p_1(x;X_2,z_2;X_1,z_1)dX_1\,dX_2\,dz_1\,dz_2,$$

where $\tilde{\pi}(X,z)$ is as in Problem III.3 and

$$p_1(x;X_2,z_2;X_1,z_1) = \int_{\mathbf{R}} p(x;X_2,z_2,1;X_1,z_1,\eta)\eta\,d\eta,$$

p being the transition probability of the Markov process $(X(x),\zeta(x),\eta(x))$ specified by the equations $\zeta' = \zeta^2 + \lambda + q$ and $\eta' = \zeta\eta$. The function p_1 can be found from the equations

$$\frac{\partial p_1}{\partial x} = -\frac{\partial}{\partial z}((z^2+\lambda-Q(X))p_1) + zp_1 + \mathfrak{A}^* p_1$$

and $p_1\big|_{x=0} = \delta(X_1,X_2)\delta(z_1-z_2)$. Show that (3.4), together with the expression above for the expectation of $e(0,x,\lambda)$, with $x=0$, leads to the result of Problem II.23(a).

27. Define

$$A(x,\Delta) = \lim_{\varepsilon\downarrow 0} \frac{\varepsilon}{\pi} \int_{\Delta} \mathsf{E}\{|G(0,x,\lambda+i\varepsilon)|^2\}\, d\lambda,$$

where $G(x,x',z)$ is the Green's function of the Schrödinger operator. By Problem I.14 (see also Section 18), from the positiveness of $A(0,\Delta)$, it follows that the spectrum of Δ has, with probability 1, a point component, and if $\int_{\mathbf{R}} A(x,\Delta)\, dx = N(\Delta)$ then, with probability 1, the spectrum of Δ is pure point. Show that $A(x,\Delta) = \int_{\Delta} \tilde{a}(x,\lambda)\, d\lambda$ and that $\tilde{a}(x,\lambda)$ has a representation similar to the one obtained in the previous problem for $\mathsf{E}\{e(0,x,\lambda)\}$, but including also the quantity $\xi_a^{(\alpha)}(x,\lambda) = \partial\zeta_a^{(\alpha)}(x,\lambda)/d\lambda$ (cf. (15.12)) which, together with $X(a)$, $\zeta_a^{(\alpha)}(x)$ and $\eta_a(x)$, forms a four-component Markov process.

28. The conductivity of an ideal Fermi gas to alternating current of frequency ν, at temperature 0 and Fermi energy λ, is [Lifshitz et al. 1982b]

$$\sigma(\lambda,\nu) = \int dx\, \mathsf{E}\{(\partial_1 - \partial_2)(\partial_3 - \partial_4) e(x_1,x_2,\lambda) e(x_3,x_4,\lambda+\nu)\}\Big|_{\substack{x_1=x_4=0\\ x_2=x_3=x}},$$

where $\partial_j = \partial/\partial x_j$ and $e(x,x',\lambda)$ is as in the previous problem. Show that the right-hand side of this formula, for the potential (15.18), is given by

$$2V^2(K) \int_{\mathbf{R}^2} (z_2 - z_1)\pi_0(X,-z_1,-z_2)\pi_1(X,z_1,z_2)\, dX\, dz_1\, dz_2,$$

where

$$\pi_\delta(X,z_1,z_2) = \int_{\mathbf{R}} w^\delta \tilde{\pi}(X,z_1,z_2,w)\, dw$$

for $\delta = 0,1$, and $\tilde{\pi}(X,z_1,z_2,w)$ is the stationary density of the Markov process

$$(X(x), \zeta_a^{(\alpha)}(x,\lambda_1), \zeta_a^{(\alpha)}(x,\lambda_2), w_a^{(\alpha)}(x,\lambda_1,\lambda_2))$$

for $\lambda_1 = \lambda$ and $\lambda_2 = \lambda + \nu$, where the last component is defined as

$$w_a^{(\alpha)}(x,\lambda_1,\lambda_2) = \frac{1}{y_a^{(\alpha)}(x,\lambda_1) y_a^{(\alpha)}(x,\lambda_2)} \int_a^x W(y_a^{(\alpha)}(t,\lambda_1), y_a^{(\alpha)}(t,\lambda_2))\, dt,$$

$W(y_1,y_2)$ being the Wronskian (12.3).

For more information on other physical quantities that can be represented in a similar way and computed approximately in some cases, see [Lifshitz et al. 1982b; Gor'kov et al. 1983] and the references therein.

29. Show that the quantity $\mu(\delta)$ from (15.36) is a convex function of δ, and vanishes for $\delta = 0, 1$.

Hint. For convexity, use Cauchy–Schwarz. The equality $\mu(0) = 0$ follows from the definition of $Z_\delta(x, u)$. Next, the operator on the right in (15.35) coincides with the adjoint of (6.53). By considering, along with that equation, the one conjugate to it, and denoting its fundamental solution by $q(x; u_1; u_2)$ and using Green's formula and the condition $Z_\delta(0,0) = 1$ (cf. Problem III.23(a)), show that

$$Z_1(x, u) = \int_{K \times S^1} q(x; u', u)\, du'.$$

Then use the fact that $q(x; u_1, u_2)$, after the replacement of $u = (X, \phi)$ by $\bar{u}(X, -\phi)$, coincides with the transition density of the Markov process $u(x)$, and therefore, by Theorem 6.12, $\lim_{x\to\infty} Z_1(x, u) = \pi(\bar{u})V(K)$, that is, $\mu(1) = 0$.

30. Show, proceeding from (15.16c), that for the potential (15.18),

$$\mathsf{E}\{\delta_{a,b}^{(\alpha,\beta)}(\lambda) \mid \mathcal{F}_0^b\} = V(K) \int_K dX\, p(a; \pi, \alpha; X, \phi_b^{(\beta)}(0, \lambda)).$$

Using this formula and Theorem 6.12, check that the analogue of Theorem 13.7 for the Schrödinger operator is true. This yields another proof of Theorem 15.4.

31. Obtain the formula

$$\int_0^\pi \delta_{a,b}^{(\alpha,\beta)}(\lambda)\, d\beta = (r_a^{(\alpha)}(b))^{-2},$$

where $\delta_{a,b}^{(\alpha,\beta)}$ is the generalized density of the measure (12.8).

Hint. Use Problems V.17–18 and Corollary 15.2.

32. Obtain the following formula, relating the spectral measures ρ and $\rho^{(\alpha)}$ of the operators H and $H_{-\infty,\infty}^{(\alpha)}$:

$$\rho(d\lambda) = \lim_{b\to\infty} \int_0^\pi \frac{\rho^{(\alpha)}(d\lambda)}{(r_b^{(\alpha)}(0))^2}\, d\alpha.$$

Hint. Use Corollary 15.2 and the result of the preceding problem.

33. Use the above problem to obtain another proof for Theorem 15.8.

Hint. Use the fact that $\int_0^\pi (r_b^{(\beta)}(x))^{-2}\, d\beta = \pi$.

34. Obtain an analogue of Problem VI.30 for the one-dimensional divergence operator (6.19) and (6.20), where the coefficient is a Markov-type process (15.18). Based on this, prove that the spectrum of this operator is pure point and that the eigenfunctions decay with rate equal to the Lyapunov exponent.

35. Find analogues for the formulas in Problems VI.25–28 and VI.30–31 for the discrete Schrödinger operator h.

36. Let A be the operator (11.94) and $\mathrm{dist}(\lambda, \sigma(A)) = A > 0$. Show that

$$|(A-z)^{-1}(x,y)| \leqslant \frac{2}{a} e^{-\delta|x-y|},$$

where $\delta > 0$ is specified by the equation $4\delta d e^{\delta} = a$.

Hint. Consider the operator $A(p) = U(p)AU^{-1}(p)$, where $(U(p)\psi)(x) = e^{px}\psi(x)$, and prove that $A(p) = A + B(p)$, with $\|B(p)\| \leqslant 2d|p|e^{|p|}$. Then use the relations

$$(A-z)^{-1}(x,y) = \big(A(p)-z\big)^{-1}(x,y)e^{-p(x-y)},$$
$$|(A(p)-z)^{-1}(x,y)| \leqslant \big(a - \|B(p)\|\big)^{-1},$$

and choose $p \in \mathbf{C}^d$ appropriately. For a detailed discussion of this technique, see [Reed and Simon IV; Spencer 1984; Martinelli and Scoppola 1986].

37. Consider the Schrödinger operator $H = -\Delta + q_1(x-a_1) + q_2(x-a_2)$ on $L_2(\mathbf{R}^d)$, where $q_j(x)$, for $j = 1, 2$, are nonpositive functions with compact support (potential wells). Assume that the operators $H_j = -\Delta + q_j(x - a_j)$, for $j = 1, 2$, have one negative eigenvalue each. Then the respective eigenfunctions $\psi_j^{(0)}(x-a_j)$ decrease exponentially with rate $\delta_j^{(0)} = (\lambda_j^{(0)})^{1/2}$ as $|x-a_j| \to \infty$ [Glazman 1965]. Prove the following facts:

(a) If $a = |a_1 - a_2|$ is large enough, H has exactly two eigenvalues λ_j, for $j = 1, 2$, and that $|\lambda_1^{(0)} - \lambda_2| > \frac{1}{2}\alpha$, where $\alpha = e^{-\kappa\sqrt{a}}$ for some $\kappa > 0$.

(b) If $|\lambda_1^{(0)} - \lambda_2^{(0)}| > \alpha$, each perturbed eigenfunction $\psi_j(x)$, for $j = 1, 2$, is exponentially localized near a_j, that is, near its potential well.

(c) If $d = 1$ and $q_j(x) = k\delta(x)$, for $k > 0$, the spectrum of H, for a arbitrarily large, has two negative eigenvalues λ_1 and λ_2, with $\lambda_1 - \lambda_2 = O(e^{-\delta a})$, where $\delta = O(k)$. The eigenfunctions, up to exponentially small terms, are given by

$$\psi_{1,2}(x) = \frac{1}{\sqrt{2}}\big(\psi_1^{(0)}(x-a_1) \pm \psi_2^{(0)}(x-a_2)\big).$$

Remark. Case (b) corresponds to nonresonant wells, while (c) corresponds to the resonance tunneling of quantum mechanics.

Hint. For (a), apply the variational principle [Glazman 1965]. For (b), set $R_0 = (-\Delta - \lambda_2)^{-1}$ and $R_j = (-\Delta + q_j - \lambda_2)^{-1}$. By the resolvent identities,

$$\psi_2 = (-R_0 + R_0 q_1 R_0 - R_0 q_1 R_1 q_1 R_0) q_2 \psi_2.$$

Since $G_0(x, y, \lambda_2)$ decreases exponentially (for example, for $d = 3$ it equals $(4\pi|x - y|)^{-1}e^{-\delta|x-y|}$, where $\delta^2 = -\lambda_2$), the first two terms are bounded from above by $e^{-\delta'|x-y|}$, for $\delta' < \delta$. To prove the third term is, too, use the bound $\|R_1\| \leqslant 1/\operatorname{dist}(\lambda_2, \sigma(H_1)) = 2/\alpha$, which here is just a result of the existence of small denominators. Therefore

$$\left|(R_0 q_1 R_1 q_1 R_0 q_2 \psi_2)(x)\right| \leqslant 2e^{-\delta|x-a_1|+\kappa\sqrt{a_1-a_2}-\delta|a_1-a_2|} \leqslant e^{-\delta'|x-a_1|},$$

for $\delta' < \delta$. The same argument applies to show that ψ_1 is localized near the center a_1 of its well.

38. Consider a discrete Schrödinger operator on $\ell^2(\mathbf{Z}^d)$ whose potential satisfies the conditions of Theorem 15.11, and let

$$(1) \qquad P_l(\lambda) = P\Big\{\max_{|x|\leqslant l/4} \sum_{y\in\Lambda_l} |G_{\Lambda_l(0)}(x, y, \lambda + i0)| \geqslant 2l^{-2}\Big\}.$$

Prove that there exists $p_0 > 0$ (small enough) such that, if $P_l(\lambda) < p_0$ for some (large enough) l, we have $\lim_{L\to\infty} P_L(\lambda) = 0$.

Remark. The event in (1) defines a somewhat different notion of singularity than the one in (15.54); we call it an l-singularity. The statement above gives a weaker form of the finite-volume criterion for a pure point spectrum than the one in Theorem 15.14. Based on this criterion, one can show that $|G(0, x, \lambda + i0)|$ is summable in x with probability 1 and, by the results of Delyon et al. [1985a] (see also Theorem 15.20) and Simon et al. [1985], the discrete Schrödinger operator has a pure point spectrum near λ, with summable eigenfunctions.

Hint. The key estimate is

$$(2) \qquad P_L(\lambda) \leqslant \big(P_l(\lambda)\big)^{R^d} + C_1 P_l(\lambda) R^{2d} + C_2 f_0 L^d l^{-R/4}$$

for $L = Rl$ and some C_1 and C_2 depending only on d. It is easy to see that for large R independent of l—for example, $R = 10d$—we have $P_{R^i l} \to 0$ as $i \to \infty$.

To prove (2), we begin by covering $\Lambda_L(0)$, which has edge length L, with all the subcubes $\Lambda_l(y)$ of edge $l = L/R$ and center y, where the coordinates of y are multiples of $l/2$. The family $\mathcal{F}$ of such cubes has at most $c_d R^d$ elements, and for any point x in $\Lambda_L(0)$ there exists some $L_l(y) \in \mathcal{F}$ containing x and such that $|x - y| \leqslant l/4$. We now consider three cases.

If all the cubes in $\mathcal{F}$ are l-regular in the sense of (1), $\Lambda_L(0)$ is L-regular in the same sense. To prove this, use the block resolvent expansion for a chain of cubes connecting an arbitrary $x \in \Lambda_L(0)$ to the origin. This gives the bound $(2l^{-2})^{R/4} \leqslant L^{-(d+2)}$ for large R, the exponent $R/4$ arising from the fact that there are at least $R/4$ factors of the form $G_{\Lambda_l(\cdot)}$ in the block resolvent expansion for a chain that goes all the way to the boundary of $\Lambda_L(0)$. (Some details about the form of the expansion near the boundary must be proved, as in Theorem 15.14.) The probability that all cubes in $\mathcal{F}$ are l-regular is at most $(P_l(\lambda))^{R^d}$.

If there are two or more disjoint l-singular cubes in $\mathcal{F}$, we get the second term in (2). The probability of this event is less than $C_1 R^{2d} P_l^2(\lambda)$ for some C_1 depending only on d: see the proof of (15.66).

If there are several l-singular cubes in $\mathcal{F}$, no two of which are disjoint, let U be the union of them. Then U is contained in a cube of side $2l$, and each cube in $\mathcal{F} \cap A$, where $A = \Lambda_L(0) \setminus U$, is l-regular. The block resolvent expansion gives the bound

(3) $$|G_A(x, y, \lambda + i0)| \leqslant (2l^{-2})^{2r},$$

where $r = |x - y|/l \geqslant 2$. Now the double iteration of the resolvent identity for the pair of domains $\Lambda = \Lambda_L(0)$ and A gives

(4) $$G_\Lambda(x, y) = (R_A + R_A \Gamma_U R_A + R_A \Gamma_U R_\Lambda \Gamma_U R_A)(x, y).$$

Consider first the third term in parentheses. It can be expressed as a double sum over the pairs (z_1, z_1') and (z_2, z_2') belonging to ∂U. Use (3) with $r_1 = |x - z_1|/l$, $r_2 = |z_2' - y|/l$ and $r_1 + r_2 \geqslant R/4$ to estimate the G_A's, and use Theorem 15.11 with $W = l^{R/4}$ to estimate G_Λ (compare the formula for ψ_2 in the hint to the previous problem, and the proof of Lemma 15.15). Thus the third term, for R large, is bounded from above by

$$\text{const}\, l^{2(d-1)} l^{-2r_1} l^{R/4} l^{-2r_2} \leqslant \text{const}\, L^{-(d+2)}.$$

The first two terms in (4) are estimated similarly. Hence $\Lambda_L(0)$ is L-regular, and, by Theorem 15.11, its probability is given by the third term in (2).

Chapter VII

Almost-Periodic Operators

As we discussed in Example 1.15(g), almost-periodic functions can be regarded as realizations of metrically transitive processes, and therefore differential and finite-difference operators with almost-periodic coefficients are metrically transitive operators. But since the "randomness" of almost-periodic functions is very weak, the spectral properties of almost-periodic operators differ in many respects from those of the random operators that we saw in Sections 6, 14 and 15.

The techniques used in the spectral analysis of almost-periodic operators are also distinctive. Instead of probabilistic, they are based largely on analytic methods, and in particular on methods for approximating almost-periodic coefficients by periodic ones.

Since almost-periodic processes are in many ways similar to periodic ones, one would expect the properties of almost-periodic and periodic operators to be also similar. This, however, is not the case. The difference, roughly speaking, is that the spectral properties of almost-periodic operators, and in particular the character of the spectrum, depends essentially on the coefficients, and can change qualitatively with their amplitude, smoothness, and so on. The spectrum of periodic operators, on the other hand, is always absolutely continuous and responds only quantitatively to changes in the coefficients.

More specifically, the spectrum of almost-periodic operators is, as a rule, a nowhere dense (Cantor) set, even when it is absolutely continuous (as is the case for limit-periodic potentials, small and smooth quasiperiodic potentials, or potentials that can be approximated well by periodic ones). Further, the spectrum of almost-periodic operators can be singular continuous and even pure point. This phenomenon has been studied well for discrete one-dimensional operators with a large enough amplitude or a singular enough potential. (For a general discussion of the features of these spectra, see [Avron and Simon 1981b].)

Thus, almost-periodic operators, even in one dimension, offer a wide variety of spectral behaviors, unlike both periodic and random operators (we showed in Sections 13 and 15 that the spectrum of random one-dimensional operators is, as a rule, pure point).

This chapter is organized as follows. Section 16 lays down various facts from the spectral theory of (mainly one-dimensional, second-order) differential and finite-difference operators whose coefficients are sufficiently simple functions such as trigonometric polynomials, or sometimes piecewise-constant functions. These facts are disparate in nature and are obtained by a variety of methods, so for several of them (especially in 16.B, 16.D and 16.E) only the merest sketch of a proof is given.

The purpose of Section 16 is to present a detailed enough picture of the spectral properties of almost-periodic operators, in particular of their nowhere dense absolutely continuous spectrum and of their pure point spectrum. The next two sections are devoted to two classes of almost-periodic operators, each demonstrating clearly one of the characteristic properties discussed above.

Namely, Section 17 deals with the one-dimensional Schrödinger operator with a limit-periodic potential that can be approximated very well (super-exponentially) by periodic functions on the whole axis. The spectrum here is absolutely continuous for any amplitude and degree of smoothness, and generally nowhere dense (precise conditions are relatively easy to state). One can get fairly complete information on the spectrum of these operators, and one can even solve the inverse problem of spectral analysis, which cannot be done for other types of metrically transitive operators. An important technical tool here is the study of the dependence of various spectral characteristics on the period, as the period increases toward infinity. This study comprises the bulk of the proofs.

Section 18 considers the Schrödinger operator and similar ones (discrete, continuous, multidimensional) whose coefficients are determined by the quasiperiodic sequence $g \tan \pi(\alpha x + \omega)$, for $x \in \mathbf{Z}$, $g \in \mathbf{R}$, $\omega \in [0,1)$ and α an irrational number. The spectrum of such operators, when α is poorly approximated by rational numbers, is pure point, and their eigenfunctions decay exponentially at infinity. The picture here is similar to that of random operators in the fluctuation region (Sections 9 and 15). The specific form of the potential results in essentially explicit formulas for the resolvent of the operators considered, and in a pretty complete spectral analysis of them, leading in particular to examples where the spectrum is mixed (pure point in one region and singular in another: see 18.C) or to a point spectrum dense in $\mathbf{R}$ coexisting with an absolutely continuous spectrum coinciding with $[\rho, \infty)$, for some $\rho > 0$ (Theorem 18.17).

16 Smooth Quasi-Periodic Potentials

Most of this section is devoted to the spectral analysis of one-dimensional differential and finite-difference operators (mainly the Schrödinger operator), with fairly smooth quasiperiodic coefficients. However, we will also discuss operators with continuous, and even discontinuous, coefficients.

We start in 16.A with the geometry of the spectrum, giving reasons and conditions for it to be a nowhere dense set. We will see that this generic property of one-dimensional almost-periodic operators is intimately related to the existence of gaps in the spectrum of periodic operators (Theorem 16.2).

In 16.B we consider the Schrödinger operator with an analytic (or sufficiently smooth) quasiperiodic potential of small amplitude, or for large values of λ. We explain, using perturbation theory, that if the frequencies are incommensurable, the absolutely continuous spectrum contains a big nowhere dense component (Theorem 16.6). But, as is well known, even in the periodic case a small potential causes the spectrum to change, with the appearance of gaps. The mechanism for this phenomenon is similar to that of the destruction of invariant tori in nonlinear mechanics, and is known in quantum mechanics as resonance tunneling, being associated with the small denominators appearing in the formal series of perturbation theory. Therefore, the approach we use is based on KAM (Kolmogorov–Arnold–Moser) theory, specifically on its variant developed for the problem of reduction of systems of differential equations with quasiperiodic coefficients to equations with constant coefficients, where excessively strong resonances are eliminated by imposing conditions of incommensurability between the frequencies.

In 16.C we look at conditions under which an almost-periodic potential gives an operator having no eigenvalues. This is done using recurrence properties of the solutions of the corresponding equation (in particular, the possibility of their taking nonzero values on a divergent sequence of points). These properties arise when long enough stretches of the potential match each other closely enough, either because the potential is well approximated over several periods by periodic functions of increasingly large periods (Theorems 16.7 and 16.10), or because it takes on only a finite number of values (Problems VII.9–10).

Thus, Subsections 16.A through 16.C discuss the properties of almost-periodic operators that are, in some sense, similar to those of periodic ones. In 16.D, by contrast, we prove a property that has no counterpart in the periodic case, namely, the fact that the the Lyapunov exponent is positive on the spectrum of a discrete operator of order two whose potential is a trigonometric polynomial (Theorems 16.13 through 16.15). In view of Theorem 12.1, this means that the spectrum has no absolutely continuous component. By combining this with the results of 16.C, we construct examples of almost-periodic operators with a singular continuous spectrum (Theorem 16.17).

Subsection 16.E discusses the point spectrum. It is natural to search for the point spectrum first in the discrete case, where the second-difference operator responsible for the continuous spectrum is bounded, and therefore, for a large potential, can be regarded as a small perturbation of the operator of multiplication by the potential. (By Example 2.13, this unperturbed operator has a pure point spectrum.)

However, an attempt to use perturbation theory, considering the operator of second difference, rather than the potential, to be the perturbation, as we do in 16.B, runs into effects of resonance tunneling, that is, small denominators in the formal series for the resolvent, that are due to matching stretches of the potential. The simplest way to overcome this hurdle is to consider potentials that only take on identical values at points very far apart: see (16.81). This condition turns out to be very strong, excluding all but very irregular almost-periodic sequences. The resulting almost-periodic operators, in one or more dimensions, do have a pure point spectrum, whose closure might be the interval $[0, 1]$, or a Cantor set of zero measure. This is the subject of the first part of 16.E.

In the second part of 16.E we discuss briefly a modification of perturbation theory, in which the conditions of absence of strong resonance (coincidence of potential values) are imposed not on the whole axis at a time, but on a sequence of bounded, expanding intervals. In this way it becomes possible to prove that the spectrum is pure point and the eigenfunctions decay exponentially (as in the case of the random operators of Chapter VI) for potentials of the form $q(x) = gQ(\alpha x + \omega)$, for $x \in \mathbf{Z}$, where g is sufficiently large, α is an irrational number poorly approximated by rational numbers, and $Q(t)$ is a function of class C^2, periodic of period 1, independent of g and having a single, nondegenerate, maximum and minimum in a period. The simplest example of such a potential is $q(x) = 2g \cos 2\pi(\alpha x + \omega)$, and the corresponding order-two operator is known as the almost-Mathieu operator. This example is important because it displays a variety of spectral behaviors and is very rich in interesting properties, including a type of symmetry, called Aubry duality, thanks to which a lot of additional information on the spectrum can be derived. We therefore devote Subsection 16.F to a summary of its spectral properties.

In 16.B, and especially in 16.E, we will often limit ourselves to brief sketches of proofs.

16.A The Integrated Density of States and the Gap Labeling Theorem

We start by recalling the main results from the spectral theory of the continuous or discrete one-dimensional Schrödinger operator with a periodic potential. For details on the continuous case, see [Titchmarsh 1958; Marchenko 1986; Reed and Simon IV]. The theory of the discrete case is largely similar; see [Toda 1981; van Moerbeke 1976].

Consider the second-order equation (6.1) or (11.4) with a periodic potential $q(x)$, for $x \in \mathbf{R}$, of period a. For any $\lambda \in \mathbf{R} \setminus S$, where $S = \{\nu_0, \nu_1^\pm, \nu_2^\pm, \ldots\}$ is a countable, discrete set, this equation has two linearly independent *Floquet–Bloch solutions*

$$\psi_\pm(x,\lambda) = e^{\pm ik(\lambda)x} u_\pm(x,\lambda), \tag{16.1}$$

where the $u_\pm(x,\lambda)$ are periodic of period a and the function $k(\lambda)$, called the *quasimomentum*, is real on the *stability intervals* or *allowed bands*

$$\Delta_1 = [\nu_0, \nu_1^-], \quad \Delta_2 = [\nu_1^+, \nu_2^-], \quad \ldots, \tag{16.2}$$

whose endpoints

$$-\infty < \nu_0 < \nu_1^- \leqslant \nu_1^+ < \nu_2^- \leqslant \nu_2^+ < \cdots, \tag{16.3}$$

form the set S. These endpoints are given by the roots of the equation $A^2(\lambda) = 1$, for

$$A(\lambda) = \tfrac{1}{2}\big(c(a,\lambda) + s'(a,\lambda)\big), \tag{16.4}$$

where $c(x,\lambda)$ and $s(x,\lambda)$ is a fundamental system of solutions of the equation determined by conditions (12.11). The function $A(\lambda)$ is called the *Lyapunov function* or *Hill discriminant*, and is related to the quasimomentum by the formula

$$A(\lambda) = \cos k(\lambda)a. \tag{16.5}$$

In the complement of the allowed bands, whose components

$$\Delta_0' = (-\infty, \nu_0), \quad \Delta_1' = (\nu_1^-, \nu_1^+), \quad \ldots \tag{16.6}$$

are called *gaps* (or *lacunas*, or *forbidden bands*), $k(\lambda)$ takes on the values

$$\frac{\pi m}{a} + ih \tag{16.7}$$

for $m \in \mathbf{Z}$ and $h \in \mathbf{R}$.

Since the spectrum of a differential or finite-difference operator is the closure of the set of $\lambda \in \mathbf{R}$ for which the corresponding equation has a polynomially bounded solution (see [Berezanskii 1968], [Simon 1982] and 12.A), it follows from (16.1) that the spectrum σ is the set of allowed bands,

$$\sigma = \bigcup_m \Delta_m \tag{16.8}$$

If $\nu_m^- = \nu_m^+$ for some m, we say that the m-th gap is closed. If there are only finitely many gaps, we are dealing with a so-called *finite-band potential*. In general, the length of the gaps tends to zero at a rate that depends on the smoothness of the potential: the smoother the potential, the higher the rate. Also, the gaps are asymptotically close to the points

$$\lambda_m = \left(\frac{\pi m}{a}\right)^2. \tag{16.9}$$

The Floquet–Bloch solutions (16.1) exist also for complex values z of the spectral parameter. In this case, the function $k(z)$ near the ends of the open gaps $\nu_m^\pm$ behaves as

$$k(z) = \frac{\pi m}{a} + \mathrm{const}\sqrt{z - \nu_m^\pm}. \tag{16.10}$$

It is analytic on the upper half-plane $\mathbf{C}_+$ and has a positive imaginary part there. Therefore, for $z \in \mathbf{C}_+$, the solutions (16.1), when normalized so that $u(0,z) = 1$, coincide with the Weyl solutions (12.12). It follows that the spectrum is absolutely continuous, has multiplicity two, and the solutions (16.1), for $\lambda \in \bigcup_m \Delta_m$, form a complete set of (generalized) eigenfunctions.

In view of definitions (6.11) and (11.25), this implies that

$$N(\lambda) = \frac{1}{\pi}\,\mathrm{Re}\,k(\lambda), \qquad \gamma(\lambda) = \mathrm{Im}\,k(\lambda) \tag{16.11}$$

and

$$\sigma = \{\lambda \in \mathbf{R} : \gamma(\lambda) = 0\}. \tag{16.12}$$

Thus, for periodic operators, the spectrum is not only the support of the measure $N(d\lambda)$, as it is for all metrically transitive operators, by Theorem 3.1, but also the set where the Lyapunov exponent vanishes. This also follows from Theorem 12.6, since the spectrum of periodic operators is absolutely continuous.

From (16.10), or from a direct calculation based on (16.1) with λ complex, it follows that $ik(z)$ coincides with the function $w(z)$ of (12.54). It also follows from (16.5) and (16.11) that $N(\lambda)$ and $\gamma(\lambda)$ are real analytic for all λ outside the set S of band edges (16.3). In the neighborhood of band edges, (16.10) shows that $\gamma(\lambda)$ satisfies a Hölder condition with exponent $\frac{1}{2}$.

There is a different approach to the Floquet–Bloch solutions (16.1) in spectral theory, consisting of taking the quasimomentum k as an independent parameter varying, say, in the interval $[-\pi/a, \pi/a]$. One defines functions $\lambda_m(k)$, for $|k| \leqslant \pi/a$, so that the m-th allowed band Δ_m is exactly the set of values of the function $\lambda_m(k)$. The advantage of this approach, quite common in theoretical physics, is that it yields in a natural way all the main facts of spectral theory (the Parseval equality, the representation of the Green's functions, etc.) in the case of multidimensional periodic operators. It was introduced to the spectral theory by Gel'fand [1950]. Its modern form, employing the techniques of direct integrals in Hilbert spaces, is presented in [Reed and Simon IV].

We start our discussion of almost-periodic operators with the simplest object in the spectral theory of metrically transitive operators, the integrated density of states. We restrict ourselves to the continuous case.

As explained in Example 1.15(g), every almost-periodic function $q_0(x)$ generates a metrically transitive structure $(\Omega, \mathcal{F}, \mathsf{P}, T_x)$, where the space of realizations Ω is the hull Ω_{q_0} of q_0. Thus the proofs of the existence and of various properties of $N(\lambda)$, given in Sections 3 and 6 for general metrically transitive operators, are equally applicable for almost-periodic operators. But since, in the almost-periodic case, there is a fairly strong topology on the space of realizations, namely, the topology of uniform convergence on the whole axis, and since $N(\lambda)$ is continuous as a functional of the potential in this topology, for every fixed $\lambda \in \mathbf{R}$ (see Theorems 4.14 and 5.24), the existence of $N(\lambda)$ (that is, of the limits (4.11), (5.10) and (6.10)) is guaranteed not only with probability 1, but for all $\omega \in \Omega$.

As shown by the example of Theorem 5.27, $N(\lambda)$ is also continuous in a much weaker sense. For bounded metrically transitive potentials, including uniform almost-periodic functions, it is true that if a metrically transitive potential $q_n(x)$ converges pointwise to $q(x)$, the integrated density of states $N_n(\lambda)$ of converges to $N(\lambda)$ for all λ. This can be proved using (5.14), for example.

Now every almost-periodic function $q_0(x)$ gives rise to a generalized Fourier series

$$\sum_{n \geqslant 1} q_{0,n} e^{2\pi i \alpha_n x},$$

and to a module $\mathcal{M}_{q_0}$ formed by integral combinations of frequencies α_n, for $n \geqslant 1$. The set of almost-periodic functions with frequencies belonging to a prescribed *module* $\mathcal{M}$ will be denoted by $\mathcal{A}(\mathcal{M})$. Clearly $\mathcal{A}(\mathcal{M})$ is a Banach algebra with respect to the ordinary operations of addition and multiplication of functions in the topology of uniform convergence on the whole axis, so that $\mathcal{A}(\mathcal{M}) \subset C(\mathbf{R})$. The next proposition gives the relationship between the prescribed function q_0 and the corresponding metrically transitive structure.

(16.1) Proposition [Levitan 1953]. *Set $\mathcal{M} = \mathcal{M}_{q_0}$ and let $g(x)$ be an almost-periodic function. Then $g \in \mathcal{A}(\mathcal{M})$ if and only if for every uniformly convergent sequence $\{q_0(\cdot + t_n)\}$, for $t_n \in \mathbf{R}$, the sequence $\{g(\cdot + t_n)\}$, converges as well. Moreover, the Banach algebra $\mathcal{A}(\mathcal{M})$ is isomorphic to the Banach algebra $C(\Omega_{q_0})$ of continuous functions on Ω_{q_0}, and if $G \in C(\Omega_{q_0})$ is the image of $g \in \mathcal{A}(\mathcal{M}_{q_0})$ under this canonical isomorphism, we have*

$$g(x) = G(q_0(\cdot + x)) \tag{16.13}$$

for $x \in \mathbf{R}$, and

$$\lim_{L\to\infty} \frac{1}{L}\int_0^L g(x)\,dx = \int_{\Omega_{q_0}} G(\omega)\,d\omega, \tag{16.14}$$

where $d\omega$ is the Haar measure on Ω_{q_0} (see Example 1.15(g)). □

Equation (16.13) is crucial for the construction of the canonical isomorphism between $C(\Omega_{q_0})$ and $\mathcal{A}(\mathcal{M}_{q_0})$, because, given an almost-periodic function $g \in \mathcal{A}(\mathcal{M}_{q_0})$, it determines G on the set $\{q_0(\cdot + x) : x \in \mathbf{R}\}$, which is dense in Ω_{q_0}.

Thus, the operator $H_0 = -d^2/dx^2 + q_0$ belongs to the family of metrically transitive operators $H(\omega) = -d^2/dx^2 + q(x,\omega)$, for $\omega \in \Omega_{q_0}$, all of which have integrated density of states $N(\lambda)$. The set of growth points of $N(\lambda)$ is, by Theorem 3.1, the spectrum of all operators in this family. In particular, $N(\lambda)$ is constant in the gaps of the spectrum, with a different constant for each gap (since $N(\lambda)$ is monotonic in λ), so each gap can naturally be labeled with the value of $N(\lambda)$ there.

This labeling, of course, can be carried out for any metrically transitive operator; but in the case of a periodic Schrödinger operator, it follows from (16.11) and the properties (16.7) and (16.11) of $k(\lambda)$ that the value of $N(\lambda)$ on the n-th gap is m/a, for $m \in \mathbf{Z}_+$, where a is the period of the potential. If we introduce the frequency $\alpha = 1/a$, more convenient in the almost-periodic case, and note that the module of a function of period a is $\alpha\mathbf{Z}$, this observation can be written as follows:

$$N\left(-\frac{d^2}{dx^2} + q, \lambda\right) \in \mathcal{M}_q \cap [0,\infty) = \mathcal{M}_q^+ \tag{16.15}$$

for $\lambda \in \mathbf{R} \setminus \sigma$. This is true for almost-periodic operators as well, as was first pointed out by Claro and Wannier [1979]:

(16.2) Theorem (gap labeling) [Johnson and Moser 1982]. *If $q(x)$, for $x \in \mathbf{R}$, is a uniform almost-periodic function with module $\mathcal{M}_q$, equation (16.15) holds.*

Proof. By the results in 12.A, the Schrödinger equation

$$-\psi''(x) + q(x)\psi(x) = z\psi(x), \tag{16.16}$$

where z belongs to the resolvent set of the operator $H = -d^2/dx^2 + q$, has a square-integrable Weyl solution $\phi_+(x, z)$ that is unique up to a multiplicative factor. In particular, such a solution also exists for real $z = \lambda$ in a gap of the spectrum, where $\phi_+(x, \lambda)$ is real. From now on we consider only such values of z.

Introduce the function

$$g(x, \lambda) = \frac{\phi'_+(x, \lambda) + i\phi_+(x, \lambda)}{\sqrt{(\phi'_+(x, \lambda))^2 + \phi^2(x, \lambda)}} = e^{i\theta(x,\lambda)},$$

where $\theta(x, \lambda)$ is a continuous function. We need some auxiliary results.

(16.3) Lemma. *$g^2(x, \lambda)$ is an almost-periodic function and $g^2 \in \mathcal{M}_q$.*

Proof. By Proposition 16.1, it suffices to prove that whenever the sequence of potentials $\{q(\cdot + t_n)\}$ converges in the uniform metric, so does the sequence $\{g^2(\cdot + t_n)\}$. If $\phi_t(x)$ denotes the solution of (16.16) with potential $q_t(x) = q(x + t)$, normalized by the conditions $\int_0^\infty \phi_t^2(x)\, dx = 1$ and $\phi_t(0) > 0$, it follows from uniqueness that $\phi_t(x) = c_t\phi_t(x + t)$ for some $c_t \in \mathbf{R}$. Thus the sequence $\{g^2(\cdot + t_n)\}$ will converge if the solutions ϕ_{t_n} converge to a nontrivial solution.

By the normalization conditions and (16.16), the norms $\|\phi_t''\|$ are uniformly bounded in t, and simple embedding theorems imply that the family $\{\phi_t : t \in \mathbf{R}\}$ is precompact both in the Sobolev space $H^1(\mathbf{R}_+)$ and in $C(\mathbf{R})$. If we assume that $\{q(\cdot + t_n)\}$ converges to some potential q^* in $C(\mathbf{R})$, it clearly follows that there is a nonzero function $\phi^*(x)$ that is the limit of $\phi(x + t_n)$ and satisfies equation (16.16) for $q = q^*$. This proves the lemma. □

(16.4) Proposition. *Let $\tilde{g}(x)$, for $x \in \mathbf{R}$, be a complex-valued almost-periodic function in $\mathcal{A}(\mathcal{M})$, with $|\tilde{g}(x)| > 0$. Then*

$$\tilde{g}(x) = |\tilde{g}(x)| e^{i(\beta x + \chi(x))},$$

where $|\tilde{g}|$, $e^{i\beta x}$ and χ belong to $\mathcal{A}(\mathcal{M})$, and therefore $\beta/2\pi \in \mathcal{M}$.

Proof. See [Levitan 1953], for example. □

We can now return to the proof of Theorem 16.2. Lemma 16.3 says that $g^2(x)$ is an almost-periodic function, and we know that $|g^2(x)| = 1$. By Proposition 16.4,

$$2\theta(x,\lambda) = \beta(\lambda)x + \chi(x,\lambda), \tag{16.17}$$

where $\beta(\lambda)/2\pi \in \mathcal{M}_q$ and $\chi \in \mathcal{A}(\mathcal{M}_q)$. Theorem 6.4 and equation (16.17) lead to $N(H,\lambda) = \beta(\lambda)/2\pi \in \mathcal{M}_q$, completing the proof. □

(16.5) Corollary. *For a one-dimensional Schrödinger operator whose potential is a quasiperiodic function* (1.37), *the values of* $N(\lambda)$ *on the gaps of the spectrum lie in the set*

$$\{l_1\alpha_1 + \cdots + l_n\alpha_n : l_1, \ldots, l_n \in \mathbf{Z}\} \cap [0,\infty). \qquad \square$$

Remarks. 1. Statements similar to Theorem 16.2 and Corollary 16.5 also hold for the discrete operator (1.4), if the coefficients are uniform (Bohr) almost-periodic functions on $\mathbf{Z}$ [Delyon and Souillard 1984]. The only difference in the statements is that the module of the potential is a subset of the unit circle $\mathbf{R}/\mathbf{Z}$. (For general Jacobi matrices (1.4) the module is the arithmetic sum of the modules of the coefficients $s(x)$ and $q(x)$). The proof is essentially the same as the one given above for Theorem 16.2, and even technically simpler. (This common proof differs from the original proofs by Johnson and Moser for the continuous case and by Delyon and Souillard for the discrete case.)

2. The closure (16.15) can be strict even in the class of periodic potentials (to say nothing of $q(x) = \text{const}$). Indeed, by using the solution to the inverse problem of spectral analysis for periodic Schrödinger operators [Marchenko 1986], one can construct periodic potentials whose spectrum has only a prescribed set of gaps, which can even be finite (finite-band potentials). A similar result for limit-periodic potentials that are well approximated by periodic ones is derived in Section 17. A regular method for constructing finite-band quasiperiodic potentials has been developed in the theory of integration of nonlinear Korteveg–de Vries equation [Marchenko 1986, 1980; Levitan 1984; Zakharov et al. 1985].

3. However, all these examples are pretty special, and it is natural to think that, generically, the set of values of $N(\lambda)$ on gaps coincides with the set $\mathcal{M}_q^+$ of (16.15), that is, there are no closed gaps. One can easily see, for example, that the set of periodic potentials of period a for which the m-th gap is present is open and dense in the space of summable functions on $[0,a)$.

Since the module of an almost-periodic function can be a dense set in $\mathbf{R}$, when all gaps are present the spectrum can turn out to be a nowhere dense set. Examples of such almost-periodic potentials, which are limit-periodic, are given in [Moser 1981; Avron and Simon 1981a; Chulaevskii 1981]. If we associate to each sequence $\{q_n\} \in \ell_1(\mathbf{Z}_+)$ the limit-periodic potential

(16.18) $$q(x) = \sum_{n=0}^{\infty} q_n \cos 2^{-n}x,$$

the set of sequences for which the spectrum of the corresponding Schrödinger operator is nowhere dense is a dense G_δ-set in $\ell_1(\mathbf{Z}_+)$. Thus a nowhere dense spectrum is typical (in the Baire category sense) for potentials of the form (16.18). In Section 17 we will study systematically the spectral properties of the Schrödinger operator with this type of potential, in the case when the numbers q_n decrease very rapidly. In particular, we will describe the structure of the dense G_δ-set mentioned above.

Bellissard and Simon [1983] proved that the spectrum of the discrete Schrödinger operator (11.4) with potential

(16.19) $$q(x) = 2g\cos 2\pi(\alpha x + \omega),$$

for $x \in \mathbf{Z}$ and $\omega \in [0,1)$, is nowhere dense for pairs (g,α) in a dense G_δ-set in $\mathbf{R}^2$.

4. Thus, generically (at least in the Baire category sense), for one-dimensional differential and finite-difference operators whose coefficients are uniform almost-periodic functions, the set of values of $N(\lambda)$ on gaps coincides with the module of the coefficients. It may seem that the presence of all the gaps is the main analytic condition for these two sets to coincide, whereas the condition that the coefficients belong to the class of uniform (Bohr) almost-periodic potentials is a technical one. For periodic potentials, this is indeed so, because Theorem 16.2 is true for them even under the assumption that the potential is square-integrable over a period, and potentials for which all gaps are present form a dense G_δ-set. A similar situation obtains for the limit-periodic potentials considered in Section 17, which are square-integrable only locally.

However, there are examples of quasiperiodic potentials that show that the assumption of uniform almost-periodicity for the coefficients of Theorem 16.2 may be important. For example, in Section 18 we will consider the finite-difference operator with potential

(16.20) $$q(x) = g\tan\pi(\alpha x + \omega),$$

where α is an irrational number, $\omega \in [0,1)$ and $\omega \neq \alpha n + \frac{1}{2} \pmod 1$, for $n \in \mathbf{Z}$. If we consider a continuous function $Q(t)$ of period 1, instead of $g\tan\pi t$, the sequence

(16.21) $$q(x) = Q(\alpha x + \omega),$$

for $x \in \mathbf{Z}$, $\omega \in [0,1)$ and α irrational, is a uniform almost-periodic function on $\mathbf{Z}$, with module

$$\alpha\mathbf{Z} \pmod 1, \tag{16.22}$$

and its hull is the interval $[0, 1)$. But, by Theorem 18.1, the spectrum of the finite-difference operator with potential (16.20) is the whole real axis, and $N(\lambda)$ is a real analytic function whose derivative is strictly positive on the whole axis.

Notice also that, for a periodic potential, the spectrum has no gaps only if the potential is identically constant [Marchenko 1986]; see also Theorem 12.15.

Here is another example to show that it is likely to be important for the potential to belong to the class of uniform almost-periodic functions. Consider (16.21) with

$$Q(t) = g\chi_{I_\beta}(t), \tag{16.23}$$

where I_β is the arithmetic sum of $\mathbf{Z}$ with the interval $(0, \beta) \subset (0, 1)$, and β is a number that is $\mathbf{Q}$-linearly independent of α. For this potential, Bellissard and Scoppola [1982] proved that the labeling of gaps requires the bigger module generated by α and β, rather than the module (16.22) needed for the potential (16.21) with a continuous $Q(t)$, as guaranteed by Theorem 16.2. Similar examples were constructed by Craig [1983] (see 16.E and Problem VII.22).

A profound analysis of the causes of this phenomenon was carried out by Bellissard and Testard [1985], using techniques from C^*-algebras and K-theory. This approach allows one to extend Theorem 16.2 to a wide class of pseudo-differential operators in $L_2(\mathbf{R}^d)$ and matrix operators in $\ell_2(\mathbf{Z}^d)$ with quasiperiodic coefficients. It is based on a formula, found by Shubin [1975, 1979], representing the integrated density of states for such operators as the relative trace of their resolution of the identity in the C^*-algebra constructed from the coefficients. These algebras were introduced by Coburn et al. [1973] to generalize the index theorem to pseudo-differential operators with almost-periodic coefficients; an example of their application will be considered in Section 18. (For a discussion of their use in the general metrically transitive case, see [Fedosov and Shubin 1978].)

Bellissard and Testard (see also [Bellissard 1985]) proved a general gap-labeling theorem, in terms of these algebras and their topological characteristics, which applies to pseudo-differential operators whose coefficients are functions on an ergodic dynamical system that is not necessarily an irrational flow on a torus, as is the case of quasiperiodic functions. For example, let Γ be a discrete subgroup of $\mathrm{SL}(2, \mathbf{Z})$ with a compact fundamental domain, Ω the compact space $\mathrm{SL}(2, \mathbf{Z})/\Gamma$, and T a diffeomorphism of Ω with every orbit dense in Ω. The theorem then implies that there are no gaps in the spectrum of the discrete one-dimensional Schrödinger operator (11.4) with potential $q(x,\omega) = Q(T^x\omega)$, where $Q(\omega)$ is a continuous function on Ω.

We have discussed, in general terms, the geometric structure of the spectrum and the behavior of $N(\lambda)$ for almost-periodic operators. We now discuss briefly certain questions that are more traditional in the spectral theory of metrically transitive operators, in particular the asymptotic behavior of $N(\lambda)$ near the edges of the spectrum. As we saw in Chapters III and IV, this behavior can be determined in great details for random operators. A first step in this study was to locate the edges. For a broad class of random operators, this can be done using Theorems 4.22 and 5.33, which is very similar to the Weyl criterion, well-known in spectral theory. Roughly speaking, these theorems are based on the fact that, because of weak correlation, the coefficients can assume values arbitrarily close to their minimum or maximum, in arbitrarily large regions of space. Therefore, spectral boundaries of random operators generally coincide with the boundaries for the same operators with constant or periodic coefficients (although the relationship can be more complicated, as in Theorem 5.35), and can often be found simply and explicitly in general. For example, for the potential (6.25), in which $k_0 > 0$ and the points x_j have a Poisson distribution, the left edge of the spectrum $\nu_0 = \inf \sigma$ equals zero, the same value it has for the Schrödinger operator with a zero potential.

For almost-periodic operators, the situation is more complicated, because, by definition, almost-periodic function display a strictly fixed behavior on sufficiently long intervals. Thus, even the bottom of spectrum ν_0 is hard to find explicitly. For the same potential (6.25), with $k_0 > 0$ and periodically distributed points x_j, we always have $\nu_0 > 0$. Now consider points x_j such that $x_{j+1} - x_j = f(\alpha j)$, where α is an irrational number and $f(t)$, for $t \in \mathbf{R}$, is a nonnegative function of period 1. Here ν_0 can be found if, for example, $f(t)$ increases monotonically in the interval $[0,1)$ and $f(t) = c(1-t)^{-1/a}(1+o(1))$, as $t \uparrow 1$ and $a > 1$ is fixed. Then the distance between the x_j can be arbitrarily large and, using arguments similar to those in the proofs of Theorem 5.33 and 5.34, one can show that $\nu_0 = 0$. The asymptotic behavior of $N(\lambda)$ as $\lambda \downarrow 0$ can also be found in this case, and, in contrast with the exponential behavior of (6.29), it has the polynomial form (6.29a) in the case of Poisson points x_j. See [Gredeskul and Pastur 1985], and Problem VII.6, where other similar potentials are considered, some very smooth.

Another situation in which one can find the bottom of the spectrum and the asymptotic behavior of $N(\lambda)$ near it is that of small or almost constant almost-periodic coefficients. Consider, for example, a Schrödinger operator with a small and smooth enough quasiperiodic potential whose frequencies satisfy the incommensurability conditions (16.32) below. Kozlov [1983] showed that there is a small number ν_0 that is the bottom of the spectrum and a positive quasiperiodic function with a frequency module the same as that of the potential, such that for $\lambda \downarrow \nu_0$ the integrated density of states has the asymptotic behavior (8.4), where λ should be replaced by $\lambda - \nu_0$ and $a_{ij}(x)$ by $\delta_{ij}p^2(x)$, where $p(x)$ is the sum of a constant with a small almost-periodic function A similar statement for the one-dimensional discrete case was obtained by Sinai [1985]. These results are obtained by KAM perturbation theory, which, as we will see later in this section, is a very efficient technique in the study of the spectral properties of almost-periodic operators.

16.B Absolutely Continuous Spectrum

We start by noticing that the main results of Section 12—in particular, Lemma 12.4 and Theorems 12.5 and 12.6—are true, in the case of almost-periodic operators, for all, not just almost all, $\omega \in \Omega$. These results become individual, rather than statistical: they apply to every one-dimensional operator with almost-periodic coefficients. This can be proved, as was done in Section 5 for the results on $N(\lambda)$, by verifying that the property or quantity in question depends continuously on the coefficients, provided that the convergence of the coefficients is uniform over the whole axis (see, for example, [Kotani 1984]).

Now, it seems intuitively obvious that almost-periodic functions are the metrically transitive fields most similar to periodic ones. In particular, they are deterministic. Therefore, in view of Theorem 12.10 and the fact that the spectrum of periodic differential and finite-difference operators is absolutely continuous, it is natural to think that the absolutely continuous spectrum of almost-periodic operators cannot be empty, although, as we know from 16.A, it can be a nowhere dense set.

There are some other heuristic arguments in favor of this conjecture, based on formal perturbation theory. Take first the one-dimensional periodic case. Assuming the generalized eigenfunctions to have the Floquet–Bloch form (16.1) (this follows for any periodic operator from the fact that the operator commutes with the operator of shift by a period) and substituting them into the Schrödinger equation, we get the following boundary problem:

$$(16.24) \qquad \left(i\frac{d}{dx} + k\right)^2 u + qu = \lambda u, \quad u(0) = u(a), \quad u'(0) = u'(a),$$

with the quasimomentum k as the parameter, $|k| \leqslant \pi/a$. For $q(x) \equiv 0$, we have

$$(16.25) \quad u_m^{(0)} = e^{2ik_m x}/\sqrt{a}, \ \lambda_{2m}^{(0)} = \left(|k| - k_{2m}\right)^2, \ \lambda_{2m+1}^{(0)} = \left(|k| + k_{2m}\right)^2,$$

where $m \in \mathbf{Z}_+$ and $k_m = \pi m/a$. Therefore the spectrum coincides with $\mathbf{R}_+$ and is absolutely continuous. Now if the periodic potential is small, or λ is large, the correction to the eigenvalues, calculated by Rayleigh–Schrödinger perturbation theory [Reed and Simon IV], is

$$\delta\lambda_m = \sum_{l \neq m} \frac{|q_{l-m}|^2}{\lambda_m^{(0)} - \lambda_l^{(0)}}, \tag{16.26}$$

where the q_m, for $m \in \mathbf{Z}$, are the Fourier coefficients of the potential, and we assume for simplicity that $q_0 = 0$. This correction will not be small if the quasimomentum is very close to $k_m = \pi m/a$, for then the denominator $4\pi(m-l)/a(k + \pi(m+l)/a)$ can be very small. These exceptional values, known in physics as *Bragg points*, correspond to electronic waves reflected by periodic cells, which are exactly in antiphase and therefore completely damp each other. Thus the particle energies

$$\lambda_m = k_m^2 = \nu_m^+ \big|_{q \equiv 0} \tag{16.27}$$

are forbidden: there must be gaps in their neighborhood. Indeed, by modifying slightly the formulas of perturbation theory so as to take into account the coincidence of the eigenvalues in (16.26) as degeneracies, or by considering simple examples of potential, one can see that such gaps (ν_m^-, ν_m^+) do indeed appear, that $\nu_m^+ - \nu_m^- = O(q_m^2)$, and that the following asymptotic formulas hold:

$$k(\lambda) = k_m \pm \sqrt{|\nu_m^{\mp} - \lambda|}$$

for λ approaching ν_m^- from below or ν_m^+ from above, respectively.

So the Bragg points are the ones we must watch out for in terms of gap creation, where the quasimomentum equals half the Fourier exponent of the potential

$$k = k_m = \pi m \alpha, \tag{16.28}$$

for $m \in \mathbf{Z}$, where $\alpha = 1/a$ is the frequency. For values of λ well away from points of the form (16.27), there are still Floquet–Bloch solutions with $k(\lambda)$ an $N(\lambda)$ essentially the same as for $q \equiv 0$, so the spectrum remains absolutely continuous for such values.

For rigorous proofs of all these facts, see [Titchmarsh 1958] and [Reed and Simon IV].

We now consider *quasiperiodic* potentials, that is, those that can be represented as

$$q(x) = Q(\alpha_1 x, \ldots, \alpha_n x), \tag{16.29}$$

with

$$Q(\xi) = \sum_{m \in \mathbf{Z}^n} q_m e^{2\pi i (m,\xi)} \tag{16.29a}$$

for $\xi \in \mathbf{R}^m$ (see (1.37)). If all components of the frequency vectors α are of the form $\alpha_j = \alpha_0 b_j / a$, where $\alpha_0 \in \mathbf{R}$ and b_j and a are integers, the function (16.29) is periodic of period a/α_0, and we conclude, as above, that gaps will appear in the neighborhood of points

$$\lambda_m = k_m^2, \qquad k_m = \pi(m, \alpha) \tag{16.30}$$

for $m \in \mathbf{Z}^n$. Therefore, it is reasonable to think that when the components of α are incommensurable, these same points will be "centers" of gaps. But then the set (16.30) is dense in $\mathbf{R}$, and the condition forbidding the values of $k(\lambda)$ from lying within this set can only be a condition that these values be poorly approximated by number of the form (16.28), that is, for example, a *Diophantine condition*

$$|k - \pi(m, \alpha)| \geqslant C|m|^{-\beta} \tag{16.31}$$

for $m \neq 0$ and C, β positive constants. In particular, for $k = 0$,

$$|(m, \alpha)| \geqslant \frac{C}{\pi}|m|^{-\beta} \tag{16.32}$$

for $m \neq 0$. This latter equation is the condition of strong incommensurability of the frequency vector $(\alpha_1, \ldots, \alpha_n)$ of the quasiperiodic potential (16.29) and, when $\beta > n$, is valid for Lebesgue-almost $\alpha \in \mathbf{R}^n$ [Sprindzhuk 1979].

Now, in order for the series obtained from perturbation theory, a good example of which is given by (16.26), to converge, the Fourier coefficients q_m of $Q(x)$ have to decrease very fast, to compensate for the smallness of the denominators (16.32). This makes the conditions in the following result reasonable:

(16.6) Theorem [Dinaburg and Sinai 1975; Belokolos 1975]. *Consider the Schrödinger operator for a quasiperiodic potential* (16.29) *such that the frequencies* $\alpha_1, \ldots, \alpha_n$ *satisfy* (16.32), *and such that the Fourier coefficients* q_m *of the function* $Q(x)$ *satisfy*

$$|q_m| \leqslant Ce^{-\rho|m|} \tag{16.33}$$

for some C *and* $\rho > 0$, *so that* $Q(x)$ *has an analytic extension to the region*

$$\Pi_\rho = \{\zeta \in \mathbf{C}^n : |\operatorname{Im} \zeta_j| < \rho \text{ for } 1 \leqslant j \leqslant n\}. \tag{16.34}$$

Then, for any $\delta > 0$, *there exist positive constants* C_1, C_2, C_3 *and* C_4 *such that every neighborhood*

$$O_m = \{\lambda > 0 : \big|\sqrt{\lambda} - |k_m|\big| \leqslant C_1(1 + |k_m|)^{-1}\}$$

of every point (16.30) *contains a neighborhood*

$$\tilde{O}_m = \{\lambda > 0 : |\sqrt{\lambda} - k'_m| \leqslant C_2 \exp(-|m|(\ln|m|)^{-(1+\delta)})\}$$

such that

$$\text{meas}\Big\{[\lambda_0, \infty) \cap \bigcup_m \tilde{O}_m\Big\} \leqslant C_3 \exp\Big(\frac{-C_4\sqrt{\lambda_0}}{\ln^{1+\delta}\lambda_0}\Big) \tag{16.35}$$

for large enough λ_0. *Now set*

$$\sigma_1 = \Big(\sigma\Big(-\frac{d^2}{dx^2} + q\Big) \setminus \bigcup_{m \in \mathbf{Z}^n} \tilde{O}_m\Big) \cap [\lambda_0, \infty). \tag{16.36}$$

Then:

(i) *For* $\lambda \in \sigma_1$, *the Schrödinger equation has two linearly independent solutions* (*cf*. (16.1))

$$\psi_+(x, \lambda) = e^{ik(\lambda)}u(x, \lambda), \qquad \psi_-(x, \lambda) = \overline{\psi_+(x, \lambda)}, \tag{16.37}$$

where the function $k(\lambda)$ *on* σ_1 *is uniformly continuous and satisfies* (16.31) *and* $|k(\lambda) - \sqrt{\lambda}| \leqslant \text{const}/\sqrt{\lambda}$, *and* $u(x, \lambda)$ *can be written as*

$$u(x, \lambda) = U(\alpha_1 x, \ldots, \alpha_n x, \lambda) \tag{16.37a}$$

with $U(x_1, \ldots, x_n, \lambda)$ *analytic in* $\Pi_{\rho/2}$ (*in particular*, $u(x, \lambda)$ *is quasi-periodic with same frequency module as the potential*).

(ii) *The spectrum is absolutely continuous on* σ_1, *with multiplicity* 2.

(iii) *The integrated density of states* $N(\lambda)$ *is given by* (16.11) *on* σ_1.

(iv) *If* $E(d\lambda)$ *is the resolution of the identity and* $\phi(x)$ *is a bounded function with compact support*,

$$(\phi, E(d\lambda)\phi) = \frac{|\Phi_+(\lambda)|^2}{2\pi\sqrt{\lambda}(1 + f(\lambda))} d\lambda$$

for $d\lambda \in \sigma_1$, *where*

$$\Phi_+(\lambda) = \int_{\mathbf{R}} \psi_+(x, \lambda)\phi(x)\, dx$$

for $\lambda \in \sigma_1$, *and the function* $f(\lambda)$ *does not depend on* ϕ *and tends to zero as* $\lambda \to \infty$.

This theorem is asymptotic, for its conditions are satisfied either for large enough λ, or for small enough potentials: the criterion of applicability is $\sup_{x\in\mathbf{R}}|q(x)|/\lambda \ll 1$. Notice that, since the solutions (16.37) are bounded, the fact that almost all $\lambda \in \sigma_1$ with respect to the spectral measure (12.35) belongs to the spectrum follows from the criterion (12.28).

Proof. The proof is based on KAM theory. This is natural in view of the difficulty caused by small denominators, as explained above.

To state the problem in a form convenient for the application of KAM techniques, we write the Schrödinger equation with potential (16.29) as a system of first-order differential equations whose right-hand side does not depend explicitly on x (cf. (11.12) and (11.13)):

$$Y' = A(\xi)Y, \qquad \xi' = \alpha, \tag{16.38}$$

where $Y = (y', y)$, $\xi = (\xi_1, \dots, \xi_n) \in \mathbf{R}^n$ and $A(\xi) = \bar{A} + S(\xi)$, with

$$\bar{A} = \begin{pmatrix} 0 & -\lambda \\ 1 & 0 \end{pmatrix}, \qquad S(\xi) = \begin{pmatrix} 0 & Q(\xi) \\ 0 & 0 \end{pmatrix}.$$

Let us diagonalize $\bar{A} = A\big|_{Q=0}$ by performing the linear change of variables

$$Y = B\Psi_0, \qquad B = \begin{pmatrix} -i\sqrt{\lambda} & i\sqrt{\lambda} \\ 1 & 1 \end{pmatrix}.$$

As a result, the system becomes

$$\frac{d}{dx}\Psi_0 = iA_0\Psi_0 + S_0(\xi)\Psi_0, \qquad \frac{d\xi}{dx} = \alpha, \tag{16.39}$$

with

$$A_0 = \begin{pmatrix} -\sqrt{\lambda} & 0 \\ 0 & \sqrt{\lambda} \end{pmatrix}, \qquad S_0(\xi) = \frac{Q(\xi)}{2\sqrt{\lambda}} \begin{pmatrix} i & i \\ i & -i \end{pmatrix}.$$

Now assume that, for a given λ, there is an invertible, C^1 change of variables

$$\Psi_0 = V(\xi)\Psi \tag{16.40}$$

bringing the system (16.39) into the form

$$\frac{d\Psi}{dx} = iK\Psi, \qquad \frac{d\xi}{dx} = \alpha, \tag{16.41}$$

where K is a diagonal matrix with entries $\pm k(\lambda)$ independent of x, and $V(\xi)$ is a periodic function of the variables $\xi_1, \dots, \xi_n$ with period 1. It is easy to see that the Schrödinger equation with this λ will have solutions of the "quasi-Bloch" form (16.37). Thus, the proof of part (i) of the theorem reduces to the proof that (16.39) can be reduced to the form (16.41). The values for which this can be done form the set σ_1 of allowed values of λ. The possibility of this reduction is studied using the KAM technique of accelerated convergence, in a form thoroughly developed by Bogolyubov et al. [1976].

Notice first that the matrix S_0 in (16.39), for sufficiently small values of $\sup_{\xi\in\mathbf{R}^n} Q(\xi)/\lambda$, can be regarded as a small perturbation of A_0. Thus, we should seek the change of variables (16.40) in the form

$$\Psi_0 = \big(I + W_1(\xi)\big)\Psi_1, \tag{16.42}$$

where $W(\xi)$ is a periodic matrix-valued function of the same order of smallness as $S_0(\xi)$. Subtract from $S_0(\xi)$ the constant diagonal part, writing

$$S_0(\xi) = D_0 + \tilde{S}_0(\xi), \tag{16.43}$$

with $(D_0)_{ij} = \delta_{ij}\int_{\mathbf{T}^n}(S_0)_{ij}(\xi)\,d\xi$ for $i,j = 1,2$. Substituting (16.42) into (16.39), we obtain, after some simple manipulations,

$$\Psi_1' = (iA_0 + D_0)\Psi_1 + \big(i(A_0W_1 + W_1A_0) + \tilde{S}_0 - (\lambda, \nabla W_1)\big) + S_1,$$

where $S_1 = (I + W_1)^{-1}S_0W_1$ has a higher order of smallness than the other terms. If we now choose W_1 using the condition

$$i(A_0W_1 + W_1A_0) + \tilde{S}_0 - (\alpha, \nabla W_1) = 0, \tag{16.44}$$

we arrive again at an equation of the form (16.39), with the diagonal matrix $A_1 = A_0 - iD_0$ and the matrix S_1 of a higher order of smallness than S_0. Indeed, since the matrices $\tilde{S}_0(\xi)$ and $W(\xi)$ are periodic with period 1 in $\xi \in \mathbf{R}^n$, (16.44) shows that their Fourier transforms $S_0^{(m)}$ and $W^{(m)}$, for $m \in \mathbf{Z}^n$, are related as

$$W_{11}^{(0)} = W_{22}^{(0)} = 0 \tag{16.45a}$$

$$W_{jl}^{(m)} = \frac{(\tilde{S}_0^{(m)})_{jl}}{\pi i\big((m,\alpha) + (j-l)k_0(\lambda)\big)^{-1}} \tag{16.45b}$$

for $m \neq 0$, where

$$-2k_0(\lambda) = (A_0)_{11} + (A_0)_{22} = -2\sqrt{\lambda}. \tag{16.46}$$

Since the Fourier coefficients $\tilde{S}_0^{(m)}$ are proportional to the Fourier coefficients q_m of the function (16.29a), we conclude, by virtue of (16.32) and (16.33), that the coefficients $W_{jl}^{(m)}$ with $j = l$ will also satisfy (16.33) with an exponent $\rho_1 < \rho$. As for the off-diagonal entries of $W^{(m)}$, their exponential decay requires only that $(m, \alpha) \pm k_0(\lambda)$ decrease polynomially. This condition (cf. (16.31)) determines, to a first approximation, the forbidden bands (gaps) inside which we cannot even start to reduce (16.39) to the form (16.41).

Now introduce, for matrix-valued functions analytic on Π_ρ, the norm

$$\|S(\xi)\|_\rho = \max_l \sum_{j=1}^{2} \max_{\zeta\in\Pi_\rho} |S_{lj}(\zeta)|.$$

By making some estimates we find that if $\|S_0\|_\rho$ is small enough,

$$\|S_1\|_{\rho_1} \leqslant \|S_0\|_\rho^{1+\kappa} \tag{16.47}$$

for some $\kappa > 0$ and $\rho_1 < \rho$.

We have described the first step of the reduction procedure. The exact representation (16.40) is constructed by iterating this step, each time making the substitution $\Psi_l = (I + W_l)\Psi_{l+1}$ into the equation $\Psi_l' = iA_l\Psi_l + S_l\Psi_l$, with the diagonal matrix A_l independent of ξ and having entries $\pm k_l(\lambda)$. Requiring that the equation for Ψ_{l+1} should have the same form, we obtain $S_{l+1} = (I + W_{l+1})^{-1}S_l W_{l+1}$, $A_{l+1} = A_l + iD_l$, and $\tilde{S}_l = S_l - D_l$, where D_l is the diagonal and constant part of the matrix-valued function $S_l(\xi)$, and

$$i(A_l W_{l+1} - W_{l+1}A_l) + \tilde{S}_l - (\alpha, \nabla W_{l+1}) = 0.$$

From the latter relation, called a homological equation, the periodic matrix-valued function $W_{l+1}(\xi)$ can be found. If

$$\big|(m, \alpha) \pm k(\lambda)\big| \geqslant M_l^{1-\kappa/4}|m|^{-\beta},$$

where $M_l = \|S_l\|_{\rho_l}$, the matrix $W_{l+1}(\xi)$ will be small, $\|W_{l+1}\| \leqslant \frac{1}{10}M_l^\kappa$, and analytic in the region $\Pi_{\rho_{l+1}}$, for $\rho_{l+1} = \rho_l - 2\delta_l$, where $\delta_l = \delta_0(1+l)^{-\alpha}$ for some $\alpha > 1$ and δ_0 sufficiently small. We then find $|k_{l+1} - k_l| \leqslant \frac{1}{2}M_l$ and $M_{l+1} \leqslant M_l^{1+\kappa}$ (cf. (16.47)).

This latter inequality, a consequence of the absence of terms of lowest order of smallness at each iterative step, is the chief feature of the accelerated convergence method. It ensures that the influence of small denominators is compensated for. The width ρ_l of the domain of analyticity of the functions decreases during this procedure; however, it turns out that $\lim_{l\to\infty} \rho_l \geqslant \rho/2 > 0$ (see the papers referred to above and [Arnold 1979]). This has other important consequences: the total Lebesgue measure of the set of values of λ for which the incommensurability condition $\big|(m, \alpha) \pm k(\lambda)\big| \geqslant M_l^{1-\kappa/4}|m|^{-\beta}$ is not fulfilled is exponentially small (cf. (16.35)); the sequence of transformations $V_l = \prod_0^l (I + W_s)$ converges to a limit $V(\xi)$ (cf. (16.40)) uniformly in $\Pi_{\rho/2}$; and $k_l(\lambda)$ tends to a uniformly continuous function $k(\lambda)$ on σ_1, as $l \to \infty$.

Those are the basic steps in the proof of statement (i) in Theorem 16.6. This statement clearly implies that $\gamma(\lambda) = 0$ on σ_1, which in turn implies part (ii) of the theorem, because of the remark at the beginning of this section that, for almost-periodic potentials, Theorem 12.4 is valid for every function in Ω_q.

To prove part (iv), it is enough to show that

$$\lim_{\varepsilon\downarrow 0} \frac{1}{\pi} \operatorname{Im}\big(R(\lambda+i\varepsilon)\phi,\phi\big) = \frac{|\Phi_+(\lambda)|^2}{2\pi\sqrt{\lambda}\big(1+f(\lambda)\big)}, \tag{16.48}$$

where $R(z)$ is the resolvent, and use Theorem A.11 from the Appendix. But, by (12.29), the kernel of the resolvent, that is, the Green's function of the operator, can be expressed in terms of the Weyl solutions $\phi_\pm(x,z)$, which are square-integrable over the respective semiaxes, for z nonreal. Thus the problem is reduced to proving that these solutions converge uniformly to the solutions (16.37) for x in any finite interval and for $z = \lambda + i\varepsilon$, with $\lambda \in \sigma_1$ and $\varepsilon > 0$. This is done by generalizing to the case of complex $z = \lambda + i\varepsilon$ the reduction procedure for the system (16.39). Namely, we want to put this system in the form

$$\Psi' = i\begin{pmatrix} k_1(z) & 0 \\ 0 & k_2(z)\end{pmatrix}\Psi, \qquad \xi' = \alpha,$$

(where z takes the place of λ), through a change of variables $\Psi_0 = (I+W_z)\Psi$ analogous to (16.42). Repeating, with some minor changes, the calculations carried out for real λ, we see that, for $\lambda \in \sigma_1$ and $0 < \varepsilon < \varepsilon_0$, where ε_0 and λ_0^{-1} are small enough, there is such a matrix W_z, analytic in $\Pi_{\rho/2}$ and such that $\|W_z\|_{\rho/2} \leqslant \frac{1}{2}$, the limit $W = \lim_{\varepsilon\downarrow 0} W_{\lambda+i0}$ exists, $\lim_{\lambda\to\infty} \|W\|_{\rho/2} = 0$,

$$\lim_{\varepsilon\downarrow 0}\sup_{\lambda\in\sigma_1} \|W_{\lambda+i\varepsilon} - W\|_{\rho/2} = 0,$$

and

$$\sup_{\lambda\in\sigma_1} \big|k_j(\lambda+i\varepsilon) - (-1)^{j+1}k(\lambda)\big| \leqslant \text{const}\cdot\varepsilon$$

for $j = 1, 2$, where $\varepsilon/\sqrt{\lambda} \leqslant \operatorname{Im} k_j(\lambda+i\varepsilon) \leqslant 3\varepsilon/\sqrt{\lambda}$.

From this it follows that the solutions $\phi_\pm(x,z)$ obtained from the relation

$$\begin{pmatrix} \phi_+' & \phi_-' \\ \phi_+ & \phi_- \end{pmatrix} = B(I+W_z)\begin{pmatrix} e^{ik_1(z)x} & 0 \\ 0 & e^{-ik_2(x)z}\end{pmatrix}$$

for $\operatorname{Im} z > 0$, decay exponentially as $x \to \pm\infty$, that is, they are Weyl solutions, and that the limit (16.48) exists. As a result, we obtain statement (iv) for $1 + f(\lambda) = \det(I + W_{\lambda+i0})$ (for details, see [Dinaburg and Sinai 1975]).

Part (iii) follows from (16.37) and from arguments similar to those used to derive (16.11) from (16.1). □

Remarks. 1. We emphasize again that Theorem 16.6 is obtained by perturbation theory, and therefore has an asymptotic character: it applies either for large λ, or for potentials of small amplitude, such that $\sup_{x\in\mathbf{R}}|q(x)|\lambda^{-1} \ll 1$.

2. The neighborhoods $\tilde{O}_m$, for $m \in \mathbf{Z}^n$, that appear in the conditions of the theorem and are treated as gaps in the proof, do not necessarily coincide with the gaps in general: they just indicate their possible location and size (see also below). We see that the centers of these intervals are somewhat displaced from the points (16.30), and it is difficult to find them explicitly using the procedure above, although certainly k'_m tends to k_m as $\sup_{x\in\mathbf{R}}|q(x)|\lambda^{-1}$ tends to 0. Rüssmann [1978] improved the technique of the above proof and in particular the procedure for the asymptotic calculation of the quasimomentum $k(\lambda)$, for $\lambda \in \sigma_1$, as the solution of the equation

$$k = \sqrt{\lambda} - a(k, 1/\sqrt{\lambda}), \tag{16.49}$$

in which the function $a(k,t)$, for values of the parameter k satisfying condition (16.31) and for t small enough, is real analytic in k and can be represented as a uniformly convergent series in powers of $\sqrt{t}$. Here $k_m = k(k'_m)$, in agreement with Theorem 16.2 and part (iii) of Theorem 16.6, which says that $N(\lambda)$ is related to the quasimomentum by (16.11), as in the periodic case.

In [Belokolos 1976] there is an investigation of several properties of the quasimomentum.

3. Theorem 16.6 was extended by Parasyuk [1978] in a different direction. Instead of hypothesis (16.33) guaranteeing the analyticity of the potential, he assumed that for some $l > 2(n+1)$ the potential has $[l]$ continuous derivatives, where the $[l]$-th derivative satisfies the Hölder condition with exponent $l-[l]$. If, in addition, the Diophantine condition (16.32) is satisfied, the length of the "gaps" $\tilde{O}_m$ is of order $O\big(|m|^{-(l-1)/2+\delta}\big)$, for $\delta > 0$, in agreement with already known asymptotic estimates of the lengths of distant gaps in the spectrum of a periodic Schrödinger operator depending on the smoothness of the potential [Marchenko 1986]. This result was obtained by the Moser–Nash smoothing technique, known in small denominator theory [Bogolyubov et al. 1976; Arnold 1980].

4. A statement similar to Theorem 16.6 is true also in the discrete case, for the second-order operator h in $\ell_2(\mathbf{Z})$ given by (11.4) with potential (16.21), where $Q(t)$ is analytic in the domain $|\operatorname{Im} t| \leqslant \rho$ of the complex plane and the frequency α satisfies

(16.50)
$$\left|\alpha - \frac{b}{a}\right| \geqslant C|a|^{-2-\delta}$$

for some $C, \delta > 0$ and all $a, b \in \mathbf{Z}^n$ with $a \neq 0$. This was proved by Bellissard et al. [1983b], by extending the techniques from [Dinaburg and Sinai 1975] and [Belokolos 1975] to the discrete case. Namely, if $g = \sup_{|\operatorname{Im} t| \leqslant \rho} |Q(t)|$ is small enough, for every $\omega \in [0, 1)$ the absolutely continuous spectrum contains a nowhere dense set σ_1, independent of ω, of positive Lebesgue measure. This set is obtained as the image in $\mathbf{R}$ of the set

(16.51) $K_0 = \{k \in [0, \frac{1}{2}) : |\alpha m + k| \geqslant C|m|^{-1-\delta} \text{ for } m \in \mathbf{Z} \text{ and } m \neq 0\}$

under the analytic map on $|\operatorname{Im} k| \leqslant \rho_1 < \rho$ given by

(16.51a)
$$\Lambda(k, g) = 2\cos 2\pi k + \Lambda_1(k, g),$$

where $\Lambda_1(k, g)$ is constructed by a rapidly converging iterative process and $\Lambda_1(k, 0) = 0$ (recall that the inverse function to $\Lambda(k, 0) = 2\cos 2\pi k$ gives the dependence of the quasimomentum on λ for the unperturbed operator $h|_{g=0}$). In the continuous case, the analogous statement follows from (16.49).

On the set $\sigma_1 = \Lambda(K_0, g)$, the generalized eigenfunctions of h have the "quasi-Bloch" form

(16.52)
$$\psi_+(x, \lambda) = e^{2\pi i k(\lambda) x} u(x, \lambda), \qquad \psi_- = \bar{\psi}_+,$$

(cf. (16.37)), where

(16.52a)
$$u(x, \lambda) = U(\alpha x + \omega, \lambda)$$

and $U(t)$, for $t \in [0, 1)$, is a periodic function of period 1, analytic on the strip $|\operatorname{Im} t| \leqslant \rho_1$.

The authors used the technique from [Rüssman 1978], which allowed them to consider other variants of the lower bound (16.50), for example, using the function $\Omega(a)$ instead of $Ca^{-2-\delta}$, where

$$\Omega(t) = \begin{cases} C\exp(-t\ln^{-1-\delta} t) & \text{for } t \geqslant t_0, \\ \Omega(t_0) & \text{for } 0 \leqslant t \leqslant t_0, \end{cases}$$

where $C, \delta > 0$. Using this technique they were also able to estimate the measure of σ_1 (see 16.G).

Sinai [1985], under the same condition (16.50) and by a modification of the technique used in the proof of Theorem 16.6, proved that for small enough g there exists $\lambda_0(g, \delta) > 0$ such that, for any $0 < \lambda_1 < \lambda_0$, the set of $\lambda \in [-2, -2 + \lambda_1]$ for which the solutions (16.52) exist has measure no less than

$$\lambda_1 - \exp(-\lambda_1^{1/2+2\delta}(-\ln \lambda_1)^{\alpha_1}),$$

for $\alpha_1 = \alpha_1(g, \delta) > 0$, and the spectrum of the almost-Mathieu operator (the discrete Schrödinger operator with potential (16.19): see 16.F) is absolutely continuous on this set. Thus, in this case the absolutely continuous spectrum exists in any small enough right half-neighborhood of $\lambda = -2$, this being the left boundary of the spectrum of h_{AM} for $g = 0$.

This result is closely associated with the existence, for small g, of a quasiperiodic ground state of the nonlinear equation

$$-u(x+1) - u(x-1) + 2u(x) + g \sin 2\pi u(x) = 0$$

for $x \in \mathbf{Z}$, that is, of a solution having the minimum energy [Aubry and Le Daeron 1983].

Though the set σ_1 of Theorem 16.6 is nowhere dense by construction, neither this theorem nor its generalizations state that the spectrum is a Cantor set, because the proof does not give any information on the spectrum outside σ_1. In particular, Theorem 16.6 does not say that $\sigma = \sigma_1$. At the end of the previous subsection, for example, we discussed finite-band potentials. Such potentials are always analytic in a domain of the form (16.34), and generically quasiperiodic, with basis frequencies satisfying the incommensurability conditions (16.32). This is evidence that under the hypotheses of Theorem 16.6 the spectrum can even be finite-band.

Nonetheless, one is led to expect, based on the proof of Theorem 16.6, on numerical calculations [Hofstadter 1976; de Lange and Janssen 1981] and on various heuristic arguments [Sokoloff 1985] that a nowhere dense (Cantor) absolutely continuous spectrum is typical of a wide class of one-dimensional almost-periodic operators.

To support this conjecture, we cite examples of almost-periodic potentials with an absolutely continuous spectrum that is a Cantor set, constructed by Levitan and Savin [1984]. The authors proceed from quasiperiodic finite-band potentials, for which the density of the spectral measure can be computed explicitly, and use the technique of approximating a certain class of infinite-band potentials by finite-band ones [Levitan 1984] to prove that, if $q(x)$ is an infinite-band potential of this class, there exists for any $\varepsilon > 0$ a quasiperiodic potential $q_n(x)$ with n basis frequencies ($n \to \infty$ as $\varepsilon \to 0$), such that $\sup_{x \in \mathbf{R}} |q(x) - q_n(x)| < \varepsilon$, and the Schrödinger operator with potential $q_n(x)$ has an absolutely continuous, nowhere dense spectrum.

Here are corollaries of this statement, obtained by "closure" of results of the spectral theory of the Schrödinger operator with a finite-band potential:

(a) In any ε-neighborhood of a potential periodic of period a, and for any number a_1 incommensurable with a, there exists a quasiperiodic potential with frequency basis (a^{-1}, a_1^{-1}) and an absolutely continuous spectrum that is a Cantor set.

(b) For any module of real numbers that is not isomorphic to **Z**, there exists an almost-periodic potential having this module as its frequency module, and other spectral properties similar to (a).

16.C Lower Bounds of Solutions and Absence of a Point Spectrum

We now consider some results, mainly from [Gordon 1976, 1986], that guarantee that several operators—the discrete or continuous one-dimensional Schrödinger operator, a few more general ones—with an almost-periodic potential have no decreasing solutions and thus no point spectrum. In addition to their intrinsic interest, these results make it possible, together with those of the preceding subsection, to construct almost-periodic operators with a pure singular continuous spectrum (Theorem 16.17).

We start with the simplest case, the discrete Schrödinger operator

$$(h\psi)(x) = -\psi(x+1) - \psi(x-1) + q(x)\psi(x) \tag{16.53}$$

for $x \in \mathbf{Z}$. For each natural number a, set

$$\delta_q(x) = \inf_{q_a} \sup_{|x| \leqslant 2a} \left| q(x) - q_a(x) \right|, \tag{16.54}$$

where the infimum is taken over the set of periodic functions of period a.

(16.7) Theorem [Gordon 1976]. *If a bounded potential* $q(x)$, *for* $x \in \mathbf{Z}$, *is such that*

$$\liminf_{a\to\infty} e^{Ca}\delta_q(a) = 0$$

for all $C > 0$, *the spectrum of the corresponding operator* h (16.53) *has no point component.*

In [Gordon 1976] this was actually proved for the continuous case (see Theorem 16.10 below). The discrete version is, if anything, simpler.

Theorem 16.7 shows how far from the class of periodic operators the property of absence of eigenvalues is still to be found. Its proof is based on the following elegant lemma.

(16.8) Lemma. *If* B *is an invertible* 2×2 *matrix,*

$$\max_{m=\pm1,\pm2} \|B^m\xi\| \geqslant \tfrac{1}{2}\|\xi\|$$

for all $\xi \in \mathbf{C}^2$.

Proof. Let $p(t) = b_0 t^2 + b_1 t + b_2$ be the characteristic polynomial of B, so that $p(B) = 0$. Choosing k_0 so that $|b_{k_0}| = \max |b_k|$, we have

$$\xi = -\sum_{k \neq k_0} b_k b_{k_0}^{-1} B^{k-k_0} \xi,$$

which clearly implies the lemma. □

Proof of Theorem 16.7. Assume that h has an eigenvalue λ, that is, $h\psi = \lambda\psi$ for some $\psi \in \ell_2(\mathbf{Z})$. We write the solution of this equation in a form similar to (11.17):

$$\Psi(x) = \begin{pmatrix} \psi(x+1) \\ \psi(x) \end{pmatrix} = \mathcal{T}(x,0)\Psi(0).$$

Let $q_k(x)$ be a sequence of potentials, each periodic of period a_k, with $a_k \to \infty$, and $\mathcal{T}_k(x,0)$ and $\Psi_k(x)$ the matrices and vectors obtained by replacing $q(x)$ by $q_k(x)$ in (11.28) and (11.15), with the condition $\Psi_k(0) = \Psi(0)$. It is easy to see that

$$\max_{|x| \leqslant 2a} \|\Psi(x) - \Psi_k(x)\| \leqslant e^{C_1 a_k} \sup_{|x| \leqslant 2a_k} |q(x) - q_k(x)|,$$

where C_1 depends only on q and λ. From this and from the hypotheses of the theorem it follows that

$$\lim_{k \to \infty} \max_{|x| \leqslant 2a_k} \|\Psi(x) - \Psi_k(x)\| = 0.$$

But since $q_k(x)$ is periodic, Lemma 16.8 gives

$$\max_{m=\pm 1, \pm 2} \|\Psi_k(ma_k)\| = \max_{m=\pm 1, \pm 2} \|\mathcal{T}^m(a_k, 0)\Psi(0)\| \geqslant \tfrac{1}{2}\|\Psi(0)\|.$$

But this contradicts the fact that $\psi(x) \in \ell_2(\mathbf{Z})$, concluding the proof. □

A simple example of an almost-periodic potential satisfying the conditions of Theorem 16.7 is the limit-periodic function

$$q(x) = \sum_{l=1}^{\infty} q_l \cos \frac{2\pi x}{2^l}, \tag{16.55}$$

where

$$\limsup_{l \to \infty} e^{C 2^l} q_l = 0$$

for all $C > 0$. Clearly we can take the partial sums in (16.55) as approximating periodic potentials.

Here is another example, discovered by Avron and Simon [1982]:

(16.9) Corollary. *Let $q(x)$ be a quasiperiodic potential of the form* (16.29), *with $n \geqslant 1$, $Q(\xi)$ a function on the torus $\mathbf{T}^n$ satisfying a Hölder condition with exponent β, and $\alpha = (\alpha_1, \ldots, \alpha_n)$ satisfying*

$$\liminf_{k\to\infty} \bigl(\|k\alpha_1\| + \cdots + \|k\alpha_n\|\bigr)^{1/k} = 0,$$

where $\|\cdot\|$ denotes the distance to the nearest integer. Then the operator (16.53) *has no eigenfunctions, that is, $\sigma_{\mathrm{P}}(h) = \varnothing$.*

An example of α satisfying this hypothesis, for $n = 1$, is $\alpha = \sum_{i=1}^{\infty} 2^{-k_i}$, where k_i is defined by $k_1 = 1$ and $k_{i+1} = 3^{k_i}$.

The proof of the corollary is based on checking the conditions of Theorem 16.7. It is convenient to use

$$\delta_q^*(a) = \max_{-2a\leqslant x\leqslant a} \bigl|q(x) - q(x+a)\bigr| \tag{16.56}$$

as a measure of the proximity of the potential to periodic potentials, instead of the quantity $\delta_q(a)$ of (16.54). It is easy to see that

$$\tfrac{1}{4}\delta_q^*(a) \leqslant \delta_q(a) \leqslant 4\delta_q^*(a). \tag{16.57}$$

The quantity $\delta_q^*(a)$ is interesting because it highlights the fact that the potential repeats itself almost exactly on several adjacent intervals of the same length, as their lengths grow indefinitely. This is the property that implies, together with Lemma 16.8, that all solutions recur in successive intervals, a behavior that is obviously incompatible with decay at infinity.

In the examples above, this recurrence behavior was guaranteed by the fact that the potential is very well approximated, on four consecutive periods, by a periodic function whose period we can take as large as desired. A different way to guarantee that the potential has repeated stretches is exemplified in [Delyon and Petritis 1986] (see also Problems VII.9–10), where it is shown that if the function $Q(t)$ takes on only finitely many values (and is, as a consequence, discontinuous), the quasiperiodic potential (16.21) gets repeated exactly, in the case $n = 1$. The simplest example of such a function is (16.23) with $\beta < 1$.

We now consider the continuous version of Theorem 16.7.

(16.10) Theorem [Gordon 1976]. *Consider a one-dimensional Schrödinger operator H with a locally summable potential $q(x)$ such that*

$$\lim_{a\to\infty} \frac{1}{2a}\int_{-a}^{a} \bigl|q(x)\bigr|\,dx \leqslant M < \infty$$

and

$$\liminf_{a\to\infty} \delta_q(a)e^{Ca} = 0$$

for all $C > 0$, *where*

$$\delta_q(a) = \inf_{q_a} \int_{-2a}^{2a} |q(x) - q_a(x)|\, dx, \tag{16.58}$$

the infimum being taken over all real, summable potentials $q_a(x)$ *periodic of period* a. *Then the spectrum of* H *has no point component*: $\sigma_{\mathrm{P}}(H) = \varnothing$.

The proof uses the same arguments as that of Theorem 16.7, with help from the following technical lemma:

(16.11) Lemma. *Consider the Schrödinger operator with potential* $q(x)$ *such that*

$$\int_{\Delta_k} |q(x)|\, dx \leqslant C < \infty$$

for every k, *where the* Δ_k *are intervals of equal length* l *that tend to infinity. If* $y(x)$ *is the solution of the corresponding equation and*

$$\lim_{|x|\to\infty} \int_x^{x+1} |y^2(x)|\, dx = 0,$$

and if $r(x)$ *is as in* (11.1), *we have*

$$\lim_{k\to\infty} \sup_{x\in\Delta_k} r(x) = 0.$$

Proof. Set $A_k = \sup_{x\in\Delta_k} |y(x)|$. There exists a point $x_k \in \Delta_k$ at which $|y'(x_k)| \leqslant 2A_k/l$, so that, by the Schrödinger equation and the hypotheses of the lemma,

$$\sup_{x\in\Delta_k} |y'(x)| \leqslant C_1 A_k$$

for $k \in \mathbf{Z}_+$. Now let $|y(u_k)| = \sup_{x\in\Delta_k} |y(x)|$ and let Δ'_k be a half-neighborhood of u_k that lies entirely within Δ_k and has length $\delta = \min\{(2C_1)^{-1}, l/2\}$. Then

$$\int_{\Delta_k} y^2\, dx \geqslant \int_{\Delta'_k} y^2\, dx \geqslant \tfrac{1}{4} A_k^2 \delta.$$

By the hypotheses of the lemma, $\lim_{k\to\infty} A_k = 0$; from this and the preceding inequality, the lemma follows. □

Proof of Theorem 16.10. Under the hypotheses of the theorem, the Schrödinger operator defined on the linear manifold of a function $\psi(x)$, having compact support and an absolutely continuous derivative such that $-\psi''(x)+$

$q(x)\psi(x) \in L_2(\mathbf{R})$, is essentially self-adjoint [Dunford and Schwartz 1963]. Assume that this operator has an eigenfunction $\psi(x) \in L_2(\mathbf{R})$, and introduce the vector function $(\psi'(x), \psi(x))$ (cf. (11.12)). In Lemma 16.11, choose $\Delta_k = [m/a_k, ma_k]$ with $= 1, 2$, and $[ma_k, m/a_k]$ with $m = -1, -2$, where a_k is a sequence such that $\lim_{k\to\infty} a_k = \infty$. Then the conditions of the lemma are fulfilled, because for any periodic potential of period q_k and, say, $m = 1$,

$$\int_{\Delta_k} |q(x)|\, dx \leqslant \int_0^1 |q(x)|\, dx + \delta_q(a_k) < \infty.$$

Thus the lemma implies

$$\lim_{k\to\infty} \|\Psi(ma_k)\| = 0$$

for $m = \pm 1, \pm 2$. On the other hand, arguments like those in the proof of Theorem 16.7 show that there is a sequence a_k tending to ∞ and such that

$$\lim_{k\to\infty} \max_{m=\pm 1,\pm 2} \|\Psi(ma_k)\| \geqslant \tfrac{1}{2}\|\Psi(0)\|.$$

Since this contradicts the previous limit, the theorem is proved. □

One can easily check that the conditions of Theorem 16.10 hold for limit-periodic potentials

$$q(x) = \sum_{l=1}^{\infty} \sum_{k\in\mathbf{Z}} q_{l,k} e^{2\pi i x k/a_l}$$

for $x \in \mathbf{R}$ (cf. (16.55)), where

$$\lim_{l\in\infty} e^{Ca_l} \sum_{k\in\mathbf{Z}} |q_{l,k}| = 0$$

for all $C > 0$, and a_l is an increasing and diverging sequence of positive numbers such that the ratio a_{l+1}/a_l is a natural number. They also hold for the quasiperiodic potential (16.29), where the function $Q(\xi)$, for $\xi \in \mathbf{T}^n$, satisfies a Hölder inequality with exponent $\beta > 0$, and the frequencies satisfy the condition

$$\liminf_{l\to\infty} \left(\left\| l\frac{\alpha_2}{\alpha_1} \right\| + \cdots + \left\| l\frac{\alpha_n}{\alpha_1} \right\| \right)^{\frac{1}{l}} = 0 \tag{16.59}$$

(compare the hypotheses of Corollary 16.9). To verify this latter fact, it is convenient to use the quantity

$$\delta_q^*(a) = \int_{-2a}^{a} |q(x+a) - q(x)|\, dx$$

(cf. (16.56)), instead of $\delta_q(a)$ from (16.58). These two quantities are related by inequalities similar to (16.57).

In particular, the point spectrum of the Schrödinger operator will be empty if the potential is

$$q(x) = q_1 \cos 2\pi x + q_2 \cos 2\pi\alpha x, \tag{16.60}$$

where the irrational number α is such that, for an increasing sequence of natural numbers a_k, we have

$$\lim_{k\to\infty} \|a_k\alpha\|^{1/a_k} = 0. \tag{16.61}$$

Remarks. 1. The Hölder condition for $Q(\xi)$ in the representation (16.29) of the quasiperiodic potential is generally necessary, because one can construct [Gordon 1976; Johnson and Moser 1982] quasiperiodic potentials of the form (16.29) with continuous functions $Q(\xi)$ and any set of basis frequencies whose corresponding Schrödinger operator has eigenvalues at prescribed points, even a countable number of them forming a dense set. This is done using the phase formalism. This might appear to contradict the results of Section 2 to 5 to the effect that metrically transitive operators have no isolated or immobile (independent of ω) eigenvalues of finite multiplicity (Theorem 2.12) and that for second-order operators the left endpoint of the spectrum is not an eigenvalue, with probability 1 (for one can prescribe an eigenvalue at $\inf \sigma(H)$). But, in fact, the location and even the existence of these eigenvalues depend essentially on the realization $\omega \in \Omega = \mathbf{T}^n$ of the potential: isolated eigenvalues, for example, exist only for a set of measure zero of realizations, maybe only for a single realization.

2. Theorems 16.7 and 16.10 admit various extensions. One of them, to one-dimensional operators of higher order, is covered in Problems VII.11–12 and VII.15. Another one, to the multidimensional discrete Schrödinger operator, was carried out by Gordon [1986]. Unlike the one-dimensional case, where the absence of eigenvalues is implied by the condition of Theorem 16.7 that the potential be superexponentially approximated, along four periods, by periodic functions, in this case the approximation has to take place in all of $\mathbf{Z}^d$. More exactly, set

$$\bar{\delta}_q(a) = \inf_{q_a} \sup_{x\in\mathbf{Z}^d} \left|q(x) - q_a(x)\right| \tag{16.62}$$

for any $a = (a_1, \dots, a_d)$ with each $a_j > 0$, the infimum being taken over the set of periodic potentials on $\mathbf{Z}^d$ with period a (that is, $q_a(x_1, \dots, x_j + a_j \dots, x_d) = q_a(x_1, \dots, x_d)$ for any j). Then the point spectrum is empty if

(16.63) $$\liminf_{|a|\to\infty} \frac{\ln \bar{\delta}_q(a)}{|a|^{2d-1}\ln^2|a|} = -\infty.$$

This is proved by replacing the partial-difference equation by a system of ordinary equations in a slab whose cross-section is determined by the periods of the approximating potential, and then using results similar to those of Problem VII.12.

16.D Lower Bounds for the Lyapunov Exponent and Absence of an Absolutely Continuous Spectrum in the Discrete Case

We now consider the one-dimensional operator (16.53) with a potential of form (16.29), where the frequencies $\alpha_1, \dots, \alpha_n$ are incommensurate and the function $Q(\xi)$, for $\xi \in \mathbf{T}^n$, is a real trigonometric polynomial

(16.64) $$Q(\xi) = \sum_{m\in\mathcal{L}} q_m e^{2\pi i(m,\xi)},$$

$\mathcal{L}$ being a finite subset of $\mathbf{Z}^n$. More general operators are covered in Problems VII.17–19.

Being a quasiperiodic function on $\mathbf{Z}$, this potential gives a metrically transitive field on $\mathbf{Z}$ (see Example 1.15(g)), whose space of realizations is $\mathbf{T}^n$ with the Lebesgue measure. The simplest example of such a potential is given by (16.19), and the corresponding operator is called the almost-Mathieu operator and denoted by h_{AM}.

We will show that, for potentials of the type above, the Lyapunov exponent $\gamma(\lambda)$ of (11.25) has a lower bound that is a function of the coefficients q_m of (16.64). When the bound is positive, we can use Theorem 12.1 to conclude that the spectrum of the operator (16.53) has no absolutely continuous component.

The estimate we are going to discuss was discovered by André and Aubry [1980] for the potential (16.19), for which it has the form

(16.65) $$\gamma(\lambda, g) \geqslant \ln|g|.$$

The authors obtained this bound by using the Herbert–Jones–Thouless formula (11.79) and identities for $N(\lambda)$ that follow from the special symmetry of h_{AM}. This symmetry, known as duality, will be discussed in more detail in 16.F (see also Problem VII.5). A general interpretation of it in terms of Fourier transformation in certain C^*-algebras was given in [Bellissard and Testard 1985; Bellissard 1985].

Herman [1983] proposed a simple method, described below, to obtain estimates like (16.65) for many interesting cases. This method, like many other results of the spectral theory of almost-periodic and particularly quasiperiodic operators, makes essential use of the fact that the space of realizations of this class of metrically transitive operators is the torus $\mathbf{T}^n$, and therefore has a much richer analytic structure than in the general case. In particular, the estimation of the Lyapunov exponent depends essentially on the possibility of extending the fundamental matrix $\mathcal{T}(x,\omega)$ of (11.17) and (11.25) from $\mathbf{T}^n$ to the polydisk

$$\mathbf{D}^n = \{(\zeta_1,\dots,\zeta_n) \in \mathbf{C}^n : |\zeta_j| \leqslant 1 \text{ for } 1 \leqslant j \leqslant n\}, \tag{16.66}$$

and on the subsequent use of the subharmonicity of $\ln\|\mathcal{T}\|$ (cf. Lemma 11.4). Here we are considering $\mathbf{T}^n$ as a subset of $\mathbf{C}^n$ by identifying each factor circle with the unit circle in the corresponding factor of $\mathbf{C}^n$.

(16.12) Lemma [Herman 1983]. *Let f be an analytic map of the polydisk (16.66) into itself that fixes the origin, takes the torus $\mathbf{T}^n \subset \mathbf{D}^n$ into itself, and preserves the Lebesgue measure $d\omega$ when restricted to $\mathbf{T}^n$. Let $\tilde{A}(\zeta)$ be a $m \times m$ matrix holomorphic on $\mathbf{D}^n$, and set*

$$\tilde{\mathcal{T}}(x,\zeta) = \tilde{A}(f^x\zeta)\dots\tilde{A}(\zeta). \tag{16.67}$$

Then the quantity

$$\Gamma = \lim_{x\to\infty} \frac{1}{x}\int_{\mathbf{T}^n} \ln\|\tilde{\mathcal{T}}(x,\omega)\|d\omega$$

satisfies

$$\Gamma \geqslant \ln R(\tilde{A}(0)),$$

where $R(A)$ is the spectral radius of the matrix A, that is,

$$R(A) = \lim_{l\to\infty} \|A^l\|^{1/l} = \inf_{l\geqslant 1} \|A^l\|^{1/l}.$$

Proof. By arguments similar to those used to prove Lemma 11.4, we conclude that $\ln\|\tilde{\mathcal{T}}(x,\zeta)\|$, for fixed $x \in \mathbf{Z}^d$, is a plurisubharmonic function on $\mathbf{D}^n$ [Hayman and Kennedy 1980], and thus satisfies the multidimensional analogue of (11.38). Since f preserves the origin, this implies that the integral in the definition of Γ is bounded below by $\ln\|\tilde{\mathcal{T}}(x,0)\| = \ln\|(\tilde{A}(0))^n\|$. The proposition follows. □

(16.13) Theorem [Herman 1983; Pastur 1987]. *Let h be the operator* (16.53) *with potential* (16.64), *and m^* the point in $\mathbf{Z}^n$ with coordinates $m_j^* = \max_{m\in\mathcal{L}} m_j$, for $j = 1,\dots,n$. Then the Lyapunov exponent $\gamma(\lambda)$ of the equation $h\psi = \lambda\psi$ satisfies*

$$\gamma(\lambda) \geqslant \ln|q_{m^*}|. \tag{16.68}$$

Proof. As we have mentioned, h is metrically transitive with $\Omega = \mathbf{T}^n$, P the Lebesgue measure on $\mathbf{T}^n$, and group $\{T^x : x \in \mathbf{Z}\}$, where

$$(T\omega)_j = \omega_j + \alpha_j \pmod 1 \tag{16.69}$$

for $1 \leqslant j \leqslant n$. By definition (11.25),

$$\gamma(\lambda) = \lim_{x\to\infty} \frac{1}{x} \int_{\mathbf{T}^n} \ln\|\mathcal{T}(x,\omega)\|\, d\omega.$$

Now since each factor of $\mathbf{T}^n$ is the unit circle $\{e^{it} : 0 \leqslant t < 2\pi\}$, we can write T in the form

$$(T\zeta)_j = e^{2\pi i\alpha_j}\zeta_j \tag{16.70}$$

for $1 \leqslant j \leqslant n$. Recalling (16.64), we get

$$q(x,\omega) = \tilde{Q}(T^x\zeta) \tag{16.71}$$

if $\zeta_j = e^{2\pi i\omega_j}$ and

$$\tilde{Q}(\zeta) = \sum_{m\in\mathcal{L}} q_m\zeta^m$$

for $\zeta^m = \zeta_1^{m_1}\ldots\zeta_n^{m_n}$. Hence, the matrix $A(x,\omega)$ of (11.15) and (11.18) can be written as $A(x,\omega) = A(T^x\zeta)$, where

$$A(\zeta) = \begin{pmatrix} \tilde{Q}(\zeta) - \lambda & 1 \\ 1 & 0 \end{pmatrix}.$$

Since, by hypothesis, $\tilde{Q}(\zeta)$ is a polynomial, $\tilde{A}(\zeta) = \zeta^{m^*}A(\zeta)$ depends polynomially on $\zeta \in \mathbf{C}^n$, so (11.28) implies that $\|\mathcal{T}(x,\omega)\| = \|\tilde{\mathcal{T}}(x,\zeta)\|$ for $\zeta = e^{2\pi i\omega}$, where $\tilde{\mathcal{T}}(x,\zeta)$ is as in (16.67). But since $\tilde{A}(\zeta)$ is a polynomial in ζ, the matrix and the map T from (16.69) can be analytically continued to the polydisk (16.66), and

$$\tilde{A}(0) = \begin{pmatrix} q_{m^*} & 0 \\ 0 & 0 \end{pmatrix}.$$

Hence, by applying Lemma 16.12, we obtain the theorem. □

16.14. Examples

(a) Let $n = 1$ in (16.64), so that

$$q(x) = \sum_{|m|\leqslant M} q_m e^{2\pi i m(\alpha x+\omega)}$$

with $M < \infty$ and α an irrational number. Then Theorem 16.13 gives

(16.72) $$\gamma(\lambda) \geqslant \ln |q_M|.$$

This was first proved by Herman [1983]. Inequality (16.65), proved by André and Aubry, is clearly a particular case of (16.72), for $M = 1$.

(b) Consider the potential (16.64) with an arbitrary finite number of basis frequencies. Theorem 16.13 gives a trivial bound for the potential

$$2\sum_{j=1}^{n} g_j \cos 2\pi(\alpha_j x + \omega_j),$$

when $q_{m^*} = q_{(1,\ldots,1)} = 0$, but a nontrivial one for the equally simple-looking potential

(16.73) $$2^n g \prod_{j=1}^{n} \cos 2\pi(\alpha_j x + \omega_j),$$

for which $q_{m^*} = q_{(1,\ldots,1)} = g$. Thus for (16.73), as for (16.19), we arrive at (16.65). We can regard (16.73) as an analogue of (16.19) for an arbitrary number of basis frequencies.

Here is an example of Lemma 16.12 being used to treat a potential more complicated than almost-periodic ones.

(16.15) Theorem [Pastur 1987]. *Let the potential in* (16.53) *be of the form*

(16.74) $$q(x) = \sum_{|m|\leqslant M} q_m e^{2\pi i m p_n(x)},$$

where $q_m^* = q_{-m}$ *and* $p_n(t) = \alpha_0 t^n + \cdots + \alpha_n$ *is a polynomial of degree* n *with real coefficients, at least one of which is irrational. Then the Lyapunov exponent of the equation* $h\psi = \lambda\psi$ *satisfies* (16.72).

Proof. Assume first that α_0 is irrational. We will use the fact [Cornfeld et al. 1980] that for any polynomial $p_n(x)$ as in the statement of the theorem, there are natural numbers $p_{j,k}$, for $1 \leqslant k < j \leqslant n$, and an irrational number α, such that the map $T : \mathbf{T}^n \to \mathbf{T}^n$ given by

$$T\omega = (\omega_1 + \alpha, \omega_2 + p_{2,1}\omega_1, \ldots, \omega_n + p_{n,1}\omega_1 + \cdots + p_{n,n-1}\omega_{n-1}) \pmod 1$$

satisfies $p_n(x) = (T^x\omega_0)_n \pmod 1$ for some $\omega_0 \in \mathbf{T}^n$. The map T preserves Lebesgue measure and is metrically transitive; it is called the complex skew shift on $\mathbf{T}^n$. For example, in the case $n = 2$ and $p_2(x) = \alpha x^2 + (2\omega_1 - \alpha)x + \omega_2$, where α is irrational, we have

(16.45) $$T^x\omega = \bigl(\omega_1 + \alpha x, \omega_2 + 2x\omega_1 + x(x-1)\alpha\bigr).$$

Choosing $\alpha = \alpha_0$, $\omega_1 = \frac{1}{2}\alpha_1 + \alpha$ and $\omega_2 = \alpha_2$, we obtain $p_2(x)$.

We will assume from now on that $n = 2$, to avoid cumbersome formulas, but the case of higher n is in every way similar.

To obtain (16.72), note that the map T specified by (16.75) can be analytically continued from $\mathbf{T}^2$, regarded as a manifold in $\mathbf{C}^2$, to the whole of $\mathbf{C}^2$:

$$T\zeta = (e^{2\pi i\alpha}\zeta_1, e^{2\pi i\alpha}\zeta_1^2\zeta_2).$$

From this it follows that, setting

$$\tilde{Q}(\zeta) = \sum_{|m|\leqslant M} q_m \zeta_1^{2m}\zeta_2^m,$$

the potential (16.74) can be expressed in a form similar to (16.71), for $\zeta_j = e^{2\pi i\omega_j}$, for $j = 1, 2$. The rest of the argument is just as in the proof of Theorem 16.13; it leads to (16.72) if the leading coefficient in the polynomial $p_n(x)$ in (16.74) is irrational. The general case can be proved similarly. □

Here is a direct corollary of Theorems 16.13 (or 16.15) and 12.1:

(16.16) Theorem. *Let h be the metrically transitive operator* (16.53) *with potential $q(x)$, for $x \in \mathbf{Z}$, satisfying the conditions of Theorem* 16.13 (*or* 16.15), *where* $|q_{m^*}| > 1$ (*or* $|q_M| > 1$). *Then the spectrum of h for Lebesgue-almost all $\omega \in \mathbf{T}^n$ has no absolutely continuous component.* □

The right-hand sides of all the lower bounds obtained above for the Lyapunov exponent are the same: they represent logarithms of the absolute value of the leading coefficient of the trigonometric polynomial that determines the potential, and are independent of λ. Problems VII.17–19 will treat more complicated operators, for which the estimates are in terms of λ and other constants determining the coefficients.

Note that the conditions of Corollary 16.9, guaranteeing that the operator h with quasiperiodic potential (16.29) has no point spectrum, can be satisfied also for the trigonometric polynomials considered in this subsection, because here we don't impose any arithmetic conditions on the frequencies. Therefore, combining this corollary with Theorem 16.16, we obtain the following statement, first proved for the important case of the potential (16.19) by Avron and Simon [1983]. The statement about multiplicity is a consequence of Theorem 12.9.

(16.17) Theorem. *For a quasiperiodic potential satisfying the conditions of Corollary* 16.9 *and Theorem* 16.16, *the spectrum of the operator* (16.53) *is singular continuous and has multiplicity* 1 *for Lebesgue-almost $\omega \in \mathbf{T}^n$.* □

Thus, we have obtained a class of metrically transitive operators with a pure singular continuous spectrum and a positive Lyapunov exponent. Below, in Section 18, we will also consider h with nonsmooth quasiperiodic potentials (16.20) having a similar property. By Theorem 13.6, in such a situation, the singular continuous spectrum is unstable under perturbations of finite rank, being turned into a point spectrum, which, according to Section 15, is more typical of differential and finite-difference metrically transitive operators of order two with positive Lyapunov exponent and random coefficients.

In spite of their instability, however, such examples are important, for several reasons. First, they display the nontrivial role of sets of measure zero in the proof that the typical spectrum is pure point, as discussed in Section 13. Secondly, they show that the role of generalized eigenfunctions corresponding to the singular continuous spectrum may be played not only by solutions that tend to zero at infinity but are not square-summable, but also by solutions that do not tend to zero at infinity (in the proofs of Theorems 16.7 and 16.10, $\limsup_{|x|\to\infty} r(x) \geqslant \frac{1}{2} r(0)$, where $r(x)$ is the envelope of the solution: see (11.1) and (11.5)). Other examples [Bellissard et al. 1982] and numerical and heuristic calculations [Prange et al. 1984] show that these solutions are characterized by more or less chaotic behavior, taking place at an infinite hierarchy of scales.

This type of behavior for generalized eigenfunctions, which seems characteristic of metrically transitive operators with a singular continuous spectrum, makes it possible to use techniques from the modern theory of dynamical systems (dynamic chaos) and statistical physics (renormalization groups) in the spectral analysis of such operators [Kohmoto and Oono 1984; Ostlund and Pandit 1984; Kalugin et al. 1986]. The almost-periodic functions employed as potentials by these authors are considered as reasonable models [Katz and Dunneau 1986] of the effect of the recently discovered quasicrystalline structures [Schechtman et al. 1984] on the spectrum of elementary excitations (electrons, phonons, etc.) in the structure.

The simplest almost-periodic potential used to model quasicrystals has the form (16.20), where $Q(t)$ is the characteristic function of some interval in $[0, 1)$. Since $Q(t)$ is not a trigonometric polynomial, we cannot prove the absence of an absolutely continuous spectrum by the method developed above, based on Herman's lemma (Lemma 16.12). Nevertheless, in all such cases for irrational α and almost all $\omega \in [0, 1)$ we have $\sigma_{\text{ac}} = \varnothing$ (see [Kotani 1989] and Problem V.48). One more criterion for the absence of the absolutely continuous spectrum was given by Simon and Spencer [1989]. In particular, for the one-dimensional Schrödinger operator h we have $\sigma_{\text{ac}} = \varnothing$ if the potential $q(x)$ (not necessarily metrically transitive) is unbounded, that is, $\lim_{x\to\infty}|q(x)| = \lim_{x\to-\infty}|q(x)| = \infty$. Thus, in the case of the potential (16.20) we have $\sigma_{\mathrm{ac}} = \varnothing$ if $Q(t)$ is an unbounded periodic function of period 1.

16.E Point Spectrum of Almost-Periodic Operators

In 16.C we mentioned examples, due to Gordon [1976], of one-dimensional Schrödinger operators with quasiperiodic potentials and having eigenvalues (for a simple version of a similar example considered in [Johnson and Moser 1982], see Problem VII.16). Refinements of these examples lead to almost-periodic one-dimensional Schrödinger operators with a countable, and even dense, discrete spectrum. But it is unclear if the spectrum is pure point in these examples.

A potential for which the spectrum of the Schrödinger operator is pure point was given by Molchanov and Chulaevskii [1984]. The potential $q(x)$ is a uniform limit-periodic function approximable by random periodic functions $q_l(x)$ whose period a_l increases very rapidly: $a_{l+1} \geqslant a_l \exp(a_l^{1+\delta})$, for $\delta > 0$. But the rate of approximation is not very high: $\sup_x |q(x) - q_l(x)| \geqslant a_l^{-2+\delta}$, for $\delta > 0$. (In Section 17 we will see that if the rate of approximation is high enough, $\sup_x |q(x) - q_l(x)| \leqslant \exp(-a_{l+1}^{1+\delta})$, for $\delta > 0$, the spectrum is purely absolutely continuous, though possibly nowhere dense, no matter how fast the periods increase.) In this situation, the spectrum is pure point and nowhere dense, the eigenfunctions decrease as $|x| \to \infty$ more rapidly than any power of $|x|^{-1}$, but the Lyapunov exponent is zero on the spectrum.

Subsequently, it was shown by Bellissard et al. [1985b] that the almost-Mathieu operator (16.19), (16.53), for g large enough and α Diophantine, has a big point component, whose measure tends to the measure of the whole spectrum as $g \to \infty$: more exactly, $\operatorname{meas} \bar{\sigma}_{\mathrm{p}} = \big(4 - o(1)\big)g$ as $g \to \infty$. This was proved using the discrete version of Theorem 16.6 and the duality property of the almost-Mathieu operator, which allows one to relate to each other solutions for large and small g. We will discuss this result in detail in 16.F.

As for Gordon's examples and their extensions, their construction begins with the construction of eigenfunctions, which are then used to find the potential. This path from the spectral data to the potential is characteristic of inverse problems of spectral theory.

It turns out that, for problems about the point spectrum of discrete almost-periodic operators, this inverse approach can be pushed further. It yields operators with a pure point spectrum concentrated on a wide variety of sets: from the whole interval $[0, 1]$ to the Cantor set in that interval. A procedure for this, implemented by Craig [1983] and Pöschl [1983], is essentially a version of perturbation theory, where the discrete Laplacian is regarded as small. The procedure starts with the almost-periodic sequence $\lambda(x)$, for $x \in \mathbf{Z}^d$, that should be the set of eigenvalues of the desired operator:

$$(h\psi)(x) = -\varepsilon \sum_{|x-y|=1} \psi(y) + q(x)\psi(x) = ((h_0 + Q)\psi)(x). \tag{16.76}$$

We wish to find, for $\varepsilon > 0$ small enough, an almost-periodic sequence $q(x)$ and a unitary operator V such that

$$V^{-1}(h_0 + Q)V = \Lambda, \tag{16.77}$$

where Λ is the diagonal operator in $\ell_2(\mathbf{Z}^d)$ defined by the sequence $\lambda(x)$.

The method of solution, borrowed from KAM theory, is similar in many ways to the one used in Theorem 16.6. We construct V and Q recursively, the following condition being true at the l-th step:

$$V_l^{-1}(h_0 + \Lambda + \delta Q_l)V_l = \Lambda + \delta\Lambda_l, \tag{16.78}$$

where

$$V_l = \prod_{s=0}^{l}(I + W_s), \qquad \delta Q_l = \sum_{s=1}^{l} Q^{(s)}. \tag{16.79}$$

The norms of W_l, $Q^{(l)}$ and $\delta\Lambda_l$ (defined appropriately) is of order ε^{2^l}; also, $Q^{(l)}$ is diagonal, and the (x, y)-entries of W_l and $\delta\Lambda_l$ are bounded above by $C_l e^{-\rho_l|x-y|}$, for some $\rho_l > 0$. For $l = 0$, we assume $W_0 = \delta Q_0 = 0$ and $\delta\Lambda_0 = h_0$. To go from step l to $l + 1$, we demand, following the basic idea of the KAM method, that there should be no terms linear in the small operators. This leads to

$$\begin{aligned} [W_{l+1}, \Lambda] &= \delta\Lambda_l + V_l^{-1}Q^{(l+1)}V_l, \\ \delta\Lambda_{l+1} &= (I + W_{l+1})^{-1}(\delta\Lambda_l + V_l^{-1}Q^{(l+1)}V_l)W_{l+1}, \end{aligned} \tag{16.80}$$

where $[A, B] = AB - BA$. The first of these relations requires that the sum on its right-hand side have no diagonal entries, so it determines the diagonal operator $Q^{(l+1)}$. We can then find the operator W_{l+1} with zero main diagonal and all other entries containing denominators $(\lambda(x) - \lambda(y))^{-1}$ (cf. (19.45)). As in the ordinary KAM method [Arnold 1980], if the nonresonance condition

$$|\lambda(x) - \lambda(y)| \geqslant C|x - y|^{-\beta} \tag{16.81}$$

is satisfied for some $C, \beta > 0$ and all $x \neq y$ (cf. (16.30)), the presence of these denominators results in a lower rate ρ_{l+1} of exponential decay for the entries of W_{l+1}, compared to those of W_l. A standard analysis in KAM theory then shows that the quadratic convergence characteristic of KAM results in $\rho_\infty \geqslant \rho/2$ in the limit $l \to \infty$ and therefore in this limit, (16.77) follows from (16.78). For details, see [Craig 1983; Pöschl 1983].

An important feature of this spectral procedure is that the diagonal matrix Λ in equation (16.80), usually called *homological*, must not be modified from one step to another, as it can be in the more traditional *direct problem*, where we have Q and seek V and Λ. An example of such a modification was given in 16.B, where we single out the diagonal part of the zeroth Fourier coefficient of the small operator in order to ensure the solvability of the homological equation (16.44).

Here, for the operator (16.76), a similar modification is needed, for essentially the same reason. If we were using KAM to solve the direct problem, we would assume that $V_l^{-1}(h_0 + Q)V_l = \Lambda_l$ at the l-th step, where V_l is of the form (16.79) and $\Lambda_l - Q$ is small in ε and tends to a diagonal operator as $l \to \infty$. Then, requiring that in the transition $l \to l+1$ there be no terms linear in W_{l+1} and $\Lambda_l + Q$, and that the homological equation arising from this requirement should be solvable, we would arrive at the relations

$$(16.82)\qquad \begin{aligned} [W_{l+1}, \Lambda_l'] &= \Lambda_l'', \\ \Lambda_{l+1}' &= \Lambda_l' + \left((I + W_{l+1})^{-1}\Lambda_l'' W_{l+1}\right)_d, \\ \Lambda_{l+1}'' &= \left((I + W_{l+1})^{-1}\Lambda_l'' W_{l+1}\right)_{od}, \end{aligned}$$

where A_d and A_{od} are the diagonal and off-diagonal parts of the matrix operator A (cf. (16.80)). As in Theorem 16.6, the modification at each step of the diagonal operator in the homological equation, designed to ensure that the equation is solvable, has the result that the nonresonance condition should be satisfied not only by the unperturbed matrix Λ, as in the inverse problem, but by all the modified Λ_l' as well. In the proof of Theorem 16.6, this was taken care of by excluding more intervals (possibly gaps) from consideration at each step, so that the set σ_1 of values of λ for which the reduction was accomplished was nowhere dense. But in the current case we must require that (16.81) be satisfied not only by $q(x)$, but by all sufficiently close sequences. This requirement that the nonresonance condition be stable forces the exclusion of all but very singular sequences. Thus, according to Bellissard et al. [1983a] and Pöschl [1983], the direct spectral problem (16.77) can be solved by this method if

$$(16.83)\qquad q(x) = Q\big((\alpha, x) + \omega\big),$$

for $x \in \mathbf{Z}^d$ and $\omega \in [0, 1)$, where $\alpha = (\alpha_1, \ldots, \alpha_d)$ is made up of independent frequencies satisfying (16.31), and the real function $Q(t)$, periodic of period 1, is monotone on the period and can be analytically continued to some strip in the complex plane. Thus, $Q(t)$ should have poles on the real axis. A typical example of such a potential is (16.20). (However, the spectral analysis of this potential and of its multidimensional analogues can be carried out fairly extensively and explicitly, not only for ε small but for all ε: see Section 18.)

Returning to the inverse problem, let's formulate more exactly the results by Craig [1983] and Pöschl [1983]. Let $\mathfrak{A}$ be the Banach algebra of real-valued sequences $f(x)$, for $x \in \mathbf{Z}^d$, with a norm invariant under shifts: if $(U_a f)(x) = f(x+a)$ and $f \in \mathfrak{A}$, then $U_a f \in \mathfrak{A}$ and $\|U_a f\|_{\mathfrak{A}} = \|f\|_{\mathfrak{A}}$. Then the analogue of (16.81) takes the form

$$\|(\lambda - U_a\lambda)^{-1}\|_{\mathfrak{A}} \leqslant C^{-1}|a|^{\beta} \tag{16.84}$$

for all $a \neq 0$ and some $C, \beta > 0$. If the sequence $\lambda(x)$ satisfies this condition and Λ is the diagonal operator defined by it, then for every $0 \leqslant \varepsilon \leqslant \varepsilon_0$, where ε_0 is small enough, there exist a sequence $q \in \mathfrak{A}$ and a unitary operator V such that (16.77) is satisfied, $\|q - \lambda\|_{\mathfrak{A}} \leqslant (\varepsilon/\varepsilon_0)^2$, and the entries $w(x,y)$ of $W = I - V$ satisfy

$$\|w(\,\cdot\,,\,\cdot\,+y)\|_{\mathfrak{A}} \leqslant \operatorname{const} \varepsilon e^{-b|y|}$$

for $y \in \mathbf{Z}^d$, where $b = 1 + \ln(\varepsilon_0/\varepsilon)$. Thus, the operator (16.76) with this potential has the prescribed pure point and simple spectrum:

$$\sigma = \bar{\sigma}_{\mathrm{p}}, \qquad \sigma_{\mathrm{p}} = \{\lambda(x) : x \in \mathbf{Z}^d\}, \tag{16.85}$$

and its eigenfunctions decay exponentially at infinity, with a high enough rate.

This spectral structure seems natural enough for almost-periodic operators of the form (16.76) with ε small (or a large potential), in view of the heuristic arguments presented in Sections 6, 9 and 13. Although those arguments, especially in Sections 6 and 9, concerned the fluctuation spectrum of random operators, they apply wherever the spectrum is due mainly to independent quantization in various potential wells (fluctuations of the potential), and resonance tunneling effects are suppressed and only smear slightly the eigenfunctions outside the wells, so that they decay rapidly at infinity.

This picture of the fluctuation spectrum of random operators, given in 6.B and in Sections 9, 10 and 15, obtains because the potential wells are exceedingly rare (in Section 9 we saw that as $\lambda \downarrow 0$, where 0 is the bottom of the spectrum, the wells are of size $1/\sqrt{\lambda}$, and are separated by about $N^{-1/d}(\lambda) \sim \exp(\operatorname{const} \lambda^{-d/2})$), and are almost certainly of various shapes. In the present case, and especially in Section 18, the insignificant role of resonance tunneling (which is the mechanism of delocalization of solutions) and the resulting point spectrum are due to the small value of the multiplier of the Laplacian, which is responsible for the tunneling, and to condition (16.81), which guarantees large enough distances between almost identical (and thus almost resonant) wells.

(16.18) Examples

(a) For $l = 1, 2, \ldots$, let $X_l \subset \mathbf{Z}$ be defined by

$$X_l = 2^l \mathbf{Z} + \begin{cases} [0, 2^{l-1}) & \text{if } l \text{ is even,} \\ [2^{l-1}, 2^l) & \text{if } l \text{ is odd,} \end{cases}$$

where + denotes arithmetic sum. The characteristic function χ_{X_l} is periodic of period 2^l, and for any integer L, $(\chi_{X_1}, \ldots, \chi_{X_L})$ gives a one-to-one map from $\mathbf{Z}/2^L\mathbf{Z}$ onto $\{0,1\}^L$. Further, if $\chi_{X_{l+m}}(x) = 1$ for $m = 1, 2, 3$ and if $|x - y| \leqslant 2^l$, we have $\chi_{l+3}(x) = 1$. Now set

$$\lambda_1(x) = \sum_{l=1}^{\infty} 2^{-l} \chi_{X_l}(x),$$

and choose as the algebra $\mathfrak{A}$ the closure, in the norm of $\ell_\infty(\mathbf{Z})$, of the set of periodic sequences of period 2^l, for all l. Then $\lambda_1 \in \mathfrak{A}$ satisfies condition (16.81) for $C = \frac{1}{16}$ and $\beta = 1$, and since the partial sum of the series with l summands is a periodic function of period 2^l taking on the values $n2^{-l}$, for $0 \leqslant n \leqslant 2^l$, we see that $\lambda_1(x)$ is a uniform limit-periodic function on $\mathbf{Z}$, and that its image is dense in $[0, 1]$. This gives an example of the operator (16.76) with a limit-periodic potential and a pure point spectrum coinciding with $[0, 1]$.

(b) Set

$$\lambda_2(x) = 2 \sum_{l=1}^{\infty} 3^{-l} \chi_{X_l}(x),$$

where X_l and $\mathfrak{A}$ are as in part (a). Then $\lambda_2 \in \mathfrak{A}$ satisfies (16.81) with $C = \frac{1}{3}$ and $\beta = \log_2 3$, and its image is dense in the standard Cantor set of real numbers in $[0, 1]$ whose base-three expansion contains only zeros and twos; this set is the point spectrum in this example.

(c) Consider the algebra of real periodic functions of period 1 and bounded variation over a period. Let the norm be $\|\Lambda\|_{L_\infty(\mathbf{R})} + \operatorname{Var} \Lambda$, and let α be a real vector satisfying (16.32). Set

$$\lambda(x) = \Lambda\big((\alpha, x)\big) \tag{16.86}$$

for $x \in \mathbf{Z}^d$: it is easy to see that this sequence cannot satisfy (16.81) if $\Lambda(t)$ is continuous, because, by the mean value theorem, there would exist for every noninteger (α, x) some number t such that $\Lambda(t) = \Lambda\big(t + (\alpha, x)\big)$. Thus, here one cannot obtain the sequence $\lambda(x)$, which is uniform almost-periodic, but only a sequence belonging to the Besicovitch space b_p, with norm

$$\|f\|_{b_p} = \lim_{V\to\infty}\left(\frac{1}{V}\sum_{x\in\Lambda}|f(x)|\right)^{\frac{1}{p}}, \tag{16.87}$$

where Λ is a cube in $\mathbf{Z}^d$ centered at the origin and V is the number of points in it.

The simplest example of a $\Lambda(t)$ satisfying these conditions is the extension of the function t on $[0,1)$ to a periodic function of period one (or, in general, the extension of any bounded monotone function). By the remarks above, in this case there is a periodic function $Q(t)$ of period 1 which gives rise, via (16.83), to a quasiperiodic potential for the operator (16.76) so that the spectrum (16.85) is pure point and that $\|q-\lambda\|_{b_p} = O(\varepsilon^2)$.

It is easy to calculate $N(\lambda)$ for the operator (16.53) in this example (see Problem II.2):

$$N(\lambda) = \operatorname{meas}\{t\in[0,1) : \Lambda(t)\leqslant\lambda\}. \tag{16.88}$$

It is clear that this construction will yield operators with a great variety of sets as spectra, including nowhere dense sets of positive and zero measure (for example, if $\Lambda(t)$ is the "devil's staircase," the function inverse to the standard Cantor function). One can also obtain integrated densities of state for which the gap labeling theorem 16.2 or the estimate (11.83) are not valid: see Problems VII.22-23.

Thus, we have a class of discrete almost-periodic operators with pure point spectrum that displays the variety of interesting phenomena possible within the general spectral theory of metrically transitive operators. But the operators in this class are given by rather irregular functions: see especially Example 16.18(c). In the direct spectral problem (16.77), for example, $Q(t)$ must have poles on the real axis (see (16.20)); this is what ensures that the nonresonance condition (16.81) holds, and therefore allows us to use more or less standard tools of KAM theory.

In order to consider other metrically transitive potentials, we need other methods. For example, while studying the one-dimensional random operators of order two in Sections 6 and 11–15, we used the phase formalism and a probabilistic Markov technique. In Section 17, the one-dimensional continuous Schrödinger operator with limit-periodic potential will be treated by means of a version of the phase formalism and the technique of approximation by periodic potentials.

For multidimensional operators of the form (16.76) with a random potential and small ε (or arbitrary ε if we consider only the fluctuation region), one can use a method developed in [Fröhlich and Spencer 1983; Jona-Lasinio et al. 1984; Fröhlich et al. 1985], ultimately deriving from KAM theory. The idea is to impose the nonresonance conditions on the values of the potential not in the whole space at once, but on a sequence of bounded, increasing regions, which eventually exhaust space. One proves the existence of exponentially decaying solutions in certain subregions of these regions, and then takes into account the influence of the rest of space (tunneling effects) at the cost of a decreasing rate of decay. Further, the probability that the nonresonance condition is fulfilled can be estimated using Theorem 4.21 (see Problem II.1), and one can show that the spectrum is pure point using the Borel–Cantelli lemma. (For variations on this last step and for the proof that the spectrum is pure point based on the existence of exponentially decaying solutions, see [Delyon et al. 1985a; Simon and Wolf 1986].) Thus the method owes something not only to KAM theory, but to the renormalization group techniques [Wilson and Kogut 1975] widely employed in theoretical and mathematical physics to study exactly those problems that involve an infinite hierarchy of spatial or temporal scales.

This scheme allows the detailed study of the fluctuation spectrum of a wide class of random operators: see 15.C and the surveys [Spencer 1985; Martinelli and Scoppola 1986]. A version of it was used by Sinai [1987] as the basis for the study of the point spectrum of certain almost-periodic operators.

Sinai considered (16.76) in one dimension with potential (16.83), where $Q(t)$ is periodic of period 1, twice differentiable, and has exactly one maximum and one minimum in the period (as in (16.19), for example), and where the continued fraction expansion $[k_1, \dots, k_l, \dots]$ of α satisfies $k_l \leqslant \text{const}\, l^2$ and the limit $\lim_{l\to\infty} l^{-1} \ln a_l$ exists and is finite, where a_l is the denominator of the truncated partial fraction $[k_1, \dots, k_l]$. The set of values of α with this property has full Lebesgue measure [Khintchine 1971].

Under these conditions, he proved that, if ε is small enough in (16.76), the spectrum is pure point for Lebesgue-almost all $\omega \in [0,1]$ in (16.21), and that, again for almost all ω, each eigenfunction $\psi(x)$ decays exponentially outside a bounded set $Z(\psi)$, called the essential support of the function. Furthermore, the set of all eigenvalues and the basis of eigenfunctions can be written as

(16.89a) $$\sigma_{\mathrm{p}}(\omega) = \sum_{n\in\mathbf{Z}} \mathcal{L}(T^n\omega),$$

(16.89b), $$\ell_2(\mathbf{Z}) = \overline{\langle \textstyle\bigcup_{n\in\mathbf{Z}} U^{-n}\Phi(T^n\omega)\rangle}$$

where $\mathcal{L}(\omega)$ is a set of numbers depending on ω, $T^n\omega = \alpha n + \omega \pmod 1$, $\Phi(\omega)$ is the set of functions $\psi(x,\omega)$ that decay exponentially outside a finite interval $Z_0(\psi) \subset \mathbf{Z}$ containing the origin, U is the shift operator on $\ell_2(\mathbf{Z})$, and $\langle \cdot \rangle$ denotes the linear span.

Clearly (16.89) holds for the almost-periodic operator of multiplication by the sequence $Q(\alpha x + \omega)$ if $Q(t)$ is periodic of period 1 and monotone in the period (see (16.20), for example). In this case $\mathcal{L}(\omega)$ is the point $Q(\omega)$ and $\Phi(\omega)$ is the function concentrated at 0, that is, δ_{x_0}.

One can write down similar representations for the pure point spectrum of metrically transitive operators that are more complicated than the multiplication operator. For the operator of Section 18, for example (see equations (18.1)–(18.4) and (18.21), and Theorem 18.2), $\mathcal{L}(\omega)$ is the point $\Lambda(\omega)$, where Λ is the function inverse to the integrated density of states (18.8), and $\Phi(\omega)$ consists of the single function $\chi(\cdot, \Lambda(\omega))$, where $\chi(x, \lambda)$ is defined by (18.27).

In the case considered by Sinai, the set $\Phi(\omega)$ also consists, roughly speaking, of the eigenfunctions ϕ of the operator h, whose normalization is contributed to mainly by a certain neighborhood $Z_0(\psi)$ of the origin; and $\mathcal{L}(\omega)$ is the set of eigenvalues corresponding to these eigenfunctions. Here, as in the case of the operator of multiplication by a monotone periodic function or by the operators (16.77) and (18.1)–(18.4), $\mathcal{L}(\omega)$ is the set of values of a certain function $\Lambda(\omega)$. It is measurable, although, generally speaking, multiple-valued. It is related to the point spectrum by relations like (16.85) and (16.86), and to the integrated density of states $N(\lambda)$ by (16.88). Further, $N(\lambda)$ has a singular continuous component, and the spectrum h has an infinite number of gaps.

To prove these facts, Sinai developed an iterative procedure to analyze the operators $h^{(l)}$ obtained from (16.76) if $\alpha = [k_1, \dots, k_l, \dots]$ is replaced by the truncated continued fraction $[k_1, \dots, k_l] = b_l/a_l = \alpha_l$ and periodic boundary conditions are imposed at the endpoints of the interval $[1, a_l]$. He showed that for each l there exist functions $\phi_j^{(l)}(x, \omega)$, for $j = 1, \dots, a_l$, which are approximate eigenfunctions of $h^{(l)}$ in the sense that for some numbers $\mu_j^{(l)}(\omega)$, whose construction he gives, the function $\big(h^{(l)} - \mu_j^{(l)}(\omega)\big)\phi_j^{(l)}$ is of order $a_l^{-2+\delta_1}$, for some $\delta_1 > 0$, when $0 \leqslant x \leqslant O(\ln a_l)$. These functions are orthonormal to within $O(a_l^{-2/3 - C/\ln \varepsilon})$, and for each of them there exists a set $Z(\phi_l^{(l)})$, called its essential support, of size $O(\ln a_l / \ln \varepsilon^{-1})$, outside of which the function decays exponentially: $\big|\phi_j^{(l)}(x)\big| \leqslant C \exp\big(\ln \varepsilon \operatorname{dist}(x, Z(\phi_j^{(l)}))\big)$.

All these results were obtained using perturbation theory in the small parameter ε. As we have emphasized, the formalism of this theory and the results it yields depend essentially on how close the eigenvalues of the unperturbed problem are, because, if they are very close to one another, the eigenfunctions change a lot even under small perturbations. Now the unperturbed operator $h_0^{(l)} = h^{(l)}\big|_{\varepsilon=0}$ is here the operator of multiplication by $Q(\alpha_l x + \omega)$, where the periodic function $Q(t)$ is continuous and therefore necessarily non-monotone. Then, for those $\omega \in [0,1]$ such that $Q(\alpha_l x + \omega) = Q(\alpha_l y + \omega)$ for different points $x, y \in \mathbf{Z}$, there are coinciding eigenvalues. Thus we must consider the sets $\Pi_{xy} = \{\omega : Q(\alpha_l x + \omega) = Q(\alpha_l y + \omega)\} \subset [0,1]$, and take the union over $x, y \in \mathbf{Z}$ of small neighborhoods of these sets. Outside this union, the eigenvalues of h_0^l are far enough apart for standard perturbation theory to apply.

For small enough ε, this gives a_l eigenvalues $\lambda^{(l)}(x,\omega)$ and eigenfunctions $\psi_x^{(l)}(y,\omega)$, for $1 \leqslant x \leqslant a_l$, such that $|\psi_x(y,\omega) \leqslant \text{const}\, e^{|x-y| \ln \varepsilon}$. Thus, in this case every eigenfunction of $h^{(l)}$ is concentrated on a small neighborhood of the point at which the unperturbed eigenfunction $\delta_{x,y}$ was concentrated. This enables us to define unambiguously, for such ω, the sets $\mathcal{L}^{(l)}(\omega)$ and $\Phi^{(l)}(\omega)$, which, as in the case of $Q(t)$ monotonic, can be taken to consist of a single element.

The study of the small neighborhoods of the sets $\Pi_{x,y}$ where there is quantum-mechanical resonance (for the functions $Q(t)$ of the type under discussion, such as (16.19), these sets have two points) requires a modification of perturbation theory, using especially chosen unperturbed eigenfunctions. This technique leads to the splitting of the close, unperturbed eigenvalues $Q(\alpha_l x + \omega)$ and to (approximate) eigenfunctions equally concentrated near x and y, which, by (16.50), are the further apart the closer together the unperturbed eigenvalues are.

Therefore, $\mathcal{L}^{(l)}(\omega)$ and $\Phi^{(l)}(\omega)$ look very different here compared with values of ω away from the sets $\Pi_{x,y}$. In particular, in the plot of the dependence of the approximate eigenvalue $\lambda^{(l)}(x,\omega)$ on ω, there appear discontinuities and forbidden intervals (future gaps in the spectrum of h) containing the values of λ at which the unperturbed eigenvalues coincide. But these changes can be studied in sufficient detail for large l and small ε.

With the information thus obtained on the spectrum of h_l, one can take the next iterative step. This is aided by the fact that the difference $h^{(l+1)} - h^{(l)}$ is small, of order $O(a^{-2+\delta})$, for $\delta > 0$, in the range $|x| \leqslant \ln a_l$. One can then find out what happens to the structure of the spectrum of $h^{(l)}$ as $l \to \infty$—in particular, that the eigenfunctions decay exponentially in intervals of order $\ln a_l$—and hence prove that the limit operator h has a pure point spectrum. For details, see [Sinai 1987].

16.F The Almost-Mathieu Operator

We compile here some spectral facts about the *almost-Mathieu Operator*

$$(h_{AM}\psi)(x) = -\psi(x+1) - \psi(x-1) + 2g\cos 2\pi(\alpha x + \omega)\psi(x). \tag{16.90}$$

This operator is interesting for two reasons: because it arises in a variety of problems of theoretical physics [Sokoloff 1985], and because it is the simplest example of a finite-difference operator having the important property of duality. This property, first used explicitly by André and Aubry [1980], allows one to obtain many spectral results in addition to those known for the operator (16.53), (16.21) with more general functions $Q(t)$, as discussed in the preceding subsections.

(i) The spectrum of h_{AM} is a Cantor set, hence nowhere dense, for pairs (α, g) forming a dense G_δ-set [Bellissard and Simon 1982]. This result, implying that a Cantor set spectrum is categorically typical for the almost-Mathieu operator, can be regarded as a partial solution to the following problem, ascribed to M. Kac: prove that, for all $g \neq 0$, $\omega \in [0,1)$ and α irrational, all the gaps of the spectrum are open (in the sense of Theorem 16.2, for example). Another result in this direction, obtained by Elliot [1982], is that the set of frequencies α for which h_{AM} is nowhere dense is uncountable.

(ii) If α is rational, say $\alpha = b/a$, the potential in 16.90 is periodic of period a, and therefore (see 16.A) its spectrum is absolutely continuous and the generalized eigenfunctions have the Floquet–Bloch form (16.52) for any g and ω.

(iii) If α is irrational, Theorem 16.13 says that the Lyapunov exponent satisfies (16.65), and therefore is positive if $|g| > 1$. By Theorem 12.1, this implies that for α irrational and $|g| > 1$ the spectrum has no absolutely continuous component for almost all $\omega \in [0,1]$: $\sigma_{\text{ac}}(h_{AM}) = \varnothing$. Inequality (16.65) can also be obtained as a consequence of the duality property of h_{AM}, rather than from Theorem 16.13 (see Problem VII.21).

(iv) If α is irrational but well approximated by rationals, so that (16.61) is satisfied (we say that α is a *Liouville number*), Corollary 16.9 implies that h_{AM} has no eigenvalues for almost all ω and g: $\sigma_{\text{P}}(h_{AM}) = \varnothing$. Theorem 16.18 is an improvement of this result for $|g| < 1$.

(v) Combining (iii) and (iv) we conclude that, for almost $\omega \in [0,1]$ and $|g| > 1$ and for α a Liouville number, the spectrum is singular continuous: $\sigma(h_{AM}) = \sigma_{\text{sc}}(h_{AM})$.

(vi) Now let α be irrational and badly approximated by rational numbers, so that (16.50) is fulfilled. Then there exists $g_0 > 0$ such that, for all $|g| \leqslant g_0$, there is a nowhere dense set $\sigma_1 \subset \mathbf{R}$ of positive Lebesgue measure and independent of ω such that, on σ_1, the spectrum is absolutely continuous and the eigenfunctions have the quasi-Bloch form (16.52). If, in addition, the continued fraction expansion $[k_1, \ldots, k_l, \ldots]$ of α satisfies $\lim_{l\to\infty} k_l = \infty$, we have $\operatorname{meas} \sigma_1 = 4 - o(1)$ as $g \to 0$. Since $4(1 + |g|)$ is evidently an upper bound for the measure of the whole spectrum, we see that in this case $\operatorname{meas}\sigma(h_{AM}) = \operatorname{meas}\sigma_{\mathrm{ac}}(h_{AM}) + o(1)$ as $g \to 0$. These results were obtained by Bellissard et al. [1983b] as extensions of Theorem 16.6 to the discrete case.

Numerical calculations [André and Aubry 1980] seem to indicate that $4(1-|g|)$ is the Lebesgue measure of the spectrum; according to [Thouless 1983], it is a lower bound for it.

(vii) Let α be irrational and $\lim_{l\to\infty} k_l \geqslant 10$. Then, for $|g| \geqslant g_0^{-1}$, where g_0 is the constant in (vi), and for almost all $\omega \in [0,1]$, the set of eigenvalues of h_{AM} is infinite, its closure has positive Lebesgue measure, and the eigenfunctions decrease exponentially. This set is

$$K_1 = K_0 \cap \{2\omega + 2\alpha\mathbf{Z}\}, \tag{16.91}$$

where K_0 is the set (16.51) of allowed values of the quasimomentum, similar to the set (16.31) for the continuous case. If $\lim_{l\to\infty} k_l = \infty$, we have $\operatorname{meas} K_1 = \big(4 - o(1)\big)g$ as $g \to \infty$.

These results, from Bellissard et al. [1983b], are weaker than those discussed in 16.E, due to Sinai [1987], which imply that, under essentially the same conditions, the spectrum of h_{AM} is pure point for g large enough. But we quote them here for two reasons: they were historically the first results on the existence of a big point spectrum for almost-periodic finite-difference operators; and they were obtained using duality, a property that helps with many spectral problems for the almost-Mathieu operator. To explain how this property is employed here, we write the equation corresponding to h_{AM}:

$$-y(x+1) - y(x-1) + 2g\cos 2\pi(\alpha x + \omega)y(x) = \lambda y(x). \tag{16.92}$$

For λ in the set σ_1 of property (vi), we substitute the generalized eigenfunction (16.52) for $y(x)$ in this equation. Setting $x = 0$ we get

$$-e^{2\pi i k}U(\omega+\alpha) - e^{-2\pi i k}U(\omega-\alpha) + 2g\cos 2\pi\omega\, U(\omega) = \lambda U(\omega), \tag{16.93}$$

which may be regarded as a difference equation on the one-dimensional torus $[0,1)$. Using Fourier coefficients

$$\psi(x) = \int_0^1 e^{-2\pi i\omega x}U(\omega)\,d\omega,$$

we obtain, after a simple calculation,

$$-\psi(x+1) - \psi(x-1) + \frac{2}{g}\cos 2\pi(\alpha x + k)\psi(x) = -\frac{\lambda}{g}\psi(x) \tag{16.94}$$

for $x \in \mathbf{Z}$. Since the function $U(t)$ of (16.52a) is analytic on a strip $|\operatorname{Im} t| \leqslant \rho$, its Fourier coefficients decay exponentially as $|x| \to \infty$. Thus, if $\lambda \in \sigma_1(h_{AM}(g,\omega))$, we have $-\lambda g^{-1} \in \sigma_{\mathrm{p}}(h_{AM}(g^{-1}, k(\lambda)))$.

Taking into account the obvious fact that $h_{AM}(g,\omega)$ and $h_{AM}(g,\omega + m\alpha)$, for $m \in \mathbf{Z}$, are unitarily equivalent, we can express this property in the following way: whenever $k+m\alpha$ is in the set (16.51) of allowed values for the quasimomentum, the expression $g^{-1}\Lambda(k+m\alpha, g)$ of (16.51a) is an eigenvalue of $h_{AM}(g^{-1}, k)$. To show property (vii), therefore, we must show that for almost all $k \in [0, \frac{1}{2}]$ the closure of (16.91) has a positive Lebesgue measure. This requires sharpening somewhat the estimates made in the proof of the discrete version of Theorem 16.6, as was done by Bellissard et al. [1983b] for α satisfying the conditions above.

(viii) We consider one more application of duality. Theorem 16.6, and its discrete version, do not give any information on the structure, or even the existence, of the spectrum outside σ_1. In particular, one cannot exclude the possibility that there might be eigenvalues outside this set. Using Theorem 16.7, we can guarantee their absence only for the Liouville frequencies α, for all g and ω. The following statement gives another condition for the absence of eigenvalues.

(16.19) Theorem [Delyon 1986]. *The point spectrum of h_{AM} is empty for any ω if α is irrational and $|g| < 1$.*

Proof. Assuming the opposite, replace $y(x)$ in (16.90) by the corresponding normalized eigenfunction $\psi(x) \in \ell_2(\mathbf{Z})$, multiply by e^{2ipx}, for $p \in [0,1)$, and sum over $x \in \mathbf{Z}$. The result is the following equation for the Fourier transform $\hat{\psi}(p)$ of $\psi(x)$:

$$-e^{2\pi i\omega}\hat{\psi}(p+\alpha) - e^{-2\pi i\omega}\hat{\psi}(p-\alpha) + \frac{2}{g}\cos 2\pi p\,\hat{\psi}(p) = -\frac{\lambda}{g}\hat{\psi}(p),$$

which differs from (16.93) only in the notation. Setting $p = x\alpha+\tau$, for $x \in \mathbf{Z}$ and $\tau = [0,1)$, and $\chi_\tau(x) = e^{2\pi i\omega x}\hat{\psi}(\alpha x + \tau)$, we find that this sequence, which is in $\ell_2(\mathbf{Z})$ for almost all $\tau \in [0,1)$, satisfies (16.94) with χ in place of ψ and τ in place of κ.

Since $\hat{\psi}(p)$ is defined as a square-integrable function on the unit circle and it has unit $L^2[0,1)$-norm, this leads to

$$\begin{aligned}\int_0^1 d\tau \sum_{x\in\mathbf{Z}} \frac{|\chi_\tau(x)|^2}{1+|x|^{1+\delta}} &= \sum_{x\in\mathbf{Z}} \frac{1}{1+|x|^{1+\delta}} \int_0^1 |\hat{\psi}(\alpha x+\tau)|^2\, d\tau \\ &= \|\psi\|^2 \sum_{x\in\mathbf{Z}} \frac{1}{1+|x|^{1+\delta}} < \infty\end{aligned}$$

for any $\tau > 0$. Therefore, for Lebesgue-almost all $\tau \in [0,1)$, there exists a finite quantity $C(\tau,\delta)$ such that

$$|\chi_\tau(x)|^2 \leqslant C(\tau,\delta)\left(1+|x|^{1+\delta}\right)$$

for all $x \in \mathbf{Z}$, and, under the same conditions,

$$\lim_{|x|\to\infty} \frac{1}{2|x|} \ln\left(|\chi_\tau(x)|^2 + |\chi_\tau(x+1)|^2\right) = 0.$$

But, by Theorem 11.3, this limit is exactly $\pm\gamma(-\lambda/g, 1/g)$. hence the Lyapunov exponent of (16.94) is zero for $|g| < 1$, contradicting (16.65). This proves the theorem. □

(ix) Similar arguments, also based on duality and on the extensive analysis and results in [Sinai 1987], discussed at the end of 16.E, allow one to show [Chulaevskii and Delyon 1989] that, for small g and frequencies satisfying the condition of Sinai's theorem, the spectrum of h_{AM} is purely absolutely continuous for almost all $\omega \in [0,1)$. Furthermore, on this set of values of g, α and ω, the generalized eigenfunctions have the quasi-Bloch form (16.52), where the amplitude $u(x)$ is an ℓ^2-quasiperiodic function. Thus, for h_{AM} in the small coupling regime we have more complete results than for the general case discussed in 16.B.

(x) The case of $|g| = 1$ and α Diophantine is of special interest. Numerical calculations [Hofstadter 1976; André and Aubry 1980] and theoretical physics arguments [Suslov 1982; Wilkinson 1984] lead one to think that the spectrum is singular continuous and that, at $|g| = 1$, the pure absolutely continuous spectrum for $|g| < 1$ changes to pure point for $|g| > 1$ (cf. the metal-insulator transition or the Anderson transition [Lifshitz et al. 1982b]). But none of this has been rigorously proved. It is only known that for $g = 1$ the spectrum cannot be pure point [Bellissard and Testard 1985], that the eigenvalues, if any, form a set independent of ω where $\gamma(\lambda) = 0$, and that the corresponding eigenfunctions cannot decay fast enough to belong to $\ell_1(\mathbf{Z})$ [Delyon 1986]. The proofs of these facts are based on the self-duality of h_{AM} for $g = 1$.

17 Limit-Periodic Potentials

In Section 16 we saw that almost-periodic operators can have a nowhere dense absolutely continuous spectrum. This property and related ones, such as the Bloch form of the eigenfunctions, solvability of the inverse problem of spectral analysis, and others, can be studied quite completely if the operator coefficients are limit-periodic functions. Recall that we defined a uniform, or Bohr, limit-periodic function as an almost-periodic function that is a uniform limit of periodic functions. By using other metrics we can define Stepanov, Besicovitch, and other types of limit-periodic functions (see [Levitan 1953], relation (17.4) and the beginning of 17.B). Here we deal mainly with Stepanov and Besicovitch limit-periodic function, which are the most convenient for spectral theory and especially for the solution of the inverse problem.

A simple, but typical, example of a limit-periodic function is given by (16.55) and its continuous counterpart. Here, if the series $\sum_l |q_l|$ converges, $q(x)$ is uniform limit-periodic; if the series $\sum_l |q_l|^2$ converges, $q(x)$ is Besicovitch limit-periodic, and so on.

An important property of limit-periodic functions is that the periods a_l of the periodic functions that converge to a limit-periodic function are related by integer ratios: $a_{l+1}/a_l = m_l$ (Problem VII.25). This gives a natural way to construct the spectral theory of limit-periodic operators, as the "closure" of the spectral theory of periodic operators under the limit of infinitely increasing periods. This transition was also applied in Section 16 in various ways: see 16.C and 16.E. In this section we carry out such a closure when the rate of approximation by periodic functions is superexponential, that is, the error is less than e^{-ba_l} for any $b > 0$: see condition (17.6). This high rate of approximation results from the fact that, in the spectral theory of periodic operators, the changes in the various spectral quantities caused by a perturbation of period a are generally estimated by e^{ba}, for some constant b: see Theorem 17.6.

The derivation of these estimates is the technical pivot of the theory presented below. Its main result is the absolutely continuity of the spectrum and the Bloch form of the generalized eigenfunctions (Theorems 17.4 and 17.5) for any amplitude of the potential, not just for a small potential, as in Theorem 16.6. Also, this method allows us to establish a one-to-one correspondence between potentials in this class and a certain set of spectral data (Theorem 17.1 and 17.3), that is, to solve the inverse problem of spectral analysis for such potentials as completely as for potentials that vanish at infinity and for periodic potentials [Marchenko 1986]. One can prove the uniqueness of the potential reconstructed from the spectral data—the so-called conditional inverse problem—and parametrize those data using independent parameters, which thus form a complete set of independent spectral data. In particular, there exist limit-periodic Schrödinger operators and Jacobi matrices whose spectrum is nowhere dense in any given interval (and, for example, having no gaps in the complement of the same interval).

The results of this section, like those of previous sections that referred to the absolutely continuous spectrum of almost-periodic operators, are true for every, not just almost every, almost-periodic potential in the hull.

17.A Basic Results

The first results on limit-periodic potentials obtained by approximation with periodic functions are due to Moser [1981], Avron and Simon [1981] and Chulaevskii [1981]. These authors proved that arbitrarily close to any continuous periodic potential of period a, in the $\|\cdot\|_\infty$ metric, there exists a limit-periodic potential with frequency basis $(a2^l)^{-1}$, for $l = 0, 1, \ldots$, whose corresponding spectrum is nowhere dense. In other words, if we regard almost-periodic functions as forming a metric space $\mathcal{Q}$ with the uniform metric, the spectrum of the Schrödinger operator is nowhere dense for potentials in a dense G_δ subset of $\mathcal{Q}$. Thus a nowhere dense spectrum is typical, at least in the Baire category sense.

This seems a natural result in view of the perturbation theory arguments of 16.B. A small change in those arguments, and in particular in (16.26), shows that adding a small potential of period $2a$ to a potential of period a results, generically, in a gap inside every allowed band. The proofs given in the three papers just mentioned for the result on limit-periodic potentials represent formalizations of this simple observation.

In [Avron and Simon 1981] and Chulaevskii [1981] it is also shown that for a dense set of potentials in $\mathcal{Q}$ (though not necessarily a G_δ) the spectrum of the Schrödinger operator is nowhere dense and absolutely continuous.

Explicit estimates of the rate of approximation of the limit-periodic potential by periodic ones are announced by Chulaevskii [1981] for a potential of the form

$$q(x) = \sum_{l=1}^{\infty} q_l u_l(x/a_l) \tag{17.1}$$

(cf. Problem VII.25), where the functions $u_l(x)$ are periodic of period 1, of class C^2, and bounded by 1 (in particular, all the $u_l(x)$ can be the same, as in (16.55)), and the periods a_l are such that a_{l+1}/a_l is an integer $\geqslant 2$. The estimates have the form

$$\lim_{l\to\infty} e^{ba_{l+1}}|q_l| = 0$$

for all $b > 0$.

All three papers above used fairly simple facts from the spectral theory of the periodic Schrödinger operator. However, in recent years, this theory has made rapid progress, in connection with the problem of integration of nonlinear evolution equations. In particular, the inverse problem for periodic Schrödinger operators was solved [Marchenko and Ostrovskii 1975, 1980; Marchenko 1986]. Based on these advances, Pastur and Tkachenko [1984, 1989] constructed a fairly complete picture of the spectral behavior of Schrödinger operators with limit-periodic potential very well approximated by periodic functions. But before we get to that, we cite the main results of Marchenko and Ostrovskii.

Let $q(x)$ be a periodic function of period a, square-integrable in a period. Denote by $C(x,\mu)$ and $S(x,\mu)$ the fundamental system of solutions (12.11) of the Schrödinger equation with potential $q(x)$, where $\lambda = \mu^2$, and by $A(a,\mu)$ the Hill discriminant (16.4). The eigenvalues $\nu_0 < \nu_2^- \leqslant \nu_2^+ < \cdots$ and $\nu_1^- \leqslant \nu_1^+ < \nu_3^- \leqslant \cdots$ of the periodic and antiperiodic problems on the period are the roots of the equations $A(a,\sqrt{\lambda}) = \pm 1$, and give the edges of the spectrum (16.2) of the Schrödinger operator on $L_2(\mathbf{R})$ with this potential. The eigenvalues $\lambda_1 < \lambda_2 < \cdots$ of the Dirichlet problem with the same potential are the roots of $S(a,\sqrt{\lambda}) = 0$, and lie in the gaps (16.6) of the spectrum. Let γ_m and $\beta_m = \sqrt{\lambda_m - \nu_0}$ be, in increasing order, the positive roots of the functions

$$\dot{A}\big(a,\sqrt{\mu^2+\nu_0}\big) \qquad \text{and} \qquad S\big(a,\sqrt{\mu^2+\nu_0}\big),$$

where the dot stands for differentiation with respect to μ.

Consider the Nevanlinna function

$$\theta(a,\mu) = \arccos A\big(a,\sqrt{\mu^2+\nu_0}\big) = \kappa(a,\mu)a, \tag{17.2}$$

where $\kappa(a,\mu) = k(a,\mu^2+\nu_0)$ (k is the quasimomentum: see (16.1)). Consider also the sequence of complex numbers $\theta = \{\theta_m : m = 0,1,\ldots\}$ specified by

$$\begin{aligned} \theta_0 &= \nu_0, \\ |\theta_m| \equiv h_m &= \operatorname{Im}\theta(a,\gamma_m), \\ \operatorname{sign}\operatorname{Im}\theta_m &= \operatorname{sign}\big(C^2(a,\beta_m) - S'^2(a,\beta_m)\big), \\ \operatorname{sign}\operatorname{Re}\theta_m &= \operatorname{sign}(\beta_m - \gamma_m), \end{aligned} \tag{17.3}$$

where the prime indicates differential with respect to x. The function $\theta(a,\mu)$ maps the upper half-plane $\mathbf{C}_+$ conformally onto the region $\Pi_+ = \mathbf{C}_+ \setminus \cup_{m\neq 0}[\pi m, \pi m + ih_{|m|}]$, and the sequence θ belongs to the space w_1^2 of complex-valued sequences with norm

$$\|\theta\|_a = |\theta_0| + \frac{\pi^2}{a^2}\Big(\sum_{m\geqslant 1} m^2\,|\theta_m|^2\Big)^{\frac{1}{2}}$$

and forms a complete set of independent spectral data for the periodic Schrödinger operator in $L_2(\mathbf{R})$ with potential $q(x)$. This means that each sequence $\theta = \{\theta_m : m = 0, 1, \ldots\} \in w_1^2$ uniquely determines a periodic potential $q(x) \in L_2[0, a)$ of period a, and consequently an operator H_q.

We now describe the class of limit-periodic potentials and the spectral data that are the main subject of this section. Let S^2 be the space of almost-periodic *Stepanov functions* with the norm

$$\|f\|_{S^2} = \sup_x \left(\int_x^{x+1} |f(t)|^2 \, dt \right)^{\frac{1}{2}}, \tag{17.4}$$

and let $\mathfrak{A} = \{a_n : n = 1, 2, \ldots\}$ be an increasing sequence of positive numbers such that a_{n+1}/a_n is an integer $\neq 1$. Denote by $\mathfrak{R}_n$ the set of real numbers $r = \pi m / a_n$, for $m \in \mathbf{Z}$, and set

$$\mathfrak{R}(\mathfrak{A}) = \bigcup_{n \geqslant 1} \mathfrak{R}_n. \tag{17.5}$$

Each $q \in S^2$ determines a set of Fourier coefficients $\hat{q}_r$, for $r \in \mathfrak{R}(\mathfrak{A})$, such that the norm $\left(\sum_{r \in \mathfrak{R}} |\hat{q}_r|^2\right)^{1/2}$ is finite. We denote by $\mathcal{Q}_\infty(\mathfrak{A})$ the set of $q \in S^2$ such that $\hat{q}_r = 0$ for $r \notin \mathfrak{R}(\mathfrak{A})$, and

$$\lim_{n \to \infty} e^{b a_{n+1}} \sum_{r \in \mathfrak{R}(\mathfrak{A}) \setminus \mathfrak{R}_n} |\hat{q}_r|^2 = 0 \tag{17.6}$$

for all $b > 0$. For $q \in \mathcal{Q}_\infty(\mathfrak{A})$, consider the sequence of periodic potentials q_n of period a_n given by

$$q_n(x) = \sum_{r \in \mathfrak{R}_n} \hat{q}_r e^{2irx}.$$

This sequence converges in the metric S^2 to $q(x)$, and it follows from (17.6) that

$$\lim_{n \to \infty} e^{b a_{n+1}} \|q - q_n\|_{S^2} = 0 \tag{17.7}$$

for all $b > 0$. Now introduce the set $\mathcal{K}(\mathfrak{A})$ of complex-valued functions $\kappa = \{\kappa_r : r \in \mathfrak{R}(\mathfrak{A})\}$, with the norm

$$\|\kappa\|_{\mathcal{K}(\mathfrak{A})} = \sup_{n \geqslant 1} \left(|\kappa_0| + \left(\sum_{r \in \mathfrak{R}_n} r^2 |\kappa_r|^2 \right)^{1/2} \right) \tag{17.8}$$

(the supremum also being the limit as $n \to \infty$), and let $\mathcal{K}_\infty(\mathfrak{A})$ be the set of κ such that $\operatorname{Im} \kappa_0 = 0$, $\kappa_r = \kappa_{-r}$ and

$$\lim_{n \to \infty} e^{b a_{n+1}} \sum_{r \in \mathfrak{R}(\mathfrak{A}) \setminus R_n} r^2 |\kappa_r|^2 = 0 \tag{17.9}$$

for all $b > 0$.

For a fixed potential $q \in \mathcal{Q}_\infty(\mathfrak{A})$, consider the periodic functions $q_n(x)$ of period a_n, which satisfy (17.7). Let $\theta_{(j)} = \{\theta_{m,j}\}$ be the Marchenko–Ostrovskii complete set of independent spectral data for the Schrödinger operator with potential q_j: see (17.3). Introduce the function $\kappa_{(j)} = \{\kappa_{r,j}\} \in \mathcal{K}_\infty(\mathfrak{A})$ given by

$$(17.10)\quad \kappa_{r,j} = \begin{cases} \theta_{0,j} & \text{if } r = 0; \\ \theta_{|m|,j}/a_j & \text{if } r = \pi m/a_j \in \mathfrak{R}_j \text{ for some integer } m \neq 0; \\ 0 & \text{if } r \notin \mathfrak{R}_j. \end{cases}$$

The first main result of this section gives a parametric description of the class of potentials $\mathcal{Q}_\infty(\mathfrak{A})$, similar to the one that we have mentioned for the periodic case [Marchenko and Ostrovskii 1975, 1980]. In particular, it solves the inverse problem of spectral analysis for this class of potentials.

(17.1) Theorem [Pastur and Tkachenko 1984]. *The sequence of functions $\kappa_{(n)} \in \mathcal{K}_\infty(\mathfrak{A})$ generated by the potential $q \in \mathcal{Q}_\infty(\mathfrak{A})$ has a limit κ in $\mathcal{K}(\mathfrak{A})$ in the metric* (17.8), *and this limit belongs to $\mathcal{K}_\infty(\mathfrak{A})$. The operator T_∞ that takes $q \in \mathcal{Q}_\infty(\mathfrak{A})$ into $\kappa \in \mathcal{K}_\infty(\mathfrak{A})$ is a one-to-one map between these two sets.*

Thus, the function $\kappa \in \mathcal{K}_\infty(\mathfrak{A})$ is a complete set of independent spectral data for the limit-periodic potential $q \in \mathcal{Q}_\infty(\mathfrak{A})$. The sets $\mathcal{K}_\infty(\mathfrak{A})$ and $\mathcal{Q}_\infty(\mathfrak{A})$ can be easily made into Frechet spaces in such a way that T_∞ is an isomorphism of Frechet spaces.

The main part of the proof of Theorem 17.1, presented in 17.C, consists in relating the L_2-norm of the difference between two periodic potentials to a certain norm of the difference of their spectral data, to show that each norm is bounded by a multiple of the other (Theorem 17.6). The continuity in the L_2-metric of the quantities β_m, γ_m, and so on, as functionals of the potential, was mentioned in [Marchenko and Ostrovskii 1980]. But for our purposes it is necessary to know the exponent in the continuity moduli of these functions and to be able to estimate the multiplicative factors in formulas such as (17.13) and (17.14) as functions of the period and the potential.

The parametric description of Schrödinger operators with periodic potential is closely associated with the quasimomentum $k(\mu^2 + \nu_0) = \theta(a, \mu)/a$, which is a Nevanlinna function. A similar fact is valid in the limit-periodic case:

(17.2) Theorem [Pastur and Tkachenko 1984]. *Let $q \in \mathcal{Q}_\infty(\mathfrak{A})$ and let $q_n(x)$ be periodic potentials of period a_n satisfying* (17.7). *If $\theta_n(a_n, \mu)$ is the Nevanlinna function* (17.2) *for the Schrödinger operator with potential q_n, the func-*

tions $\kappa_n(\mu) = \theta_n(a_n, \mu)/a_n$ *converge uniformly on the closed half-plane* $\mathbf{C}_+$ *to a Nevanlinna function* $\kappa(\mu)$. *This limit function maps* $\mathbf{C}_+$ *conformally onto the domain*

$$\Pi_\infty^+(\mathfrak{A}) = \mathbf{C}_+ \setminus \bigcup_{r \in \mathfrak{R}(\mathfrak{A}) \setminus \{0\}} [r, r + i|\kappa_r|],$$

so that $\kappa(0) = 0$ *and* $\lim_{y\to\infty} \kappa(iy)/(iy) = 1$. *Furthermore,* $\kappa(\mu)$ *is absolutely continuous on the real axis and has derivative* $\dot{\kappa} \in L_p^{\mathrm{loc}}(\mathbf{R})$, *for* $p \in (1,2)$.

(17.3) Theorem [Pastur and Tkachenko 1984]. *A closed set* $\sigma \subset \mathbf{R}$ *is the spectrum of an operator* H_q, *for* $q \in \mathcal{Q}_\infty(\mathfrak{A})$, *if and only if we can write*

$$\sigma = \{\lambda \geqslant \nu_0 : \operatorname{Im} \kappa(\sqrt{\lambda - \nu_0}) = 0\} \tag{17.11}$$

(*cf*. (16.12)), *where* ν_0 *is a real number and* $\kappa(\mu)$ *is the conformal map from the upper half-plane* $\mathbf{C}^+$ *onto* $\Pi_\infty^+(\mathfrak{A})$ *uniquely determined by the condition* $\lim_{y\to\infty} \kappa(iy)/(iy) = 1$.

Theorem 17.3 allows one to construct a limit-periodic Schrödinger operator, given the geometric properties of its spectrum. For example, let $\kappa \in \mathcal{K}_\infty(\mathfrak{A})$ be a function that is nonzero at all points of $\mathfrak{R}(\mathfrak{A}) \cap \Delta$, where Δ is a fixed interval. Then we obtain an operator whose spectrum is a Cantor set, and in particular nowhere dense, in the interval $(\kappa^{-1}(\Delta))^2 + \nu_0$. Examples of such potentials for $\Delta = \mathbf{R}$ are given, somewhat implicitly, by Moser [1981], Avron and Simon [1981], and Chulaevskii [1981]. In our terms, such potentials $q \in \mathcal{Q}_\infty(\mathfrak{A})$ correspond to those and only those functions $\kappa \in \mathcal{K}_\infty(\mathfrak{A})$ that have a dense support in $\mathfrak{R}(\mathfrak{A})$. The set of such functions is a dense G_δ in $\mathcal{K}(\mathfrak{A})$.

Note that $\pi^{-1} \operatorname{Re} \kappa(\sqrt{\lambda - \nu_0})$ coincides with the integrated density of states $N(\lambda)$ of the operator H_q: cf. (16.11). Therefore, by Theorem 13.1, the spectrum of H_q is the set of growth points of $\operatorname{Re} \kappa(\sqrt{\lambda - \nu_0})$. Further, $\pi N(\lambda)$ is constant on spectral gaps, and takes values on $\mathfrak{R}(\mathfrak{A})$, in conformity with Theorem 16.2. In our case, the set of values taken by $N(\lambda)$ on gaps coincides with the support of $\kappa \in \mathcal{K}_\infty(\mathfrak{A})$, which parametrizes the potential $q(x)$.

The quantity $\operatorname{Im} \kappa(\sqrt{\lambda - \nu_0})$ is the Lyapunov exponent $\gamma(\lambda)$ of H_q. We have the following improvement for Theorems 12.6 and 12.9:

(17.4) Theorem [Pastur and Tkachenko 1984]. *The spectrum of a Schrödinger operator with potential* $q \in \mathcal{Q}_\infty(\mathfrak{A})$ *is absolutely continuous and has multiplicity two.*

Thus the spectrum of a Schrödinger operator with a limit-periodic potential of the type considered here has the same properties as that of an operator with periodic potential. Theorem 17.4 follows from the next result, which also has an analogue in the periodic case, and states that, for almost all $\mu > 0$, the equation $H_q\psi = \mu^2\psi$ has solutions of Floquet–Bloch type (16.1).

(17.5) Theorem [Pastur and Tkachenko 1984]. *If $q \in \mathcal{Q}_\infty(\mathfrak{A})$, the corresponding equation has, for almost all $\lambda \in \mathbf{R}$ with respect to the Lyapunov measure, linearly independent solutions*

$$\psi_+(x,\mu) = e^{i\kappa(\mu)x}u(x,\mu), \qquad \psi_-(x,\mu) = \overline{\psi_+(x,\mu)}$$

(cf. (16.1) and (16.37)), where $\lambda = \mu^2 + \nu_0$ and $u(\,\cdot\,,\mu) \in S^2$. In addition, there is a sequence $u_n(x,\mu)$, for $n \in \mathbf{Z}_+$, of periodic functions of period a_n, such that, for any $\Delta \subset \mathbf{R}_+$ and any $b > 0$,

$$\lim_{n\to\infty} e^{ba_{n+1}} \int_\Delta \|u(\,\cdot\,,\mu) - u_n(\,\cdot\,,\mu)\|_{S^2}^p \, d\mu = 0 \tag{17.12}$$

for every $p \in (1,2)$. The set of functions $\psi_\pm(x,\sqrt{\lambda-\nu_0})$ for Lebesgue-almost all λ that belong to the spectrum σ of the operator H_q represent a complete system of generalized eigenfunctions of H_q, so that the resolution of the identity E_Δ of H_q is

$$(E(\Delta)f)(x) = \frac{1}{\pi}\int_{\sigma\cap\Delta} \operatorname{Re}\bigl(F_+(\sqrt{\lambda-\nu_0})\psi_-(x,\sqrt{\lambda-\nu_0})\bigr)\, d\lambda,$$

where $F_\pm(\mu)$ is the generalized Fourier transform of the function $f \in L_2(\mathbf{R})$:

$$F_\pm(\mu) = \int_{\mathbf{R}} \psi_\pm(x,\mu)\, f(x)\, dx.$$

Results similar to Theorems 17.1–5 are also true in the discrete case, if the coefficients $s(x)$ and $q(x)$ in (1.4) are limit-periodic sequences very closely (superexponentially) approximated by periodic ones. They also apply to a Dirac operator whose matrix of potentials consists of limit-periodic functions satisfying (17.6). For details, see [Egorova 1986, 1987].

The main assumption that guarantees the validity of all these results is condition (17.6). As applied to the limit-periodic potential (17.1), this condition means that for a high rate of decrease of the amplitude q_l with respect to the period a_l, the terms of (17.1) with large l can only make the spectrum a Cantor set, without changing the fact that it is absolutely continuous. But it is reasonable to think that if the amplitudes don't decay too fast, other spectral types can arise, and in particular point spectra.

This possibility was considered by Molchanov and Chulaevskii [1984]. It is realized when the sequence $\{u_n(t)\}$, for $t \in [0,1)$, in (17.1) is taken as a family of Markov processes, independent for different values of n, and each leading to a point spectrum: see Section 15, and in particular Theorem 15.4. In this case, we should expect, based on the results of Chapters V and VI, that if the amplitudes decrease moderately fast and the periods grow very rapidly, a point spectrum will appear.

And indeed, the authors claim that if u_l and q_l exceed $a_{l+1}^{2+\delta}$ for some $\delta > 0$ and a_{l+1}/a_l exceeds $\exp(a_l^{1+\delta})$, the spectrum is pure point with probability 1. However, the Lyapunov exponent is zero, so all we can say about the eigenfunctions is that they decrease faster than any power of $|x|^{-1}$, as $|x| \to \infty$.

17.B Spectral Data for Periodic Potentials of Increasing Period

In order to state the next theorem, which is the main analytic result that we will use to prove Theorem 17.1, we define the so-called *Besicovitch norm* $\|\cdot\|_B$ by

$$\|q\|_B = \lim_{L\to\infty}\left(\frac{1}{L}\int_0^L |q(x)|^2\,dx\right)^{\frac{1}{2}}.$$

Clearly, if (17.7) is satisfied for the norm $\|\cdot\|_{S^2}$ of (17.4) it is also satisfied for $\|\cdot\|_B$, and vice versa; but it turns out to be more convenient to use $\|\cdot\|_B$ than $\|\cdot\|_{S^2}$.

(17.6) Theorem [Pastur and Tkachenko 1984]. *Let $q_1(x)$ and $q_2(x)$ be periodic potentials of period a and lying in $L_2[0,a]$, and $\theta_{(1)}$ and $\theta_{(2)}$ the complete set of spectral data associated with them by* (17.3). *Then*

$$\|\theta_{(2)} - \theta_{(1)}\|_a \leqslant Ca^{-1}\sqrt{\|q_2 - q_1\|_B}\, e^{Ca\sqrt{Q}}, \tag{17.13}$$

$$\|q_2 - q_1\|_B \leqslant Ca^{-1}\sqrt{\|\theta_{(2)} - \theta_{(1)}\|_a}\, e^{C(\Theta_2 + \sqrt{\Theta_1})}, \tag{17.14}$$

where C is a constant, $Q = \max\|q_j\|_B$, $\Theta_1 = \max\bigl(\sum_{k\geqslant 1}|\theta_{k,j}|^2k^2\bigr)^{1/2}$, and $\Theta_2 = \sup_{j,k\geqslant 1}|\theta_{k,j}|$.

Proof. We will need some auxiliary statements. For the sake of simplicity, we assume $a = \pi$; we will switch to an arbitrary period at the end. All the constants to be derived in the subsequent calculation will be denoted by C.

Let $E(x,\mu)$ be the solution of the Schrödinger equation satisfying the initial conditions $E(0,\mu) = 1$ and $E'(0,\mu) = \mu$. It is known [Marchenko 1986] that for every potential $q \in L_2(0,\pi)$ there exist transformation operators with kernels $K_e(x,t)$, $K_c(x,t)$ and $K_s(x,t)$ realizing the representation

$$y(x,\mu) = y_0(x,\mu) + \int_{x_y}^{x} K_y(x,t)y_0(t,\mu)\,dt, \tag{17.15}$$

for $y = e, s, c$, where $e_0(x, \mu) = e^{i\mu x}$, $s_0(x, \mu) = \mu^{-1} \sin \mu x$, $c_0(x, \mu) = \cos \mu x$, and $x_e = -x$, $x_s = x_c = 0$.

(17.7) Lemma. *Set* $\tilde{Q} = \|q\|_{L_2[0,\pi]}$, $\mathcal{M}(t) = (\partial/\partial x)K_s(\pi, t) - (\partial/\partial t)K_c(\pi, t)$, $\mathcal{N}(t) = (\partial/\partial t)K_c(\pi, t)$, $\bar{\mathcal{N}}(t) = \mathcal{N}(t) - \frac{1}{4}q(\frac{1}{2}(\pi + t)) - \frac{1}{4}q(\frac{1}{2}(\pi - t))$ *and let* $R(x,t)$ *be one of* $K_e(x,t)$, $K_s(x,t)$ *or* $K_c(x,t)$. *Further, let* $\tilde{Q}_j$, R_j, $\mathcal{M}_j$ *and* $\mathcal{N}_j$, *for* $j = 1, 2$, *be the corresponding quantities for the potentials* q_j, *periodic of period* π, *and set* $\hat{Q} = \max \tilde{Q}_j$. *Then*

$$|R(x,t)| \leqslant C\tilde{Q}e^{\sqrt{\tilde{Q}}}; \tag{17.16a}$$

$$|\mathcal{M}(t)| + |\bar{\mathcal{N}}(t)| \leqslant C\tilde{Q}^2 e^{C\sqrt{\tilde{Q}}}; \tag{17.16b}$$

$$|R_2(x,t) - R_1(x,t)| \leqslant C\|q_2 - q_1\|_{L_2[0,\pi]} e^{C\sqrt{\hat{Q}}}; \tag{17.16c}$$

$$|\mathcal{M}_2(t) - \mathcal{M}_1(t)| + |\bar{\mathcal{N}}_2(t) - \bar{\mathcal{N}}_1(t)| \leqslant C\hat{Q}\|q_2 - q_1\|_{L_2[0,\pi]} e^{C\sqrt{\hat{Q}}}. \tag{17.16d}$$

Proof. We know [Marchenko 1986] that the kernel K_e satisfies the integral equation

$$K_e(x,t) - \int_0^{\frac{x+t}{2}} d\alpha \int_0^{\frac{x-t}{2}} d\beta\, K_e(\alpha + \beta, \alpha - \beta) = \frac{1}{2}\int_0^{\frac{x+t}{2}} q(\alpha)\, d\alpha, \tag{17.17}$$

which, by successive approximations, leads to the representation

$$K_e(x,t) = \sum_{n \geqslant 0} K_n\left(\frac{x+t}{2}, \frac{x-t}{2}\right), \tag{17.18}$$

where the summand is recursively defined by $K_0(u,v) = \frac{1}{2}\int_0^u q(\alpha)\, d\alpha$ and $K_n(u,v) = \int_0^u d\alpha \int_0^v d\beta K_{n-1}(\alpha, \beta) q(\alpha + \beta)$. It is easy to check by induction that, for all $n \geqslant 0$,

$$|K_n(u,v)| \leqslant \frac{\sqrt{\pi}}{2}\frac{u^n}{n!^2}\left(\int_0^{u+v} |q(\alpha)|\, d\alpha\right)^n \tilde{Q}, \tag{17.19}$$

which results in the inequality

$$K_e(x,t) \leqslant \frac{\sqrt{\pi}}{2}\tilde{Q}\sum_{n \geqslant 0}\frac{(\sqrt{\pi}\tilde{Q})^n}{n!^2} \leqslant C\tilde{Q}e^{C\sqrt{\tilde{Q}}}.$$

To prove (17.16a), we just have to use the identities

$$K_c(x,t) = K_e(x,t) + K_e(x,-t), \quad K_s(x,t) = K_e(x,t) - K_e(x,-t). \tag{17.20}$$

To obtain (17.16b), we differentiate (17.17) and apply (17.16a); to get (17.16c,d), we write (17.17) for $R_1 - R_2$, then apply (17.16a). □

Remark. The exponents for the corresponding estimates in [Marchenko 1986] and [Marchenko and Ostrovskii 1980] include expressions quadratic in the period, such as $\int_0^x (x-t)q(t)\,dt$. With those estimates one must limit oneself to limit-periodic potentials satisfying a version of (17.6) with a_{n+1}^2 instead of a_{n+1}, that is, the rate of approximation of $q(x)$ by periodic potentials must be higher.

(17.8) Corollary. *If $y_j(x,\mu)$ is the solution of the Schrödinger equation with potential $q_j(x)$, for $j=1,2$, and we have $y_1(0,\mu)=y_2(0,\mu)$ and $y_1'(0,\mu)=y_2'(0,\mu)$, the following inequalities are satisfied for all $x\in[0,\pi]$:*

$$\begin{aligned}&\bigl|y_j(x,\mu)\bigr|+\bigl|y_j'(x,\mu)\bigr|\leqslant C\bigl(|y_j(0,\mu)|+|y_j'(0,\mu)|\bigr)e^{C\sqrt{\tilde{Q}_j}+\pi|\operatorname{Im}\mu|};\\&\bigl|y_2(x,\mu)-y_1(x,\mu)\bigr|+\bigl|y_2'(x,\mu)-y_1'(x,\mu)\bigr|\\&\qquad\qquad\leqslant C\bigl(|y_1(0,\mu)|+|y_1'(0,\mu)|\bigr)\|q_2-q_1\|_{L_2[0,\pi]}e^{C\sqrt{\hat{Q}}+\pi|\operatorname{Im}\mu|}.\end{aligned}$$

(17.9) Lemma. *If the potential $q\in L_2[0,\pi]$ is such that $\nu_0=0$, the numbers β_m and γ_m of* (17.3), *for all $m\geqslant 1$, can be represented as*

$$\beta_m=m+\frac{\langle q\rangle}{2m}+\frac{(-1)^{m+1}}{\pi m}\int_0^\pi \mathcal{N}(t)\cos mt\,dt+\frac{\omega_m}{m^2},\tag{17.21a}$$

$$\gamma_m=m+\frac{\langle q\rangle}{2m}+\frac{(-1)^{m+1}}{2\pi m^2}\int_0^\pi \bigl(t\mathcal{M}'(t)+2\mathcal{M}(t)\bigr)\sin\mu t\,dt+\frac{\rho_m}{m^2}\tag{17.21b}$$

where $\langle q\rangle=\pi^{-1}\int_0^\pi q(t)\,dt$ and ω_m and ρ_m satisfy

$$|\omega_m|\leqslant C\tilde{Q}\biggl(\frac{1}{m}+\biggl|\int_0^\pi t\mathcal{N}(t)\sin mt\,dt\biggr|\biggr)e^{C\sqrt{\tilde{Q}}},\tag{17.22a}$$

$$|\rho_m|\leqslant\frac{C\tilde{Q}}{m}e^{C\sqrt{\tilde{Q}}}.\tag{17.22b}$$

Proof. This is a strengthened version of well-known asymptotic formulas for β_m and γ_m, proved by the techniques developed by Marchenko [1986]. □

(17.10) Lemma. *Let $\hat{Q}$, $\mathcal{M}_j$ and $\mathcal{N}_j$, for $j=1,2$, be as in Lemma* 17.7, *and assume that $\nu_{0,1}=\nu_{0,2}=0$. Then, for all $m\geqslant 1$,*

$$\begin{aligned}\beta_{\mathrm{m},2}-\beta_{\mathrm{m},1}&=\frac{\langle \mathrm{q}_2-\mathrm{q}_1\rangle}{2\mathrm{m}}\\&\quad+\frac{(-1)^{m+1}}{\pi m}\int_0^\pi\bigl(\mathcal{N}_2(t)-\mathcal{N}_1(t)\bigr)\cos mt\,dt+\frac{\Omega_m}{m^2}\end{aligned}\tag{17.23a}$$

and

(17.23b) $$\gamma_{m,2}-\gamma_{m,1}=\frac{\langle q_2-q_1\rangle}{2m}$$
$$+\frac{(-1)^{m+1}}{2\pi m^2}\int_0^\pi \big(t(\mathcal{M}_2'(t)-\mathcal{M}_1'(t))+2(\mathcal{M}_2(t)-\mathcal{M}_1(t))\big)\sin mt\,dt+\frac{\mathcal{P}_m}{m^2},$$

where

(17.24a)
$$|\Omega_m|\leqslant C\left(\frac{\|q_2-q_1\|_{L_2[0,\pi]}}{m}+\left|\int_0^\pi t(\mathcal{N}_2(t)-\mathcal{N}_1(t))\sin mt\,dt\right|\right)e^{C\sqrt{\hat{Q}}},$$

(17.24b) $$|\mathcal{P}_m|\leqslant C\|q_2-q_1\|_{L_2[0,\pi]}e^{C\sqrt{\hat{Q}}}.$$

Proof. The estimates (17.24a,b) are proved separately for large and small m. We start with the estimate of $\beta_{m,2}-\beta_{m,1}$. Integrating (17.5) for $y=s$ by parts, and taking into account that $K_s(\pi,\pi)=\frac{1}{2}\pi\langle q\rangle$ and $K_s(\pi,0)=0$ by (17.17) and (17.20), we see that the numbers β_m, which are zeros of the entire function $S(\pi,\mu)$, satisfy the equation

(17.25) $$\sin n\beta=\frac{\pi}{2\beta}\langle q\rangle\cos\pi\beta-\frac{1}{\beta}\int_0^\pi \mathcal{N}(t)\cos\beta t\,dt.$$

Assume that $m\geqslant l=2(10^{10}\pi^2(1+\hat{Q})(1+Ce^{C\sqrt{\hat{Q}}}))+1$, where C is the constant from (17.16a). Subtracting the two equations of the form (17.25) for q_1 and q_2, and setting $\sigma_m=\beta_{m,2}-\beta_{m,1}$, we find that σ_m is a solution of

(17.26) $$\zeta_m(z)=\Phi_m,$$

where

$$\zeta_m(z)=\frac{2}{\pi}\sin\frac{\pi z}{2}\cos\pi(\beta_m-m+\frac{z}{2})+\Psi_m(z),$$
$$\Psi_m(z)=\frac{(-1)^{m+1}}{2}\langle q_1\rangle\frac{\cos(\beta_m(1)+u)}{\beta_{m,1}+u}\Bigg|_{u=0}^{z}+\frac{(-1)^m}{\pi}\int_0^\pi \mathcal{N}_2(t)\,dt\frac{\cos(\beta_m(1)+u)t}{\beta_{m,1}+u}\Bigg|_{u=0}^{z},$$
$$\Phi_m=\frac{(-1)^m}{2\beta_{m,2}}\langle q_2-q_2\rangle\cos\pi\beta_{m,2}-\int_0^\pi(\mathcal{N}_2(t)-\mathcal{N}_1(t))\frac{\cos\beta_{m,2}t}{\beta_{m,2}}dt.$$

Applying the implicit function theorem to (17.26), we obtain inequality (17.24a) in the case $m>l$ [Marchenko 1986]. For the complementary case $m\leqslant l$, we will show that

(17.27) $$|\dot{S}(\beta_m,\pi)|\geqslant C^{-1}e^{-C\sqrt{\hat{Q}}}.$$

Indeed, the function $\dot{S}(x,\mu)$ is a solution of the equation

$$-\frac{d^2y}{dx^2} + qy = \mu^2 y + 2\mu S,$$

with $y(0) = y'(0) = 0$. Therefore,

$$\dot{S}(x,\mu) = 2\mu \int_0^x \big(C(t,\mu)S(x,\mu) - S(t,\mu)C(x,\mu)\big)S(t,\mu)\,dt.$$

Since $S(\pi,\beta_m) = 0$, we have $\dot{S}(\pi,\beta_m) = -2\beta_m C(\pi,\beta_m)\int_0^\pi S^2(t,\beta_m)\,dt$. But $C(\pi,\beta_m)S'(\pi,\beta_m) = 1$, and, by Corollary 17.8, $S'(\pi,\beta_m) \leqslant Ce^{C\sqrt{\hat{Q}}}$. Therefore, using the representation

$$\dot{S}(x,\mu) = x + \int_0^x (x-t)\big(\mu^2 - q(t)S(t,\mu)\big)\,dt,$$

we obtain

$$\begin{aligned}
\int_0^\pi t^2\,dt &\leqslant 2\int_0^\pi S^2(t,\beta_m)\,dt + 2\int_0^\pi dt\left|\int_0^t (t-u)\big(\beta_m^2 - q(u)\big)S(u,\beta_m)\,du\right| \\
&\leqslant 2\left(1 + \int_0^\pi dt\int_0^t (t-u)^2\big(\beta_m - q(u)\big)^2\,du\right)\int_0^\pi S^2(t,\beta_m)\,dt \\
&\leqslant C(m^4 + \hat{Q}^2)\int_0^\pi S^2(t,\beta_m)\,dt \\
&\leqslant C\int_0^\pi S^2(t,\beta_m)\,dt\, e^{C\sqrt{\hat{Q}}}.
\end{aligned}$$

Therefore

$$\big|\dot{S}(\pi,\beta_m)\big| = \frac{2\beta_m}{\big|S'(\pi,\beta_m)\big|}\int_0^\pi S^2(t,\beta_m)\,dt \geqslant C^{-1}|\beta_1|e^{-C\sqrt{\hat{Q}}}. \tag{17.28}$$

Let $\alpha_m^\pm = \operatorname{sign} m(\nu_{|m|}^\pm)^{1/2}$ for $m \neq 0$, and $\alpha_0^\pm = 0$. Then the following intermittence takes place, for all $m \geqslant 1$ [Marchenko 1986]:

$$\alpha_m^- \leqslant \beta_m \leqslant \alpha_m^+, \qquad \alpha_m^- \leqslant \gamma_m \leqslant \alpha_m^+. \tag{17.29}$$

We now use Bernstein's theorem [Levin 1972], which says that if $f(z)$ is an entire function of exponential type σ, we have

$$\sup_{x\in\mathbf{R}}\big|f'(x)\big| \leqslant \sigma \sup_{x\in\mathbf{R}}\big|f(x)\big|. \tag{17.30}$$

This, and the fact that $A(\pi,\alpha_m^\pm) = \pm 1$, imply that

$$\beta_1 \geqslant C^{-1}e^{-C\sqrt{\hat{Q}}}, \qquad |\beta_{m+1} - \beta_m| \geqslant C^{-1}e^{-C\sqrt{\hat{Q}}}. \tag{17.31}$$

Similarly,

$$|\alpha_m^- - \alpha_{m-1}^+| \geqslant C^{-1} e^{-C\sqrt{\hat{Q}}}. \tag{17.32}$$

The first inequality in (17.31), combined with (17.28), yields (17.27), while the second shows that every circle $\mathcal{D}_m = \{\mu : |\mu - \beta_{m,1}| \leqslant (4C)^{-1} e^{-C\sqrt{\hat{Q}}}\}$ contains exactly one root of $S_1(\pi, \mu)$.

We now need a lower bound for $S_1(\pi, \mu)$ on circles of small radius, centered at $\beta_{m,1}$. To get this bound, we must take into account that, if an analytic function $f(z)$ on the circle $|z| \leqslant r$ has no zeros inside the circle and takes the value 1 at the origin,

$$\ln|f(z)| \geqslant -\frac{2r_1}{r - r_1} \ln \max_{|z| \leqslant r} |f(z)| \tag{17.33}$$

for $|z| \leqslant r_1 < r$ [Levin 1972]. Applying this to the function

$$\frac{S_1(\pi, \mu)}{\dot{S}_1(\pi, \beta_{m,1})(\mu - \beta_{m,1})}$$

we conclude that $|S_1(\pi, \mu)| \geqslant C^{-1} e^{-C\sqrt{\hat{Q}}}$ for $|\mu - \beta_{m,1}| = \frac{1}{2} r$.

If $\|q_2 - q_1\|_{L_2[0,\pi]} \geqslant (2C)^{-2} e^{-C\sqrt{\hat{Q}}}$, where C is taken from the preceding lower bound for $S_1(\pi, \mu)$, Lemma 17.9, together with the trivial inequality $|\beta_{m,2} - \beta_{m,1}| \leqslant |\beta_{m,2} - m| + |\beta_{m,1} - m|$, gives (17.24a). Therefore, without loss of generality, we can assume from now on that $\|q_2 - q_1\|_{L_2[0,\pi]} \leqslant (2C)^{-2} e^{-2C\sqrt{\hat{Q}}}$. From Corollary 17.8 we get

$$|S_2(\pi, \mu)| \geqslant |S_1(\pi, \mu)| - |S_2(\pi, \mu) - S_1(\pi, \mu)| \geqslant 2C^{-1} e^{-C\sqrt{\hat{Q}}}.$$

Taking into account that

$$\begin{aligned} \beta_{m,2} - \beta_{m,1} = \frac{1}{2\pi i} \int_{|\mu - \beta_{m,1}| = r/2} \mu \frac{\dot{S}_1(\pi, \mu) - \dot{S}_2(\pi, \mu)}{S_1(\pi, \mu)} \, d\mu \\ + \frac{1}{2\pi i} \int_{|\mu - \beta_{m,1}| = r/2} \mu \dot{S}_2(\pi, \mu) \frac{S_2(\pi, \mu) - S_1(\pi, \mu)}{S_1(\pi, \mu) S_2(\pi, \mu)} \, d\mu, \end{aligned}$$

and applying Corollary 17.8 again, we get $|\beta_{m,2} - \beta_{m,1}| \leqslant C \|q_2 - q_1\|_{L_2[0,\pi]} \times e^{C\sqrt{\hat{Q}}}$. Setting

$$m^{-2} \Omega_m = \beta_{m,2} - \beta_{m,1} - \frac{1}{2m} \langle q_2 - q_1 \rangle + \frac{(-1)^m}{\pi m} \int_0^m (\mathcal{N}_2(t) - \mathcal{N}_1(t)) \cos mt \, dt$$

and recalling that $m \leqslant l \leqslant C e^{C\sqrt{\hat{Q}}}$, we obtain (17.24a) from (17.16d).

The estimate of $\beta_{m,2} - \beta_{m,1}$ for $m \leqslant l$ could also have been obtained from the variational principle. But this method doesn't work for $\gamma_{m,2} -$

$\gamma_{m,1}$, because we cannot write down a boundary-value problem for which the numbers γ_m are eigenvalues.

The proof of representation (17.23b) and of estimate (17.24b) is based on the analysis of the following equation, whose solutions are the γ_m, for $m \geqslant 1$:

$$\sin \pi\gamma = \frac{\pi}{2\gamma}\langle g\rangle \cos \pi\gamma + \left(\frac{\pi^2}{8}\langle q^2\rangle - \frac{\langle q\rangle}{2}\right)\frac{\sin \pi\gamma}{\gamma^2} - \frac{1}{2\pi}\int_0^\pi \left(t\mathcal{M}'(t) + 2\mathcal{M}(t)\right)\frac{\sin \gamma t}{\gamma^2}\,dt. \tag{17.34}$$

Now the difference $\sigma_m = \gamma_{m,2} - \gamma_{m,1}$ is again the solution of an equation of the form (17.26), where

$$\zeta_m(z) = (-1)^m \frac{2}{\pi}\sin\frac{\pi z}{2}\cos\pi\left(\gamma_{m,1} + \tfrac{1}{2}z\right) + \Psi_m(z),$$

$$(-1)^m\Psi_m(z) = \frac{\langle q_1\rangle}{2}\frac{\cos\pi u}{u}\Bigg|_{u=\gamma_{m,1}}^{\gamma_{m,2}} + \left(\frac{\pi\langle q_1\rangle^2}{8} - \frac{\langle q_1\rangle}{2\pi}\right)\frac{\sin \pi u}{u^2}\Bigg|_{u=\gamma_{m,1}}^{\gamma_{m,2}} + \frac{1}{2\pi^2}\int_0^\pi \left(t\mathcal{M}_1'(t) + 2\mathcal{M}_1(t)\right)\frac{\sin ut}{u^2}\Bigg|_{u=\gamma_{m,1}}^{\gamma_{m,2}}\,dt,$$

$$\begin{aligned}(-1)^m\Phi_m = {} & \frac{1}{2\gamma_{m,2}}\langle q_2 - q_1\rangle \cos\pi\gamma_m(2)\\ & + \left(\frac{\pi}{8}\left(\langle q_2\rangle^2 - \langle q_1\rangle^2\right) - \frac{1}{2\pi}\langle q_2 - q_1\rangle\right)\frac{\sin\gamma_m(z)}{\gamma_m(z)}\\ & - \frac{1}{2\pi^2}\int_0^\pi \frac{\sin\gamma_m(z)t}{\gamma_m^2(z)}\left(t(\mathcal{M}_2'(t) - \mathcal{M}_1'(t)) + 2(\mathcal{M}_2(t) - \mathcal{M}_1(t))\right)dt.\end{aligned}$$

The estimation of σ_m for $m > l$ is carried out as in the proof of (17.24a), using the methods from [Marchenko 1986]. For the case $m \leqslant l$, recall that $\dot\theta(\pi, \gamma_m) = 0$ and that $\dot\theta(\pi, \mu)$, being a derivative a the map from the upper half-plane onto the domain Π_+, can be represented by the Christoffel–Schwarz formula. Moreover, since $A(\pi,\mu) = \cos\theta(\pi,\mu)$, we have, for $\gamma_{-m} = \gamma_m$:

(17.35)

$$\left|\ddot A(\pi,\gamma_m)\right| = \pi\frac{\sinh h_m}{\sqrt{|\gamma_m - \alpha_m^-|\,|\gamma_m - \alpha_m^+|}}\prod_{n\neq m}\frac{|\gamma_m - \gamma_n|}{\sqrt{|\gamma_m - \alpha_n^-|\,|\gamma_m - \alpha_n^+|}}.$$

We now use the inequalities $|\gamma_{m+1} - \gamma_m| \geqslant C^{-1}e^{-C\sqrt{\hat Q}}$, which is similar to (17.31), and $\left|\ddot A(t,\mu)\right| \leqslant Ce^{C\sqrt{\hat Q}}$, which follows from (17.30) and (17.31). Repeating the arguments used in the proof of inequality (3.4.38) of [Marchenko 1986], we arrive at the estimate

$$\frac{h_m}{\sqrt{|\gamma_m - \alpha_m^-|\,|\gamma_m - \alpha_m^+|}} \geqslant C^{-1} e^{-C\sqrt{\hat{Q}}}.$$

It follows from this and from (17.35) that $|\ddot{A}(\pi, \gamma_m)| \geqslant C^{-1} e^{-C\sqrt{\hat{Q}}}$. We now apply (17.33) to the function

$$\frac{A_1(\pi, \mu)}{\ddot{A}_1(\pi, \gamma_{m,1})(\mu - \gamma_{m,1})},$$

with μ in a circle of small radius r around the point $\gamma_{m,1}$. We get $\dot{A}_1(\pi, \mu) \geqslant C^{-1} e^{-C\sqrt{\hat{Q}}}$ for $|\mu - \gamma_{m,1}| = \varepsilon r$, for $\varepsilon \in [\frac{1}{2}, 1]$. From this it follows that $|\gamma_{m,2} - \gamma_{m,1}| \leqslant C\|q_2 - q_1\|_{L_2[0,\pi]} e^{C\sqrt{\hat{Q}}}$, yielding (17.24b). □

(17.11) Lemma. *Let $\hat{Q}$ and $\tilde{Q}_j$, for $j = 1, 2$, be as in Lemma 17.7, and let $\nu_{0,j}$, for $j = 1, 2$, be the smallest eigenvalues of the periodic boundary problem with potential $q(x)$, periodic of period π. Then*

(17.36a) $$\left|\nu_{0,j}\right| \leqslant C(1 + \tilde{Q}_j)\tilde{Q}_j,$$

(17.36b) $$\left|\nu_{0,2} - \nu_{0,1}\right| \leqslant C(1 + \hat{Q})\|q_2 - q_1\|_{L_2[0,\pi]}.$$

Proof. This would follow from the variational principle if the we were using the uniform norm, but here we must use instead the resolvent identity and the explicit form of the resolvent of the operator with $q = 0$. □

We now resume the proof of Theorem 17.6, starting with inequality (17.13). Assume, for simplicity, that m is even, so, by (17.2),

$$|\theta_m| = \ln\bigl(A(\pi, \gamma_m) + \sqrt{A^2(\pi, \gamma_m) - 1}\bigr).$$

Then, from the representation

(17.37) $$A(\pi, \mu) = \cos \pi\mu + \frac{\pi}{2\mu}\langle q\rangle \sin \pi\mu + \frac{1}{2}\int_0^\pi \mathcal{M}(t)\frac{\sin \mu t}{\mu}\,dt,$$

which is derived just like (17.15), based on equality (17.16b), we get

$$|\theta_m| \leqslant C\sqrt{A(\pi, \gamma_m) - 1}\, e^{C\sqrt{\hat{Q}}}.$$

Therefore, $\left|\theta_{m,2} - \theta_{m,1}\right|$ equals

$$\ln\left(1 + \frac{A_2(\pi, \gamma_{m,2}) - A_1(\pi, \gamma_{m,1}) + \sqrt{A_2^2(\pi, \gamma_{m,2}) - 1} - \sqrt{A_1^2(\pi, \gamma_{m,1}) - 1}}{A_1(\pi, \gamma_{m,1}) + A_2(\pi, \gamma_{m,2})}\right),$$

and since $A_j(\pi, \gamma_m) \geqslant 1$ and $|\sqrt{a} - \sqrt{b}| \leqslant \sqrt{|a - b|}$ for all $a, b \geqslant 0$, we have

$$(17.38)\qquad \big||\theta_{m,2}| - |\theta_{m,1}|\big| \leqslant C\sqrt{|A_2(\pi,\gamma_{m,2}) - A_1(\pi,\gamma_{m,1})|}\, e^{C\sqrt{\hat{Q}}}.$$

Similarly, we have

$$(17.39)\qquad \big||\operatorname{Im}\theta_{m,2}| - |\operatorname{Im}\theta_{m,1}|\big| \leqslant C\sqrt{|A_2(\pi,\beta_{m,2}) - A_1(\pi,\beta_{m,1})|}\, e^{C\sqrt{\hat{Q}}}.$$

To estimate the right-hand side of these inequalities, we integrate by parts the representation (17.37) for $A(\pi,\mu)$, obtaining

(17.40)

$$A(\pi,\mu) = \cos\pi\mu + \frac{\pi}{2\mu}\langle q\rangle \sin\pi\mu - \frac{\pi^2}{8}\langle q\rangle^2 \frac{\cos\pi\mu}{\mu^2} + \frac{1}{2}\int_0^\pi \mathcal{M}'(t)\frac{\cos\mu t}{\mu^2}\,dt.$$

From (17.21a) and (17.23a), we have

$$\begin{aligned}\big|\cos\pi\beta_{m,2} - \cos\pi\beta_{m,1} + \tfrac{1}{2}\pi^2(\beta_{m,2}-\beta_{m,1})(\beta_{m,2}+\beta_{m,1}-2m)\big|&\\ \leqslant Cm^{-1}\|q_2-q_1\|_{L_2[0,\pi]}e^{C\sqrt{\hat{Q}}}&,\\ \big|\sin\pi\beta_{m,2} - \sin\pi\beta_{m,1} - \pi(\beta_{m,2}-\beta_{m,1})\big| \leqslant Cm^{-3}\|q_2-q_1\|_{L_2[0,\pi]}e^{C\sqrt{\hat{Q}}}&.\end{aligned}$$

Now we use the representation from Lemma 17.7,

$$(17.41)\quad \mathcal{M}(t) = \int_{\frac{\pi-t}{2}}^{\pi} d\alpha\, K(\alpha,\alpha-\pi+t)q(\alpha) - \int_{\frac{\pi+t}{2}}^{\pi} d\alpha\, K(\alpha,\alpha-\pi-t)\,q(\alpha),$$

to obtain

$$\mathcal{M}'(t) = \frac{1}{4}\int_{\frac{\pi-t}{2}}^{\pi} q(u)q\Big(u-\frac{\pi-t}{2}\Big)\,du + \frac{1}{4}\int_{\frac{\pi+t}{2}}^{\pi} q(u)q\Big(u-\frac{\pi+t}{2}\Big)\,du + W(t),$$

where the function $W(t)$ is absolutely continuous in the interval $[0,\pi]$ and its derivative is summable in it. Therefore, for any integer $m \geqslant 1$,

$$\begin{aligned}(17.42)\quad &\int_0^\pi \mathcal{M}'(t)\cos mt\,dt\\ &= \frac{1}{4}\int_0^\pi \cos mt\,dt \int_{\frac{\pi-t}{2}}^{\pi} q(u)q\Big(u-\frac{\pi-t}{2}\Big)\,du\\ &\quad + \frac{1}{4}\int_0^\pi \cos mt\,dt \int_{\frac{\pi+t}{2}}^{\pi} q(u)q\Big(u-\frac{\pi+t}{2}\Big)\,du - \int_0^\pi W'(t)\frac{\sin mt}{m}\,dt\\ &= \frac{1}{2}\int_0^\pi \cos 2mv\,dv \int_v^\pi q(u)q(u-v)du - \int_0^\pi W'(t)\frac{\sin mt}{m}\,dt.\end{aligned}$$

Also, it is easy to see that

$$\int_0^\pi \cos 2mv\,dv \int_v^\pi q(u)q(u-v)\,du = \frac{1}{2}\left(\int_0^\pi q(u)\cos 2mu\,du\right)^2 + \frac{1}{2}\left(\int_0^\pi q(u)\sin 2mu\,du\right)^2.$$

Setting $\delta_m = \beta_m - m$, we have from (17.40) and the preceding estimates:

$$\begin{aligned}
&|A_2(\pi,\beta_{m,2}) - A_1(\pi,\beta_{m,1})| \\
&\leqslant \Big| -\frac{\pi^2}{2}(\beta_{m,2}-\beta_{m,1})(\delta_{m,2}+\delta_{m,1}) + \frac{\pi^2}{2m}\big(\langle q_2 - q_1\rangle\delta_{m,2} + \langle q_1\rangle(\delta_{m,2}-\delta_{m,1})\big) \\
&\quad - \frac{\pi^2}{8m^2}\big(\langle q_2\rangle^2 - \langle q_1\rangle^2\big) + \frac{1}{2}\int_0^\pi (\mathcal{M}_2'(t) - \mathcal{M}_1'(t))\frac{\cos mt}{m^2}\,dt\Big| \\
&\leqslant \frac{C}{m^3}\left(\frac{1}{m} + \left|\int_0^\pi t\mathcal{M}_1'(t)\sin mt\,dt\right|\right)\|q_2 - q_1\|_{L_2[0,\pi]}e^{C\sqrt{\hat{Q}}}.
\end{aligned}$$

Using (17.21a), (17.22a), (17.23a), (17.24a) and (17.42), we find

$$m^2|A_2(\pi,\beta_{m,2}) - A_1(\pi,\beta_{m,1})| \leqslant \Gamma_m, \tag{17.43}$$

where

$$\begin{aligned}
\Gamma_m = &\left|\left(\int_0^\pi \mathcal{N}_2(t)\cos mt\,dt\right)^2 - \left(\int_0^\pi \mathcal{N}_1(t)\cos mt\,dt\right)^2\right| \\
&+ \left|\left(\int_0^\pi q_2(t)\cos 2mt\,dt\right)^2 - \left(\int_0^\pi q_1(t)\cos 2mt\,dt\right)^2\right| \\
&+ \left|\left(\int_0^\pi q_2(t)\sin 2mt\,dt\right)^2 - \left(\int_0^\pi q_1(t)\sin 2mt\,dt\right)^2\right| \\
&+ Cm^{-1}\|q_2 - q_1\|_{L_2[0,\pi]}\left(\frac{1}{m} + \left|\int_0^\pi t\mathcal{M}_1'(t)\sin mt\,dt\right|\right. \\
&\qquad \left. + \left|\int_0^\pi t\mathcal{N}_2(t)\sin mt\,dt\right| + \left|\int_0^\pi \mathcal{N}_1(t)\sin mt\,dt\right|\right) \\
&+ Cm^{-1}\left(\left|\int_0^\pi (\mathcal{N}_2(t) - \mathcal{N}_1(t))\cos mt\,dt\right|\right. \\
&\qquad \left. + \left|\int_0^\pi (\mathcal{N}_2(t) - \mathcal{N}_1(t))\sin mt\,dt\right| + \left|\int_0^\pi (W_2'(t) - W_1'(t))\sin mt\,dt\right|\right).
\end{aligned} \tag{17.44}$$

Similarly, from (17.21b), (17.22b), (17.23b) and (17.24b), we find

$$m^2|A_2(\pi,\gamma_{m,2}) - A_1(\pi,\gamma_{m,1})| \leqslant \Xi_m,$$

where Ξ_m has a representation similar to that of Γ_m.

Now consider the function $B(\pi,\mu) = \frac{1}{2}\big(C(\pi,\mu) - S'(\pi,\mu)\big)$, which, in view of (17.3), determines the sign of $\operatorname{Im}\theta_m$. If we introduce

$$Y(t) = \frac{1}{2}\left(\int_{\frac{\pi+t}{2}}^{\pi} d\alpha\, K(\alpha,\pi+t-\alpha)q(\alpha) - \int_{\frac{\pi-t}{2}}^{\pi} K(\alpha,\pi-t-\alpha)q(\alpha)d\alpha\right),$$

it is easy to see that

$$B(\pi,\mu) = \frac{1}{2}\int_0^{\pi} q(t)\frac{\sin\mu(\pi-2t)}{\mu}dt - \int_0^{\pi} Y'(t)\frac{\cos\mu t}{\mu^2}dt.$$

In the same way as in the proof of (17.43), we arrive at the estimate

$$|B_2(\pi,\beta_{m,2}) - B_1(\pi,\beta_{m,1})| \leqslant \Lambda_m, \tag{17.45}$$

where
(17.46)

$$\Lambda_m = C\left(\frac{1}{2m}\left|\int_0^{\pi}(q_2(t)-q_1(t))\sin 2mt\,dt\right| + \frac{1}{m^2}\|q_2-q_1\|_{L_2[0,\pi]}\right)e^{C\sqrt{\hat{Q}}}.$$

If $|B_1(\pi,\beta_{m,1})| \geqslant 3\Lambda_m$, (17.45) implies that $B_1(\pi,\beta_{m,1})$ and $B_2(\pi,\beta_{m,2})$ have the same sign. But then, from (17.39) and (17.43), we have

$$m^2(\operatorname{Im}\theta_{m,2}-\operatorname{Im}\theta_{m,1})^2 = m^2\big(|\operatorname{Im}\theta_{m,2}|-|\operatorname{Im}\theta_{m,1}|\big)^2 \leqslant C\Gamma_m e^{C\sqrt{\hat{Q}}}. \tag{17.47}$$

If, on the contrary, $|B_1(\pi,\beta_{m,1})| \leqslant 3\Lambda_m$, it follows from (17.45) that $|B_2(\pi,\beta_{m,2})| \leqslant 4\Lambda_m$. Using the inequality $A^2(\pi,\beta_m) = B^2(\pi,\beta_m)+1$, we get

$$\begin{aligned} |\operatorname{Im}\theta_{m,1}| &= \ln\big(|B_1(\pi,\beta_{m,1})| + \sqrt{1+B_1^2(\pi,\beta_{m,1})}\big) \\ &\leqslant |B_1(\pi,\beta_{m,1})|\big(1+\tfrac{1}{2}|B_1(\pi,\beta_{m,1})|\big) \leqslant 5\Lambda_m(1+\Lambda_m). \end{aligned}$$

Estimating $|\operatorname{Im}\theta_{m,2}|$ in a similar way, we get

$$m^2|\operatorname{Im}\theta_{m,2}-\operatorname{Im}\theta_{m,1}|^2 \leqslant 2m^2\big(|\operatorname{Im}\theta_{m,1}|+|\operatorname{Im}\theta_{m,2}|\big) \leqslant 50\Lambda_m^2(1+\Lambda_m)^2m^2.$$

Together with (17.47), this leads, for all $m \geqslant l$, to

$$m^2|\operatorname{Im}\theta_{m,2}-\operatorname{Im}\theta_{m,1}|^2 \leqslant C(\Gamma_m + m^2\Lambda_m^2)e^{C\sqrt{\hat{Q}}}. \tag{17.48}$$

We now consider the quantities $\operatorname{Re}(\theta_{m,2}-\theta_{m,1})$. If we set $f(t) = \ln\big(A(\pi,t)+\sqrt{A^2(\pi,t)-1}\big)$ for $A(\pi,t) \geqslant 1$, we have $|\theta_m| = f(\gamma_m)$, $|\operatorname{Im}\theta_m| = f(\beta_m)$, and

$$\begin{aligned}|\operatorname{Re}\theta_m|^2 &= |\theta_m|^2 - |\operatorname{Im}\theta_m|^2 = f^2(\gamma_m) - f^2(\beta_m)\\ &= 2\int_{\beta_m}^{\gamma_m} f(t)f'(t)\,dt = 2\int_{A(\pi,\beta_m)}^{A(\pi,\gamma_m)} \frac{\ln(u+\sqrt{u^2-1})}{\sqrt{u^2-1}}\,du\\ &\leqslant 4\big|A(\pi,\gamma_m) - A(\pi,\beta_m)\big|.\end{aligned}$$

Since $\dot{A}(\pi,\gamma_m) = 0$, we have $A(\pi,\beta_m) = A(\pi,\gamma_m) + \int_{\gamma_m}^{\beta_m}(\beta_m - t)\ddot{A}(\pi,t)\,dt$, whence

$$|\operatorname{Re}\theta_m|^2 \leqslant C|\beta_m - \gamma_m|^2 e^{C\sqrt{\hat{Q}}}. \tag{17.49}$$

In view of (17.23a,b), we have

$$|\gamma_{m,2} - \beta_{m,2} - \gamma_{m,1} + \beta_{m,1}| \leqslant \Delta_m, \tag{17.50}$$

where

$$\begin{aligned}\Delta_m = \frac{1}{\pi m}\left|\int_0^\pi (\mathcal{N}_2(t) - \mathcal{N}_1(t))\cos mt\,dt\right| + \frac{1}{m^2}(\Omega_m + \mathcal{P}_m)\\ + \frac{1}{2\pi m^2}\left|\int_0^\pi \big(t(\mathcal{M}_2'(t) - \mathcal{M}_1'(t)) - 2(\mathcal{M}_2(t) - \mathcal{M}_1(t))\big)\sin mt\,dt\right|.\end{aligned} \tag{17.51}$$

If, for some fixed $m \geqslant 1$, we have $|\gamma_{m,1} - \gamma_{m,2}| \geqslant 3\Lambda_m$, we deduce from (17.50) that $\gamma_{m,1} - \beta_{m,1}$ and $\gamma_{m,2} - \beta_{m,2}$ have the same sign, and therefore so do $\operatorname{Re}\theta_{m,1}$ and $\operatorname{Re}\theta_{m,2}$. In this case,

$$\begin{aligned}&m^2|\operatorname{Re}\theta_{m,2} - \operatorname{Re}\theta_{m,1}|^2\\ &\quad\leqslant m^2\big||\operatorname{Re}\theta_{m,2}|^2 - |\operatorname{Re}\theta_{m,1}|^2\big|\\ &\quad\leqslant 2m^2\left|\left(\int_{A_2(\pi,\beta_{m,2})}^{A_2(\pi,\gamma_{m,2})} - \int_{A_1(\pi,\beta_{m,1})}^{A_1(\pi,\gamma_{m,1})}\right)\frac{\ln(u+\sqrt{u^2-1})}{\sqrt{u^2-1}}\,du\right|\\ &\quad\leqslant 2m^2\big(|A_2(\pi,\gamma_{m,2}) - A_1(\pi,\gamma_{m,1})| + |A_2(\pi,\beta_{m,2}) - A_1(\pi,\beta_{m,1})|\big)\\ &\quad\leqslant 2(\Gamma_m + \Xi_m).\end{aligned}$$

If, contrariwise, $|\gamma_{m,1} - \gamma_{m,2}| \leqslant 3\Lambda_m$, we deduce from (17.50) that $|\gamma_{m,2} - \beta_{m,2}| \leqslant 4\Delta_m$. Then, from (17.49), we get

$$m^2|\operatorname{Re}\theta_{m,2} - \operatorname{Re}\theta_{m,1}|^2 \leqslant 2m^2\big(|\operatorname{Re}\theta_{m,2}|^2 + |\operatorname{Re}\theta_{m,1}|^2\big) \leqslant Cm^2\Delta_m^2 e^{C\sqrt{\hat{Q}}}.$$

Thus, for any $m \geqslant 1$,

$$m^2|\operatorname{Re}\theta_{m,2} - \operatorname{Re}\theta_{m,1}|^2 \leqslant C(\Gamma_m + \Xi_m + m^2\Delta_m^2)e^{C\sqrt{\hat{Q}}}.$$

Taking (17.48) into account, we get

(17.52) $$\sum_{m\geqslant 1} m^2|\theta_{m,2}-\theta_{m,1}|^2 \leqslant C\sum_{m\geqslant 1}(\Gamma_m+\Xi_m+m^2\Delta_m^2+m^2\Lambda_m^2)e^{C\sqrt{\hat{Q}}}.$$

From (17.44), (17.46) and (17.51), and using Cauchy's inequality and Parseval's equality, we arrive at the estimate (17.13) for $\theta_{0,2}=\theta_{0,1}=0$ and $a=\pi$. In the case of arbitrary $\theta_{0,j}$, we consider the shifted potentials $\tilde{q}_j=q_j-\theta_{0,j}$. The parameters for the new potentials are

$$\tilde{\theta}_{0,j}=0, \qquad \tilde{\theta}_{m,j}=\theta_{m,j} \quad \text{for } m\geqslant 1.$$

From this and from Lemma 17.11, we obtain (17.13) for $a=\pi$.

Now if the potentials $q_j(x)$, for $j=1,2$, are periodic of period $a>0$, consider the potentials $q_{a,j}(x)=(a/\pi)^2q_j(ax/\pi)$, which have period π. The functions $\theta_j(a,\mu)$ and $\theta_j(\pi,\mu)$ corresponding to the potentials q_j and $q_{a,j}$ are related by

$$\theta_j(a,\mu)=\theta_j(\pi,a\mu/\pi).$$

Since we already know that (17.13) holds for $a=\pi$, we have

$$\begin{aligned}\|\theta_{(2)}-\theta_{(1)}\|_a &= \frac{\pi^2}{a^2}\|\tilde{\theta}_{(2)}-\tilde{\theta}_{(1)}\|_\pi+|\theta_{0,2}-\theta_{0,1}|\\ &\leqslant \frac{C}{a^2}\|q_{a,2}-q_{a,1}\|^{1/2}_{L_2[0,\pi]}e^{C\sqrt{\hat{Q}_a}},\end{aligned}$$

where $\hat{Q}_a=\max_{j=1,2}\|q_{a,j}\|_{L_2[0,\pi]}=\pi^{-3/2}a^2\hat{Q}$ and $\|q_{a,2}-q_{a,1}\|_{L_2[0,\pi]}=\pi^{-3/2}a^2\|q_2-q_1\|_B$. This completes the proof of (17.13).

To prove (17.14), we use Lemma 17.7 and repeat almost literally the argument of the proof of Theorem 2 of [Marchenko and Ostrovskii 1980]. This concludes the proof of Theorem 17.6. □

(17.12) Lemma. *For a potential q, periodic of period a, let*

$$U(a,\mu)=\begin{pmatrix} C(a,\mu) & S(a,\mu)\\ C'(a,\mu) & S'(a,\mu)\end{pmatrix}$$

be the monodromy matrix, that is, the matrix $\mathcal{T}(a,0)$ from (11.17). The monodromy matrix $\mathcal{T}(ma,0)=U(ma,\mu)=U^m(a,\mu)$ for the same potential, regarded as having period ma, is

$$\frac{1}{\sin\theta}\begin{pmatrix} C\sin m\theta-\sin(m-1)\theta & S\sin m\theta\\ C'\sin m\theta & -C\sin m\theta-\sin(m+1)\theta\end{pmatrix},$$

where $\theta=\theta(a,\mu)$, $S=S(a,\mu)$ and $C=C(a,\mu)$.

Proof. Reduce $U(a,\mu)$ to diagonal form. □

(17.13) Lemma. *The functions $\theta_{(j)}(a,\mu)$, corresponding to the potentials $q_j(x)$, for $j=1,2$, satisfy*

$$\theta_{(2)}(a,\mu)-\theta_{(1)}(a,\mu)\leqslant Ca\sqrt{\|q_2-q_1\|_B}e^{C\sqrt{Q}a}, \tag{17.53}$$

where C is a function of μ and $Q=\max_j\|q_j\|_B$.

Proof. Consider first the case $a=\pi$, $\nu_{0,1}=\nu_{0,2}=0$. Using the same arguments as in the proof of Lemma 17.10, we see that, for all $m\geqslant 1$,

$$|\alpha_{m,2}{}^{\pm}-\alpha_{m,1}{}^{\pm}|\leqslant Cm^{-1}\|q_2-q_1\|_{L_2[0,\pi]}e^{C\sqrt{Q}}. \tag{17.54}$$

Denote by τ_m, for $m\geqslant 1$, the successive positive roots of $A(\tau,\mu)$. Since $|A(\pi,\tau_m)-A(\pi,\alpha_m^-)|=|A(\pi,\tau_m)-A(\pi,\alpha_m^+)|$, we can assume that

$$|\tau_m-\alpha_{m-1}^+|\geqslant C^{-1}e^{-C\sqrt{Q}},\quad |\tau_m-\alpha_{m-1}^-|\geqslant C^{-1}e^{-C\sqrt{Q}}, \tag{17.55}$$

with the same absolute constant C as in (17.54). Suppose first that q_2-q_1 is so small that

$$\|q_2-q_1\|_{L_2[0,\pi]}\leqslant(2C)^{-2}e^{-2C\sqrt{Q}}. \tag{17.56}$$

Introducing the family of potentials $q_t=(t-1)q_2-(2-t)q_1$, for $t\in[1,2]$, we have

$$\|q_t\|_{L_2[0,\pi]}\leqslant\pi Q,\qquad \|q_t-q_1\|_{L_2[0,\pi]}\leqslant\|q_2-q_1\|_{L_2[0,\pi]},$$

and as t increases continuously from 1 to 2, the numbers $\alpha_{m,t}^{\pm}$ corresponding to H_{q_t} change continuously from $\alpha_{m,1}^{\pm}$ to $\alpha_{m,2}{}^{\pm}$. By virtue of (17.54) and (17.56), we have, for each value of t,

$$|\alpha_{m,t}^{\pm}-\alpha_{m,1}^{\pm}|\leqslant C\|q_t-q_1\|_{L_2[0,\pi]}e^{C\sqrt{Q}}\leqslant(4C)^{-1}e^{-C\sqrt{Q}},$$

whence

$$\tau_{m,1}\leqslant\alpha_{m,t}^-\leqslant\alpha_{m,t}^+\leqslant\tau_{m+1,1}. \tag{17.57}$$

But then the intervals $[\tau_{m,1},\alpha_{m,1}^-]$ and $[\alpha_{n-1,2}^+,\alpha_{n,2}^-]$ can only intersect when $n=m$ or $n=m+1$, and the intervals $[\alpha_{m-1,1}^+,\tau_{m,1}]$ and $[\alpha_{n-1,2}^+,\alpha_{n,2}^-]$ can only intersect when $n=m$ or $n=m-1$.

There are three possible cases for a fixed value $\mu\geqslant 0$, depending on whether μ is in a gap of both H_{q_1} and H_{q_2}, in a gap of only one of them, or in the spectrum of both operators. Consider the latter case, for example. Let m be the number of the band of the spectrum of H_{q_1} to which μ belongs. Then $\theta_1(\pi,\mu)=\pi m-\arccos A_1(\pi,\mu)$. If $\mu\in[\alpha_{m-1,2}^+-\alpha_{m,2}^-]$, we have $\theta_2(\pi,\mu)=\pi m-\arccos A_2(\pi,\mu)$ and $|\theta_2(\pi,\mu)-\theta_1(\pi,\mu)|\leqslant 4\sqrt{A_2(\pi,\mu)-A_1(\pi,\mu)}$.

Thus, in view of (17.37) and Lemma 17.7, we arrive at (17.53). If the case $\mu \notin [\alpha^+_{m-1,2} - \alpha^-_{m,2}]$, we must have $\mu \in [\alpha^+_{m,2}, \tau_{m+1,2}]$, and therefore $\alpha^-_{m,2} \leqslant \mu \leqslant \alpha^+_{m,1}$. Now $\theta_2(\pi,\mu) = \pi\mu + \arccos A_2(\pi,\mu)$ and $|\theta_2(\pi,\mu) - \theta_1(\pi,\mu)| \leqslant \arccos A_1(\pi,\mu) + \arccos A_2(\pi,\mu)$. Hence

$$
\begin{aligned}
|\arccos A_1(\pi,\mu)| &\leqslant 4\sqrt{|A_1(\pi,\mu) - A_1(\pi,\alpha^-_{m,1})|} \\
&\leqslant C\sqrt{\mu - \alpha^-_{m,1}}\,e^{C\sqrt{Q}} \leqslant C\|q_2 - q_1\|_{L_2[0,\pi]} e^{C\sqrt{Q}}.
\end{aligned}
$$

The estimation of $\arccos A_2(\pi,\mu)$ is similar and again leads to (17.53). The other two possibilities for the position of μ relative to the spectra of H_{q_1} and H_{q_2} are treated similarly.

It remains now to consider the case when (17.56) does not hold. Then, if $\operatorname{Im}\theta(\pi,\mu) \neq 0$, we have $\operatorname{Im}\theta(\pi,\mu) \leqslant 2A(\pi,\mu) \leqslant Ce^{C\sqrt{Q}}$, so that

$$
|\operatorname{Im}\theta_2(\pi,\mu) - \operatorname{Im}\theta_1(\pi,\mu)| \leqslant C\|q_2 - q_1\|_{L_2[0,\pi]}
$$

for all $\mu \in \mathbf{R}$. If $\operatorname{Im}\theta(\pi,\mu) = 0$, we have $|A(\pi,\mu)| \leqslant 1$ and for some $k \geqslant 1$ such that $\alpha^+_{k-1} \leqslant \mu \leqslant \alpha^-_k$, we have $\operatorname{Re}\theta(\pi,\mu) = \mu\pi - \arccos A(\pi,\mu)$. Therefore, $\operatorname{Re}\theta(\pi,\mu) = \mu\pi + \sigma(\mu)$, where $\sigma(\mu) = \mu(k - \beta_k) + \pi(\beta_k - \mu) + \arccos A(\pi,\mu)$. But it follows from the inequality $|\beta_k - \mu| \leqslant \beta_{k+1} - \beta_k$ and (17.21a) that $|\sigma(\mu)| \leqslant Ce^{C\sqrt{Q}}$, and this inequality holds for $\operatorname{Im}\theta(\pi,\mu) \neq 0$ as well. But then $|\operatorname{Re}(\theta_2(\pi,\mu) - \theta_1(\pi,\mu))| \leqslant Ce^{C\sqrt{Q}}$, so (17.53) follows also when (17.56) does not hold.

As in the proof of Theorem 17.6, the case of general a, $\nu_{0,1}$ and $\nu_{0,2}$ is easily reduced to the case of $a = \pi$ and $\nu_{0,1} = \nu_{0,2} = 0$, which we have just considered. This concludes the proof of Lemma 17.13. □

(17.14) Lemma. *Let q_1 and q_2 be periodic potentials of period a. Assume $\nu_{0,1} = \nu_{0,2}$, set $Q = \max\|q_j\|_B$, and let $\alpha^\pm_0 = 0$ and $\alpha^\pm_{m,j} = \operatorname{sign} m\sqrt{\nu^\pm_{|m|,j}}$, for $m \neq 0$, be the solutions of* (17.29). *Suppose the numbers $\sigma_{m,j}$ and $\sigma_{-m,j}$ satisfy*

$$
\alpha^-_{m,j} \leqslant \sigma_{m,j} \leqslant \alpha^+_{m,j} \tag{17.58}
$$

for $m = \pm1, \pm2, \ldots$, and

$$
|\sigma_{m,2} - \sigma_{m,1}| \leqslant Ca|m|^{-1}\|q_2 - q_1\|_B e^{C\sqrt{Q}}. \tag{17.59}
$$

Define the functions $\Phi_j(\mu)$, for $j = 1, 2$, by

$$
\Phi_j(\mu) = \prod_{m\neq0} \frac{\mu - \sigma_{m,j}}{\sqrt{(\mu - \alpha^+_{m,j})(\mu - \alpha^-_{m,j})}}; \tag{17.60}
$$

these functions are analytic on the half-plane $\mathbf{C}_+$, *and are real and have the same sign for small real values of* μ.

Then, for every finite interval $\Delta \in \mathbf{R}$ *and ever* $p \in (1,2)$, *there exists a constant* $C(p)$ *such that*

$$\|\Phi_2 - \Phi_1\|_{L_p(\Delta)} \leqslant C(p)|\Delta|^{1/p} a^{2\nu} \|q_2 - q_1\|_B^{\nu} e^{Ca\sqrt{Q}}, \tag{17.61}$$

where $\nu = \min\{\frac{1}{p} - \frac{1}{2}, \frac{1}{4}\}$ *and* $|\Delta| = \operatorname{meas}\Delta$. *In addition, for any* $p > 1$,

$$\|\arg\Phi_2 - \arg\Phi_1\|_{L_p(\Delta)} \leqslant C(p)|\Delta|^{1/p}|q_2 - q_1\|_B^{1/p} e^{Ca\sqrt{Q}}. \tag{17.62}$$

Proof. As above, we restrict ourselves to the case $a = \pi$. Moreover, since the functions Φ_j are even, we can take $\Delta = [0,h]$ for $h > 0$. We start by estimating $\|\Phi_j\|_{L_2[0,h]}$. Each factor in (17.60) has two square-root singularities, at the points $\alpha^{\pm}_{m,j}$. In view of (17.58), for $\mu \in [\alpha^{+}_{m+1,j}, \alpha^{-}_{m,j}]$ we have

$$|\Phi_j(\mu)| \leqslant C\sqrt{\frac{\mu - \alpha^{-}_{m,j}}{\mu - \alpha^{+}_{m-1,j}}}\, e^{C\sqrt{Q}}.$$

Since

$$0 \leqslant \alpha^{-}_{m} - \alpha^{+}_{m-1} \leqslant \beta_m - \beta_{m-1} \leqslant Ce^{C\sqrt{Q}}$$

by Lemma 17.9, this implies that

$$\begin{aligned}\int_{\alpha^{+}_{m-1,j}}^{\alpha^{-}_{m,j}} |\Phi_j(\mu)|^p\, d\mu &\leqslant C\int_{\alpha^{+}_{m-1,j}}^{\alpha^{-}_{m,j}} |\mu - \alpha^{+}_{m-1,j}|^{-p/2}|\mu - \alpha^{-}_{m,j}|^{-p/2}\, d\mu\, e^{C\sqrt{Q}} \\ &\leqslant C(p)e^{C\sqrt{Q}}.\end{aligned}$$

Similarly, $|\Phi_j(\mu)| \leqslant Ce^{C\sqrt{Q}}|\mu - \alpha^{-}_{m,j}|^{-1/2}$ for $\mu \in [\alpha^{-}_{m,j}, \sigma_{m,j}]$, and $|\Phi_j(\mu)| \leqslant Ce^{C\sqrt{Q}}|\mu - \alpha^{+}_{m,j}|^{-1/2}$ for $\mu \in [\sigma_{m,j}, \alpha^{+}_{m,j}]$. Therefore,

$$\begin{aligned}&\int_{\alpha^{-}_{m,j}}^{\alpha^{+}_{m,j}} |\Phi_j(\mu)|^p\, d\mu \\ &\qquad \leqslant C\Big(\int_{\alpha^{-}_{m,j}}^{\sigma_{m,j}} |\mu - \alpha^{-}_{m,j}|^{-p/2}\, d\mu + \int_{\sigma_{m,j}}^{\alpha^{+}_{m,j}} |\mu - \alpha^{+}_{m,j}|^{-p/2}\, d\mu\Big) e^{C\sqrt{Q}} \\ &\leqslant C(p)e^{C\sqrt{Q}}.\end{aligned}$$

In view of (17.32), the interval $[0,h]$ contains no more than $Che^{C\sqrt{Q}}$ intervals of the form $[\alpha^{+}_{m-1,j}, \alpha^{-}_{m,j}]$, whence

$$\|\Phi_j\|_{L_p[0,h]} \leqslant C(p)h^{1/p}e^{C\sqrt{Q}}. \tag{17.63}$$

Thus, if the functions $\Phi_j(\mu) \in L_p[0,h]$ are considered as the result of the application of a nonlinear operator to the potential $q_j \in L_2[0,\pi]$, it follows from (17.63) that this operator is a bounded map from $L_2[0,\pi]$ into $L_p[0,h]$. Inequality (17.61) says that this operator satisfies a Hölder condition with exponent ν and multiplicative constant Ce^{CQ}.

Before proving (17.61), we note that, by (17.63), we can assume that $\|q_2 - q_1\|_{L_2[0,\pi]}$ is small, at that (17.56), for example, is satisfied.

The essence of the arguments below is the verification that the L_p-norm of the difference between individual factors in (17.60) with same m, corresponding to the potentials q_1 and q_2, can be estimated by the right-hand side of (17.61). We divide $[0,h]$ into a finite number of intervals with endpoints of the form $\alpha^{\pm}_{m,j}$ and $\sigma_{m,j}$. Within each of these subintervals $[b_1,b_2]$, the main contribution to the product (17.60) comes from the values of $\alpha^{\pm}_{m,j}$ closest to b_1 and b_2. When $[b_1,b_2]$ is contained in $[\alpha^+_{m-1,1},\alpha^-_{m,1}] \cap [\alpha^+_{m-1,2},\alpha^-_{m,2}]$ or $[\alpha^-_{m,1},\alpha^+_{m,1}] \cap [\alpha^-_{m,2},\alpha^+_{m,2}]$, estimating $\|\Phi_1 - \Phi_2\|_{L_p}$ reduces to estimating the difference between the leading factors

$$\int_0^\infty \left| \frac{1}{\mu} - \frac{1}{\mu+\delta} \right|^{\frac{p}{2}} d\mu \leqslant \operatorname{const} \delta^{1-p/2},$$

which gives (17.61). In all other cases, since $\|q_2 - q_1\|_{L_2[0,\pi]}$ is small, it is enough to use the trivial inequality

$$\|\Phi_2 - \Phi_1\|_{L_p[b_1,b_2]} \leqslant \|\Phi_1\|_{L_p[b_1,b_2]} + \|\Phi_2\|_{L_p[b_1,b_2]}.$$

To carry out this procedure for proving (17.61), it is convenient to introduce the characteristic functions $\chi_{m,j}$ and $\psi_{m,j}$ of the intervals $[\alpha^+_{m-1,j},\alpha^-_{m,j}]$ and $[\alpha^-_{m,j},\alpha^+_{m,j}]$, respectively, and to set $\chi_j = \sum_m \chi_{m,j}$ and $\psi_j = \sum_m \psi_{m,j}$. As we showed in the proof of Lemma 17.13, if inequalities (17.54), (17.55) and (17.56) (with C instead of $2C$) are fulfilled, we have $\chi_{m,1}\chi_{n,2} = 0$ for $|n-m| > 1$. Clearly, then,

$$\begin{aligned}(17.64)\quad \Phi_2 - \Phi_1 = \chi_2\chi_1(\Phi_2 - \Phi_1) + \chi_1\psi_2(\Phi_2 - \Phi_1)\\ + \chi_2\psi_1(\Phi_2 - \Phi_1) + \psi_2\psi_1(\Phi_2 - \Phi_1)\end{aligned}$$

and

$$(17.65)\quad \begin{aligned}\chi_2\chi_1 &= \sum_m \chi_{m,2}(\chi_{m,1} + \chi_{m+1,1} + \chi_{m-1,1}),\\ \chi_1\psi_2 &= \sum_m \chi_{m,2}(\psi_{m,1} + \psi_{m-1,1}).\end{aligned}$$

If a_m and b_m are the endpoints of $[\alpha^+_{m-1,2},\alpha^-_{m,2}] \cap [\alpha^+_{m-1,1},\alpha^-_{m,1}]$, the functions $\Phi_1(\mu)$ and $\Phi_2(\mu)$ are real and have the same sign for $\mu \in [a_m,b_m]$. Therefore,

$$(17.66)\quad \left\|\chi_{m,2}\chi_{m,1}(\Phi_2 - \Phi_1)\right\|_{L_p[0,\pi]} \leqslant \left(\int_{a_m}^{b_m} \left|\Phi_2^2(\mu) - \Phi_1^2(\mu)\right|^{p/2} d\mu\right)^{1/p} = \|\Phi_2^2 - \Phi_1^2\|_{L_p[0,\pi]}.$$

We estimate the integral in (17.66) over the interval $[\tau_{m,1}, b_m]$, where $\tau_{m,1}$ is the root of the function $A_1(\pi,\mu)$ in the interval $[\alpha^+_{m-1,1}, \alpha^-_{m,1}]$; the integral over $[a_m, \tau_{m,1}]$ can be considered similarly. For $j = 1, 2$, let

$$u_{n,j} = \frac{\mu - \sigma_{n,j}}{(\mu - \alpha^-_{n,j})(\mu - \alpha^+_{n,j})}$$

and $v_{n,j} = \Phi_j^2(\mu)/u_{n,j}(\mu)$. Then

$$\sqrt{|\Phi_2^2 - \Phi_1^2|} \leqslant \sqrt{|u_{m,2} - u_{m,1}|\,|v_{m,1}|} + \sqrt{|v_{m,2} - v_{m,1}|\,|u_{m,2}|}.$$

Since $\alpha^+_{m-1,j} \leqslant a_m \leqslant \tau_{m,1} \leqslant \mu \leqslant b_m \leqslant \alpha^-_{m,j}$, and since

$$|\tau_{m,1} - a_m| \geqslant Ce^{-C\sqrt{Q}}, \qquad |\tau_{m,1} - b_m| \geqslant Ce^{-C\sqrt{Q}}$$

by (17.55), we obtain

$$(17.67)\qquad |v_{m,1}(\mu)| = \left|\prod_{n\neq m} \frac{(\mu - \sigma_{n,1})^2}{(\mu - \alpha^-_{n,1})(\mu - \alpha^+_{n,1})}\right| \leqslant \frac{1}{|\mu - \alpha^-_{m+1,1}|\,|\mu - \alpha^+_{m-1,1}|} \leqslant Ce^{C\sqrt{Q}}.$$

Moreover,

$$(17.68)\qquad \int_{\tau_{m,1}}^{b_m} \left|u_{m,2}(\mu)\right|^{p/2} d\mu \leqslant |\sigma_{m,2} - \tau_{m,1}|^{p/2} \int_{\tau_{m,1}}^{b_m} |\mu - \alpha^-_{m,2}|^{-p/2}\, d\mu \leqslant C(p)e^{C\sqrt{Q}}.$$

For $j = 1, 2$ and $n = \pm 1, \pm 2, \ldots$, set $r^{\pm}_{n,j}(\mu) = (\mu - \alpha^{\pm}_{n,j})^{-1}$ and

$$\kappa^{\pm}_{n,j} = \frac{(\alpha^{\pm}_{n,j} - \sigma_{n,j})^2}{\alpha^-_{n,j} - \alpha^+_{n,j}}.$$

Then

$$(17.69)\quad u_{n,2}(\mu) - u_{n,1}(\mu) = \kappa^-_{n,2}\big(r^-_{n,2}(\mu) - r^-_{n,1}(\mu)\big) - \kappa^+_{n,2}\big(r^+_{n,2}(\mu) - r^+_{n,1}(\mu)\big) + (\kappa^-_{n,2} - \kappa^-_{n,1})r^-_{n,1}(\mu) - (\kappa^+_{n,2} - \kappa^+_{n,1})r^+_{n,1}(\mu).$$

Now, if we set $\delta^{\pm}_n = |\alpha^{\pm}_{n,2} - \alpha^{\pm}_{n,1}|$, we have, by (17.54):

$$(17.70)\qquad \int_{\tau_{m,1}}^{b_m} \left|r^{\pm}_{m,2}(\mu) - r^{\pm}_{m,1}(\mu)\right|^{p/2} d\mu \leqslant \int_0^{\infty} \left|\frac{1}{t} + \frac{1}{t+\delta^{\pm}_m}\right|^{p/2} dt$$
$$\leqslant C(p)\|q_2 - q_1\|^{1-p/2}_{L_2[0,\pi]} e^{C\sqrt{Q}},$$

and, for $n \neq m$ and $\mu \in [a_m, b_m]$,

$$(17.71)\qquad \left|r^{\pm}_{n,2}(\mu) - r^{\pm}_{n,1}(\mu)\right| \leqslant \delta^{\pm}_n \max_{a_m \leqslant \mu \leqslant b_m} \left|r^{\pm}_{n,2}(\mu) - r^{\pm}_{n,1}(\mu)\right|$$
$$\leqslant C\|q_2 - q_1\|_{L_2[0,\pi]}|n-m|^2 e^{C\sqrt{Q}}.$$

Moreover,

$$(17.72)\qquad \int_{\tau_{m,1}}^{b_m} \left|r^{\pm}_{m,1}(\mu)\right|^{-p/2} d\mu \leqslant \int_0^{b_m - \tau_{m,1}} t^{-p/2}\, dt \leqslant C(p) e^{C\sqrt{Q}},$$

and, for $n \neq m$ and $m \in [a_m, b_m]$,

$$(17.73)\qquad \left|r^{\pm}_{n,1}(\mu)\right| \leqslant C|n-m|^{-1} e^{C\sqrt{Q}}.$$

If, for the value of n under consideration,

$$(17.74)\qquad |\alpha^+_{n,1} - \alpha^-_{n,1}| \geqslant 8C|n|^{-1}\|q_2 - q_1\|^{1/2}_{L_2[0,\pi]} e^{-C\sqrt{Q}},$$

where C is the constant from (17.54), we conclude from (17.54) that

$$|\alpha^+_{n,2} - \alpha^-_{n,2}| \geqslant |\alpha^+_{n,1} - \alpha^-_{n,1}| - |\alpha^+_{n,2} - \alpha^+_{n,1}| - |\alpha^-_{n,2} - \alpha^-_{n,1}|$$
$$\geqslant C|n|^{-1}\|q_2 - q_1\|^{1/2}_{L_2[0,\pi]} e^{-C\sqrt{Q}},$$

so that

$$(17.75)\quad |\kappa^{\pm}_{n,2} - \kappa^{\pm}_{n,1}| \leqslant \frac{\left(|\alpha^{\pm}_{n,2} - \alpha^{\pm}_{n,1}| + |\sigma_{n,2} - \sigma_{n,1}|\right)\left(|\alpha^{\pm}_{n,2} - \sigma_{n,2}| + |\alpha^{\pm}_{n,1} - \sigma_{n,1}|\right)}{|\alpha^-_{n,2} - \alpha^+_{n,2}|}$$
$$+ |\alpha^{\pm}_{n,1} - \sigma_{n,1}|^2 \frac{|\alpha^-_{n,2} - \alpha^-_{n,1}| + |\alpha^+_{n,2} - \alpha^+_{n,1}|}{|\alpha^-_{n,2} - \alpha^+_{n,2}|\,|\alpha^-_{n,1} - \alpha^+_{n,1}|}$$
$$\leqslant C|n|^{-1}\|q_2 - q_1\|^{1/2}_{L_2[0,\pi]} e^{C\sqrt{Q}}.$$

If, on the contrary, (17.74) does not hold, we again write, using (17.54),

$$|\alpha^+_{n,2} - \alpha^-_{n,2}| \leqslant C|n|^{-1}\|q_2 - q_1\|^{1/2}_{L_2[0,\pi]} e^{C\sqrt{Q}}.$$

Then

$$(17.76)\qquad |\kappa^{\pm}_{n,2} - \kappa^{\pm}_{n-1}| \leqslant |\alpha^+_{n,2} - \alpha^-_{n,2}| + |\alpha^+_{n,1} - \alpha^-_{n,1}|$$
$$\leqslant C|n|^{-1}\|q_2 - q_1\|^{1/2}_{L_2[0,\pi]} e^{C\sqrt{Q}}.$$

From the estimates (17.67), (17.70), (17.72), (17.75), (17.76) and the representation (17.69), we have

(17.77)
$$\Big(\int_{\tau_{m,1}}^{b_m} |u_{m,2}(\mu)-u_{m,1}(\mu)|^{p/2}|v_{m,1}(\mu)|^{p/2}\,d\mu\Big)^{\frac{1}{p}}$$
$$\leqslant C(p)\big(\|q_2-q_1\|_{L_2[0,\pi]}^{1/p-1/2}+\|q_2-q_1\|_{L_2[0,\pi]}^{1/4}\big)e^{C\sqrt{Q}}$$
$$\leqslant 2C(p)\|q_2-q_1\|_{L_2[0,\pi]}^{\nu}e^{C\sqrt{Q}}.$$

We now express the difference $v_{m,2}-v_{m,1}$ in the form

(17.78)
$$v_{m,2}(\mu)-v_{m,1}(\mu)=\sum_{n\neq 0}{}'\sum_{s<r}{}'u_{s,1}(\mu)\prod_{s>n}{}'u_{s,2}(\mu)\big(u_{n,2}(\mu)-u_{n,1}(\mu)\big),$$

where the primes in the sums and product mean the m-th term is absent. As before, we assume that $\mu\in[a_m,b_m]$, so that

$$\prod_{s<n}{}'u_{s,1}(\mu)\prod_{s>n}{}'u_{s,2}(\mu)\leqslant Ce^{C\sqrt{Q}}.$$

Finally, from (17.71), (17.73), (17.75), (17.76) and (17.69), we get

$$\sum_{n\neq 0}{}'|u_{n,2}(\mu)-u_{n,1}(\mu)|$$
$$\leqslant C\sum_{n\neq 0}{}'\Big(\frac{1}{|n-m|^2}+\frac{1}{|n-m|\,|m|}\Big)\|q_2-q_1\|_{L_2[0,\pi]}^{\nu}e^{C\sqrt{Q}}$$
$$\leqslant C\|q_2-q_1\|_{L_2[0,\pi]}^{\nu}e^{C\sqrt{Q}}.$$

Hence

$$\Big(\int_{\tau_{m,1}}^{b_m}|v_{m,2}(\mu)-v_{m,1}(\mu)|^{p/2}|u_{m,2}(\mu)|^{p/2}\,d\mu\Big)^{\frac{1}{p}}\leqslant C(p)\|q_2-q_1\|_{L_2[0,\pi]}^{1/2}e^{C\sqrt{Q}}.$$

By combining this with (17.77) and estimating the integral over $[a_m,\tau_{m,1}]$, we get

$$\|\chi_{m,2}\chi_{m,1}(\Phi_2-\Phi_1)\|_{L_p[0,h]}\leqslant C(p)\|q_2-q_1\|_{L_2[0,\pi]}^{\nu}e^{C\sqrt{Q}}.$$

We now estimate $\|\chi_{m,2}\chi_{m+1,1}(\Phi_2-\Phi_1)\|_{L_p[0,h]}$; the estimation of the quantity $\|\chi_{m,2}\chi_{m-1,1}(\Phi_2-\Phi_1)\|_{L_p[0,h]}$ is entirely similar. If $\chi_{m,2}\chi_{m+1,1}$ does not vanish identically, we have $\operatorname{supp}\chi_{m,2}\chi_{m+1,1}=[\alpha_{m,1}^{+},\alpha_{m,2}^{-}]$. But then

$$0\leqslant\alpha_{m,2}^{-}-\alpha_{m,1}^{+}\leqslant\alpha_{m,2}^{-}-\alpha_{m,1}^{-}\leqslant C\|q_2-q_1\|_{L_2[0,\pi]}e^{C\sqrt{Q}},$$

and it is enough to use the obvious fact that $\|\chi_{m,2}\chi_{m+1,1}(\Phi_2-\Phi_1)\|_{L_p[0,h]}$ is bounded above by

$$\|\chi_{m,2}\chi_{m+1,1}\Phi_2\|_{L_p[0,h]}+\|\chi_{m,2}\chi_{m+1,1}\Phi_1\|_{L_p[0,h]}.$$

The first of these norms is estimated by

$$\begin{aligned}\left(\int_{\alpha^+_{m,1}}^{\alpha^-_{m,2}}|\Phi_2(\mu)|^p\,d\mu\right)^{\frac{1}{p}} &\leqslant C\left(\int_{\alpha^+_{m,1}}^{\alpha^-_{m,2}}\sqrt{|u_{m,2}(\mu)|}\,d\mu\right)^{\frac{1}{p}}e^{C\sqrt{Q}}\\ &\leqslant C\left(\int_{\alpha^+_{m,1}}^{\alpha^-_{m,2}}\frac{d\mu}{\sqrt{|\mu-\alpha^-_{m,2}|}}\right)^{\frac{1}{p}}e^{C\sqrt{Q}}\\ &\leqslant C(p)|\alpha^-_{m,2}-\alpha^+_{m,1}|^{1/p-1/2}e^{C\sqrt{Q}}\\ &\leqslant C(p)\|q_2-q_1\|^{\nu}_{L_2[0,\pi]}e^{C\sqrt{Q}}.\end{aligned}$$

A similar estimate holds for the integral of $|\Phi_1|^p$, whence

$$\begin{aligned}\|\chi_{m,2}\chi_{m+1,1}(\Phi_2-\Phi_1)\|_{L_p[0,h]}+\|\chi_{m,2}\chi_{m-1,1}(\Phi_2-\Phi_1)\|_{L_p[0,h]}&\\ \leqslant C(p)\|q_2-q_1\|^{\nu}_{L_2[0,\pi]}e^{C\sqrt{Q}}&.\end{aligned}$$

Hence

$$\begin{aligned}\|\chi_2\chi_1(\Phi_2-\Phi_1)\|_{L_p[0,h]}&\\ =\Big\|\sum_m\chi_{m,2}(\chi_{m,1}+\chi_{m+1,1}+\chi_{m-1,1})(\Phi_2-\Phi_1)\Big\|_{L_p[0,h]}&\\ \leqslant C(p)h^{1/p}\|q_2-q_1\|^{\nu}_{L_2[0,\pi]}e^{C\sqrt{Q}}&.\end{aligned}\tag{17.79}$$

The other terms on the right-hand side of (17.64) are estimated similarly, and the result is inequality (17.61).

To prove (17.62), we use an identity similar to (17.64), with Φ_j replaced by $\arg\Phi_j$. On the supports of $\chi_{m,2}\chi_{m,1}$ and $\psi_{m,2}\psi_{m,1}$, the arguments of Φ_1 and Φ_2 coincide, so

$$\big\|\chi_{m,2}\chi_{m,1}(\arg\Phi_2-\arg\Phi_1)\big\|_{L_p[0,h]}=\big\|\psi_{m,2}\psi_{m,1}(\arg\Phi_2-\arg\Phi_1)\big\|_{L_p[0,h]}=0.$$

On the supports of $\chi_{m,2}\chi_{m+1,1}$ and $\chi_{m,2}\chi_{m-1,1}$, the arguments differ by π. Therefore

$$\begin{aligned}\big\|\chi_{m,2}\chi_{m+1,1}(\arg\Phi_2-\arg\Phi_1)\big\|_{L_p[0,h]}&\leqslant\pi|\alpha^+_{m,2}-\alpha^+_{m,1}|^{1/p},\\ \big\|\chi_{m,2}\chi_{m-1,1}(\arg\Phi_2-\arg\Phi_1)\big\|_{L_p[0,h]}&\leqslant\pi|\alpha^+_{m-1,2}-\alpha^+_{m-1,1}|^{1/p}.\end{aligned}$$

Similarly,

$$\left\|\chi_{m,1}\psi_{m,2}(\arg\Phi_2-\arg\Phi_1)\right\|_{L_p[0,h]}\leqslant\tfrac{1}{2}\pi|\alpha^-_{m,2}-\alpha^-_{m,1}|^{1/p},$$
$$\left\|\chi_{m+1,1}\chi_{m,2}(\arg\Phi_2-\arg\Phi_1)\right\|_{L_p[0,h]}\leqslant\tfrac{1}{2}\pi|\alpha^+_{m,2}-\alpha^+_{m,1}|^{1/p}.$$

This proves (17.62), and Lemma 17.14. □

This proof also provides the following fact, which we will use below:

(17.15) Lemma. *For the periodic potentials $q_1(x)$ and $q_2(x)$ of period a, with $\nu_{0,1}=\nu_{0,2}=0$, let $\chi_j(\mu)$ be the characteristic function of the set $\bigcup_m[\alpha^+_{m,j},\alpha^-_{m+1,j}]$, for $j=1,2$. If $Q=\max_j\|q_j\|_B$ and Δ is an interval of length $|\Delta|$, we have*

$$\int_\Delta|\chi_2(\mu)-\chi_1(\mu)|\,d\mu\leqslant C|\Delta|\|q_2-q_1\|_Be^{Ca\sqrt{Q}}. \qquad \square$$

Given a periodic potential q of period a such that the bottom of the spectrum ν_0 equals 0 for H_q, we consider the functions

$$r(\mu)=\sqrt{\frac{S(a,\mu)}{\sqrt{1-A^2(a,\mu)}}},\qquad v(\mu)=\sqrt{-\frac{C'(a,\mu)}{\sqrt{1-A^2(a,\mu)}}},$$
$$t(\mu)=\frac{C(a,\mu)-A(a,\mu)}{\sqrt{S(a,\mu)\sqrt{1-A^2(a,\mu)}}}.$$

These functions are holomorphic in the upper half-plane, and take on real values of the same sign for small real values of μ. We extend them to the real axis by taking their limit from the upper half-plane.

These functions, as we will see in the next subsection, determine the entries of the spectral matrix (12.33). Using Lemma 17.12, it is easy to see that they do not change if a is replaced by a multiple of itself. Finally, it is immediate that

$$t^2(\mu)=v^2(\mu)-r^{-2}(\mu). \tag{17.80}$$

The proof of the next lemma is similar to that of Lemma 17.14, but rather cumbersome, so we omit it. For details, see [Pastur and Tkachenko 1989].

(17.16) Lemma. *Let $q_1(x)$ and $q_2(x)$ be periodic potentials of period a, and assume that $\nu_{0,1}=\nu_{0,2}=0$. If $f(\mu)$ is one of the functions $r_2(\mu)-r_1(\mu)$, $r_2^{-1}(\mu)-r_1^{-1}(\mu)$, $v_2(\mu)-v_1(\mu)$ or $t_2(\mu)-t_1(\mu)$, we have, for every interval Δ,*

(17.81) $$\|f\|_{L_p(\Delta)} \leqslant C(p)|\Delta|^{1/p} a^{\varepsilon(p)} \|q_2 - q_1\|_B^{\nu(p)} e^{Ca\sqrt{Q}},$$

where $Q = \max_j \|q_j\|_B$ *and* $C(p)$, $\varepsilon(p)$ *and* $\nu(p)$, *for* $p \in (1,2)$, *are positive, finite constants not depending on* a, Δ, q_1 *or* q_2. □

17.C Proof of the Main Theorems

Proof of Theorem 17.1. Suppose $q \in \mathcal{Q}_\infty(\mathfrak{A})$ and q_n, for $n \geqslant 1$, satisfy the condition (17.7), and let $\kappa^{(n)} \in \mathcal{K}_\infty(\mathfrak{A})$ be defined by (17.10). Then $Q = \sup_{n\geqslant 1} \|q_n\|_B$ is finite and the number b in (17.7) can be taken so large that $b > 4C\sqrt{Q}$, where C is taken from (17.13). Let a_n be the period of q_n. For each n, we introduce two sequences: $\theta_{(n)} = \{\theta_{m,n}\}$, which parametrizes the spectrum of the corresponding Schrödinger operator H_{q_n}, and $\tilde\theta_{(n)} = \{\tilde\theta_{m,n}\}$, which parametrizes the same operator, but seen as having a potential of period a_{n+1}. It follows from Lemma 17.12 that

(17.82) $$\tilde\theta_{m,n} = \begin{cases} \theta_{0,n} & \text{for } m = 0, \\ \theta_{m,n} l_n & \text{for } m = kl_n, \\ 0 & \text{for } m \neq kl_n, \end{cases}$$

where $l_n = a_{n+1}/a_n$ is a natural number and $k \in \mathbf{Z}$. By (17.10), $\theta_{(n)}$ and $\tilde\theta_{(n)}$ lead to the same function $\kappa^{(n)} \in \mathcal{K}(\mathfrak{A})$.

Based on the estimate (17.13), we have

$$\begin{aligned} \|\kappa_{(n+1)} - \kappa_{(n)}\|_{\mathcal{K}(\mathfrak{A})} &= |\theta_{0,n+1} - \theta_{0,n}| + 2\Big(\sum_{m\geqslant 1} \frac{m^2}{a_{n+1}^2} |\theta_{m,n+1} - \theta_{m,n}|^2\Big)^{\frac12} \\ &\leqslant 2\|\theta_{(n+1)} - \tilde\theta_{(n)}\|_{a_{n+1}} \leqslant C a_{n+1}^{-1} e^{(C\sqrt{Q} - b/2)a_{n+1}} \\ &\leqslant C_1 e^{-ba_{n+1}/5}, \end{aligned}$$

with the constant C_1 independent of n. But then, for $m > n$,

$$\|\kappa_{(m)} - \kappa_{(n)}\|_{\mathcal{K}(\mathfrak{A})} \leqslant 2C_1 e^{-ba_{n+1}/5},$$

so that the sequence $\kappa_{(n)}$ of elements of $\mathcal{K}(\mathfrak{A})$ has a limit $\kappa \in \mathcal{K}(\mathfrak{A})$.

To check that $\kappa \in \mathcal{K}_\infty(\mathfrak{A})$, notice that $\kappa_{r,n} = 0$ for $r \notin \mathfrak{R}_n$, by (17.10). Therefore,

$$\sum_{r\in\mathfrak{R}(\mathfrak{A})\setminus\mathfrak{R}_n} r^2|\kappa_r|^2 = \sum_{r\in\mathfrak{R}(\mathfrak{A})\setminus\mathfrak{R}_n} r^2|\kappa_r - \kappa_{r,n}|^2 \leqslant \|\kappa - \kappa^{(n)}\|_{\mathcal{K}(\mathfrak{A})} \leqslant 4C_1^2 e^{-2ba_{n+1}/5};$$

since b is arbitrary, we get (17.9).

Now suppose, conversely, that a function $\kappa \in \mathcal{K}_\infty(\mathfrak{A})$ is given. For each $n \geqslant 1$, define the sequence $\theta_{(n)}$ of spectral data as

$$\theta_{m,n} = \begin{cases} \kappa_0 & \text{if } m = 0, \\ a_n \kappa_r & \text{if } m = a_n r/\pi. \end{cases} \tag{17.83}$$

Let $q_n(x)$ be the periodic potential of period a_n described by these data, according to the correspondence discussed above [Marchenko and Ostrovskii 1975, 1980]. If we consider $q_n(x)$ as having period a_{n+1} and pass from $\theta_{(n)}$ to $\tilde{\theta}_{(n)}$ according to (17.82), we have, by (17.14),

$$\begin{aligned} \|q_{n+1} - q_n\|_B &\leqslant C a_{n+1}^{-1} \|\theta_{(n+1)} - \tilde{\theta}_{(n)}\|_{a_{n+1}}^{1/2} e^{C(\Theta_{2,n} + \sqrt{\Theta_{1,n}})} \\ &= C a_{n+1}^{-1} \Bigl(\sum_{r \in \mathfrak{R}_{n-1} \setminus \mathfrak{R}_n} r^2 |\kappa_r|^2 \Bigr)^{\frac{1}{4}} e^{C(\Theta_{2,n} + \sqrt{\Theta_{1,n}})}, \end{aligned}$$

where

$$\begin{aligned} \Theta_{1,n} &= \max\Bigl\{ \Bigl(\sum_{m \geqslant 1} m^2 |\theta_{m,n+1}|^2 \Bigr)^{\frac{1}{2}}, \Bigl(\sum_{m \geqslant 1} m^2 |\tilde{\theta}_{m,n}|^2 \Bigr)^{\frac{1}{2}} \Bigr\} \\ &= \Bigl(\sum_{m \geqslant 1} m^2 |\theta_{m,n+1}|^2 \Bigr)^{\frac{1}{2}}, \\ \Theta_{2,n} &= \max\{ \sup_{m \geqslant 1} |\theta_{m,n+1}|, \sup_{m \geqslant 1} |\tilde{\theta}_{m,n}| \} \leqslant a_{n+1} \sup_{r \in \mathfrak{R}(\mathfrak{A})} |\kappa_r|. \end{aligned}$$

But, for an arbitrary $b > 0$, we have

$$\begin{aligned} \sup_{r \in \mathfrak{R}(\mathfrak{A})} |\kappa_r| &\leqslant \kappa_0 + \sup_{n \geqslant 1} \sup_{r \in \mathfrak{R}_{n+1} \setminus \mathfrak{R}_n} |\kappa_r| \leqslant |\kappa_0| + \sup_{n \geqslant 1} a_{n+1} \Bigl(\sum_{r \in \mathfrak{R}(\mathfrak{A}) \setminus \mathfrak{R}_n} r^2 |\kappa_r|^2 \Bigr)^{\frac{1}{2}} \\ &\leqslant |\kappa_0| + \sup_{n \geqslant 1} a_{n+1} e^{-b a_{n+1}} \leqslant C_b < \infty. \end{aligned}$$

Consequently, there is for any $\kappa \in \mathcal{K}_\infty(\mathfrak{A})$ a constant $C_{\kappa,1}$ such that $\Theta_{2,n} \leqslant C_{\kappa,1} a_{n+1}$. Moreover, since

$$\Theta_{1,n} = a_{n+1}^2 \Bigl(\sum_{r \in \mathfrak{R}_{n+1}} r^2 |\kappa_r|^2 \Bigr)^{\frac{1}{2}} \leqslant a_{n+1}^2 \|\kappa\|_{\mathcal{K}(\mathfrak{A})},$$

there is another constant $C_{\kappa,2}$ such that $\Theta_{1,n}^{1/2} \leqslant C_{\kappa,2} a_{n+1}$. Therefore,

$$\|q_{n+1} - q_n\|_B \leqslant \left(\sum_{r \in \Re(\mathfrak{A}) \setminus \Re_n} r^2 |\kappa_r|^2 \right)^{\frac{1}{4}} e^{C_{\kappa,3} a_{n+1}}.$$

Now since $\kappa \in \mathcal{K}_\infty(\mathfrak{A})$ the sum above does not exceed $C_{\kappa,4} e^{-b a_{n+1}}$, for some $C_{\kappa,4}$ and any $b > 0$. Therefore the sequence of potentials q_n converges in the Besicovitch metric to a limit q, which evidently belongs to $\mathcal{Q}_\infty(\mathfrak{A})$. By construction, $\kappa \in \mathcal{K}_\infty(\mathfrak{A})$ is a complete set of spectral data for q. This proves Theorem 17.1. □

Proof of Theorem 17.2. Since the function $\theta(a, \mu)$ does not change if a constant is added to the potential q, we can assume, without loss of generality, that $\nu_0 = 0$. Let $\{q_n : n = 1, 2, \ldots\}$ be a sequence of approximating periodic potentials, of period a_n, for the potential $q \in \mathcal{Q}_\infty(\mathfrak{A})$. Now consider the potential q_n as a function of period a_{n+1}. By Lemma 17.2, it gives rise to a function $\theta_n(a_{n+1}, \mu) = (a_{n+1}/a_n)\theta(a_n, \mu)$. Thus $\kappa_n(\mu) = \theta_n(a_n, \mu)/a_n = \theta_n(a_{n+1}, \mu)/a_{n+1}$, and from (17.53) and (17.7), with $Q = \sup_{n \geqslant 1} \|q_n\|_B$, we get

$$(17.84) \quad \begin{aligned} |\kappa_{n+1}(\mu) - \kappa_n(\mu)| &= a_{n+1}^{-1} |\theta_{n+1}(a_{n+1}, \mu) - \theta_n(a_{n+1}, \mu)| \\ &\leqslant C \|q_{n+1} - q_n\|_B^{1/2} e^{C a_{n+1} \sqrt{Q}} \leqslant C(b) e^{-b a_{n+1}} \end{aligned}$$

for all $\mu \in \mathbf{R}$. This implies that the sequence $\kappa_n(\mu)$ converges uniformly on the whole axis. Since each of these functions is related to the quasimomentum $k_n(z)$ by $\kappa_n(\mu) = k_n(\mu^2 + \nu_{0,n})$, we can write, by Problem VII.3,

$$\kappa_n(\mu) = -i w_n(\mu^2 + \nu_0)$$

for $\mu \in \mathbf{C}_+$, where $w_n(z)$ is the function (12.54a) for the operator H_{q_n}. By (12.54a) and Problem III.5, this implies that $|\kappa_n(\mu) - \kappa_n(\mu)| \leqslant c_n / \operatorname{Im} \mu$ for $\mu \in \mathbf{C}_+$. But, by (17.84), we have $|\kappa_n(\mu) - \kappa_n(\mu)| \leqslant C(b)$ on the real axis, so by the Phragmen–Lindelöf principle [Levin 1972] this same estimate is valid on the closed upper half-plane $\bar{\mathbf{C}}_+$. We conclude that the sequence of Nevanlinna functions $\kappa_n(\mu)$ converges to a limit Nevanlinna function $\kappa(\mu)$ uniformly on $\bar{\mathbf{C}}_+$, and that

$$(17.85) \qquad \lim_{t \to \infty} i t \kappa(it) = 1, \qquad \kappa(\mu) = -i w(\mu^2 + \nu_0),$$

where $w(z)$ is the function (12.54a) for the operator H_q.

In the sequel we will need the relations

$$\dot{\theta}(a, \mu) = \frac{\dot{A}(a, \mu)}{\sqrt{1 - A^2(a, \mu)}} = a \prod_{m \neq 0} \frac{\mu - \gamma_m}{\sqrt{(\mu - \alpha_m^-)(\mu - \alpha_m^+)}}.$$

The first equality follows directly from the definition of $\theta(a,\mu)$, and the second can be obtained either by the Christoffel–Schwarz formula or by the representation of $A(a,\mu)$ and $1-A^2(a,\mu)$ by canonical products. The functions $\Phi_1(\mu)=\dot{\theta}_{n+1}(a_{n+1},\mu)$ and $\Phi_2(\mu)=\dot{\theta}_n(a_{n+1},\mu)$ satisfy the conditions of Lemma 17.14; by (17.61), this leads to

$$\|\dot{\kappa}_{n+1}(\mu)-\dot{\kappa}_n(\mu)\|_{L_p(\Delta)}\leqslant C(p,\Delta,b)e^{-ba_{n+1}}$$

for every $p\in(1,2)$ and every interval Δ. From this we conclude that $\dot{\kappa}\in L_p^{\text{loc}}$ for $p\in(1,2)$, and that $\kappa(\mu)$ is absolutely continuous. □

To prove Theorem 17.3, it is important to know that the resolvents of H_{q_n} converge uniformly to the resolvent of H_q:

(17.17) Lemma. *Let $\{q_n : n=1,2,\ldots\}$ be a sequence of functions belonging to the Stepanov space S^2 of (17.4), and converging in the metric of S^2 to $q\in S^2$. Then the resolvents $R^{(n)}=(H_{q_n}-zI)^{-1}$ of the corresponding Schrödinger operators converge in norm to the resolvent R of H_q, for every $z\in\mathbf{C}$ such that $\liminf_{n\to\infty}d_n(z)>0$, where $d_n(z)$ is the distance from z to the spectrum of H_{q_n}.*

Proof. Consider first the case $z=i\varepsilon$, for $\varepsilon\geqslant 1$. Using the identity (4.27) for the pairs H_q, H_{q_n} and H_q, H_0, where $H_0=-d^2/dx^2$, we find that

$$R-R^{(n)}=(I-Rq)R_0p_nR^{(n)},\tag{17.86}$$

where $p_n=q-q_n$. Hence, since $\|R^{(n)}\|\leqslant\varepsilon^{-1}$, the lemma will follow if we prove that

$$\|Rq\|\leqslant\infty,\tag{17.87}$$

$$\lim_{n\to\infty}\|R_0p_n\|=0.\tag{17.88}$$

But (4.27), applied to H_q and H_0, implies that

$$\|Rq\|\leqslant\|R_0q\|\left(1-\|R_0q\|\right)^{-1}.$$

Therefore, (17.87) is true if

$$\|R_0q\|\leqslant\delta<1.\tag{17.89}$$

Relations (17.88) and (17.89) are, in turn, consequences of the inequality

$$\|R_0q\|\leqslant C\|q\|_{S^2}\varepsilon^{-3/4},\tag{17.90}$$

which holds for $\varepsilon\geqslant 1$ and some absolute constant C. This inequality is easily obtained by direct calculation, using the explicit form of the kernel $G_0(x,x',z)$ of the resolvent R_0:

$$G(x,x',z)=\frac{e^{i\sqrt{z}|x-x'|}}{2i\sqrt{z}}. \tag{17.91}$$

For an arbitrary, nonreal z, we represent $R^{(n)}$ as a norm-convergent series in powers of $(z-i\varepsilon)R^{(n)}\big|_{z=i\varepsilon}$, where $i\varepsilon$ is the center of a circle lying entirely in the upper half-plane and containing z. Since $\left\|(R^{(n)}-R)\big|_{z=i\varepsilon}\right\|$ tends to 0 as n tends to ∞, and $\left\|R^{(n)}\big|_{z=i\varepsilon}\right\|\leqslant\varepsilon^{-1}$, we can take the limit under the expansion.

A similar argument proves the lemma for $z=\lambda$, where λ lies on the real axis, in an interval contained in the resolvent set of all H_{q_n} for n large enough. This concludes the proof. □

Proof of Theorem 17.3. For $q\in\mathcal{Q}_\infty(\mathfrak{A})$, let $\kappa(\mu)$ be the corresponding Nevanlinna function given by Theorem 17.2, $k(\lambda)=\kappa(\sqrt{\lambda-\nu_0})$ the quasi-momentum, and $\kappa\in\mathcal{K}_\infty(\mathfrak{A})$ the spectral data parametrizing q, given by Theorem 17.1. We show first that the bottom of the spectrum ν_0 coincides with κ_0. Let $\{q_n\}$, for $n=1,2,\ldots$, be a sequence of periodic functions of period a_n satisfying (17.6). By Theorem 17.1, $k_0=\lim_{n\to\infty}\kappa_{0,n}$, where $\kappa_{0,n}=\theta_{0,n}$ is the bottom of the spectrum of H_{q_n}. Since $\liminf_{n\to\infty}d_n(\lambda)>0$ for every $\lambda<\kappa$, it follows from Lemma 17.17 that $\kappa_0\leqslant\nu_0$. Now, for every $n\geqslant 1$, there exists a normalized function $\psi_n\in L_2(\mathbf{R})$ for which

$$\left\|(R^{(n)}\big|_{z=i}-(\kappa_{0,n}-i)^{-1})\psi_n\right\|\leqslant\frac{1}{n}.$$

But then

$$\begin{aligned}\left\|(R\big|_{z=i}-(\kappa_0-i)^{-1})\psi_n\right\|&\leqslant\left\|(R-R^{(n)})\big|_{z=i}\psi_n\right\|\\&\quad+\left\|(R^{(n)}\big|_{z=i}-(\theta_{0,n}-i)^{-1})\psi_n\right\|+|(\theta_{0,n}-i)^{-1}+(\kappa_0-i)^{-1}|,\end{aligned}$$

As $n\to\infty$, the first term on the right tends to zero by Lemma 17.17, the second by our choice of ψ_n, and the third because $\kappa_{0,n}$ converges to κ_0. Thus, $(\kappa_0-i)^{-1}$ is in the spectrum of $R\big|_{z=i}$, and κ_0 is in the spectrum of H_q. Therefore, $\kappa_0=\nu_0$ and $(-\infty,\nu_0]$ is contained in the resolvent set $\rho(H_q)$ of H_q.

Now let $\lambda_0\in\mathbf{R}$ be a point where $\operatorname{Im}k(\lambda_0)>0$. If $\lambda_0<\nu_0$, we have seen that $\lambda_0\in\rho(H_q)$. To prove this inclusion for $\lambda_0>\nu_0$, notice that the uniform convergence of the Nevanlinna functions $\kappa_n(\mu)$ to $\kappa(\mu)$, proved in Theorem 17.2, implies that there exist positive number ε and δ and an integer n_0 such that $\operatorname{Im}\kappa_n(\mu)\geqslant\varepsilon$ for all μ in the interval $\{\mu:|\mu-\sqrt{\lambda_0-\nu_0}|\leqslant\delta\}$ and for all $n>n_0$. Then, for some $\delta_1>0$, the interval $\{\lambda:|\lambda-\lambda_0|\leqslant\delta_1\}$ lies entirely in the resolvent set $\rho(H_{q_n})$ for all $n\geqslant n_0$. By Lemma 17.17, this means that $\lambda_0\in\rho(H_q)$, so we conclude that the spectrum $\sigma(H_q)$ is contained in the set of λ such that $\operatorname{Im}k(\lambda)=0$. On the other hand, (17.85),

(12.55) and Theorem 12.7 imply that $\sigma_{\text{ac}}(H_q) = \{\lambda : \operatorname{Im} k(\lambda) = 0\}$, so the whole spectrum is absolutely continuous and coincides with the set σ given by (17.11).

If, conversely, we have a set σ of the form (17.11) for ν_0 real and $\kappa(\mu)$ a conformal mapping from $\mathbf{C}_+$ onto the region $\Pi_\infty^+(\mathfrak{A})$, so that $\kappa(0) = 0$ and $\lim_{y\to\infty} iy\kappa(iy) = 1$, we can use Theorem 17.1 to construct a potential $q(x)$—and even a family of such potentials with infinitely many parameters—whose spectrum is σ. This concludes the proof of Theorem 17.3. □

Proof of Theorem 17.5. As we have seen, for a Schrödinger equation with a periodic potential and for values of λ in the resolvent set, the Floquet–Bloch solutions (16.1), normalized so that $u_\pm\big|_{x=0} = 1$, coincide with the Weyl solutions (12.12). Therefore, by comparing (12.12) and (16.1), we arrive at the following expressions for the Weyl functions $m_\pm(z)$:

$$m_\pm(\mu^2) = \pm\frac{e^{\pm i\kappa(\mu)a} - C(a,\mu)}{S(a,\mu)}. \tag{17.92}$$

Here we will use the normalization of the Floquet–Bloch solutions in which their Wronskian (12.3) equals $2i$. Then, from (12.12) and (17.92), it follows that

$$\psi_\pm(x,\mu) = r(\mu)C(x,\mu) - \big(t(\mu) \pm ir^{-1}(\mu)\big)S(x,\mu), \tag{17.93}$$

where $r(\mu)$ and $t(\mu)$ are defined just before Lemma 17.16.

Take a sequence of approximating periodic potentials q_n for q, having period a_n, and consider the functions $r_n(\mu)$, $v_n(\mu)$ and $t_n(\mu)$, corresponding to the potential q_n, and the Floquet–Bloch solutions $\psi_{\pm,n}(x,\mu)$. By Lemma 17.16, the sequences $r_n(\mu)$, $v_n(\mu)$ and $r_n^{-1}(\mu)$ converge in $L_p(\Delta)$ for all $p \in (0,1)$ and $\Delta \subset \mathbf{R}$. Moreover, $C_n(x,\mu)$ and $S_n(x,\mu)$ converge to $C(x,\mu)$ and $S(x,\mu)$ uniformly for (x,μ) on any compact set. Therefore, for almost all μ there exist limit functions $\psi_\pm(x,\mu)$, which are solutions for the potential q.

By Theorem 17.2, the sequence $\kappa_n(\mu)$ converges uniformly on $\mathbf{C}_+$, and $Q = \sup_{n\geqslant 1}\|q_n\|_B$ is finite. Therefore

$$\begin{aligned}
&\|u_{\pm,n+1}(\,\cdot\,,\mu) - u_{\pm,n}(\,\cdot\,,\mu)\|_B \\
&\quad = \left(\frac{1}{a_{n+1}}\int_{a_n}^{a_{n+1}} \left|\psi_{\pm,n+1}(x,\mu)e^{\mp i\kappa_{n+1}(\mu)x} - \psi_{\pm,n}(x,\mu)e^{\mp i\kappa_n(\mu)x}\right|^2 dx\right)^{\frac{1}{2}} \\
&\quad \leqslant C\big\|\psi_{\pm,n+1}(\,\cdot\,,\mu) - \psi_{\pm,n}(\,\cdot\,,\mu)\big\|_{L_2[0,a_{n+1}]} \\
&\qquad + \big\|\psi_{\pm,n}(\,\cdot\,,\mu)\big\|_{L_2[0,a_{n+1}]} \sup_{\mu\in\mathbf{R}}\big|\kappa_{n+1}(\mu) - \kappa_n(\mu)\big| e^{Ca_{n+1}\sqrt{Q}}.
\end{aligned}$$

Using Lemmas 17.13 and 17.16, Corollary 17.8 and the condition (17.7) of superexponential approximation, we arrive at

$$\left(\int_{\Delta} \|u_{n+1}(\,\cdot\,,\mu) - u_n(\,\cdot\,,\mu)\|_B^p \,d\mu\right)^{\frac{1}{p}} \leqslant C(p,\Delta,q)e^{-ba_{n+1}},$$

with some constant b and $p \in (1,2)$.

We now construct the spectral expansion for H_q. The expansion (12.30) for the operator with potential q_n can be readily rewritten by means of the functions $\psi_{\pm,n}(x,\sqrt{\lambda-\kappa_{0,n}})$ (see also [Titchmarsh 1958]):

$$(17.94)\quad f(x) = \frac{1}{2\pi}\int_{\mathbf{R}} \big(F_{+,n}(\sqrt{\lambda-\kappa_{0,n}})\psi_{-,n}(x,\sqrt{\lambda-\kappa_{0,n}}) + F_{-,n}(\sqrt{\lambda-\kappa_{0,n}})\psi_{+,n}(x,\sqrt{\lambda-\kappa_{0,n}})\big)\,\tfrac{1}{2}\chi_n(\lambda)\,d\lambda,$$

where $F_{\pm,n}(\mu) = \int_{\mathbf{R}} f(x)\psi_{\pm,n}(x,\mu)\,dx$ and $\chi_n = \chi_{\sigma(H_{q_n})}$ is the characteristic function of the spectrum. Parseval's equality, applied to the generalized Fourier transforms with respect to these functions, reads as follows:

(17.95)

$$\|f\|^2_{L_2(\mathbf{R})} = \frac{1}{2\pi}\int_{\mathbf{R}} \big(|F_{+,n}(\sqrt{\lambda-\kappa_{0,n}})|^2 + |F_{-,n}(\sqrt{\lambda-\kappa_{0,n}})|^2\big)\,\tfrac{1}{2}\chi_n(\lambda)\,d\lambda.$$

Take $f \in C_0^\infty(\mathbf{R})$. Lemma 17.17 shows that the sequences $F_{\pm,n}(\mu)$, as $n \to \infty$, converge in the $L_p(\Delta)$-norm for any $p \in (1,2)$ and any interval $\Delta \subset \mathbf{R}_+$, and that the limit functions $F_\pm(\mu)$ belong to the same space. As a result,

$$(17.96)\qquad F_\pm(\mu) = \int_{\mathbf{R}} f(x)\psi_\pm(x,\mu)\,dx.$$

By Lemma 17.15, the sequence of characteristic functions $\chi_n(\lambda)$ converges in the norm of $L_1(\Delta)$. Therefore, we can take the limit in Parseval's equality (17.95), obtaining Bessel's inequality for H_q:

$$(17.97)\quad \frac{1}{2\pi}\int_{\mathbf{R}} \big(|F_+(\sqrt{\lambda-\kappa_0})|^2 + |F_-(\sqrt{\lambda-\kappa_0})|^2\big)\,\tfrac{1}{2}\chi(\lambda)\,d\lambda \leqslant \|f\|^2_{L_2(\mathbf{R})}.$$

The relation $F_\pm(\sqrt{\lambda-\kappa_0}) \in L_2(\Delta,\chi(\lambda)d\lambda)$, which follows from (17.97), can also be proved directly, by the arguments used to prove Lemmas 17.14 and 17.16.

From (17.97), by a standard technique of spectral analysis [Titchmarsh 1958], we arrive at the Parseval equality

$$(17.98)\quad \frac{1}{2\pi}\int_{\mathbf{R}} \big(|F_+(\sqrt{\lambda-\kappa_0})|^2 + |F_-(\sqrt{\lambda-\kappa_0})|^2\big)\,\tfrac{1}{2}\chi(\lambda)d\lambda = \|f\|^2_{L_2(\mathbf{R})},$$

which is true for any $f \in L_2(\mathbf{R})$. Equation (17.96) can also be extended to any $f \in L_2(\mathbf{R})$. After this, we see that the generalized Fourier transform $F_\pm(\sqrt{\lambda - \kappa_0})$ belongs to $L_2(\mathbf{R}_+, \chi(\lambda) d\lambda)$, and the family of operators

$$(17.99) \quad (E(\Delta)f)(x) = \frac{1}{2\pi} \int_\Delta \big(F_+(\sqrt{\lambda - \kappa_0})\psi_-(x, \sqrt{\lambda - \kappa_0}) + F_-(\sqrt{\lambda - \kappa_0})\psi_+(x, \sqrt{\lambda - \kappa_0})\big) \tfrac{1}{2}\chi(\lambda) d\lambda$$

defines the resolution of the identity of H_q. This concludes the proof of Theorem 17.5. □

Proof of Theorem 17.4. The absolute continuity of the spectrum follows from Theorems 17.3 and 12.5. Given this, Theorem 12.9 implies that the spectrum has multiplicity two. Both the absolute continuity and the multiplicity also follow from the representation (17.99). □

18 Unbounded Quasiperiodic Potentials

Most of this section is devoted to the study of the matrix metrically transitive operator of the form (1.8), the sum of a Toeplitz and a diagonal operator, the latter being defined by the quasiperiodic function $g \tan\big(\pi((\alpha, x) + \omega)\big)$, for $x \in \mathbf{Z}^d$, where $\alpha = (\alpha_1, \ldots, \alpha_d) \in \mathbf{R}^d$ is the frequency vector. We derive the spectral quantities for this operator in an essentially explicit form, thus obtaining a fairly complete picture of its spectral properties.

We will show that, so long as the values of (α, x) do not come too close to the set of integers (see (18.21)), the spectrum is pure point and has multiplicity 1, and the eigenfunctions decay exponentially as $|x| \to \infty$. This is true in arbitrary dimension. In other words, such operators exhibit Anderson localization (15.C) for all λ, just as random operators (Sections 13 and 15) did in one dimension for all λ and in several dimensions in the pure point vicinity of a fluctuation boundary.

Concentrating on the one-dimensional case, we find various spectral types, depending on the rate of approximation of α by rational numbers: pure point, if the rate of approximation is polynomial; mixed (pure point for large $|\lambda|$ and singular continuous for small $|\lambda|$), if the rate is exponential; purely singular continuous, if the rate is superexponential; and purely absolutely continuous, if α is rational.

This latter is the only situation where we can get an absolutely continuous spectrum: we will show that for α irrational there is no absolutely continuous component. For an operator of order two, this can be shown as in Section 16, using Theorem 12.1, because, as we will see, the Lyapunov exponent is positive for all $\lambda \in \mathbf{R}$. For a Toeplitz operator of arbitrary—even infinite—order, we approximate the resolvent by the resolvents of operators with rational α: see Lemma 18.12.

In 18.D we consider the Schrödinger operator on $L_2(\mathbf{R}^d)$ having the nonlocal quasiperiodic potential (18.116), equal to the sum of the orthogonal projections generated by the translates of a rapidly decreasing function, multiplied by $\tan\big(\pi((\alpha,n)+\omega)\big)$. This metrically transitive operator is a generalization of the matrix operator of (18.1)–(18.4). For an appropriate choice of parameters, it exhibits an interesting new phenomenon: the coexistence of a point and an absolutely continuous component on the semiaxis $[\rho,\infty)$, for $\rho \leqslant \frac{1}{4}$.

18.A General Results and the Integrated Density of States

We consider the metrically transitive operator $A(\omega)$ on $\ell_2(\mathbf{Z}^d)$ given by the matrix

$$a(x,y,\omega) = a_0(x-y) - g\tan\alpha(x)\delta_{x,y}, \tag{18.1}$$

where $a_0(x)$, for $x \in \mathbf{Z}^d$, is a sequence of complex numbers satisfying

$$a_0^*(x) = a_0(-x), \qquad \big|a_0(x)\big| \leqslant Ce^{-\rho|x|}, \tag{18.2}$$

for C and ρ positive; g is a real constant, taken positive for definiteness; and $\alpha(x)$ is the sequence given by

$$\alpha(x) = \pi\big((\alpha,x)+\omega\big), \tag{18.3}$$

where $\alpha \in \mathbf{R}^d$, $\omega \in [0,1)$, and $(\alpha,x) = \sum_{j=1}^d \alpha_j x_j$. The sequence a_0 defines a bounded, self-adjoint Toeplitz operator A_0 on $\ell_2(\mathbf{Z}^d)$. In order to exclude infinite values of $\tan\alpha(x)$ for finite values of x (such values are irrelevant to our subsequent analysis), we assume that

$$\omega \neq \tfrac{1}{2} - (\alpha,x) \pmod 1 \tag{18.4}$$

for all $x \in \mathbf{Z}^d$.

The operator A is obviously periodic if all the components of α are rational numbers. Its spectrum is then absolutely continuous, by general principles [Reed and Simon IV]; in addition, for almost all λ with respect to the spectral measure, the generalized eigenvalues are of Bloch form, an obvious generalization of (16.1) to arbitrary dimension. These results can also be obtained directly, from our knowledge of the special structure of A. (It is also good to avoid resorting to the general theory because its application to A requires special justification, in view of the unboundedness of the potential.)

When at least one component of α is irrational, the set of numbers of the form $(\alpha, x) \pmod 1$, for $x \in \mathbf{Z}^d$, is dense in $[0,1)$, and therefore, for any continuous function $Q(t)$, periodic of period 1, the hull of the function $f(x) = Q\big((\alpha, x)\big)$, that is, the set of translates $\{f(x+a) : a \in \mathbf{Z}^d\}$, is the circle of circumference 1. This set, as explained in Example 1.15(g) and in 16.A, can be seen as the space of realizations Ω of the metrically transitive field on $\mathbf{Z}^d$ of the form $f(x, \omega) = Q\big((\alpha, x) + \omega\big)$, for $\omega \in [0,1)$, the group action being given by the maps $T_x\omega = \omega + (\alpha, x) \pmod 1$, for $x \in \mathbf{Z}^d$, and the probability measure P being the Lebesgue measure on $[0,1)$.

The function $\tan t$ is not continuous on Ω, so that our potential is not a uniformly continuous (Bohr) quasiperiodic function on $\mathbf{Z}^d$. But, since the set of points satisfying (18.4) has full P-measure in Ω, the function $g \tan\big(\pi((\alpha, x) + \omega)\big)$ defines a metrically transitive, although unbounded, field on $\mathbf{Z}^d$. Thus the operator Q of multiplication by this function is self-adjoint in $\ell_2(\mathbf{Z}^d)$. Since the Toeplitz operator A_0 generated by the function $a_0(x)$ of (18.2) is bounded, it follows that $A = A_0 + Q$ is self-adjoint, being a sum of two self-adjoint operators, one of them bounded.

The operator A defined by (18.1)–(18.4) is a natural extension to several dimensions of a one-dimensional operator first proposed by Grempel et al. [1982]. These authors discovered this remarkable operator while considering the problem of quantum chaos (stability of solutions of the Schrödinger equation with a time-periodic potential). They showed that, if α is approximated poorly enough by rationals (see (18.21) below), the eigenfunctions can be explicitly found for a countable, dense set of simple eigenvalues.

However, the method they used was very indirect, even in the one-dimensional case. In the discussion below, based on [Figotin and Pastur 1984b], we develop a convenient representation of the resolvent of A, using only fairly elementary means. This representation allows a detailed spectral analysis for various values of $a_0(x)$, d and α. Similar results were obtained by Simon [1985], whose technique was an extension of the methods of Grempel et al. [1982].

Denote by V the unitary operator on $\ell_2(\mathbf{Z}^d)$ given by

(18.5) $$(V\psi)(x) = e^{2\pi i(\alpha, x)}\psi(x).$$

Using the elementary formula $\tan t = i(1 - e^{-2it})/(1 + e^{2it})$, we have, for $\operatorname{Im} z > 0$,

(18.6)
$$A - z = (A_0 - z - ig)\big(1 + (A_0 - z + ig)(A_0 - z - ig)^{-1}e^{2\pi i\omega}V\big)\big(1 + e^{2\pi i\omega}V\big)^{-1},$$

where we denote the operator zI simply by z. If we now set

(18.7) $$B = (A_0 - z - ig)^{-1}, \quad C = -(A_0 - z + ig)(A_0 - z - ig)^{-1}, \quad \kappa = e^{2\pi i\omega},$$

we find that the resolvent $R(z) = (A - z)^{-1}$ of A can be written as

$$R(z) = (I + \kappa V)(I - \kappa CV)^{-1}B. \tag{18.8}$$

In view of (18.2) and (18.7), $\|C\| < 1$ for $\operatorname{Im} z > 0$, so (18.8) implies that $R(z)$, for $\operatorname{Im} z > 0$, can be written as the following series, which converges in the operator norm:

$$\begin{aligned} R(z) &= \sum_{n \geqslant 0} (I + \kappa V)(CV)^n B \kappa^n \\ &= B - 2ig \sum_{n \geqslant 1} B(VC)^{n-1} V B \kappa^n. \end{aligned} \tag{18.9}$$

In what follows we shall need the Fourier transform, which maps the vector $\psi(x) \in \ell_2(\mathbf{Z}^d)$ to the square-integrable function $\hat{\psi}(\eta)$ on the d-dimensional torus $\mathbf{T} = \mathbf{T}^d = \{\eta \in \mathbf{C}^d : |\eta_j| = 1 \text{ for } j = 1, \dots, d\}$ given by

$$\hat{\psi}(\eta) = \sum_{x \in \mathbf{Z}^d} \psi(x) \eta^x, \tag{18.10}$$

where $\eta^x = \eta_1^{x_1} \dots \eta_d^{x_d}$; the inverse transform is given by

$$\psi(x) = (2\pi i)^{-d} \int_{\mathbf{T}} \hat{\psi}(\eta) \eta^{x-1}\, d\eta. \tag{18.11}$$

This transformation maps the Toeplitz operators A_0, B and C of (18.1) and (18.7) to operators of multiplication by the functions

$$\hat{a}_0(\eta) = \sum_{x \in \mathbf{Z}^d} a_0(x) \eta^x, \qquad \hat{B}(\eta) = (\hat{a}_0(\eta) - z - ig)^{-1} \tag{18.12}$$

$$\hat{C}(\eta) = -\big(\hat{a}_0(\eta) - z + ig\big)\big(\hat{a}_0(\eta) - z - ig\big)^{-1}, \tag{18.13}$$

and the operator V of (18.5) to the shift operator

$$(\widehat{V\psi})(\eta) = \hat{\psi}(\gamma\eta), \tag{18.14}$$

where

$$\gamma\eta = (\gamma_1\eta_1, \dots, \gamma_d\eta_d), \qquad \gamma = (e^{2\pi i \alpha_1}, \dots, e^{2\pi i \alpha_d}). \tag{18.15}$$

Now any operator Φ of multiplication by a bounded function $\Phi(\eta)$ satisfies the commutation relation

$$V^n \Phi(\eta) = \Phi(\gamma^n \eta) V^n, \tag{18.16}$$

with $n \in \mathbf{Z}$ and $\gamma^n = (\gamma_1^n, \dots, \gamma_d^n)$. Thus (18.9) can be written as

(18.17) $$R = B - 2ig \sum_{n \geqslant 1} B \prod_{s=0}^{n-1} \hat{C}(\gamma^s \eta) \hat{B}(\gamma^n \eta) V^n \kappa^n.$$

Relations (18.8), (18.9) and (18.17) provide the basis for the subsequent analysis.

As we have mentioned, $A(\omega)$ is a metrically transitive operator, so, by Theorem 4.4, it has an integrated density of states, whose form is given by the following result:

(18.1) Theorem [Grempel et al. 1982; Figotin and Pastur 1984b; Simon 1984]. *If A is a metrically transitive operator of the form* (18.1)–(18.4), *where at least one component of $\alpha \in \mathbf{R}^d$ is irrational, the integrated density of states $N(A, \lambda)$ is absolutely continuous and its density $n(\lambda)$ is given by*

(18.18) $$n(\lambda) = \frac{1}{\pi} \frac{1}{(2\pi i)^d} \int_{\mathbf{T}} \frac{g}{\left(\lambda - \hat{a}_0(\eta)\right)^2 + g^2} \frac{d\eta}{\eta}.$$

Proof. From equation (A.4) of the Appendix, it follows that the resolvent $R(z)$ and the resolution of the identity $E(d\lambda)$ of a self-adjoint operator are related by

(18.19) $$\lim_{\varepsilon \downarrow 0} \frac{\varepsilon}{\pi} \int_{\Delta} \left\| R(\lambda + i\varepsilon)\psi \right\|^2 d\lambda = \left(E(\Delta)\psi, \psi\right),$$

where ψ is a vector in the corresponding Hilbert space and Δ is an interval of the real axis neither endpoint of which is an eigenvalue.

Since $N(A, \lambda)$ satisfies representation (4.12), we conclude from (18.17) that

(18.20) $$\int_{\mathbf{R}} \frac{N(A, d\lambda)}{\lambda - z} = \int_0^1 \left(R(z)e_0, e_0\right) d\omega = \left(R_0(z + ig)e_0, e_0\right),$$

where $e_0(x) = \delta_{0,x}$ for $x \in \mathbf{Z}^d$, and we took into account the fact that B in (18.7) is the resolvent of A_0 at $z + ig$.

Separating the imaginary part in (18.20) and taking the limit $\varepsilon \downarrow 0$, as prescribed by (18.19), we obtain (18.18). □

Remarks. 1. The operator (18.1) has an integrated density of states of the same form for a much wider class of almost-periodic potentials, and even potentials of other type, as in (16.75). See Problems VII.30–32.

2. Under the condition of Theorem 18.1, the one-dimensional distribution $F(dq)$ of the metrically transitive potential $q(x) = g\tan(\pi((\alpha, x) + \omega))$ is obviously given by

$$F(dq) = \frac{g}{\pi(g^2 + q^2)} dq,$$

and so coincides with the Cauchy distribution (7.15). Now consider another metrically transitive potential, consisting of independent, identically distributed variables $q(x)$, for $x \in \mathbf{Z}^d$, all with the same Cauchy distribution as above. According to Problem II.24, for the operator (1.8) with this potential, the integrated density of states also has the form (18.18). (The derivation of (18.18) in Problem II.24 uses (4.51), which allows one to average successively over each $q(x)$, for $x \in \mathbf{Z}^d$. But here (18.18) is also an immediate consequence of the representation (18.17) for the resolvent. It is enough to note that if a Cauchy-distributed random variable q is written as $g \tan \pi\alpha$, the random variable α is uniformly distributed in $[0, 1)$.)

Thus, although the statistical properties of these two metrically transitive potentials are very different—the first, being almost-periodic, has no mixing properties, while the second, being independent for distinct x, has the strongest possible mixing properties—the integrated densities of state of the corresponding operators coincide.

3. Formula (18.18) can be written as

$$N(A_0 + Q, \lambda) = \int_{\mathbf{R}} N(A_0, \lambda - \mu) C(g, d\mu).$$

According to Problem II.37, the integrated density of states of the operator (1.8) is given by the same expression for any metrically transitive potential of the form $q(x) = Q(x/R)$, in the limit $R \to \infty$, if $C(g, dQ)$ is the one-dimensional distribution of the metrically transitive field $Q(x)$. Thus, the two metrically transitive operators of the previous remark share their integrated density of states with a wide class of metrically transitive operators with a slowly varying potential.

18.B The Case of Strongly Incommensurate Frequencies

Suppose the frequency vector α of (18.3) satisfies

$$\big|(\alpha, x) - m\big| \geqslant C|x|^{-\beta} \tag{18.21}$$

for all $x \in \mathbf{Z}^d \setminus 0$ and $m \in \mathbf{Z}$, and some $C, \beta > 0$ (compare with (16.31)). This is often called a *Diophantine condition*, and implies that the behavior of the corresponding quasiperiodic function is very irregular ("chaotic"). For $\beta > d$, (18.21) holds for almost all α with respect to the Lebesgue measure on $\mathbf{R}^d$ [Sprindzhuk 1977] (but the set of vectors with linearly independent components that do not satisfy this condition is dense in $\mathbf{R}^d$).

The main result of this subsection is that the spectrum of (18.1)–(18.4) is pure point if (18.21) is satisfied. This is very natural from the quantum-mechanical point of view, at least in one dimension. For in this case the potential $\tan\bigl(\pi((\alpha,x)+\omega)\bigr)$, for $x \in \mathbf{Z}^d$, represents a very irregular system of peaks whose height increases, although in an irregular way, as we go off to infinity. It is no wonder that a quantum-mechanical particle gets tangled up in such a chaotic forest, which becomes denser and taller as the distance from the origin increases.

It is interesting to notice that, in the multidimensional case, this forest is also sufficiently dense, tall and chaotic to prevent a particle from going to infinity, although one might think that in any direction perpendicular to α the motion of a particle is practically unimpeded. That this is not so is due to the discreteness of the problem (see also [Prange et al. 1984]).

Before we formulate and prove our result rigorously, let's discuss the property of A that lies at the basis of the proof.

Assume that $\lambda \in \mathbf{R}$ is an eigenvalue of A. Setting $z = \lambda$ in (18.6) and (18.7), we get

$$(A-\lambda) = (A_0 - \lambda - ig)(I - \kappa CV)(I+\kappa V)^{-1}. \tag{18.22}$$

Now $A_0 - \lambda - ig$ is invertible when $g \neq 0$, and $(I+\kappa V)^{-1}$, where V is defined in (18.5), is well-defined by virtue of (18.21). Thus there exists $v \in \ell_2(\mathbf{Z}^d)$ such that

$$(I - \kappa CV)v = 0. \tag{18.23}$$

Taking the Fourier transform (18.10) of (18.23), we get

$$\hat{v}(\eta) = \kappa \hat{C}(\eta)\hat{v}(\gamma\eta). \tag{18.24a}$$

From this and from (18.14), it follows readily that the vectors $\eta^x \hat{v}(\eta)$, for $x \in \mathbf{Z}^d$, are eigenvectors of CV, with $\kappa CV \eta^x \hat{v} = \gamma^x \eta^x \hat{v}$.

Now consider the function $\hat{v}(\eta)$ as the operator of multiplication by v in $L_2(\mathbf{T}^d)$. Then, since the functions η^x, for $x \in \mathbf{Z}^d$, form a complete system of eigenvectors of (18.5), with eigenvalues γ^x, we can write

$$\kappa CVv = vV. \tag{18.24b}$$

But $\bigl|\hat{C}(\eta)\bigr| = 1$ for λ real by (18.7), so (18.24a) implies that $\bigl|\hat{v}(\eta)\bigr| = \bigl|\hat{v}(\gamma\eta)\bigr|$. From (18.21) and the measurability of $\hat{v}(\eta)$ it follows that the absolute value of this function is a constant, which we may as well assume is 1. This means that the operator v is invertible, and then (18.22) and (18.24) imply that the eigenvalues of A exist if and only if there exists a Toeplitz operator v that makes κCV and V unitarily equivalent. For this operator, $\hat{v}(\eta)$ can be written as $\eta^y e^{t(\eta)}$, for $y \in \mathbf{Z}^d$; here $t(\eta)$ satisfies the so-called *homological equation*

$$-t(\eta) + \ln \hat{C}(\eta) + t(\gamma\eta) + 2\pi i\omega + (y, \ln\gamma) = 0 \tag{18.25}$$

[Arnold 1980], and the factor η^y fixes the branch of $t(\eta)$. By integrating (18.25) over the torus, we find that a necessary condition for it to be solvable is that

$$\int_{\mathbf{T}} \ln \hat{C}(\eta, \lambda) \frac{d\eta}{\eta} = -2\pi i \omega - (y, \ln \gamma). \tag{18.26}$$

Under the conditions (18.2), the function $\ln \hat{C}(\eta)$ is analytically continuable into the polycylinder $T_\rho = \{\eta : e^{-\rho} \leqslant |\eta_j| \leqslant e^\rho \text{ for } j = 1, \ldots, d\}$. This, together with (18.21), implies that (18.26) is not only necessary, but also sufficient, for (18.28) to have a solution in the class of functions analytic on T_ρ. Therefore, if λ_y satisfies (18.26) for a given $y \in \mathbf{Z}^d$ (we'll see later that y defines λ_y uniquely), λ_y is an eigenvalue of A. By (18.22), (18.25) and (18.26), the corresponding normalized eigenfunction $\psi_y(x)$ is

$$\psi_y(x) = \chi(x - y, \lambda_y), \tag{18.27}$$

where

$$\chi(x, \lambda) = \frac{1}{(2\pi i)^d \sqrt{4\pi g n(\lambda)}} \int_{\mathbf{T}} \frac{e^{t(\eta,\lambda)}}{\hat{a}_0(\eta) - \lambda - ig} \eta^{x-1}\, d\eta.$$

Since $t(\eta)$ and $\hat{a}_0(\eta)$ are analytic in T_ρ, the function $\chi(x, \lambda)$ decreases exponentially as $|x| \to \infty$. Its rate of decay depends on the absolute value of the maximal root η_0 of the equation

$$\hat{a}_0(\eta) - \lambda - ig = 0 \tag{18.28}$$

inside the unit disk; more precisely, the rate is given by

$$\gamma_-(\lambda, g) = -\ln |\eta_0|. \tag{18.29}$$

For a one-dimensional operator of order two, this coincides with the Lyapunov exponent of (11.4) with potential $g \tan(\pi(\alpha x + \omega))$ (Problem VII.34).

An important question that is not settled by the arguments above is whether the system of eigenfunctions thus described is complete. It turns out that, even for complex values of the spectral parameter, there still exists a Toeplitz operator that makes $\kappa C V$ equivalent to a diagonal operator on $\ell_2(\mathbf{Z}^d)$. Indeed, assume such an operator v exists, and let $\hat{v}(\eta)$ be its symbol. Then

$$\kappa C V = v D v^{-1},$$

with D diagonal. By (18.16), we can rewrite this as

$$\kappa C v v_\gamma^{-1} = D V^{-1},$$

where v_γ is the Toeplitz operator with symbol $\hat{v}(\gamma\eta)$. Now the operator on the left is Toeplitz, and the one on the right is diagonal, so they must both equal the identity times some number c_0, that is, $D = c_0 V$ and $\kappa C = c_0 v v_\gamma^{-1}$. These relations are analogous to (18.24) and (18.25). We see that the problem is again reduced to the factorization of $\hat{C}(\eta)$, that is, to the solution of a homological equation similar to (18.25), but involving complex values of the spectral parameter, and the factor $c_0(z)$ (which does not depend on η). The existence of this factor makes the problem uniquely solvable for all $z \in \mathbf{C}$ (Lemma 18.5), and, together with (18.8), leads to a convenient, essentially diagonal, representation of the resolvent of A, which in turn allows the basic spectral properties of A to be obtained.

(18.2) Theorem [Figotin and Pastur 1984b]. *The spectrum of the metrically transitive operator A defined by* (18.1)–(18.4) *and* (18.21) *coincides with* $\mathbf{R}$, *is pure point, and has multiplicity* 1. *All eigenvalues can be found from the equation*

$$N(A, \lambda) = \omega + (\alpha, y) \pmod 1$$

for $y \in \mathbf{Z}^d$, where $N(A, \lambda)$ is given by (18.18). *For each $y \in \mathbf{Z}^d$ there exists a unique eigenvalue λ_y, and the corresponding eigenfunction $\psi_y(x)$ has the form* (18.27), *where $t(\eta, \lambda)$ is the solution of the homological equation* (18.25). *The eigenfunction decays exponentially as $|x| \to \infty$, and, in the one-dimensional case,*

$$-\lim_{|x|\to\infty} |x|^{-1} \ln|\psi_y(x)| = \gamma_-(\lambda_y, g),$$

where $\gamma_-(\lambda, g)$ is defined in (18.29), *and the limit is understood to be taken over some subsequence, since $\psi_y(x)$ can have zeros.*

The proof will require several auxiliary results. We start by rewriting the Diophantine equation (18.21) in terms of (18.10) and (18.15):

(18.30) $$|1 - \gamma^x| \geqslant C|x|^{-\beta}$$

for $x \in \mathbf{Z}^d \setminus 0$.

(18.3) Lemma. *Let $T_{\rho,r}$ be the domain $T_\rho \times \{\xi : e^{-r} < |\xi| < e^r\} \subset \mathbf{C}^{d+1}$, and let $s(\eta, \zeta)$, for $\eta \in \mathbf{C}^d$ and $\zeta \in \mathbf{C}$, be analytic in $T_{\rho,r}$. Let $s_n(\eta)$, for $n \in \mathbf{Z}$, denote the coefficients of the Laurent expansion of s in the variable ζ (and the operators of multiplication by these coefficients), so that*

(18.31) $$s(\eta, \zeta) = \sum_{n\in\mathbf{Z}} s^n(\eta)\zeta^n;$$

and consider the operator

(18.32) $$s(\eta, \kappa V) = \sum_{n\in\mathbf{Z}} s_n(\eta)(\kappa V)^n,$$

where V is the shift operator (18.14), *and $\kappa = e^{2\pi i\omega}$. The action of $s(\eta, \kappa V)$ on the basis vectors is given by*

$$s(\eta, \kappa V)\eta^x = s(\eta, \kappa\gamma^x)\eta^x \tag{18.33}$$

for $x \in \mathbf{Z}^d$.

Proof. This follows immediately from the equation

$$V^n\eta^x = (\gamma^x)^n\eta^x. \tag{18.34}$$

□

The operators constructed in this way form an algebra that will be considered at the end of this subsection.

(18.4) Lemma. *If $\mathcal{F}_0$ be the set of analytic functions $\hat{\phi}(\eta)$ on T_ρ such that*

$$\phi(0) \equiv \frac{1}{(2\pi i)^d}\int_{\mathbf{T}} \hat{\phi}(\eta)\frac{d\eta}{\eta} = 0,$$

the operator $I - V$ is a one-to-one map of $\mathcal{F}_0$ into itself, satisfying

$$((I-V)^{-1}\phi)(x) = \begin{cases} 0 & \text{for } x = 0, \\ (1-\gamma^x)^{-1}\phi(x) & \text{for } x \neq 0. \end{cases} \tag{18.35}$$

Proof. Recall that T_ρ, for $\rho > 0$, denotes the polycylinder

$$T_\rho = \{\eta \in \mathbf{C}^d : e^{-\rho} \leqslant |\eta_j| \leqslant e^\rho \text{ for } j = 1, \ldots, d\}.$$

By (18.2), the function $\hat{a}_0(\eta)$ defined by (18.12) is analytic in T_ρ. In view of (18.12), (18.13) and the fact that $\hat{a}_0(\eta)$ takes on real values on the torus $\mathbf{T} \subset T_\rho$, the functions $\hat{B}(\eta)$ and $\hat{C}(\eta)$ are analytic in T_ρ. On the other hand, an arbitrary analytic function $\hat{\psi}(\eta)$ on T_ρ can be represented there by the Laurent series (18.10), with coefficients $\psi(x)$ defined by (18.11). These coefficients decrease as

$$|\psi(x)| \leqslant \text{const}\, e^{-\rho_1|x|},$$

for some positive $\rho_1 < \rho$. This, together with (18.30), implies the lemma. □

Now the function $\hat{C}(\eta, z)$ defined by (18.13) is evidently analytic in η and z in the region $T_\rho \times L_g$, where $\rho > 0$ and L_g is the strip

$$L_g = \{z \in \mathbf{C} : |\operatorname{Im} z| \leqslant g/2\}. \tag{18.37}$$

The function $\hat{C}(\eta, z)$ is continuous up to the boundary of $T_\rho \times L_g$, and obviously cannot be nonpositive there. Therefore, taking the standard branch of the log function, with a cut along the negative semiaxis, we obtain the following analytic function on $T_\rho \times L_g$:

$$f(\eta, z) = \ln \hat{C}(\eta, z). \tag{18.38}$$

Using Lemma 18.4, we find the function $t(\eta, z)$ from this equation:

$$f(\eta, z) = f_0(z) + t(\eta, z) - (Vt)(\eta, z), \tag{18.39}$$

$$f_0(z) = \frac{1}{(2\pi i)^d} \int_{\mathbf{T}} f(\eta, z) \frac{d\eta}{\eta}. \tag{18.40}$$

As a result, we have

$$t(\eta, z) = (I - V)^{-1}\big(f(\eta, z) - f_0(z)\big), \tag{18.41}$$

and $t(\eta, z)$ and $f_0(z)$ have the following properties:

(18.5) Lemma. *The function $f_0(z)$ is analytic on the strip L_g, and we have*

$$\operatorname{Re} f_0(z) < 0 \textit{ if } \operatorname{Im} z > 0; \tag{18.42}$$

$$\operatorname{Re} f_0(\lambda) = 0 \textit{ for } \lambda \in \mathbf{R}; \tag{18.43}$$

$$m(\lambda) = \operatorname{Im} f_0(\lambda) = 2 \int_{\mathbf{T}} \arctan\left(\frac{\hat{a}_0(\eta) - \lambda}{g}\right) \frac{d\eta}{(2\pi i)^d \eta}. \tag{18.44}$$

The function $m(\lambda)$ is a one-to-one map from $\mathbf{R}$ onto $[-\pi, \pi]$, and satisfies

$$2\pi N(A, \lambda) = \pi - m(\lambda). \tag{18.45}$$

The function $t(\eta, z)$ is analytic in $T_\rho \times L_g$, and

$$\operatorname{Re} t(\eta, \lambda) = \operatorname{Re} f(\eta, \lambda) = 0 \textit{ for } \lambda \in \mathbf{R}. \tag{18.46}$$

Proof. The properties of $f_0(z)$ and $m(\lambda)$ follow immediately from (18.38), (18.40), (18.13) and Theorem 18.1. The analyticity of $t(\eta, z)$ follows from its representation by equality (18.41) and from Lemma 18.4.

It follows from (18.41) that $\operatorname{Re} t(\eta, z) = \operatorname{Re}\big(f(\eta, z) - f_0(z)\big)$. But when $\operatorname{Im} z = 0$, we evidently have $\operatorname{Re} f(\eta, z) = 0$ from (18.38) and (18.13). Now (18.46) follows from (18.43). □

From equality (18.39) and the commutation relations (18.16) we deduce the following equality, which will be useful later:

$$C(z)V = e^{f_0(z)} e^{t(z)} V e^{-t(z)} \tag{18.47}$$

for $z \in L_g$, where we explicitly indicate the dependence of C and t on z. From this and from (18.6), we get

$$A - z = B^{-1}e^{t}(I - e^{f_0(z)}\kappa V)e^{-t}(I + \kappa V)^{-1}. \tag{18.48}$$

(18.6) Lemma. *The set of eigenvalues of A is the set of solutions of the equation*

$$m(\lambda) + 2\pi\omega + 2\pi(\alpha, y) = 0 \pmod{2\pi} \tag{18.49}$$

for $y \in \mathbf{Z}^d$. For any y, there exists a unique solution λ_y, and $\lambda_{y_1} \neq \lambda_{y_2}$ if $y_1 \neq y_2$. The set of eigenvalues $\{\lambda_y : y \in \mathbf{Z}^d\}$ is dense in $\mathbf{R}$. The eigenfunction corresponding to λ_y is

$$u_y(\eta) = (I + \kappa V)e^{t(\eta,\lambda_y)}\eta^y; \tag{18.50}$$

if we normalize $u_y(\eta)$ so as to have norm 1 as an element of $\ell_2(\mathbf{Z}^d)$, we get the function (18.27), which decays exponentially as $|x| \to \infty$.

Proof. From (18.48) it follows that u is an eigenvalue of A if and only if

$$(I - e^{f_0(\lambda)}\kappa V)\big(e^{-t}(I + \kappa V)^{-1}u\big) = 0. \tag{18.51}$$

If we set

$$v = e^{-t}(I + \kappa V)^{-1}u, \tag{18.52}$$

we can rewrite (18.51) in the form

$$Vv = \kappa^{-1}e^{-f_0(\lambda)}v. \tag{18.53}$$

Since the set of vectors $\{\eta^y \in L_2(\mathbf{T}) : y \in \mathbf{Z}^d\}$ is a complete set of eigenvectors of the unitary operator V, the set of vectors v satisfying (18.53) is the same as the set $\{\eta^y : y \in \mathbf{Z}^d\}$, and the eigenvalues are the solutions of (18.49). The properties of the set of eigenvalues listed in the lemma follows directly from the properties of the function $m(\lambda)$ of Lemma 18.5. The expression for the eigenvector $\hat{u}_y(\eta)$ is obtained from (18.52) by the substitution $v = \eta^y$, and (18.27) is obtained from $\hat{u}_y(\eta)$ by taking the inverse Fourier transform and normalizing, taking (18.18) into account. □

Proof of Theorem 18.2. All that remains to show is that the system of eigenvectors of A is complete. We now depart from the proof in [Figotin and Pastur 1984b], which is incomplete.

It is convenient to consider the representations of the eigenvectors in the space $L_2(\mathbf{T})$. The plan for the proof is the following: We first prove the existence of an analytic function $s(\eta, \zeta)$ in the domain $T_{\rho,r}$ defined in Lemma 18.3, so that the vectors

$$\tilde{u}_y(\eta) = s(\eta, \kappa\gamma^y)\eta^y \tag{18.54}$$

for $y \in \mathbf{Z}^d$, where $\kappa = e^{2\pi i\omega}$, form a complete system of normalized eigenfunctions of A. Together with Lemma 18.3, this implies that

$$\hat{u}_y(\eta) = S(\kappa)\eta^y, \qquad S(\kappa) = s(\eta, \kappa V). \tag{18.55}$$

This means that the operator S transforms the orthonormal basis $\{\eta^y : y \in \mathbf{Z}^d\}$ of $L_2(\mathbf{T})$ into an orthonormal system of vectors $\{u_y : y \in \mathbf{Z}^d\}$, and thus is at least an isometric operator. We then show that

$$S(\kappa)S^*(\kappa) = I, \tag{18.56}$$

that is, S is a unitary operator; this concludes the proof of completeness.

To simplify the argument, we assume that $g = 1$. From (18.50) and (18.39) we obtain the following representation for the unnormalized eigenvector:

$$u_y(\eta) = -2i\big(\hat{a}_0(\eta) - \lambda_y - i\big)^{-1} e^{t(\eta,\lambda_y)}\eta^y. \tag{18.57}$$

The normalized eigenvector of A is defined up to a factor of unit absolute value, and (18.54) will take place if only if this factor is chosen properly.

We introduce the change of variables

$$\lambda = -i\frac{1-\xi}{1+\xi}, \tag{18.58}$$

so that values of ξ on the unit circle S^1 correspond to real values of λ. We now consider the functions $\hat{C}(\eta, \lambda)$, $f(\eta, \lambda)$, $f_0(\lambda)$ and $t(\eta, \lambda)$ as functions of the variable ξ, and, to simplify the notation, we write $f(\eta, \xi)$ instead of $f(\eta, \lambda(\xi))$, and so on. We will do the same for the variable $\zeta \in S^1$ that appears below.

(18.7) Lemma. (i) *We have*

$$f(\eta, \xi) = \ln(-\xi^{-1}) + \ln\frac{\hat{a}_0 + 2i}{\hat{a}_0 - 2i} + \ln\frac{1 + \hat{a}_0(\hat{a}_0 + 2i)^{-1}\xi}{1 + \hat{a}_0(\hat{a}_0 - 2i)^{-1}\xi^{-1}} + c(\xi),$$

where $\hat{a}_0 = \hat{a}_0(\eta)$, we take a continuous branch of the log with $\ln 1 = 0$, and $c(\xi)$ is a multiple of $2\pi i$.

(ii) *The function $e^{f_0(\xi)}$ is analytic in the annulus $K_r = \{\xi : e^{-r} \leqslant |\xi| \leqslant e^r\}$ for $r > 0$ small enough, and there is some $c > 0$ such that*

$$c^{-1}|\xi_1 - \xi_2| \geqslant |e^{f_0(\xi_1)} - e^{f_0(\xi_2)}| \geqslant c|\xi_1 - \xi_2| \tag{18.59}$$

for $\xi, \xi_2 \in K_r$.

(iii) *There exists a function l on K_r that is inverse to $e^{f_0(\xi)}$, that is,*

$$\zeta = e^{f_0(\xi)} \text{ if and only if } \xi = l(\zeta); \tag{18.60}$$

this function is analytic in some neighborhood of the circle, and satisfies inequalities similar to (18.59).

(iv) *The functions* $t(\eta,\xi)$ *and* $t(\eta,\zeta)$ *are analytic in some neighborhood of* S^1, *and* $\operatorname{Re} t = 0$ *if* $\eta \in \mathbf{T}$ *and* $\xi, \zeta \in S^1$.

Proof. Part (i) follows from the identity

$$C(\eta,\xi) = -\xi^{-1}\frac{\hat{a}_0 + 2i}{\hat{a}_0 - 2i}\,\frac{1 + \hat{a}_0(\hat{a}_0 + 2i)^{-1}\xi}{1 + \hat{a}_0(\hat{a}_0 - 2i)^{-1}\xi^{-1}},$$

if one takes into account that both sides of the equation to be shown are obviously continuous in η, that is, the difference between them does not depend on η.

The analyticity of $e^{f_0(\xi)}$ follows immediately from (i). To show (18.59), note that (i) leads to

$$\frac{\partial}{\partial \xi} e^{f_0(\xi)} = \int_{\mathbf{T}}^{\exp(f_0(\xi))} \xi^{-1} \frac{1 - |h(\eta)|^2}{\left|1 + h(\eta)\xi\right|^2} \hat{d}\eta,$$

where $\hat{d}\eta = (2\pi i)^{-d}\, d\eta/\eta$ and $h(\eta) = \hat{a}_0(\eta)/\big(\hat{a}_0(\eta) + 2i\big)$. Thus there is some $c_1 > 0$ such that

$$c_1^{-1} \geqslant \left|\frac{\partial}{\partial \xi} e^{f_0(\xi)}\right| \geqslant c_1. \tag{18.61}$$

Now set $\xi = e^{i\phi} \in S^1$. In view of (18.43)–(18.45), $F(\phi) = i^{-1} f_0(e^{i\phi})$ can be defined as a real-valued, continuous, monotonically decreasing function of the real variable ϕ. Further, $F(\phi + 2\pi n) = F(\phi) - 2\pi n$ for $n \in \mathbf{Z}$. Since $\partial/\partial\phi = i\xi\partial/\partial\xi$ and F is monotonic, (18.61) implies that

$$c_1^{-1} \geqslant -\frac{\partial F}{\partial \phi} \geqslant c_1.$$

Together with (18.61), this shows that (18.59) is valid for $\xi_1, \xi_2 \in S^1$ and $|\xi_1 - \xi_2|$ small in the annulus K_r. Because $e^{f_0(\xi)}$ is analytic, it follows that (18.59) holds for any $\xi_1, \xi_2 \in K_r$ if r is small enough.

Part (iii) clearly follows from (ii), and (iv) from (i) and (iii). □

(18.8) Lemma. *The normalized eigenfunctions* $\tilde{u}_y$ *of* A *can be represented in the form* (18.54), *where* $s(\eta,\zeta)$ *is analytic in* $T_{\rho,r}$ *and*

$$\int_{|\zeta|=1} \left|s(\eta,\zeta)\right|^2 \hat{d}\zeta = 1 \tag{18.62}$$

for $\eta \in \mathbf{T}$, *where* $\hat{d}\zeta = (2\pi i)^{-1} d\zeta/\zeta$.

Proof. Introduce the function

$$u_y(\eta,\lambda) = -2i\big(\hat a_0(\eta) - \lambda - i\big)^{-1} e^{t(\eta,\lambda)}\eta^y.$$

Since $|\xi| = 1$ for real λ in (18.58), we get, setting $\hat a_0 = \hat a_0(\eta)$:

$$\begin{aligned}
\text{(18.63)}\quad & \frac{u_y(\eta,\lambda)\eta^{-y}}{\|u_y\|} \\
&= \frac{-i(\hat a_0 - \lambda - i)^{-1}}{\Big(\int_{\mathbf{T}}\big((\hat a_0-\lambda)^2+1\big)^{-1}\hat{d\eta}\Big)^{1/2}} e^{t(\eta,\lambda)} \\
&= \tau\,\frac{1 + i\hat a_0(i\hat a_0+2)^{-1}\xi^{-1}}{i\hat a_0 + 2} e^{t(\eta,\xi)} \\
&\quad\times\left(\int_{\mathbf{T}} \frac{\hat{d\eta}}{2(\hat a_0^2+2)}\left(1 + \frac{(\hat a_0^2 - 2i\hat a_0)\xi + (\hat a_0^2 + 2i\hat a_0)\xi^{-1}}{2(\hat a_0^2+2)}\right)^{-1}\right)^{-\frac12},
\end{aligned}$$

where τ does not depend on η and $|\tau| = 1$. Further, for $\eta \in \mathbf{T}$ and $\xi \in S^1$ we have

$$\text{(18.64a)}\qquad \left|\frac{(\hat a_0^2 - 2i\hat a_0)\xi + (\hat a_0^2 + 2i\hat a_0)\xi^{-1}}{2(\hat a_0^2+2)}\right| \leqslant \frac{|\hat a_0|}{\sqrt{\hat a_0^2+1}} \leqslant c_2 \leqslant 1,$$

$$\text{(18.64b)}\qquad \big|i\hat a_0(i\hat a_0+2)^{-1}\big| \leqslant c_2.$$

This latter inequality implies that for ρ and r small enough and $(\eta,\xi) \in T_{\rho,r}$, the integrand in the right-hand side of (18.63) takes values in the region $\operatorname{Re} z \geqslant c_2'$, where $c_2' > 0$ is a constant. For the square root of the integral, choose the standard branch, with $\sqrt{1} = 1$. Then (18.64) implies that the right-hand side of (18.63), which we denote by $s(\eta,\xi)$, is an analytic function of η and ξ in $T_{\rho,r}$. By (18.49) the eigenvalues λ_x satisfy $e^{f_0(\lambda)} = \kappa\gamma^x$. Introducing the variable ζ of (18.60) and using (18.57), we arrive at (18.54).

To check (18.62), note that

$$\hat{d\zeta} = \frac{1}{2\pi i}\frac{\partial}{\partial\lambda} f_0(\lambda)\,d\lambda.$$

This, together with (18.44), (18.63) and Lemma 18.7(iv), gives

$$\begin{aligned}
\int_{|\zeta|=1} |s(\eta,\zeta)|^2 \hat{d\zeta} &= \frac{1}{2\pi}\int_{-\infty}^{\infty} \frac{|m'(\lambda)|\,|\hat a_0 - \lambda - i|^{-2}}{\int_{\mathbf{T}}\big((\hat a_0-\lambda)^2+1\big)^{-1}\hat{d\eta}}\,d\lambda \\
&= \frac{1}{\pi}\int_{-\infty}^{\infty} |\hat a_0 - \lambda - i|^{-2}\,d\lambda = 1
\end{aligned}$$

for $\eta \in \mathbf{T}$. This concludes the proof. □

Since we have proved (18.54), the proof of Theorem 18.2 will be complete if we show that the operator S defined in (18.55) satisfies (18.56). The function s is analytic, so S depends continuously on κ, and SS^* is an orthogonal projection, that is, $SS^* \leqslant I$; thus, it is enough to show that

$$\int_{|\kappa|=1} SS^* \hat{d}\kappa = I. \tag{18.65}$$

In view of (18.54) and (18.62), we have, for $\phi \in L_2(\mathbf{T})$,

$$\begin{aligned}\int_{|\kappa|=1} \|S\phi\|^2 \hat{d}\kappa &= \sum_{y\in\mathbf{Z}^d} \int_{|\kappa|=1} \left| \int_{\mathbf{T}} s(\eta, \kappa\gamma^y)\eta^y \phi^*(\eta)\hat{d}\eta \right|^2 d\kappa \\ &= \sum_{y\in\mathbf{Z}^d} \int_{|\kappa|=1} \left| \int_{\mathbf{T}} s(\eta, \kappa)\eta^y \phi^*(\eta)\hat{d}\eta \right|^2 d\kappa \\ &= \int_{|\kappa|=1} \int_{\mathbf{T}} \left|s(\eta,\kappa)\right|^2 \left|\phi(\eta)\right|^2 \hat{d}\eta \hat{d}\kappa = \int_{\mathbf{T}} \left|\phi(\eta)\right|^2 \hat{d}\eta.\end{aligned} \tag{18.66}$$

This clearly implies (18.65), and with it Theorem 18.2. □

Remark. Our arguments here are a variation of Problem I.15.

In this connection there arises naturally an algebra, containing the resolvents of operators of the form (18.1)–(18.4), that is useful in generalizing Theorem 18.2 to a wider class of operators. (For physical applications, in particular the study of the asymptotic properties of the low-frequency conductivity associated with A, see [Figotin and Pastur 1984b]. Similar algebras were introduced to the theory of differential and pseudo-differential operators in [Coburn et al. 1973], and analyzed in detail in [Shubin 1975] and [Bellissard and Testard 1985].)

Consider the set $\mathcal{F}_{\rho,r}$ of functions analytic in the polycylinder $T_{\rho,r}$ defined in the statement of Lemma 18.3. Such functions are represented by the Laurent series

$$\hat{u}(\eta, \xi) = \sum_{\substack{x\in\mathbf{Z}^d \\ k\in\mathbf{Z}}} u(x,k)\eta^x \xi^k, \tag{18.67}$$

with

$$u(x,k) = \frac{1}{(2\pi i)^{-d-1}} \int_{|\eta_j|=|\xi|=1} \hat{u}(\eta,\xi)\eta^{-x-1}\xi^{-k-1}\, d\eta\, d\xi, \tag{18.68}$$

$$\left|u(x,k)\right| \leqslant \text{const}\, e^{-\rho_1|x| - r_1|k|} \tag{18.69}$$

for $0 < \rho_1 < \rho$ and $0 < r_1 < r$.

A function $\hat{u} \in \mathcal{F}_{\rho,r}$ generates the operator

(18.70) $$\hat{u}(\eta, V) = \sum_{\substack{x \in \mathbf{Z}^d \\ k \in \mathbf{Z}}} u(x,k)\eta^x V^k.$$

Since η and V do not commute, the order of the multiplication in (18.70) is essential.

We call the function $u(x,k)$, for $x \in \mathbf{Z}^d$ and $k \in \mathbf{Z}$, the kernel of the operator $\hat{u}(\eta, V)$, and denote by $\mathcal{A}_{\rho,r}$ the set of operators defined by (18.70), for $\hat{u} \in \mathcal{F}_{\rho,r}$. Clearly, $\mathcal{A}_{\rho,r}$ is an algebra, and by the commutation relation (18.16) and the equality $V\eta^x = \gamma^x \eta^x$, we have

(18.71) $$(u+v)(x,k) = u(x,k) + v(x,k),$$

(18.72) $$(uv)(x,k) = \sum_{\substack{x_1 \in \mathbf{Z}^d \\ k_1 \in \mathbf{Z}}} u(x - x_1, k - k_1)v(x_1, k_1)\gamma^{(k-k_1)x}.$$

If the components of α, together with 1, form a rationally independent set (for $d = 1$ this amounts to saying that α is irrational), it is easy to see that

(18.73) $$\hat{u}(\eta, V) = 0 \iff u(x,k) = 0 \text{ for all } x \in \mathbf{Z}^d \text{ and } k \in \mathbf{Z}.$$

Since we will not be interested in the particular values of ρ and r, it is convenient to consider the algebra $\mathcal{A} = \bigcup_{\rho,r} \mathcal{A}_{\rho,r}$, for all ρ and r. The elements of $\mathcal{A}$ are generated by functions $\hat{u} \in \mathcal{F} = \bigcup_{\rho,r} \mathcal{F}_{\rho,r}$, by (18.70).

We will say that a complex-valued function $Q(\xi)$, for $\xi \in \mathbf{C}$, belongs to class $\mathcal{P}$ if $Q(\xi)$ is a meromorphic function on a neighborhood of the circle C^1, and there exists $C > 0$ such that

(18.74) $$|Q(\xi e^{ia}) - Q(\xi)| \geqslant C|e^{ia} - 1|$$

for all $a \in \mathbf{R}$. Functions of class $\mathcal{P}$ have a variety of interesting properties [Bellissard et al. 1983a]: for example, if $Q \in \mathcal{P}$ takes on real values on S^1, it has a single pole on S^1, and takes on all real values on S^1. If $w(\xi)$ is an analytic function in the neighborhood of S^1, we have $Q + \varepsilon w \in \mathcal{P}$ for all ε small enough.

(18.9) Theorem [Bellissard et al. 1983a]. *Let A be the operator on $\ell_2(\mathbf{Z}^d)$ given by*

(18.75) $$A = A_0 + gQ(V),$$

where $A_0 \in \mathcal{A}$, $Q \in \mathcal{P}$ and g is a real constant. There exists a positive constant g_0 such that, for all $|g| \geqslant g_0$, there exists $\tilde{Q}_g \in \mathcal{P}$ and $U_g \in \mathcal{A}$ such that $U_g^{-1} \in \mathcal{A}$ and

(18.76) $$U_g A U_g^{-1} = U_g\big(A_0 + gQ(V)\big)U_g^{-1} = \tilde{Q}_g(V).$$

If A_0 is a self-adjoint operator and $\tilde{Q}_g$ takes on real values on the circle, U_g is a unitary operator, the spectrum of A is pure point, simple and dense on $\mathbf{R}$, *and the eigenfunctions of A decrease exponentially.*

Proof. The proof uses a version of the KAM method, adapted to the algebraic setting described above, where the diagonal operator $Q(V)$ in $\ell_2(\mathbf{Z}^d)$ is perturbed by the operator $g^{-1}A$ (see also 16.E). □

It is easy to see that if $Q(\xi) = i(\xi - 1)(\xi + 1)^{-1}$, the operator A defined by (18.75) coincides with the operator (18.1) for $\omega = 0$. But Theorem 18.2, unlike Theorem 18.9, is true for all $g \neq 0$, not just for g big enough.

Theorems 18.2 and 18.9 have the following consequence:

(18.10) Corollary [Figotin and Pastur 1984b]. *Let $w(\xi)$ be an analytic function on a neighborhood of S^1, taking real values on S^1. There exists a positive constant δ_0 such that, for all real δ with $|\delta| \leqslant \delta_0$, the statement of Theorem* 18.2 *holds for the operator $\tilde{A} = A + \delta W$, where W is the operator of multiplication by the function $w(e^{2\pi i((\alpha,x)+\omega)})$, for $x \in \mathbf{Z}^d$.* □

18.C The One-Dimensional Case

Operators of the form (18.1)–(18.4) with $d = 1$ are of particular interest because for them one can analyze in considerable detail how the spectrum depends on the arithmetic properties of α, and in particular on how well α is approximated by rationals.

We studied in 16.F the almost-Mathieu operator (16.90). For the operator (18.1)–(18.4), the results are even more complete; many of them do not depend on the operator being of order two (they are even true for operators of infinite order satisfying (18.2)), and are obtained by a unified approach based on the representation (18.9) for the resolvent.

Here is how the structure of the spectrum depends on the arithmetic properties of α:

(i) If $\alpha = b/a$ is rational, the potential is periodic of period a (assuming a and b are relatively prime), so the spectrum is absolutely continuous [Reed and Simon IV].

(ii) If α is irrational, the absolutely continuous spectrum is empty for almost all $\omega \in [0, 1)$. To see this, consider first an operator of order two, with $a_0(x) = 0$ for $|x| > 1$ in (18.1). From Theorem 18.1 and Problem II.24, we know that $N(\lambda)$ for this operator is the same as that for an operator of order two whose potential consists of independent, identically Cauchy-distributed random variables (the Lloyd model: see [Lloyd 1969]). But equation (11.78b), which is valid by Theorem 11.9 because (11.16b) is clearly satisfied, implies that the Lyapunov exponent of the two operators coincide. Now we found the Lyapunov exponent of the Lloyd model in Problem V.14:

$$\inf_{\lambda \in \mathbf{R}} \gamma(\lambda, g) = \gamma(0, g), \qquad \sinh \gamma(0, g) = \tfrac{1}{2}|g| > 0,$$

so the Lyapunov exponent is strictly positive for all $\lambda \in \mathbf{R}$. (This also follows from Theorem 14.2). Therefore, by Theorem 12.1, the spectrum has no absolutely continuous component if α is irrational and $a_0(x) = 0$ for $|x| > 1$.

The general case, when A_0 satisfies only condition (18.2), will be considered in Theorem 18.11 below (see also [Simon and Spencer 1989]).

To gain more information, we introduce the following measure of how well α is approximated by rationals [Simon 1985]:

$$L(\alpha) = -\liminf_{n \to \infty} \frac{1}{|n|} \ln |1 - e^{2\pi i \alpha n}| \geqslant 0. \tag{18.77}$$

The set of numbers α for which $L(\alpha) = 0$ obviously includes the Diophantine numbers satisfying condition (18.21) for $d = 1$, and therefore has full Lebesgue measure in $\mathbf{R}$. For $L(\alpha) > 0$, the rate of approximation is exponential, that is, for some sequence of rational numbers b_n/a_n we have

$$\left| \alpha - \frac{b_n}{a_n} \right| = O(e^{-L(\alpha) a_n}).$$

Similarly, the Liouville numbers of Corollary 16.9 correspond to $L(\alpha) = \infty$. This set is of second Baire category (a dense G_δ). By using continued fraction theory [Khintchine 1963], one can see that the set $\{\alpha \in \mathbf{R} : L(\alpha) = L_0\}$ is uncountable and dense for any $L_0 > 0$.

(iii) Now assume that, for almost all $\lambda \in \mathbf{R}$, the root η_0 of (18.28) of maximal absolute value $|\eta_0| \leqslant 1$ is simple and is the only one on the circle of radius $|\eta_0|$. Set

$$\Sigma_1 = \{\lambda \in \mathbf{R} : \gamma(\lambda, g) > L(\alpha)\}, \tag{18.78}$$

$$\Sigma_2 = \{\lambda \in \mathbf{R} : \gamma(\lambda, g) < \tfrac{1}{2} L(\alpha)\}, \tag{18.79}$$

where the quantity $\gamma(\lambda, g) = \gamma_-(\lambda, g)$ is defined in (18.29). Then the spectrum is pure point and simple on Σ_1 and is purely singular continuous on Σ_2:

$$\sigma(A) \cap \Sigma_1 = \bar{\sigma}_{\mathrm{p}}(A) \cap \Sigma_1 = \Sigma_1, \tag{18.80}$$

$$\sigma(A) \cap \Sigma_2 = \sigma_{\mathrm{sc}}(A) \cap \Sigma_2 = \Sigma_2. \tag{18.81}$$

These relations are obviously valid for operators of order two, when (18.28) takes the form $a_0(0) + a_0(1)(\eta + \eta^{-1}) = \lambda + ig$. Another example is the operator of infinite order given by $a_0(x) = e^{-\rho|x|}$, for $\rho > 0$. Problem VII.33 gives the value of $\gamma(\lambda, g)$ in this case.

Thus, if $L(\alpha) > 2\inf_\lambda \gamma(\lambda, g)$, the operator A has mixed spectral type: pure point for large $|\lambda|$ and singular continuous for small $|\lambda|$. For it is easy to see that $\gamma(\lambda, g) \to \infty$ as $\lambda \to \infty$, since $\eta_0 \to 0$. (Notice also that for order-two operators, where $\gamma(\lambda, g)$ is given in Problem V.14, the condition $L(\alpha) > 2\inf_\lambda \gamma(\lambda, g)$ is equivalent to $L(\alpha) > 2\,\mathrm{arcsinh}(g/2)$, which is certainly satisfied if $L(\alpha) > g$.)

In 18.D we will encounter another example of a metrically transitive operator (this time multidimensional) of mixed spectral type, but having a different structure.

Relation (18.80) is a strengthened version of Theorem 18.2. Indeed, the key point in the proof of that theorem was the construction of a solution $t(\eta)$ of the homological equation (18.25) that can be analytically continued into some annulus, because the existence of this solution shows that the operators κCV and V are similar. The formal solution to this equation, obtained by Fourier series, is

(18.82)
$$t(\eta) = \sum_{n \neq 0} \frac{f(n)}{1 - e^{2\pi i \alpha n}} \eta^n,$$

where $f(n)$, for $n \in \mathbf{Z}$, are the Fourier coefficients of the function $\ln \hat{C}(\eta)$ of (18.13). Under our conditions on $\hat{a}_0(\eta)$ these coefficients have the property that

(18.83)
$$\limsup_{|n| \to \infty} \frac{1}{|n|} \ln|f(n)| = -\gamma(\lambda, g)$$

(cf. (18.29)). Therefore, if $\gamma > L(\alpha)$, the series (18.82) defines functions analytic in the annulus $e^{-\rho_1} \leqslant |\eta| \leqslant e^{\rho_1}$, where $\rho_1 = \gamma - L > 0$, and the proof of the Theorem 18.2 can be carried through.

A relation similar to (18.80) also holds in several dimensions, if we define $L(\alpha)$ and $\gamma(\lambda, g)$ by

(18.84)
$$L(\alpha) = -\liminf_{|x| \to \infty} \frac{1}{|x|} \ln\left|1 - e^{2\pi i(\alpha, x)}\right|,$$

(18.85)
$$\gamma(\lambda, g) = -\limsup_{|x| \to \infty} \frac{1}{|x|} \ln|f(x)|,$$

where $f(x)$, for $x \in \mathbf{Z}^d$, are the Fourier coefficients of the function $\ln \hat{C}(\eta)$, for $\eta \in \mathbf{T}^d$.

Relation (18.81) will be proved below (Theorem 18.14 and remark following).

If α is a Liouville number, $L(\alpha)=\infty$, it follows from (18.79) and (18.81) that the whole spectrum is singular continuous. For an equation of order two ($a_0(x)=0$ for $|x|>1$) this can also be deduced from Theorem 16.6, as was done in 16.F for the almost-Mathieu equation (16.90). Also, since for a second-order equation the quantity $\gamma(\lambda,g)$ defined by (18.29) and (18.85) coincides with the Lyapunov exponent, we get from (11.25) and (11.28) that

$$\gamma(\lambda,g)\leqslant\int_0^1 \ln\|A(0)\|\,d\omega=\tilde{\gamma}(\lambda,g)$$

(cf. (11.27b)), where $A(x)$ is defined by formula (11.15), with $q(x)=-g\tan(\pi\alpha x+\omega)$. Thus the spectrum must be singular continuous in the region $\{\lambda\in\mathbf{R}:\tilde{\gamma}<\frac{1}{2}L(\alpha)\}$. This result was also proved by Simon [1985] in a different way.

A fairly detailed study of the behavior of the generalized eigenfunctions of the operator A in this case is presented in [Prange et al. 1984], based on numerical calculations and heuristic, but persuasive, arguments.

The rest of this subsection is devoted to the statement and proof of Theorems 18.11 and 18.14, mentioned above.

(18.11) Theorem [Figotin and Pastur 1984b]. *Let A be an operator of the form* (18.1)–(18.4), *with $d=1$. If α is irrational, the spectrum of A has no absolutely continuous component, that is, it is singular.*

Proof. If we have, say,

$$\left|\alpha-\frac{b}{a}\right|\geqslant a^{-20} \tag{18.86}$$

for all integers b and a, Theorem 18.2 says that the spectrum is pure point, and there is no absolutely continuous component. (We chose the number 20 just to avoid fractional powers of a in the subsequent argument.) Now assume that this condition is not fulfilled, that is, there are sequences of natural numbers a_n and b_n such that $\lim_{n\to\infty}b_n/a_n=\alpha$ and

$$\left|\alpha-\frac{b_n}{a_n}\right|\leqslant a_n^{-20}. \tag{18.87}$$

We will show that for any ω satisfying (18.4) the absolutely continuous component of the spectrum is empty. By Theorem A.10(i) of the Appendix, it suffices to show that, for every $x\in\mathbf{Z}$ and Lebesgue-almost every $\lambda\in\mathbf{R}$,

$$\lim_{\varepsilon\downarrow 0}\operatorname{Im}\big((\eta^x,R(\lambda+i\varepsilon)\eta^x)\big)=0, \tag{18.88}$$

where one should take the representation of the resolvent R over the space $L_2(\mathbf{T}^1) = L_2(S^1)$.

Our proof of (18.88) is organized as follows. The operator A and its resolvent R depend on α, that is, $R(\lambda + i\varepsilon) = R(\lambda + i\varepsilon, \alpha)$. Condition (18.87) says that α is well approximated by rationals, so one can, using representation (18.17) for the resolvent, reduce the proof of (18.88) to verifying that, for Lebesgue-almost $\lambda \in \mathbf{R}$,

$$\lim_{n\to\infty} \operatorname{Im}\big((\eta^x, R(\lambda + i\varepsilon_n, b_n/a_n)\eta^x)\big) = 0$$

for every $x \in \mathbf{Z}$, where $\varepsilon_n = a_n^{-5}$. This we prove by an estimate based on the representation (18.17) or (18.9).

As before, it is convenient to consider the representation of the operators over the space $L_2(S^1)$, and also to change from the variable $\eta \in S^1$ to the variable $p \in \mathbf{R}$, by the substitution

$$\eta = e^{2\pi i p}. \tag{18.89}$$

With this substitution, (18.16) becomes

$$(V\psi)(p) = \psi(p+\alpha), \qquad V^n\Phi(p) = \Phi(p+n\alpha)V^n \text{ for } n \in \mathbf{Z}. \tag{18.90}$$

From this and from (18.9), we get

$$R(z) = \sum_{n\geqslant 0} R_n(p)\kappa^n V^n, \tag{18.91}$$

with $R_0(p) = \hat{B}(p)$ and $R_n(p) = -2ig\hat{B}(p)\prod_{s=0}^{n-1}\hat{C}(p+s\alpha)\hat{B}(p+n\alpha)$. Now fix an arbitrary finite interval $\Delta \subset \mathbf{R}$, and set

$$L_\Delta = \{z \in \mathbf{C} : \operatorname{Re} z \in \Delta \text{ and } 0 \leqslant \operatorname{Im} z \leqslant g/4\}. \tag{18.92}$$

From the definition of $\hat{B}$ and $\hat{C}$ ((18.12) and (18.13)), we have the following estimates:

$$\left|\hat{B}(p, \lambda + i\varepsilon)\right| \leqslant c_1, \tag{18.93}$$

$$\left|\hat{C}(p, \lambda + i\varepsilon)\right| \leqslant e^{-c_2\varepsilon}, \tag{18.94}$$

$$\left|R_n(p, \lambda + i\varepsilon)\right| \leqslant c_1 e^{-c_2\varepsilon}, \tag{18.95}$$

$$\left|\frac{\hat{B}(p+\Delta p, \lambda + i\varepsilon)}{\hat{B}(p, \lambda + i\varepsilon)} - 1\right| \leqslant c_1|\Delta p|, \tag{18.96}$$

$$\left|\frac{\hat{C}(p+\Delta p, \lambda + i\varepsilon)}{\hat{C}(p, \lambda + i\varepsilon)} - 1\right| \leqslant c_1|\Delta p|, \tag{18.97}$$

where c_1 and c_2 are positive constants independent of $z \in L_\Delta$ and α.

(18.12) Lemma. *Let $\alpha_1, \alpha_2 \in \mathbf{R}$, $z = \lambda + i\varepsilon \in L_\Delta$. There exist positive constants c_3 and c_4, independent of $z \in L_\Delta$, such that, for any natural number k,*

$$\left|(\eta^x, R(\lambda+i\varepsilon,\alpha_1)\eta^x) - |(\eta^x, R(\lambda+i\varepsilon,\alpha_2)\eta^x)\right| \leqslant \frac{c_3}{\sqrt{|\alpha_2-\alpha_1|}} \int_0^{c_3 k\sqrt{|\alpha_2-\alpha_1|}} (e^{\tau^2}-1)\,d\tau + c_3\varepsilon^{-1} e^{-c_4 k\varepsilon}.$$

Proof. From (18.91), it follows that

$$R(z,\alpha)\eta^x = \sum_{n\geqslant 0} \tilde{R}_{n,x}(p,\alpha)\eta^x, \tag{18.98}$$

with $\tilde{R}_{n,x} = R_n(p,\alpha)(\kappa e^{2\pi i\alpha x})^n$. Using (18.93) to (18.97), we get for $n \geqslant 0$, where $\Delta\alpha = |\alpha_2 - \alpha_1|$:

$$\begin{aligned}\left|\frac{\tilde{R}_{n,x}(p,\alpha_2)}{\tilde{R}_{n,x}(p,\alpha_1)} - 1\right| &\leqslant 2\pi\,\Delta\alpha\, n|x| + (1+nc_1\,\Delta\alpha)\prod_{s=1}^{n}(1+sc_1\,\Delta\alpha) - 1\\ &\leqslant 2\pi\,\Delta\alpha\, n|x| + \exp\left(nc_1\,\Delta\alpha + \sum_{s=1}^{n} sc_1\Delta\alpha\right) - 1\\ &\leqslant e^{c_1'\,\Delta\alpha\, n^2} - 1,\end{aligned}$$

where $c_1' = 2c_1 + 2\pi|x|$. This, together with (18.95), (18.91), (18.98) and the obvious identity

$$\tilde{R}_{n,x}(p,\alpha_2) - \tilde{R}_{n,x}(p,\alpha_1) = \left(\frac{\tilde{R}_{n,x}(p,\alpha_2)}{\tilde{R}_{n,x}(p,\alpha_1)} - 1\right)\tilde{R}_{n,x}(p,\alpha_1), \tag{18.99}$$

gives

$$\begin{aligned}&\left|(\eta^x, R(\lambda+i\varepsilon,\alpha_2)\eta^x) - (\eta^x, R(\lambda+i\varepsilon,\alpha_1)\eta^x)\right|\\ &\qquad\leqslant \sum_{n\geqslant 0}\int_{-\frac{1}{2}}^{\frac{1}{2}} dp\,\left|\tilde{R}_{n,x}(p,\alpha_2) - \tilde{R}_{n,x}(p,\alpha_1)\right|\\ &\qquad\leqslant \sum_{n=0}^{k-1}(e^{c_1'\,\Delta\alpha\, n^2}-1)c_1 e^{-c_2 n\varepsilon} + \sum_{n\geqslant k} 2c_1 e^{-c_2 n\varepsilon}\\ &\qquad\leqslant c_1\int_0^k d\tau\,(e^{c_1'\,\Delta\alpha\,\tau^2}-1) + 2c_1 e^{-c_2 k\varepsilon}(1-e^{c_2\varepsilon})^{-1}.\end{aligned}$$

This clearly implies the lemma. □

Now notice that it is enough to show (18.98) for any sequence ε_n that tends to zero. Thus, consider a sequence b_n/a_n that approximates α at the rate specified by (18.87), and use Lemma 18.12, with $\alpha_1 = \alpha$, $\alpha_2 = b_n/a_n$, $k = a_n^6$ and $\varepsilon_n = a_n^{-5}$. Then $k_n\sqrt{|\alpha - b_n/a_n|}$ approaches 0 as $n \to \infty$, and, using (18.88), we have

$$\begin{aligned}\big|(\eta^x, R(\lambda + i\varepsilon_n, \alpha)\eta^x) - (\eta^x, R(\lambda + i\varepsilon_n, b_n/a_n)\eta^x)\big|& \\ \leqslant 2c_3^4 k_n^3 |\alpha - b_n/a_n| + c_3\varepsilon_n^{-1} e^{-c_4 k_n \varepsilon_n}&.\end{aligned}$$

Since this obviously goes to zero, it is enough to prove that for $\varepsilon_n = a_n^{-5}$ and Lebesgue-almost all $\lambda \in \Delta$ we have

$$\lim_{n\to\infty} \operatorname{Im}\big((\eta^x, R(\lambda + i\varepsilon_n, b_n/a_n)\eta^x)\big) = 0. \tag{18.100}$$

Let us investigate the expression $R(z, \alpha)$ given by (18.91) and (18.98) for rational α. Assume $\alpha = b/a$, where b and a are relatively prime and $a > 0$. Introduce the following auxiliary functions:

$$C_a(p) = \prod_{s=1}^{a} \hat{C}(p + sb/a), \tag{18.101}$$

$$D_k(p) = -2ig\hat{B}(p)\hat{B}\big(p + (k+1)b/a\big) \prod_{s=1}^{k} \hat{C}(p + sb/a). \tag{18.102}$$

Then, if $\kappa_a = \kappa e^{2\pi i (b/a)x}$, we have $\kappa_a^a = 1$, and for $n \geqslant 1$ and $n - 1 = la + k$, for $0 \leqslant k \leqslant a - 1$, we have

$$\tilde{R}_{n,x}(p) = \big(C_a(p)\kappa^a\big)^l D_k(p)\kappa_a^{k+1}. \tag{18.103}$$

From (18.98) and the identity $\operatorname{Im}(\eta^x, R(\lambda + i\varepsilon)\eta^x) = \varepsilon\big\|R(\lambda + i\varepsilon)\eta^x\big\|^2$, we get

$$\operatorname{Im}\big(\eta^x, R(\lambda + i\varepsilon, b/a)\eta^x\big) = \varepsilon \int_{-\frac{1}{2}}^{\frac{1}{2}} dp\,\Big|R_0(p) + \sum_{n\geqslant 1} \tilde{R}_{n,x}(p)\Big|^2.$$

This, together with (18.95), reduces the proof of (18.100) to the proof of the following lemma:

(18.13) Lemma. *For Lebesgue-almost all $\lambda \in \Delta$,*

$$\lim_{n\to\infty} \varepsilon_n \int_{-\frac{1}{2}}^{\frac{1}{2}} dp \Big|\sum_{m\geqslant 1} \tilde{R}_{m,x}(p, \lambda + i\varepsilon_n, b_n/a_n)\Big|^2 = 0, \tag{18.104}$$

where $\varepsilon_n = a_n^{-5}$.

Proof. From (18.103), we get

$$\sum_{m\geqslant 1} \tilde{R}_{m,x}(p) = \left(\sum_{k=0}^{a-1} D_k(p)\kappa_a^{k+1}\right)\left(1 - C_a(p)\kappa^a\right)^{-1}. \tag{18.105}$$

Together with (18.94) and (18.95), this shows there exists a constant $c > 0$ such that, for all $z \in L_\Delta$,

$$\left|\sum_{m\geqslant 1} \tilde{R}_{m,x}(p, z, b/a)\right| \leqslant ca\left|1 - C_a(p)\kappa^a\right|^{-1}. \tag{10.106}$$

Now notice that, in view of (18.38), the function $C_a(p)$ from (18.101) can be written as

$$C_a(p) = \exp\left(af_0 + a\sum_{k\neq 0} f_{ka}e^{2\pi ipka}\right). \tag{18.107}$$

Since, as we showed earlier, $f(\eta, z)$ is analytic in $T_\rho \times L_g$, there exist positive constants c_1 and c_2 such that

$$\left|f_m(z)\right| \leqslant c_1 e^{-c_2|m|} \tag{18.108}$$

for $m \in \mathbf{Z}$. From this and from (18.107) we conclude that, for some positive constants c_3 and c_4 and all $z \in L_\Delta$,

$$\left|C_a(p, z) - e^{af_0(z)}\right| \leqslant c_3 e^{-c_4 a}.$$

Using Lemma 18.5, we conclude that, for $z = \lambda + i\varepsilon \in L_\Delta$ with $\varepsilon a \leqslant 1$,

$$\left|e^{af_0(\lambda+i\varepsilon)} - e^{iam(\lambda)}\right| \leqslant c_5\varepsilon a.$$

Setting $a = a_n$ and $\varepsilon_n = a_n^{-5}$ in the last two inequalities, we get

$$\left|C_{a_n}(p, \lambda + i\varepsilon_n) - e^{ia_n m(\lambda)}\right| \leqslant c_6 a_n^{-4},$$

where $c_6 > 0$; for $\kappa = e^{2\pi i\omega}$ it follows that

$$\left|C_{a_n}(p, \lambda + i\varepsilon_n)\kappa^{a_n} - e^{ia_n(m(\lambda)+2\omega)}\right| \leqslant c_6 a_n^{-4}.$$

Since $m(\lambda)$ is continuous and monotonous by (18.44), it follows [Khintchine 1963] that, for Lebesgue-almost all $\lambda \in \mathbf{R}$ and all a large enough,

$$|1 - e^{ia(m(\lambda)+2\omega)}| \geqslant c_7 a^{-1}\ln^{-2} a. \tag{18.109}$$

This, together with the preceding inequality and (18.107), shows that for almost all $\lambda \in \Delta$ and a_n large enough,

$$\left|\sum_{m\geqslant 1} \tilde{R}_{m,x}(p, \lambda + i\varepsilon_n, b_n/a_n)\right|^2 \leqslant 2c_8 a_n^{-4} \ln^4 a_n.$$

This clearly implies (18.104), and with it (18.100) and Theorem 18.11. □

Theorem 18.11 can also be derived from the results in [Simon and Spencer 1990]. But the technique developed in the proof given here allows us to establish the following addition result, which is in some sense a converse:

(18.14) Theorem. *Let A be an operator of the form* (18.1)–(18.4), *with $d = 1$, $a_0(x) = 0$ for $|x| > 1$, and α satisfying* (18.71). *If $\gamma(\lambda, g)$ is the Lyapunov exponent of the corresponding order-two equation* (Problem V.14), *the spectrum is continuous on the set Σ_2 of* (18.79) *for Lebesgue-almost all $\omega \in [0, 1)$.*

Proof. Without loss of generality, we can take $a_0(0) = 0$ and $a_0(1) = 1$. In view of Problem I.14(b), it is enough to show that, for any finite interval $\Delta \subset \Sigma_2$,

$$\lim_{\varepsilon\downarrow 0} \int_\Delta d\lambda \int_0^1 \varepsilon\left|R_{0,0}(\lambda + i\varepsilon, \alpha)\right|^2 d\omega = 0, \tag{18.110}$$

where $R_{0,0}(z) = \big(e_0, R(z,\alpha)e_0\big)$ and $R(z, \alpha)$ is the resolvent of A. But by representation (18.91), the inner integral in (18.110) is

$$\varepsilon \sum_{n\geqslant 0} \int_{-\frac{1}{2}}^{\frac{1}{2}} \left|R_n(p, \alpha)\right|^2 dp,$$

where the functions $R_n(p, \alpha)$ are given by (18.91). Using arguments similar to those in the proof of Lemma 18.1, we conclude that the integrand in (18.110), for $\varepsilon = \varepsilon_n \to 0$, coincides with the limit of the same expression with α replaced by b_n/a_n, provided that there is a sequence of integers k_n such that $\lim_{n\to\infty} k_n = \infty$, $\lim_{n\to\infty} \varepsilon_n k_n = 0$ and

$$\lim_{n\to\infty} \varepsilon_n k_n^3 \left|\alpha - b_n/a_n\right| = 0.$$

In view of (18.77), we can choose $k_n = a_n \varepsilon_n^{-1}\big(1 + o(1)\big)$, where

$$\varepsilon_n = e^{-L' a_n} \text{ with } L' \in \left(\tfrac{1}{2}L, \inf_{\lambda\in\Delta}\{\gamma(\lambda, g)\}\right). \tag{18.111}$$

Since, by (18.95) and (18.98),

$$\varepsilon \int_0^1 \left|R_{0,0}(\lambda + i\varepsilon, b_n/a_n)\right|^2 d\omega \leqslant \text{const}$$

as $\varepsilon \downarrow 0$, this reduces the proof of (18.110) to checking the relation

$$\lim_{n\to\infty} \varepsilon_n \int_0^1 \left|R_{0,0}(\lambda + i\varepsilon_n, b_n/a_n)\right|^2 d\omega = 0. \tag{18.112}$$

But, according to (18.101) and (18.102),

$$\begin{aligned} &\int_0^1 \left|R_{0,0}(\lambda + i\varepsilon, b/a)\right|^2 d\omega \\ &\quad = \left|\int_{-\frac{1}{2}}^{\frac{1}{2}} dp\, \hat{B}(p)\right|^2 + \sum_{k=1}^{a-1} \iint_{-\frac{1}{2}}^{\frac{1}{2}} \frac{D_k(p_1)D_k^*(p_2)}{1 - C_a(p_1)C_a^*(p_2)} dp_1\, dp_2 \\ &\quad \leqslant c_1 + c_2 a \int_{-\frac{1}{2}}^{\frac{1}{2}} \frac{1}{\left|1 - C_a(p_1)C_a^*(p_2)\right|} dp_1\, dp_2, \end{aligned} \tag{18.113}$$

where the inequality follows from (18.93), (18.94) and (18.102). To estimate this last integral, remember that we are dealing with a second-order equation with $\hat{a}_0(p) = 2\cos 2\pi p$. By calculating the Fourier coefficients f_m of $\ln \hat{C}(p)$, where $\hat{C}(p)$ is defined in (18.7), we find

$$f_m = -\frac{2}{i|m|} r^{|m|} \sin |m|\phi, \tag{18.114}$$

where $re^{i\phi} = \eta_0$ is the root of the equation $\eta + \eta^{-1} = \lambda + ig$ in the unit disk. We consider only those $\lambda \in \mathbf{R}$ such that, for all integers a large enough,

$$|e^{ia\phi} - 1| \geqslant \text{const}\, |a|^{-1} \ln^{-2} |a| \tag{18.115}$$

(cf. (18.109)). Since η_0, and hence ϕ, depend analytically on λ, the set of such λ has full Lebesgue measure in $\mathbf{R}$ [Khintchine 1963]. Divide the last integral in (18.113) into two, over the regions $|\cos p_1 - \cos p_2| \geqslant \delta$ and $|\cos p_1 - \cos p_2| < \delta$. In the first region, we have

$$\begin{aligned} \left|1 - C_a(p_1)C_a^*(p_2)\right| &\geqslant \left|1 - e^{-4i \sin a_n\phi}(\cos p_1 - \cos p_2) + O(a_n r^{2a_n})\right| \\ &\geqslant \text{const}\, a_n^{-1} \ln^{-2} a_n\, r^{a_n} \delta, \end{aligned}$$

by (18.107), (18.114) and (18.115), so the integral is bounded below by $\text{const}\, \delta^{-1} a_n \ln^2 a_n r^{-a_n}$. To estimate the integral over the region $|\cos p_1 - \cos p_2| < \delta$, use the bound $|C_a(p)| \leqslant e^{-c_9 a\varepsilon}$, which follows from (18.88) and (18.101); the result does not exceed $\text{const}\, \delta^{1/2}(1 - e^{-c_9 a_n \varepsilon_n})$. Therefore the right-hand side of (18.113) is bounded by

$$\text{const}\left(1 + \frac{a_n^2 \ln^2 a_n}{\delta r^{a_n}} + \frac{\sqrt{\delta} a_n}{1 - e^{-c_9 a_n \varepsilon_n}}\right).$$

Recalling that $\ln r = -\gamma(\lambda, g)$ and that ε_n has the form (18.111), we conclude that the limit as $n \to \infty$ of this bound times ε_n is a constant times $\sqrt{\delta}$. But δ is arbitrary, so the limit is in fact zero, and (18.110) is proved. This concludes the proof of Theorem 18.14. □

Remark. In this proof the fact that the operator has order two was only used in deriving (18.114) for the coefficients f_m, in which the phase ϕ of the root of (18.28) is an analytic function of λ. Clearly, then, Theorem 18.14 is still true for any Toeplitz operator A_0 such that a formula similar to (18.114) holds at least asymptotically as $m \to \infty$, and that ϕ is a piecewise analytic function of λ (this latter property ensures the validity of (18.115), which is used in estimating (18.113)).

As an example, consider the operator of infinite order defined by $a_0(x) = e^{-\rho|x|+i\beta x}$, where $\rho > 0$ and $\beta \in S^1$. Then $a_0(\eta) = b\eta(1-b\eta)^{-1} + b(\eta - b)^{-1}$, with $b = e^{-\rho+i\beta}$, and the unique root of (18.28) lies in the unit disk, just as for an operator of order two. Here, then, formula (18.114) is true, and the phase ϕ is analytic in λ.

Another example is given by an operator A_0 of finite order n, where $\eta^n a_0(\eta)$ is a polynomial of degree $2n$. In this case it is enough to require that, for Lebesgue-almost all $\lambda \in \mathbf{R}$, the root of (18.28) of highest absolute value be unique and simple. Then (18.114) holds asymptotically, and since the set of $\lambda \in \mathbf{R}$ for which the roots of (18.28) have this property is open, we conclude that on each interval making up this set the phase ϕ is piecewise analytic.

18.D The Schrödinger Operator with a Nonlocal Quasiperiodic Potential

We consider here the continuous counterpart of the operators of the form (18.1), that is, a certain class of multidimensional, almost-periodic operators on $L_2(\mathbf{R}^d)$, whose spectral analysis can be carried out in a fair amount of detail [Figotin and Pastur 1990]. These operators differs from their discrete counterparts in having overlapping absolutely continuous and point spectra (Theorem 18.17). This is somewhat unusual, in view of the hypothesis, widely accepted in the physics literature, that the pure point and absolutely continuous components of the spectrum of typical disordered structures are separated by a so-called *mobility edge*, which is the only point belonging to both. The operators we discuss are not, of course, counterexamples to the hypothesis, but rather examples by means of which one might try to understand what the "proper" conditions of existence of the mobility edge must be.

Let H be the symmetric operator on $L_2(\mathbf{R}^d)$ defined on smooth functions with compact support by the expression

$$H = -\frac{1}{(2\pi)^2}\Delta + \sum_{n\in\mathbf{Z}^d} \tau_n |\phi_n\rangle\langle\phi_n|, \tag{18.116}$$

where Δ is the Laplacian,

$$\tau_n = -\tan\pi\big((n,\alpha)+\omega\big) \tag{18.117}$$

for $\omega \in [0,1)$, $\alpha \in \mathbf{R}^d$ satisfies (18.21), ω the condition (18.4), and $|\phi_n\rangle\langle\phi_n|$ is the one-dimensional operator acting according to the formula

$$\big(|\phi_n\rangle\langle\phi_n|\,\psi\big)(x) = \phi(x+n)\int_{\mathbf{R}_d}\phi(y+n)\psi(y)\,dy.$$

Definition. *A function $\phi(x)$, for $x \in \mathbf{R}^d$, is said to satisfy condition Φ if its Fourier transform*

$$\hat\phi(p) = \int_{\mathbf{R}^d} e^{-2\pi i(p,x)}\phi(x)\,dx,$$

for $p \in \mathbf{R}^d$, has the representation

$$\hat\phi(p) = a\big(|p|\big),$$

where the real-valued function $a(\mu)$, for $\mu \geqslant 0$, is infinitely differentiable on the semiaxis $[0,\infty)$ and satisfies

$$a(\mu) > 0 \textit{ for } 0 \leqslant \mu < \rho,$$
$$a(\mu) = 0 \textit{ for } \mu \geqslant \rho,$$

for some $\rho \in (0,\infty]$; if $\rho = \infty$, this should be interpreted to mean that the function $a(\mu)$ and all its derivatives decay superpolynomially as $\mu \to \infty$.

The operator (18.116) is not a Schrödinger operator, because its "potential" is not an operator of multiplication; it is a sum of projections onto functions localized in neighborhoods of points of $\mathbf{Z}^d$. Nevertheless, this operator seems interesting and useful as a model for more realistic almost-periodic Schrödinger operators. Denote by $E_0(d\lambda)$ and $E(d\lambda)$ the resolutions of the identity of the operators $-(2\pi)^{-2}\Delta$ and H, respectively, and set

$$\mathcal{H}_{<\rho} = E_0\big((-\infty,\rho)\big)L_2(\mathbf{R}^d), \qquad \mathcal{H}_{>\rho} = E_0\big((\rho,\infty)\big)L_2(\mathbf{R}^d).$$

(18.15) Theorem. *The operator H is essentially self-adjoint on the Schwartz space S. The subspaces $\mathcal{H}_{<\rho}$ and $\mathcal{H}_{>\rho}$ reduce H, and the restriction of H to $\mathcal{H}_{>\rho}$ coincides with the restriction of $(-2\pi)^{-2}\Delta$, so that if $d\lambda \in (\rho,\infty)$ we have*

$$E(d\lambda)\mathcal{H}_{>\rho} = E_0(d\lambda)\mathcal{H}_{>\rho}. \tag{18.118}$$

Furthermore, the spectrum H has an absolutely continuous component on the whole semiaxis (ρ, ∞), that is, $\sigma_{ac}(H) \supset (\rho, \infty)$.

(18.16) Theorem. *Suppose the function ϕ defining the operator H in (18.116) satisfies condition Φ with $\rho > \frac{1}{4}$. Then, in addition to the statements of Theorem 18.15, there is a positively number ρ_1 such that the spectrum of H is pure point and simple in $(-\infty, \rho_1)$, the set of eigenvalues is dense in $(-\infty, \rho_1)$, the corresponding eigenfunctions belong to S.*

(18.17) Theorem. *Suppose the function ϕ defining the operator H in (18.116) satisfies condition Φ with $\rho < \frac{1}{4}$. Then, in addition to the statements of Theorem 18.15, the spectrum of H is exhausted by the point and absolutely continuous components. The pure point spectrum coincides with $\mathbf{R}$, is simple, and the corresponding eigenfunctions belong to S. The orthogonal projections onto the pure point and absolutely continuous components of H coincide with the projections onto the subspaces $\mathcal{H}_{<\rho}$ and $\mathcal{H}_{>\rho}$, respectively.*

(18.18) Theorem. *The integrated density of states of the operator (18.116) is absolutely continuous for $d \leqslant 3$, and its derivative $n(\lambda)$ satisfies*

$$n(\lambda) = \frac{1}{\pi} \int_{\mathbf{T}} \frac{W_1^2(p, \lambda)}{W^2(p, \lambda) + 1} dp + \chi_{(0,\infty)}(\lambda - \rho) n_0(\lambda), \tag{18.119a}$$

where $\mathbf{T} = [-\frac{1}{2}, \frac{1}{2})^d$ is the d-dimensional torus, $W(p, \lambda)$ is given by formula (18.123) below, $W_1(p, \lambda)$ is given by the same formula with $\hat{\phi}^2(p)$ replaced by $\hat{\phi}(p)$, and $n_0(\lambda) = N_0'(\lambda)$ is the density of states of the operator $-(2\pi)^{-2}\Delta$. When $\rho < \frac{1}{4}$, formula (18.119a) is modified as follows:

$$n(\lambda) = \frac{1}{\pi} \int_{p^2 < \rho} \frac{\hat{\phi}^2(p)}{\hat{\phi}^4(p) + (p^2 - \lambda)^2} dp + \chi_{(0,\infty)}(\lambda - \rho) n_0(\lambda). \tag{18.119b}$$

Remark. In view of the spectral theory of the Schrödinger operator, the most interesting case is when $\rho = \infty$, that is, $\phi(x)$ is a generalized function, concentrated at a single point. For $d = 1$ this means that $\phi(x) = g\delta(x)$, and we recover the potential (1.31), with $u_a(x) = g \tan \pi(\alpha a + \omega)\delta(x)$, for $a \in \mathbf{Z}$. (See 6.B for the meaning of the δ-function as a potential.) The spectral analysis of the resulting operator can be reduced to the analysis of its discrete analogue, with the same potential (to within a multiplier) by a simple procedure: see [Lifshitz et al. 1982b] and the proof to Theorem 14.3. Therefore the results of the preceding subsections can be extended to this case, essentially without change. In particular, for a Diophantine number α, the spectrum is pure point.

As is well-known in quantum mechanics [Baz' et al. 1971], the Schrödinger operator with a point potential can also be properly defined for $d = 2, 3$—for example, as a strong resolvent limit of operators of the type we are discussing, the support of $\phi(x)$ shrinking to a point while the amplitude increases infinitely. Other variants of this definition can be found in [Berezin and Fadeev 1961; Grossman et al. 1980]. However, there is no known procedure in dimension $d > 1$ analogous to the one we mentioned to reduce the one-dimensional Schrödinger operator to a discrete operator; and the method we use for the spectral analysis of (18.116) for $\rho < \infty$ proves to be of little help when $\rho = \infty$. These difficulties are not merely technical; the spectrum of (18.116) for $\rho = \infty$ may in fact be absolutely continuous for large enough values of λ. This will be justified, though not with rigorous arguments, at the end of this subsection.

As in the case of the operators (18.1)–(18.4) and (18.21), the study of the spectral properties of (18.116) is based on a suitable representation of the resolvent $R(z) = (H - z)^{-1}$. We will not attempt a full-fledged discussion here, but merely highlight the most important ways in which this operator differs from its discrete counterpart. For details, see [Figotin and Pastur 1990].

As in 18.C, it is convenient to pass from the functions $\psi(x) \in L_2(\mathbf{R}^d)$ to their Fourier transforms by the formulas

$$\hat{\psi}(p) = \int_{\mathbf{R}^d} e^{-2\pi i(p,x)}\psi(x)\,dx,$$

$$\psi(x) = \int_{\mathbf{R}^d} e^{2\pi i(p,x)}\hat{\psi}(p)\,dp.$$

Introduce the operator Q on $\ell_2(\mathbf{Z}^d)$ given by $(Q\psi)_n = \tau_n^{-1}\psi_n$, for $n \in \mathbf{Z}^d$. We will look at the Fourier transform of functions $\psi_n \in \ell_2(\mathbf{Z}^d)$ as well:

$$\hat{\psi}(p) = \sum_{n\in\mathbf{Z}^d} e^{2\pi i(p,n)}\psi_n,$$

$$\psi_n = \int_{\mathbf{T}} e^{-2\pi i(pn,n)}\hat{\psi}(p)\,dp,$$

where $n \in \mathbf{Z}^d$ and $p \in \mathbf{T} = [-\frac{1}{2}, \frac{1}{2})^d$. The functions $\hat{\psi} \in L_2(\mathbf{T})$ will often be regarded as periodic and defined on the whole of $\mathbf{R}^d$. Finally, introduce the linear operator $(\,\cdot\,)_{\mathbf{T}}$ that takes a function $\psi(p)$, for $p \in \mathbf{R}^d$, to the periodic function $(\psi)_{\mathbf{T}}(p) = \sum_{n\in\mathbf{Z}^d}\psi(p+n)$.

With this notation, we have the following representation for H:

$$(H\psi)(p) = p^2\psi(p) + \hat{\phi}(p)Q^{-1}(\hat{\phi}\psi)_{\mathbf{T}}(p). \tag{18.120}$$

Setting $g = (H - z)\psi$, we get

$$(p^2 - z)^{-1} g = \psi + (p^2 - z)^{-1} \hat{\phi} Q^{-1} (\hat{\phi}\psi)_{\mathbf{T}}. \tag{18.121}$$

Multiplying both sides by $\hat{\phi}$ and applying the operation $(\,\cdot\,)_{\mathbf{T}}$, we obtain

$$\left(\frac{\hat{\phi} g}{p^2 - z}\right) = (\hat{\phi}\psi)_{\mathbf{T}} + W_z Q^{-1} (\hat{\phi}\psi)_{\mathbf{T}}, \tag{18.122}$$

where W_z denotes the operator of multiplication by the function

$$W_z(p) = \left(\frac{\hat{\phi}^2(p)}{p^2 - z}\right)_{\mathbf{T}} \tag{18.123}$$

in $L_2(\mathbf{T})$. Solving for $(\hat{\phi}\psi)_{\mathbf{T}}$ in (18.122) and plugging the result into (18.120), we obtain, after simple manipulations:

$$\big(R(z)g\big)(p) = \frac{g(p)}{p^2 - z} - \frac{\hat{\phi}(p)}{p^2 - z} (Q + W_z)^{-1} \left(\frac{\hat{\phi} g}{p^2 - z}\right)_{\mathbf{T}} (p). \tag{18.124}$$

Now (18.124) (or even (18.120)) imply that the subspaces $\mathcal{H}_{<\rho}$ and $\mathcal{H}_{>\rho}$ reduce H, and that the restriction of H to $\mathcal{H}_{>\rho}$ coincides with the restriction of $-(2\pi)^{-2}\Delta$, so that (18.118) must hold.

We now determine the eigenfunctions $\hat{\psi}_\lambda(p)$ and the corresponding eigenvalues λ. From (18.120) we have

$$(p^2 - \lambda)\hat{\psi}_\lambda + \hat{\phi} Q^{-1} (\hat{\phi}\hat{\psi}_\lambda)_{\mathbf{T}} = 0.$$

Hence, the eigenfunctions $\hat{\psi}_\lambda$ must be

$$\hat{\psi}_\lambda(p) = \frac{\hat{\phi}(p)}{p^2 - \lambda} u_\lambda(p), \tag{18.125}$$

where $u_\lambda(p) = u_\lambda(p + n)$ is periodic and, as an element of $L_2(\mathbf{T})$, satisfies

$$(Q + W_\lambda) u_\lambda = 0. \tag{18.126}$$

A given λ will be an eigenvalue of H if and only if (18.126) admits a nontrivial solution u_λ. Now note that the operator on the left-hand side of (18.126) is in fact of the same form as the operator (18.1)–(18.4). Therefore, it satisfies the following representation, corresponding to (18.5)–(18.7):

$$Q + W_\lambda = (W_\lambda + i)\left(I - \frac{W_\lambda - i}{W_\lambda + i}\kappa V\right)(I - \kappa V)^{-1}. \tag{18.127}$$

As in the discrete case, introduce the functions $f(p, \lambda)$, $f_0(\lambda)$ and $t_\lambda(p)$, corresponding to (18.3) and (18.38):

$$f(p,\lambda) = \ln C(p,\lambda), \qquad C(p,\lambda) = \frac{W_\lambda(p) - i}{W_\lambda(p) + i}; \tag{18.128}$$

$$f_0(\lambda) = \int_{\mathbf{T}} f(p,\lambda)\,dp = im(\lambda), \qquad m(\lambda) \in \mathbf{R}; \tag{18.129}$$

$$t_\lambda(p) = (I - V)^{-1}\big(f(p,\lambda) - f_0(\lambda)\big). \tag{18.130}$$

If $f(p,\lambda)$ is differentiable infinitely often in p, so is $t_\lambda(p)$. This is easy to check by considering f and t_λ as elements of $\ell_2(\mathbf{Z}^d)$ and using the important condition (18.21) for the vector α.

We now explain the role of the parameter ρ in the formation of the spectrum of H. From definition (18.123) we get

$$W_\lambda(p) = \sum_{n\in\mathbf{Z}^d} \frac{\hat{\phi}^2(p+n)}{(p+n)^2 - \lambda}$$

for $p \in \mathbf{T}$. For $\rho < \frac{1}{4}$, the sum on the right has only one nonzero term, corresponding to $n = 0$. Therefore, for any real λ, the set of values of the function $W_\lambda(p)$ has a gap: $W_\lambda(p) \notin \Delta_\lambda$, where Δ_λ is a certain interval in $\mathbf{R}$. This means that the branch of the log function can be chosen so that $f(p,\lambda)$ in (18.128) is smooth in p, just as in the discrete case (18.1)–(18.4).

If $\rho > \frac{1}{4}$ and $d \geqslant 2$, and λ is sufficiently large, the set of values of $W_\lambda(p)$ is $\mathbf{R} \cup \infty$. To check this, we introduce some notation: for $O \subset \mathbf{R}^d$, set

$$(O)_{\mathbf{T}} = \{p \in \mathbf{T} : p + n \in O \text{ for some } n \in \mathbf{Z}^d\};$$
$$O_\rho = \{p \in \mathbf{R}^d : |p^2| = \rho\};$$
$$\tilde{O}_\rho = \bigcup_{n\in\mathbf{Z}^d} \big(O_\rho \cap (O_\rho + n)\big)_{\mathbf{T}}.$$

Then, for $\lambda > 0$ large enough, in any neighborhood of any point of $\tilde{O}_p$, the function $f(p,\lambda)$ takes on all values in $\mathbf{R}\cup\infty$. This means that $f(p,l)$ will not be continuous everywhere in $\tilde{O}_p$, and we have no guarantee that there exists a function $t_\lambda(p)$ measurable in p that satisfies the homological equation and is therefore correctly specified by (18.130). Even if $t_\lambda(p)$ exists, it is quite irregular and assumes all its values in the neighborhood of any point $p \in \mathbf{T}$. This difficulty in finding a solution of the homological equation (18.39) for $d \geqslant 2$ and large λ is closely associated with the absence of a point component in this part of the spectrum.

However, in the one-dimensional case the function $f(p,\lambda)$ is smooth for all ρ, including $\rho = \infty$. For if we set $\mu = \frac{1}{2} - \left|\{\lambda\} - \frac{1}{2}\right|$, where $\{\lambda\}$ is the fractional part of λ, we find that $W_\lambda(p)$, for $p \in \mathbf{T} = [-\frac{1}{2}, \frac{1}{2})$, is an even function, smooth when $|p| \neq \mu$. Moreover, $\left|W_\lambda(\pm\mu)\right| = \infty$, and on the intervals $(-\mu, \mu)$ and $\mathbf{T} \setminus [-\mu, \mu]$ the function $W_\lambda(p)$ takes on values in the intervals $(\pm\infty, C_1)$ and $(\mp\infty, C_2)$, respectively, where C_1 and C_2 are finite.

Together with (18.128), this shows that the branch of the log can be chosen so that $f(p, \lambda)$ is smooth. This means that, in the one-dimensional case, for $\rho = \infty$ (and for $\phi(x) = g\delta(x)$), the spectrum of H is the same as that of the discrete operator (18.1), and in particular it is pure point.

Returning now to the analysis of the point component of the spectrum of (18.116), we obtain from (18.127)–(18.130) the following representation similar to (18.48):

$$Q + W_\lambda = (W_\lambda + i)e^{t_\lambda}(I - e^{f_0(\lambda)}\kappa V)e^{-t_\lambda}(I - \kappa V)^{-1}. \tag{18.131}$$

If we define the function $v_\lambda(\rho)$ by a relation similar to (18.52), it follows from (18.131) that (18.126) can be transformed into an equation similar to (18.53). From this and from (18.129) it is already clear that the eigenvalues of H are the solutions of (18.49). One can check that for every $n \in \mathbf{Z}^d$ this equation has the unique solution λ_n, which means that the point spectrum of H is simple. For each such λ_n, we evidently have $v_{\lambda_n} = e^{2\pi i(n,p)}$, so (18.52) gives

$$u_{\lambda_n}(p) = (I - \kappa V)e^{t_{\lambda_n}(p)}e^{2\pi i(n,p)}.$$

Using (18.130), we can represent u_{λ_n} as

$$u_{\lambda_n}(p) = -\frac{2i}{W_{\lambda_n}(p) - i}e^{t_{\lambda_n}(p)}e^{2\pi i(n,p)}.$$

By substituting either of these expressions for u_{λ_n} into (18.125), we arrive at the desired eigenfunctions.

The representation (18.119a) for $n(\lambda)$ is obtained from the equality (18.124) for the resolvent. Indeed, formula (5.11)—whose use is justified for this operator—together with the results of the Appendix gives

$$n(\lambda) = \lim_{\varepsilon \downarrow 0} \frac{1}{\pi} \int_0^1 \operatorname{Im}(\psi, R(\lambda + i\varepsilon)\psi)\, d\omega \bigg|_{\hat{\psi}(p)=1}. \tag{18.132}$$

From (18.127), with λ replaced by $z = \lambda + i\varepsilon$, we get

$$\int_0^1 (Q + W_z)^{-1}\, d\omega = (W_z + i)^{-1}.$$

Thus, in view of (18.124),

$$\int_0^1 \hat{\psi}^*(p)\big(R(z)\hat{\psi}\big)(p)\, d\omega \bigg|_{\hat{\psi}(p)=1} = \frac{1}{p^2 - z} - \frac{\hat{\phi}(p)}{p^2 - z}\frac{1}{W_z(p) + i}\left(\frac{\hat{\phi}(p)}{p^2 - z}\right)_{\mathbf{T}}.$$

This, together with (18.132), gives (18.119a) after simple transformations.

For more details on the proofs of Theorems 18.15 and 18.18, as well as related results, see [Figotin and Pastur 1990].

Problems

1. Let $\mathcal{A}$ be the metric space of bounded Hermitian operators on some Hilbert space, with the metric $\rho(A, B) = \|A - B\|$. Prove that the set $\mathcal{N} \subset \mathcal{A}$ of operators whose spectrum is nowhere dense is a G_δ (a countable intersection of dense, open sets). Prove a similar fact for the Schrödinger operator $-\Delta + q$, where $q \in C(\mathbf{R}^d)$ and $\|q\| = \sup_x |q(x)|$.

Hint. Write $\mathcal{N}$ as the intersection, over all intervals (a, b) with rational endpoints, of the set of operators whose resolvent set is disjoint from (a, b).

2. For a periodic potential, the set of zeros of the sine solution (12.11) of the Schrödinger equation or its discrete analogue coincides with the spectrum of the Dirichlet problem on the period.

3. Prove that, for a periodic potential, the function $w(z)$ of (12.54a) is related to the quasimomentum by $w(z) = ik(z)$.

Hint. Compare (16.1) and (12.12) and show that

$$m_\pm(z) = ik(z) \pm \left.\frac{u'_\pm}{u_\pm}\right|_{x=0};$$

then use Problem V.26.

4. Prove the existence of the Floquet–Bloch solutions (16.1) for $\operatorname{Im} z \neq 0$ for a periodic potential.

Hint. Integrate (12.46) with respect to x.

5. Suppose given an $L \times M$ matrix $\alpha : \mathbf{R}^L \to \mathbf{R}^M$; points $\omega = (\omega_1, \ldots, \omega_M) \in \mathbf{T}^M$ and $\tilde{\omega} = (\tilde{\omega}_1, \ldots, \tilde{\omega}_L) \in \mathbf{T}^L$; and two functions $A(s)$ and $B(q)$ on $\mathbf{Z}^L$ and $\mathbf{Z}^M$, respectively, with finite supports $S = \operatorname{supp} A$ and $Q = \operatorname{supp} B$, and such that $A^*(s) = A(-s)$ and $B^*(q) = B(-q)$. Consider the finite-difference operators

$$(h_{AB}u)(x) = \sum_{s \in S} A(s)u(x+s) + \sum_{q \in Q} e^{2\pi i(q, \alpha x + \omega)} B(q)u(x),$$

$$(\tilde{h}_{BA}v)(y) = \sum_{q \in Q} B(q)v(y+q) + \sum_{s \in S} A(s)e^{2\pi i(s, \alpha^T y + \tilde{\omega})} v(s)$$

for $x \in \mathbf{Z}^L$ and $y \in \mathbf{T}^M$, where α^T is the transpose of α. Prove that h_{AB} and $\tilde{h}_{BA}$ are metrically transitive operators with the same integrated density of states.

Hint. Approximate α with a matrix of rational entries, and use the form (16.1) of the Floquet–Bloch generalized eigenfunctions in the resulting periodic operator so as to obtain the desired relation for periodic operators. Then use Corollary 4.19 to pass to the limit of the original matrix α with rationally independent elements. For another method, see [Bellissard and Testard 1985].

6. Take an almost-periodic potential of the form (6.25), where $k_0 > 0$ and the points x_j are constructed by the rule $x_j = f(\alpha j)$, for $j \in \mathbf{Z}$, for α irrational and $f(t)$ a nonnegative periodic function of period 1, monotonically increasing in the period and satisfying $f(t) \sim C(1-t)^{-1/\delta}$ as $t \uparrow 1$. Prove that:

(a) the bottom of the spectrum is 0;
(b) if $\delta > 1$, and we set $a = \lim_{n\to\infty} n^{-1} \sum_{j=1}^{n} f(\alpha j)$, we have

$$N(\lambda) = a\left(\frac{C\sqrt{\lambda}}{\pi}\right)^{\delta} \zeta(\delta)(1 + o(1))$$

as $\lambda \downarrow 0$, where $\zeta(\delta)$ is the Riemann zeta function;
(c) if $0 < \delta < 1$, we have $N(\lambda) = N_0(\lambda) = \pi^{-1}\sqrt{\max\{\lambda, 0\}}$, the integrated density of states of the operator $-d^2/dx^2$.

Hint. Use Theorem 6.7.

7. Prove that under the conditions of Theorem 16.6, the Floquet–Bloch solutions (16.37) exist for all $\operatorname{Im} z \neq 0$.

Hint. Use the relation between the Weyl functions (12.46) and the Green's function of the Neumann problem (see (12.18)), and prove that $m_{\pm}(z)$, with $z = \lambda + i\varepsilon$, is an analytic function in the domain (16.34) for $\rho = \rho(\varepsilon)$, with $\lim_{\varepsilon\to 0} \rho(\varepsilon) = 0$. Then argue as in Problem VII.4.

8. Let h be a discrete Schrödinger operator (16.53) on $\ell_2(\mathbf{Z})$ whose potential q satisfies $q(x \pm a_n) = q(x)$, for $0 \leqslant x \leqslant a_n$, where $\{a_n\}$ is an infinitely increasing sequence of natural numbers. Prove that, if $y(x)$ is a solution of the equation $hy = \lambda y$ and we set $Y(x) = \big(y(x+1), y(x)\big)$, we have

$$\max\{\|Y(\pm a_n)\|, \|Y(2a_n)\|\} \geqslant \tfrac{1}{2}\|Y(0)\|.$$

Hint. Use Lemma 16.8.

9. Consider the operator h of the form (16.53) with a potential q given by (16.21), the function $Q(t)$ being equal to $\sum_{n\in\mathbf{Z}} \chi_{(t_1,t_2)}(t+n)$, where $(t_1, t_2) \subset (0,1)$. Show that the set of α for which the point spectrum of h is empty for Lebesgue-almost all $\omega \in [0,1)$ has full Lebesgue measure.

Hint. If $\alpha = (\alpha_1, \alpha_2, \ldots)$ is the continued-fraction expansion of α, let $b_n/a_n = (\alpha_1, \ldots, \alpha_n)$ be the n-th convergent, and $X_j(n)$, for $j = 1, 2$, the set of $\omega \in [0, 1)$ for which $|(\alpha n + \omega - t_j) \pmod 1| \geqslant |\alpha a_n - b_n|$. For $\omega \in X_1(n) \cap X_2(n)$, apply the preceding problem for the sequence of denominators $\{a_n\}$. Next, verify that $\operatorname{meas} X_j(n) \geqslant 1 - 2a_n |\alpha a_n - b_n|$. Then use the relations $|\alpha a_n - b_n| \leqslant a_{n+1}$ and $a_{n+1} \geqslant \alpha_n a_n$, and the fact that the set of α such that $\limsup_{n\to\infty} \alpha_n = \infty$ has full Lebesgue measure [Khintchine 1963].

10. Extend the result of the preceding problem to the case of a function $Q(t)$, periodic of period one and taking on only a finite number of values.

11. Show that Lemma 16.8 can be extended to an invertible matrix B or arbitrary order n, as follows:

$$\max_{m=\pm 1, \ldots, \pm n} \|B^m \xi\| \geqslant \frac{1}{n} \|\xi\|,$$

for $\xi \in \mathbf{C}^n$. Verify that the constant $1/n$ in this inequality cannot be improved on.

12. Let A_n be a finite-difference operator on $\ell_2(\mathbf{Z})$,

$$(A_n \psi)(x) = \sum_{k=-l}^{n} q_k(x) \psi(x + k)$$

for $x \in \mathbf{Z}$, where $l, n > 0$. Suppose that $\sup_x |q_k(x)| < \infty$, and that $\inf_x q_k(x) > 0$ for $k = -l, n$. As in (16.54), set

$$\delta_q(a) = \inf_{q_a} \sup_{|x| \leqslant (l+n)a} \|q(x) - q_a(x)\|_{\mathbf{C}^{n+l-1}},$$

where $q(x) = (q_{-l}(x), \ldots, q_n(x))$, and the infimum is taken over the set of all vector functions $q_a(x)$ on $\mathbf{Z}$ periodic of period a. Prove that Theorem 16.7 remains valid if

$$\liminf_{a\to\infty} \frac{1}{a} \ln \delta_q(a) = -\infty.$$

13. Show that the quantity $\delta_q(a)$ of the previous problem and the quantity

$$\delta_q^*(a) = \max_{-(n+l)a \leqslant x \leqslant (n+l-1)a} |q(x) - q(x + a)|$$

satisfy

$$C^{-1} \leqslant \frac{\delta_q^*(a)}{\delta_q(a)} \leqslant C,$$

where the constant $C > 0$ does not depend on q.

14. Show that, if we replace the condition

$$\lim_{a\to\infty} \frac{1}{2a} \int_{-a}^{a} |q(x)|\, dx \leqslant M < \infty$$

in Theorem 16.10 by the condition that there exists a sequence $a_k \to \infty$ such that

$$\delta_q(a_k) \leqslant -A_k a_k - \frac{1}{2} \int_{|x|\leqslant 2a_k} |q(x)|\, dx,$$

where A_k is defined in the proof of Lemma 16.11, the theorem remains valid.

15. Let A_n be a differential operator on $L_2(\mathbf{R})$:

$$A_n = \left(\frac{d}{dx}\right)^n + q_{n-1}(x)\left(\frac{d}{dx}\right)^{n-1} + \cdots + q_0(x),$$

where $q_{n-1}(x), \ldots, q_0(x)$ are measurable, locally summable, complex-valued functions (concerning such operators, see [Naimark 1969]). As in (16.58), set

$$\delta_q(a) = \inf_{q_a} \int_{-na}^{na} \|q(x) - q_a(x)\|_{\mathbf{C}^n}\, dx,$$

where $q = \{q_{n-1}, \ldots, q_0\}$ and the infimum is taken over the set of vector functions q_a periodic of period a. Prove that Theorem 16.10 still holds if:

(1) $\limsup_{a\to\infty} (2a)^{-1} \int_{-a}^{a} |q_j(x)|\, dx < \infty$ for $j = 0, \ldots, n-1$;
(2) $\sup_{x\in\mathbf{R}} \int_x^{x+1} |q_{n-1}(x')|^p\, dx' < \infty$ for some $p > 1$;
(3) $\liminf_{a\to\infty} a^{-1} \ln \delta_q(a) = -\infty$.

16. Let Q be the function on the torus $\mathbf{T}^2$ given by

$$Q(\xi_1, \xi_2) = \sum_{n=1}^{\infty} Q_n \sin(j_n\xi_1 - k_n\xi_2),$$

where $j_n = (n-1)!$, $k_n = j_n \sum_{m=0}^{n-1} (-1)^m/m!$, and

$$Q_n = (j_n - k_n\alpha)\varepsilon_n^\delta$$

for $\alpha = e^{-1}$, $\delta \in (0,1)$ and $\varepsilon_n = |j_n - k_n\alpha|$. Show that the quasiperiodic function $q(x) = Q(x, \alpha x)$ is real analytic and that

$$g(x) = \int_0^x q(t)\, dt \geqslant C|x|^{1-\delta}$$

for $|x| \geqslant 1$ and some $C > 0$. Use this to verify that $\psi(x) = e^{-g(x)}$ is an $L^2(\mathbf{R})$-solution of the Schrödinger equation with real analytic potential $f(x)$. In other words, $q(x)$ is an eigenfunction of the Schrödinger operator with this potential, although the function $Q(\xi_1, \xi_2)$ is just continuous, not smooth, in the representation (16.29).

17. Using Lemma 16.12, prove the following estimates for the Lyapunov exponent of equation (6.69):

(a) If $s(x) = g_1 + 2g_3\cos\bigl(2\pi(x-\frac{1}{2})\alpha+\omega\bigr)$ and $q(x) = 2g_2\cos(2\pi\alpha x+\omega)$, we have

$$\gamma(\lambda) \geqslant \frac{g_2+\sqrt{g_2^2-4g_3^2}}{g_1+\sqrt{g_1^2-4g_3^2}}.$$

(b) If $s(x) = 2g_2\cos(2\pi\alpha x+\omega)$ and $q(x) = 2g_1\cos\bigl(2\pi(x-\frac{1}{2})\alpha+\omega\bigr)$, we have

$$\gamma(\lambda) \geqslant \ln\frac{g_1}{g_2}.$$

These two sets of coefficients correspond to the motion of an electron in a two-dimensional crystal with a homogeneous magnetic field perpendicular to it, in the case of a square anisotropic and a triangular anisotropic lattices, respectively [Thouless 1983].

18. Prove, using Lemma 16.12, that the Lyapunov exponent of the equation

$$-\psi(x+1)-\psi(x-1)+2\psi(x) = \lambda m(x)\psi(x)$$

with $m(x) = 1+\varepsilon\cos(2\pi\alpha x+\omega)$, for $|\varepsilon|<1$, satisfies $\gamma(\lambda) \geqslant \ln|\lambda\varepsilon|$.

19. Prove that the Lyapunov exponent of the equation

$$-s(x+1)\bigl(\psi(x+1)-\psi(x)\bigr)-s(x)\bigl(\psi(x)-\psi(x-1)\bigr) = \lambda\psi(x),$$

where $s(x) = 1+\varepsilon\cos(2\pi\alpha x+\omega)$ for $|\varepsilon|<1$, satisfies

$$\gamma(\lambda) \geqslant \ln\bigl(2\varepsilon(1+\sqrt{1-\varepsilon^2})^{-1}\bigr).$$

Hint. Define the Lyapunov exponent by writing $r^2(x) = \Delta^2(x)+\psi^2(x)$, where $\Delta(x) = s(x)\bigl(\psi(x)-\psi(x-1)\bigr)$, and use Lemma 16.12.

20. Prove that the Lyapunov exponent of the Schrödinger operator with potential

$$q(x) = \sum_{n\in\mathbf{Z}} 2g\cos\bigl(2\pi(n\alpha+\omega)\bigr)\delta(x-n),$$

for $x\in\mathbf{R}$, where $g>0$, is positive for all $\lambda\neq\pi^2n^2$.

Hint. Use (14.22) and Theorem 16.13.

21. Obtain a lower bound for the Lyapunov exponent of discrete, order-two operators with nondiagonal disorder like those of Problems VII.17 and VII.19, by using the Herbert–Jones–Thouless formula (11.82) and the duality considered in Problem VII.5.

22. Find the integrated density of states of the quasiperiodic operator of order two considered in Example 16.18(c), in the case where $\Lambda(t)$ is a periodic function of period 1, equal to t for $0 < t < \beta < 1$ and to $t+1$ for $\beta \leqslant t < 1$. Show that there exists $\beta \in (0,1)$ such that Theorem 16.2 (gap labeling) is not valid.

23. Construct an example of a quasiperiodic operator with an isolated eigenvalue. Verify that such an operator exists only on a subset of of measure zero of the potential's hull, so that, as expected, the probability of such an event is zero.

Hint. For $d = 1$, consider the function $\Lambda(t)$ of the preceding problem, and redefine it for $t = \beta$, so that $\beta < \Lambda(\beta) < \beta + 1$ with $\beta = \alpha n$, for $n \in \mathbf{Z}$.

24. The papers by Craig [1983] and Pöschel [1983] treat a wider class of sequences $\lambda(x)$ than we do in Section 16, including some sequences satisfying $|\lambda(x) - \lambda(y) \geqslant \Omega(|x-y|)$, where

$$\Omega(t) = \begin{cases} e^{-Ct/(\ln t)^{1+\beta}} & \text{for } t > e, \\ e^{-Ce} & \text{for } 0 \leqslant t \leqslant e, \end{cases}$$

with $\beta > 0$. Construct, with the help of Example 16.18(c), an operator with a quasiperiodic potential such that $N(\lambda)$ does not satisfy (11.83).

Hint. Take a nondecreasing function $N(\lambda)$, with $N(0) = 0$ and $N(1) = 1$ and satisfying

$$C_1 \exp\bigl(-C_0 |N(\lambda) - N(\lambda')|^{-1} (-\ln|N(\lambda) - N(\lambda')|)^{-1-\beta}\bigr) \leqslant |\lambda - \lambda'|$$

for $|\lambda - \lambda'| \leqslant \frac{1}{2}$, and verify that the function $\Lambda(t)$ corresponding to this $N(\lambda)$ by (16.88) leads to

$$\bigl|\lambda(x) - \lambda(y)\bigr| \geqslant \text{const}\, \Omega\bigl(|x-y|\bigr).$$

25. Show that:

(a) if $f_1(x)$ and $f_2(x)$, for $x \in \mathbf{R}$, are continuous periodic functions with incommensurate periods, we have

$$\Bigl|\sup_{x \in \mathbf{R}} |f_1(x) - f_2(x)|\Bigr| = \max\bigl\{|\max_x f_1(x) - \min_x f_2(x)|, |\min_x f_1(x) - \max_x f_2(x)|\bigr\};$$

(b) a function $f(x)$, for $x \in \mathbf{R}$, is limit-periodic if and only if its module M_f is contained in $\beta\mathbf{Q}$, where β is a real number;

(c) if $f(x)$, for $x \in \mathbf{R}$, is a limit-periodic function, there exist periodic functions $f_l(x)$ of period 1 and natural numbers $n_1, n_2, \ldots$ such that $\|f_l\|_\infty \leqslant 2^{-(l-2)} \|f\|_\infty$ and

$$f(x) = \sum_{l=1}^{\infty} f_l \left(\frac{x}{n_1 \cdots n_l} \right),$$

where the series converges uniformly on the whole axis;

(d) if a function is both limit-periodic and quasiperiodic, it is periodic.

26. Prove Lemma 17.17 for the d-dimensional Schrödinger operator under the condition

$$\|q\|_{S^p} = \sup_{x \in \mathbf{R}^d} \int_{C+x} |q(x')|^p \, dx' < \infty,$$

where C is the unit cube in $\mathbf{R}^d$ centered at zero, and $p = 2$ for $d \leqslant 3$, $p > 2$ for $d = 4$, and $p = \frac{1}{2}d$ for $d > 4$.

Hint. (i) Extend the proof of (17.90) to the multidimensional case by using the explicit form of the Green's function of the Laplacian. (ii) Use the inequality

$$\|q\psi\|^2 \leqslant \|q\|_{S^p}^2 (\varepsilon \|\Delta\psi\|^2 + C(\varepsilon)\|\psi\|^2)$$

[Reed and Simon IV, Theorem XIII.96], which is valid for any $\varepsilon > 0$ and some constant $C(\varepsilon)$.

27. Prove that the Floquet–Bloch solution (17.12) exists, for nonreal z, if the limit-periodic potential $q(x)$ is approximable in the Besicovitch metric (Theorem 17.6) by a_n-periodic functions $q_n(x)$ at the rate $\|q - q_n\|_B \leqslant \text{const}\, a_n^{-2-\delta}$, for $\delta > 0$.

Hint. By arguments similar to those used to prove Lemma 17.17, estimate the difference $|m^{(n+1)} - m^{(n)}|$ of the Weyl functions for the potentials q_{n+1} and q_n in terms of $\|q_{n+1} - q_n\|_B$, and proceed as in Problems VII.4 and VII.7.

28. Let W be the class of function $w(z)$ such that w, $-iw$ and w' are Nevanlinna functions, and let $w_1, w_2 \in W$. Assume that

$$-iw_j(z)\sqrt{z} = z - \frac{\sigma_j^2}{2z} + O(|z|^{-1})$$

as $\operatorname{Im} z \to \infty$, for $j = 1, 2$. Show that $\sigma_1^2 \geqslant \sigma_2^2$ if $w_1(\mathbf{C}_+) \subset w_2(\mathbf{C}_+)$.

Hint. Use Problem V.22 and consider the functions $-u_j(z) = w_j^2(z)$ for which $u_j(z) = z - \sigma_j^2/z + o(|z|^{-1})$. Therefore $u(z) = u_2^{-1}(u_1(z))$ is a Nevanlinna function, and $u(z) = z - \delta z^{-1} + o(|z|^{-1})$, for $\delta = \sigma_1^2 - \sigma_2^2$. Then use the representation (A.12) for Nevanlinna functions.

29. Let $w \in W$ (see previous problem) be such that

$$-iw(z)\sqrt{z} = z - \frac{\sigma^2}{2z} + o(|z|^{-1})$$

as $\operatorname{Im} z \to \infty$. Show that there exists a sequence of functions $w_n \in W$ converging to w uniformly on compacts of $\mathbf{C}_+$, such that the boundary of $w_n(\mathbf{C}_+)$ consists of a finite number of cuts parallel to the real axis and that

$$-iw_n(z)\sqrt{z} = z - \frac{\sigma_n^2}{2z} + o(|z|^{-1}),$$

with $\sigma_n^2 \leqslant \sigma^2$.

Hint. Let a_n, for $n \geqslant 1$, be an arbitrary dense set in $i\mathbf{R}$, and D_n a simply connected domain in the quadrant $\{w \in \mathbf{C} : \operatorname{Im} w > 0, \operatorname{Re} w < 0\}$, whose boundary consists of line segments parallel to the x-axis and connecting a_n with the boundary $w(\mathbf{C}_+)$. Take w_n as a conformal map from $\mathbf{C}_+$ onto D_n, so that $w_n(\mathbf{C}_+) = D_n \supset w(\mathbf{C}_+)$. Then use Problmes V.25 and VII.28 and Caratheodory's theorem [Duren 1983] on the convergence of conformal maps (see [Kotani 1987b] for details).

30. Let T be an ergodic automorphism of the circle $[0, 1)$ whose invariant measure is the Lebesgue measure. Show that the integrated density of states of the operator $A = A_0 + Q$, where A_0 is a Toeplitz operator and Q the operation of multiplication in $\ell^2(\mathbf{Z})$ by the sequence $q(x) = g \tan \pi T^x \omega$, for $x \in \mathbf{Z}$ and $\omega \in [0, 1)$, coincides with (18.18).

31. Extend (18.18) for $d = 1$ to the case where the potential is

$$q(x) = \sum_{j=1}^{n} g_j \tan(\alpha_j x + \omega_j)\pi,$$

the numbers $1, \alpha_1, \ldots, \alpha_n$ being rationally independent. More precisely, show that

$$N(A, \lambda) = \int_{\mathbf{R}} N(A_0, \lambda - \mu) \frac{g}{\pi(g^2 + \mu^2)} d\mu,$$

where $g = |g_1| + \cdots + |g_n|$.

32. Let $q(x) = \sum_{a \in \mathbf{Z}^d} \tan \pi\big((\alpha, a) + \omega\big) u(x - a)$, for $x \in \mathbf{Z}^d$, where $u(x) \geqslant 0$ and $\sum_{x \in \mathbf{Z}^d} u(x) < \infty$. Show that $N(\lambda)$ for the operator of Problem VII.30 has the same form as that of Theorem 18.1, with g replaced by $\sum_{x \in \mathbf{Z}^d} u(x)$ (cf. Problem II.25).

33. Show that, if in the operator (18.1) with $d = 1$ we have $a_0(x) = a^{|x|}$ for $0 < a < 1$, the quantity $\gamma_-(\lambda, g)$ of (18.29) and Theorem 18.2 (which coincides, for an order-two operator, with the Lyapunov exponent $\gamma(\lambda, g)$ of Problem V.14) is given by

$$\gamma_-(\lambda, g) = \Gamma\left(-\frac{1+a^2}{a} + \frac{1-a^2}{a(1-z)}\right),$$

with $z = \lambda + ig = 2\cos\alpha$ and $\Gamma(z) = \operatorname{Im}\alpha > 0$, whereas $\gamma(\lambda, g) = \Gamma(\lambda + ig)$.

34. Prove that the Lyapunov exponent of the one-dimensional order-two operator (18.1) is the rate of exponential decrease of the mathematical expectation of the operator's Green's function. (According to [Lifshitz et al. 1982b], these two quantities can differ by as much as a factor of four for random operators.)

Appendix A: Nevanlinna Functions

We start by recalling the properties of the Stieltjes transform of (generally complex-valued) measures m on the real axis that satisfy

$$\text{(A.1)} \qquad |m(dt)| \leqslant m_+(dt), \qquad \int_{\mathbf{R}} \frac{m_+(dt)}{1+|t|} < \infty.$$

Throughout this appendix we assume that (A.1) holds.

(A.1) Definition. *The Stieltjes transform of the measure $m(dt)$ is the function*

$$\text{(A.2)} \qquad f(z) = \int_{\mathbf{R}} \frac{m(dt)}{t-z}$$

which, by (A.1), *is defined for all nonreal z.*

(A.2) Theorem (The Stieltjes-Perron inversion formula [Akhiezer 1964]). *If Δ is an open interval whose endpoints are not atoms of $m(d\lambda)$,*

$$\text{(A.3)} \qquad m(\Delta) = \lim_{\varepsilon \downarrow 0} \frac{1}{\pi} \int_{\Delta} \big(f(\lambda + i\varepsilon) - f(\lambda - i\varepsilon)\big) \frac{d\lambda}{2i}.$$

If the measure m is real, formula (A.3) can be written as

$$\text{(A.4)} \qquad m(\Delta) = \lim_{\varepsilon \downarrow 0} \frac{1}{\pi} \int_{\Delta} \operatorname{Im} f(\lambda + i\varepsilon)\, d\lambda.$$

If, in addition, the measure has a density $m(dt) = m'(t)\,dt$ satisfying the Hölder condition $|m'(dt_1) - m'(dt_2)| \leqslant C|t_1 - t_2|^{\alpha}$, the limit

$$\text{(A.5)} \qquad f(\lambda \pm i0) = \lim_{\varepsilon \downarrow 0} f(\lambda \pm i\varepsilon)$$

exists for all $\lambda \in \mathbf{R}$, and we have the Sokhotski–Plemelj formula [Muskhelishvili 1962]

$$\text{(A.6)} \qquad f(\lambda \pm i0) = \pm i\pi m'(\lambda) + \int_{\mathbf{R}} \frac{m'(t)}{t-z}\, dt,$$

where the integral is in the sense of the Cauchy principal value. In this case, the inversion formulas (A.3) *and* (A.4) *become local.* □

The next result follows from the inversion formula:

(A.3) Theorem uniqueness. *If the Stieltjes transforms f_1 and f_2 of two measures m_1 and m_2 coincide for all nonreal z, the measures are equal.* □

To formulate an important improvement of this theorem, we need the following result, due to Fatou.

(A.4) Theorem [Koosis 1984]. *The limiting values* (A.5) *of the Stieltjes transform exist and are finite for almost all values $\lambda \in \mathbf{R}$ with respect to Lebesgue measure:*

$$\left|f(\lambda + i0)\right| < \infty. \tag{A.7}$$ □

(A.5) Theorem. *If f_1 and f_2 are the Stieltjes transforms of the measures m_1 and m_2, and $f_1(\lambda + i0) = f_2(\lambda + i0)$ for all λ in a set of positive Lebesgue measure, we have $m_1 = m_2$. In particular, if $f \not\equiv 0$, we have*

$$f(\lambda + i0) \neq 0 \tag{A.8}$$

for almost all $\lambda \in \mathbf{R}$. □

This is a corollary of the well-known Luzin–Privalov uniqueness theorem [Privalov 1950; Koosis 1984] and Theorem A.3.

(A.6) Theorem (continuity).

(i) *Suppose a sequence of measures m_n, for $n \geqslant 1$, satisfying* (A.1) *and*

$$\lim_{c \to \infty} \sup_{n \geqslant 1} \int_{|t| \geqslant c} \frac{m_n^+(dt)}{|t|} dt = 0, \tag{A.9}$$

converges weakly to a measure m. Then the sequence $f_n(z)$, for $n \geqslant 1$, of their Stieltjes transforms converges, for every nonreal z, to the Stieltjes transform of m.

(ii) *Suppose a sequence of measures m_n satisfying* (A.1) *and* (A.9), *is such that the Stieltjes transforms f_n converge to a function $f(z)$ for every nonreal z. Then the measures m_n converge weakly to a limiting measure m that satisfies* (A.1) *and* (A.9), *and f is the Stieltjes transform of m.*

□

For nonnegative measures we have a simple characterization of the set of Stieltjes transforms.

(A.7) Definition. *A function $f(z)$ of a complex variable is called a* Nevanlinna function *if $f(z)$ is analytic for nonreal z, $f^*(z) = f(z^*)$, and $\operatorname{Im} z \operatorname{Im} f(z) > 0$ for $\operatorname{Im} z \neq 0$. The set of Nevanlinna functions is denoted by $\mathcal{N}$.*

(A.8) Theorem [Akhiezer 1964; Kac and Krein 1968]. *A function $f(z)$ defined for $\operatorname{Im} z \neq 0$ is the Stieltjes transform of a nonnegative measure m satisfying*

$$\int_{\mathbf{R}} \frac{m(dt)}{1+|t|} < \infty \tag{A.10a}$$

or

$$m(\mathbf{R}) < \infty \tag{A.10b}$$

if and only if f is a Nevanlinna function and satisfies the conditions

$$\int_1^\infty \operatorname{Im} f(it) \frac{dt}{t} < \infty \tag{A.11a}$$

or

$$\sup_{t>0} t \operatorname{Im} f(it) < \infty, \tag{A.11b}$$

respectively. In addition,

$$\sup_{t>0} t \operatorname{Im} f(it) = \lim_{t\to\infty} t f(it) = m(\mathbf{R}). \tag{A.12}$$ □

For us it is important to be able to characterize the properties of m in terms of the limiting values of its Stieltjes transform. One such characterization is based on classical results of Fatou and de la Vallée–Poussin.

(A.9) Definition. *A measure is said to be concentrated on a set $X \subset \mathbf{R}$ if* $m(\mathbf{R} \setminus X) = 0$. (Notice that X is not necessarily the support of m.)

Recall also that any measure on $\mathbf{R}$ can be decomposed into its absolutely continuous, singular continuous and point components, which we denote by m_{ac}, m_{sc} and m_p.

(A.10) Theorem [Reed and Simon IV]. (i) *The absolutely continuous component of the measure m is concentrated on the set*

$$X_{ac} = \{\lambda \in \mathbf{R} : f(\lambda + i0) \text{ exists, is finite, and } \operatorname{Im} f(\lambda + i0) \neq 0\}.$$

In addition, $m_{ac}(dt) = (1/\pi)\,\mathrm{Im}\, f(t+i0)\, dt$.

(ii) *The singular component* $m_{sc} + m_p$ *is concentrated on the set*

$$X_s = \{\lambda \in \mathbf{R} : f(\lambda + i0 = \infty)\} \qquad \square$$

(A.11) Theorem. *Let* $f(z) \not\equiv 0$ *be a Nevanlinna function. Then:*

(i) *If* $\mathrm{Re}(\lambda + i0) = 0$ *for all* λ *in a set* X *of positive measure, we have* $\mathrm{meas}\{\sigma_{ac} \cap X = \mathrm{meas}\, X\}$, *where* σ_{ac} *is the support of* m_{ac}.

(ii) *If the sequence* $f_n(z)$, *for* $n \geqslant 1$, *of Nevanlinna functions converges in the upper half-plane* $\mathrm{Im}\, z > 0$ *to a function* $f(z)$ *satisfying (A.11a), and* $\mathrm{Re}\, f_n(\lambda+i0) = 0$ *for* λ *is a set* X *of positive Lebesgue measure, we have* $\mathrm{Re}\, f(\lambda + i0) = 0$ *for almost all* $\lambda \in X$, *and* $m_{ac}(X) > 0$.

(iii) *If at all points of an open interval* Δ *the limit* $\mathrm{Im}\, f(\lambda + i0)$ *exists and is finite, the measure* m *is absolutely continuous on this interval, and* $m(dt) = (1/\pi)\,\mathrm{Im}\, f(t + i0)\, dt$ *for Lebesgue-almost all* $\lambda \in \Delta$.

(iv) *If* $\mathrm{Re}\, f(\lambda + i0) = 0$ *for Lebesgue-almost all* $\lambda \in \Delta$, *the measure* m *is absolutely continuous on* Δ *and its density is strictly positive and real analytic on* Δ.

Proof. Part (i) is a corollary of Theorems A.10(i) and A.5; part (ii) [Kotani 1985] follows from the readily proved weak convergence of $\mathrm{Re}\, f_n(\lambda + i0)$ to $\mathrm{Re}\, f(\lambda + i0)$; part (iii) follows from Theorem A.10, and part (iv) from the same theorem and the Schwartz symmetry principle. $\square$

(A.12) Theorem [Akhiezer 1964]. *For an analytic function on the upper half-plane to be a Nevanlinna function it is necessary and sufficient for it to have a representation*

$$f(z) = \alpha + \beta z + \int_{\mathbf{R}} \frac{1+tz}{t-z} \frac{m(dt)}{1+t^2},$$

with $\alpha \in \mathbf{R}$, $\beta \geqslant 0$ *and* m *a nonnegative measure on* $\mathbf{R}$ *satisfying*

$$\int_{\mathbf{R}} \frac{m(dt)}{1+t^2} < \infty. \qquad \square$$

Appendix B: Distribution of Eigenvalues of Large Random Matrices

We discuss here the asymptotic properties of eigenvalues of $L \times L$ symmetric random matrices with independent entries, in the limit $L \to \infty$. Such matrices, which arise in many mathematical and physical problems [Wigner 1965, Mehta 1967, Pastur 1973, Brody et al. 1981, Girko 1988] differ from the matrices studied in this book in the following respect: their entries all have, roughly speaking, the same order of magnitude (being, for example, independent, identically distributed random variables), instead of tending to zero away from the main diagonal.

This has two important consequences. First, their treatment requires a certain normalization as a function of L, as in formulas (B.1), (B.29), (B.38) and others. Secondly, one can find more or less explicit formulas for a variety of spectral characteristics: in particular, the integrated density of states obeys the so-called semicircle law (B.4). This was discovered by Wigner for matrices whose entries have a Gaussian distribution (such a matrix is known as a Gaussian orthogonal ensemble, or GOE). It was later proved under much weaker, and quite natural, conditions (B.33), similar to those in the hypothesis of the central limit theorem of probability theory: see [Wigner 1965, Mehta 1967, Pastur 1973, Brody et al. 1981, Girko 1988] and references therein.

We also discuss here a more general distribution called the deformed semicircle law, and describe several asymptotic procedures that allow us to obtain this distribution as the limit of the integrated density of states of certain metrically transitive operators. In the conclusion we formulate some results on the repulsion of eigenvalues (see 15.B) for a Gaussian orthogonal ensemble.

B.1. The Semicircle Law. *Let v_L be a symmetric $L \times L$ real matrix of the form*

(B.1) $$v_L(x, y) = L^{-1/2} v(x, y),$$

where $x, y = 1, \dots, L$ and the $v(x, y)$, are Gaussian random variables, independent for different pairs (x, y) (except for the symmetry condition), and such that

(B.2)

$$\mathsf{E}\{v(x,y)\} = 0, \qquad \mathsf{E}\{v^2(x,x)\} = 2, \qquad \mathsf{E}\{v^2(x,y)\} = 1 \quad \textit{for } x \neq y.$$

If we define $N_L(\lambda)$, *as everywhere in this book, as* $1/L$ *times the number of eigenvalues of* v_L *that do not exceed* λ, *we have, with probability* 1:

(B.3)
$$\lim_{L\to\infty} N_L(\lambda) = N_W(\lambda)$$

(*cf.* (3.1)), *where*

(B.4)
$$N'_W(\lambda) \equiv n_W(\lambda) = \frac{1}{2\pi}\chi_{(-2,2)}(\lambda)\sqrt{4-\lambda^2}.$$

Equation (B.4) is known as the semicircle law or Wigner's Law.

Proof. There exist a number of proofs of this important result: see, for example, [Girko 1988] and references therein. We give here a new method that can be applied to a wide range of related problems.

Let $R_L = (v_L - z)^{-1}$ for $z = \lambda + i\varepsilon$, with $\varepsilon \neq 0$, and consider the sequence $r_L(z) = \{r_L^{(l)}(z)\}$ for $l = 1, 2, \ldots$, where

(B.5)
$$r_L^{(l)}(z) = \mathsf{E}\{(L^{-1}\operatorname{Tr} R_L)^l\}.$$

We will need two elementary facts:

(a) If ξ is a Gaussian random variable with $\mathsf{E}\{\xi\} = 0$ and $\mathsf{E}\{\xi^2\} = 1$, and $f(x)$ is a bounded differentiable function, we have

(B.6)
$$\mathsf{E}\{\xi f(\xi)\} = \mathsf{E}\{f'(\xi)\}.$$

(b) If v is an $L \times L$ symmetric real matrix with entries $v(x,y)$ and $R = (v-z)^{-1}$ has entries $R(x,y)$, we have

(B.7)
$$\frac{\partial R(x,y)}{\partial V(t,u)} = -R(x,t)R(u,t) - R(x,u)R(t,y).$$

Using these relations and the identity $R_L = (1/z)(-1 + v_L R_L)$, we find

(B.8)
$$r_L^{(1)} = -\frac{1}{z} - \frac{1}{z} r_L^{(2)} - \delta_L^{(1)},$$

where $\delta_L^{(1)} = \operatorname{Tr} R_L^2/(L^2 z)$. Since

(B.9)
$$|L^{-1}\operatorname{Tr} A| \leqslant \|A\|$$

for any $L \times L$ matrix A, and

(B.10)
$$\|(V-z)^{-1}\| \leqslant |\operatorname{Im} z|^{-1}$$

for any symmetric matrix V, we have the bound

$$|\delta_L^{(1)}| \leqslant \frac{1}{L\varepsilon^3}. \tag{B.11}$$

Similar arguments give

$$r_L^{(l)} = -\frac{1}{z} r_L^{(l-1)} - \frac{1}{z} r_L^{(l+1)} + \delta_L^{(l)} \tag{B.12}$$

for any $l = 2, 3, \ldots$, where

$$|\delta_\lambda^{(l)}| \leqslant \frac{Cl}{L\varepsilon^{l+2}}. \tag{B.13}$$

Together, (B.5), (B.9) and (B.10) imply that $|r_L^{(l)}| \leqslant |\operatorname{Im} z|^{-l}$.

Now consider the Banach space $\mathcal{L}$ of sequences $f_l(z)$, for $l = 1, 2, \ldots$, with the norm

$$\|f\| = \sup_{l \geqslant 1} \sup_{|\operatorname{Im} z| \geqslant 3} |f_l(z)|.$$

Since $|\operatorname{Im} z|^{-l} \leqslant 3^{-l}$ if $|\operatorname{Im} z| \geqslant 3$, the sequence r_L belongs to $\mathcal{L}$. By (B.11) and (B.13), the sequence $\delta_L = \{\delta_L^{(l)}\}$ also belongs to $\mathcal{L}$, and

$$\|\delta_L\| \leqslant CL^{-1}, \tag{B.14}$$

where C does not depend on L. Therefore we can rewrite (B.8) and (B.12) as the following equation in $\mathcal{L}$:

$$r_L = \mathcal{A} r_L + b_L, \tag{B.15}$$

where the operator $\mathcal{A}$ is defined by the relations

$$(\mathcal{A}f)_1 = -\frac{1}{z} f_2, \qquad (\mathcal{A}f)_l = -\frac{1}{z} f_{l-1} - \frac{1}{z} r_{l+1} \quad \text{for } l \geqslant 2, \tag{B.16}$$

$b_L = b + \delta_L$, and

$$b_l = -\frac{1}{z} \delta_{l1}. \tag{B.17}$$

It is easy to see that $\|\mathcal{A}\| \leqslant \frac{2}{\varepsilon} \leqslant \frac{2}{3} \leqslant 1$. Thus the equation $f = \mathcal{A}f + \phi$ is uniquely solvable in $\mathcal{L}$, and $\|f\| \leqslant \frac{1}{3}\|\phi\|$. From this and from (B.15) it follows that the sequence r_L converges to the unique solution $r = \{r^{(l)}\}$ of the equation

$$r = \mathcal{A} r + b \tag{B.18}$$

uniformly in z, for $|\operatorname{Im} z| \geqslant 3$. In particular,

(B.19) $$\sup_{|\operatorname{Im} z| \geqslant 3} \left| r_L^{(l)}(z) - r^{(l)}(z) \right| \leqslant \frac{\text{const}}{L}.$$

But, in view of (B.16) and (B.17), the solution of (B.18) is

(B.20) $$r(z) = r_1^l(z),$$

where r_1 is the solution of the quadratic equation

(B.21) $$r_1^2 + zr_1 + 1 = 0.$$

According to (B.5) and the spectral theorem,

(B.22) $$r_L^{(1)}(z) = \int_{\mathbf{R}} \frac{\mathsf{E}\{N_L(d\lambda)\}}{\lambda - z},$$

that is, $r_L^{(1)}(z)$ is a Nevanlinna function. By (B.19), $r_1(z)$ has the same property, and, in view of (B.21),

(B.23) $$r_1(z) = \frac{1}{z}(-z + i\sqrt{4 - z^2}),$$

where the branch of the square root is taken to be analytic outside the cut $(-2, 2)$ and to have a positive imaginary part just above the cut.

From this and from Theorems A.2 and A.6, we find that

(B.24) $$\lim_{L \to \infty} \mathsf{E}\{N_L(\lambda)\} = N_W(\lambda).$$

We now show that

(B.25) $$\lim_{L \to \infty} \left(\mathsf{E}\{N_L^2(\lambda)\} - \mathsf{E}^2\{N_L(\lambda)\} \right) = 0,$$

which implies the convergence of (B.3) in probability.

Consider the sequence

(B.26) $$r_L^{(l,m)} = \mathsf{E}\{L^{-l-m} (\operatorname{Tr} R_L)^l (\operatorname{Tr} R^*)^m\}.$$

Obviously, $r_L^{(l,0)} = r_L^{(l)}$ and $r_L^{(0,m)} = (r_L^{(m)})^*$. Arguments similar to those used in deriving (B.8) and (B.12) give, for $l \geqslant 1$ and $m \geqslant 0$,

(B.27) $$r_L^{(l,m)} = -\frac{1}{z} r_L^{(l-1,m)} - \frac{1}{z} r_L^{(l+1,m)} + \delta_L^{(l,m)},$$

where for $l = 1$ and $m = 0$ the first term on the right is z^{-1} (see (B.17)), and

$$|\delta_L^{(l,m)}| = O\left(\frac{l + m}{L \varepsilon^{l+m+2}} \right).$$

Now, repeating almost literally the arguments leading to (B.19), we find that

(B.28)
$$\sup_{|\operatorname{Im} z|\geqslant 3} \left|r_L^{(l,m)}(z) - r_1^l(z)(r_1^*(z))^m\right| = O\left(\frac{1}{L}\right).$$

Combining this with (B.19) we obtain (B.25), and the convergence of (B.3) in probability. To prove convergence with probability 1 one must iterate (B.12) and (B.27) once, to find explicity the terms of order L^{-1}. This allows one to prove that these terms cancel in the differences $r_L^{(l,m)} - r_L^{(l)}(r_L^{(m)})^*$, and consequently that the right-hand side of (B.25) has order $O(L^{-2})$. By the Borel–Cantelli lemma, this is equivalent to (B.3) and (B.4). □

The same fact also shows that the rate of convergence in our case is higher than in the probability theory and the spectral theory of metrically transitive operators, where this rate is usually $O(L^{-1})$: see, for example, [Feller 1966] and [Reznikova 1981]. One more manifestation of this higher rate of convergence (the strong self-averaging property) is the relation (B.34) below.

B.2. The Deformed Semicircle Law and Related Topics. We now consider a more general ensemble or random matrix

(B.29)
$$h_L = h_L^{(0)} + v_L,$$

where v_L is given by (B.1)–(B.2) and $h_L^{(0)}$ is some "unperturbed" matrix having a limiting density of states. In particular, $h_L^{(0)}$ may be the restriction to the interval $[1, L]$ of the matrix of some metrically transitive operator acting on $\ell^2(\mathbf{Z})$ and independent of v_L. We studied such operators in Sections 3 and 4.

For the matrix ensemble (B.29) we also have the relation (cf. (B.3))

(B.30)
$$\lim_{L\to\infty} N_L(\lambda) = N(\lambda)$$

with convergence in probability, but instead of (B.4) we now have the functional equation

(B.31)
$$r_1(z) = r_0\big(z + r_1(z)\big)$$

for the Stieltjes transform

$$r_1(z) = \int_{\mathbf{R}} \frac{N(d\lambda)}{\lambda - z}$$

of the limiting integrated density of states, and r_0 is the same transform of the "unperturbed" integrated density of states $N_0(\lambda)$.

Equation (B.31) is uniquely solvable for $\operatorname{Im} z \neq 0$ in the class of Nevanlinna functions, and if $f_0(z) = -1/z$, that is, $N_0'(\lambda) = \delta(\lambda)$, this equation is obviously equivalent to (B.4).

The proof of (B.30) and (B.31) is given in [Pastur 1972b] for $h_0^{(L)}$ diagonal. The proof for arbitrary $h_0^{(L)}$, which will be published elsewhere, is based on the same technique of a finite system of equations for some set of moments of the resolvent. In this case this set consists of the moments

$$r_L^{(l)}(x,y) = \mathsf{E}\{R_L(x,y)(L^{-1}\operatorname{Tr} R_L)^l\},$$

for $l = 0, 1, 2, \ldots$. The corresponding system is again solved with error $O(L^{-1})$ by the Factorization Ansatz (cf. (B.21))

$$r^{(l)}(x,y) = r(x,y)r_1^l, \tag{B.32}$$

where $r(x,y)$ satisfies the equation

$$r(x,y) = R_L^{(0)}(x,y) = r_1 \sum_{t=1}^{L} r(x,t) R_L^{(0)}(t,y),$$

in which

$$R_L^{(0)}(x,y) = (h_L^{(0)} - z)^{-1}(x,y), \qquad r_1 = \frac{1}{L}\sum_{t=1}^{L} r(t,t).$$

Thus,

$$r(x,y) = (h_L^{(0)} - z - r_1)^{-1}(x,y).$$

Setting $x = y$ in this relation, applying the operation $L^{-1}\sum_{x=1}^{L}$ and taking into account our hypothesis on the existence of $N_0(d\lambda)$, we obtain (B.30) and (B.31) in the limit $L \to \infty$.

Till now we have assumed that the entries $v_L(x,y)$ in (B.1) and (B.29) have a Gaussian distribution. In fact, the results (B.3), (B.4) and (B.30), (B.31) are valid for a wide class of distributions. Namely, let $F_{xy}(dv)$ be the distribution of $v(x,y)$ in (B.1) or in (B.29). If

$$\lim_{L\to\infty} \frac{1}{L^2} \sum_{x,y=1}^{L} \int_{|v|>\tau\sqrt{L}} v^2 F_{xy}(dv) = 0, \tag{B.33}$$

for any $\tau > 0$, the limiting integrated density of states can again be found from (B.31). This result, for a diagonal $h_L^{(0)}$ in (B.29), was found by Pastur and Girko (see [Girko 1988] and references therein). For the case of a pure GOE, that is, when $h_L^{(0)} = 0$, it is shown in [Girko 1988] that (B.33) is also a necessary condition for the semicircle law (B.4) to hold.

Condition (B.33) is the natural counterpart of the well-known Lindeberg condition, which is necessary and sufficient for the central limit theorem to apply.

Thus, the limiting form (B.31) of the distribution of eigenvalues of the random matrices (B.29) with independent entries is valid under conditions of essentially the same generality as for the central limit theorem. In other words, the deformed semicircle law is a fairly universal form of such distributions.

We have already mentioned that the rate of convergence to the semicircle distribution is high, and in particular that the variance of $L^{-1}\operatorname{Tr} R_L$ has order $O(L^{-2})$. One more evidence of this fact is the relation

$$\text{(B.34)} \qquad \lim_{L\to\infty} L^{-1}\mathsf{E}\{\ln\det(h_L - z)\} = \lim_{L\to\infty} L^{-1}\ln\mathsf{E}\{\det(h_L - z)\}$$

for $\operatorname{Im} z \neq 0$, which was found by Berezin [1973] for Gaussian orthogonal ensembles. We will prove this relation by explicitly computing both of its sides.

The left-hand side follows from (B.3) and (B.4), since

$$\text{(B.35)} \quad \int_{\mathbf{R}} \ln(z-\lambda) N_W(d\lambda) = -\frac{1}{2} + \frac{z + iz\sqrt{4-z^2}}{4} + \ln\frac{-z + i\sqrt{4-z^2}}{2}$$

To calculate the right-hand side, we use the representation of the determinant of a symmetric matrix A as the integral over the Grassman variables $\psi(x), \psi^*(x)$ [Berezin 1966, Efetov 1983], which satisfy the anticommutation relations $\{\psi(x),\psi(y)\} = \{\psi(x),\psi^*(y)\} = \{\psi^*(x),\psi^*(y)\} = 0$, where $\{a,b\} = ab + ba$. The representation is

$$\text{(B.36)} \qquad \det A = \int e^{-(A\psi,\psi)} D\psi,$$

where

$$(A\psi,\psi) = \sum_{x,y=1}^{L} A(x,y)\psi(x)\psi^*(y), \qquad D\psi = \prod_{x=1}^{L} d\psi(x)\, d\psi^*(x),$$

and $d\psi(x)$ and $d\psi^*(x)$ anticommute with one another and with all the $\psi(x)$ and $\psi^*(x)$. Plugging into (B.36) the random matrix $v_L - z$, where v_L is given by (B.1), and taking the mathematical expectation, we find that

$$\text{(B.37)} \qquad \mathsf{E}\{\det(v_L - z)\} = \int e^{z(\psi,\psi)-(1/2L)(\psi,\psi)^2} D\psi.$$

Using the identity

$$e^{-u^2/2L} = \sqrt{\frac{L}{2\pi}} \int_{-\infty}^{\infty} e^{Lt^2/2 - itu} dt$$

and formula (B.36) once more, we can rewrite (B.37) as

$$\mathsf{E}\{\det(v_L - z)\} = i^l \sqrt{\frac{L}{2n}} \int_{-\infty}^{\infty} e^{Lz^2/2 - Lt^2/2 - itzL}\, t^L \, dt.$$

Now it is sufficient to calculate this integral for large L by the steepest descent metrod, and to make sure that the logarithm of the resulting expression, after division by L tends to (B.35) as $L \to \infty$.

For other applications of the Grassman integral in the spectral theory of random operators its applications, see [Campanino and Klein 1986, Efetov 1983] and references therein.

B.3. The Deformed Semicircle Law as a Limiting Behavior for Metrically Transitive Operators. We saw in Sections 3 and 4 thta the integrated density of states is at the same time a characteristic of the limiting eigenvalue distribution of the matrices A_Λ as Λ grows to exhaust all of $\mathbf{Z}^d$ or $\mathbf{R}^d$, and a spectral characteristic of the metrically transitive operator A acting on $\ell^2(\mathbf{Z}^d)$, that is, of the corresponding "limiting" object. These two properties of the integrated density of states are reflected in (3.1) and (3.2), respectively.

For the random matrices considered in this appendix, there is no limiting object. Therefore, we cannot interpret the deformed semicircle law as the integrated density of states of a metrically transitive operator. Nevertheless, as we explain now, the law can be obtained as the limiting form of the integrated density of states of certain metrically transitive operators as a result of certain additional limiting transitions. We consider three classes of such operators, analogous to three types of Hamiltonians in statistical physics: those having a large interaction radius, a large dimensionality and a large number of field components, respectively.

We define these operators by explicitly writing their random infinite matrices (cf. 4.A). The first is given by

$$h_R^{(1)}(x,y) = h_0(x-y) + R^{-d/2}\phi\left(\frac{x-y}{R}\right)v(x,y), \tag{B.38}$$

where $h_0^*(x) = h_0^*(-x)$ and $|h(x)| \leqslant Ce^{-\rho|x|}$ for all $x \in \mathbf{Z}^d$ and some positive C and ρ; $\phi(x)$ is a real-valued, continuous function on $\mathbf{R}^d$ with compact support; and $v(x,y)$, for $x, y \in \mathbf{Z}^d$, is a family of independent random variables (apart from the symmetry condition) such that the restricted matrices $L^{-d/2}v(x,y)$, for $x, y \in \Lambda$, with $|\Lambda| = L^d$, satisfy the semicircle law.

The second operator is given by

$$h_d^{(2)}(x,y) = h_0(x-y) + \frac{1}{\sqrt{d}} v_1(x,y), \tag{B.39}$$

where $h_0(x) = d^{-1/2}$ for $|x| = 1$ and $h_0(x) = 0$ for $|x| > 1$ (that is, $h_0(x-y)$ is the d-dimensional discrete Laplacian); and $v_1(x,y)$ is the same as $v(x,y)$ in (B.38) for $|x-y| \leqslant 1$, and $v_1(x,y) = 0$ for $|x-y| > 1$.

By the results of Section 4, $h_R^{(1)}$ and $h_d^{(2)}$ are essentially self-adjoint metrically transitive operators having an integrated density of states. It turns out that for these operators, the integrated density of states tends, as $R \to \infty$ and $d \to \infty$, to the deformed semicircle law (B.31), in which $N_0(\lambda)$ can be calculated from (3.12) for $h_R^{(1)}$, and $N_0'(\lambda) = (4\pi)^{-1/2} e^{-\lambda^2/4}$ for $h_d^{(2)}$.

The third class of metrically transitive operators acts on the space $\ell^2(\mathbf{Z}^d) \times \mathbf{C}^n$ and is defined by the matrices

$$h_{ik}^{(3)}(x,y) = h_0(x,y)\delta_{ik} + \frac{1}{\sqrt{n}} v_{ik}^{(n)}(x)\delta_{xy}, \tag{B.40}$$

for $i,k = 1,2,\ldots,n$, where $h_0(x)$ is as in (B.38); and the matrices $v_{ik}^{(n)}(x)$ are identically distributed and independent for distinct $x \in \mathbf{Z}^d$, and satisfy the conditions: $v_{ik}^{(n)}(x) = v_{ki}^{(n)}(x)$ for each x, relations (B.2) hold, and $\mathsf{E}\{|v_{ik}^{(n)}|^3\} \leqslant C < \infty$ for $i,k = 1,\ldots,n$ uniformly in n.

In this case the integrated density of states also exists (see the remark after Theorem 4.13), and it can be shown that as $n \to \infty$ its limit is given by (B.31), with $N_0(d\lambda)$ calculated from (3.12).

B.4. Repulsion of Eigenvalues. Our discussion here follows mainly the book by Mehta [1967]. Denote by $\lambda_1,\ldots,\lambda_L$ the eigenvalues of matrices (B.1)–(B.2), and by $S_L(\mu_1,\mu_2)\,d\mu_1\,d\mu_2$ the probability that an interval (μ_1,μ_2) contains no eigenvalues and that the infinitesimal neighborhoods $(\mu_1,\mu_1+d\mu_1)$ and $(\mu_2,\mu_2+d\mu_2)$ contain at least one eigenvalue each. Simple arguments show that

$$S_L(\mu_1,\mu_2) = \frac{\partial^2 R_L(\mu_1,\mu_2)}{\partial\mu_1\partial\mu_2},$$

where $R_L(\mu_1,\mu_2)$ is the probability that an interval (μ_1,μ_2) does not contain eigenvalues. If $p_L(\lambda_1,\ldots,\lambda_L)$ is the joint probability density of all eigenvalues, we have

$$R_L(\mu_1,\mu_2) = \int_{-\infty}^{\infty} p_L(\lambda_1,\ldots,\lambda_L) \prod_{i=1}^{L} \chi_{(-\infty,\mu_1)\cup(\mu_2,\infty)}(\lambda_i)\,d\lambda_i.$$

Note that, in terms of $p_L(\lambda_1,\ldots,\lambda_L)$,

$$\mathsf{E}\{N_L'(\lambda)\} = \int_{-\infty}^{\infty} p_L(\lambda,\lambda_2,\ldots,\lambda_L)\,d\lambda_2,\ldots,d\lambda_L. \tag{B.41}$$

To avoid some technical complications, it is convenient to consider Hermitian complex-valued matrices of the form (B.1), (that is, $v(x,y) = v^*(x,y)$), with Gaussian-distributed entries, and satisfying

$$\begin{aligned} &\mathsf{E}\{v(x,y)\} = \mathsf{E}\{v^2(x,y)\} = 0, \\ \text{(B.42)} \qquad &\mathsf{E}\{v^2(x,x)\} = 2, \\ &\mathsf{E}\{|v(x,y)|^2\} = 1, \quad \text{for } x \neq y. \end{aligned}$$

This ensemble is known as the Gaussian unitary ensemble (GUE). Its joint probability density is

$$\text{(B.43)} \qquad C_{L1} e^{-L/2 \operatorname{Tr} VV^*} \prod_{x=1}^{L} dv(x,x) \prod_{x<y} d\operatorname{Re} v(x,y)\, d\operatorname{Im} v(x,y),$$

where C_{L1} is a normalizing factor. This, in turn, gives

$$p_L(\lambda_1, \ldots, \lambda_L) = C_{L2} e^{-L/2 \sum_1^L \lambda_i^2} \Delta^2(\lambda_1, \ldots, \lambda_L),$$

where

$$\Delta(\lambda_1, \ldots, \lambda_L) = \prod_{i<j} (\lambda_i - \lambda_j)$$

is the Vandermonde determinant. Using Gram's theorem, one can show that

$$\text{(B.44)} \qquad p_L(\lambda_1, \ldots, \lambda_L) = \frac{1}{L!} \left(\frac{L}{2}\right)^{L/2} \left(\det |\phi_{k-1}(\lambda_i)|_{i,k=1}^{L}\right)^2,$$

where $\phi_n(x)$ is the n-th orthonormalized eigenfunction of the harmonic oscillator, given by $\phi_n(x) = e^{-x^2/2} H_n(x)$, where $H_n(x)$ is the n-th Hermite polynomial.

From this and from (B.41) we find

$$\text{(B.45)} \qquad \mathsf{E}\{N_L'(\lambda)\} = \frac{1}{L} \sum_{k=0}^{L-1} \phi_k\left(\lambda \sqrt{\frac{L}{2}}\right),$$

and, since

$$\text{(B.46)} \qquad \phi_{2m}(x) = (-1)^m m^{1/4} \pi^{-1/4} \cos 2\sqrt{m} x \,(1 + o(1))$$

as $m \to \infty$ and $\phi_{2m+1}(x)$ has the same asymptotic with the cosine replaced by sine, we obtain from (B.45) the semicircle law in the limit $L \to \infty$. Similarly, again by Gram's theorem,

$$R_L(\mu_1, \mu_2) = \det \left| \delta_{ij} - \int_{\mu_1}^{\mu_2} \phi_i(t) \phi_j(t)\, dt \right|_{i,j=1}^{L-1}$$

for $t = \lambda\sqrt{L/2}$. This determinant is the Fredholm determinant of the integral equation

$$f(t) = \int_{\mu_1}^{\mu_2} K_L(t,t')f(t')\,dt'$$

for $t, t' = \lambda, \lambda'\sqrt{L/2}$, and

$$K_L(t,t') = \sum_{i=0}^{L-1} \phi_i(t)\phi_i(t').$$

From this and from (B.46) it can be shown that, if $\mu_i = \pi\theta_i/L$, for $i = 1, 2$, the limit of $R_L(\mu_1, \mu_2)$ as $L \to \infty$ equals the Fredholm determinant of the integrral equation on the interval (θ_1, θ_2) with kernel

$$\frac{1}{2}\left(\frac{\sin\pi(t-s)}{t-s} + \frac{\sin\pi(t+s)}{t+s}\right).$$

This fact allows us to prove that, for $\theta_2 - \theta_1 = \theta$ approaching zero, R_L behaves like a constant times θ^4, so that $S_L(\mu_1, \mu_2)$ is a constant times θ^2. By definiton (see 15.B), this implies that repulsion occurs for matrices of the form (B.1)–(B.2). The calculations take on the simplest form if $\theta_1 = -\theta_2$, when the integral equation above is closely related to the differential equation for spheroidal functions, so that one can use known asymptotic formulas for these functions.

Bibliography

N. I. Akhiezer (1964). The Classical Moment Problem. Oliver and Boyd, London.

B. L. Altshuler, A. G. Aronov, D. E. Khmelnitskii and A. I. Larkin (1982). Coherent effects in disordered conductors. In Quantum Theory of Solids, edited by I. M. Lifshitz, 130–237. Mir, Moscow.

P. W. Anderson (1958). Absence of diffusion in certain random lattices. Phys. Rev. **109**, 1492-1501.

G. Andre and S. Aubry (1980). Analyticity breaking and Anderson localization in incommensurate solids. Ann. Isr. Soc. **3**, 133–142.

V. V. Anshelevich, K. M. Khanin and Ya. G. Sinai (1982). Symmetric random walks in random environments. Commun. Math. Phys. **85**, 449–461.

T. N. Antsygina, L. A. Pastur and V. A. Slyusarev (1981). Localization of states and kinetic properties of one-dimensional disordered systems (in Russian). Fiz. Nizkikh Temp. **7**, 5–46.

L. Arnold, G. Papanicolaou and V. Wihstutz (1986). Asymptotic analysis of the Lyapunov exponent and rotation number of a random oscillator and applications. SIAM J. Appl. Math. **46**, 427–450.

L. Arnold and V. Wihstutz (editors) (1986). Lyapunov Exponents. Springer, Berlin (Lecture Notes in Mathematics, 1186).

V. I. Arnold (1980). Chapitres supplémentaires de la théorie des équations différentielles ordinaires. Mir, Moscow.

N. Aronszajn (1957). On a problem of Weyl in the theory of singular Sturm–Liouville equations. Am. J. Math. **79**, 597–610.

F. V. Atkinson (1964). Discrete and Continuous Boundary Problems. Academic Press, New York and London.

S. Aubry and P. V. Le Daeron (1983). The discrete Frenkel–Kontorova model and its extensions. I: Exact results for the ground states. Physics D **7**, 240–258.

J. Avron, W. Craig and B. Simon (1983). Large coupling behavior of the Lyapunov exponent for tight-binding one-dimensional random systems. J. Phys. A **16**, L209–L211.

J. Avron and B. Simon (1981a). Almost periodic Schrödinger operators. I: Limit-periodic potentials. Commun. Math. Phys. **82**, 101–120.

J. Avron and B. Simon (1981b). Cantor sets and Schrödinger operators: transient and recurrent spectrum. J. Func. Anal. **43**, 1–31.

J. Avron and B. Simon (1982). Singular continuous spectrum of a class of almost periodic Jacobi matrices. Bull. (new series) Am. Math. Soc. **6**, 81–87.

J. Avron and B. Simon (1983). Almost periodic Schrödinger operators. II: The integrated density of states. Duke Math. J. **50**, 369–391.

N. S. Bakhvalov and G. P. Panasenko (1984). Averaging of Processes in Periodic Media: Mathematical Problems of the Mechanics of Composite Materials (in Russian). Nauka, Moscow.

A. I. Baz', Ya. I. Zeldovich and A. M. Perelomov (1971). Scatterings, Reactions and Decays in Nonrelativistic Quantum Mechanics (in Russian). Nauka, Moscow.

J. Bellissard (1985). K-theory of C^*-algebras in solid-state physics. In Statistical Mechanics and Field Theory, Mathematical Aspects, edited by T. C. Dorlas, 99–156. Springer, Berlin.

J. Bellissard, D. Bessis and P. Moussa (1982). Chaotic states of almost-periodic Schrödinger operators. Phys. Rev. Lett. **49**, 701–704.

J. Bellissard, R. Lima and E. Scoppola (1983a). Localization in v-dimensional incommensurable structures. Commun. Math. Phys. **88**, 465–477.

J. Bellissard, R. Lima and D. Testard (1983b). A metal-insulator transition for the almost-Mathieu model. Commun. Math. Phys. **88**, 207–234.

J. Bellissard and D. Testard (1985). Almost-periodic Schrödinger operators. In Mathematics and Physics, Lectures on Recent Results, vol. I, edited by L. Streit, 1–64. World Scientific, Singapore and Philadelphia.

J. Bellissard and E. Scoppola (1982). The density of states for almost-periodic Schrödinger operators: a counterexample. Commun. Math. Phys. **85**, 301–308.

J. Bellissard and B. Simon (1982). Cantor spectrum for the almost-Mathieu equation. J. Funct. Anal. **49**, 191–213.

E. P. Belokolos (1975). Quantum particle in a one-dimensional deformed lattice: Estimates of the gaps in the spectrum (in Russian). Teor. Mat. Fiz. **25**, 344–357.

E. P. Belokolos (1976). Quantum particle in a one-dimensional deformed lattice: Dependence of the energy on the quasimomentum (in Russian). Teor. Mat. Fiz. **26**, 35–40.

Yu. K. Belyaev (1959). Analytic random processes (in Russian). Teor. Ver. Primen. **4**, 437–444.

M. M. Benderskii and L. A. Pastur (1969). Calculations of the average number of states in a model problem. Soviet Phys. JETP **30**, 158–162.

M. M. Benderskii and L. A. Pastur (1970). On the spectrum of the one-dimensional Schrödinger equation with a random potential (in Russian). Mat. Sb. **82**, 273–284.

M. M. Benderskii and L. A. Pastur (1975). On the asymptotics of the solutions of second-order equations with random coefficients (in Russian). Teoria Funkcii, Func. Anal. i Priloz. (Kharkov University) **N 22**, 3–14.

F. Bentosela, R. Carmona, P. Duclos, B. Simon, B. Souillard and R. Weder (1983). Schrödinger operators with electric field and random or deterministic potential. Commun. Math. Phys. **88**, 387–397.

Y. M. Berezanskii (1968). Expansions in Eigenfunctions of Self-Adjoint Operators. Am. Math. Soc., Providence (Transl. of Math. Mon., 17).

F. A. Berezin (1966). The Method of Second Quantization. Academic Press, New York.

F. A. Berezin (1973). Several remarks on the Wigner distribution (in Russian). Teor. Mat. Fiz. **17**, 305–318.

F. A. Berezin and L. D. Fadeev (1961). Remark on the Schrödinger operator with a singular potential (in Russian). Dokl. Acad. Nauk SSSR **137**, 1011-1014.

I. Bernasconi and W. P. Schneider (1983). Diffusion in random one-dimensional systems. J. Stat. Phys. **30**, 355-362.

N. N. Bogolyubov and Yu. A. Mitropolski (1974). Asymptotic Methods in the Theory of Nonlinear Oscillations (in Russian). Nauka, Moscow.

N. N. Bogolyubov, Yu. A. Mitropolski and A. M. Samoilenko (1976). Methods of Accelerated Convergence in Nonlinear Mechanics. Springer, Berlin.

T. T. Bogorodskaya (1979). Density of states of multidimensional integral operators (in Russian). Funkc. Anal. i Priloz. **13**, 79–80.

T. T. Bogorodskaya and M. A. Shubin (1986). The variational principle and the asymptotic behavior of the density of states for random pseudodifferential operators (in Russian). In Trudy sem. Petrovskii (Moscow University) **14**, 98–114.

V. L. Bonch-Bruevich, I. P. Zvyagin, A. G. Mironov, R. Kaiper, R. Enderdain and B. Esser (1981). Introduction to the Electron Theory of Disordered Semiconductors (in Russian). Nauka, Moscow.

R. E. Borland (1963). Nature of the electronic states in disordered one-dimensional systems. Proc. Roy. Soc. London A **274**, 529–531.

P. Bougerol and J. Lacroix (1985). Products of Random Matrices with Applications to Schrödinger Operators. Birkhäuser, Boston.

E. N. Bratus', S. A. Gredeskul, L. A. Pastur and V. S. Shumeiko (1986). Nonlinear resonant interaction between acoustic impulses and electrons in superconductors. Fiz. Nizkikh Temp. **12**, 322-325.

T. A. Brody, J. Flores, J. B. French, P. A. Mello, A. Pandey and S. S. M. Wong (1981). Random matrix physics: spectrum and strength fluctuations. Rev. Mod. Physics **53**, 385–480.

B. F. Bylov, R. E. Vinograd, P. M. Brovman and V. V. Nemytskii (1966). Theory of Lyapunov Exponents (in Russian). Nauka, Moscow.

M. Campanino and A. Klein (1986). A supersymmetric transfer matrix and differentiability of the density of states in the one-dimensional Anderson model. Commun. Math. Phys. **104**, 227-241.

M. Campanino and A. Klein (1990). Anomalies in the one-dimensional Anderson model at weak disorder. Commun. Math. Phys. **130**, 441–456.

R. Carmona (1982). Exponential localization in one-dimensional disordered systems. Duke Math. J. **49**, 191–213.

R. Carmona (1983). One-dimensional Schrödinger operators with random and deterministic potential: new spectral types. J. Funct. Anal. **51**, 229–258.

R. Carmona (1984a). One-dimensional Schrödinger operators with random potential. Physica A **124**, 181–188.

R. Carmona (1984b). Random Schrödinger operators. In Ecole d'Eté de Probabilités de Saint-Flour, edited by P. L. Hennequin. Springer, Berlin (Lecture Notes in Mathematics, 1180).

R. Carmona, A. Klein and F. Martinelli (1987). Anderson localization for Bernoulli and other random potentials. Commun. Math. Phys. **108**, 41–67.

R. Carmona and S. Kotani (1987). On the inverse problem for random Jacobi matrices. J. Stat. Phys. **46**, 1091–1114.

G. Casati (1985). Chaotic Behavior in Quantum Systems. Plenum, New York.

A. Casher and J. Lebowitz (1971). Heat flow in regular and disordered harmonic chains. J. Math. Phys. **12**, 1701–1711.

J. Chayes and L. Chayes (1986). Percolation in random media. In Phénomènes critiques, systèmes aléatoires, théories de jauge, edited by K. Osterwalder and R. Stora. North-Holland, Amsterdam and New York, 1001-1138.

V. A. Chulaevski (1981). On perturbation of a Schrödinger operator with periodic potential. Russian Math. Surv. **36(5)**, 143–144.

V. A. Chulaevski (1984). Inverse spectral problem for limit-periodic Schrödinger operators (in Russian). Funkc. Anal. i Priloz. **18**, 63-66.

V. A. Chulaevski and F. Delyon (1989). Purely absolutely continuous spectrum of the almost-Mathieu operator. J. Stat. Phys. **55**, 1279–1284.

F. Claro and G. Wannier (1979). Magnetic subband structure of electrons in hexagonal lattices. Phys. Rev. B **19**, 6068–6074.

L. A. Coburn, R. P. Moyer and I. M. Singer (1973). C^*-algebras of almost-periodic pseudodifferential operators. Acta Math. **138**, 279–307.

E. A. Coddington and N. Levinson (1955). Theory of Ordinary Differential Equations. McGraw–Hill, New York.

F. Constantinescu, J. Fröhlich and T. Spenser (1984). Analyticity of density of states and replica method for random Schrödinger operators on a lattice. J. Stat. Phys. **34**, 371–396.

I. P. Cornfeld, S. V. Fomin and Ya. G. Sinai (1982). Ergodic Theory. Springer, Berlin.

R. Courant and D. Hilbert (1968). Methoden der mathematischen Physik I. Springer, Berlin.

W. Craig (1983). Pure point spectrum for discrete Schrödinger operators. Commun. Math. Phys. **88**, 113–131.

W. Craig and B. Simon (1983a). Log-Hölder continuity of the integrated density of states for stochastic Jacobi matrices. Commun. Math. Phys. **90**, 207–218.

W. Craig and B. Simon (1983b). Subharmonicity of the Lyapunov index. Duke Math. J. **90**, 389–411.

P. Dean (1961). Vibrational spectra of diatomic chain. Proc. Roy. Soc. London A **260**, 263–271.

P. Deift and B. Simon (1983). Almost-periodic Schrödinger operators. III: The absolutely continuous spectrum in one dimension. Commun. Math. Phys. **90**, 389–411.

C. de Lange and T. Janssen (1981). Incommensurability and recursivity: lattice dynamics of modulated crystals. J. Phys. C **14**, 5269–5292.

F. Delyon (1985). Apparition of purely singular continuous spectrum in a class of random Schrödinger operators. J. Stat. Phys. **40**, 421–630.

F. Delyon (1987). Absence of localization in the almost-Mathieu equation. J. Phys. A **20**, L21–L23.

F. Delyon and P. Foulon (1986). Adiabatic invariants and asymptotic behavior of Lyapunov exponents of the Schrödinger equation. J. Stat. Phys. **45**, 41–48.

F. Delyon, H. Kunz and B. Souillard (1983). One-dimensional wave equation in random media. J. Phys. A **16**, 25–42.

F. Delyon, Y. Levy and B. Souillard (1985a). Anderson localization for multidimensional systems of large disorder or large energy. Commun. Math. Phys. **100**, 463–470.

F. Delyon, Y. Levy and B. Souillard (1985b). Anderson localization for one- and quasi-one-dimensional systems. J. Stat. Phys. **41**, 375–388.

F. Delyon, Y. Levy and B. Souillard (1985c). Approach à la Borland to multidimensional localization. Phys. Rev. Lett. **55**, 618–621.

F. Delyon and D. Petritis (1986). Absence of localization in a class of Schrödinger operators with quasiperiodic potential. Comm. Math. Phys. **103**, 441–445.

F. Delyon, B. Simon and B. Souillard (1985). From power pure-point to continuous spectrum in disordered systems.. Ann. Inst. H. Poincaré Phys. Theor. **42**, 283–309.

F. Delyon, B. Simon and B. Souillard (1987). Localization for off-diagonal disorder and for continuous Schrödinger operators. Commun. Math. Phys. **109**, 157–165.

F. Delyon and B. Souillard (1983). The rotation number for finite-difference operators and its properties. Commun. Math. Phys. **89**, 415–426.

F. Delyon and B. Souillard (1984). Remark on the continuity of the density of states of ergodic finite-difference operators. Commun. Math. Phys. **94**, 289–291.

Y. Derriennic (1975). Sur le théorème ergodique sous-additif. Comptes Rendues Acad. Sc. Paris A **281**, A985.

P. Devillard and B. Souillard (1986). Polynomially decaying transmission for the nonlinear Schrödinger equation in a random medium. J. Stat. Phys. **43**, 423–430.

E. I. Dinaburg and Ya. G. Sinai (1975). The one-dimensional Schrödinger equation with a quasiperiodic potential. Funct. Anal. Appl. **9**, 279–289.

W. Donoghue (1965). On the perturbations of spectra. Commun. Pure Appl. Math **18**, 559–579.

M. D. Donsker and S. R. S. Varadhan (1975a). Asymptotic evaluation of certain Markov process expectations for large time, I. Commun. Pure Appl. Math. **28**, 279–301.

M. D. Donsker and S. R. S. Varadhan (1975b). Asymptotic evaluation of certain Markov integrals for large time.. In Functional Integration and its Applications, edited by A. M. Arthurs, 15–33. Clarendon, Oxford.

M. D. Donsker and S. R. S. Varadhan (1975c). Asymptotics for the Wiener sausage. Commun. Pure Appl. Math. **28**, 525–565.

M. D. Donsker and S. R. S. Varadhan (1976). Asymptotic evaluation of certain Markov process expectations for large time, III. Commun. Pure Appl. Math. **29**, 389–461.

M. D. Donsker and S. R. S. Varadhan (1983). Asymptotic evaluation of certain Markov process expectations for large time, IV. Commun. Pure Appl. Math. **36**, 183–212.

J. L. Doob (1952). Stochastic Processes. Wiley, New York.

H. von Dreifus (1987). On the effect of randomness in ferromagnetic models and Schrödinger operators. Ph. D. thesis, New York University.

H. von Dreifus and A. Klein (1989). A new proof of localization in the Anderson tight binding model. Commun. Math. Phys. **124**, 285–299.

N. Dunford and J. Schwartz (1958). Linear Operators, I. Interscience, New York.

N. Dunford and J. Schwartz (1963). Linear Operators, II. Interscience, New York.

P. Duren (1983). Univalent Functions. Springer, Berlin.

V. V. Dyakin and S. I. Petrukhnovskii (1982). Some geometrical properties of the Fermi surface (Russian). Dokl. Akad. Nauk SSSR **264**, 1117–1119.

A. M. Dykhne (1970). Conductivity of two-dimensional two-phase systems. Sov. Phys. JETP **32**, 63–71.

E. B. Dynkin (1965). Markov Processes, vols. I and II. Springer, Berlin.

F. J. Dyson (1953). The dynamics of a disordered linear chain. Phys. Rev. **92**, 1331–1338.

F. Dyson (1962). Statistical theory of the energy levels of complex systems. J. Math. Phys. **3**, 140–165.

M. S. Eastham, W. D. Evans and J. B. McLead (1976). The essential self-adjointness of Schrödinger-type operators. Arch. Rat. Mech. Anal. **60**, 53–127.

J. T. Edwards and D. J. Thouless (1971). Regularity of density of states in Anderson's localized electron model. J. Phys. C **4**, 453–457.

K. Efetov (1983). Supersymmetry and theory of disordered systems. Adv. in Phys. **32**, 53–127.

A. L. Efros and B. I. Shklovski (1984). Electronic Properties of Doped Semiconductors. Springer, Berlin.

I. E. Egorova (1986). Spectral properties of the Dirac operator with a limit-periodic potential (in Ukranian). Dopovidi Akad. Nauk Ukr. SSR 1986(5), 10–13.

I. E. Egorova (1987). Spectral analysis of a limit-periodic Jacobi matrix (in Ukranian). Dopovidi Akad. Nauk Ukr. SSR 1987(3), 7–9.

S. D. Eidel'man and S. D. Ivasishen (1970). Investigation of the Green's matrix of a homogeneous parabolic boundary problem (in Russian). In Trudy Mosk. Mat. Obz. **23**, 179–234.

G. A. Elliot (1982). Gaps in the spectrum of an almost-periodic Schrödinger operator. C. R. Math. Ref. Acad. Sci. Canada **4**, 255-261.

M. Endrullis and H. Englisch (1987). Special energies and special frequencies. Commun. Math. Phys. **108**, 591–604.

H. Englisch (1983). Schrödinger operator with ergodic potential. Zeit. Anal. und Anwend. **2**, 411-416.

H. Englisch and K. Kursten (1983). Infinite representability of Schrödinger operators with ergodic potential. Zeit. Anal. und Anwend. **3**, 329–350.

W. G. Faris (1986). A localization principle for multiplicative perturbations. J. Funct. Anal. **67**, 105–114.

W. G. Faris (1987). Localization estimates for a random discrete wave equation at high frequency. J. Stat. Phys. **46**, 477–491.

B. V. Fedosov and M. A. Shubin (1978). Index of random operators (in Russian). Math. Sb. **106**, 455–483.

W. Feller (1966). An Introduction to Probability Theory and its Applications. Wiley, New York.

X. Fernique (1964). Continuité des process gaussiens. C. R. Acad. Sc. Paris **258**, 6058–6060.

P. Figari, E. Orlandi and G. Papanicolaou (1982). Diffuse behavior of a random walk in a random medium. In Stochastic Analysis, edited by K. Ito, 105–119. North-Holland, Amsterdam.

A. L. Figotin (1979). Asymptotic behavior for large time of some Wiener integrals, I and II. Teoria Funkcii, Funkc. Anal. i Priloz. (Kharkov University) **32**, 88–91; **33**, pp. 132–135.

A. L. Figotin (1980a). On the exponential growth of solutions of finite-difference equations with random coefficients (in Russian). Dokl. Akad. Nauk Uzb. SSR 1980(2), 9–11.

A. L. Figotin (1980b). Spectral properties of some classes of random self-adjoint operators (in Russian). Dissertation, Tashkent University.

A. L. Figotin (1981). Eigenvalue distribution of random Schrödinger equations and asymptotic behavior of certain Wiener integrals (in Ukrainian). Dopovidi Akad. Nauk Ukr. SSR 1981(6), 27–29.

A. L. Figotin (1983a). Essential self-adjointness and ergodic properties of the Schrödinger operator with a random potential (in Ukrainian). Dopovidi Akad. Nauk Ukr. SSR 1983(8), 18–20.

A. L. Figotin (1983b). On the absence of an absolutely continuous spectrum for random Jacobi matrices (in Russian). In Functional Analysis and Applied Mathematics, 200-207, edited by V. A. Marchenko. Naukova Dumka, Kiev.

A. L. Figotin (1987). Ergodic properties and essential self-adjointness of random matrix operators (in Russian). In Operators in Functional Spaces and Topics of Function Theory, edited by V. A. Marchenko. Naukova Dumka, Kiev.

A. L. Figotin and L. A. Pastur (1983). Ergodic properties of the eigenvalue distribution of certain classes of random self-adjoint operators. Selecta Math. Sov. **3**, 69–86.

A. L. Figotin and L. A. Pastur (1984a). The positivity of the Lyapunov exponent and the absence of the absolutely continuous spectrum for the almost-Mathieu equation. J. Math. Phys. **25**, 774–777.

A. L. Figotin and L. A. Pastur (1984b). An exactly solvable model of a multidimensional incommensurate structure. Commun. Math. Phys. **95**, 401–425.

A. L. Figotin and L. A. Pastur (1990). A Schrödinger operator with a nonlocal potential whose absolutely continuous and point spectra coexist. Commun. Math. Phys. **130**, 357–380.

M. A. Freidlin and A. D. Wentzell (1984). Random Perturbations of Dynamical Systems. Springer, Berlin (Grundlehren der mathematischen wissenschaften, 260).

A. Friedman (1964). Partial Differential Equations of Parabolic Type. Prentice-Hall, Englewood Cliffs, NJ.

H. L. Frisch and S. P. Lloyd (1960). Electron levels in a one-dimensional random lattice. Phys. Rev. **120**, 1179–1194.

J. Fröhlich and T. Spencer (1983). Absence of diffusion in the Anderson tight-binding model for large disorder or low energy. Commun. Math. Phys. **88**, 151–189.

J. Fröhlich, T. Spencer and P. Wittwer (1990). Localization for a class of one-dimensional quasiperiodic Schrödinger operators. Commun. Math. Phys. **132**, 5–26.

J. Fröhlich, F. Martinelli, B. Scoppola and T. Spencer (1985). Constructive proof of localization in the Anderson tight-binding model. Commun. Math. Phys. **101**, 21–46.

M. Fukushima (1974). On the spectral distribution of a disordered system and the range of a random walk. Osaka Jour. Math. **11**, 73–85

M. Fukushima, M. Nagai and S. Nakao (1975). On the asymptotic properties of spectra of a random difference operator. Proc. Japan Acad. **51**, 100-102.

M. Fukushima and S. Nakao (1977). On the spectra of the Schrödinger operator with white-noise potential. Zeit. Wahrscheinlichkeitstheorie verw. Gebiete **37**, 267–274.

H. Furstenberg (1963). Noncommuting random products. Trans. Am. Math. Soc. **108**, 377–428.

H. Furstenberg and H. Kesten (1960). Product of random matrices. Ann. Math. Statist. **31**, 457–469.

H. Furstenberg and V. Kifer (1983). Random matrix products and measures on projective spaces. Israel J. of Math. **46**, 12–32.

I. Gel'fand (1950). Expansion in series of eigenfunctions of an equation with periodic potential (in Russian). Dokl. Akad. Nauk SSSR **73**, 1117–1120.

R. Giachetti and R. Johnson (1984). Spectral theory of second-order almost-periodic differential operators and its relation to classes of nonlinear evolution equations. Nuovo Cimento B **82**, 125–168.

I. I. Gihman and A. V. Skorohod (1974–1979). Theory of Stochastic Processes, vols. I–III. Springer, Berlin.

V. L. Girko (1988). Spectral Theory of Random Matrices (in Russian). Nauka, Moscow.

I. M. Glazman (1965). Direct methods of qualitative spectral analysis of singular differential operators. Israeli Progr. Scient. Transl., Jerusalem.

I. Ya. Goldsheidt (1976). The law of large number in certain functional spaces (in Russian). Usp. Mat. Nauk **31**, 211-212.

I. Ya. Goldsheidt (1980). Asymptotic properties of the product of random matrices depending on a parameter. In Multicomponent systems. Advances in Probability **8**, 239–289.

I. Ya. Goldsheidt (1981). The structure of the spectrum of the Schrödinger random difference operator. Sov. Math. Dokl. **22**, 670–674.

I. Ya. Goldsheidt and G. A. Margulis (1987). Conditions for the simplicity of the Lyapunov exponent spectrum (in Russian). Dokl. Akad. Nauk SSSR **293**, 297–301.

I. Ya. Goldsheidt, S. Molchanov and L. Pastur (1977). A pure-point spectrum of the stochastic one-dimensional Schrödinger equation. Funct. Anal. Appl. **11**, 1–10.

A. Ya. Gordon (1976). On the point spectrum of the one-dimensional Schrödinger operator (in Russian). Usp. Math. Nauk **31**, 257–258.

A. Ya. Gordon (1979). On the continuous spectrum of the one-dimensional Schrödinger operator (in Russian). Funkc. Anal. i Priloz. **13**, 77–78.

A. Ya. Gordon (1986). Sufficient conditions for the continuity of the spectrum of the discrete Schrödinger equation (in Russian). Funkc. Anal. i Priloz. **20**, 70–71.

L. P. Gor'kov, O. N. Dorokhov and F. V. Prygara (1983). Structure of wave functions and conductivity in disordered one-dimensional conductors (in Russian). Zh. Eksp. Teor. Fiz. SSSR **85**, 1470–1488.

L. P. Gor'kov and G. M. Eliashberg (1965). Small metallic particles in an electromagnetic field (in Russian). Zh. Teor. Eksp. Fiz. SSSR **48**, 1407–1418.

S. A. Gredeskul and L. A. Pastur (1975). Behavior of the density of states in one-dimensional disordered systems near the spectrum edges (in Russian). Teor. Mat. Fiz. **23**, 132–139.

S. A. Gredeskul and L. A. Pastur (1978). Density of states in a one-dimensional disordered system in the two-band approximation (in Russian). Zh. Exp. Teor. Fiz. **75**, 1444-1459.

S. A. Gredeskul and L. A. Pastur (1985). Density of states near the spectrum edge of one-dimensional incommensurate structures (in Russian). Teor. Mat. Fiz. **62**, 316–319.

D. R. Grempel, S. Fishman and R. E. Prange (1982). Localization in an incommensurate potential: an exactly soluble model. Phys. Rev. Lett. **49**, 833-835.

U. Grenander and G. Szegő (1958). Toeplitz Forms and Their Applications. Univ. Calif. Press, Berkeley.

L. G. Grenkova (1981). On essential self-adjointness of the Schrödinger operator with a random potential (in Russian). Usp. Mat. Nauk **36**, 211-212.

A. Grossmann, R. Hoegh-Krohn and M. Mebkhout (1980). A class of explicitly soluble, local, many-center Hamiltonians for one-particle quantum mechanics in two and three dimensions. J. Math. Phys. **21**, 2376–2380.

Y. Guivarc'h (1984). Exposants caracteristiques des produits de matrices aléatoires en dépendance markovienne. In Probability Measure on Groups, edited by H. Heyer, 161–181. Springer, Berlin (Lecture Notes in Mathematics, 1064).

Y. Guivarc'h and A. Raugi (1985). Frontière de Furstenberg, propriétés de contraction et théorème de convergence. Zeit. Wahrscheinlichkeitstheorie verw. Gebiete **69**, 187–242.

A. I. Gusev (1977). Density of states and other spectral invariants of self-adjoint elliptic operators with random coefficients (in Russian). Mat. Sb. **104**, 207–226.

P. Hall and C. C. Heyde (1980). Martingale Limit Theory and its Application. Academic Press, New York.

B. Halperin (1965). Green's functions for a particle in a one-dimensional random potential. Phys. Rev. A **139**, 104–107.

B. Halperin (1967). Properties of a particle in a one-dimensional random potential. Adv. Chem. Phys. **13**, 123–177.

B. Halperin and M. Lax (1967). Impurity-band tails in the high-density energy limit. Phys. Rev. **153**, 802-814.

P. Hartman (1964). Ordinary Differential Equations. Wiley, New York.

P. Hartman and C. R. Putnam (1950). The gaps in the essential spectra of wave equations. Am. J. Math. **72**, 849–892.

W. K. Hayman and P. B. Kennedy (1976). Subharmonic Functions. Academic Press, New York.

D. Herbert and R. Jones (1971). Localized states in disordered systems. J. Phys. C **4**, 1145–1150.

I. W. Herbst and J. S. Howland (1981). The Stark ladder and other one-dimensional external field problems. Commun. Math. Phys. **80**, 23–42.

M. Herman (1983). Une méthode pour minorer les exposants de Lyapounov et quelques examples montrant le caractère local d'Arnold et de Moser sur le tore de dimension 2. Comment. Math. Helv. **58**, 453–502.

D. R. Hofstadter (1976). Energy levels and wave functions of Bloch electrons in a random or irrational magnetic field. Phys. Rev. B **14**, 2239–2245.

L. Hörmander (1967). Hypoelliptic differential equation of the second order. Acta Math. **119**, 147–171.

J. Howland (1987). Perturbation theory of dense point spectrum. J. Funct. Anal. **74**, 52–80.

K. Ichihara and H. Kunita (1974). A classification of second-order degenerate elliptic operators and its probabilistic characterization. Zeit. Wahrscheinlichkeitstheorie verw. Gebiete **30**, 235–254.

K. Ishii (1973). Localization of eigenstates and transport phenomena in one-dimensional disordered systems. Progress Theor. Phys. Suppl. **53**, 77–118.

K. Ishii and H. Matsuda (1970). Localization of normal models and energy transport in the disordered harmonic chain. Progress Theor. Phys. Suppl. **45**, 56–86.

R. Johnson and J. Moser (1982). The rotation number for almost-periodic potentials. Commun. Math. Phys. **84**, 403–438.

G. Jona-Lasinio, F. Martinelli and E. Scoppola (1985). A quantum particle in a hierarchical potential with tunneling over arbitrarily large scales. Ann. Inst. H. Poincaré Phys. Theor. **42**, 73–108.

R. Joynt and R. E. Prange (1984). Conditions for the quantum Hall effect. Phys. Rev. B **29**, 3303–3317.

I. S. Kac (1963). Spectrum multiplicity and eigenfunction expansion of second-order differential operators (in Russian). Izv. Akad. Nauk SSSR Mat. **27**, 1081–1112.

I. S. Kac and M. G. Krein (1968). R-analytic functions mapping the upper half-plane into itself. Am. Math. Soc. Transl. (series 2) **103**, 1–18.

P. A. Kalugin, A. Yu. Kitaev and L. S. Levitov (1986). Electronic spectrum of a one-dimensional quasicrystal (in Russian). Zh. Eksp. Teor. Fiz. SSSR **91**, 692–701.

T. Kato (1966). Perturbation Theory of Linear Operators. Springer, Berlin.

A. Katz and M. Dunneau (1986). Quasiperiodic patterns and icosahedral symmetry. J. Physique **47**, 181–196.

H. Kesten (1982). Percolation Theory for Mathematicians. Birkhäuser, Boston.

A. Ya. Khintchine (1963). Continued Fractions. Nordhoff.

D. N. Khomskii (1966). Edges of the spectrum of disordered systems. Fiz. Tverd. Tela **8**, 1592–1598.

W. Kirsch (1985). Random Schrödinger operators and the density of states. In Stochastic Aspects of Classical and Quantum Systems, 68–102, edited by S. Albeverio, P. Combe and M. Sirugue Collin. Springer, Berlin (Lecture Notes in Mathematics, 1109).

W. Kirsch, S. Kotani and B. Simon (1985). Absence of absolutely continuous spectrum for some one-dimensional random but deterministic Schrödinger operators. Ann. Inst. H. Poincaré Phys. Theor. **42**, 383–406.

W. Kirsch and F. Martinelli (1982a). On the ergodic properties of the spectrum of general random operators. J. Reine Angew. Math. **334**, 141–156.

W. Kirsch and F. Martinelli (1982b). On the spectrum of Schrödinger operators with a random potential. Commun. Math. Phys. **85**, 329–350.

W. Kirsch and F. Martinelli (1982c). On the density of states of Schrödinger operators with a random potential. J. Phys. A **15**, 2139–2156.

W. Kirsch and F. Martinelli (1983a). Large deviations and Lifshitz singularity of the integrated density of states of random Hamiltonians. Commun. Math. Phys. **89**, 27–40.

W. Kirsch and F. Martinelli (1983b). On the essential self-adjointness of stochastic Schrödinger operators. Duke Math. J. **50**, 1255–1260.

W. Kirsch and B. Simon (1986). Lifshitz tails for periodic plus random potential. J. Stat. Phys. **42**, 799-808.

A. Klein, F. Martinelli and J. F. Peresz (1986). A rigorous replica trick approach to Anderson localization in one dimension. Commun. Math. Phys. **106**, 623–634.

M. Kohmoto and Y. Oono (1984). Cantor spectrum for an almost-periodic Schrödinger equation and a dynamical map. Phys. Lett. A **102**, 145–148.

P. Koosis (1980). Introduction to H_p-spaces. Cambridge University Press, London.

S. Kotani (1976). On asymptotic behavior of the spectrum of a one-dimensional Hamiltonian with a certain coefficient. Publ. Res. Inst. Math. Sci. Kyoto Univ. **12**, 447–492.

S. Kotani (1984). Lyapunov indices determine absolutely continuous spectra of stationary random one-dimensional Schrödinger operators. In Stochastic Analysis, edited by K. Ito, 225–247. North-Holland, Amsterdam.

S. Kotani (1985a). Support theorem for random Schrödinger operators. Commun. Math. Phys. **97**, 443–452.

S. Kotani (1985b). On an inverse problem for random Schrödinger operators. In Particle Systems, Random Media and Large Deviations, edited by R. Durrett, 270–280. American Mathematical Society, Providence, RI (Contemporary Mathematics, 41).

S. Kotani (1986). Lyapunov exponents and spectra for one-dimensional random Schrödinger operators. In Random Matrices and Their Applications, edited by J. E. Cohen, H. Kesten and C. M. Newman, 277–286. American Mathematical Society, Providence, RI (Contemporary Mathematics, 50).

S. Kotani (1987a). Link between periodic potential and random potential in one-dimensional Schrödinger operators.. In Differential Equations and Mathematical Physics, edited by I. W. Knowles and Y. Saito, 256–269. Springer, Berlin (Lecture Notes in Mathematics, 1285).

S. Kotani (1987b). One-dimensional Schrödinger operators and Herglotz functions. In Probabilistic Methods in Mathematical Physics, edited by K. Ito and N. Ikeda, 219–250. Academic Press, Boston.

S. Kotani (1989). Jacobi matrices with random potentials taking finitely many values. Review in Math. Phys. **1**, 129–133.

S. Kotani and B. Simon (1987). Localization in general one-dimensional systems. II: Continuous Schrödinger operators. Commun. Math. Phys. **112**, 103–119.

S. Kotani and N. Ushiroya (1988). One-dimensional Schrödinger operators with random decaying potentials. Commun. Math. Phys. **115**, 247–266.

S. M. Kozlov (1979a). Averaging of random structures (in Russian). Math. Sb. **109**, 188–202.

S. M. Kozlov (1979b). Conductivity of two-dimensional media (in Russian). Usp. Math. Nauk **34**, 193–199.

S. M. Kozlov (1982). Eigenvalue distribution of differential operators in large domains (in Russian). Usp. Math. Nauk **37**, 185–186.

S. M. Kozlov (1983). Ground states of quasiperiodic operators (in Russian). Dokl. Akad. Nauk SSSR **271**, 532–534.

S. M. Kozlov (1985). Averaging method and random walk in inhomogeneous media (in Russian). Usp. Math. Nauk **40**, 61–120.

S. M. Kozlov and S. A. Molchanov (1984). On the conditions for applicability of the central limit theorem to random walks on a lattice. Sov. Math. Dokl. **30**, 410–413.

P. Koosis (1980). Introduction to H_p-Spaces. Cambridge Univ. Press, London.

H. Kramer and M. Leadbetter (1967). Stationary and Related Stochastic Processes. Wiley, New York.

M. A. Krasnosel'skii (1962). Positive Solutions of Operator Equations (in Russian). Fizmatgiz, Moscow.

H. Kunz and B. Souillard (1980). Sur le spectre des opérateurs aux différences finies aléatoires. Commun. Math. Phys. **79**, 201–249.

H. Kunz and B. Souillard (1983a). The localization transition of the Bethe lattice. J. Physique Lett. **44**, L411-L414.

H. Kunz and B. Souillard (1983b). On the upper critical dimension and the critical exponents of the localization transition. J. Physique Lett. **44**, L503–L506.

J. Lacroix (1983). Singularités du spectre de l'opérateur de Schrödinger aléatoire dans un ruban ou un demi-ruban. Ann. Inst. H. Poincaré Phys. Théor. **38**, 385–399.

J. Lacroix (1984). Localization pour l'opérateur de Schrödinger aléatoire dans un ruban. Ann. Inst. H. Poincaré Phys. Theor. **40**, 97–116.

L. D. Landau and E. M. Lifshitz (1965). Quantum Mechanics. Pergamon, Oxford.

L. D. Landau and Ya. A. Smorodinskii (1957). Lectures on the Theory of Atomic Nuclei (in Russian). GITTL, Moscow.

M. Lax and J. Phillips (1958). One-dimensional impurity bands. Phys. Rev. **110**, 41–48.

P. A. Lee and T. V. Ramakrishnan (1985). Disordered electronic systems. Rev. Modern Phys. **57**, 287–337.

E. le Page (1984). Repartition d'états pour les matrices de Jacobi à coefficients aléatoires. In Probability Measure on Groups, edited by H. Heyer, 309–367. Springer, Berlin (Lecture Notes in Mathematics, 1064).

B. Ya. Levin (1972). Distribution of Zeros of Entire Functions. Am. Math. Soc., Providence, RI (Transl. Math. Monographs, 5).

B. M. Levitan (1953). Almost-Periodic Functions (in Russian). GITTL, Moscow.

B. M. Levitan (1984). Inverse Sturm–Liouville Problems (in Russian). Nauka, Moscow.

B. M. Levitan and I. S. Sargsyan (1976). Introduction to Spectral Theory. Am. Math. Soc., Providence, RI (Transl. Math. Monographs, 39).

B. M. Levitan and A. V. Savin (1984). Examples of Schrödinger operators with almost-periodic potential and nowhere dense absolutely continuous spectrum. Dokl. Akad. Nauk SSSR **276**, 539–542.

I. M. Lifshitz (1964). Energy spectrum structure and quantum states of disordered quantum systems (in Russian). Usp. Fiz. Nauk **83**, 617–663.

I. M. Lifshitz (1967). Theory of fluctuation levels in disordered systems. Sov. Phys. JETP **26**, 462–479.

I. M. Lifshitz, S. A. Gredeskul and L. A. Pastur (1982a). On the transmission of waves and particles through random inhomogeneous media (in Russian). Zh. Eksp. Teor. Fiz. **83**, 2362–2376.

I. M. Lifshitz, S. A. Gredeskul and L. A. Pastur (1982b). Introduction to the Theory of Disordered Systems (in Russian). Nauka, Moscow. (English version, 1988, Wiley, New York).

P. Lloyd (1969). Exactly solvable model of electronic states in a three-dimensional disordered Hamiltonian: Non-existence of localized states. J. Phys. C **2**, 1717–1725.

W. Loeve (1960). Probability Theory. Van Nostrand, Princeton, NJ.

J. M. Luttinger and H. K. Sy (1973). Low-lying energy spectrum of a one-dimensional disordered system. Phys. Rev. A **7**, 701–712.

R. S. Maier (1986). The density of states of random Schrödinger operators. In Random Matrices and Their Applications, edited by J. E. Cohen, H. Kesten and C. M. Newman, 287–294. American Mathematical Society, Providence, RI (Contemporary Mathematics, 50).

V. A. Marchenko (1986). Sturm–Liouville Operators and Applications. Birkhäuser, Boston.

V. A. Marchenko and I. V. Ostrovskii (1975). A characterization of the spectrum of Hill's operator. Math. USSR Sb. **26**, 493–554.

V. A. Marchenko and I. V. Ostrovskii (1980). Approximation of periodic potentials by finite-zone potentials (in Russian). Vestnik Kharkov. Univ. **205**, 4–39. (English version in Selecta Math. Sov. **6** (1987), pp. 101–136).

V. A. Marchenko and L. A. Pastur (1967). Distribution of eigenvalues in some ensembles of random matrices (in Russian). Mat. Sb. **72**, 507–536.

A. V. Marchenko and L. A. Pastur (1985). Wave-transmission coefficient for a one-dimensional random medium (in Russian). Usp. Mat. Nauk **40**, 213–214.

F. Martinelli and H. Holden (1984). On the absence of diffusion near the bottom of the spectrum of a random Schrödinger operator on $L^2(\mathbf{R}^d)$. Commun. Math. Phys. **93**, 197–217.

F. Martinelli and L. Micheli (1985). On the large coupling-constant behaviour of the Lyapunov exponent of a binary alloy. J. Stat. Phys. **48**, 1–18.

F. Martinelli and E. Scoppola (1985). Remark on the absence of the absolutely continuous spectrum for d-dimensional Schrödinger operator with random potential for large disorder or low energy. Commun. Math. Phys. **97**, 465–471.

F. Martinelli and E. Scoppola (1987). Introduction to the mathematical theory of Anderson localization. Nuovo Cimento **10**, 2–90.

H. Matsuda (1964). Spectral frequencies in the vibrational spectra of disordered chains. Progr. Theor. Phys. **31**, 161–162.

D. Mattis and E. Lieb (1966). Mathematical Physics in One Dimension. Academic Press, New York.

H. P. McKean (1969). Stochastic Integrals. Academic Press, New York.

M. L. Mehta (1967). Random Matrices and the Statistical Theory of Energy Levels. Academic Press, New York.

G. A. Mezincescu (1985). Bounds on the integrated density of electronic states for disordered Hamiltonians. Phys. Rev. B **32**, 6272–6277.

G. A. Mezincescu (1986). Internal Lifshitz singularities of disordered finite-difference Schrödinger operators. Commun. Math. Phys. **103**, 167–176.

I. B Mikhailovskaya (1987). On the spectrum of random multidimensional Jacobi matrices (in Russian). Vest. Mosk. Univ. 1987(1), 55–58.

V. M. Millionsčikov (1969). Criterion of stability of the probabilistic spectrum of linear systems of differential equations with recurrent coefficients and criterion of reducibility with almost-periodic coefficients (in Russian). Mat. Sb. **78**, 179–202.

N. Minami (1986). An extension of Kotani's theory to generalized Sturm–Liouville operators. Commun. Math. Phys. **103**, 387–402.

R. A. Minlos and A. Ya. Povzner (1968). On the thermodynamic limit of entropy (in Russian). In Trudy Mosk. Mat. Obz. **17**, 243–272.

P. van Moerbeke (1976). The spectrum of Jacobi matrices. Invent. Math. **37**, 45–81.

S. A. Molchanov (1978). The structure of eigenfunctions of one-dimensional disordered structures. Math. USSR Izv. **12**, 69–101.

S. A. Molchanov (1981). The local structure of the spectrum of the one-dimensional Schrödinger operator. Commun. Math. Phys. **78**, 429–446.

S. A. Molchanov and V. A. Chulaevskii (1984). Structure of the spectrum of lacunary limit-periodic Schrödinger operators (in Russian). Funkc. Anal. i Priloz. **18**, 91–92.

S. A. Molchanov, A. A. Ruzmaikin and D. D. Sokolov (1985). Kinematic dynamo in a random flow (in Russian). Usp. Fiz. Nauk **145**, 593–628.

S. A. Molchanov and H. Seidel (1982). Spectral properties of the general Sturm-Liouville equation with random coefficients, I.. Math. Wachz. **109**, 57–79.

S. A. Molchanov and H. Seidel (1982). Spectral properties of the general Sturm-Liouville equation with random coefficients, II.. Preprint Univ. Dresden.

S. A. Molchanov and A. V. Stepanov (1979). Peaks of a Gaussian random field above a high level. Dokl. Akad. Nauk SSSR **249**, 294–297.

S. A. Molchanov and V. N. Tutubalin (1984). Linear model of hydrodynamical dynamo and random matrix products (in Russian). Teor. Ver. Primen. **29**, 234–247.

J. Morrison (1962). On the number of electronic levels in a one-dimensional random lattice. J. Math. Phys. **3**, 1023–1027.

J. Moser (1981). An example of a Schrödinger equation with almost-periodic potential and nowhere dense spectrum. Comment. Math. Helv. **56**, 198–215.

N. Mott and W. D. Twose (1961). The theory of impurity conduction. Adv. in Phys. **10**, 107–155.

N. I. Muskhelishvili (1962). Singular Integral Equations (in Russian). Fizmatgiz, Moscow.

S. V. Nagaev (1957). Some limit theorems for stationary Markov chains. Teor. Ver. Primen. **11**, 339–416.

M. A. Naimark (1969). Linear Differential Operators (in Russian). Nauka, Moscow.

Th. M. Nieuwenhuizen and J. M. Luck (1985). Singular behaviour of the density of states and the Lyapunov coefficient in binary random harmonic chains. J. Stat. Phys. **41**, 745.

Th. M. Nieuwenhuizen, J. M. Luck, J. Canisius, L. van Hemmen and W. J. Ventevogel (1986). Spectral frequencies and Lifshitz singularities in binary harmonic chains. J. Stat. Phys. **45**, 395–418.

G. Obermair (1983). Bloch electrons in rational and irrational magnetic fields. Helvet. Phys. Acta **56**, 245–254.

A. J. O'Connor and J. L. Lelowitz (1973). Heat conduction and sound transmission in isotopically disordered harmonic crystals. J. Math. Phys. **15**, 692–703.

V. Oseledec (1968). The multiplicative ergodic theorem: Lyapunov characteristic numbers for dynamical systems (in Russian). In Trudy Mosk. Mat. Obz. **19**, 197–231.

S. Ostlund and R. Pandit (1984). Renormalization-group analysis of a discrete quasiperiodic Schrödinger equation. Phys. Rev. B **29**, 1394–1397.

G. Papanicolaou and S. R. S. Varadhan (1982). Diffusion with random coefficients. In Statistics and Probability: Essays in Honor of C. R. Rao, 547–552. North-Holland, Amsterdam.

I. S. Parasyuk (1978). On instability zones of the Schrödinger equation with a quasiperiodic potential (in Russian). Ukr. Mat. Zh. **30**, 70–78.

L. A. Pastur (1971a). On Schrödinger equations with a random potential (in Russian). Teor. Mat. Fiz. **6**, 415–424.

L. A. Pastur (1971b). Self-averageness of the number of states of the Schrödinger operator with a random potential. In Matem. Fiz., Funkc. Anal. Trudy FTINT Akad. Nauk Ukr. SSR **2**, 111–116. Ukr. Acad. Sci., Kharkov.

L. A. Pastur (1972a). On the spectrum of random matrices (in Russian). Teor. Mat. Fiz. **10**, 102–112.

L. A. Pastur (1972b). On the distribution of the eigenvalues of the Schrödinger equation with a random potential. Funct. Anal. Appl. **8**, 163–165.

L. A. Pastur (1973). Spectra of random self-adjoint operators. Russ. Math. Surveys **28**, 1–67.

L. A. Pastur (1974a). On the spectrum of random Jacobi matrices and the Schrödinger equation on the whole axis with a random potential (in Russian). Preprint FTINT Ukr. Akad. Sci., Kharkov.

L. A. Pastur (1974b). On the eigenvalue distribution of the Schrödinger equation with a random potential (in Russian). In Matem. Fiz., Funkc. Anal. Trudy FTINT Akad. Nauk Ukr. SSR **5**, 141–146. Ukr. Acad. Sci., Kharkov.

L. A. Pastur (1976). The density of states in the fluctuation region of the spectra for the Schrödinger equation with a random potential. In Proc. of the VI Intern. Conf. on Amorphous and Liquid Semiconductors, edited by B. Kolomiets, 145–146. Nauka, Moscow.

L. A. Pastur (1977). Behavior of some Wiener integrals as $t \to \infty$ and the density of states of the Schrödinger equation with a random potential (in Russian). Teor. Mat. Fiz. **32**, 88-95.

L. A. Pastur (1980). Spectral properties of disordered systems in the one-body approximation. Commun. Math. Phys. **75**, 179–196.

L. A. Pastur (1982). Disordered spherical model. J. Stat. Phys. **27**, 119–151.

L. A. Pastur (1985). On the pure-point spectrum of the one-dimensional Anderson model with the Gaussian potential. Preprint Karl-Marx-Univ., Leipzig.

L. A. Pastur (1987). Lower estimates for the Lyapunov exponent of certian finite-difference equations with quasiperiodic coefficients (in Russian). In Operators in Functional Spaces and Topics of Function Theory, edited by V. A. Marchenko, 115–127. Naukova Dumka, Kiev.

L. A. Pastur and E. P. Feldman (1975). Wave transmission for a thick layer of a randomly inhomogeneous medium. Sov. Phys. JETP **40**, 241–243.

L. A. Pastur and V. A. Tkachenko (1984). On the spectral theory of the one-dimensional Schrödinger operator with a limit-periodic potential (in Russian). Dokl. Akad. Nauk SSSR **279**, 1051–1054.

L. A. Pastur and V. A. Tkachenko (1989). Spectral theory of a class of one-dimensional Schrödinger operator with limit-periodic potentials (in Russian). Trans. Moscow Math. Soc. **51**, 115–166.

J. Piepenbrink and P. Rejto (1974). Some singular Schrödinger operators with deficiency indices (n^2, n^2). Duke Math. J. **41**, 593–605.

J. Pöschel (1983). Examples of discrete Schrödinger operators with pure-point spectrum. Commun. Math. Phys **88**, 447-463.

R. E. Prange, D. R. Grempel and S. Fishman (1984). Solvable model of quantum motion in an incommensurate potential. Phys. Rev. B **29**, 6500-6512.

I. I. Privalov (1956). Boundary Properties of Analytic Fuctions (in German). Deutscher Verlag, Berlin.

M. S. Raghunathan (1979). A proof of Oseledec's multiplicative ergodic theorem. Isr. J. Math. **32**, 356–362.

M. Reed and B. Simon (1972–1979). Methods of Modern Mathematical Physics. I: Functional Analysis; II: Fourier Analysis, Self-Adjointness; III: Scattering Theory; IV: Analysis of Operators. Academic Press, New York.

A. Reznikova (1981). The central limit theorem for the spectrum of the random one-dimensional Schrödinger operator. J. Stat. Phys. **25**, 291–302.

M. Romerio and W. Wreszinski (1979). On the Lifshitz singularity and the tailing in the density of states for random lattice systems. J. Stat. Phys. **21**, 169–179.

G. Royer (1980). Croissance exponentielle de produits markoviens de matrices aléatoires. Ann. Inst. H. Poincaré A **16**, 46–62.

D. Ruelle (1969a). A remark on bound states in potential scattering. Nuovo Cimento A **61**, 655-661.

D. Ruelle (1969b). Statistical Mechanics. Benjamin, New York.

D. Ruelle (1978). Analyticity properties of the characteristic exponents of random matrix products. Adv. Math. **32**, 68–91.

D. Ruelle (1979). Ergodic theory of differentiable dynamical systems. Publ. Math. I.H.E.S. **50**, 275–306.

H. Rüssman (1978) On the one-dimensional Schrödinger equation with quasi-periodic potential. Annals of the New York Acad. Sci. **357**, 90–107.

S. Saks (1937). Theory of Integral. Warsaw.

E. Sanchez-Palencia (1980). Non-Homogeneous Media and Vibration Theory. Springer, Berlin (Lecture Notes in Physics, 127).

V. V. Sazonov and V. N. Tutubalin (1966) Probability distributions on topological groups. Theor. Probab. Appl. **13**, 1–45.

D. Schechtman, I. Blech, D. Cratias and J. V. Cahn (1984). Metallic phase with long-range oriental order and no translation symmetry. Phys. Rev. Lett. **53**, 1951.

H. Schmidt (1958). Disordered one-dimensional crystals. Phys. Rev. **105**, 411–425.

I. E. Sch'nol (1954). On the behavior of eigenfunctions (in Russian). Dokl. Akad. Nauk SSSR **94**, 389–392.

M. A. Shubin (1975). Elliptic almost-periodic operators and van Neumann algebras (in Russian). Funkc. Anal. Priloz. **9**, 89–90.

M. A. Shubin (1978). Density of states for self-adjoint elliptic operators wiht almost-periodic coefficients (in Russian). In Trudy sem. Petrovskii (Moscow University) **3**, 243–281.

M. A. Shubin (1979). Spectral theory and index of elliptic operators with almost-periodic coefficients. Russ. Math. Surveys **34**, 109–157.

B. Simon (1982). Schrödinger semigroups. Bull. Am. Math. Soc. **7**, 447–526.

B. Simon (1983). Kotani theory for one-dimensional stochastic Jacobi matrices. Commun. Math. Phys. **89**, 227–234.

B. Simon (1985a). Lifshitz tails for the Anderson model. J. Stat. Phys. **38**, 65–76.

B. Simon (1985b). Almost-periodic Schrödinger operators. IV: The Maryland model. Ann. Phys. New York **159**, 157–182.

B. Simon (1985c). Localization in genreal one-dimensional random systems. I: Jacobi matrices. Commun. Math. Phys. **102**, 327–336.

B. Simon (1987). Internal Lifshitz tails. J. Stat. Phys. **46**, 911–918.

B. Simon and T. Spencer (1989). Trace-class perturbations and the absence of absolutely continuous spectra. Comm. Math. Phys. **125**, 113–125.

B. Simon and M. Taylor (1985). Harmonic analysis on SL(2, **R**) and smoothness of the density of states in the one-dimensional Anderson model. Commun. Math. Phys. **101**, 1–19.

B. Simon, M. Taylor and T. Wolff (1985). Some rigorous results for the Anderson model. Phys. Rev. Lett. **54**, 1589–1592.

B. Simon and T. Wolff (1986). Singular continuous spectrum under rank-one perturbations and localization for random Hamiltonians. Commun. Pure Appl. Math. **39**, 75–90.

Ya. G. Sinai (1985). On the spectrum structure of the finite-difference Schrödinger operator with an almost-periodic potential near the left edge. (in Russian). Funkc. Anal. Priloz. **19**, 42–48.

Ya. G. Sinai (1987). Anderson localization for one-dimensional difference Schrödinger operators with quasiperiodic potentials. J. Stat. Phys. **46**, 861–909.

A. V. Skorohod (1984). Random Linear Operators. D. Reidel, Dordrecht and Boston.

M. M. Skriganov (1984). Structure of the spectrum of the multidimensional Schrödinger operator with a periodic potential (in Russian). Dokl. Akad. Nauk SSSR **262**, 846–850.

I. M. Slivnyak (1960). On the spectrum of the Schrödinger operator with a random potential (in Russian). Zhurn. Vychisl. Matem. i Matem. Fysiki **6**, 1104–1108.

J. B. Sokoloff (1985). Unusual band structure, wave functions and electrical conductance in crystals with incommensurate periodic potentials. Phys. Reports **126**, 189–244.

T. Spencer (1984). The Schrödinger equation with a random potential: a mathematical review. In Phénomènes critiques, systèmes aléatoires, théories de jauge, edited by K. Osterwalder and R. Stora, 895–944. North-Holland, Amsterdam and New York.

T. Spencer (1988). Localization for random and almost-periodic potentials. J. Stat. Phys. **51**, 1009–1017.

V. G. Sprindzhuk (1979). Metric Theory of Diophantine Approximations. Wiley, New York.

R. L. Stratonovich (1961). Selected Topics on Fluctuation Theory in Radio Technique (in Russian). Sov. Radio, Moscow.

I. M. Suslov (1982). Localization in one-dimensional incommensurate systems (in Russian). Zh. Eksp. Teor. Fiz. **83**, 1079–1086.

G. Temple (1928). The theory of Rayleigh's principle as applied to continuous systems. Proc. Roy. Soc. London A **119**, 276–293.

W. Thirring (1981) A Course of Mathematical Physics. III: Quantum Mechanics of Atoms and Molecules. Springer, Berlin.

L. Thomas (1973). Time-dependent approach to scattering from impurities in a crystal. Commun. Math. Phys. **33**, 335–343.

L. Thomas and E. Wayne (1986). On the stability of dense point spectra for self-adjoint operators. J. Math. Phys. **27**, 71–75.

D. Thouless (1972). A relation between the density of states and range of localization for one-dimensional systems. J. Phys. C **5**, 77–81.

D. Thouless (1981). Localization and the two-dimensional Hall effect. J. Phys. C **14**, 3475–3480.

D. Thouless (1983). Bandwidth for a quasiperiodic tight-binding model. Phys. Rev. B **28**, 4272–4276.

E. C. Titchmarsh (1946). Eigenfunction expansions associated with second-order differential equations. I. Clarendon Press, Oxford.

E. C. Titchmarsh (1958). Eigenfunction expansions associated with second-order differential equations. II. Clarendon Press, Oxford.

M. Toda (1981). Theory of Nonlinear Lattices. Springer, Berlin.

V. N. Tutubalin (1965). On limit theorems for products of random matrices. Theor. Probab. Appl. **10**, 15–27.

V. N. Tutubalin (1968). Convergence of probability measures in variation and products of random matrices (in Russian). Teor. Ver. Primen. **13**, 63–81.

V. K. Vardazaryan (1977). Spectral theory of the one-dimensional Dirac system. Dokl. Akad. Nauk Arm. SSR **64**, 264–270.

O. A. Veliev (1987). Asymptotic formulas for the eigenvalues of the periodic Schrödinger operator and the Bethe–Sommerfeld hypothesis (in Russian). Func. Anal. i Priloz. **21**, 1–15.

A. D. Virtser (1979). On products of random matrices and operators. Theory Prob. Appl. **24**, 367–377.

A. D. Virtser (1983). On the simplicity of the spectrum of the Lyapunov characteristic indices of a product of random matrices. Theory Prob. Appl. **28**, 122–136.

F. Wegner (1979). Disordered systems with n orbitals per site: the limit $n = \infty$. Phys. Rev. B **19**, 783–792.

F. Wegner (1981). Bounds on the density of states in disordered systems. Zeit. für Physik B **44**, 9–15.

E. Wigner (1967). Random matrices in physics. SIAM Review J. **9**, 1–23.

V. Wihstutz (1985). Analytic expansion of the Lyapunov exponent associated to the Schrödinger operator with random potential. Stochast. Anal. and Appl. **3**, 93–118.

M. Wilkinson (1984). Critical properties of electron eigenstates in incommensurate systems. Proc. Roy. Soc. London A **391**, 306–350.

K. G. Wilson and J. Kogut (1974). The renormalization group and ε-expansions. Phys. Rep. **12C**, 79–199.

V. V. Yurinskii (1982) On averaging nondivergent second-order equations with random coefficients (in Russian). Sib. Mat. Zh. **23**, 176–188.

V. E. Zakharov, S. V. Manakov, S. P. Novikov and L. P. Pitaevskii (1985). Theory of Solitons: The Inverse Problem Method (in Russian). Nauka, Moscow.

G. M. Zaslavskii (1984). Stochasticity of Dynamical Systems (in Russian). Nauka, Moscow.

Ya. B. Zel'dovich, S. A. Molchanov, A. A. Ruzmaikin and D. D. Sokolov (1987). Intermittency in random media. Sov. Phys. Usp. **30**, 353–369.

V. V. Zhikov (1967). On inverse Sturm–Liouville problems on a finite interval. Uzv. Akad. Nauk SSSR ser. mat. **31**, 965–981.

V. V. Zhikov, S. M. Kozlov and O. A. Oleinik (1982). Averaging of parabolic operators (in Russian). In Trudy Mosk. Mat. Ob. **45**, 11–58.

V. V. Zhikov, S. M. Kozlov, O. A. Oleinik and M. T. Nguyen (1979). Averaging and G-convergence of differential operators (in Russian). Usp. Mat. Nauk **34**, 65–133.

V. V. Zhikov and M. M. Sirazhudinov (1981). Averaging of nondivergent elliptic second-order operators and the stabilization of the solutions of the Cauchy problem (in Russian). Mat. Sb. **116**, 166–186.

List of Symbols

B^c complement of set B
$\bar{B}$ closure of B
∂B boundary of B
$B_1 \triangle B_2$ symmetric difference between B_1 and B_2
z^* complex conjugate of z
$\mathbf{R}$ set of real numbers
$\mathbf{R}_+$ set of positive numbers
$\mathbf{Z}$ set of integers
$\mathbf{Z}_+$ set of natural numbers
$\mathbf{C}$ set of complex numbers
$\mathbf{C}_+$ set of complex numbers with positive imaginary part
$\mathbf{R}^d, \mathbf{Z}^d, \mathbf{C}^d$ d-fold cartesian product of $\mathbf{R}$, $\mathbf{Z}$, $\mathbf{C}$
$\mathbf{T}^d$ d-dimensional torus
S^1 circle of length π, that is, $[0, \pi)$
$(\cdot\,,\,\cdot)$ scalar product in Hilbert space
I identity operator
$\mathcal{D}(A)$ domain of operator A
$A\big|_{\mathcal{D}}$ restriction of A to $\mathcal{D}$
$\bar{A}$ closure of A
A^* adjoint of A
$\|A\|$ norm of A
$\mathrm{Tr}\, A$ trace of A
$E(A, d\lambda), E(d\lambda)$ resolution of the identity of A
$\sigma, \sigma(A)$ spectrum of A
$\sigma_{\mathrm{p}}, \sigma_{\mathrm{p}}(A)$ point spectrum (set of eigenvalues) of A
$\sigma_{\mathrm{ac}}, \sigma_{\mathrm{ac}}(A)$ absolutely continuous spectrum of A
$\sigma_{\mathrm{sc}}, \sigma_{\mathrm{sc}}(A)$ singular continuous spectrum of A
$N(A, d\lambda), N(\lambda)$ integrated density of states of A
$-\Delta$ Laplacian
$H = -\Delta + q$ Schrödinger operator
h discrete analogue of the Schrödinger operator
$\mathrm{meas}(\,\cdot\,)$ Lebesgue measure on $\mathbf{R}$, $\mathbf{R}^d$, $\mathbf{T}^d$
$\mathrm{supp}\,\mu$ topological support of μ
δ_x Dirac delta function
$\delta_{x,y}$ Kronecker delta
χ_B characteristic function of B
$C(G)$ set of continuous functions of G
$C_0^\infty(\mathbf{R}^d)$ space of smooth functions with compact support
ω space of realizations
$\mathcal{F}$ σ-algebra of events
P probability measure
$(\omega, \mathcal{F}, \mathsf{P})$ probability space
$\mathsf{E}\{\,\cdot\,\}$ mathematical expectation
$\mathsf{E}\{\,\cdot \mid \mathcal{F}_1\}$ conditional expectation with respect to σ-algebra $\mathcal{F}_1$
$\{T_g : g.\ InG\}$ metrically transitive group of automorphisms

Index

Grundlehren der mathematischen Wissenschaften

A Series of Comprehensive Studies in Mathematics

A Selection

190. Faith: Algebra: Rings, Modules, and Categories I
191. Faith: Algebra II, Ring Theory
192. Mal'cev: Algebraic Systems
193. Pólya/Szegö: Problems and Theorems in Analysis I
194. Igusa: Theta Functions
195. Berberian: Baer*-Rings
196. Arthreya/Ney: Branching Processes
197. Benz: Vorlesungen über Geometrie der Algebren
198. Gaal: Linear Analysis and Representation Theory
199. Nitsche: Vorlesungen über Minimalflächen
200. Dold: Lectures on Algebraic Topology
201. Beck: Continuous Flows in the Plane
202. Schmetterer: Introduction to Mathematical Statistics
203. Schoeneberg: Elliptic Modular Functions
204. Popov: Hyperstability of Control Systems
205. Nikol'skiĭ: Approximation of Functions of Several Variables and Imbedding Theorems
206. André: Homologie des Algébres Commutatives
207. Donoghue: Monotone Matrix Functions and Analytic Continuation
208. Lacey: The Isometric Theory of Classical Banach Spaces
209. Ringel: Map Color Theorem
210. Gihman/Skorohod: The Theory of Stochastic Processes I
211. Comfort/Negrepontis: The Theory of Ultrafilters
212. Switzer: Algebraic Topology - Homotopy and Homology
215. Schaefer: Banach Lattices and Positive Operators
217. Stenström: Rings of Quotients
218. Gihman/Skorohod: The Theory of Stochastic Processes II
219. Duvant/Lions: Inequalities in Mechanics and Physics
220. Kirillov: Elements of the Theory of Representations
221. Mumford: Algebraic Geometry I: Complex Projective Varieties
222. Lang: Introduction to Modular Forms
223. Bergh/Löfström: Interpolation Spaces. An Introduction
224. Gilbarg/Trudinger: Elliptic Partial Differential Equations of Second order
225. Schütte: Proof Theory
226. Karoubi: K-Theory. An Introduction
227. Grauert/Remmert: Theorie der Steinschen Räume
228. Segal/Kunze: Integrals and Operators
229. Hasse: Number Theory
230. Klingenberg: Lectures on Closed Geodesics
231. Lang: Elliptic Curves: Diophantine Analysis
232. Gihman/Skorohod: The Theory of Stochastic Processes III
233. Stroock/Varadhan: Multidimensional Diffusion Processes
234. Aigner: Combinatorial Theory
235. Dynkin/Yushkevich: Controlled Markov Processes
236. Grauert/Remmert: Theory of Stein Spaces
237. Köthe: Topological Vector Spaces II
238. Graham/McGehee: Essays in Commutative Harmonic Analysis
239. Elliott: Probabilistic Number Theory I
240. Elliott: Probabilistic Number Theory II
241. Rudin: Function Theory in the Unit Ball of C^n

242. Huppert/Blackburn: Finite Groups II
243. Huppert/Blackburn: Finite Groups III
244. Kubert/Lang: Modular Units
245. Cornfeld/Fomin/Sinai: Ergodic Theory
246. Naimark/Štern: Theory of Group Representations
247. Suzuki: Group Theory I
248. Suzuki: Group Theory II
249. Chung: Lectures from Markov Processes to Brownian Motion
250. Arnold: Geometrical Methods in the Theory of Ordinary Differential Equations
251. Chow/Hale: Methods of Bifurcation Theory
252. Aubin: Nonlinear Analysis on Manifolds. Monge-Ampère Equations
253. Dwork: Lectures on p-adic Differential Equations
254. Freitag: Siegelsche Modulfunktionen
255. Lang: Complex Multiplication
256. Hörmander: The Analysis of Linear Partial Differential Operators I
257. Hörmander: The Analysis of Linear Partial Differential Operators II
258. Smoller: Shock Waves and Reaction-Diffusion Equations
259. Duren: Univalent Functions
260. Freidlin/Wentzell: Random Perturbations of Dynamical Systems
261. Bosch/Güntzer/Remmert: Non Archimedian Analysis – A Systematic Approach to Rigid Analytic Geometry
262. Doob: Classical Potential Theory and Its Probabilistic Counterpart
263. Krasnosel'skiĭ/Zabreĭko: Geometrical Methods of Nonlinear Analysis
264. Aubin/Cellina: Differential Inclusions
265. Grauert/Remmert: Coherent Analytic Sheaves
266. de Rham: Differentiable Manifolds
267. Arbarello/Cornalba/Griffiths/Harris: Geometry of Algebraic Curves, Vol. I
268. Arbarello/Cornalba/Griffiths/Harris: Geometry of Algebraic Curves, Vol. II
269. Schapira: Microdifferential Systems in the Complex Domain
270. Scharlau: Quadratic and Hermitian Forms
271. Ellis: Entropy, Large Deviations, and Statistical Mechanics
272. Elliott: Arithmetic Functions and Integer Products
273. Nikol'skiĭ: Treatise on the Shift Operator
274. Hörmander: The Analysis of Linear Partial Differential Operators III
275. Hörmander: The Analysis of Linear Partial Differential Operators IV
276. Liggett: Interacting Particle Systems
277. Fulton/Lang: Riemann-Roch Algebra
278. Barr/Wells: Toposes, Triples and Theories
279. Bishop/Bridges: Constructive Analysis
280. Neukirch: Class Field Theory
281. Chandrasekharan: Elliptic Functions
282. Lelong/Gruman: Entire Functions of Several Complex Variables
283. Kodaira: Complex Manifolds and Deformation of Complex Structures
284. Finn: Equilibrium Capillary Surfaces
285. Burago/Zalgaller: Geometric Inequalities
286. Andrianov: Quadratic Forms and Hecke Operators
287. Maskit: Kleinian Groups
288. Jacod/Shiryaev: Limit Theorems for Stochastic Processes
289. Manin: Gauge Field Theory and Complex Geometry
290. Conway/Sloane: Sphere Packings, Lattices and Groups
291. Hahn/O'Meara: The Classical Groups and K-Theory
292. Kashiwara/Schapira: Sheaves on Manifolds
293. Revuz/Yor: Continuous Martingales and Brownian Motion
294. Knus: Quadratic and Hermitian Forms over Rings
295. Dierkes/Hildebrandt/Küster/Wohlrab: Minimal Surfaces I
296. Dierkes/Hildebrandt/Küster/Wohlrab: Minimal Surfaces II
297. Pastur/Figotin: Spectra of Random and Almost-Periodic Operators
298. Berline/Getzler/Vergne: Heat Kernels and Dirac Operators